Complex product development model

Technical foundation

by Christer Sandahl

Complex product development model (*Cpdm*) ™

www.cpdm.com

Sandahl Consulting AB

Tegelvaegen 7, 227 30 Lund, Sweden
+46 733 736008, +46 708 152538
christer@sandahlconsulting.se, gunnel@sandahlconsulting.se

ISBN 978-91-984195-0-4

To my dear family
for taking proper care of me
when I fade into book coma

Copyright 2017 by Christer Sandahl
—First Edition—

In cooperation with

PO Box 221974 Anchorage, Alaska 99522-1974
books@publicationconsultants.com—www.publicationconsultants.com

and

Copyediting Services, Marthy W. Johnson, editor
Anchorage, Alaska, USA
mjces@gci.net

Manufactured in the United States of America

Brief table of contents

Brief table of contents 3

Detailed table of contents 5

CHAPTER 1   *Thank mentors*   13

CHAPTER 2   *Reveal background*   17

CHAPTER 3   *Explain book*   25

CHAPTER 4   *Start development*   37

CHAPTER 5   *Identify value chain*   65

CHAPTER 6   *Utilize processes*   81

CHAPTER 7   *Capture staffing & requirements*   175

CHAPTER 8   *Predetermine solutions & suppliers*   281

CHAPTER 9   *Design architectures*   333

CHAPTER 10   *Finalize design & requisites*   435

CHAPTER 11   *Realize & integrate white-box*   593

CHAPTER 12   *Verify black- & white-box*   659

Watch references 781

Track indexes 783

Share glossary 787

Detailed table of contents

Brief table of contents 3

Detailed table of contents 5

CHAPTER 1 *Thank mentors   13*

    1.1 My technical mentors   14
    1.2 The author   16

CHAPTER 2 *Reveal background   17*

    2.1 What has gone wrong?   18
    2.2 Mastering complexity   20
    2.3 How can I help you?   24

CHAPTER 3 *Explain book   25*

    3.1 Why another book about this?   26
    3.2 Organization of this book   31

CHAPTER 4 *Start development   37*

    4.1 About start development   38
    4.2 Study strategy   40
    4.3 Study concepts   41
    4.4 Plan portfolio   42
    4.5 Study business   44
    4.6 Prototype details   45
    4.7 Research market   46
    4.8 Plan products   47
    4.9 Capture staffing & requirements   48
    4.10 Predetermine solutions & suppliers   49
    4.11 Design architectures   51
    4.12 Finalize design & requisites   52
    4.13 Realize & integrate white-box   58
    4.14 Verify black- & white-box   59
    4.15 Validate prototype   61
    4.16 Promote campaigns   62
    4.17 Launch product   63
    4.18 Prepare manufacturing   63
    4.19 Finalize product   64

CHAPTER 5   *Identify value chain   65*

5.1 About value chain   66
5.2 Different ways to illustrate a value chain?   69
5.3 Elaborate concurrence in a value chain   74
5.4 Value chain roles   78
5.5 Document value chain by a process   80

CHAPTER 6   *Utilize processes   81*

6.1 About processes   82
6.2 Process schedule constituents   83
6.3 Partition, variant, terms, and govern   85
6.4 Symbols used throughout this book   95
6.5 Cpdm process   97
6.6 P. Cpdm generic development schedule   99
6.7 Pt:1. Cpdm generic elementary technical schedule   101
6.8 Pt:2. Cpdm generic advanced technical schedule   103
6.9 Pt:2. Cpdm generic advanced technical schedule   106
6.10 Pt:2. Cpdm generic advanced technical schedule   108
6.11 Pt:2.0. Map environment and develop interfaces to the environment   109
6.12 Pt:2.n. Develop outermost black-/white-box and each inward embedded black-/white-box   128
6.13 EXAMPLE: Tailored process overall schedules   160
6.14 Schedules seen as programs   172
6.15 Some other process models   172

CHAPTER 7   *Capture staffing & requirements   175*

7.1 About requirements   176
7.2 About explore restrictions   177
7.3 About environment staffing allocation   179
7.4 EXAMPLE House environment: Explore environment restrictions and ensure environment staffing   180
7.5 About black-box requirements   184
7.6 About development staffing   185
7.7 EXAMPLE House: Refine house requirements from environment restriction requirements, and ensure house development staffing   187
7.8 Requirements refinement hierarchy   192
7.9 EXAMPLE House rooms: Refine room requirements from house requirements, and ensure room development staffing   193
7.10 EXAMPLE House rooms machinery: Refine machinery requirements from rooms requirements and ensure machinery development staffing   198
7.11 Restriction requirements   203

7.12 Behavior requirements   206

7.13 Requirements refinement and relation to architecture   207

7.14 EXAMPLE Multiplication toy environment: Explore environment restrictions and ensure environment staffing   211

7.15 EXAMPLE Multiplication toy: Refine product requirements from environment restriction requirements and ensure black-/white-box development staffing   215

7.16 Use-case behavior requirements   221

7.17 Requirements formalism   224

7.18 Formal use-case requirements   227

7.19 Scenario behavior requirements   230

7.20 Wait-state-machine   232

7.21 EXAMPLE Calc_logic environment (the pocket calculator firmware): Explore environment restrictions and ensure environment staffing   237

7.22 EXAMPLE Calc_logic environment (the Windows calculator application): Explore environment restrictions and ensure environment staffing   240

7.23 EXAMPLE Calc_logic for reuse: Refine product requirements from environment restriction requirements and ensure black-/white-box development staffing   243

7.24 EXAMPLE Formal and complete requirements for calculator   250

7.25 Uncorrelated stimuli on different interfaces   255

7.26 Capturing data structures   259

7.27 EXAMPLE Phonebook environment: Explore environment restrictions and ensure environment staffing   261

7.28 EXAMPLE Phonebook: Refine product requirements from environment restriction requirements and ensure black-/white-box development staffing   264

7.29 Huge networks of scenarios   275

7.30 Road maps   276

7.31 Frequently asked questions about requirements   277

CHAPTER 8   *Predetermine solutions & suppliers*   *281*

8.1 About predetermine solutions & suppliers   282

8.2 Demarcate product by environment solutions with possible supplier opportunities (environment nesting level = 0)   287

8.3 EXAMPLE House environment: Demarcate product by environment solutions with possible supplier opportunities   290

8.4 Predetermine solutions with sourcing options (nesting levels n ≥ 1)   294

8.5 EXAMPLE House: Predetermine solutions with sourcing options   300

8.6 EXAMPLE House rooms: Predetermine solutions with sourcing options   305

8.7 EXAMPLE House rooms machinery: Predetermine solutions with sourcing options   309

8.8 EXAMPLE Multiplication toy environment: Demarcate product by environment solutions with possible supplier opportunities   313

8.9 EXAMPLE Multiplication toy: Predetermine solutions with sourcing options   316

8.10 EXAMPLE Calc_logic environment (the pocket calculator firmware): Demarcate product by environment solutions with possible supplier opportunities   319

8.11 EXAMPLE Calc_logic environment (the Windows calculator application): Demarcate product by environment solutions with possible supplier opportunities   323

8.12 EXAMPLE Calc_logic for reuse: Predetermine solutions with sourcing options   326

8.13 EXAMPLE Phonebook environment: Demarcate product by environment solutions with possible supplier opportunities   328

8.14 EXAMPLE Phonebook: Predetermine solutions with sourcing options   330

CHAPTER 9   *Design architectures   333*

9.1 About architecture   334

9.2 EXAMPLE House environment: Constitute environment architecture   339

9.3 Architecture white-box   343

9.4 Architecture interfaces   346

9.5 Transforming requirements into architecture ingredients   354

9.6 EXAMPLE House: Satisfy house requirements by decomposing house black-box into white-box design containing room black-boxes   357

9.7 EXAMPLE House rooms: Satisfy room requirements by decomposing room black-boxes into white-box design containing machinery black-boxes   365

9.8 EXAMPLE House rooms machinery: Satisfy machinery requirements by decomposing machinery black-boxes into white-box design not containing black-boxes   371

9.9 Managed interfaces   378

9.10 Illustrate interactions between elements   380

9.11 EXAMPLE Multiplication toy environment: Constitute environment architecture   382

9.12 EXAMPLE Multiplication toy commonalities: Satisfy refined requirements by decomposing product black-box into white-box design containing no black-boxes   386

9.13 EXAMPLE Hard-wired multiplier: Further satisfy refined requirements by decomposing product black-box into white-box design containing no black-boxes   391

9.14 EXAMPLE: Typical microcontroller architecture   394

9.15 Low-level programs   396

9.16 EXAMPLE Microcontroller multiplier: Further satisfy refined requirements by decomposing product black-box into white-box design containing no black-boxes   397

9.17 Multi-threads and interrupts   400

9.18 Usage of threads   403

9.19 EXAMPLE: Architecture ID numbering used throughout this book   406

9.20 EXAMPLE Calc_logic environment (the pocket calculator firmware): Constitute environment architecture   408

9.21 EXAMPLE Central Processing Unit (CPU)   412

9.22 High-level program architecture   414

9.23 EXAMPLE Calc_logic environment (the Windows calculator application): Constitute environment architecture   415

9.24 EXAMPLE Calc_logic for reuse: Satisfy refined requirements by decomposing product black-box into white-box design containing no black-boxes   419

9.25 Object-oriented program architecture   424

9.26 EXAMPLE Phonebook environment: Constitute environment architecture   425

9.27 EXAMPLE Phonebook: Satisfy refined requirements by decomposing product black-box into white-box design containing no black-boxes   429

9.28 Large architecture documentation   434

CHAPTER 10   *Finalize design & requisites   435*

10.1 About finalize design & requisites   436

10.2 About finalize design items   440

10.3 About finalize product requisites   443

10.4 EXAMPLE: House kitchen machinery: Finalize design of machinery with product requisites   445

10.5 EXAMPLE House rooms: Finalize design of rooms with product requisites   452

10.6 EXAMPLE House: Finalize design of house with product requisites   465

10.7 EXAMPLE House environment: Finalize design of environment interfaces with product requisites   477

10.8 EXAMPLE Specifics for multiplication toy commonalities: Finalize design items of product white-box with product requisites   482

10.9 EXAMPLE Specifics for multiplication toy environment commonalities: Finalize design of environment interfaces with product requisites   493

10.10 EXAMPLE Specifics for multiplication toy with hard-wired multiplier: Finalize design items of product white-box with product requisites   499

10.11 EXAMPLE Multiplication toy environment (hard-wired multiplier variant): Finalize design of environment interfaces with product requisites   509

10.12 Enter into computer programs   512

10.13 Machine program   513

10.14 Assembler programs   516

10.15 EXAMPLE Specifics for multiplication toy with microcontroller multiplier: Finalize design items of product white-box with product requisites   518

10.16 EXAMPLE Multiplication toy environment (microcontroller multiplier variant): Finalize design of environment interfaces with product requisites   526

10.17 Structured programs   529

10.18 EXAMPLE Hosted calc_logic for reuse: Finalize design items of product white-box with product requisites   533

10.19 EXAMPLE Calc_logic environment (the Windows calculator application): Finalize design of environment interfaces with product requisites   545

10.20 EXAMPLE Calc_logic environment (the pocket calculator firmware) development board and pocket calculator: Finalize design of environment interfaces with product requisites   549

10.21 Cpdm definition   553

10.22 Object-oriented programs   554

10.23 Accounting in yet finer granularity   560

10.24 EXAMPLE Phonebook: Finalize design items of product white-box with product requisites   561

10.25 EXAMPLE Phonebook environment: Finalize design of environment interfaces with product requisites   585

10.26 Scaling up programs   590

CHAPTER 11   *Realize & integrate white-box*   *593*

11.1 About realize & integrate white-box   594

11.2 EXAMPLE House environment: Realize interfaces to environment and insert/await outermost white-box   601

11.3 EXAMPLE House: Procure house items (and await invading room black-boxes) in order to build house   607

11.4 EXAMPLE House rooms: Procure room items (and await invading machinery black-boxes) in order to build rooms   613

11.5 EXAMPLE House kitchen machinery: Procure machinery items in order to install machinery   618

11.6 Completeness of prototypes   623

11.7 EXAMPLE Multiplication toy with hard-wired multiplier: Procure items and linkable programs in order to realize white-box   624

11.8 EXAMPLE Multiplication toy environment (hard-wired multiplier variant): Realize interfaces to environment and insert/await outermost white-box   628

11.9 Integrate assembler programs in a white-box   631

11.10 EXAMPLE Multiplication toy with microcontroller multiplier: Procure items and linkable programs in order to realize white-box   632

11.11 EXAMPLE Multiplication toy environment (microcontroller multiplier variant): Realize interfaces to environment and insert/await outermost white-box   637

11.12 Integrate all kind of elements in a white-box   640

11.13 Reuse and porting   641

11.14 EXAMPLE: Hosted calc_logic for reuse: Procure items and linkable programs in order to realize white-box   644

11.15 EXAMPLE: Hosted calc_logic for reuse: Porting to Windows application, development board, and pocket calculator   646

11.16 EXAMPLE Calc_logic environment (the Windows calculator application): Realize interfaces to environment and insert/await outermost white-box   646

11.17 EXAMPLE Calc_logic environment (the development board fixture firmware): Realize interfaces to environment and insert/await outermost white-box   649

11.18 EXAMPLE Calc_logic environment (the pocket calculator firmware): Realize interfaces to environment and insert/await outermost white-box   653

11.19 EXAMPLE Phonebook: Procure items and linkable programs in order to realize white-box   655

11.20 EXAMPLE Phonebook environment: Realize interfaces to environment and insert/await outermost white-box   657

CHAPTER 12   *Verify black- & white-box   659*

12.1 About verification   660

12.2 EXAMPLE Lake: Size of meshes in fishing nets   665

12.3 Troubleshooting   665

12.4 EXAMPLE House environment: Partially verify interfaces to environment with house fake   673

12.5 EXAMPLE House: Partially verify black-/white-box with embedded fakes   675

12.6 EXAMPLE House kitchen: Partially verify black-/white-box with embedded fakes   677

12.7 EXAMPLE: House rooms machinery: Fully verify machinery black-/white-boxes   678

12.8 EXAMPLE House kitchen: Fully verify room black-/white-box   684

12.9 EXAMPLE House: Fully verify house black-/white-box   693

12.10 EXAMPLE House environment: Fully verify prototype in environment   699

12.11 Verification depends on type of requirements   704

12.12 EXAMPLE Multiplication toy with hard-wired multiplier: Verify black-/white-box   706

12.13 EXAMPLE Multiplication toy environment (hard-wired multiplier variant): Verify prototype in environment   712

12.14 Prepare verification during manufacturing   713

12.15 Making verification efficient   713

12.16 EXAMPLE Multiplication toy with microcontroller multiplier: Verify black-/white-box   717

12.17 EXAMPLE Multiplication toy environment (microcontroller multiplier variant): Verify prototype in environment   725

12.18 EXAMPLE Lake: Which fish to catch first ?   729

12.19 Usage profile modeling   730

12.20 EXAMPLE Hosted calc_logic for reuse: Verify black-/white-box   733

12.21 EXAMPLE Calc_logic environment (the Windows calculator application): Verify prototype in target environment   758

12.22 EXAMPLE Calc_logic environment (the pocket calculator firmware) on development board: Verify prototype in environment   761

12.23 EXAMPLE Calc_logic environment (the pocket calculator firmware): Verify prototype in environment   767

12.24 EXAMPLE Lake: Fishing in same lake with changed environment   769

12.25 About human-decided verification coverage   769

12.26 EXAMPLE Phonebook: Verify black-/white-box   770

12.27 EXAMPLE Phonebook environment: Verify prototype in environment   780

Watch references 781

Track indexes 783

Share glossary 787

## CHAPTER 1

Thank mentors

You do not understand how lucky you are
until your companions are not at hand anymore.

Impressions for a lifetime

1.1 My technical mentors

During a professional lifetime in product development, I have, of course, had a lot of knowledgeable contacts. It is not humanly possible to enumerate them all here, but some of them have been essential when navigating the crossroads and making decisions for life. The important influences below must be acclaimed.

I cannot enumerate all my excellent contacts, but some important influences must be acclaimed.

Enok Sandahl. This was my grandfather, acting as my second father. He had a small carpentry in Åsafors in the countryside in Småland, where I spent all my free time up to six years of age, and much time during later school holidays. I was the chief designer and he was my carpenter and butler. We built small machines for everything needed in my life, such as a machine for playing with cats, for spanking my little brother, for automatically passing nails when hammering, and so forth. He was old already at that time and is no longer alive.

Kjell Andersson. Kjell is an engineer married to one of my father's sisters. Every time I met him at family reunions, I jumped up on his knee and discussed engineering until he got tired and almost apathetic. He was more formally educated than my grandfather, so I could discuss even some electronics with him.

Per-Olof Ewers. We met in the Värnamo Gymnasium School. He was an amateur radio operator and constructed his own colossal radio transmitters. I was impressed and started to construct colossal audio amplifiers. At that time all our constructions contained radio tubes, which were fatal for a fumbling teenager, but we obviously had guardian angels. Per-Olof now has his own successful company, helping customers with EMC (electromagnetic compatibility) problems.

Bertil Lindberg and other student friends. At the Chalmers University of Technology in Gothenburg, we were a group of students from the surroundings of the city of Värnamo, who among others constructed electronics. Bertil was always my helpful genius, and we attended many classes and took many labs together.

Ingvar Sandblom. This was a unique bohemian and autodidact from Skillingaryd, the small village where I grew up. He was a radio and television repairman and served customers from the whole Småland province. During my summer vacations, I practiced in his company, and we traveled around and repaired truck communication equipment. Later he became a close friend to me. Unfortunately, he smoked a lot and died of lung cancer.

Erik Dahlbergs Gymnasium colleagues. For a couple of years in the early eighties I was a teacher in electric power and control engineering, and met truly brainy people at this school. I developed an enjoyable intellectual relationship with the crazy and egocentric adventurer **Ludvig Bergström**, and I worked on a silicon custom design education with the ultradynamic and friendly **Anders Lindberg**.

Gunnel Sandahl. I met Gunnel during her education in program development at the University in Lund. Apart from sorting out a lot of technical problems together, we fell in love and married, and we now have a sweet family together with our sons **Sebastian** and **Jonathan**.

Axis Communication colleagues. I worked at Axis during their startup period in the late eighties, when there were only three development groups. The genius **Per Zander** worked in the same group as I, and was shockingly skilled in constructing computer boards for complex printers. The Axis manager, **Mikael Karlsson**, was talented and incredibly promising, but sadly got a fatal cancer some years after I had left Axis. The director for new products, **Martin Gren**, was a spontaneous genius, which is quite the reverse of my own inclination, and despite our problems in communication, we always aimed to respect each other. At Axis I also met the technician entrepreneur **Bernt Böhmer**, and together we collected wines of unreal quality, and we still meet for super dinners with special wines and technical discussions.

Q-Labs colleagues. A dynamic Norwegian **Geir Fagerhus** set up this consultant company during the economically harsh times in Sweden in the nineties. He succeeded in finding dozens of extraordinarily dynamic and skillful people, such as **Henrik Cosmo** and **Anders Gustavsson**, to mention two of them. Q-Labs collaborated with many universities and several international product-developing corporations, and used the Capability Maturity Model (CMM) of **Bill Curtis** at SEI, and the Cleanroom methodology by **Harlan Mills** in Florida.

Sony Ericsson colleagues. I have been in various positions at Ericsson and Sony Ericsson for more than a dozen years. I began as a mobile phone technical manager under the highly talented **Jan Svensson**, who afterwards had a long, successful career ending at the absolute top of the development departments of both Ericsson and Sony Ericsson. After some years, I shifted to operational development and together with **Mats Pettersson**, another talented manager, got the yearly award from the hands of the president, Miles Flint. A great deal of this success was attributable to **Jan Sjunnesson**, who was the knowledgeable consultant mentor during this period. During my operational development work I met many technical geniuses, such as **Even Andre Karlsson** and **Magnus Augustinsson,** to mention some. My ever present social mentor was **Per Göran Ohlsson**.

Cooperation with the academic world. At many of the companies where I have worked, there was a tight connection with the technical universities. I got to know professor **Claes Wohlin** during the Q-Labs era; I came in contact with professor **Boris Magnusson** during Java development at Sony Ericsson; and as operational developer in verification methodology I cooperated with professor **Per Runesson.**

Thord Sandahl. This is my brother, who has a degree in civil and environmental engineering from Chalmers University of Technology. He is a full-fledged entrepreneur, and is the successful managing director of our large family transportation corporation. He has been indispensable during our Hungary wine production adventure.

Bakó Ambrus. I met this microbiologist in Badacsony in Hungary while setting up our own winery, called Villa Sandahl. He established his own oenology consulting company to assist us, and we have developed some of the best white dry wines produced in Hungary. He is a living lexicon and has answers to almost anything one may want to know. He is now replaced with the worthy successor **Palkó Zsolt**. We also had invaluable help from wine producer **Fabien Stirn** in Alsace.

1.2 The author

Christer Sandahl. I was born and grew up in the Gnosjö area, the well-known region of entrepreneurship in Småland in southern Sweden. My family holds a large transportation company, which was founded by my grandfather, expanded by my father, and is now managed by my brother.

I have been a passionate engineer for over 40 years. Beginning with photo and chemistry labs in my early teenage years, to electrical engineering at the university, and computer and program design as a professional.

Several times in my early career, I constructed large computer systems all by myself, including mechanics, electronics, and programs. When computers grew larger and complex, for long periods I managed programming groups in successful local companies, as well as in large worldwide corporations, such as Axis Communications, Ericsson, and Sony Ericsson.

I also manage my brother's and my winery in Hungary, which nowadays is considered to be a quality breakthrough in Hungary's dry white wine production.

I have always been an engineer in the front lines of technology, and although I am very fond of management matters, I have not jumped into a formal career to get more and more people under me. Neither am I an academic longing for a formal scientific career, presenting an impressive list of references to show my formal education. This book is more a collection of unique and creative ideas from long experience in combination with curiosity about theories in technical and methodology fields. To conclude, I am an artist in technology, loving to express myself.

## CHAPTER 2

Reveal background

"There is a lot of noise in the jungle;
you must only be aware of the dangerous."

What went wrong?

2.1 What has gone wrong?

Product development is nowadays in many respects an established and ordinary business. For example, house and bridge development are several thousand years of age. Other fields of *product development* are much younger; for example, dealing with programming began in the 1960s.

However, in all development fields, large numbers of *products* are still being created which fail to satisfy *end users*. In some newer fields, such as programming, trouble looks to be the standard. For example, the editor I use to write this book has crashed over 200 times and caused me several weeks of extra work. And note, there is no conspiracy behind this—no *supplier* likes to disappoint an *end user*. So what is the problem?

Uncontrolled complexity is likely to emerge when a business is forced to scale up.

Digging deeper into the development topic in order to find the root cause of why *products* fail to satisfy *end users*, reveals the most common reason to be an exploding number of market opportunities and technical possibilities, which in turn causes competency and growth problems when organizing development of related *products*.

Two thousand years ago, the Pantheon building in Rome was the ultimate *complex* building construction at the very edge of development knowledge at that time, but it would nowadays be a rather modest target for a midsized construction company. A *complex* building of today is a skyscraper many hundred meters high, which needs high-tech solutions and sophisticated calculations for strength of construction materials, and hundreds of workers within many disciplines who are well organized to erect the building. *Complexity* of *products* has increased over time, but so has our ability to cope with greater and greater *complexity*.

To analyze *complexity* a bit further, imagine for a given time point the two diametrically opposed ends of prevailing *complexity*, the "ordinary low end," and the "extraordinary high end."

2.1.1 The ordinary low end

Houses have been constructed with success for a very long time. Handy persons (with some drive) can, for example, expand their private house with some new rooms. They may have to contact experts to sort out problems beyond their competency and hire specialists to help them build, but on the whole, this is not too *complex* for them to lead the design work and to also take part in the craftsmanship. The extended space will be of desired cost and *quality*, and will function mainly as planned. Whatever possibly fails can most often be repaired at modest cost later.

This scene is true for many ordinary-scaled businesses in our modern society (leaving aside for now the new *complexity* currently faced in the home construction business because of massive energy-saving *requirements*).

2.1.2 The extraordinary high end

There might be airplane crashes and medicines with severe side effects, but to travel by air or follow a doctor's prescription is generally very safe. In these cases, the high *complexity* of developing aircraft or medicine is undoubtedly handled with success. Obviously, in these fields, the mastering of *complexity* has worked out very well, even if not totally free from disasters. One can object that a lot of money is behind the mastery of that *complexity*, but nevertheless, the *complexity* is, in fact, handled with success.

2.1.3 Transition from low complexity to high

However, not all businesses and companies have managed to make a proper transition. Computers often crash and spoil large amounts of work, electronics fall into pieces and must be expensively repaired, cars barely keep together until warranties expire, and so forth.

Some companies maintain success, and others fail when forced into complexity.

In these unsuccessful cases there are, of course, a lot of extenuating circumstances, such as: Everything must be developed in a rush because the market changes quickly; testing is not given enough time and is passed on to the *end users*; money is spent on commercials rather than development, and so forth. And in the programming discipline, typically one after another page of *programs* is added to *programs*, and the scaled-up *complexity* creeps in unnoticed and all of a sudden has forever ruined the structure of the *product*.

Many unsuccessful companies might argue that it is not really their problem if they fail to deliver satisfying *complex products*. Who hasn't heard that "*customers* simply get what they pay for"? But most often an economic analysis would have shown that poor *products* cost more than they save for both producer and *customer*. *Products* in the poor end of the *quality* scale would, in fact, have been much more profitable if developed better right from the beginning (at least when considering the full *product life cycle*). Producers of indefensible *quality* risk being killed by a more *customer-oriented* competitor all the time.

2.1.4 Is there any catch when scaling up?

Scaling up is like 1 + 1 = 3. At some point more than the size has increased; a paradigm shift has occurred.

It has long been recognized that when things dramatically change in scale, concepts do not stay the same. It becomes more like 1 + 1 = 3. Sometimes this is referred to as "At some point when quantity increases, there is also a change in *quality*," or in our case, "At some point during the *product complexity* growth there must be a change in development approach." Such a dramatic change is sometimes referred to as a *paradigm shift*.

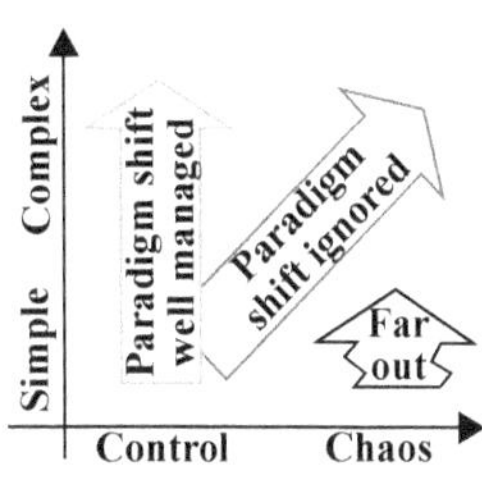

If the *paradigm shift* has occurred, the old methodology and approach must be replaced, and a radical new way of thinking must be applied. Small and concrete examples of *paradigm shift* are when too many stairways in a house must be replaced by one elevator; when growing *programs* must be *partitioned* into smaller pieces separated by clear *interfaces*; when slide rules are replaced with digital calculators; when keyhole surgery is far more efficient than big open wounds, and so forth. The world is full of (smaller and bigger) *paradigm shifts*.

It may sound a bit self-aggrandizing, but exchanging the traditional messy *product development* with the consistent and understandable *Cpdm* may be a *paradigm shift* for development companies.

2.1.5 Complexity dead end

Let's look at the solar system. When Copernicus placed the sun in the middle and the earth to orbit it, Isaac Newton was able to describe this system with his "laws of motion." This is rather ordinary mathematics, referred to as the "n-body problem," which can very well be analytically handled.

But *complexity* reappears in this example. It was rather simple to solve the n-body problem for n = 2; for instance, two planets like the sun and the earth being alone in the universe. It took several hundred years to solve it for n = 3, that is, three planets like the sun, the earth, and the moon being the only planets. For n greater than 3 the question is still not completely analytically solved, but the challenge has led to a lot of chaos research.

This is, in short, what happens with growing *complexity*. Very soon the system parts form a total system that might possibly be described, but gets hard to analyze and predict. Often such systems are referred to as systems in chaos. Of course, the system itself doesn't know that it is in chaos; it is our understanding that is not sufficient.

Cpdm definition of complexity: Something currently not sufficiently understood and thereby not predictable enough, in severe cases denoted as chaos. *Paradigm shifts* may accelerate understandability.

Cpdm definition of unmanageable complexity: Something whose further development is unpredictable, because is has turned into something that can no longer be sufficiently understood.

2.2 Mastering complexity

The general way to master *complexity* is to spend effort on investigation and research in order to adequately understand the *complexity*. If still too *complex* to be handled, some mitigation may be tried. One way may be to limit the degree of freedom and accept a lesser accuracy of understanding, for example, by approximations (the moon has no gravitational influence on the sun), or to freeze some relations (the sun can be assumed to be fixed in the center although the planets cause it to move a little).

Have you ever reflected on why houses preferably have right angles between most building *elements*? Do you get the point? Simply because this lowers the *complexity* and makes a house easier to understand, predict, and build. The new Beijing Bird's Nest, not having two similar angles anywhere, has such great *complexity* that it would have been impossible to handle in the era of slide rules, but could be mastered through design with powerful computers.

When developing *products* it is important to not end up in the *complexity* dead end, where the *products* get beyond salvation. Time and money are spent, and only bad market response and uncertain profitability can be expected.

The discussion so far boils down to how to stay away from the *complexity* dead ends. *Complexity* shows in many different ways, which need some different cures to be mastered. Below, some examples are given of *complexity* areas, with explanations of how *Cpdm* gets away from the *product development* dead ends.

There are certainly many areas, apart from *product development*, that contain *complexity* dead ends, such as *line management* and *project management*. There are also many *activities* trying to minimize *complexity*, such as *quality* assurance and *process* improvement. This book focuses mostly on technical foundations, but I will extend *Cpdm* with a second book on technical overhead.

2.2.1 EXAMPLE: Artificial complexity being promoted

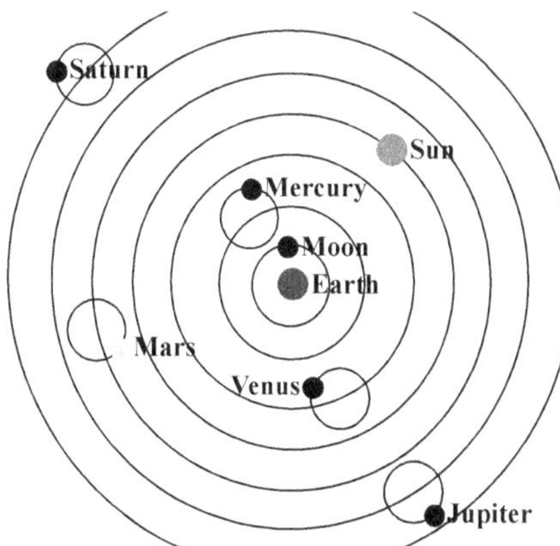

One of the most illustrative examples of this, even if far from *product development*, is the geocentric solar system model, with the earth in the center (see Fig. 2-1, left.)

This medieval model shows the movement of planets around the sun, assuming that they are fixed at spheres that rotate at different speeds.

FIGURE 2-1 The geocentric planet model

Some observations are really difficult to explain with this model. For example, it is not seldom seen that a planet moving across the sky suddenly reverses its direction and regresses for some time, before reversing once again and proceeding in the original direction. (This effect is caused by the fact that the earth and the planets move relative to each other.) The more irregularities of planet movement were observed, the more epicycle spheres had to be put into the geocentric solar system model.

The *complexity* by epicycle spheres was, of course, added by humans, and consequently most of it disappeared when Copernicus released his model, with the sun in the center and each planet following an oval orbit instead of a jumble of circles. Everybody knows that an explanation becomes so much simpler when the problem behind it is enough understood, and *artificial complexity* is removed.

In *product development* it is also easy to end up with too many "spheres" and everything gets more and more impossible to track. Not every model builder is as brilliant as Copernicus, but the warning bell should be ringing when *products* (or *processes*) gradually get too *complex* to be understood. It might be the consequence of a model that has evolved beyond its initial simple context; a *paradigm shift* must then upgrade the model to a new level which is more understandable.

Things should be made as simple as possible—but not simpler.

Albert Einstein

Cpdm is constructed in a fresh and modern way to explain efficient *product development* with a minimum of confusing spheres.

2.2.2 EXAMPLE: Intrinsic complexity being ignored

This is the opposite of the preceding example and might ring the warning bell as well. Modern *products* are really *complex*, and are not properly developed because of the rush to get them out on the market.

A standard example of this is construction of a modern airport. An airport might be seen as a rather simple system, only guiding people from the entrance to their aircraft. A similar system is sorting letters at a post office; letters arrive and should be channeled to a vehicle for transport. Such systems initially appear rather simple, but airports and terminals often go into completion delays and massive cost explosions, and seem to never disappear from news channels.

When beginning a development, understanding and describing the *target product* can at first look easy, and the organization is fooled into believing that the *product* has modest *complexity*. If that *product* continues to be developed and *complexity* grows, without methodology being properly scaled up to the necessary degree, understanding of the *product* will gradually get out of hand.

Again, things should be made as simple as possible—but not simpler.

Albert Einstein

Developing a *product* without losing control requires that everything is so carefully described that everybody can see it is correct. This is called to achieve *intellectual control* over the *product* to be developed. In practice, it is safer to begin with too much *intellectual control*, and to loosen up a bit if *intrinsic complexity* is found to be modest. The reverse is almost impossible, to increase the *intellectual control* when understanding has got lost.

Cpdm demands some effort to be understood, because no important aspects are omitted. If it had ignored too many important details of *product development*, *Cpdm* might have been easy to read, but useless to apply in reality. If *Cpdm* is found to be overkill for the *intrinsic complexity* of the *product* to be developed, it is easy to *tailor Cpdm* for a lesser *product complexity*.

2.2.3 EXAMPLE: Impossible principles applied

Such a competitive edge it would be for a company—succeeding in "withdrawing gravitation." But what is the chance?

A striking example of applying impossible principles is the design of the Swedish warship Wasa. Sweden was at war with Poland, and needed better firepower in their navy. Thus, there was heavy pressure from the Swedish king to equip the ships with as many cannons as possible. Consequently, the ship Wasa was equipped with too many cannons for its size and ballast, but despite this the king ordered the launch of the ship. After ten minutes in fine weather, the ship capsized and sank in 1628.

An example from a well-known development company is the start of an improvement *project* with upper management announcing "the principle of decoupling," meaning that different parts of the *product* should be developed separately from each other and handed over to smaller organizations for *integration* and assembly. Certainly it was a great idea, with the "only annoying obstacle" being that the architecture of the current *product* had been designed to be far from decoupled, and no internal *interfaces* were described, visualized, or managed. Few *managers* understood that if the *product* wasn't *partitioned* in parts, it was useless to *partition* the organization. A lot of time, energy, and money were thrown away to divide the company into a decoupled organization, but without also decoupling the technical parts.

The great advantage of *Cpdm* is that well-tried parts are integrated in a consistent whole, which fits seamlessly together. To even better illustrate that *Cpdm* is something quite different from the impossible, a lot of detailed real examples show how *Cpdm* takes care of every aspect of the *product development*. And even better, *Cpdm* is so good and so well explained in the book, that it allows experienced engineers to take *intellectual control* over it (to realize that *Cpdm* is correct by just studying it).

2.2.4 EXAMPLE: Watertight connection between mechanics, electronics, and programs

Complex products are often built from mechanics, electronics, and *programs*, which must be tightly integrated with each other. If these disciplines are isolated from each other, maybe also isolated from different parts of the organization, with special cultures and attitudes, *product integration* will, of course, suffer.

Since *programs* are immaterial and a rather new discipline, it has created attitudes that are very different from electronics and mechanics. It is much easier to hide poorly constructed *programs* than, for example, to hide an ugly wall in a house. The development community has still not transferred good habits from mechanics to *programs*, for instance, by making good *architecture* drawings, perform reviews, and work in teams.

One often hears that contrary to mechanics and electronics, the nature of *programs* is a huge flat set of *instructions*, not having any structure to illustrate it. The *program* examples of CHAPTER 9 "design architectures," page 333 and CHAPTER 10 "finalize design & requisites," page 435 of this book, show how wrong this is. The structures of mechanics, electronics, and *programs* are very similar and should be drawn together in the same *architecture* to tie together the disciplines, and to lower development *complexity*.

2.2.5 EXAMPLE: Product structure being degenerated

Product structure and *architecture* are often the most misunderstood of all development requisites, particularly for *programs*. It is very strange, because for the house construction business the *architect* is both important and well understood. In electronics development *structure* and *architecture* are fairly well understood, because printed boards of *components* are tangible and can be likened to rooms in a house.

But when it comes to *programs*, it might be totally impossible to draw analogies to rooms, apartments, floors, and so forth. A *program* construction that has been uncontrollably extended may have an *architecture* very similar to an extended summer cottage. Small rooms have been added to the house-body every summer but the body itself has never been redesigned. This results in a cottage with a large number of small rooms, nooks, and corners, but nowhere any continuous space for living. Even if more expensive in the beginning, the only long-term solution is frequent redesign and restructuring of the *product architecture*.

A program structure might be as degenerated as an ever extended summer cottage— a lot of cubicles, nooks, and corners, but nowhere space for living.

To prevent general degeneration, a diversity of illustration types is supported by *Cpdm* in order to let everybody take part in development, not the least *managers*. To maintain a sound *product structure*, *requirements* and *architectures* are preferably shown in full graphics, and all development documentations are *modularized* to be illustrated by tables.

2.2.6 EXAMPLE: Products hijacked by engineers

Many *product* markets (even for very technical *products*) are not much different from the fashion market. It is the price tag and the appearance (or behavior) of the *product* that are most important, and the rest of the *product* characteristics must only reach above a "hygiene standard."

Engineer-driven development is often said to be worthless. But to ensure that it does not happen is difficult.

If a company is *governed* by mostly technical people, the reverse might occur. Then the invisible technical systems within a *product* case become most important and heavily improved, but the user-friendliness is kept as boring as ever. User-perceived *quality* may slip because feedback from the market is ignored. A high return rate and warranty cost are often the result

A company may have a lot of market-driven people and *managers*, but despite this might still be engineering driven, because a strong channel might be missing to convey undistorted market-driven *requirements* into the center of the engineering departments. *Cpdm* provides ample *requirement management* possibilities, which can be table based or even graphic, and easy to review also for nontechnicians.

2.3 How can I help you?

Back out from the mess, and let Cpdm from this book be your new platform, so as to avoid ending up in the complexity dead end again.

What can clearly be seen from this chapter is that development might be very messy, and many of you are in the middle of the mess. If so, you have to back out of the mess, and get a chance to recover. I have been in the same mess, and have also seen a lot of failures. Don't copy them all; let me instead help you out, and save your time from a further mess.

Start learning *Cpdm* and persuade your surroundings to use it. It is as simple as that!

## CHAPTER 3

Explain book

Cpdm is straightforward but difficult for a book, suggesting the need for a chapter to explain itself.

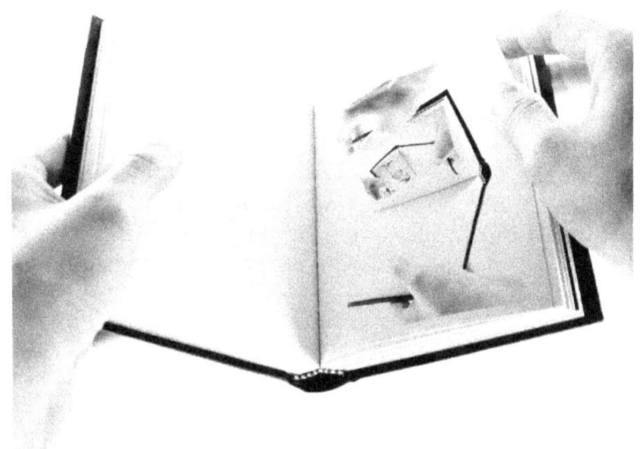

Organizing a complex topic

3.1 Why another book about this?

There are many books presenting overviews or fragments from *product development* models. Some can be good on *requirement management* (RM) and others on *architecture design* (AD). But rarely are there descriptions of how *requirements* relate to *architectures*, and even more seldom how the hierarchy of *architecture* decomposition relates to *requirement refinements* and how hierarchical *process schedules* are constructed, and how all these hierarchies relate to each other.

How many books explain requirements, architecture, and process hierarchies, as well as the relationship between them?

How many books explain tight integration of technology and processes with high quality?

Process descriptions have similar shortcomings. There are good books on *process management*, but they rarely connect to the technique being developed. It is even worse with *quality management*, which too often is seen as a set of very separate *activities*, when in fact, these must be deeply integrated with the development *process*, which in turn must be deeply integrated with technique.

My interest has always been to understand connections between theory and practice, and connections between the whole and its details. I like cross-bordering, which is to tie together and relate matters that too often are seen as independent from each other. Since few books are really holistic down to essential details, I have the ambition, and hopefully the ability, to achieve something special with this book. I admit that the examples in this book have forced me to take care of every detail, and thereby make *Cpdm* holistic to keep these details together.

This book is a compilation of experience from my whole life of *product development*, going back to my own laboratories in my early teenage years. Furthermore, I have the necessary education for building models of *complex products*, and I have worked in small and midsize companies as well as in international corporations.

How many books are a merge between advanced methodology theories, and a lifetime of experience in development with all kinds of companies?

The most fundamental concepts are defined in a formal way in this book, to help establish a solid superstructure. Many of the definitions are similar to those commonly accepted, but some may differ significantly from what is usually found, because all concepts in this book must fit together.

Despite the fact that I will explain everything very carefully, do not expect this to be an easily approachable book. It will delve deeply and broadly into *complexity*, which by definition is difficult. I have tried to simplify as much as possible, but not to the extent that the model gets diluted and misses the point in order to cure the *complexity*. You have to put in some energy when reading this book, but I promise as exciting a journey as I have had when writing it.

With this book in your hands, you will find it much easier to dig into *complexity*, because the technical postulates now are elaborated and explained. For example, before explaining *project management*, the technical *activities* and *results* must first be visualized and clarified. The same goes for *configuration management* and *quality management*. Consequently, I will follow up the fundamentals of this book with their superstructures in a future book addressing technical overhead.

How many books explain the foundation of product development, which perfectly supports a discussion of development overhead?

3.1.1 What is in this book . . .

This book focuses on explaining the foundation of product development.

The attempt was initially to document most experiences in *product development* in one book. However, after 600 pages just to define a skeleton, most of which was superstructure, I had to rethink and divide the book into several volumes. This is the first volume, covering the *complex* technical foundation, presented as basically as possible.

In my career, I have often tried to explain *complex* concepts from the development world, such as *requirements*, *architectures*, and *quality*, but always without concrete examples to point to. It is surely tedious to argue endlessly without being able to anchor in technical fundamentals, and this is why this book was completed to describe the basis of *product development*.

Even if this book explains a lot of foundations, there is still plenty of development overhead to explain.

This book contains the development foundations of mechanics, electronics, and programs.

Furthermore, different technologies have a lot to learn from each other. *Architecture* is rather easy to explain when considering mechanics such as house construction, but often impossible to apply to *programs*. In this book, the intent is to show that mechanics, electronics, and *programs* can be developed in almost the same way.

This book covers real experience and longtime active learning. It contains an almost endless number of examples from various relevant development cases. But it doesn't stop even there—it also puts all these examples into a uniform concept. And that is, in fact, what you need, but seldom get from other books.

This book contains countless detailed examples from all technology areas.

It might be said about me that I am a nerd born out of the box. True or not, much of what is in this book is very personalized, for example, usage of graphic symbols. I have chosen to invent a lot of them for this book, even if there were decent existing candidates, because I simply don't want to compromise clarity and accuracy. Already accepted concepts are often loaded with obscurity; they have many meanings and different interpretations, or they don't fit together. However, when these are explained and understood by the readers of this book, they can easily adapt these symbols to the graphical editors they normally use.

3.1.2 . . . and what is not

This book is not an enumeration of fragments, not a simplified textbook, and not a book about smart products.

This book is not about how to design the funkiest house, how to construct the most intriguing hard-wired gate circuits, or how to make a popular iPhone application. A lot of technology is used to provide examples in this book, but only for the purpose of illustrating how to develop *products* and master *complexity*. There are possibly some technical mistakes in some places in this book, but don't scrutinize them for their own sakes. If the example effectively illustrates *complex* development, please just be satisfied with that.

Furthermore, this is not an academic book that simply enumerates all plausible facts of *complex* development. Lots of books present small corners of the world that are consistent and harmonious within themselves, but have no real connection to each other. I remember that the technology university offered a lot of classes on different technologies without teaching the connection between them. Even courses on mathematics were entirely disconnected from applied topics using mathematics.

Overhead topics not included in this basic book, but possible candidates to be seamlessly integrated in the technical overhead in the next book:

Services. This book does not cover development of services, such as events, guidance, consultancy, and so forth that sometimes may be interpreted and treated as *products*.

Configuration management. This is a book about developing a *product* from scratch, and does not cover extensions or changes of existing *products*, which is a very common way to realize feature growth or cost reduction. *Configuration management* is the solution for this, and in itself a big topic for a coming book to cover.

Subsystems with various development times. At *integration* time of a *product prototype*, all subsystems must be ready to integrate. Many *complex products* contain mechanics, electronics, and *program* subsystems that have very different development times. Thus, all subsystems will have their own unique start times in order to converge in the same *integration* time, causing a challenge for holistic *requirement management*.

Incremental development. When developing a *product* by gradually extending it, the growth of functionality can be arranged in many ways. One beneficial order is to plan incremental deliveries in such a way that all deliveries become useful (or at least possible to evaluate) for the *customer*. This results in frequent strong feedback from the *customer*.

Quality aspects. *Quality* is often seen as an aspect of the work that can be handled by specialists in a separate organization. Nothing could be more wrong, because *quality* is the result of how well the *value chain* is organized, and this is a responsibility for the *line management* of the development company. However, *process* and *quality management* can be a very good support for *line management* when improving the *value chain*.

Management and change management. For most overall topics above, there is also unavoidable overhead, such as *line management* and *project management*, which must be organized through appropriate *process* support. The intention is to also cover this in a future book.

3.1.3 Is it boring to be formal?

This book may be accused of being too formal and too expensive to follow. It will be claimed that no developers can, want to, or like to be so careful with *formalism*, with documented identifiers and references and *processes*.

What *formalism* boils down to is that developers must do their best to get *intellectual control* over the *product* being developed. *Intellectual control* means that the *product* is so well described that involved persons understand just by looking at the documentation how the *product* will work and that the *product* contains no failures.

Intellectual control is a bit challenging to obtain, but is far better than *verification*, even if the *verification* is done extensively. The nature of *verification* is nothing more than spot-checking, relying on a finite set of *test-cases*. Regardless of how many *test-cases* are prepared, some will always be lacking, which in turn prevents *failure detections* of some failures. Even worse is trial-and-error methodology, where the developer tries by guessing, and is adequately satisfied the moment a fix is obtained. By this method, the developer has not learned anything about the nature of the problem, and has no idea if the remedy has a broader application.

With a good understanding of the essence of this book, the development model can be changed and tailored to fit any development company.

It is very important to stress that a company doesn't need to arrange and implement everything in the way this book determines (see ch. 6.3.12, "Tailoring a process schedule," p. 94). Development cultures are different just as *products* are, and there is no standard recipe.

It should be respectfully stated that *Cpdm* might constitute overkill for many companies that develop smaller *products*. Keep in mind the option of easily cutting away pieces of *Cpdm* that obviously are unnecessary while keeping the rest of the model intact, thus simplifying development, without the risk of throwing out the baby with the bathwater.

However, to be able to *tailor Cpdm* into something concrete, or to introduce *Cpdm* stepwise without creating a mess, there must be enough skill and competence to understand a book such as this. It means that not only some odd engineers have to understand the content of this book, but rather the opposite, that this book must be more or less possible to grasp for the whole development organization, including its technical *managers*.

3.1.4 How to illustrate and teach complexity?

Writing a book such as this is to prepare a huge minefield. For example, when trying to teach and explain general and overall principles, people will call the book academic and theoretical. On the contrary, when going into details, the other half of the people will accuse this book of spending too much time on nonessentials. Catch 22 is the recurrent feeling.

Different traps in mastering a *complex product* have already been identified in previous chapters. In addition, *product development* might create shelves of documentation binders, and it is simply impossible to squeeze all that into one book. To cover *complexity*, this book selects different smaller examples that are orthogonal to each other, thus together defining a huge volume of combinations. One example can be a *product* with a lot of details, and another example can be a small *product* of few but very intricate details. These two examples together then explain how to develop a *product* of a huge number of very intricate details.

A house example presents a huge number of concrete details, and a programs example presents abstract details in smaller numbers. Combining these examples will extend understanding of a huge abstract product.

There are a lot of educative and elaborately explained details in this book, such as unique identity numbers for everything, complicated graphic symbols with different flaps in the corners (see Fig. 6-5, Flaps and corners of symbols, p. 95), and *process schedules* to cover all possible imaginary situations. In actual work this does not need to be applied to any degree. For example, continue to use your normal text and graphic editor, but first understand *Cpdm*, because that gives you the key to use your tools in a smart way.

The unity of overall principles and technology with particular details must be taught simultaneously with addressing complexity.

Since this book is intended to be exhaustive, the simple resort is to tandem overall principles and nitty-gritty details, and most important, the connection between them. To be reliable and convincing, this book is also broad in technology, covering mechanics, electronics, and *programs*, with most of their inherent technical specialties. To realize this objective, a lot of examples are covered that are meant to exemplify principles as well as details.

It is true that this book contains a lot of detailed examples, many more than necessary to simply explain the *Cpdm* theory. But experience shows that there are many obstacles when trying to use a new method and approach. The idea is that examples rich in details will lower the comfort barriers and invite developers to start changing immediately. Many good methods are never used or are misused, because developers on the floor don't get enough help with their particular details to fit into theories of excellence.

Please read this book with a multitasking approach. Concurrently try to keep an eye on important big lines, and on the details building up the big lines. Also be open and unbiased about digging into immaterial concepts such as *value chains, processes,* and *programs.* Just as the immaterial text of this book has a structure of chapters, paragraphs, and pictures, *Cpdm* will show that things like *programs* and *processes* have their structure as well, which can also be illustrated and managed successfully.

3.1.5 Who needs this book?

Now to an essential question—who needs this book? For me the answer is very simple: Everybody who has anything to do with *complex product* development should study *Cpdm*. Targets in particular for this book are:

Managers. *Managers* who don't understand how their subordinates develop *complex products*, risk getting lost in their everyday work. They plan awkwardly, they promote the wrong people, they simply create chaos. Often *managers* are recruited engineers and have a good understanding of technology, but this is only partially useful when it comes to managing *complex* development. I strongly recommend that *managers* study the connection between technique and *processes* in order to manage *complex* development.

Students. This target group is often the most unbiased and ready to learn. Apart from technical topics, they have a positive opinion also of organizing and conducting development. This book should be an outstanding background book for technology students of many disciplines, showing how to tie together methodology and technology.

Engineers. Oddly enough, this is often the most reluctant target. Many engineers get offended because they are supposed to know everything already. It is much more convenient to claim that this book is based on faulty understanding, or that it is valid but irrelevant for them. If *managers* are prepared to buy some ideas from this book, they are in a much better position to convert and improve their engineers, beginning with the curious and open-minded. A good trick is to involve *line management* when teaching engineers.

3.2 Organization of this book

The *Complex product development model (Cpdm)* presented in this book is illustrated by Figure 3-1 below (with *Cpdm* concepts and own symbols). In the figure below, each chapter is referenced by the same symbols used throughout this book.

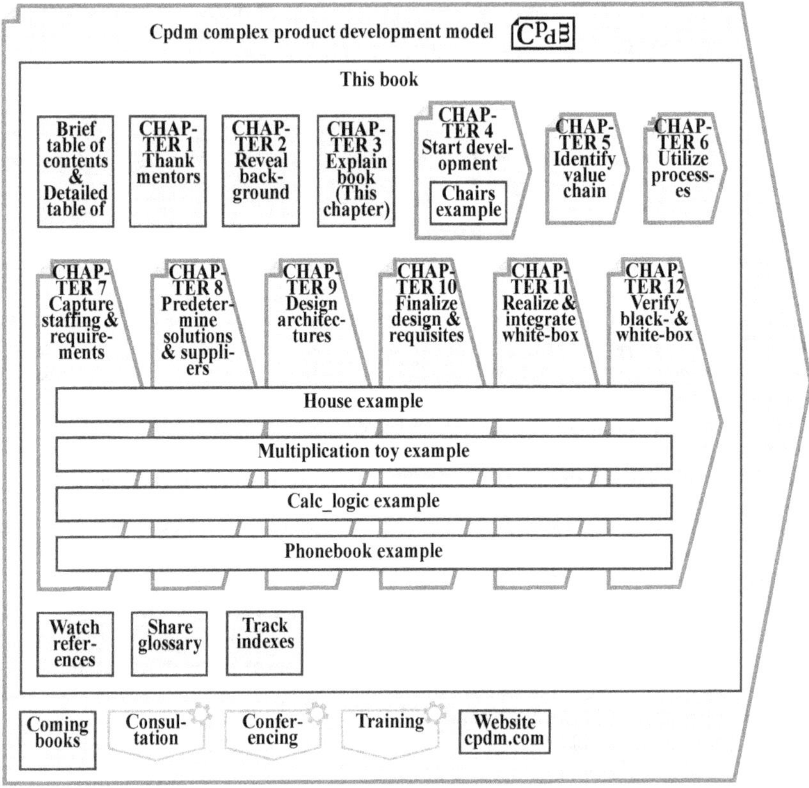

FIGURE 3-1 Cpdm explained by this book

3.2.1 The vast examples

Since this book intends to master all degrees of *complexity* of any kind of *product* technology, the development *process* unfortunately becomes a bit *complex* (but as simple as possible), which might be a challenge for new readers to follow.

To explain complex development principles, nothing is better than complex real examples.

To illustrate the way things work, an abundance of examples is the best way. Even if the examples may be somewhat strained and messy, the purpose of each example is to appear self-explanatory without too much lecturing. To gradually build up a theoretical background for the examples, all examples are preceded by general explanations.

The examples are, of course, not fully developed to the extent of real *complex products*. A *complex* building, for example, needs stacks of binders and terabytes of disk space to be adequately documented. Although many examples in this book might be perceived to contain more details than necessary, the level of details is not even close to the detail level in a real *complex product.*

3.2.2 EXAMPLE Chair: What is the purpose?

Everybody knows what a chair is, and many people have constructed something to sit on during their lifetime. The purpose of the chair example is:

- To establish an easy-to-understand example, which can be expanded to its full *complexity* in future chapters
- To introduce a broader view, from *product* idea to *product* launch, with succeeding examples focusing on the strictly technical parts of development
- To lower the risk of losing readers who are interested in this topic, but lack experience, and who have some initial gaps to fill

The example starts in chapter 4.2 on page 40.

3.2.3 Follow either the chapter topic or the process schedule

There are different ways to read the last 6 chapters in this book (see Fig. 3-2 below).

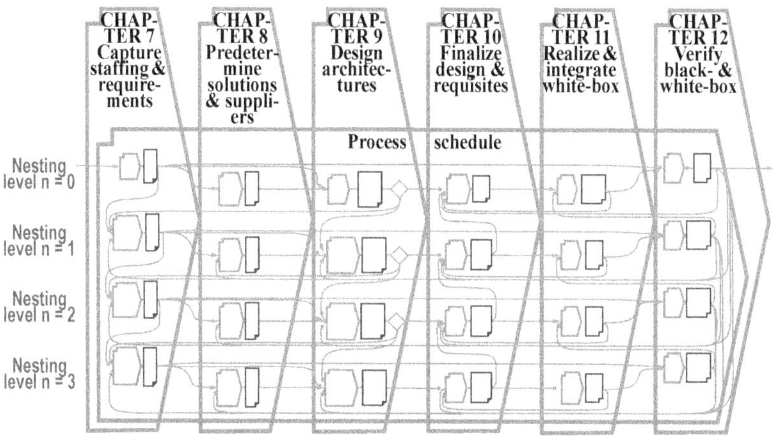

FIGURE 3-2 Follow chapters or follow example

To follow the *process schedule* for each example is the most rewarding way to read this book, but also the most demanding. The *process schedule governing* development is a big network of *tasks* and *masters*, which leads to many forks in the road, resulting in many options for how to read the book to follow an example. But this book can also be read without following an example *process schedule* at all. Thus, the book can be read in the following main ways:

1. Reading sequentially, from beginning through end, passing each example many times
2. Following a particular example along the *process schedule* on a similar *nesting level* (where possible)
3. Following a particular example along the *process schedule* by moving one *nesting level inwards* or *outwards* (where possible)

Wherever you are in these four examples and six chapters, your way of reading can alternate freely between the above alternatives.

3.2.4 EXAMPLE House: What is the purpose?

Most people, and engineers in particular, are adequately familiar with house design and how to build a house. Because of that, it is a good example to start showing real *complexity*.

- When planning to build a house, maybe writing *requirements* is not the first thing many people consider. But since using a house is easy to understand for almost everybody, it is also good for illustrating how to capture staffing & requirements.
- The most obvious in a house development is the *architecture* and layout. Everybody understands how important this is, and the *Cpdm* concept of organizing hierarchical *architectures* will be illustrated.
- A house is also a good example to clearly show how *Cpdm* keeps *requirements* and *architectures* interconnected.
- A house contains an awful lot of building *elements*, and illustrates how *Cpdm* keeps track of them by being formal enough.
- It might be a bit awkward to apply a *product prototype* concept for a house, since most houses are built as single *product prototypes*, without any intent of series manufacturing.
- Developing an ordinary house requires more documentation than what fits into this book. This example will be far from complete, but nonetheless shows the *Cpdm* principles.
- A house is a special *product* since it is so large that humans can enter into it and use it from the inside. This doesn't mean that *black-boxes* have been opened. One can expect behavior from inside a house without opening its *black-box*.
- A house is a good example to illustrate *boundary interfaces*. For example, the outer wall is a *boundary interface* of the house's *black-box*, and the same wall seen from inside belongs to a room *boundary interface*.

The example starts in chapter 7.4 on page 180.

3.2.5 EXAMPLE Multiplication toy: What is the purpose?

It is now time to extend the *product development* with an electronics example:

- This example shows that the same *Cpdm process* can be used for electronics and *programs* that was previously used for the house (but with 2 *nesting levels* instead of 4).

- This example is divided into two *realizations*, one with hard-wired gates and one with a small microcontroller with a *program*. This is to illustrate that *Cpdm* supports development of exchangeable *elements* (see Figure 3-3 below).

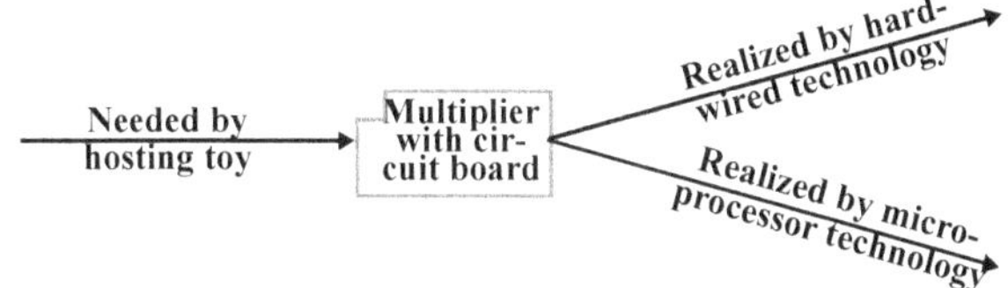

FIGURE 3-3 Development splits up in two product variants

- The house was made as a single *product prototype*, but this example is intended for series manufacturing, which creates a lot of attention to manufacturing requisites for different manufacturing series, which is also easily supported by *Cpdm*.

The example starts in chapter 7.14 on page 211.

3.2.6 EXAMPLE Calc_logic for reuse: What is the purpose?

Now it is time to move over to structured *programs*. The *product* to be developed containing the math logic for a calculator:

- What might seem simple in the beginning can be really nasty if made with pre-served *intellectual control*, because a calculator must be trusted to never fail.

- The calc_logic *element* should be *reused* by two different *realizations*, one stand-alone calculator with small microprocessor, and one graphic Windows calculator (see Figure 3-4 below).

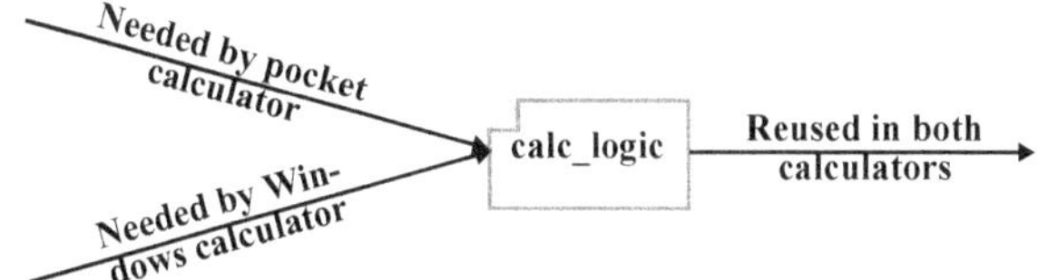

FIGURE 3-4 Development of product to fit in two environments

- Behavior of the calculator logic is specified with *use-case requirements* of different kinds, which is a strong methodology in *Cpdm*.

- The implementation of the *program* is keeping the structure of the *use-case wait-states* from the *requirements*, showing nice *traceability* between *requirements* and *architectures*.

- *Verification* is done by a powerful statistical method, based on expected usage of the calculator. From a probability distribution, *test-cases* are machine-generated and automatically executed.

- Again, the same *Cpdm process schedule* is used, only with minor modifications to suit structured *programs*.

The example starts in chapter 7.21 on page 237.

3.2.7 EXAMPLE Phonebook: What is the purpose?

The last example illustrates a *product development* of an *object-oriented program*. A simple phonebook *product prototype* will be founded for future large-scale expansion.

- The structure of the *program* is data-oriented, which must be future-proof, designed to be extended to a full phonebook implementation.

- The *requirement's* formal *use-cases* are rich in *exception use-cases*, which is often more difficult than *normal use-cases*.

- The *program* is extremely *modularized*, to be distributed in the future to a large organization with a lot of engineers. There is an illustrated time follow-up chart that can also be used as a responsibility chart for almost all people developing the phonebook.

- The generic *Cpdm process schedule* fits this development example very well also.

- The concept of executing many *test-cases* of each behavior is shown. One *test-case* may execute the clear *normal use-cases*, and one *test-case* can execute the clear exception. Two more *test-cases* are forced as close as possible to the boundary between the exceptions and normal, one of them on the normal side and one of them on the exception side.

The example starts in chapter 7.27 on page 261.

Start development

"A figure says more than a thousand words — examples from reality say even more."

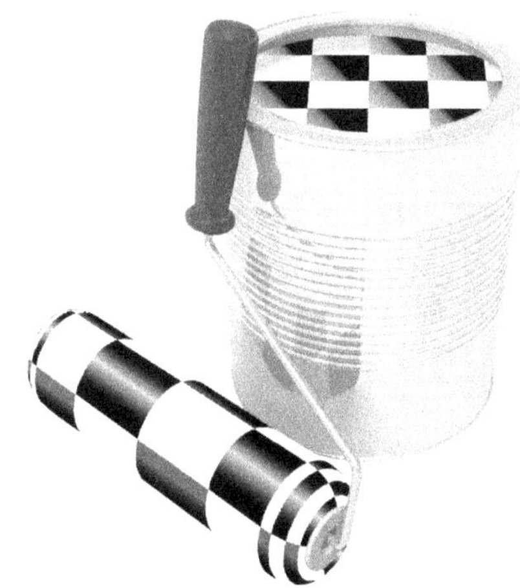

Begin with something obvious.

4.1 About start development

This chapter is meant to establish the basis and context for *product development*, and is quite simple for experienced *product* developers to understand. The *complexity* in this overview is not yet significant enough to present an obstacle to those less experienced, but the chapter serves several purposes:

- To establish an easy-to-understand example, which can be expanded to its full *complexity* in chapters to come
- To introduce a broader view, from *product* idea to *product* launch, with succeeding examples focusing on the strictly technical parts of development
- To lower the risk of losing readers who are interested in this topic, but lack experience, and who have some initial gaps to fill

The more *complex* the *product*, the greater the need to formalize its development. Subsequent chapters will elaborate on that.

4.1.1 Cpdm definition

In *Cpdm*, *product* is defined according to Table 4-1 below.

TABLE 4-1 Cpdm definition

Aspect	Product
Definition	An artifact created by somebody, from raw materials to finished goods for a market, to satisfy a need
Synonyms	Manufacture, creation, goods, artifact, produce (verb sometimes wrongly used as noun e.g. "produce of Sweden").
Symbol	Product

4.1.2 EXAMPLE Chairs: Development

A series of examples in the following chapters elaborates on the crazy idea of taking up competition with the Swedish home furnishing company IKEA. The example briefly explains developing in order to produce and sell some ordinary modern chairs, just to exemplify the basics of *product development*.

An example of developing chairs is shown throughout this chapter.

4.1.3 Definition of product life cycle

Only the development stage of the product life cycle is the main topic for this book.

The *product life cycle* encompasses the different *stages* of a *product's* life, from when a *product* demand is first identified, up to when the *product* is phased out of the market and warranties are cleared.

Revenue from the *product life cycle* is often illustrated as a curve as in Figure 4-1 below. Note that the topic for this book is primarily the first *stage*, the *product development stage*.

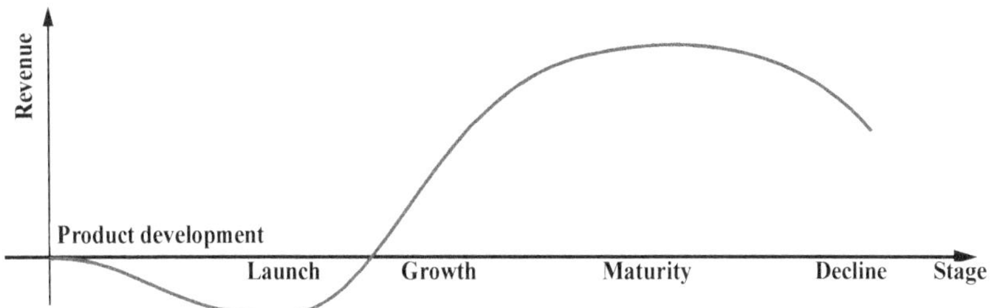

FIGURE 4-1 The stages in a product life cycle

How to administer the *product life cycle* is often called *product management* (PLM).

4.1.4 About product development

Product development includes the *stages* that occur from the time a new *product* is first being identified to the launch of that particular *product* on the market.

Product development starts with a product idea, and ends with the market launch for that product.

The *product development stage* is in turn divided into several *phases* (see the chairs example presented below). Note that in the *product development stage* there are as yet no *products* at hand, apart from *product prototypes* and *samples* for the *product prototype*. Most of the development work is documentation and descriptions of various kinds.

Many companies focus on the engineering work, and forget market aspects. Other companies are too business-oriented and have poor engineering capability. But on the whole, marketing and engineering are equally important and must be tightly integrated.

The *product development* might be driven as a separate *project*, in the *line* organization or in any other controlled form. This book doesn't concern management and organization, but a sequel will take up such topics.

4.1.5 Cpdm definition

The most important *realization* from development is the product prototype. In *Cpdm*, *product prototype* is defined according to Table 4-2 below.

TABLE 4-2 Cpdm definition

Aspect	Product prototype
Definition	Realizations made during product development, for the purpose of being verified against its requirements, and to be validated against its user profiles
Synonyms	Example, model, mockup, test item
Symbol	Product prototype

Product prototypes are most often manual *realizations* of *products* in small numbers, according to developed *requirements* of different kinds. The purposes of product prototypes are to ensure that they conform to *stakeholder* expectations (*verification*), and that the *product prototype* corresponds to market demand (validation).

Prototypes are built according to specified requirements and design, to ensure technical correctness and market capability.

4.2 Study strategy

4.2.1 About study strategy

Developing *products* and getting them out on the market may take several years. Because of this, it is very important to always perform regular strategic planning that includes at least the duration of development time plus launch time. Waiting to start development until the competitors have their offerings available on the market, is equal to being years behind the competitors.

A road map identifies requirements a couple of years ahead, to get time to adapt development organization to technology shifts..

A *road map* must be both business and technology driven. Business driven implies that the demand from the market is estimated for some years ahead. Technology driven implies that technical development opportunities are estimated that the market isn't yet aware of, but it may be quite possible to create this demand when the technical solution is presented.

Because the *road map* is an amalgam of market demand and technology opportunities, it is important to always have the *road map* documented in such a way that both engineers and market people can understand and approve it.

When starting the chair development, it is also good to ensure that necessary overall developers and marketers will be available, and can already be booked for the overall development.

4.2.2 EXAMPLE Chairs: Study strategy road map

It is not expected that fashion chairs will stay attractive for more than a few years. Style and textiles will change and new technical opportunities will see the light of day. However, to lower the risk of too short *product life cycles*, the *product portfolio* will be mixed with some durable chair models.

The manufacturing capacity at hand is estimated to be slightly more than three chairs per working day, say at most 750 chairs per year. This is also what is reasonable to sell with the current staffing. In total, during four years on the market, this will result in a maximum of 3 000 chairs. The sale price is estimated between SEK 250 and SEK 450 per chair. This implies a total income of around SEK 1 000 000 during the *product life cycle*. Half of this is assumed to be material cost, which makes around 500 000 SEK.

The concept of swivel chairs is expected to be successful on the market for many years to come. For the company developing these chairs, the technology fits well with existing competency and a retailer chain is interested in selling such chairs.

TABLE 4-3 Chairs road map

Field	Second year, first half	Second year, second half	Third year
Pa.1.Castors	Plain bearing	Ball bearing, cost kept	
Pa.2.Textile printing	No printing	By designer	By famous designer
Pa.3.Upholstery	Textile or unupholstered	Durable awning	Leather
Pa.4.Seat	Fixed	Fixed	Adjustable
Pa.5.Backrest	Fixed	Adjustable	Massage

4.3 Study concepts

4.3.1 About study concepts

During the **Study concepts** *phase*, the *road map* is examined to identify *products* interesting to be developed. Planned *products* are examined and sketched from different angles, and alternative approaches for the *product realization* are evaluated. The purpose is to identify and judge early how to realize the strategies of the *road map*, what technology is the best for proposed *products*, and estimate how the *product* will fare on the market.

Based on the road maps, concepts and technology, being candidates for near-term development, get studied in more detail.

New standards are addressed all the time by *road maps* of different areas, and these in turn affect new concepts to be applied by coming *products*. If a company wants to take the lead in new technology, some developers might be sent to standardization forums. Or if some new bright concepts are invented in-house, it may be advantageous to apply for patents.

In the study concept *phase* it is also important to judge which *components* to "make" by developing them in-house with own manufacturing capacity and what *components* might be more feasible to "buy" from *suppliers*. Some ordinary *components* might even be standard *components*, that can be bought off the shelf and don't need to be developed at all.

4.3.2 EXAMPLE Chairs: Study concepts

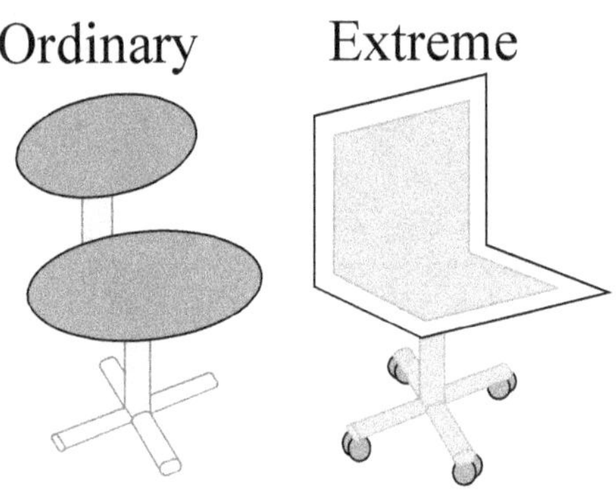

Based on the *road map* example in the previous chapter, some conceptual chairs are discussed. The idea is to develop one ordinary chair model and one extreme chair model. After some discussions, the concepts in Figure 4-2 were selected.

FIGURE 4-2 Chairs concept brief

To conclude, the concept brief is:

Pj.1.   Ordinary swivel chairs with cheap technology

By using almost only standard components from low-price regions, these chairs will be attractive for functionality and affordable prices.

Pj.2.   Extreme swivel chairs

By using mostly special design in-house components, these chairs will be attractive for their uniqueness and trendy fashion.

4.4 Plan portfolio

4.4.1 About plan portfolio

Most often, not only a single *product* is offered from a developing company. If developing a set of related *products*, described by a *product portfolio*, there will be technical and manufacturing synergies between the different *products*, and a bigger chance to impact more completely on a broader market.

To reach a wide market and many customer segments, a portfolio of products is defined.

A *product portfolio* should at least:

1.   contain an integrated offering of a range of *product* models;
2.   give higher market coverage than each *product* offered separately;
3.   distribute and balance *product* advantages to avoid *product* cannibalism.

Since cost is often the *product* differentiator, the *product portfolio* is often sorted by selling price. The margin might be very different for each of the *products*. For example, some of the *products* may have a low margin in order to tempt the *customers*, and in turn open the market for more expensive high-margin *products*.

A *product line* is an often used name for a set of different *products* that *reuse* many of their *components* from each other. A *product line* is a powerful way to reduce development costs, because fewer *components* need to be developed. It also reduces *product* price because of larger-scale manufacturing. If preferred, costs saved can instead be used for higher *quality*.

> Similar products in the portfolio are grouped together, for reuse of common components.

4.4.2 EXAMPLE Chairs: Plan portfolio

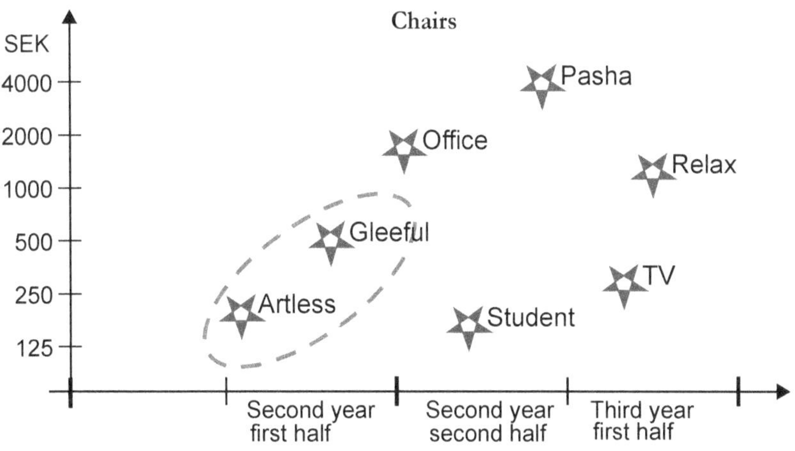

FIGURE 4-3 Chairs portfolio proposition

Description of the second-year, first-half chair *product portfolio* models (encircled):

Pca.1. The Artless chair is a cheap and simple chair.

This chair should not have anything unnecessary, only a stand, a seat, and a backrest with cheap upholstery. The market target is economical: ordinary persons who want an ordinary cheap chair for their home desk.

Pca.2. The Gleeful chair is extreme in appearance.

This chair may not be too expensive because of teenagers' limited economy. The target is teenagers who want to personalize with a colorful and fancily designed chair for their homework table.

4.4.3 EXAMPLE Chairs: Plan portfolio reuse

A *reuse plan* is prepared, based on the two chosen *products*, but later *products* in the *product portfolio* may also be considered. A *component* list will later be prepared (see Table 4-11, p. 54), to identify which *components* could be shared between the *products*.

Chairs *reuse plan*:

Pcb.1. The stand and the holder are the anonymous components, which can be reused by many chairs.

Pcb.2. Considering future products, the stand and the holder should be divided into two separate parts, having specified interfaces between them, allowing only one of these parts to be reused with the other to be chair-specific.

Pcb.3. All stands should have attachments for optional castors.

Pcb.4. All holders should have attachments for optional armrests.

4.5 Study business

4.5.1 About study business

The idea with a *business case* is to estimate a *product* accounting in advance, in order to judge and make conscious business decisions on *product* profitability before continuing to develop it. The accounting should cover the full *product life cycle*, and estimate income, expenses, and net result.

Cost and profitability estimates are performed on products in the portfolio.

Preferably, several alternative *business cases* should be prepared for discussion and selection. One alternative *business case* might even be to not develop anything new at all, but just proceed producing everything as before.

4.5.2 EXAMPLE Chairs: Study business

The following *business case* for Artless and Gleeful is calculated and presented.

TABLE 4-4 Chairs business case 1

Amounts in SEK	First year	Second year				Third year	Fourth year	Last year
		Q1	Q2	Q3	Q4			
Pf.1a. Development and prototypes	150 000							
Pf.1b. Chair units		30	45	90	180	720	480	240
Pf.1c. Material and assembly		5 000	7 000	13 000	25 000	100 000	70 000	30 000
Pf.1d. Adm. & sale	20 000	5 000	5 000	5 000	5 000	30 000	20 000	20 000

TABLE 4-4 Chairs business case 1

Amounts in SEK	First year	Second year				Third year	Fourth year	Last year
		Q1	Q2	Q3	Q4			
Pf.1e. Sale income		10 000	15 000	30 000	60 000	240 000	160 000	80 000
Pf.1f. Net	-170 000	0	3 000	12 000	30 000	110 000	70 000	30 000
Pf.1g. Accumulated net	-170 000	-170 000	-167 000	-155 000	-125 000	-15 000	55 000	85 000

The investment payback should cover the first-year development cost, such as *requirements, architectures, design items, product requisites, product prototype,* and *verification.*

4.6 Prototype details

4.6.1 About prototype details

At a **Study concepts** *phase*, some technical risks are typically identified when planning to realize a *product.* Unknown materials may be utilized, a new way to manufacturing may be planned, or *program* performance may be crucial and must be verified in advance, and so forth.

To avoid surprises from troublesome details, these may be evaluated in limited prototype projects.

In the prototype details *phase* such details are investigated. Material is ordered and examined in the laboratory, new manufacturing machinery is investigated, and some *programs* may be written to ensure that approaching crucial *functions* will be possible to realize.

4.6.2 EXAMPLE Chairs: Prototype details

The following was decided for chair *prototyping.*

- Partial prototype chairs were prepared from pressed wood board, to investigate comfortable sitting for seat height, depth, and so forth.
- Joint technique—melding versus screwing—was investigated. It is important to understand the cost of jointing, depending on investments. Screwing is expensive but needs no big investments, and welding is cheap but needs big investments in welder machines.

Chairs partial prototypes findings:

PI.1. The average height for a working chair is around 46 cm.

PI.2. The adjustable range for most adults should be from 37 cm to 53 cm. This range also covers teenagers.

PI.3. The depth of the chair seat was comfortable enough at 43 cm.

PI.4. Welding machine investments and welding work should not be done in-house.

FIGURE 4-4 Chairs partial prototypes

4.7 Research market

4.7.1 About research market

In some cases the *customers* are gathered in strong organizations and prepare their own *requirements* for their manufacturers to take care of. In many cases (like for chairs), there is no such strong *customer* organization or representation of individuals.

Market research brings extra understanding before specifying technical requirements.

Nevertheless, it is a good idea to contact planned retailers, to get their perception of *customer requirements*. The retailers have long and intense *customer* understanding and at the end of the day, it will be the retailer who will offer the product for sale to *customers*. Another way is to turn to *customers* directly through inquiries or interviews, to understand *customers* and their demands.

4.7.2 EXAMPLE Chairs: Research market

The planned retailers were invited to judge selected parts from the business case in Table 4-4, page 44 and to evaluate the partial prototypes in Figure 4-4 above. Some potential *customers* were invited at the same time. A lot of opinions were collected, and the following was documented.

TABLE 4-5 Chairs user profile

	User profile	Priority
Pn 1.	Teenagers prefer a smashing and colorful appearance over comfort.	Middle
Pn 2.	Adults prefer the opposite to above, and black is the preferred color.	Middle
Pn 3.	Armrest is preferable but not necessary.	Low
Pn 4.	Backrest is necessary.	High
Pn 5.	Teenagers can afford a maximum price of SEK 400.	High
Pn 6.	Economical ordinary adults do not spend more than SEK 300.	Middle

4.8 Plan products

4.8.1 About plan products

Now it is time to plan the *products* to be developed. That is, to select chairs from the planned chair *product portfolio* for the next period of development, and describe each of them as an offer to satisfy potential *customers*.

Each product of the portfolio must be proposed and planned.

A **Pi. Product proposition** contains at least:

- The *product's* target *customers*
- Benefits and advantages of the *product*
- Package, service, capabilities, and marketing
- Price

4.8.2 EXAMPLE Chairs: Plan products

Selected from the *product portfolio* and supported with a *business case*, the Artless *product proposition* is:

Ph.1.a. Reliable and cheap, for people thinking logically and economically. Discreet appearance.

Ph.1.b. Very stable stand construction for all kinds of situations—paperwork in the home office or hobby work in the recreation room. Adjustable height. Fully swiveling.

Ph.1.c. Delivered in four pieces in a compact box. Easy to assemble, only screwdriver required.

Ph.1.d. Price: SEK 299

FIGURE 4-5
Pr.1. Artless chair product proposition

4.8.3 EXAMPLE Chairs: Plan products

Also selected from the *product portfolio* and supported with a *business case*, the Gleeful *product proposition* is:

Ph.2.a. A fancy and personal chair for young and modern people. Ideal for any use, including TV watching as well as homework. The center of a modern teenager pad.

Ph.2.b. Fully swiveling and moving chair. Lovely rattan from birch and super metallic violet stand. Adjustable height.

Ph.2.c. Delivered in 2 pieces, assembled without tools.

Ph.2.d. Price: SEK 399

FIGURE 4-6
Pr.2. Gleeful chair product proposition

4.9 Capture staffing & requirements

4.9.1 About capture staffing & requirements

To *capture requirements* is to position yourself in the place of the *customer*, and note what he or she would demand from the *supplier* (yourself). The difficult thing with *requirements* is not so much to document them, but to afterwards analyze and understand their implications on solutions.

Based on product planning, each product gets its technical requirements specified.

If there is a huge number of *requirements*, it might also be time consuming to handle, sort, and manage them. But the biggest obstacle of them all is prioritizing every *requirement* (originating from enthusiastic *stakeholders)* in order to keep the *product* at an attractive cost.

Requirements may be derived from any *phase* executed in the past. *User profiles* are a perfect *requirement* source, but they may be more or less individual and vague. In the prototype details some nasty ones may be found, which will not be forgotten if documented as a *requirement.* The *road map*, concept brief, and *reuse plan* and all other prepared documentation may also be useful to generate valuable *requirements.*

4.9.2 EXAMPLE Chairs: Capture staffing & requirements

The *user profiles* are already documented, and are useful as the base for *requirements.* Documentation by partial prototype is also used. All other documentation from *phases* of the past are reviewed during specification, to not miss anything.

The following *requirements* were captured (see Table 4-6 below).

TABLE 4-6 Chairs requirements

	Requirements	Priority
Pl.1.	The average height for a working chair is around 46 cm.	
R 1.	The height for an average working chair is 46 cm.	High
Pl.2.	The adjustable range for most adults should be from 37 cm to 53 cm. This range also covers teenagers.	
R 2.	The adjustable range should be 37 cm to 53 cm.	Middle
Pl.3.	The depth of the chair seat was comfortable enough at 43 cm.	
R 3.	The depth of the seat is most comfortable at 43 cm.	Middle
Pn 3.	Armrest is preferable but not necessary.	
R 4.	Smooth swiveling is preferred by all customers when they buy a chair for their desks. In this case, adults need armrest and backrest.	High
R 5.	To smoothly and safely move around with the chair is highly preferred but manually lifting the chair to be moved is also acceptable.	Low
Pn 4.	Backrest is necessary..	High
R 6.	Teenagers are more fond of rolling around than adults. In this case the chair needs a backrest.	Middle
R 7.	Danger of falling off or overturning must be negligible when swiveling and rolling around.	High
R 8.	Adjustable height is a must, but does not need to be instantly changeable.	High
Pn 1.	Teenagers prefer a smashing and colorful appearance over comfort.	
Pn 2.	Adults prefer the opposite to above, and black is the preferred color.	
R 9.	Use solid-colored fabric for adult chairs and colorful solids for teenager chairs.	Middle
Pn 5.	Teenagers can afford a maximum price of SEK 400.	
Pn 6.	Economical ordinary adults do not spend more than SEK 300.	
R 10.	Material must not cost more than 150 SEK for any chair.	High

4.10 Predetermine solutions & suppliers

4.10.1 About predetermine solutions & suppliers

After the *requirements* have been captured and prioritized, it is time to convert them into design. However, it is not easy to transform all kinds of *requirements* into a designed *product* in one single step. To make the path to design step by step in order to keep it controlled instead of rushing into disorder, begin to analyze each *requirement*, and propose solutions that satisfy the *requirement*.

Analyze requirements to proposed solutions, and check if any supplier offers appropriate samples to fit these solutions, before architecture work is started.

With many *white-box solution alternatives* available, it is easier to get the total design to fit better together since many *solution alternatives* gives more possible combinations to best possible satisfy the *requirements*.

Often the *requirements* are contradictory or overly ambitious, ending up in a design that is impossible to realize. If so, it is necessary to sacrifice some *requirements* by beginning with the lowest prioritized *requirements.*

4.10.2 EXAMPLE Chairs: Predetermine solutions

The chairs *requirements* are analyzed one by one and given *solution alternatives* (see below table).

TABLE 4-7 Chairs solutions

Solution alternative	Requirement
SL.1. Optimize steel stand for this height.	R 1.The height for an average working chair is 46 cm.
SL.2. Use adjustable stand with button locking.	R 2.The adjustable range should be 37 cm to 53 cm..
SL.3. Make seats of this size.	R 3.The depth of the seat is most comfortable at 43 cm.
SL.4. Same holder with bearing for Artless and Gleeful **SL.5. Armrest for Artless**	R 4.Smooth swiveling is preferred by all customers when they buy a chair for their desks. In this case, adults need armrest and backrest.
SL.6. No castors for Artless	R 5.To smoothly and safely move around with the chair is highly preferred but manually lifting the chair to be moved is also acceptable.
SL.7. Castors for Gleeful **SL.8. Gleeful needs backrest but no armrest.**	R 6.Teenagers are more fond of rolling around than adults. In this case the chair needs a backrest.
SL.9. Make safe steel stand with 5 castor legs.	R 7.Danger of falling off or overturning must be negligible when swiveling and rolling around.
SL.10. Make cheap and robust adjustable stand.	R 8.Adjustable height is a must, but does not need to be instantly changeable.
SL.11. Black cotton fabric for Artless **SL.12. Use rattan seat and painted stand for Gleeful.**	R 9.Use solid-colored fabric for adult chairs and colorful solids for teenager chairs.
SL.13. Cheap ordinary material for Artless **SL.14. Cheap but special material for Gleeful**	R 10.Material must not cost more than 150 SEK for any chair.

4.10.3 EXAMPLE Chairs: Predetermine suppliers

When the possible solutions are considered, the next thing to do is to imagine how to realize them. When developing *products* and designing *elements* which build up the *product*, at some point it is better to buy the part from *suppliers* than to develop and manufacture them in-house. Some parts are generic and can be bought off the shelf, other parts can be ordered custom-made from external *suppliers* and bought from them during manufacturing.

White-box solution alternatives are analyzed and considered for effective *realization* (see below table).

TABLE 4-8 Chairs suppliers

Supplier opportunity	Solution alternative
SU.1. Castors are common products, which don't need to be reinvented. Buy off the shelf.	SL.7. Castors for Gleeful
SU.2. It is a competitive advantage to have a very inexpensive solution. Design this in-house. SU.3. For manufacturing, use Standit Inc to weld and mount stand and seat holder with bearing.	SL.9. Make safe steel stand with 5 castor legs.
SU.4. This is a crucial part of the chair. Ask Bearing & Co AB to get a good price for the bearing.	SL.4. Same holder with bearing for Artless and Gleeful
SU.5. Import standard seats from contacts in China. SU.6. Order special rattan seat from contacts in India.	SL.11. Black cotton fabric for Artless SL.12. Use rattan seat and painted stand for Gleeful..

4.11 Design architectures

4.11.1 About design architectures

Previous *phases* specified the *requirements* and proposed *solution alternatives* for the chairs. The trick is now to decide on a construction that *realizes* solutions as well as possible. To describe these decided solutions, it is convenient to draw *architectures*, which show the whole *product* with major *elements*, and not less important, the *interfaces* between the *elements*.

Architectures show product elements and how these are connected by interfaces.

It is very important to analyze, understand, and prepare *interface* descriptions, to allow development to proceed in parallel for each *element*. Strong definition of *interfaces* guarantees that the *elements* can be handled totally separately and without disturbing each other (see ch. 9.4.3, p. 349). Another advantage with *elements* having well-described *interfaces* is that these *components* might be easily *reused* (without much rework) for other chairs, also in future development. The *reuse plan* should already have identified some of these opportunities.

4.11.2 EXAMPLE Chairs: Design architectures

A *product architecture* is prepared for each of the chairs (see the Artless chair in Fig. 4-7). This drawing must be complete enough to identify all *elements* and *interfaces* needed for the chair. For example, an *architecture* for mechanics often contains all measurements for the *elements*. Despite the fact that such a drawing can be simple, it conveys a lot of information to an uninformed observer.

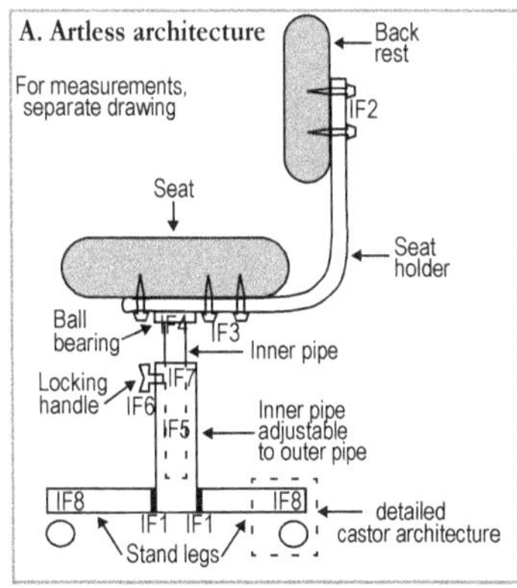

FIGURE 4-7 Artless chair architectures.

Since welding should be done by a subcontractor, he or she must agree on a dimensioned drawing before signing the order.

The available ball bearing must also be checked, to fit to the inner pipe and seat holder. Before signing the order with this *supplier*, the welding subcontractor is asked to make the exact design for the ball bearing to fit to.

The seats are fastened with ordinary screws. A check is made with *suppliers* abroad, to make sure that these *interfaces* are efficient and not ugly.

4.12 Finalize design & requisites

4.12.1 About finalize design & requisites

A *product structure* is a hierarchical *partition* of a *product*. The *product structure* is important to control during all development *activities*, and becomes crucial when approaching the start of manufacturing. In the earlier *activities* the *product structure* is often visualized by *architecture* and *component* drawings, where *elements* and the way their *interfaces* fit together are illustrated.

Design items and product requisites are finalized, which also is the base for prototype preparation.

As manufacturing gets closer, the *architectures* are detailed and finally formalized to a list of *product requisites*, which explains all *items* (including work) in detail, and specifies from which *supplier* to buy them and at what price. If a *product* is *complex*, the list of prime costs may have a huge number of columns, describing every attribute and aspect of the parts and work.

The purpose with finalize *elements* is to prepare everything needed for manufacturing. Some part of a *product* may be more *complex* than other parts. In these cases it is necessary to add detailed drawings for these *elements* and *interfaces*.

4.12.2 EXAMPLE Chairs: Finalized design

For a chair, the castor details need to be described and better understood (see Fig. 4-3, right).

This drawing shows how the castor is fixed to the jacket in the stand leg. If no castor is used, a plug should be used (even in the same hole) as a foot to the chair.

Since it has been decided to buy these castors off the shelf, minor compromises to the drawing are expected.

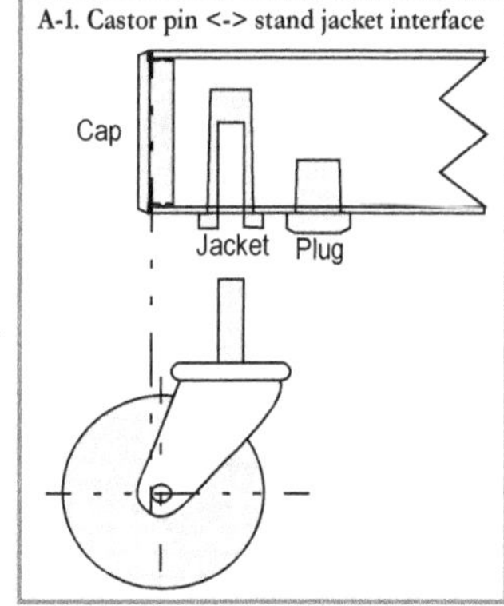

FIGURE 4-8 Chair castor finalized architectures

4.12.3 EXAMPLE Chairs: Finalize design

When all design of the *architecture* and the *components* is finalized, it might be a good idea to conclude that no *elements* or *interfaces* have been overlooked.

Since *finalize design items* is so crucial for manufacturing and *failure elimination* costs a lot, this list must be handled formally. For example, each part in the list must be carefully reviewed and approved.

Available *architectures* and drawings were scanned and the following *interfaces* were identified (see table below).

TABLE 4-9 Chairs finalized interface design

Interface	Design
IF 1. Stand legs <-> outer pipe	Arc welding
IF 2. Backrest <-> seat holder	Screws
IF 3. Seat <-> seat holder	Screws
IF 4. Seat holder <-> inner pipe	Ball bearing
IF 5. Inner pipe <-> outer pipe	Sliding with clearance H7
IF 6. Lock handle <-> outer pipe	Tread
IF 7. Lock handle <-> inner pipe	Clamping
IF 8. Castor pin <-> stand jacket	Plain bearing

All needed *elements* were identified (see table below).

TABLE 4-10 Chairs finalized element design

Item	Photo	Type	Size	Material	Cover
A.1. Castor		RR-349	Ø 7,3 cm Pin Ø 7,9 mm	Plastic wheel Steel pin	Black
A.2. Jacket		Plain bearing 525534	Ø 8mm Ø 12 mm L 33 mm	Nylon	Black
A.3. Plug		525536	Ø 12 mm	Nylon	Black
A.4. Cap		525535	4,2 cm 1,8 cm L 42 cm	Nylon	Black (planned violet)
A.5. Stand legs		Square pipe		Steel	Black or violet
A.6. Outer stand		Pipe-BV	Ø 5,2 cm	1 mm steel	
A.7. Locker		Handle 525581	M5 tread	Plastic, steel pin	Black
A.8. Inner stand		Pipe-C	Ø 5,0 cm	1 mm steel	Chrome
A.9. Swing bearing		Ball bearing with bracket and screws	Ø 6,3 cm	Steel	Black

TABLE 4-10 Chairs finalized element design

Item	Photo	Type	Size	Material	Cover
A.10. Seat holder 1		Pipe-B with brackets	Ø 3 cm	1 mm steel	Black
A.11. Seat holder 2		Pipe-B with brackets	Ø 3 cm		
A.12. Seat screws		Screw	9 x 1,5 inch	Steel	
A.13. Back-rest screws		Screw	12 x 2 inch	Steel	
A.14. Seat		Seat PE 361a	40 x 40 x 8 cm	Cotton fabric	Black with grey pattern
A.15. Back-rest		Seat PE 361b	35 x 48 x 6 cm	Polyethylene padding	
A.16. Seat with back-rest		Seat R 426	38 x 35 cm 52 x 38 cm	Rattan	Varnish

4.12.4 Finalize requisites

Now the design is finalized, and it is time to estimate and prepare requisites for everything needed for development, prototypes, and manufacturing, regarding included material, work, and equipment. Some costs have already arisen, and remaining costs in the future must be predicted as well as possible.

This is also a final chance to verify that the *business case* hasn't been destroyed by unexpected costs. If needed at this time, some *requirements* can still be sacrificed in favor of low cost.

4.12.5 EXAMPLE Chair: Finalized requisites

The finalized design was scanned and the following manufacturing materials were identified (see table below).

TABLE 4-11 Estimated chairs manufacturing material requisites

3000 Artless chairs			3000 Gleeful chairs					
Pcs	SEK/pc	SEK	Pcs	SEK/pc	SEK	Item	Product	Supplier
			5	4.40	22	Castor	RR-349	Rollon Inc.
			5	0.20	1	Jacket	Plain bearing 525534	Allo HB
5	0.12	0.60				Plug	525536	Allo HB
5	0.26	1.30	5	0.26	1.30	Cap	525535	Allo HB
5	2.2	11	5	2.2	11	Stand legs	Square pipe	Pips AB
1	3.30	3.30	1	3.60	3.60	Outer stand	Pipe-BV	Pips AB

TABLE 4-11 Estimated chairs manufacturing material requisites

3000 Artless chairs			3000 Gleeful chairs					
Pcs	SEK/pc	SEK	Pcs	SEK/pc	SEK	Item	Product	Supplier
1	5.50	5.50	1	5.50	5.50	Locker	Handle 525581	Allo HB
1						Inner stand	Pipe-C	Pips AB
1	11.70	11.70	1	11.70	11.70	Swing bearing	Ball bearing with bracket and screws	Bear-Ing AB
			1	7.60	7.60	Seat holder 1	Pipe-B with brackets	Pips AB
1	9.10	9.10				Seat holder 2	Pipe-B with brackets	Pips AB
3	14.4 / gross	0.30	3	14.4 / gross	0.30	Seat screws	Screw	Allo HB
3	14.4 / gross	0.20				Backrest screws	Screw	Allo HB
1	43	33				Seat	Seat PE 361a	Seated Inc.
1	39	29				Backrest	Seat PE 361b	Seated Inc.
			1	56	66	Seat with backrest	Seat R 426	Seated Inc.
Sum		105			130			

Manufacturing cost for assembling the chairs is then identified and estimated (see table below).

TABLE 4-12 Estimated chairs manufacturing assembly requisites

3000 Artless chairs			3000 Gleeful chairs					
Hour	SEK/hour	SEK	Hour	SEK/hour	SEK	Item	Provider	Interface
0.04	480	19.2	0.05	480	24	Welder	Standit Inc	IF 1. Stand legs <-> outer pipe IF 8. Castor pin <-> stand jacket IF 6. Lock handle <-> outer pipe
0.02	320	6.4	0.02	320	6.40	Assembler	Standit Inc	IF 4. Seat holder <-> inner pipe
		-			-	Customer		IF 5. Inner pipe <-> outer pipe IF 7. Lock handle <-> inner pipe
0.1	320	32	0.1	340	34	Assembler	in-house	IF 2. Backrest <-> seat holder IF 3. Seat <-> seat holder
Sum		57.6			64.40			

All overall development material cost must also be included in the requisites to correctly calculate all development costs (see table below).

Estimated chairs overall development material requisites

SEK		Item	Phase	Supplier
	4 400	Seats of pressed wood board	Prototype details	Seated Inc.
	6 300	Campaign material	Promote campaigns	Shadow AB
	4 000	Costs for 3 scrapped chairs	Prepare manufacturing	in-house
Sum	14 700			

All development work cost must be included in the requisites to correctly calculate all development costs (see table below).

TABLE 4-13 Estimated chairs overall development work requisites

Hour	SEK/ hour	SEK	Item	Provider	Phase
15	460	6 900	Product analyst	In-house	Study strategy Study concepts Promote campaigns Launch product
20	400	8 000	Product planner	In-house	Plan portfolio Study business Plan products
12	450	5 400	Specifier	In-house	Research market Validate prototype
24	400	9 600	Manufacturing planner	Prospective manufacturer	Prototype details Prepare manufacturing Finalize product
Sum		29 900			

Material costs for the chairs *product prototypes* are shown in table below. Note that some smaller amount of material costs arises from the *phases* prior to technical development.

Estimated cost for *product prototype* material is calculated in the table below.

TABLE 4-14 Estimated chairs prototype material requisites

SEK		Item	Supplier
	8 600	Castors	Rollon Inc.
	3 700	Jacket, Plug, Cap, Locker, Seat screws, Backrest screws	Allo HB
	10 700	Stand legs, Outer stands, Inner stands, Seat holder 1s, Seat holder 2s	Pips AB
	2 100	Swing bearings	Bear-Ing AB
	8 200	Seats, Backrests, Seat with backrest	Seated Inc.
Sum	33 300		

Estimated costs to assemble the chairs *product prototypes* are:

TABLE 4-15 Estimated chairs prototype work requisites

SEK	Item	Provider	Interface
5 000	Welder	Standit Inc	IF 1. Stand legs <-> outer pipe
18 700	Assembler	In-house	IF 2. Backrest <-> seat holder IF 3. Seat <-> seat holder IF 4. Seat holder <-> inner pipe IF 5. Inner pipe <-> outer pipe IF 6. Lock handle <-> outer pipe IF 7. Lock handle <-> inner pipe IF 8. Castor pin <-> stand jacket
Sum	**23 700**		

The estimated cost of the technical development work is then shown in the table below.

TABLE 4-16 Estimated chairs technical development requisites

Hour	SEK/ hour	SEK	Item	Provider	Phase
30	450	13 500	Specifier	In-house	Capture staffing & requirements
46	400	18 400	Designer	In-house	Predetermine solutions & suppliers
24	500	12 400	Architect	In-house	Design architectures
60	380	22 800	Designer	In-house	Finalize design & requisites
25	440	11 600	Verifier	In-house	Verify black- & white-box
Sum		**78 700**			

Sum of costs is shown in the table below. To make it simple, the development costs of the two chairs have been accounted together as one cost, and has been equally divided between the two chairs.

TABLE 4-17 :Estimated chairs total requisites

Volume / Item	Development SEK	3 000 Artless SEK/chair	3 000 Gleeful SEK/chair
Chairs manufacturing material requisites		105.00	130 00
Chairs manufacturing assembly requisites		57.60	64.40
Chairs overall development material requisites	14 700		
Chairs overall development work requisites	29 900		
Chairs prototype realization material	33 000		
Chairs prototype realization assembly	23 700		
Chairs technical development work	78 700		
Sum	180 000	162.60	194.40
Distributed across 6 000 products		30.00	30.00
Total sum		192.60	224.40

These figures correspond rather well with the much earlier estimates in Table 4-4, "Chairs business case 1" on page 44.

4.13 Realize & integrate white-box

4.13.1 About realize & integrate white-box

Flow of goods is an essential part of manufacturing. It starts up early during the building of *product prototypes* from *items* during development, continuing with *components* for *null series* during manufacturing ramp-up, and will last as long the *product* is being manufactured.

Items and components are acquired and the prototype is prepared and started.

At this point, calculate some supply backup plans, for example, second sources for crucial *items*. If received *items* don't fulfill the expectations, it might be too late to change them and reorder. Major changes beyond this time point cannot be allowed, because of the risk of delaying start of assembling the *product prototype*.

To conclude so far, from *sample catalogues* a set of *samples* has been chosen and incorporated as *architecture ingredients*, which have been further detailed to *design items* with exact *design item properties*.

When all *items* have been received, *product prototype integration* can be performed. This is the first time the full *product* is assembled from its *items* (almost authentic *components*) and all *interfaces* are finally exposed in reality. *Complex product* problems may appear, depending on interaction between *items* that have never been verified together. In the worst case, the design has to be reworked, and a second *product prototype* has to be prepared.

4.13.2 EXAMPLE Chairs: Realize & integrate white-box

The chair *product prototypes* are assembled from obtained *items*. Each *interface* is examined and documented (see table below).

TABLE 4-18 Chair realized white-box

Integration report	IdInterface or joint	Technology
IR.1a. Done by Standit Inc. approved	IF 1. Stand legs <-> outer pipe	Arc welding
IR.1b. Easy to fasten, strong enough, approved	IF 2. Backrest <-> seat holder	Screws
IR.1c. Easy to fasten, strong enough, approved	IF 3. Seat <-> seat holder	Screws
IR.1d. Delivered by Standit Inc, approved	IF 4. Seat holder <-> inner pipe	Ball bearing
IR.1e. Too tight clearance for smooth adjust, new stand ordered	IF 5. Inner pipe <-> outer pipe	Sliding with clearance H7
IR.1f. Easy to lock, easy to unlock, approved for seat load up to 180 kg	IF 6. Lock handle <-> outer pipe	Tread
IR.1g. Approved for seat load up to 180 kg	IF 7. Lock handle <-> inner pipe	Clamping
IR.1h. When mounted the jacket is fixed. Approved up to 40 kg load	IF 8. Castor pin <-> stand jacket	Plain bearing

4.14 Verify black- & white-box

4.14.1 About verify black- & white-box

Product prototype verification is to check that a *product* fulfills (and preferably surpasses) what is expected by specifications. If the *requirements* are well written and cover all important aspects from representative *stakeholders*, they are obviously used for *verification*.

To ensure correct design, the prototype is verified according to original technical requirements.

To be able to verify if the *requirements* are fulfilled by the *product prototype*, the *requirements* are most often still too informal to be accurate enough. The *requirements* are scanned through and each *requirement* is chosen to be the base for at least one *test-case*. Sometimes the most important *requirements* are personal safety, environmental considerations, and national regulations, which also need prepared *test-cases*.

4.14.2 EXAMPLE Chairs: Prepare test-cases

Requirements with *test-cases* for chair safety are important. For example, the swivel tap or the castors may not get stuck or rotate unevenly, causing the seated person to tilt over and fall to the floor. This is also an important comfort feeling. If the chair does not roll and swivel smoothly and evenly, it is considered to be of low *quality*.

TABLE 4-19 Chairs requirements test-cases

Test-case	Requirement
CH.1. Check if the chairs can easily be preset to 46 cm when assembled.	R 1. The height for an average working chair is 46 cm.
CH.2. Check that the minimum height can be set less than 37 cm and the maximum height can be greater than 53 cm.	R 2. The adjustable range should be 37 cm to 53 cm.
CH.3. Measure the seat depth with a ruler. It shall be between 41 and 43 cm.	R 3. The depth of the seat is most comfortable at 43 cm.
CH.4. Perform a swivel of 360° with a torque instrument fastened on the seat periphery. The torque must be less than 0,1 Nm, and must not vary up and down more than 0.1 Nm.	R 4. Smooth swiveling is preferred by all customers when they buy a chair for their desks. In this case, adults need armrest and backrest.
CH.5. Perform a movement of 1 meter in 5 arbitrary directions by a force instrument fastened on the seat center. The force must be less than 0.2 N, and may not vary more than 0.1 N during the movement.	R 5. To smoothly and safely move around with the chair is highly preferred but manually lifting the chair to be moved is also acceptable.
CH.6. Check that Artless chair has no castor and the Gleeful chair has both castors and backrest.	R 6. Teenagers are more fond of rolling around than adults. In this case the chair needs a backrest.
CH.7. Load the chair with 100 kg. Perform the test-cases from requirements R 4. and R 5. again. To avoid tilting, the values may not be higher than double.	R 7. Danger of falling off or overturning must be negligible when swiveling and rolling around.

TABLE 4-19 Chairs requirements test-cases

Test-case	Requirement
CH.8. Take place on the seat. Check that the height lock is easy to unlock and check that the lowering goes smoothly and stops gently at lowest height.	R 8. Adjustable height is a must, but does not need to be instantly changeable.
CH.9. Get off the chair and stand at the side of it. Check that the height lock is easy to unlock and check that the chair can be easily adjusted up and down.	
CH.10. Adjust the height to 46 cm. Load the seat with 300 kg and check that the lock keeps the seat at this height and is still working when the load is removed.	
CH.11. Check that the fabrics and solids are complying with the requirement.	R 9. Use solid-colored fabric for adult chairs and colorful solids for teenager chairs.
CH.12. Check that the product requisite costs are still valid after receiving the items.	R 10. Material must not cost more than 150 SEK for any chair.

4.14.3 EXAMPLE Chairs: Verify black- & white-box

After the *test-cases* are completed, the chair *product prototypes* can be verified. Scan through the *test-cases* one by one, and perform what they stipulate. For each *test-case*, notify whether they fail or pass, as in the table below.

TABLE 4-20 Chairs verification report 1

Artless outcome	Gleeful outcome	Requirement
Pass	Pass	R 1. The height for an average working chair is 46 cm.
Pass	Pass	R 2. The adjustable range should be 37 cm to 53 cm.
Pass	Pass	R 3. The depth of the seat is most comfortable at 43 cm.
Pass	Pass	R 4. Smooth swiveling is preferred by all customers when they buy a chair for their desks. In this case, adults need armrest and backrest.
Pass	Pass	R 5. To smoothly and safely move around with the chair is highly preferred but manually lifting the chair to be moved is also acceptable.
Pass	Pass	R 6. Teenagers are more fond of rolling around than adults. In this case the chair needs a backrest.
Pass	Pass	R 7. Danger of falling off or overturning must be negligible when swiveling and rolling around.
Pass	Pass	R 8. Adjustable height is a must, but does not need to be instantly changeable.
Pass	Pass	
VR1.1 Fail. After the test the locking handle was deformed and failed to lock.		
Pass	Pass	R 9. Use solid-colored fabric for adult chairs and colorful solids for teenager chairs.
Pass	Pass	R 10. Material must not cost more than 150 SEK for any chair.

The locking mechanism was found to be too weak, and was deformed after the first locking. A harder steel was selected, and now passes a repeated *verification*.

4.15 Validate prototype

4.15.1 About validate prototype

In the previous *phase* a *verification* was performed, which ensured that the *product prototype* was built in the right way, specified by *requirements*. If the *product prototype* passes, the *requirement stakeholders* know that it fulfills their original intentions.

In this **Validate prototype** *phase*, the *product prototype* is validated, which will ensure that it corresponds with potential *customer* demand, specified by *user profiles*. If the *product prototype* passes, the marketers know that for this *product* there is a true demand on the market.

4.15.2 EXAMPLE Chairs: Prepare quiz-cases

User profiles were documented just before the technical part of development. Based on these *user profiles*, some quiz-cases are prepared for validation (see table below). A panel of 5 teens and 5 adults is used to validate whether the *user profiles* are fulfilled (or to change the *user profiles* descriptions).

To ensure that the product will be a market success, prototypes are validated against estimated user profiles.

TABLE 4-21 Chairs user profiles quiz-cases

Inquiry	User profile
Pr.1. Provide one of each chair's prototype to some teenagers. Examine which of them they would like to use.	Pn 1. Teenagers prefer a smashing and colorful appearance over comfort.
Pr.2. Provide one of each chair's prototype to some adults. Examine which of them they used. Ask them if they would prefer another color.	Pn 2. Adults prefer the opposite to above, and black is the preferred color.
Pr.3. Let teenagers and adults use both chairs. Examine if they would spend SEK 100 extra on armrests.	Pn 3. Armrest is preferable but not necessary.
Pr.4. Let teenagers and adults use both chairs. Examine if they would like to save SEK 100 on skipping the backrest.	Pn 4. Backrest is necessary.
Pr.5. Examine if they would like to sell it for under SEK 400.	Pn 5. Teenagers can afford a maximum price of SEK 400.
Pr.6. Provide some adult the artless chair. Examine if he/she would like to sell it for under SEK 300.	Pn 6. Economical ordinary adults do not spend more than SEK 300.

4.15.3 EXAMPLE Chairs: Validate prototype

Based on the *user profiles* quiz-cases above, validate the chairs, see the table below.

TABLE 4-22 Chairs validated prototype report 1

Artless chair outcome	Gleeful chair outcome	Customer habit
Conform	Conform	Pn 1. Teenagers prefer a smashing and colorful appearance over comfort.
VAR 1.1. Mismatch. Grey is generally the preferred color.	Conform	Pn 2. Adults prefer the opposite to above, and black is the preferred color.
VAR 1.2. Mismatch. Armrest is preferred	Conform	Pn 3. Armrest is preferable but not necessary.
Conform	Conform	Pn 4. Backrest is necessary..
Conform	Conform	Pn 5. Teenagers can afford a maximum price of SEK 400.
Conform	Conform	Pn 6. Economical ordinary adults do not spend more than SEK 300.

The mismatch is not judged big enough to significantly change the profitability of the product, and it is not decided to be changed.

4.16 Promote campaigns

4.16.1 About promote campaigns

All planning for launch on the market must be prepared. This may include contacting promotion agents, preparing interesting "selling stories," preparing advertising, commercials, brochures, pamphlets, preparing websites, booking space in papers, radio and TV, and so forth.

A lot of selling is often needed before introducing a new product.

Retailers must have the campaign material, including demo *products* or *product prototypes* in place, and market people need to come to their site to be informed and trained in the selling story, prices, deductions, delivery time, and so forth.

4.16.2 EXAMPLE Chairs: Promote campaigns

The rather limited series cannot afford more than modest promotion material.

Pp.1. A brochure is planned for the retailer to hand out to interested customers.

Pp.2. One chair prototype of each model must be sent out to each retailer for demo.

Pp.3. For these chairs a floor standing poster must also be prepared.

Pp.4. Some Gleeful chair prototypes will be sent out to youth centers, schools, and hamburger restaurants.

Pp.5. The local hardware store will have one chair at its "hobby exhibition corner."

Pp.6. One sample of each chair was sent out to 9 retailers to use a week before launch.

Pp.7. A leaflet was printed to be used at exhibitions and fairs.

Pp.8. A manual was released, describing how to assemble the chair after delivery to the customer, how to wash and clean it, how to oil the bearings and so on.

4.17 Launch product

4.17.1 About launch product

Now all connections must be set up between manufacturer and *customers*, including a distribution network. Procedures are established for volume pricing, invoicing with reminders, *customers* and *suppliers* payment monitoring, and complaints with returned *products*.

The product is manufactured in a small series and sent out to preferred retailers and customers.

4.17.2 EXAMPLE Chairs: Launch product

The following market conditions were decided:

Py.1. A retailer price list was compiled, with retailer purchase price, delivery time and cost, expirations of payment periods, and proposed consumer prices.

Py.2. Deduction rates were agreed on for volume staffing.

Py.3. The manufacturer already had a carrier, and some new freight times and prices were agreed upon.

Py.4. The producer already had office systems for bookkeeping, invoicing, and payment watching, implying only the necessity of adding the two new products into these systems.

Py.5. Procedures were agreed on for how to handle chair return from dissatisfied customers.

Py.6. Some discussions were concluded about procedures to apply if retailers go bankrupt.

Py.7. A contract with each retailer was prepared by a lawyer and signed by the manufacturer.

4.18 Prepare manufacturing

4.18.1 About prepare manufacturing

This is the *phase* to adapt the verified and validated *product prototype* to series manufacturing, and to get this manufacturing tuned in. For example, *product quality* must rise to predetermined levels to obtain the expected profit.

The manufacturing site is now scaled up to planned volumes.

At this point the development *stage* is finished and the launch continues. High volume manufacturing is started after the *null series* and sale and income increase.

4.18.2 EXAMPLE Chairs: Prepare manufacturing

The below *null series* is ordered from the manufacturer:

Pw.1.   All components and assembly work are ordered for a null-series of 2 x 15 chairs.

These are produced and evaluated. All problems are evaluated, and if any severe problem is found, the *product* and/or manufacturing failures are immediately eliminated. To continue manufacturing *products* with problems is often not a way forward, although this is not seldom seen.

Pw.2.   After fixing null series problems of the 2 x 15 chairs, 27 of them can be delivered and sold as regularly produced chairs.

After the *null series* is produced and manufacturing problems are fixed, the development team gets a bottle of champagne to celebrate a good development effort.

4.19 Finalize product

4.19.1 About finalize product

A small maintenance organization is established to fix problems showing up.

Latent problems may remain that were overlooked in the *product development*, and which show after some use. Also, manufacturing is a dynamic organism, where *quality* can suffer because of new personnel, change of machinery, decreasing *component quality*, or whatever.

Despite skilled staff and excellent development, customers always pop up to report obscure failures or to claim warranty compensation.

4.19.2 EXAMPLE Chairs: Finalize product

The owner of the chair business took the responsibility of maintenance. He already knew the manufacturers and retailers, and could quickly resolve *quality* issues to calm down the *end user*.

For example, the ball bearing showed to be too weak for persons weighing more than 120 kg. This exceeded the value specified in the chair manual, but insisting on this limitation to such a *customer* may give the chair manufacturer a bad image.

The established product still lacks some development:

Pö.1.   The decision in this case is to select a bit stronger bearing with the same dimensions and only a bit more expensive.

Pö.2.   Heavy customers who still have problems with the bearing will be given one free chair with better bearing if they complain.

## CHAPTER 5

Identify value chain

The value chain is the engine in all companies, but most often scarcely fueled and maintained.

Is the value chain being valued?

5.1 About value chain

5.1.1 Cpdm definition

Throughout this book, there is a strict interpretation of a *value chain*. What sometimes makes the *value chain* difficult to understand is that it has many names. In *Cpdm*, *value chain* is defined according to Table 5-1 below.

TABLE 5-1 Cpdm definition of value chain

Aspect	Value chain
Definition	Human work ongoing in reality for the purpose of transforming source results to added value results
Synonyms	Operation, action, performance, maneuver, function, phase, event, stage
Symbol	Source result → Value chain → Added value result

In the figure, the arrow indicates in what direction the value increases.

5.1.2 Value added

Value chain and added value are well known as the principle for VAT (value added tax).

The *added value* is the difference between the value of the output *result* and the value of the input *result*. Value added tax (VAT), for example, is based on this difference.

One way to interpret *added value* is that the salary of developers is transformed by their work to add value to the output *result*. The more competent the developer is, the higher the *added value*, and the higher the salary that can then be paid (see Fig. 5-1, right).

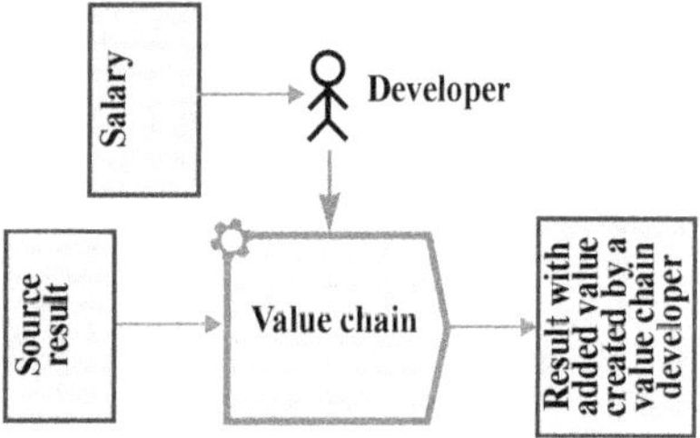

FIGURE 5-1 Salary transformed by added value

5.1.3 EXAMPLE Chair: Product development value chains

A *product development* is a kind of a *value chain*, which in turn is part of the larger *product life cycle*. The *product development value chain* is a *complex* sequence and concurrence of engineering and business *activities*, with the purpose to transform a brilliant market idea into a set of detailed manufacturing descriptions (sometimes accompanied by *product prototypes*), which becomes the base for the manufacturing of *products* for *customer* satisfaction (see Fig. 5-2 below).

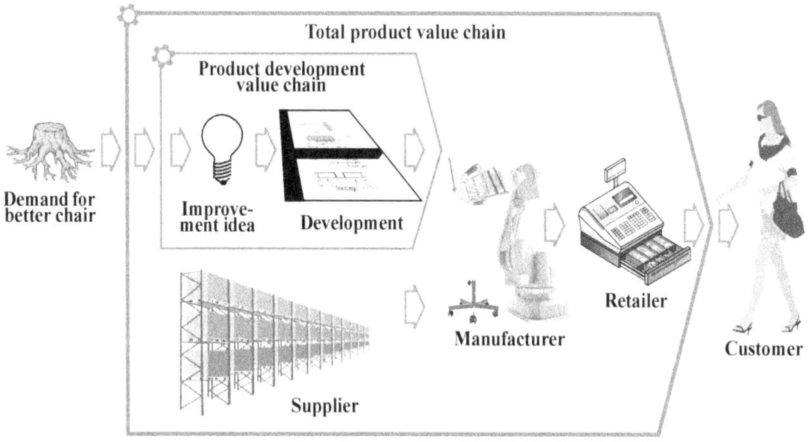

FIGURE 5-2 Activities in a product value chain

5.1.4 EXAMPLE Chair: Product development value chain

The *activities* required to develop a chair are typical for development of not overly *complex products*. This is a typical *value chain*, and might be illustrated as in Figure 5-3 below.

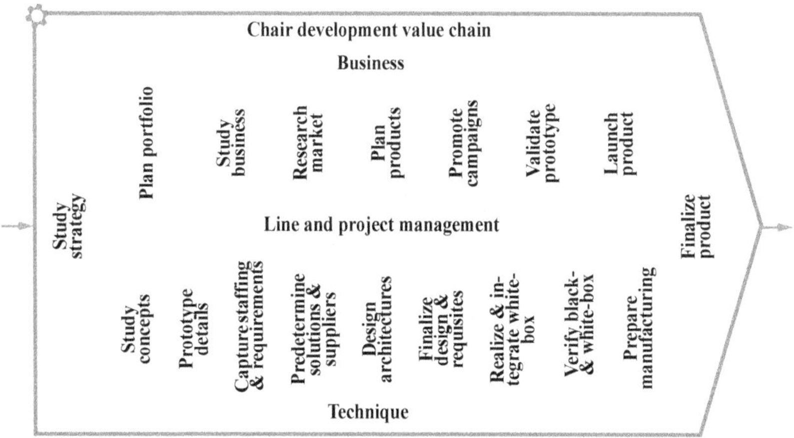

FIGURE 5-3 Typical product development value chain

5.1.5 Cpdm definitions

In *Cpdm*, *activity* and *result* are defined according to Table 5-1 below.

TABLE 5-2 Definitions

Aspect	Activity	Result
Definition	A piece of human work	Tangible output from an activity or value chain
Synonyms	Task, job, project, action, operation, undertaking	Outcome, creation, delivery, deliverable, product
Symbol	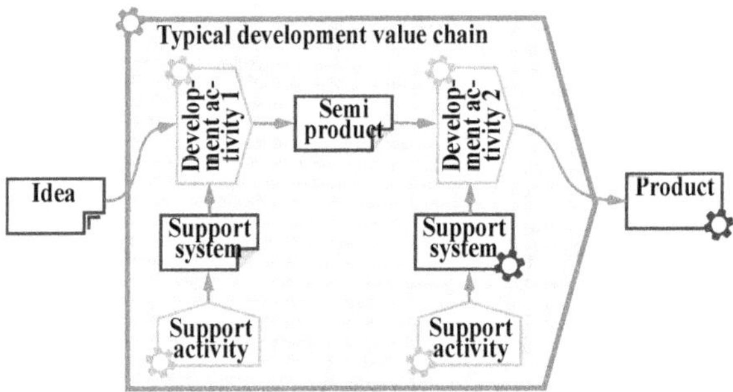	

Note that *results* can be of many different kinds, from actual tangibles to descriptions of tangibles, and even descriptions of descriptions (see ch. 6.2, "Process schedule constituents," p. 83).

5.1.6 Typical development value chain

In general, a company *value chain* acquires input from external *suppliers*, and transforms it to a more valuable output. Internally, the value increase can be made over and over again in parallel and sequential order, until the output has finally taken the shape of a final *product*, which is offered to *customers* on the market (see Fig. 5-4 below).

FIGURE 5-4 A value chain of a typical company

Generally, a company also conducts support *activities*, which are not directly involved in refining the *product*, but are indirectly needed for the *value chain* to keep on going, for example, "human resource support" and "bookkeeping support" *activities*.

5.1.7 Value chain visibility

In all organizations developing attractive products, a value chain surely exists, being the navigator for work.

Efficient *value chain*s are among the biggest nonmaterial assets existing in *product development* companies, because this *value chain* is the engine to create all *added values* finally paid for by the *customers*. For example, the *value chain* is more crucial than the *line* organization structure. Note that in any organization where productive work is done a *value chain* exists, regardless of whether it is understood or neglected, visualized or hidden. The *value chain* is handled very differently in companies, ranging from being documented and managed to going undiscovered and unexploited.

If engineers are skilled, market people are *customer-oriented*, *managers* are good decision makers, and *product* competency is in place, it is very much the *quality* of the *value chain* that makes the difference between profitable and unprofitable companies. Or to put it in reverse order, if the *value chain* gets neglected, skilled engineers feel untapped and leave, and competence decreases, *customers* disappear, and profit goes absent.

5.2 Different ways to illustrate a value chain?

There are many different interpretations of how to best understand a *value chain*. Some examples of this are illustrated in chapters below.

There are different ways to interpret a value chain.

5.2.1 EXAMPLE Chair: Development by progressing activities

Figure 5-3 above is fairly simple, but much better than not visualizing the *value chain* at all. It is still a bit vague, because it doesn't explain the preferred path when executing the *activities*.

As explained above, a *value chain* is a set of interconnected *activities*, and it would be good to know more precisely how these *activities* relate to each other. By placing arrows between the *activities*, it is easier to understand how and where the specific values are added, as in Figure 5-5 below.

The arrows between any two sequential *activities* can be interpreted as "*value chains* are continual activities from origin to end," and the approach can be named *progressing activities*.

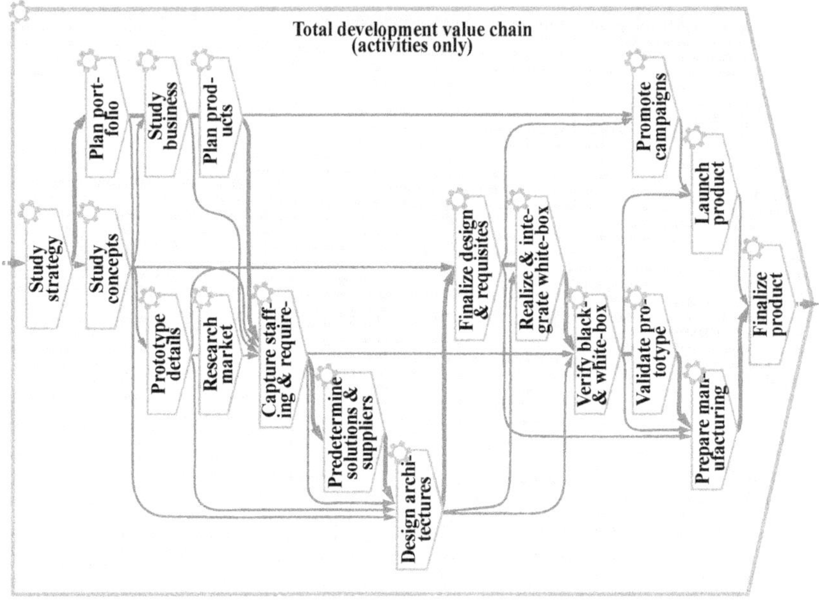

FIGURE 5-5 Chair development value chain seen as progressing activities

It should be noted that the above picture is an abstraction and approximation of the chair development *value chain*. There are many ways to organize development of chairs, and this is the *Cpdm* way, out of many other imaginable ways. For example, it may be questioned if the *activities* are chosen in the best way. And maybe it can be questioned to an even higher degree whether the interconnections between the *activities* are the best possible.

Anyway, at this point in the book, what is most important is to open up the world of development. How to support a good *value chain* by utilizing *process* descriptions will be explained and exemplified in greater detail in chapter 6, "Utilize processes," from page 81.

5.2.2 EXAMPLE Chair: Development by cascading results

It is also possible to completely disregard the *activities* used in the previous chapter, and instead use only *results* to illustrate the *value chain*. This makes the *value chain* much more *result-oriented*. The arrows between any two sequential *results* now can be interpreted as "a *value chains* are extensions from idea to *product*," and the approach can be called *cascading results* (see Fig. 5-6 below).

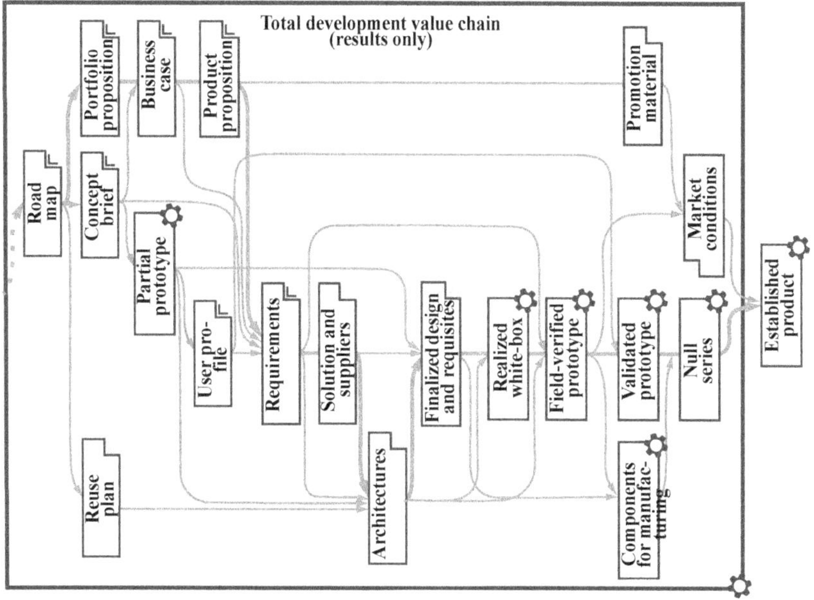

FIGURE 5-6 Chair development value chain seen as cascading results

Although this *value chain* might seem a bit oversimplified, it is very good as a base for information exercises. Such questions as "What *results* must be documented and when?" and "What documents need to be followed up?" might now find good answers.

5.2.3 EXAMPLE Chair: Development by both progressing activities and cascading results

So far, interconnections of exclusively *activities* or exclusively *results* have been taken into account.

One drawback of dealing with *activities* exclusively is that they are difficult to follow up. It is hard to monitor afterwards if a performed *activity* has created what was expected.

Much easier to check than the *activities* themselves are the *results* from these *activities*. A *result* (e.g., a document or a *product prototype*) is tangible and permanent, which is much easier to revisit than an attempt to replay the volatile *activity* itself. A *result* can be inspected, judged, and investigated, and a responsible developer can be assigned to a *result*, in order to check at any time if the *result* is sufficient (see Fig. 5-7 below).

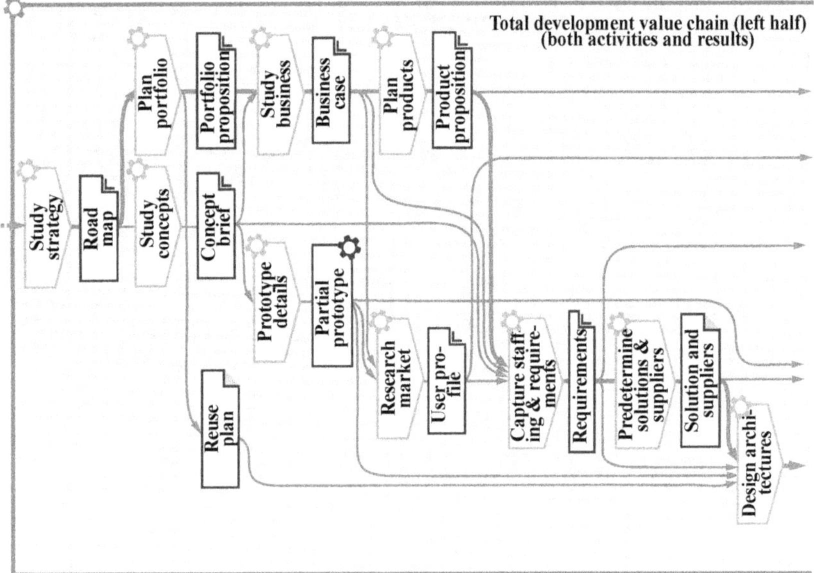

FIGURE 5-7 Chair development value chain seen as both progressing activities and cascading results (left half)

However, the *activities* are good to retain because they illustrate what must be done. Many of the methodologies are explained by the *activities* and are, of course, also important in a *value chain* illustration.

As can be seen from the development *value chain* containing *alternating activities and results*, it is now much more comprehensive and precise. According to this illustration, newly employed developers can understand much faster how to become productive, and old employees can obtain better *quality* and efficiency.

How to interpret the arrows between *alternating activities and results* is explained in 5.2.4, "Relations in a value chain," page 73, and in greater detail in chapter 6, "utilize processes," from page 81.

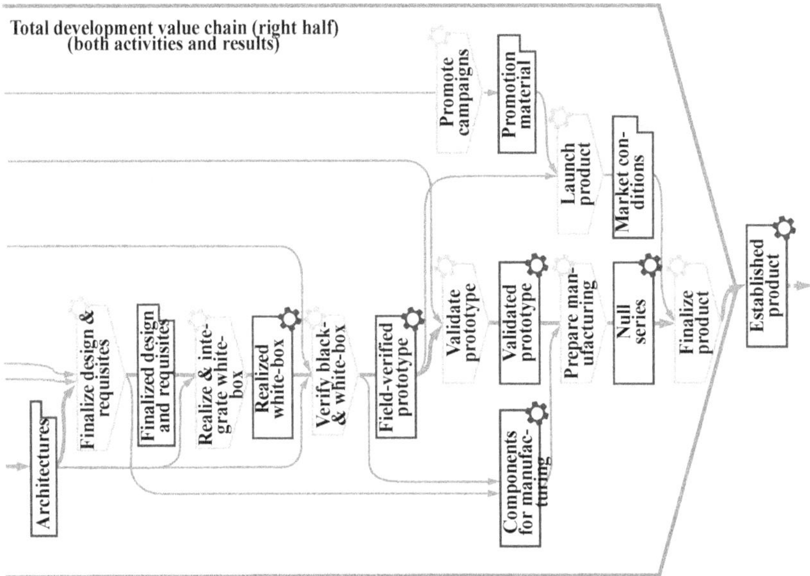

FIGURE 5-8 Chair development value chain seen as both progressing activities and cascading results (right half)

5.2.4 Relations in a value chain

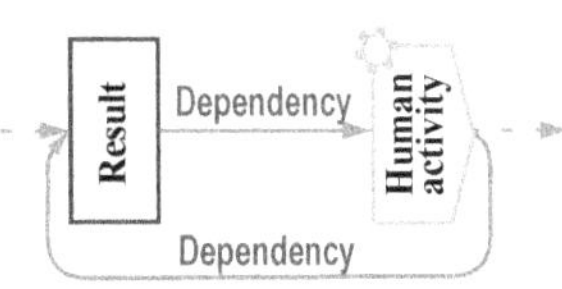

Now *activities* and *results* are defined, but what do the arrows in between represent? Generally the arrows mean that there is a dependency between these (see Figure 5-9, left).

FIGURE 5-9 General meaning of arrows in a value chain

In fact, the arrows have different meanings depending on where they depart from and where they arrive (see Figure 5-10 below).

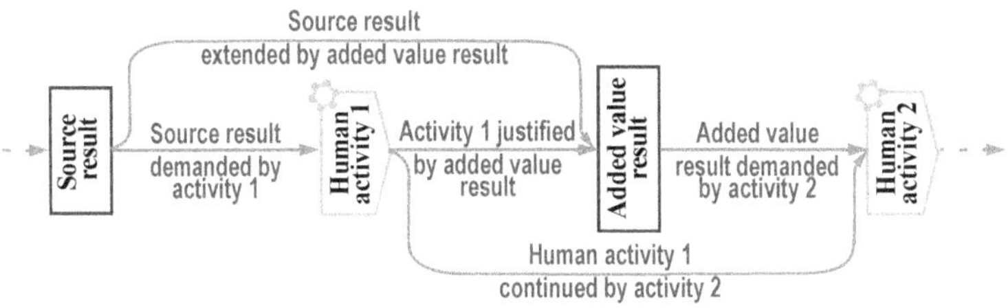

FIGURE 5-10 Specific meaning of arrows in a value chain

5.3 Elaborate concurrence in a value chain

5.3.1 Sequential activities

Executing *activities* in strict sequences is the simplest way to understand and execute. *Activity* 1 completes *result* 1, before *activity* 2 starts to develop *result* 2, and so on (see Fig. 5-11 below).

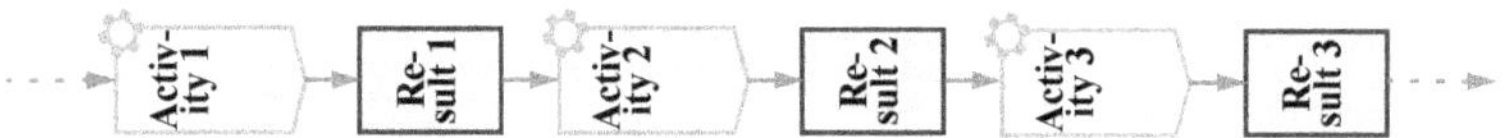

FIGURE 5-11 Truly sequential activities executed sequentially

Such a sequence is sometimes called *waterfall model* since a *value chain* execution resembles water flowing through falls.

5.3.2 Concurrent activities

Similar to previous sequential *activities*, concurrent *activities* are also very straightforward to understand and execute. Since the *results* and the *activities* are not interdependent, the *activities* can execute totally independently from each other (see Fig. 5-12 below).

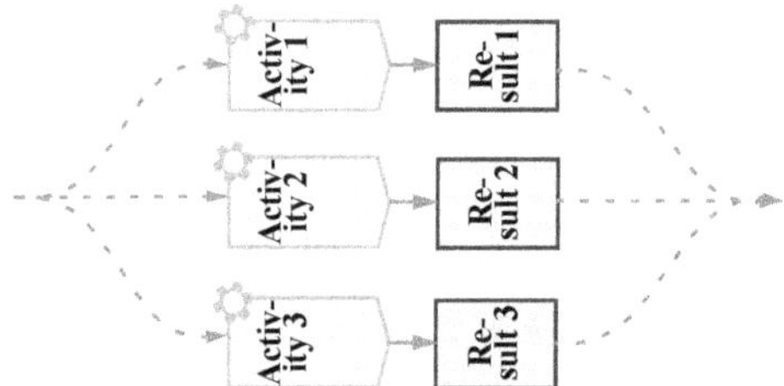

FIGURE 5-12 Truly concurrent activities executed concurrently

5.3.3 Sequential activities executed in full concurrence

In development organizations there is often a lot of stress, especially if *products* on the market evolve all the time, and it is always important to be ahead of competitors. A sequence of *activities* (as in Fig. 5-10 above) is seen as adding too much calendar time, and the consequence is that often even strictly sequential *activities* are started simultaneously in a parallel execution (see Fig. 5-13 below).

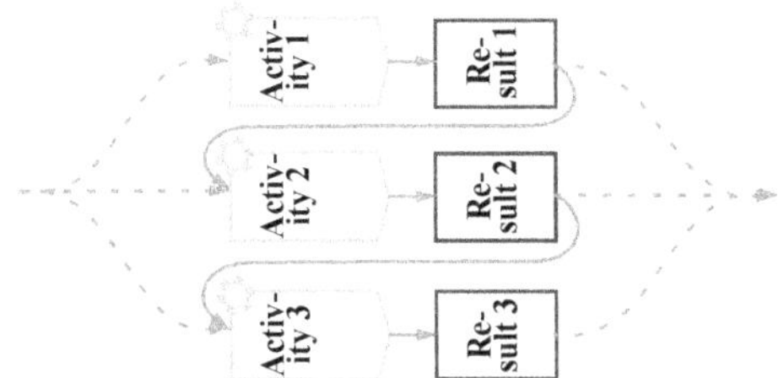

FIGURE 5-13 Sequential activities executed in concurrence

What will occur is the following:

- *Activity* 2 must have information from *result* 1 to be able to create and finish *result* 2. This is not very strange in daily life; for example, you need to get a bill before you can pay the amount specified on the bill.

- If *activity* 2 cannot yet get the demanded information from *result* 1 because it is not yet developed, but *activity* 2 is asked to develop *result* 2 anyway, *activity* 2 must more or less guess what information will be available later on in *result* 1. For example, before the bill is issued, the amount due might be approximated, and the amount that is deemed reasonable can be paid immediately. When the bill is finally issued, payment can be made once again to adjust the difference. Thus, additional work will occur, or to express it quite accurately, a waste of work occurs when the wrong amount is paid the first time.

Value chains are often mistakenly understood to be static in concept and execution. Many orthodox employees claim that no *activity* must start until its source *results* are totally completed. To bring some agility into such arguing, it is important to look more closely at consequences when executing sequences fully in parallel (see Fig. 5-14 below). The idea is to estimate how much waste will occur by this approach.

The model for estimate of waste is the following:

The level of waste in an *activity* at any point in time, caused by incomplete source *results*, is assumed to be directly proportional to the degree of completeness of that source *result*. For example, if a *result* is 50% completed when an *activity* needs to use it, the efficiency for the dependent *activity* becomes 50%, because the *activity* needs to waste time guessing at the 50% not yet found in the source *result*.

Waste in a value chain may occur in a forced activity, if previous results are not completed but have to be guessed at for a while.

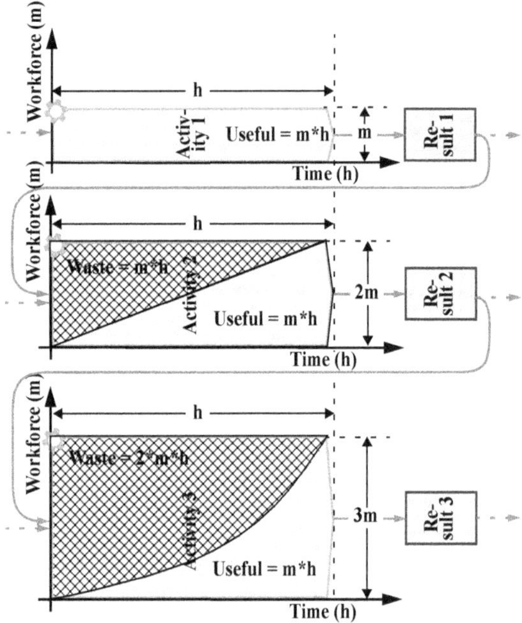

FIGURE 5-14 Waste / Usefulness ratio when executing sequential activities in full concurrence

So, the amount of waste in *activity* 2 at any point in time is assumed to be directly proportional to the degree of completeness of *result* 1. Consequently, the usefulness of *activity* 2 is 0% at the start (because *result* 1 is not yet available and *activity* 2 is forced into 100% guessing). Usefulness increases linearly ($y \sim x$) to become 100% in the end (because *result* 1 is complete and *activity* 2 needs no longer guess).

Similarly, the amount of waste in *activity* 3 at any time is assumed to be directly proportional to the degree of completeness of *result* 2. Since the usefulness of *activity* 2 is increasing linearly, the usefulness of *activity* 3 will grow quadratically ($y \sim x^2$).

Also note that workforce amount of *activity* 2 needs to be 2, and workforce amount of *activity* 3 needs to be 3, because the useful parts in total must be m * h in order to complete the assumed development.

As can be seen from Figure 5-14 above, the calendar time reduction is large, but waste is significant when executing many sequential *activities* in full concurrence. The next chapter will show a compromise of executing sequential *activities* in "half" concurrence.

5.3.4 Sequential activities executed in half concurrence

To examine executing sequential *activities* in half concurrence, make the presumptions in Figure 5-15 below.

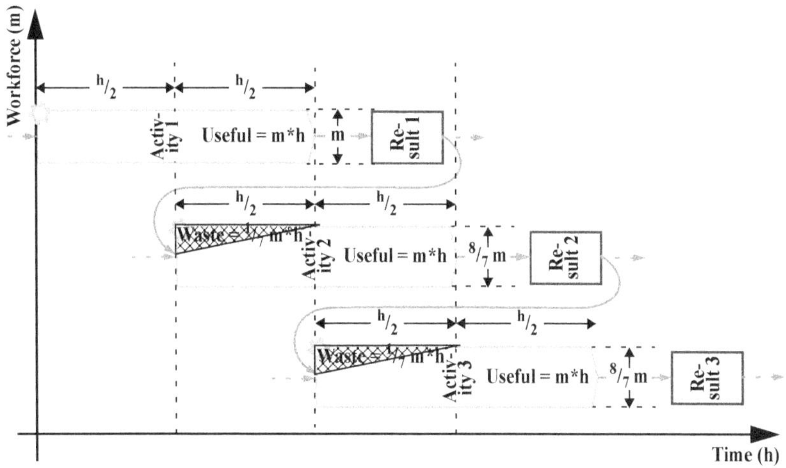

FIGURE 5-15 Assumed waste caused by sequential activities executed in concurrence

Compared to executing sequential *activities* in full concurrence, it is striking how much waste is saved by starting sequential *activities* when the preceding *result* is at least half finished. To execute all 3 *activities* takes 2 hours of calendar time, compared to the full sequential execution taking 3 hours.

Another nice benefit is that wastage doesn't increase with the number of *activities*, since waste always accumulates only from the previous *activity*.

To avoid using too much calendar time, activities in sequence must be executed in some concurrence, and waste is not significant if concurrence is kept moderate.

5.3.5 Waste depending on degree of concurrence

From the above chapters, a diagram can be made, showing approximately how waste will vary depending on degree of concurrent execution of strict sequential *activities* (see Fig. 5-16 below).

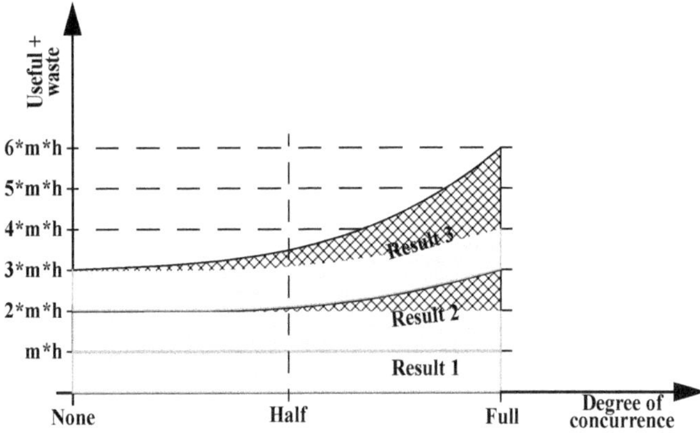

FIGURE 5-16 Correlation between value chain parallelism and waste

The conclusions from Figure 5-16 are:

- Sequential *activities* certainly don't need to be executed strictly sequentially, and subsequent *activities* can start before the needed *results* are completed without generating significant amounts of waste.
- When many *activities* are executed in too high parallelism, however, or any *activities* are executed with a high degree of parallelism, an unacceptable amount of waste is created.
- Since most of the waste occurs at the beginning of the concurrent *activities*, the waste could be somewhat prevented by lower *activity* staffing in the beginning and increasing it in the end. However, it might be difficult to continually reallocate employees in this way.
- An unpleasant side effect from a lot of waste with subsequent rework is that the consciousness of producing waste is quite demoralizing, risking a mediocre and inconsistent final *result*, which ends up in poor *product quality*.
- Concurrent execution is an extremely important and still misunderstood topic, which will be further examined in the follow-up to this book.

5.4 Value chain roles

There is much more to explain about *roles, project management*, and *line management*, but discussion of these topics will be continued together with other management overhead issues in a future book.

5.4.1 Cpdm definition

In *Cpdm*, *role* is defined according to Table 5-3 below.

TABLE 5-3 Definition of role

Aspect	Role
Definition	A set of competencies, authorities, and responsibilities of a person (typically a user, an employee, or third-party contractor)
Synonyms	Character, representative, duty, function, position
Symbol	Role

Roles are among the most important parts of *value chains*. Many *roles* are needed when executing the *value chain*, both for managing *activities* and *results*, but also for managing business gates. Note that one named person can be assigned one or many different *roles*, or none at all. The hat symbol is the role description, see chapter 6.10.1, page 108.

Roles are very useful to demarcate responsibilities and authority.

5.4.2 Cpdm definitions

In *Cpdm*, the standard roles *performer*, *preparer*, and *manager* are defined according to Table 5-4 below.

TABLE 5-4 Cpdm definition

Aspect	Performer	Preparer	Manager
Definition	The role of a human participating in a value chain activity	The role of a human contributing to a value chain result	The role of a human supervising performers and preparers in a value chain
Synonyms and examples	Operator, arranger, doer, coordinator, administrator	Creator, maker, issuer, former, approver, responsible person, releaser	Team leader, project leader, project manager, line manager
Symbol example	Developer team / Expected activity	Approver / Expected result	Project leader / Expected value chain / Expected result

In the *value chain* of a company, the employees execute all *activities* and produce all *results*. For this reason, in a *process* it is often defined who organizes *activities* and who produces the *results*. To avoid making the *process* person-dependent, *roles* instead of persons are defined to execute *activities* and produce *results*.

5.4.3 EXAMPLE: Some other roles

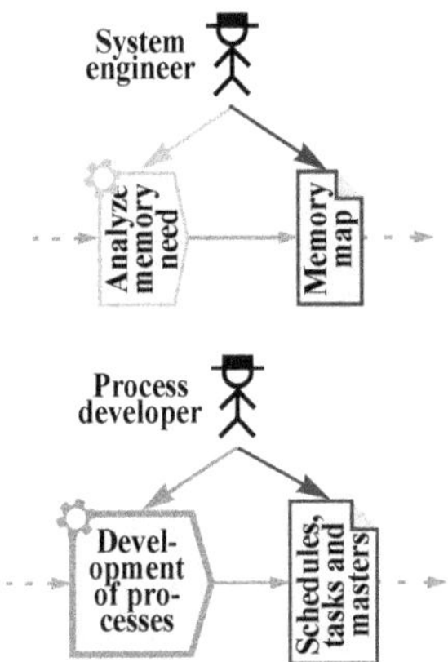

Many times in real development, it is the same *role* that is involved in an *activity* and the *result* from that *activity*. Then there is only the need for one hat, as in Figure 5-17.

FIGURE 5-17 Same role for activity and result

When actively managing *value chains* in a company, there is also a need of *process* support from *quality* organizations developing *processes*. *Process* descriptions, such as *schedules*, *tasks*, and *masters* are sometimes called *meta information* and are developed by *process* developers.

FIGURE 5-18 Process developer

5.5 Document value chain by a process

Larger complex organizations or high-*quality* organizations need to visualize work and attitudes to get more control and efficiency, which implies that the *value chain* must be documented. In chapter 6, "Utilize processes," from page 81, there are documented and visualized standard *process schedules* which *govern complex product development*.

Utilize processes

A good process never presumes a static way of working, but guarantees that nothing is overlooked when completed.

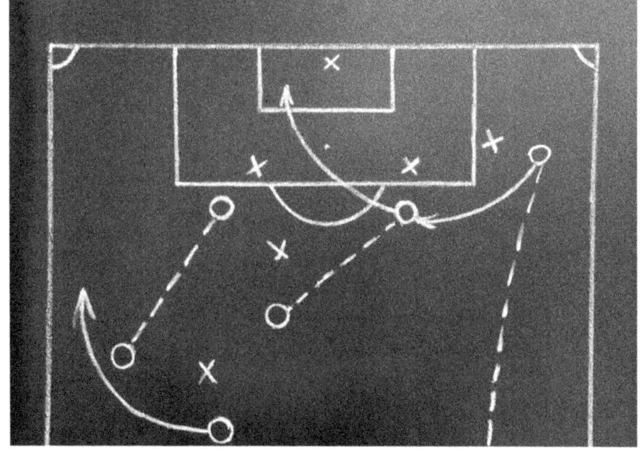

A process to score

6.1 About processes

6.1.1 Why processes

Processes are (like *architectures* or *programs*) among the most misunderstood concepts of engineering. When *processes* are viewed as efficient tools for development and not as obstacles to be sidestepped, organizations will more naturally apply and maintain them. This chapter will try to sort out the anxiety and fear that surround the use of this versatile tool.

6.1.2 EXAMPLE: Map the process from ongoing value chain

In chapter 5, "Identify value chain," page 65, the factual *activities* and *results* were examined. The *value chain* could be seen as a flow of *activities* only, as both *activities* and *results*, or as *results* only. To be able to control the *value chain* it is crucial to document it in these ways (see Fig. 6-1 below).

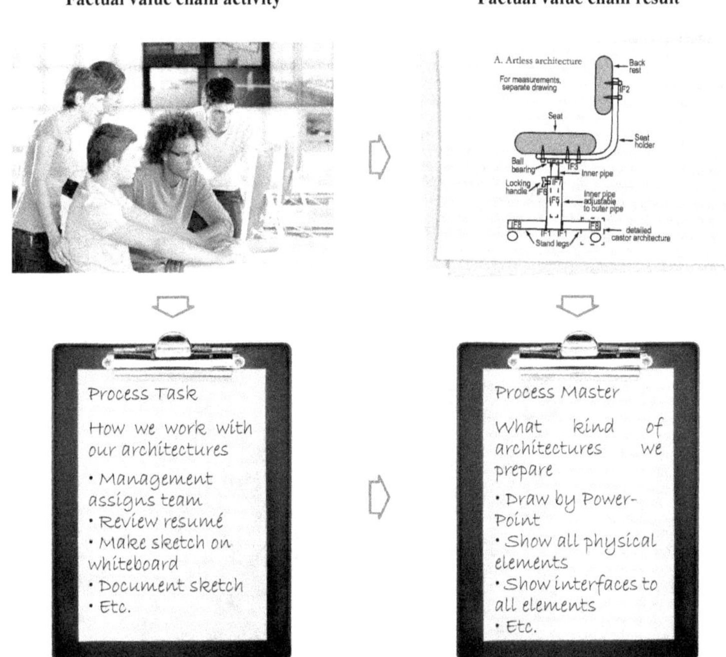

FIGURE 6-1 Factual activity and result documented as process task and master

The top half of Figure 6-1 illustrates the identified *value chain activities* and *results*. As the downward arrows illustrate, both of them might be documented as *process tasks* and *process masters*.

It is, in fact, very beneficial to distinguish between factual *activity* and the description of that *activity*, and similarly between the factual *results* and the description of those *results*. A consequence of keeping separate the factual reality and the description of it, is that there is now an opportunity to model them separately if they happen to differ.

Companies taking advantage of their value chains are documenting them by processes.

Differences between the factual reality and the documentation of it are often seen in *process* descriptions. *Managers* may like *process* descriptions that show more their desire of how developers should work, than descriptions of the messy way in which they really work.

Differences between the factual reality and the *process* description of it are not acceptable. The description should reflect the expectations of the *line management*. In the factual reality there will always be individuals who prefer to work in their own way instead of following the expected way of the *process* descriptions.

The factual work (the value chain) can differ a lot from what is prescribed in the process.

This difference is, of course, demoralizing and should be evaluated, and actions should be taken to change either the *process* or the factual *value chain*, in order to minimize this undesirable difference.

6.1.3 Process constituents

A *process* is a documented description which *governs* all aspects of a company. It is a great tool for everybody, and in particular for *line management*, to get efficiency and deliver the right *result* at the right time and of the specified *quality*. A *process* can be divided into several aspects, such as:

- A management aspect, which describes the management and organizational mechanisms. Organization maps are such examples.
- A *schedule* aspect, which *governs* the *value chain* with its *activities* and *results*.
- A *metrics* aspect, which describes measurements and follow-up performed in organizations.

Only the *schedule* aspect of a *process* is described in this book, although the other aspects might occasionally be mentioned. In a planned book "*Cpdm* technical overhead," all other *process* aspects will be central.

6.2 Process schedule constituents

Activities and *results* have already been discussed in the previous chapter as being two parts of a *value chain*. A documented *activity* is called a *task*, and a documented *result* is called a *master. Tasks* and *masters* are the con-

The process schedule may contain embedded schedules, tasks and masters.

stituents of *schedules*, which are the foundation for implementation, training, and follow-up of the *value chain*.

6.2.1 Cpdm definitions

As said before, there are a lot of ways to interpret the daily work of the *value chain*. Some developers are *activity-oriented* and see the *value chain* as *progressing activities*, and others are *result-oriented* and see the *value chain* as *cascading results*. Neither is right or wrong, but the concept is most easily understood when both the *progressing activities* and the *cascading results* are modeled.

In *Cpdm*, *task* and *master* are defined according to Table 6-1 below.

TABLE 6-1 Cpdm definitions

Aspect	Task	Master
Definition	A documented process regulation governing how to organize an activity demanding source results	A documented process regulation, governing how to organize a result to justify foregoing activity
Synonyms	Process task, activity description, work description, team instruction	Process master, result description, output explanation, expected delivery, document instruction
Symbol		

When *processes* are described formally, the purpose of arrows can be assigned proper names, which always reflect the direction from the arrow start to the arrow tip (again see Table 6-1 above).

A *task* is now an available regulative document for developers which *governs* their organized *activities*, and a *master* is another available regulative document for them which *governs* their *results*. Note the strict separation between the two, that a *master governs* only how to organize a *result* and tells nothing about how an *activity* is organized, and a *task governs* only an *activity* and tells nothing about how a *result* is organized.

The *task* document can be rather extensive, describing an extensive *activity*, if there are a lot of source *masters* to refer to, or if the *activity* is *complex*. *Results* are often parts of databases, and a *master* typically *governs* the structure of such a database. This book doesn't dig to the bottom of how *tasks* and *masters* look in detail, but a next book addresses how to organize in detail a generic document system for all documentation in a company.

6.2.2 Cpdm definition

The *schedule* is simply a package of *tasks* and *masters*, to make the *process* hierarchical in order to be scalable (see Table 6-2 below).

TABLE 6-2 Cpdm definition

Aspect	Schedule
Definition	A documented process regulation package of alternating tasks and masters, governing how to organize a value chain
Synonyms	Workflow, work model, quality manual, process map, development process, design process, business process
Symbol	

6.3 Partition, variant, terms, and govern

So far, some *process artifacts* have been defined, but little has been said about the relation between them. In the real world, the concept of *variants*, *partitions*, and *govern* are common, and in the *process* world these are also applicable. Note that architects often use the word *decompose* instead of the more general word *partition*.

To get good schedules, the concepts variant, partition, and instance must be defined and used.

6.3.1 Cpdm definitions

An *artifact* divided into *artifact partitions* is one of the most common relations between *artifacts*. In *Cpdm*, *partition* and *aggregate* are defined according to Table 6-3 below.

TABLE 6-3 Cpdm definitions

Aspect	Partition	Aggregate
Definition	Detaching parts from the whole	Merging parts to the whole
Synonyms	Divide, split, portion, decompose, fragment, separate	Merge, unify, incorporate, fuse, conflate

TABLE 6-3 Cpdm definitions

Aspect	Partition	Aggregate
General symbol (preferred in architectures; see chapter 9.3.2, page 344)		
Schedule symbols		

An artifact can be any thing whatsoever, for example a red car, an unfinished drawing, a stupid regulation, and so forth. Identities can be added to make the relation more unambiguous.

6.3.2 Hierarchy of schedules

To master large and *complex schedules*, it is beneficial to *partition* an overall *schedule* into several subordinated *schedules*, thus creating a hierarchical structure (see Fig. 6-2 below).

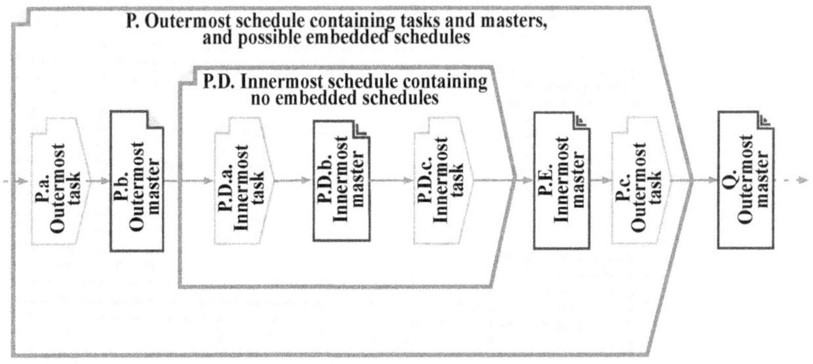

FIGURE 6-2 Schedule hierarchies

This *schedule* hierarchy has the same principle as the *product structure* with its *nesting levels* (see Fig. 9-10, p. 345). Any *nesting level* in the hierarchy can contain *schedules*, *tasks*, and *masters*, but by definition, the innermost *schedule* contains no embedded *schedules,* but only *tasks* and *masters* (like the innermost *nesting level* contains no *embedded black-boxes*). If an innermost *task* happens to be *complex,* just replace this *task* with a *schedule,* which now can be extended and detailed further *inward.*

Also note that *schedules, tasks,* and *masters* symbols are normally slightly shaded. For maximal clarity, when these symbols are contained within each other, they may be alternatingly shaded and not shaded just to distinguish them better from each other.

6.3.3 Cpdm definitions

Another important relation between *artifacts* is the *variants* of an *artifact.* In *Cpdm, variant, version,* and *view* are defined according to Table 6-4 below.

TABLE 6-4 Cpdm definitions

Aspect	Variant	Version	View
Definition	Variants are contemporary diversifications from a common artifact. A gradually extended diversification is often called version. The opposite is unification of variants to a primal artifact		Different perceptions of same artifact, depending on how the observer relates to the artifact
Synonyms	Revision, version, update, specialization, concretization		Aspect, perspective, impression, perception
General symbol	Primal artifact — Unify / diversify — Unify / diversify — Artifact : variant 1 — Artifact : variant 2		View of observer 1 — Artifact — View of observer 3 — View of observer 2
Process symbol	Architecture — Architecture :created — Extend — Architecture :draft — Extend — Architecture :reviewed — Extend — Architecture :corrected — Extend — Architecture :approved		
Architecture element symbols	Definition in chapter 10.21, "Cpdm definition," page 553		

A *view* is not the same as a *variant. Variants* are two different *artifacts* that have so much in common that it becomes meaningful to keep the same name for both, but add a *version* identifier to differentiate them from each other. Different *views* refer to one and the same *artifact,* but with different ways to observe it. For example, a *black-box* and a *white-box* are exactly the same box, which may be observed in two ways.

The direction of the arrows between *variants* needs a comment. It has become a convention for the arrowhead to turn to the common *artifact*, but sometimes it can be seen as pointing away from the common *artifact*.

6.3.4 EXAMPLE: Life cycle status

During development, *results* are most often changed and extended a number of times. Then it is very convenient to preserve the name of the *result*, but separate the degrees of completeness from each other by a *life cycle status* (see list below).

Useful *life cycle status* of *results* completeness during the *value chain* might be:

1. : Initiated (or created)
2. : Preliminary (or draft, sketch)
3. : Inspected (or reviewed)
4. : Updated (or corrected)
5. : Agreed (or accepted, approved)
6. : Change-controlled
7. : Released (or launched)
8. : Communicated
9. : Realized
10. : Verified
11. : Validated

Life cycle status is part of the *complex* topic *configuration management* and needs a thorough explanation in my next book, "*Cpdm* technical overhead."

6.3.5 Cpdm definition

In *Cpdm*, *terms* is defined according to Table 6-5 below.

TABLE 6-5 Cpdm definition

Aspect	Terms
Definition	Mutual governing regulation of a role between a development company and its employees, consultants, or hired contractors
Synonyms	Third-party contract, employment contract, role description, etc.
Symbol	Terms

Roles are among the most important parts of the *value chain*. Many *roles* are needed when performing *activities* and preparing *results*, and the way in which *roles* are *governed* is described by *terms*. Note that one named individual can be assigned one or many different *roles*, or none at all.

In a *process schedule* containing *roles*, most often in practice the *role* is named (for example "developer") and the *terms* are not well documented. It is assumed in this book that *terms* are documented for all used *performers*, *preparers*, and *managers*. In many cases, only *terms* for standard *roles* are documented, but *terms governing* any particular *roles* must be prepared when needed. Many types of *roles* will be used when accounting for *product requisites* and one of them is *role items* for costs of development work.

6.3.6 Cpdm definitions

For relations between *schedules*, *tasks*, and *masters*, see Table 6-5 below.

TABLE 6-6 Cpdm definitions

Aspect	Govern	Obey
Definition	Documented process regulation, controlling development in reality	Development in reality performed in accordance with documented process regulation
Synonyms	Regulate, rule, control, command, lead	Comply with, adhere to, follow, keep to
General symbol	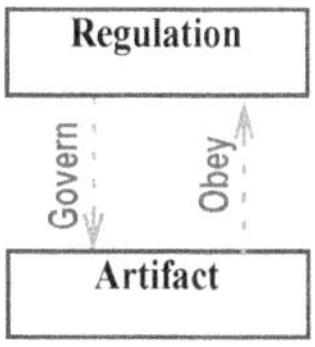	
Schedule and value chain symbol	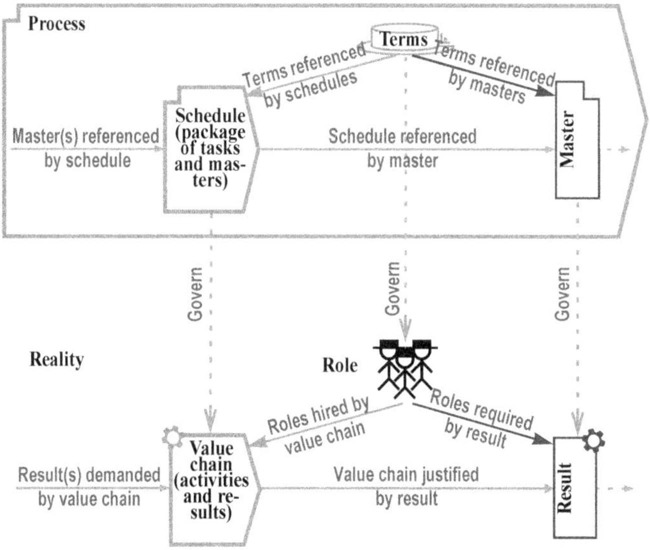	

6.3.7 Naming of process versus naming of reality

Cpdm clearly distinguishes between *process* regulation and factual reality, and assigns different names according to whether the regulation or the reality is considered. The question is how to practically handle this distinction, and how necessary it is to always point it out.

Often we don't distinguish too much between the name on a description of an artifact (regulation) and the name of the artifact itself (regulated). For example, on a map a mountain symbol can be called Matterhorn, which is the same name as used in reality. If we for some reason must distinguish between the two, we can refer to the "Matterhorn map symbol" and the "Matterhorn mountain."

The same is true for *process* regulations and the factual *value chain*. For example, **Pt.A. Design architectures** can mean the *architecture process* regulation or it can mean the *architecture* itself. If we need to distinguish, *Cpdm* now offers the possibility to refer to the **Pt.A. Design architectures** *task* regulation (with description flap in left figure) or to the **Pt.A. Design architectures** *activity* work (with cogwheel flap in left figure).

FIGURE 6-3 Example of identical naming of task and reality

6.3.8 EXAMPLE: Task, master, and terms documents

A palpable example will explain the tricky relations between above *artifacts* (see Fig. 6-4 below). Four *process* documents show how their initial parts appear, and how they reference each other, in order to build up and join to the *process* network shown in Figure 6-14, page 106, and Figure 6-15, page 107 below.

These *process* document might be rather thick, and cover everything from hard regulations to proposed recommendations. In a coming book such content, as well as how *process* documents are improved and are part of the *quality* work, will be shown in more detail,.

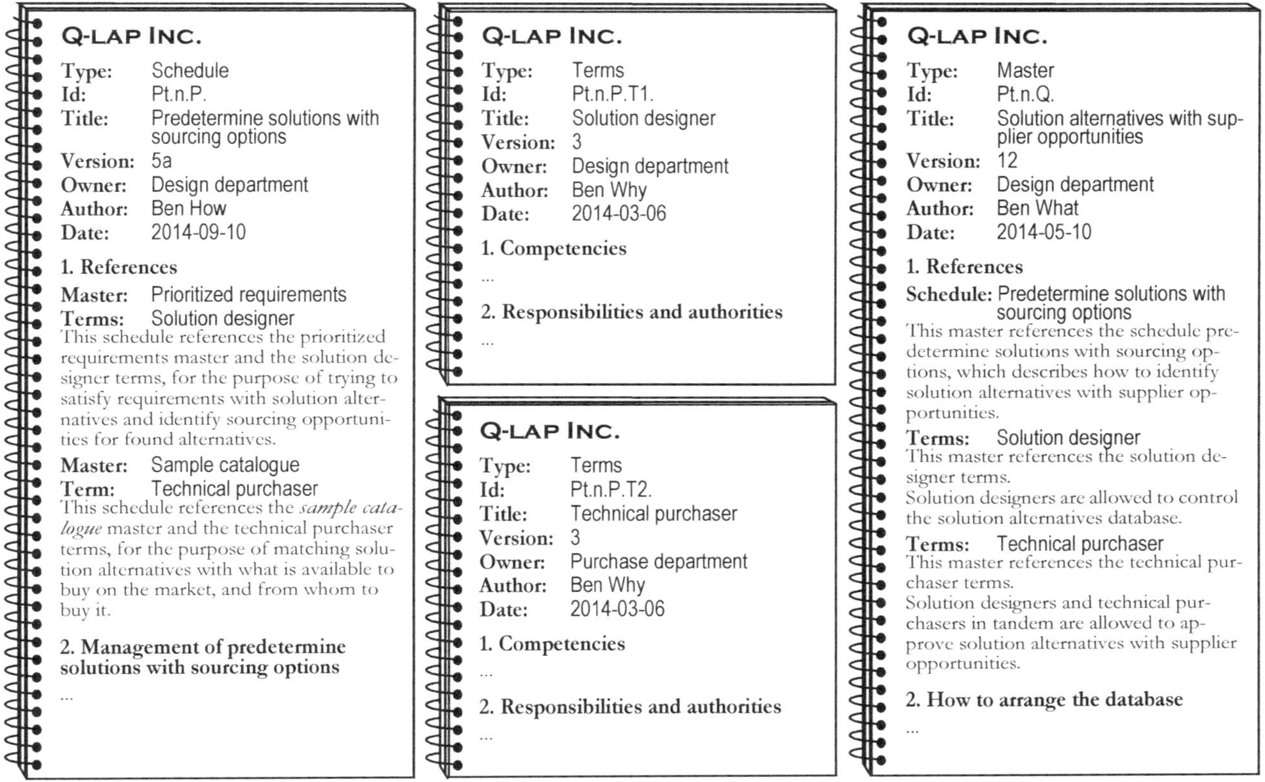

Q-LAP INC.

Type:	Schedule
Id:	Pt.n.P.
Title:	Predetermine solutions with sourcing options
Version:	5a
Owner:	Design department
Author:	Ben How
Date:	2014-09-10

1. References

Master: Prioritized requirements
Terms: Solution designer
This schedule references the prioritized requirements master and the solution designer terms, for the purpose of trying to satisfy requirements with solution alternatives and identify sourcing opportunities for found alternatives.

Master: Sample catalogue
Term: Technical purchaser
This schedule references the *sample catalogue* master and the technical purchaser terms, for the purpose of matching solution alternatives with what is available to buy on the market, and from whom to buy it.

2. Management of predetermine solutions with sourcing options

...

Q-LAP INC.

Type:	Terms
Id:	Pt.n.P.T1.
Title:	Solution designer
Version:	3
Owner:	Design department
Author:	Ben Why
Date:	2014-03-06

1. Competencies

...

2. Responsibilities and authorities

...

Q-LAP INC.

Type:	Terms
Id:	Pt.n.P.T2.
Title:	Technical purchaser
Version:	3
Owner:	Purchase department
Author:	Ben Why
Date:	2014-03-06

1. Competencies

...

2. Responsibilities and authorities

...

Q-LAP INC.

Type:	Master
Id:	Pt.n.Q.
Title:	Solution alternatives with supplier opportunities
Version:	12
Owner:	Design department
Author:	Ben What
Date:	2014-05-10

1. References

Schedule: Predetermine solutions with sourcing options
This master references the schedule predetermine solutions with sourcing options, which describes how to identify solution alternatives with supplier opportunities.

Terms: Solution designer
This master references the solution designer terms.
Solution designers are allowed to control the solution alternatives database.

Terms: Technical purchaser
This master references the technical purchaser terms.
Solution designers and technical purchasers in tandem are allowed to approve solution alternatives with supplier opportunities.

2. How to arrange the database

...

FIGURE 6-4 Example of task, terms, and master process documentation

Now all concepts are in place to support *process schedules* to describe all aspects of *value chains* in reality, which will be further presented in Figure 6-12, Pt:2. Cpdm generic advanced technical schedule (left half), page 104 and Figure 6-13, Pt:2. Cpdm generic advanced technical schedule (right half), page 105. All concepts are also in place to identify the differences between the factual *value chain* and the regulations-based *process*, which leads to measurements and *quality* discussions, a topic for the next book "*Cpdm* technical overhead" describing *quality* and *process management* overhead.

6.3.9 Schedule versus creativity

It is occasionally claimed that a strict *process schedule* kills creativity.

The traffic law is a kind of strict traffic *process* regulation. It controls what single individuals must *obey* to make traffic efficient and safe for all. As everybody knows, this is not comfortable for all, and some people violate this regulation for their own gain.

Process schedules are often misinterpreted so as to forbid personal initiatives and kill creativity.

This behavior has a lot to do with the culture, and with what kind of personality fits into the development organization. This book claims that there is no contradiction between making a rich travel experience and following the traffic rules. Companies applying *processes*, I am sure, will be strong survivors in an ever growing world competition.

It is often heard that it is impossible to work according to a *process schedule*, which produces everything in the most structured and logical way, because you learn so much during the development. It is claimed that there is no point in documenting everything every time, because after some learning it is no longer valid.

I agree with this, but with this difference: You should document everything from the start, but not ever forget to update the documentation when you have learned more. This is the only way to remember and share what was found, get it reviewed by other developers, and get it updated to be shared again.

For example, if you are good at specification and design, but hesitate to be innovative to create *white-box solution alternatives*, it can be a step forward to propose one particular *architecture* to some of the *requirements*. If you connect these *requirements* and the one *architecture* you proposed, you can probably express and document the solution you have found. When you have documented one solution with the connection between *requirements* and *architecture*, it is much easier to figure out more *white-box solution alternatives*. That might in turn lead to another *architecture* or another way to satisfy the *requirements*, which may open your mind to new *white-box solution alternatives*. But don't worry about going back and forth, just remember to update all *requirements* and *white-box solution alternatives* when the final *architecture* is firming up. And accept that this strategy consumes and even wastes a significant amount of time, but may be by far the best way to go forward in development.

6.3.10 How not to use a process schedule

Unfortunately, it is easier to exemplify how not to use a *process schedule*, than how to use it. There must always be a reasonable balance between accepting *process* guidance and deciding for yourself. To slavishly use the *schedule* creates a very rigid and paralyzed culture. On the other hand, if developers choose only what suits them best, there will be a mess of redundancy and at the same time a lot of shortage.

When using a schedule, there must be a balance between accepting guidance and deciding for yourself.

Even with a good *process, complex products* are difficult to develop. It is hard to imagine and plan everything in detail. For example, when the *product prototypes* begin to take shape, you understand a lot better how you should have done it instead. Also, the *process schedules* are only guidance for *activities* and *results*, and say nothing about how to develop the technique itself. For example, even with the best of *processes*, finding a design from *requirements* can go very wrong if technical competence is lacking.

Whenever something is better understood, go back in the process schedule and add anything necessary, and then trace downstream again, to also correct the consequences from this addition.

As concluded in chapter 5, "Identify value chain," from page 65, a sad misunderstanding is often that the *process schedule* must be followed in strict sequence. Not everything about a *product prototype* can be understood in a strict sequence, progressing strictly from strength to strength. A shortage may appear late in the development, and it is perfectly in order to trace the *process schedule upstream*, and add or change what was forgotten or not well enough understood from the start. But in this case, don't forget to trace *downstream* again, and correct consequences that have arisen from the newly made addition or change.

Another sad misunderstanding is that one *result* must be totally completed before any subsequent *activities* begin to use this *result*. Any *results* can be used at any degree of completion, but the risk for rework increases the less complete they are (see ch. 5.3, "Elaborate concurrence in a value chain," p. 74). This risk must be considered and balanced with the time available to await completion. Often, time to market is very limited, and some risk for rework must be accepted. However, there is a breaking point, when the time for rework exceeds the time saved.

A *process schedule* can even be used as a quick way of scouting. Before *product development* starts, spend some time walking through the *schedules*, with the planned *product* in mind. A lot of easily overlooked things will quickly become apparent, for example, how to manage *requirements*, when and how to find *samples*, how to document *architectures*, who will realize and verify and so forth.

A *process schedule tailored* to own organizations and *products* is also very good to use in training. Everything gets more serious when training new employees or subcontractors how to *govern* according to *process schedules*.

6.3.11 Applying a process schedule

A development *process schedule* describes what must be done before manufacturing a *product*. The good thing with a *schedule* is that it can always be used as a checklist, to follow up and see that everything is being considered during a hectic and turbulent development. Of course, the more turbulent the development, the more rework will be needed, but there is sometimes no other way, especially if a *product* is new to an organization or the first since a *paradigm shift* occurred for the whole world.

> Almost everybody can agree that a schedule is great to use at least as a checklist, to watch that nothing important gets overlooked.

For example, development of this book has resulted in much rework from the start. It was simply impossible to write with the necessary *quality* from the first page to the last in one single go, particularly when the challenge was not only to write, but also at the same time to organize my experience and develop the *Cpdm* model itself. Of course, a good *process* would have brought order and structure to the turbulence, and heightened the *quality* and decreased the development time considerably. Naturally, it was useful to know how to be the *project leader* over yourself, and to understand the basic *phases* of development, but unfortunately, as far as I know, there is no Cbdm (Complex book development model) available.

So skip the outside-in and inside-out quarrel which only leads to the chicken-or-the-egg dead end, and try a more pragmatic way. Identify what part of the *process* seems to be the most valuable for the particular *target product*. Try to keep the logic sequence as intact as possible and emphasize important *activities* with special care.

6.3.12 Tailoring a process schedule

> To be efficient, schedules must be adapted to the specific organization and product, and introduced step by step.

One *product* is not like another, and the same goes for departments. A standard *process schedule* covers all imaginable details, which may be unnecessary or unusable for a specific *product* or department.

For example, a simple *product* is developed by a few engineers. Everybody understands the *product* and it is more or less obvious how to develop it.

On the other hand, a *complex product* with mechanics, electronics, and *programs* is not entirely understood by any single mind in the organization, and the *value chain* becomes crucial for everybody to work together in an efficient way.

Cpdm contains a generic *process schedule* for development of *complex products*, and it covers everything imaginable that must be taken care of. This *schedule* can preferably be adapted to the particular *product* to be developed as well as to the specific development organization. This is often called *tailoring* the *value chain* (see ch. 6.13, p. 160), and often there is a *quality* and *process* organization to support this.

This book also uses a lot of graphic symbols to enhance educational value and comprehension when learning *Cpdm*. These *Cpdm* symbols can also be *tailored* and adapted to practical usage with editors and all other tools at hand.

6.4 Symbols used throughout this book

6.4.1 Types of symbols

Process schedules can be made somewhat simpler than shown in this book, without losing too much significance, but for the purposes of this book it is better to try to catch the *complexity* by using symbols reflecting every aspect of it.

All in all, to get a structure into the *schedule* legends, the symbols used in this book are categorized according to Figure 6-5 below.

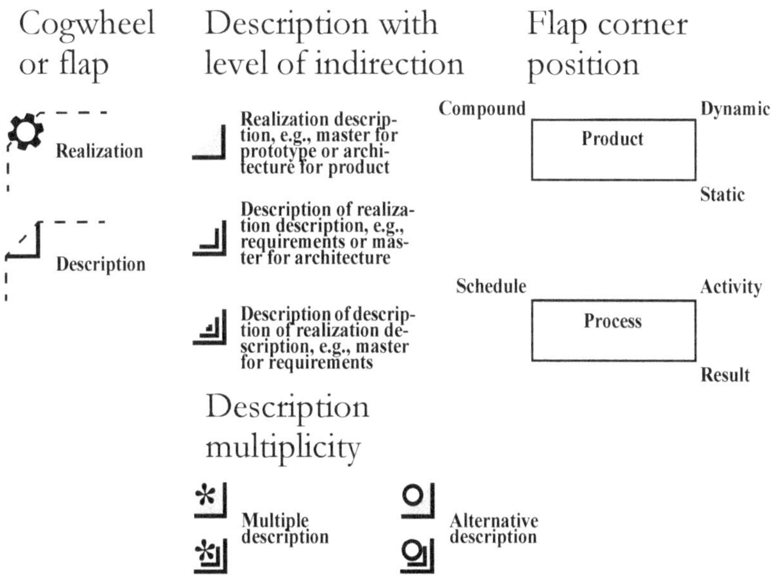

FIGURE 6-5 Flaps and corners of symbols

6.4.2 Symbol collection

Most of the symbols used in this book are shown in Table 6-6 below. How they are used will be explained more in coming chapters.

TABLE 6-7 Cpdm definitions

Aspect		Realization	Realization description	Descriptions of realization description	Description of description of realization description
Dynamic	Value chain and process	Activity in progress	Task (governs activity); Terms; Gate board		
Dynamic	Low-level program instruction	Linkable machine instructions	Conditional jump; Unconditional jump; Subroutine call; Input or output; *address label*:; Other instruction		
Dynamic	High-level program instruction		Sequence of instructions; Iteration of instructions; Selection of instructions		
Dynamic	General system	Machinery, electronics or linkable programs	Dynamic element; Dynamic interface	Behavior requirements	
Static	Value chain and process	Realization	Reference schedule or task; Terms; Master (governs realization)	Master (governs description of realization)	Master (governs description of description of realization)
Static	program variables	Linkable machine variables	High level variables; Assembler variables		
Static	General system	Mechanics realization	Static element description; Connection interface; Boundary interface	Restriction requirement	
Compound of static and dynamic	Value chain and process	Value chain in progress (activities and results)	Reference master(s); Schedule governs value chain (package of tasks and masters); Terms		
Compound of static and dynamic	Program	Executable program (instructions and variables)	Thread; Object instance	Class shape requirement	
Compound of static and dynamic	General system	Realization containing mechanics, electronics and executable programs	Black-box; Interface; White-box; Bus interface; 8; Compound element; Managed interface : revision	Compound requirement	

Note that these symbols may be rotated at any right angle to better fit into pictures in this book. Also note that development symbols are by default slightly shaded, but for clarity of *complex schedules*, these symbols might alternate between slightly shaded and clear.

6.5 Cpdm process

6.5.1 Requirements of Cpdm

There are some general *requirements* to satisfy by *Cpdm*.

- *Cpdm* shall be described and documented, in order to be available for training and implementation in an organization.

- *Cpdm* shall be as simple and comprehensible as possible. However, *Cpdm* shall not be simplified to a degree that is insufficient to explain and support the *intrinsic complexity* in *product development*.

- *Cpdm* shall be general enough to fit all kind of *products*, from static mechanics to electronics and dynamic *programs*. However, the generality shall not be at the expense of sufficient detailed explanations.

- *Cpdm* is too large to be completely covered by one book. This first book shall concentrate on basic technical parts of *product development*. After this has been described, it becomes easy to expand on the subject in the next book "*Cpdm* technical overhead," for example, discussing *quality management, configuration management*, and *project* and *line management*.

- *Cpdm* shall be formal enough to support almost any grade of *product quality*. If such high-*quality* grade is not needed, *Cpdm* shall be easy to simplify.

- *Cpdm* shall not only rely on descriptions of *schedules, tasks*, and *masters*, but shall be mainly explained through the use of examples.

Now all definitions and *requirements* are completed to settle *Cpdm* in its context (see Fig. 6-6 below). This nice figure relates the wide world, the development organization, and the manufacturing organization.

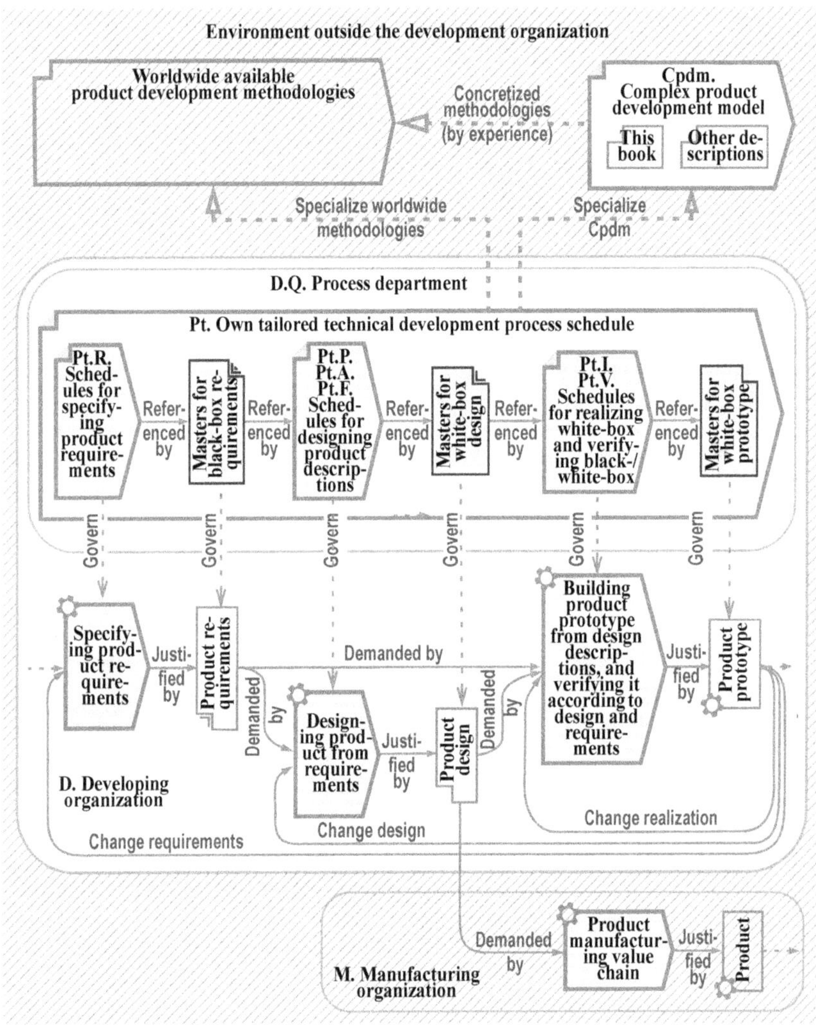

FIGURE 6-6 Connection between methodology, own tailored process, and manufacturing process

Generic *process schedules* are part of *Cpdm* later explained in this book. From these, own *tailored* development *process schedules* can be derived.

The picture above contains a lot of information and is very important for most developers to grasp, and in particular for engineers developing *process schedules*.

6.5.2 Range of Cpdm overall schedules

This book presents general *process schedules* from broadest survey to detailed *tasks*, and also some *tailored schedule versions* (see Fig. 6-7 below).

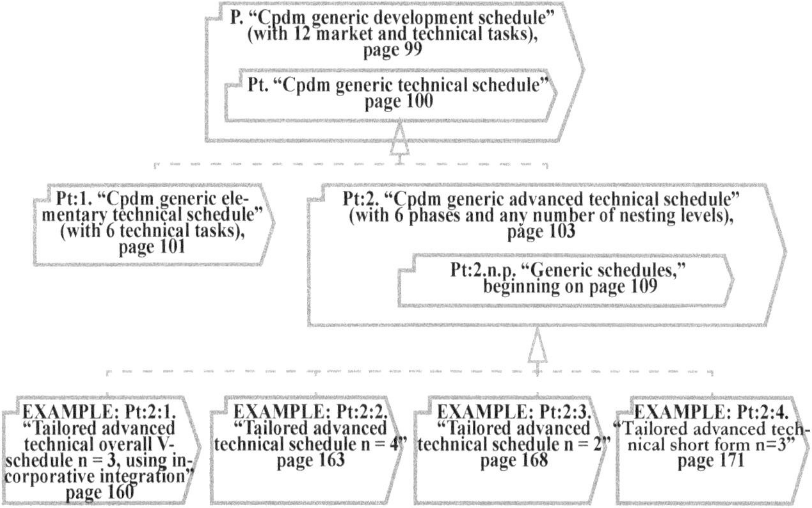

FIGURE 6-7 Cpdm process schedules explained in this book

Observe that *product development* and *prototype development* are rather similar but differ a bit in purpose. *Prototype development* results in a *product prototype* for *verification* before manufacturing starts, whereas *product development* results in *product* design descriptions for *product* manufacturing.

6.6 P. Cpdm generic development schedule

The *process schedule* **P. Cpdm generic development schedule** covering the entire development *stage*, is shown in Figure 6-8 below. The development *schedule* contains a number of *tasks* plus one embedded *schedule*, which *govern* the core technical part of the development *schedule*. Each single *task governs* one development *phase*, and the technical *schedule governs* six *phases* (illustrated later).

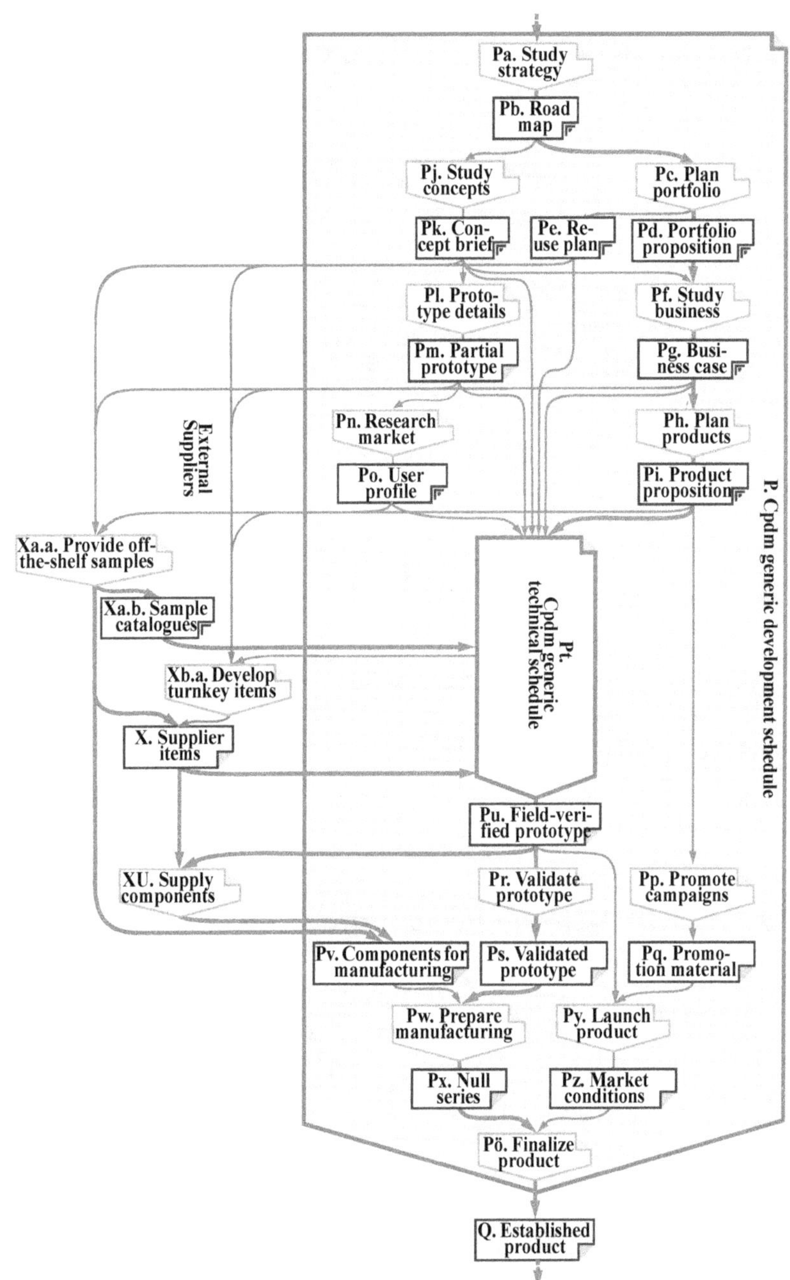

FIGURE 6-8 Cpdm generic development schedule (definition)

Note that the **P. Cpdm generic development schedule** is assigned the identification **P.** and the embedded **Pt. Cpdm generic technical schedule** is assigned the identification **Pt.**

The **P. Cpdm generic development schedule** is intended to be used for both technical and market organizations needing a *process schedule* covering all development *phases* from an idea to a launched *product*. This *schedule* was applied for the chair development starting with chapter 4.1.2, "EXAMPLE Chairs: Development," page 38.

6.7 Pt:1. Cpdm generic elementary technical schedule

The inner core of Figure 6-8, page 100 is the technical *schedule*, named **Pt. Cpdm generic technical schedule**, which consists of two *variants*, the **Pt:1. Cpdm generic elementary technical schedule** and the **Pt:2. Cpdm generic advanced technical schedule**.

The **Pt:1. Cpdm generic elementary technical schedule** is intended for *products* that are technically not overly *complex* to develop, and was consequently applied in the chair example, beginning with chapter 4.1.2, page 38.

The elementary technical *schedule* **Pt:1. Cpdm generic elementary technical schedule** is illustrated in Figure 6-9 below. Note that this *schedule* is denoted as the elementary *variant* 1 of the **Pt. Cpdm generic technical schedule**. For an alternative *view* of this *schedule*, see **Pt:1. Cpdm generic elementary technical schedule (arrow view)**, page 162.

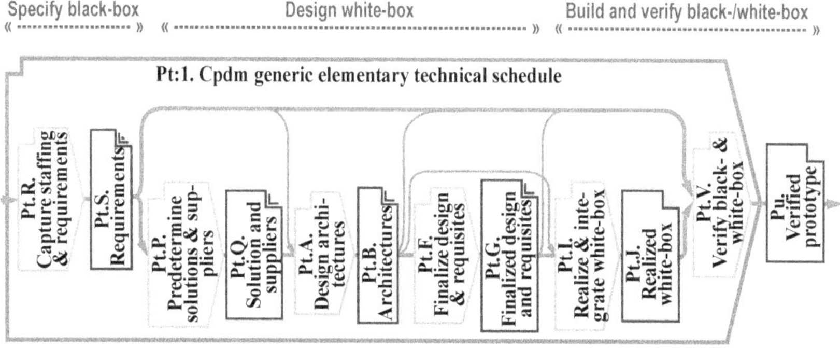

FIGURE 6-9 Cpdm generic elementary technical schedule (definition)

6.7.1 Scaling up the elementary technical schedule

As said, the *process schedule* in Figure 6-9 above is intended for *products* with limited *complexity*, and now needs to be scaled up to govern full-blown *product complexity*.

The scaling-up principle is to *decompose* the *product* into hierarchical *black-boxes* with a surrounding *environment*, which are all *governed* by separate *tasks*. The result is a hierarchical *process* (see Fig. 6-10 below) with as many layers as the *architecture* has *nesting levels* of *black-boxes* and keeps the same *phases* as the elementary *process*. Note that this *schedule* is denoted as the second *variant* **Pt:2. Cpdm generic advanced technical schedule.**

Furthermore, what was simple *tasks* in **Pt:1. Cpdm generic elementary technical schedule,** page 101, is now scaled up to *schedules*, which in turn contain *tasks* and *masters*. The *environment nesting level* n = 0 *governs* mapping of the *environment*, and development of the *environment interfaces*. All *inward nesting levels* n ≥ 1 *govern* the development of each *embedded black-box*.

For example, the **Pt.F. Finalize design & requisites,** page 101, has now scaled up to the **Pt.0.F. Finalize design of environment interfaces with product requisites,** and the multilevel **Pt.n.F. Finalize design of white-box with product requisites**.

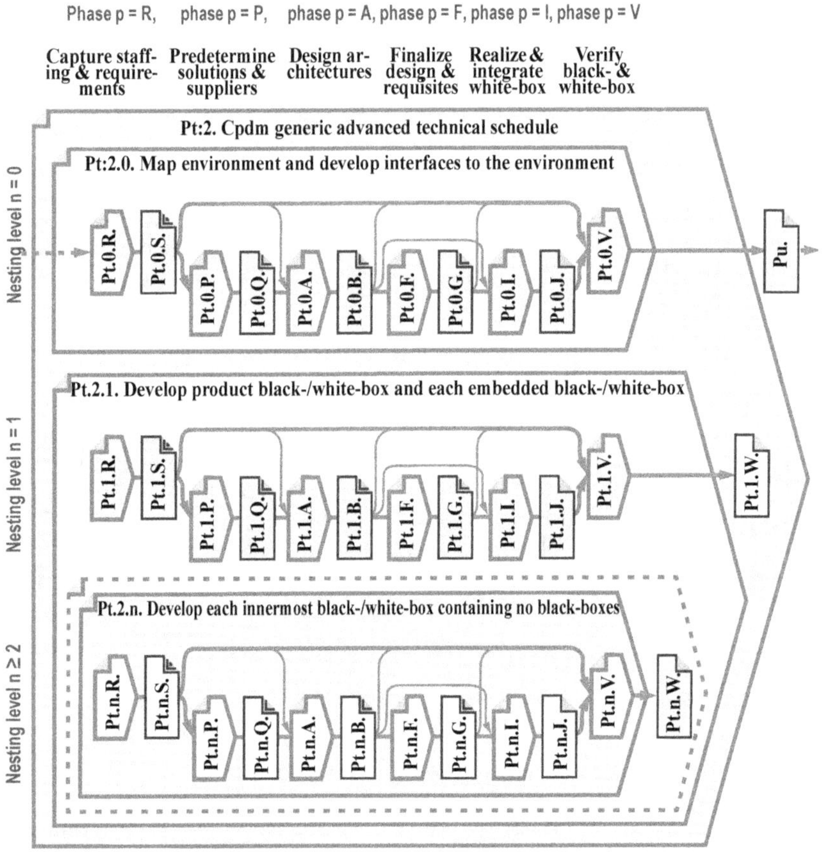

FIGURE 6-10 Scaling up the schedule by using architecture hierarchy

6.7.2 Tying together the schedule nesting levels

As said, the point with hierarchical *nesting level schedules* is to allow the *process* size to scale up at the same pace as the *product* size scales up (see Fig. 6-11). Consequently, there must be relations between *nesting levels* (see the named arrows below).

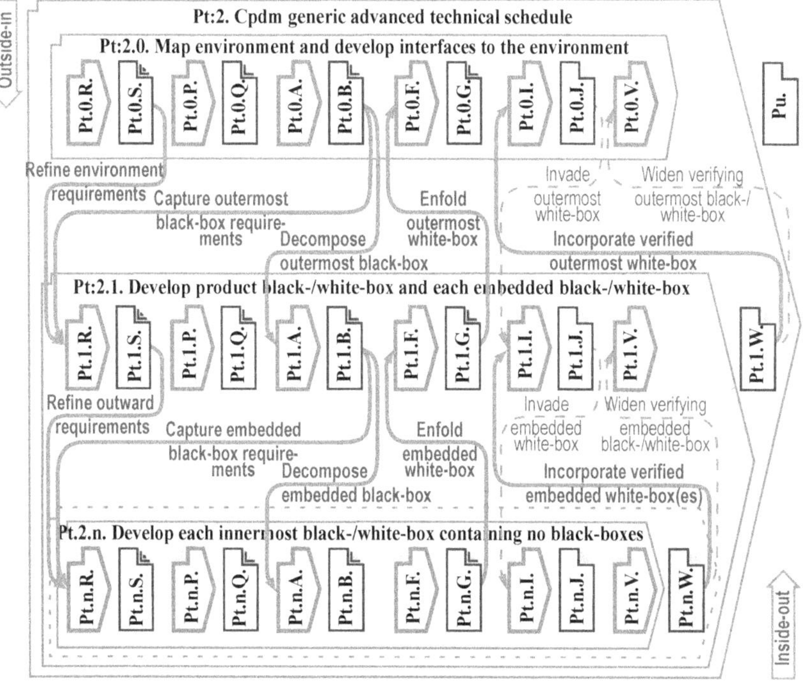

FIGURE 6-11 Scaling up the schedule by using architecture hierarchy

The normal way to realize & integrate white-boxes is denoted in *Cpdm* as *incorporative integration*. An alternative is *invasive integration* and *implicit verification*, which is marked with dashed relation − − →. A lot more about *integration* will be explained later in this book.

6.8 Pt:2. Cpdm generic advanced technical schedule

The scaled-up generic *nesting level schedule* is illustrated below by two different *views*. The definition *view* is illustrated in Figure 6-12 and Figure 6-13 and the survey *view* is illustrated in Figure 6-14 and Figure 6-15. *Environment nesting level* n = 0 is called **Pt:2.0. Map environment and develop interfaces to the environment**, and the *nesting levels* n ≥ 1 are collapsed into one multilevel *schedule* called **Pt:2.n. Develop outermost black-/white-box and each inward embedded black-/white-box**.

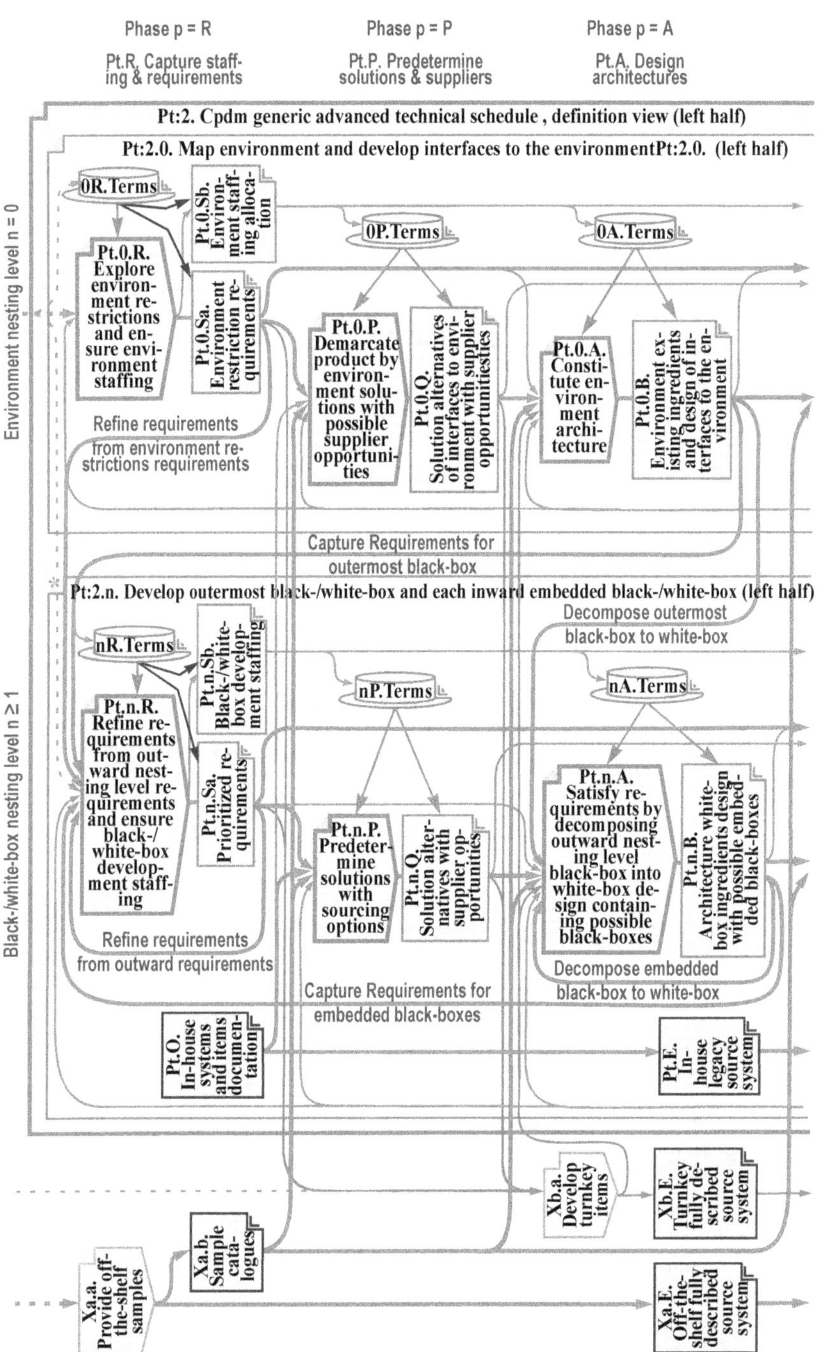

FIGURE 6-12 Pt:2. Cpdm generic advanced technical schedule (left half)

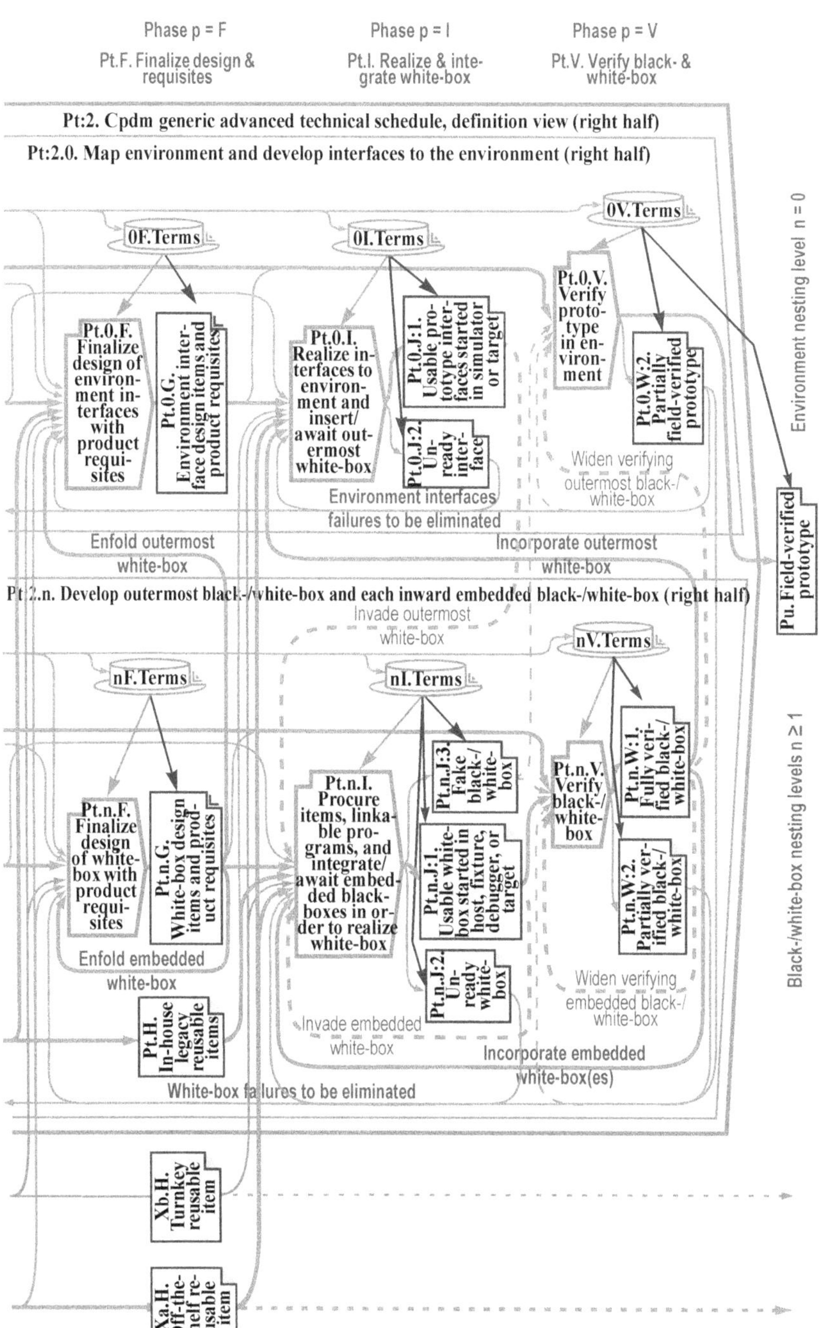

FIGURE 6-13 Pt:2. Cpdm generic advanced technical schedule (right half)

6.9 Pt:2. Cpdm generic advanced technical schedule

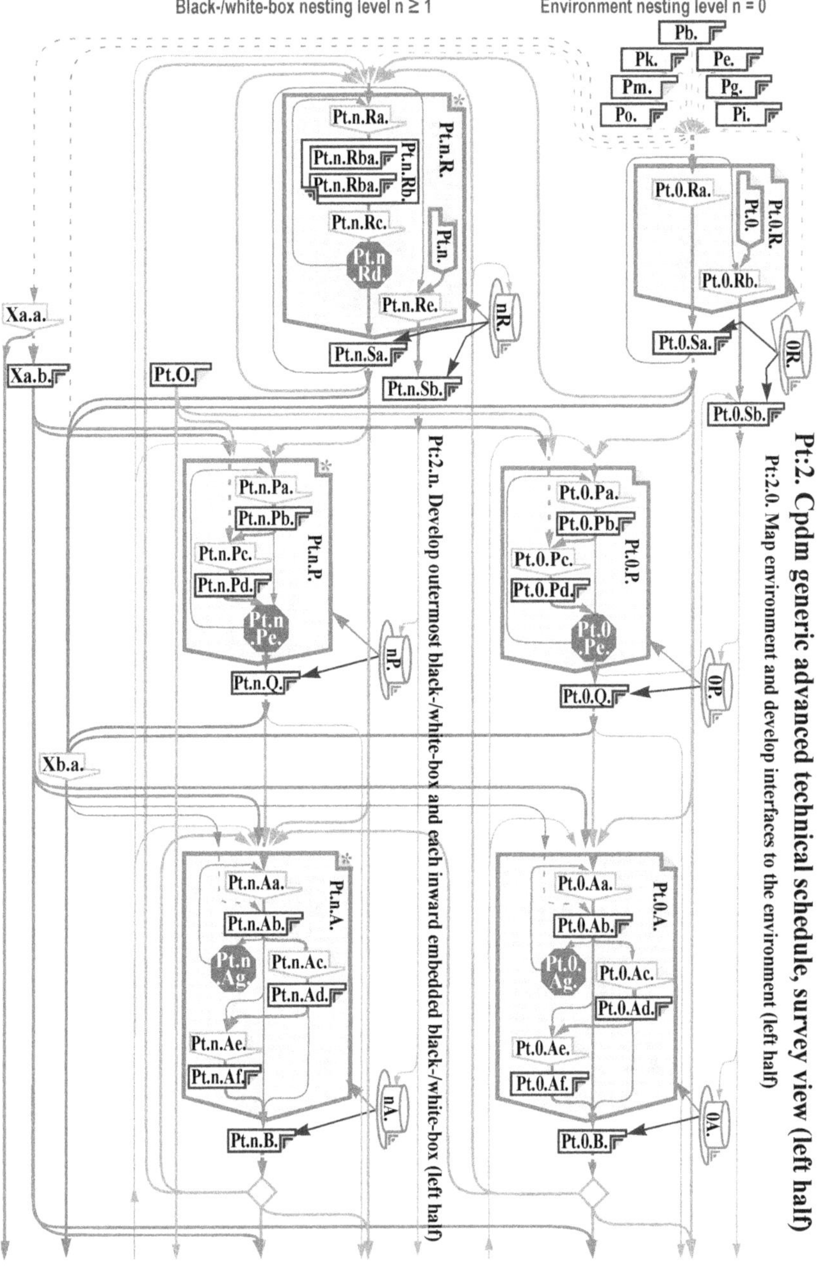

FIGURE 6-14 Pt:2. Cpdm generic advanced technical schedule, survey view (left half)

Pt:2. Cpdm generic advanced technical schedule, survey view (right half)

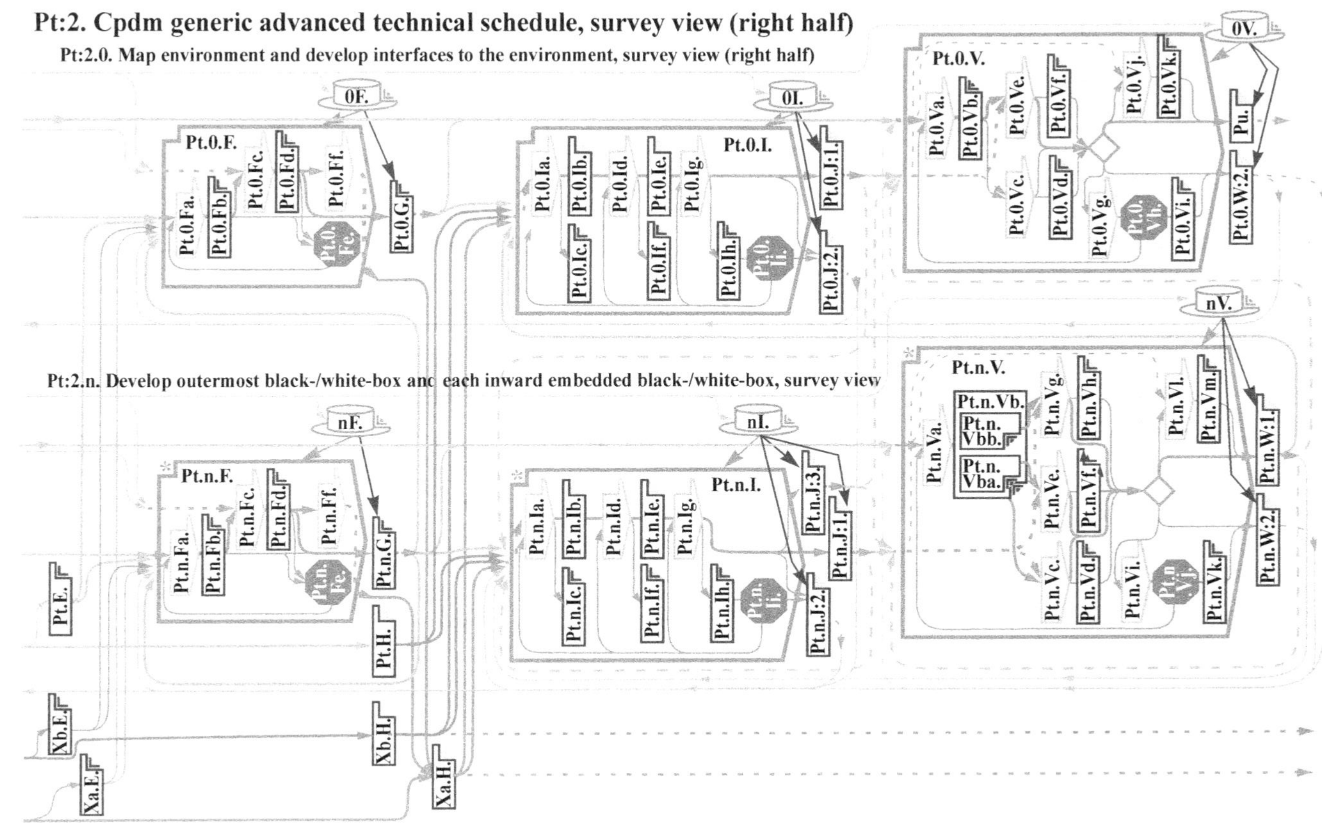

FIGURE 6-15 Pt:2. Cpdm generic advanced technical schedule, survey view (right half)

Complete wall charts and other illustrations are downloadable in high resolution from the *Cpdm* website www.cpdm.com.

To exemplify *tailoring*, this *schedule* has been *tailored* to several *schedules* illustrated below (see ch. 6.13, "EXAMPLE: Tailored process overall schedules," p. 160 below). These *tailored schedules* have in turn *governed* the 4 vast development examples of this book.

6.10 Pt:2. Cpdm generic advanced technical schedule

6.10.1 np.Terms

Pt:2. Cpdm generic advanced technical schedule

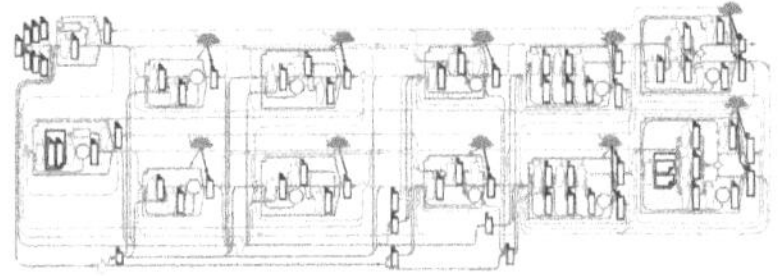

This chapter explains the *terms*:

• The *terms* are part of **Pt:2. Cpdm generic advanced technical schedule, survey view,** pages 106 and 107 (see left thumbnail).

• The *terms* explained in this chapter are highlighted in left thumbnail (the ten small hats).

All *process schedules* and *masters* are connected to *role* descriptions, in order to point out which staff competencies must be allocated to develop specific parts of the *product*.

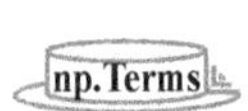

In *Cpdm role* descriptions are called *terms*, and are illustrated by a hat symbol (see Fig. 6-16, left).

FIGURE 6-16 np.Terms (nesting level n and phase p)

Staff allocation to *roles* for performing *activities* and *results* are important and necessary, since staffing is crucial for solving technological and engineering matters. Staff allocations are made in the **Pt.R. Capture staffing & requirements** *phase* (see Fig. 6-17, p. 110 and Fig. 6-23, p. 130).

Terms may be any competency *role* description, such as *specifier*, allocator, *architect, designer,* integrator, verifier, team leader, team member, reviewer, and so on. For in-house personnel, some parts of *terms* from the employment contract might be used, but for external developers a formal contract may be needed. *Terms* are not crucial for the technical *value chain*, and are introduced mostly for completeness of this basic book. More about *terms* will come in the next book "*Cpdm* technical overhead."

6.11 Pt:2.0. Map environment and develop interfaces to the environment

Pt:2. Cpdm generic advanced technical schedule

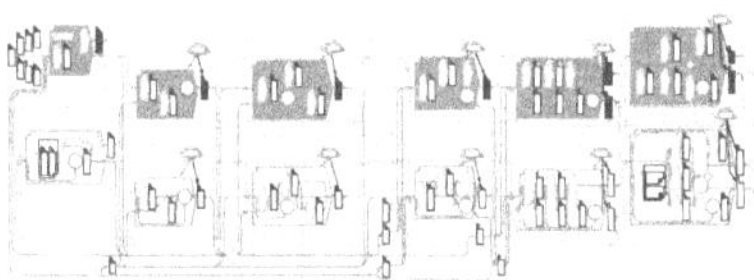

This chapter explains six *schedules*:

- These *schedules* are part of **Pt:2.0. Map environment and develop interfaces to the environment,** pages 106 and 107 (see left thumbnail).

- The six *schedules* explained in this chapter 6.11 are highlighted in left thumbnail.

This chapter contains explanations of the six *schedules* for the *environment nesting level* n = 0. Next chapter 6.12, page 128, proceeds with an explanation of the six *schedules* for all embedded *schedules* at all *nesting levels* n ≥ 1.

Some of these twelve *schedules* have some *complex tasks*, which are transferred to *schedules* counting embedded *tasks* and *masters*.

These *schedule* explanations mainly have the purpose to guide developers in their daily work. For space reasons, explanations in this book are far too short compared to what is often needed in an organization with inexperienced or newly employed developers. How to develop *schedules* with *tasks* and *masters* in detail will be covered more in a second book "*Cpdm* technical overhead".

6.11.1 Pt.0.R. Explore environment restrictions and ensure environment staffing

Pt:2. Cpdm generic advanced technical schedule

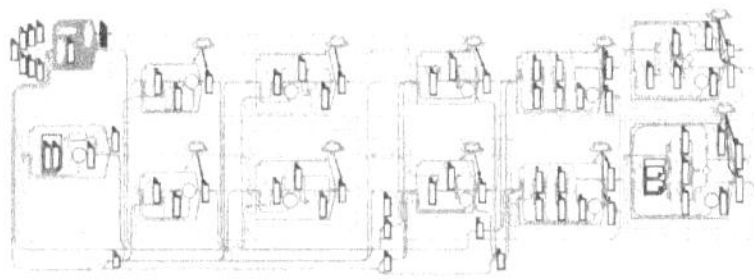

This chapter explains a *schedule*:

- The *schedule* is part of **Pt:2.0. Map environment and develop interfaces to the environment,** pages 106 and 107 (see left thumbnail).

- The *schedule* **Pt.0.R. Explore environment restrictions and ensure environment staffing** is explained below (highlighted in left thumbnail).

When starting to work according to the **Pt:2. Cpdm generic advanced technical schedule,** the first thing is to explore restrictions in the *environment* of the *product* to be developed (see Fig. 6-17 below).

Capturing *environment restriction requirements* is most probably a matter to be performed in-house by the developing organization, since this is too intimate to the *product* to be outsourced to a third party. This is the main reason why staffing allocations are also prepared in-house in this *phase*. Succeeding *phases* are easier to outsource.

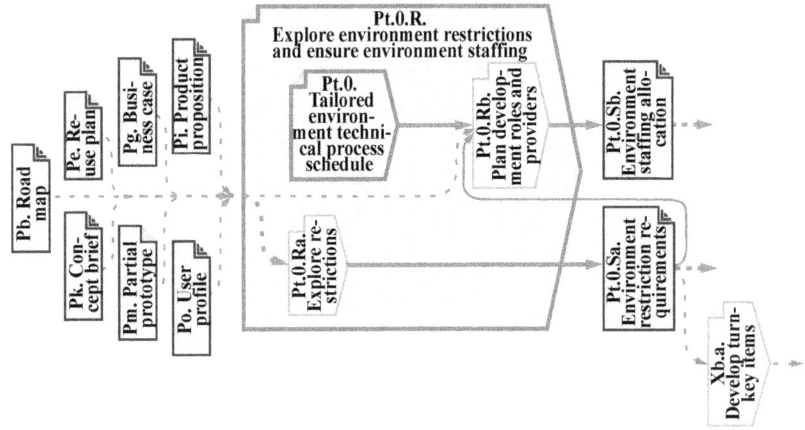

FIGURE 6-17 Explore environment restrictions and ensure environment staffing schedule

Masters referred to in this schedule

Many of the *schedules* in the **Pt:2. Cpdm generic advanced technical schedule,** page 104 (particularly the earlier *phases*) reference external *masters* in the overall **P. Cpdm generic development schedule,** page 100. These *masters* are not explicitly explained by this book, but the *results* are shown in the chair example. All of these earlier *upstream results targeting* the *environment* may now be inserted as *environment restriction requirements.*

Pb. Road map. This *master* is not explicitly described in this book but instead see description in chapter 4.2.2, page 41.

Pk. Concept brief. This *master* is not explicitly described in this book but instead see description in chapter 4.3.1, page 41.

Pe. Reuse plan. This *master* is not explicitly described in this book but instead see description in chapter 4.4.1, page 42.

Pg. Business case. This *master* is not explicitly described in this book but instead see description in chapter 4.5.1, page 44.

Pm. Partial prototype. This *master* is not explicitly described in this book but instead see description in chapter 4.6.1, page 45.

Po. User profile. This *master* is not explicitly described in this book but instead see description in chapter 4.7.1, page 46.

Pi. Product proposition. This *master* is not explicitly described in this book but instead see description in chapter 4.8.1, page 47.

Schedule content

Pt.0.Ra. Explore restrictions. The first *task* in this first *schedule* is to explore and document all kinds of restrictions in the *environment* affecting the *product* to be developed. There may be any kind of *environment restriction requirements* to capture, for example, technical restrictions in the *environment interfaces*, or there may be legal restrictions for how the *product* may appear or behave in the environment.

Pt.0. Tailored environment technical process schedule. This is a *tailored process schedule* for the *environment nesting level* = 0, adapted for the specific development organization and for the particular *product* to be developed in a particular *environment*, and should be provided for the entire development organization. Some *tailored schedules* are prepared for the examples of this book (see ch. 6.13, "EXAMPLE: Tailored process overall schedules," p. 160 below).

Pt.0.Rb. Plan development roles and providers. Among the first things to do after capturing *environment restriction requirements*, is planning which competency is needed to develop the *environment nesting level* = 0. If the *environment* is rather well known, there may be competent in-house *roles* to allocate, whereas if technology is new there may be third-party *providers* to match needed *roles*. If *terms* are missing, this is the right time to update and complete them also, especially if external *providers* are planned to be used. At this time, it may not be easy to exactly imagine all kinds of competencies and *roles* needed for the entire *environment nesting level* development, but allocations must be estimated and secured. Allocations in the most remote *phases* may be more preliminary, with some updates to be expected.

Masters referring to this schedule

Pt.0.Sa. Environment restriction requirements. This *master governs* the documentation of the *environment restriction requirements*, which must be structured and easy to read and understand for all *stakeholders* concerned with the *product*. *Environment restriction requirements* are often abundant external regulations which need not be entirely copied, but only pointed out by *environment restriction requirement* references. To be easily referable and traceable, each captured *environment restriction requirement* must be a short, distinct point, which is uniquely identifiable. For this reason, large numbers of *environment restriction requirements* may need to be kept in a database.

Pt.0.Sb. Environment staffing allocation. This *master* describes documentation of planned staffing allocations from *providers*. Each needed *role* and the *provider* of such a *role* (including in-house *roles*) constitute a *development staffing* list. Staffing is used for *product requisite* but also essential for the development *project management* (part of next book "*Cpdm* technical overhead").

Tasks referring to this schedule

Xb.a. Develop turnkey items. When ordering a *turnkey item*, the reason often is that specialist competence or capability is already in place with *suppliers*, compared to finding and hiring new specialists for in-house development. This is the *task* which *governs* how to *demand turnkey items*. The best way to inform *suppliers* about what is needed is to share own adequate *requirements* with them. Note that the third-party *turnkey suppliers* are not decided within this *phase*, and so far only their *environment interfaces solution alternatives* are asked for (also see **Xb.a.,** pp. 100, 114, 116, 133, 135, 137).

6.11.2 Pt.0.P. Demarcate product by environment solutions with possible supplier opportunities

Pt:2. Cpdm generic advanced technical schedule

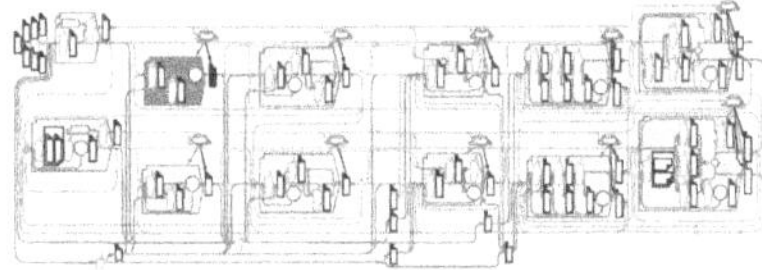

This chapter explains a *schedule*.

• The *schedule* is part of **Pt:2.0. Map environment and develop interfaces to the environment,** pages 106 and 107 (see left thumbnail).

• The *schedule* **Pt.0.P. Demarcate product by environment solutions with possible supplier opportunities** is explained below (highlighted in left thumbnail).

After *environment restriction requirements* are captured and staff allocations are in place, this *phase* analyzes these restrictions, in order to understand needed *environment interfaces*, and thereafter demarcate the *environment* from the *product* to be developed (see Fig. 6-18 below).

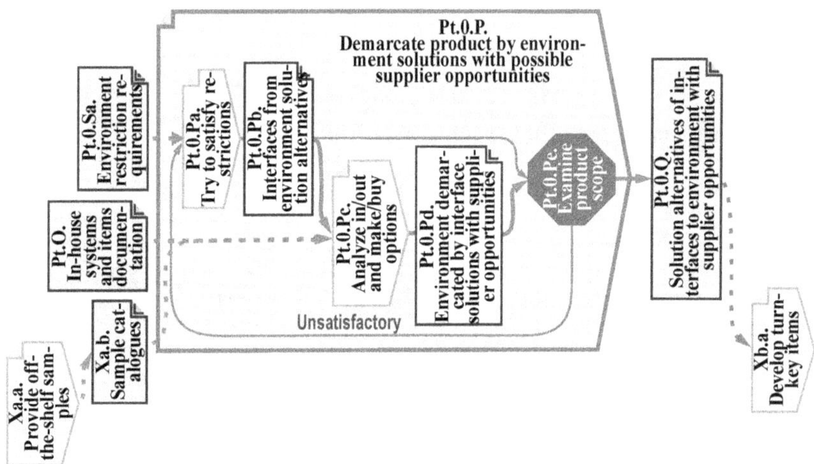

FIGURE 6-18 Demarcate product by environment solutions with possible supplier opportunities schedule

Masters referred to in this schedule

Pt.0.Sa. Environment restriction requirements. This is the *master* which *governs* the appearance of *environment restriction requirements*, which are the foundation for finding *environment interfaces solution alternatives*.

Xa.a. Provide off-the-shelf samples. This is the *task* which *governs* how to *demand off-the-shelf items*. These *suppliers* develop *samples* to be offered off the shelf, which means that they must watch market demand early, and are often proactive in cooperating with developing companies to find out their needs early. This can be used for own benefit to obtain *samples* promptly after their launch, fitting in own *products* better, and fitting competitors' in worse (also see **Xa.a.**, pp. 100, 113, 134).

Pt.O. In-house systems and items documentation. This is the *master* which *governs* how in-house *source systems* and *white-box design items* are documented and how to search and find them, a bit like browsing an in-house *sample catalogue*. Reusing known and reliable in-house and *supplier* items as *environment interfaces solution alternatives* should always be a top priority.

Xa.b. Sample catalogues. This is the *master* which *governs* how to *demand off-the-shelf supplier sample catalogues*. *Sample catalogues* play the role of enabling and catalyzing the invention of *environment interfaces solution alternatives* for the connection to the *environment*. The *environment interfaces* are seldom at their trickiest during *product development*, but for example, an easy-to-use man-machine *interface* can seem simple while often hiding a huge amount of *complexity* (also see **Xa.b.**, pp. 100, 119, 134, 137, 141).

Schedule content

Pt.0.Pa. Try to satisfy restrictions. This is the *task* which *governs* how to satisfy *environment restriction requirements* by identifying and documenting *environment interfaces solution alternatives*. For example, there might be many alternatives to arranging needed technical *environment interfaces*, or there may be alternatives to walk around legal environmental restrictions. Avoid delving deeply into the *product* to be developed, but just propose *environment interfaces solution alternatives* for the *environment interfaces*.

Pt.0.Pb. Interfaces from environment solution alternatives. This is the *master* which *governs* how to document *environment interfaces solution alternatives*, originating from *environment restriction requirements* causing these alternatives. There must be a minimum of one *solution alternative* for each *environment restriction requirement*, but there is no upper limit for *environment interfaces solution alternatives* to the same *environment restriction requirement*. *Environment restriction requirements* may also be grouped together matching one or many *environment interfaces solution alternatives*. Even as parts of a group, *environment restriction requirements* may also have more own *environment interfaces solution alternatives*. The purpose is to come up with as many (realistic) *environment interfaces solution alternatives* as possible to make the succeeding design more pluralistic. For each solution alternative, it might be helpful in a future analysis to present solution assessments, for example price, durability, *quality*, and so on.

Pt.0.Pc. Analyze in/out and make/buy options. When alternatives of *environment interfaces solution alternatives* are at hand, the next thing to figure out is which parts of the *environment* can stay unchanged, to be attached by the *environment interfaces*. This is the *task* which *governs* how to perform this. Sometimes there is no need for change in the *environment*; for example, an existing potential *user* may be part of the *environment interfaces*. In other cases, there may be a lot of parts of the *environment* that should be replaced by the *product* to be developed; for example, when replacing a lot of instances of existing manual administration by an automated computer system.

For solutions that end up as parts of the *product* to be developed, an analysis must also be made to determine if the *environment interfaces solution alternatives* can be matched by *samples*. *Sample catalogues* with data sheets, dimensions, and so forth can be *demanded* to catalyze the invention of *environment interfaces solution alternatives*. Beware of unnecessarily focusing too much on very specific *samples*, because this will limit the ability to freely optimize solutions.

Pt.0.Pd. Environment demarcated by interface solutions with supplier opportunities. This is the *master* which *governs* documentation, which must make very clear which *environment interfaces solution alternatives* will not be part of the development, since there is always a danger that more and more solutions get swallowed up enthusiastically in *product development*. It must also be documented who could possibly provide *samples* for the selected *environment interfaces solution alternatives*.

Pt.0.Pe. Examine product scope. This is a decision gate *task*, which *governs* how *managers* and technical experts act, to ensure that the *product* has a realistic demarcation to the *environment*. Information serving as the basis for gate decisions must be at hand, such as *environment interfaces solution alternatives*. If the decision gate conclusion is to not open the decision gate, developers have to reconsider this *phase* based on directives from the gate decision makers.

Masters referring to this schedule

Pt.0.Q. Solution alternatives of interfaces to environment with supplier opportunities. This is the *master* which *governs* the total *result* from this *schedule* **Pt.0.P.**.

Tasks referring to masters referring to this schedule

Xb.a. Develop turnkey items. This *task governs* how to *demand* chosen *turnkey items*, including their need to understand identified *white-box solution alternatives*, in order to incorporate their *environment interfaces solution alternatives* into all other *environment interfaces solution alternatives* (also see **Xb.a.,** pp. 100, 114, 116, 133, 135, 137).

6.11.3 Pt.0.A. Constitute environment architecture

Pt:2. Cpdm generic advanced technical schedule

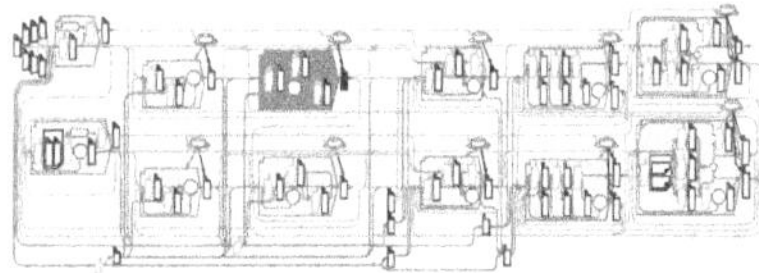

This chapter explains a *schedule*:

• The *schedule* is part of **Pt:2.0. Map environment and develop interfaces to the environment,** pages 106 and 107 (see left thumbnail).

• The *schedule* **Pt.0.A. Constitute environment architecture** is explained below (highlighted in left thumbnail).

The time has come to illustrate the *environment interfaces*, by using *architectures* of different types (see Fig. 6-19 below). If the *environment* is large or *complex*, or if the interaction between *environment* and *product* to be developed is *complex*, appropriate parts of the existing *environment* must also be properly illustrated.

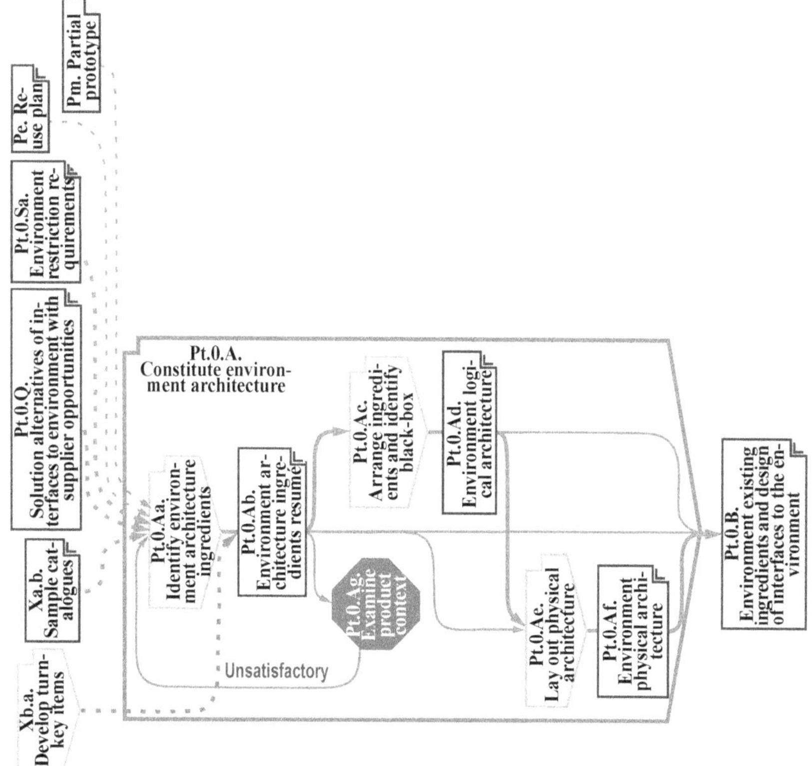

FIGURE 6-19 Constitute environment architecture schedule

Masters referred to in this schedule

Pe. Reuse plan. This *master* is not explicitly described in this book but instead see description in chapter 4.4.1, page 42.

Pm. Partial prototype. This *master* is not explicitly described in this book but instead see description in chapter 4.6.1, page 45.

Pt.0.Sa. Environment restriction requirements. This is the *master* which *governs environment restriction requirements*, which are also useful for the *architecture design*.

Pt.0.Q. Solution alternatives of interfaces to environment with supplier opportunities. This is the *master* which *governs environment interfaces solution alternatives*, which are the main input for *architecture design*.

Xa.b. Sample catalogues. In a previous *phase sample catalogues* were used to catalyze the invention of *environment interfaces solution alternatives*. This *master* also *governs* how the development now becomes more concrete and how *sample catalogues* details are *demanded*, to ensure that *supplier samples* match *architecture elements* and *interfaces* (also see **Xa.b.**, pp. 100, 119, 134, 137, 141).

Tasks referred to in this schedule

Xb.a. Develop turnkey items. This is the *task* which *governs* how to *demand turnkey items*. These *suppliers* are asked for their *environment interfaces solution alternatives*, which now supplement those provided in-house. Based on all these proposed alternatives, architects can select and optimize the most appropriate *environment interfaces* (also see **Xb.a.**, pp. 100, 114, 116, 133, 135, 137).

Schedule content

Pt.0.Aa. Identify environment architecture ingredients. This *task governs* how all *environment interfaces solution alternatives*—in-house, from *turnkey suppliers*, and from *off-the-shelf suppliers*—can now be used for *architecture* design and get documented in various *architectures*.

Pt.0.Ab. Environment architecture ingredients resumé. This is the *master* that *governs* how to specify the *architecture resumé*. All existing ingredients (such as *elements* and *interfaces*) in the *environment* are listed, as well as all ingredients for the *environment interfaces*. To be easily referable, each *element* and *interface* must be assigned a unique identifier. All *architecture ingredients* must be traceable to one or many *environment interfaces solution alternatives*. If the *environment interfaces* are built up with particular *samples* or if a specific *sample* is crucial for the *interface*, such *samples* must also be traceable from the *architecture resumé*. The *architecture resumé* is not prepared separately, but developed together with the other *architectures* explained below.

Pt.0.Ac. Arrange ingredients and identify black-box. This is the *task* which *governs* how *interface* ingredients are listed by the *architecture resumé*, and are logically arranged around the *product outermost black-box*. Everything inside the *product outermost black-box* is as yet hidden and is for the moment not subject to any attention.

Pt.0.Ad. Environment logical architecture. Two types of *architectures* will be developed from the *architecture resumé*. The first is *governed* by this *logical architecture master*, which explains how to illustrate the *environment architecture ingredients*, including the *environment interfaces*. The *product outermost black-box* acts as the hub in the *logical architecture*, and *interfaces* from this *black-box* are connected to surrounding *environment elements* or *users*. It is important that the *environment logical architecture* is wide enough to include key *users*.

Pt.0.Ae. Lay out physical architecture. A lifelike picture says more than a thousand words. Now this *master governs* how the *architecture ingredients* are illustrated as real, in order for all possible *stakeholders* to understand. Especially if potential *customers* or their representatives are available, they may be interested in following and reviewing the *physical architecture*.

Pt.0.Af. Environment physical architecture. This *master governs* how to illustrate the physical appearance of the *environment,* and how *environment interfaces* appear, including relevant *elements* and *interfaces* located in the existing *environment* or those to be developed by other organizations. Often *elements* and *interfaces* are physical artifacts that can be illustrated or silhouetted, bringing concrete and palpable information about the appearance of the *environment.* 3D simulation drawings, hand sketches, and mock-ups are also welcome to bring insight into the physical reality. For *programs, porting* the *outermost white-box* to its *target* may be the case, needing *header files* for built-in *development system libraries,* or interchangeable *header files* when *porting* to different *targets.*

Pt.0.Ag. Examine product context. This is the decision gate *task,* which *governs* how this decision gate scrutinizes the *architecture resumé* when all *architectures* have been prepared. At that time the *architecture design* work has tried out which *elements* are really needed to get the *environment* interacting well with the *product black-box,* and the *architecture resumé* is expected to be highly consistent. In this way the *architecture resumé* is the proof that right *architecture ingredients* have been decided, and this is the reason to let developers review the *architecture resumé,* and decision makers to approve it. If doubtful *architecture ingredients* are found in the *architecture resumé,* decision makers may lock the decision gate, and order an *architecture* redesign and an update of the *architecture resumé.*

Masters referring to this schedule

Pt.0.B. Environment existing ingredients and design of interfaces to the environment. This is the *master* which *governs* the total *architectures* from this *schedule* **Pt.0.A.** These *environment architectures* are the starting point for finalizing the *environment* design in the coming *phase.*

6.11.4 Pt.0.F. Finalize design of environment interfaces with product requisites

Pt:2. Cpdm generic advanced technical schedule

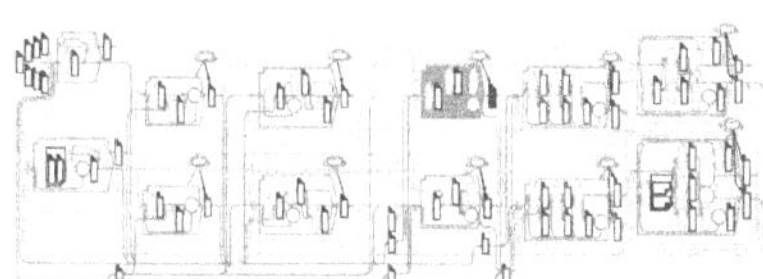

This chapter explains a *schedule:*

- The *schedule* is part of **Pt:2.0. Map environment and develop interfaces to the environment,** pages 106 and 107 (see left thumbnail).

- The *schedule* **Pt.0.F. Finalize design of environment interfaces with product requisites** is explained below (highlighted in left thumbnail).

In *Cpdm,* the **Pt.F. Finalize design & requisites** *phase* is performed inside-out, which means that the *innermost white-box* is finalized first. When progressing *outward,* the *white-boxes* being finalized will always have their *embedded white-boxes* finalized, which in the end means that when finalizing the *environment interfaces,* the *product outermost white-box* is completely finalized.

However, to win some calendar time, the *environment interfaces* can be designed while this is still a *black-box,* and not wait until the *black-box* is a finally designed and complete *white-box*. The problem with this is that it may be difficult to finalize *interfaces* to a *black-box*, without yet knowing exactly where these *interfaces* are proceeding within the *black-box*.

The **Pt:2:1. Tailored advanced technical overall V-schedule n = 3, using incorporative integration,** page 161, illustrates better the ambivalence referred to above. The question to sort out is when to perform *schedule* **Pt.0.F.** (shown in Fig. 6-20 below)—directly after the Pt.0.B. *architecture design* is ready, or when the Pt.1.G. *product outermost white-box* is finalized. A middle way is to perform as much as possible of the *architecture design* but not finish it until the *integration* is done.

Nevertheless, the *environment interfaces* must sooner or later be finalized (see Fig. 6-20 below).

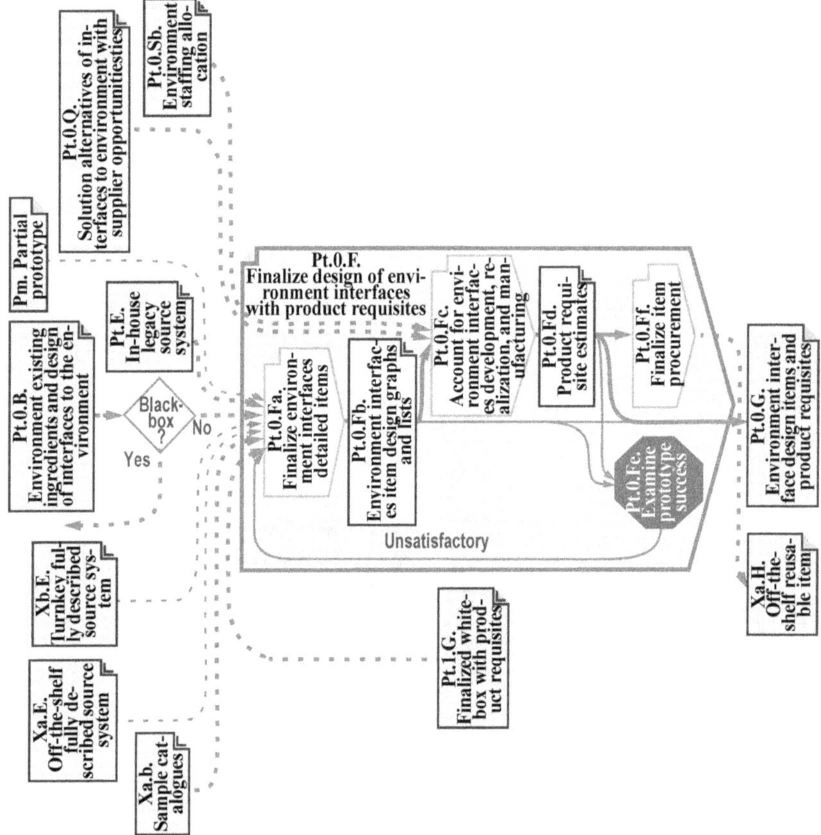

FIGURE 6-20 Finalize design of environment interfaces with product requisites schedule

Masters referred to in this schedule

Pm. Partial prototype. This *master* is not explicitly described in this book but instead see description in chapter 4.6.1, page 45.

Pt.0.Sb. Environment staffing allocation. This is the *master* that *governs* how to provide *development staffing*, which is *demanded* when accounting development work (see **Pt.0.Sb. Environment staffing allocation,** p. 111).

Pt.0.Q. Solution alternatives of interfaces to environment with supplier opportunities. This is the *master* which *governs* the *environment interfaces solution alternatives* with *suppliers*, useful when finishing and settling *environment interfaces design items*.

Pt.0.B. Environment existing ingredients and design of interfaces to the environment. This *master governs* the *architecture* package, which is the most important when detailing and finalizing design.

Pt.E. In-house legacy source system. This is the *master* that *governs* the needed appearance of in-house legacy *source systems*, when using parts from earlier *product versions* or even from other *products*. Appropriate ingredients can be recycled from such *products*, by reshaping them in the **Pt.F. Finalize design & requisites** *phase* to fit the new *environment interfaces* (also see **Pt.E.,** pp. 140, 119).

Xb.E. Turnkey fully described source system. This is the *master* which *governs* the appearance of *source systems* when buying from *turnkey suppliers*. *Turnkey items* not only hold detailed descriptions for finalizing design, but also all other documentation and descriptions that were developed to obtain these *items*, for example, mechanical drawings, electronic diagrams, construction calculations, and *programs*. During finalization of design, all *source system* documentation is integrated forever, and will in the future be treated the same way as any in-house documentation would be. Since the *source system* has been ordered turnkey for ongoing *product development*, it should need almost no adaptations by **Pt.F. Finalize design & requisites** (also see **Xb.E.,** p. 140).

Xa.E. Off-the-shelf fully described source system. This is the *master* that *governs* the needed appearance of *source systems* when buying from *off-the-shelf suppliers*. This *master* is very similar to the turnkey *source system* explained above, with the major difference that this kind of *source system* is acquired off the shelf, and may need more or less reconstruction by **Pt.F. Finalize design & requisites** (also see **Xa.E.,** pp. 140, 119).

Xa.b. Sample catalogues. This is the *master* which *governs* how to *demand sample catalogues*, now needed to be studied in detail. Maybe supplementary data sheets and *samples* must be ordered from the *suppliers* to ensure that all details of the *environment interfaces* will fit together. Physical measurements must be considered, voltage need and current consumption as well, timing sheets must fit together for logic circuits, and *function interfaces* and *porting* compatibilities must match for *programs* (also see **Xa.b.,** pp. 100, 119, 134, 137, 141).

Pt.1.G. Finalized white-box with product requisites. This *master governs* the *product outermost white-box*, to be *enfolded* by the *environment interfaces* (see **Pt.n.G. White-box design items and product requisites**, p. 142).

Schedule content

Pt.0.Fa. Finalize environment interfaces detailed items. In the developed *architectures*, the *environment interfaces* were indicated, and this is the *task* which *governs* how all details must be finalized for those *interfaces*. Observe that by far the greater part of the *product development* is to develop the *product black-/white-box* (by selecting "Yes" in Fig. 6-19 above).

Pt.0.Fb. Environment interfaces item design graphs and lists. This is the *master* which *governs* the design of *environment interfaces*, which must be fully described to be the basis for internal development or for ordering *samples* and *items* from external *suppliers*. If *environment interfaces* are *complex*, for example, a house exterior wall or a human man-machine *interface*, such *interfaces* might even be composed of several *elements*, and each *element* might need many *environment interfaces design items* to be realized. All *environment interfaces design items* needed to realize the *environment interfaces* must be identified and technically documented, including their key technical *design item properties* and parameters. Accumulating parameters such as weight, electricity, and water consumption must be finally calculated and documented. Preferably this information can reside in well-structured tables or in a database.

Pt.0.Fc. Account for environment interfaces development, realization, and manufacturing. This is the decision *task*, which *governs*—after all needed *environment interfaces design items* are identified and technically documented—an economical analysis to serve as the basis for development economy and manufacturing planning. The need to ensure profitability for the manufactured *product*, and the need to ensure the right *quality* for all *items*. Note that the **Pt.I. Realize & integrate white-box** and **Pt.V. Verify black- & white-box** *phases* have not yet been performed, and their costs are estimates only.

Pt.0.Fd. Product requisite estimates. Since finalizing design is performed inside-out, the *environment interfaces* are the last to get finalized. This is the *master* which *governs* how the costs must be presented and summarized, including all parameters accumulated from *inward embedded white-boxes*. Preferably this information can reside in well-structured tables or databases.

Pt.0.Fe. Examine prototype success. With the finalizing of design of *environment interfaces*, all documentation for manufacturing is finalized (yet with a chance for failure elimination after verifying the *product prototype*). Because of the importance of correct *product* manufacturing and economic information, this is the decision gate *task* which *governs* how to scrutinize both the technical ability to produce from developed documentation, and the economy and profitability expected from the *product*. This means that both technical documentation and accrued expenses and cost estimates must be prepared for the decision gate meeting, and that experienced manufacturing engineers should be called, as well as manufacturing and market economists.

Pt.0.Ff. Finalize item procurement. The *product requisites* are now finalized, and this *task governs* how all *environment interfaces design items* can be finally ordered from *suppliers*. Ever since the **Pt.P. Predetermine solutions & suppliers** *phase*, or even earlier, many *samples* have been evaluated off the shelf, but it is now time to exactly order what finally has been designed. Observe that *complex design items* identified very recently can be hard to quickly find off the shelf or may require some delivery times.

Masters referring to this schedule

Pt.0.G. Environment interface design items and product requisites. This is the *master* for the total documentation for manufacturing of the *product environment interfaces*. But before *product* manufacturing, this documentation will also be the basis for building and connecting the *product prototype* to the *environment*.

Xa.H. Off-the-shelf reusable item. This is the *master* which *governs* the appearance of *off-the-shelf items* bought from *off-the-shelf suppliers* (also see **Xa.H.**, pp. 144 122, 143, 144).

6.11.5 Pt.0.I. Realize interfaces to environment and insert/await outermost white-box

Pt:2. Cpdm generic advanced technical schedule

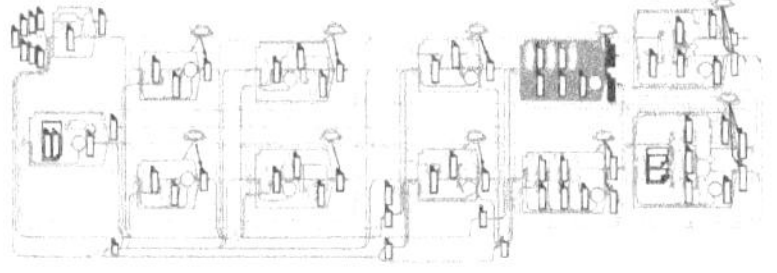

This chapter explains a *schedule*:

• The *schedule* is part of **Pt:2.0. Map environment and develop interfaces to the environment,** pages 106 and 107 (see left thumbnail).

• The *schedule* **Pt.0.I. Realize interfaces to environment and insert/await outermost white-box** is explained below (highlighted in left thumbnail).

The *integration* of the *product prototypes* into the *environment* can be made in two ways, inside-out *incorporative integration* finishing with the *environment*, or outside-in *invasive integration* beginning with the *environment* (see ch. 11.2, p. 601 and ch. 11.1.4, p. 595). In any case, sooner or later the *environment interfaces* must be realized (see Fig. 6-21 below).

Observe that the *environment interfaces*, might be anything from a simple screw to a *complex* graphic *user interface*.

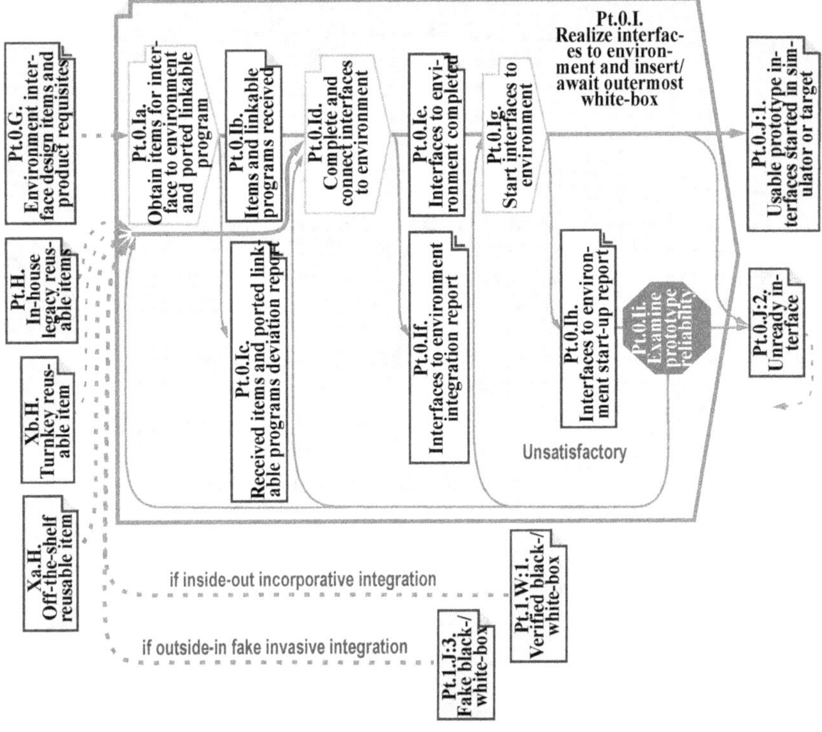

FIGURE 6-21 Realize interfaces to environment and insert/await outermost white-box schedule

Masters referred to in this schedule

Pt.0.G. Environment interface design items and product requisites. This is the *master* which *governs* the *environment interfaces design items* and *product requisite* of the *environment interfaces* prepared in a previous *phase*.

Pt.H. In-house legacy reusable items. This is the *master* which *governs* the needed appearance of in-house already completed *environment interfaces design items*. Like any other *reusable items*, such *design items* are *reused* as they are, without any reshaping or adaptations. If changes are needed, a full in-house *source system* is instead retrieved and *recycled* (see **Pt.H.**, pp. 144, 122).

Xa.H. Off-the-shelf reusable item. This is the *master* which *governs* the appearance of *off-the-shelf items* bought from *off-the-shelf suppliers*. These *off-the-shelf items* are used for immediate construction of the *environment interfaces*. *Linkable programs* are also regarded as *environment interfaces design items* and must be obtained by compiling and possibly be *ported* to the *target environment interfaces* (also see **Xa.H.**, pp. 144 122, 143, 144).

Xb.H. Turnkey reusable item. This is the *master* which *governs the appearance of turnkey items* bought from *turnkey suppliers*. These *environment interfaces design items* are now used for construction of the *environment interfaces*. *Linkable programs* are also regarded as *design items*, which may already be *ported* to the *target environment* by the *turnkey supplier* (also see **Xb.H.**, pp. 144, 123).

If inside-out incorporative integration, the intended master is Pt.1.W:1. Verified black-/white-box. This is the *master* which *governs* the *outermost white-box* to be integrated to the *environment*. When integrating inside-out, the now finished *outermost white-box* can be integrated to the *environment*, and the *environment interfaces* may get started and verified in the *target environment*.

If outside-in invasive integration, the intended master is Pt.1.J:3. Fake black-/white-box. This is the *master* which *governs* the outermost *fake black-/white-box* to be temporarily integrated to the *environment* (see **Pt.n.J:3. Fake black-/white-box,** p. 147). If the *environment interfaces* are desired to be partially verified, the entire *product outermost white-box* can now be a fake connected to the *environment*. The outermost *fake black-/white-box* enables the *environment interfaces* to be partially verified.

Schedule content

Pt.0.Ia. Obtain items for interface to environment and ported linkable program.

This is the *task* which *governs* how to receive all ordered *environment interfaces design items*. *In-house design items* and *off-the-shelf supplier design items* are checked and brought into storage waiting for the *environment interfaces* to be built. Sometimes development companies have their own stocks of approved standard *design items*, but often an order must be sent to a preferred *supplier*. *Turnkey items* have been ordered long before and these *turnkey items* are now received. Also, all finalized *programs* for the *environment interfaces* must be compiled and received together with possible *program libraries* and *program design items*.

Pt.0.Ib. Items and linkable programs received. This is the *master* which *governs* the appearance of the received *environment interfaces design items*. Acquired *design items* from *suppliers* should be expected to arrive with delivery notes, which describe what has been ordered and sent. It is also highly recommended that internal deliveries are arranged in a formal way, with documentation and delivery notes. All *environment interfaces design items* should be kept in storage, promptly available for *product prototype realizations*.

Pt.0.Ic. Received items and ported linkable programs deviation report. This is the *master* which *governs* the receiving report appearance. When *environment interfaces design items* are received, a check should be made to ensure that what has been received is what was ordered. Any deviations between what was ordered and received should be reported, for example if an *environment interfaces design item* has been damaged during transport, has sold out or expired, has been replaced by a similar *environment interfaces design item*, or if price has changed.

Pt.0.Id. Complete and connect interfaces to environment. This is the *task* which *governs* how to realize (build) the *environment interfaces* and how the *product out-*

ermost white-box integrates to the *environment*. If the *product prototype* is *complex*, it may be too big a leap to integrate and connect the *product prototype* directly in the *environment*, or if it is difficult to access the interfaces to the *environment* (there are awkward observation posts in the *environment*), a simulated *environment* may be a good intermediate step.

Pt.0.Ie. Interfaces to environment completed. This *master governs* the appearance of the realized *environment interfaces*. Often company standards are developed and *tailored*, describing what must be fulfilled by *environment interfaces realizations*.

Pt.0.If. Interfaces to environment integration report. This *master governs* the *integration* report appearance. While the *environment interfaces* are being realized, it is recommended to keep a diary of key events as a basis for the *integration* report. Most issues will concern *interface integration*, and the report should focus on problems of *interface* mismatching. Also possible *porting* problems are important to report.

Pt.0.Ig. Start interfaces to environment. The *product prototype interfaces* are now integrated to the *environment*, and this is the *task* that *governs* how the *environment interfaces* are put into operation. Sometimes there may be *target* simulators where the *product prototype* is started, to serve as an intermediate step before *porting* to the *target* and starting the *product prototype* in the real *environment*. Sometimes there are even strict *requirements* for authority approvals, and the *product prototype* must be inserted and started in a well-specified field simulator.

Pt.0.Ih. Interfaces to environment start-up report. This is the *master* which *governs* the *start-up report* appearance. Small problems at start-up can be immediately cured, but not seldom small problems disclose bigger problems emanating from earlier *phases* of development. This report focuses on how well the *environment interfaces* perform after start-up, to constitute the input for a future decision gate.

Pt.0.Ii. Examine prototype reliability. This is the decision gate *task* which *governs* how to evaluate how ready the *environment interfaces* are for *verification*. There is no point in starting *verification* if it continuously needs special care by engineers or continuously fails after restarting. If so, some disaster is already there and at least the *environment interfaces* must be more or less redesigned and realized again.

Masters referring to this schedule

Pt.0.J:2. Unready interface. This is the *master* which *governs* what kind of appearance or behavior is unacceptable for *verification* in the simulated or *target environment*, and causes the *interface* to be sent back for correction or redesign.

Pt.0.J:1. Usable prototype interfaces started in simulator or target. This is the *master* which *governs* what degree of *function* is sufficient for the start of a field test in a simulated or *target environment*.

6.11.6 Pt.0.V. Verify prototype in environment

Pt:2. Cpdm generic advanced technical schedule

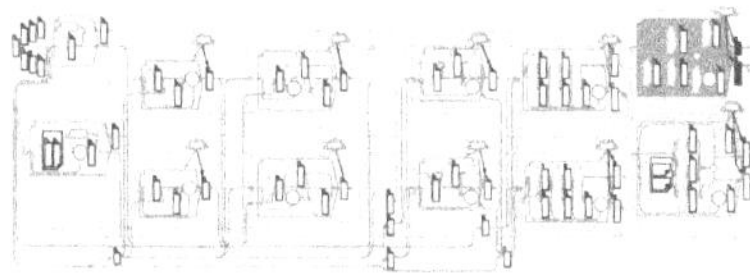

This chapter explains a *schedule*:

- The *schedule* is part of **Pt:2.0. Map environment and develop interfaces to the environment,** pages 106 and 107 (see left thumbnail).
- The *schedule* **Pt.0.V. Verify prototype in environment** is explained below (highlighted in left thumbnail).

This is the last *schedule* to apply, when developing according to the whole *Cpdm* advanced technical *schedule* (see Fig. 6-22 below).

Note that the purpose of this *schedule* is to only verify the *environment restriction requirements*. The *outermost black-/white-box* should already have been verified by the **Pt.1.V.** *schedule*.

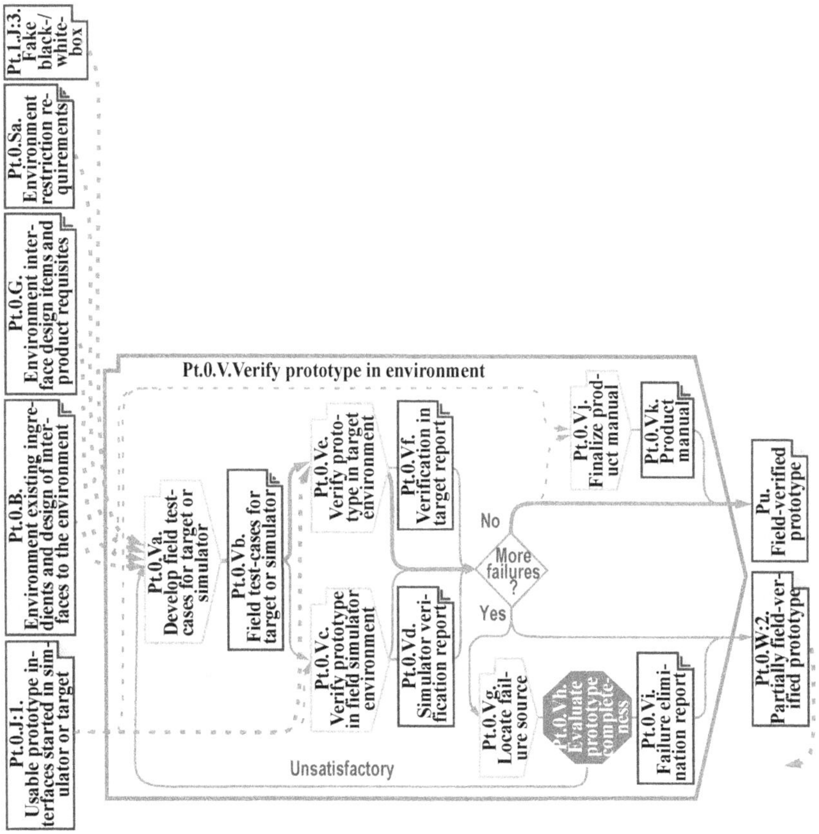

FIGURE 6-22 Verify prototype in environment schedule

Masters referred to in this schedule

Pt.0.Sa. Environment restriction requirements. This is the *master* which *governs* the appearance of *environment restriction requirements,* among others showing how they can be organized to best suit the origin for *test-cases.*

Pt.0.B. Environment existing ingredients and design of interfaces to the environment. This *master governs* the *architecture* package, which is used to develop *test-cases* to verify that *architectures* have been properly applied for the *environment interfaces* design.

Pt.0.G. Environment interface design items and product requisites. This is the *master* which *governs environment interfaces design items* and *product requisites,* demanded by *verification* of details in the *environment interfaces* (see **Pt.0.G. Environment interface design items and product requisites,** p. 122).

Pt.0.J:1. Usable prototype interfaces started in simulator or target. This is the *master* which *governs* unacceptable outcomes from start-up of the *environment interfaces* in simulator or in the *target environment* (see **Pt.0.J:1. Usable prototype interfaces started in simulator or target,** p. 124).

Schedule content

Pt.0.Va. Develop field test-cases for target or simulator. This is the *task* which *governs* how *test-cases* are developed, both for *target environment verification* and also for possible simulated *environment verification.* For equipment that is dangerous or possibly harmful to humans, an authority certificate is often mandatory to sell such a *product,* and approval for a field simulator and a set of field *test-cases* for such *verification* is often mandatory.

Pt.0.Vb. Field test-cases for target or simulator. This is the *master* which *governs* appearance of the field *test-cases.* Each *environment restriction requirement* must refer to at least one *test-case,* and each *test-case* must refer to at least one *environment restriction requirement* that should be verified. Each *test-case* contains a description of how to arrange the *verification,* and contains an expected outcome and pass-fail criteria. Each *test-case* must be traceable to the *environment restriction requirement* causing it.

Pt.0.Vc. Verify prototype in field simulator environment. This *task governs* how to verify in simulators of the real *environment.* For such *verification,* the newly started *environment interfaces* to the *product prototype* are exposed to the *test-cases* one after the other. If this is a certification needed for selling the *product,* such *verification* is often performed by an authority.

Pt.0.Vd. Simulator verification report. This is the *master* which *governs* how the *verification* outcome is documented. *Test-cases* that have passed need only minor information in the documentation, but failed *verifications* need detailed documentation of how the failure responded to the *test-cases* compared to what was expected.

Pt.0.Ve. Verify prototype in target environment. This is the *task* that *governs verification* of the *environment interfaces* to the *product prototype.* This may be a very tricky *verification,* since it may be difficult in field operation to inject specified *stimuli* into the *environment interfaces* and to detect *responses* from it. If the *product* is very expensive or will be damaged by the *verification*, most *verifications* are made in a simulated *environment*, and only limited destructive *verification* is reasonable in the *target environment*.

Pt.0.Vf. Verification in target report. This is the *master* which *governs* how *verification* outcome is documented, regardless of whether the *product* has passed *test-case* criteria or failed to do so. *Test-cases* that have passed need only minor information to be documented, but failed *verification*s need detailed documentation of how the failure responded to the *test-case* compared to what was expected.

Pt.0.Vg. Locate failure source. This is the *task* which *governs* how detected failures must be examined to find the primary failure location in the *environment interfaces* or deeper inside the *product prototype*, after the *environment interfaces* to the *product prototype* are verified. The failure must be located accurately enough to understand which organization is responsible for the failure and thus has to correct it. Observe that failures may first be identified in the *environment interfaces* or *product prototype*, but not seldom these failures are caused by failures somewhere in the developed documentation. The *failure elimination report* contains a list of all detected failures and their location in the *environment interfaces* or *product prototype* and/or in developed documentation.

Pt.0.Vh. Evaluate prototype completeness. This is the decision gate *task* which *governs* a walk-through of the *verification* documentation and judging how to cure each detected failure. Many failures may be easy to cure, but some failures may be extremely expensive to get rid of, and maybe a cheaper walk-around is sufficient. Clusters of many failures may indicate that parts of the *environment interfaces* or *product prototype* must be redeveloped. Even the absence of failures must be questioned, since this can be a sign of insufficient *test-case* coverage or too indulgent pass/fail criteria. Finally, it must be decided when the *product prototype* with its *environment interfaces* is verified enough to be passed to the following *phases*. These decisions need senior verifiers and authorized *managers* to participate in the decision gate, and the decisions must be documented in the *failure elimination report*.

Pt.0.Vi. Failure elimination report. This *master governs* the *failure elimination report*, which is sent around for actions together with the *environment interfaces* and *product prototype*, to all organizations having caused a detected failure. The organization behind the primary failure eliminates this in the *product prototype* and/or *environment interfaces* and/or in adequate documentation, adds its findings to the *failure elimination report*, and sends it all further to organizations where secondary or spin-off locations need failure elimination. When all *failure eliminations* have been made, including failure elimination of all documentation to restore *consistency*, the *product prototype* is returned for *reverification*. Observe that severe failures sometimes require the *product prototype* with its *environment interfaces* to be rebuilt in order to proceed with *verification*.

Pt.0.Vj. Finalize product manual. This is the *task* which *governs* how to develop the final *product* manual. The understanding of the *product prototype* culminates in *environment interfaces verification*, and there is no better moment than this to collect documentation and finalize the manual for it.

Pt.0.Vk. Product manual. This is the *master* which *governs* the appearance of the *product* manual, which most often is attached to the *product* at delivery to the *customer*. This means that it must be properly edited with appropriate user-friendly layout.

Masters referring to this schedule

Pt.0.W:2. Partially field-verified prototype. This *master governs environment interfaces* and *product prototype* in continued need of more *verification*. As long as *verification* leads to *failure detection*s, the *product prototype* is sent back to the organization that once has developed the location of the failure. After the *product prototype* have all failures eliminated, including all developed documentation for it, the *product prototype* is returned for *reverification*.

Pu. Field-verified prototype. This is the *master* which *governs verification* approval of the *product prototype* with *environment interfaces*. When all field *verifications* are done without remaining failures, all technical development is finished, and involved teams can celebrate this achievement with appropriate amounts of genuine champagne.

6.12 Pt:2.n. Develop outermost black-/white-box and each inward embedded black-/white-box

Pt:2. Cpdm generic advanced technical schedule

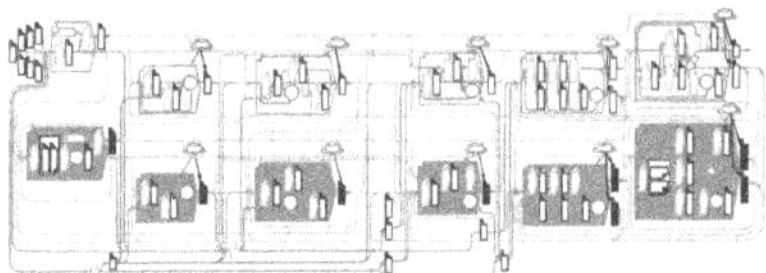

This chapter explains six *schedules.*

- These *schedules* are parts of **Pt:2.n. Develop outermost black-/white-box and each inward embedded black-/white-box (left half),** pages 106 and 107 (see left thumbnail).

- The *schedules* explained in this chapter are highlighted in left thumbnail.

This chapter contains descriptions of the six *schedules* applicable from *nesting level* 1 and *inwards* up to the innermost *nesting level*. It means that the following six *schedules* are nested to fit any *inward nesting level,* and the same *schedules* will be used for any depth in the embedded *architectures*.

Note that a single *black-box* can in turn contain many *black-boxes*, and each of them is developed according to this *black-/white-box schedule*. Also note that each *black-box* may contain different numbers of nested *embedded black-boxes*, making some *black-box* hierarchies deeper than others.

6.12.1 Pt.n.R. Refine requirements from outward nesting level requirements and ensure black-/white-box development staffing

Pt:2. Cpdm generic advanced technical schedule

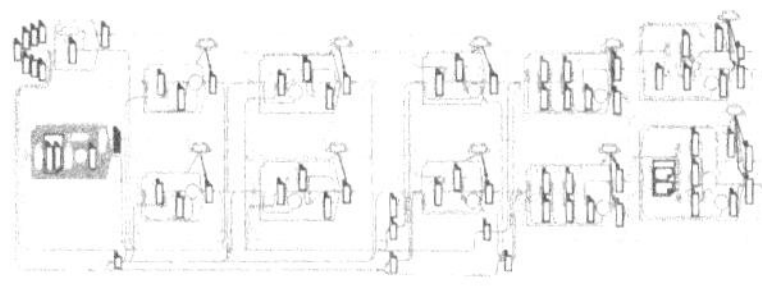

This chapter explains a *schedule*:

• The *schedule* is part of **Pt:2.n. Develop outermost black-/white-box and each inward embedded black-/white-box (left half),** pages 106 and 107 (see left thumbnail).

• The *schedule* **Pt.n.R. Refine requirements from outward nesting level requirements and ensure black-/white-box development staffing** is explained below

(highlighted in left thumbnail).

All *embedded black-boxes* (including the *outermost black-box*) must now be provided with *requirements* that are created by refining the *requirements* from one *nesting level outward* (including *environment restriction requirements*), and to these *requirements* finally add new appropriate *requirements* (see Fig. 6-23 below).

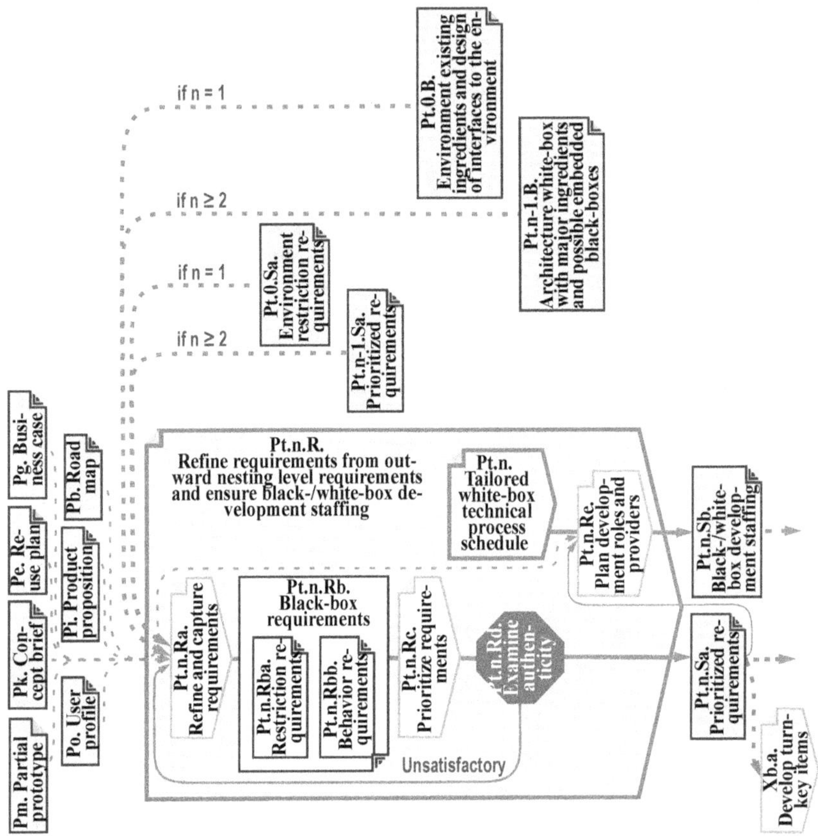

FIGURE 6-23 Refine requirements from outward nesting level requirements and ensure black-/white-box development staffing schedule

Masters referred to in this schedule

Observe that some referred *masters* below are assumed in the chair development example, which therefore is referred to (see below). All of these *upstream results targeting* the *product* on any *nesting level* ≥ 1 may be inserted as *restriction requirements* (or possibly *behavior requirements*).

Pb. Road map. This *master* is not explicitly described in this book but instead see description in chapter 4.2.2, page 41.

Pk. Concept brief. This *master* is not explicitly described in this book but instead see description in chapter 4.3.1, page 41.

Pe. Reuse plan. This *master* is not explicitly described in this book but instead see description in chapter 4.4.1, page 42.

Pg. Business case. This *master* is not explicitly described in this book but instead see description in chapter 4.5.1, page 44.

Pm. Partial prototype. This *master* is not explicitly described in this book but instead see description in chapter 4.6.1, page 45.

Po. User profile. This *master* is not explicitly described in this book but instead see description in chapter 4.7.1, page 46.

Pi. Product proposition. This *master* is not explicitly described in this book but instead see description in chapter 4.8.1, page 47.

If n = 1 the intended master is Pt.0.Sa. Environment restriction requirements. This is the *master* which *governs* the *environment restriction requirements* on *nesting level* 0, which now will be evaluated and refined for the *nesting level* 1 *outermost black-box*.

If n ≥ 2 the intended master is Pt.n-1.Sa. Prioritized requirements. This is the *master* which *governs* the *black-box requirements* for the *outermost black-box* and all *nesting levels inwards*, which will be refined for *black-box requirement* at this *nesting level*.

If n = 1 the intended master is Pt.0.B. Environment existing ingredients and design of interfaces to the environment. This is the *master* which *governs* the *outermost black-box* on *nesting level* n = 0, which in this *schedule* will be provided with refined and newly captured *requirements*.

If n ≥ 2 the intended master is Pt.n-1.B. Architecture white-box with major ingredients and possible embedded black-boxes. This is the *master* which *governs* *black-boxes* one *nesting level outward*, which now will be provided with refined and newly captured *requirements* for this *nesting level* of development (see **Pt.n.B. Architecture white-box ingredients design with possible embedded black-boxes,** p. 139).

Schedule content

Pt.n.Ra. Refine and capture requirements. This is the *task* which *governs* how to refine the *outward nesting level requirements* to *black-box(es)* at the current *nesting level*. When all possible *outward requirements* are refined and distributed to *black-box requirements*, any new *black-box requirement* may be added to any *black-boxes* of current *nesting level*.

Pt.n.Rb. Black-box requirements. This is the *master* which *governs* the appearance of the refined *black-box requirement* on this *nesting level*. Each *black-box requirement* must be traceable back to its *outward requirement* source, and thus must have a unique identifier. Each *black-box requirement* must also be traceable to the *black-box* it specifies. If a *requirement* is not a clear *restriction requirement* or *behavior requirement*, the *requirement* is a *compound requirement* (see ch. 7.5, "About black-box requirements," p. 184).

Pt.n.Rba. Restriction requirements. This is the *master* which *governs* the appearance of *restriction requirements* on this *nesting level* (see ch. 7.11, "Restriction requirements," p. 203).

Pt.n.Rbb. Behavior requirements. This is the *master* which *governs* the appearance of *behavior requirements* on this *nesting level* (see ch. 7.12, "Behavior requirements," p. 206). *Behavior requirements* can be severely *complex*, and usage of advanced requirements is an excellent resource to master *complexity*, including *use-cases* and *scenarios* (see ch. 7.16, "Use-case behavior requirements," p. 221; ch. 7.18, "Formal use-case requirements," p. 227; and ch. 7.19, "Scenario behavior requirements," p. 230).

Pt.n.Rc. Prioritize requirements. This is the *task* which *governs* how to prioritize *requirements*. *Requirements* for *complex products* are often impossible to fully satisfy, and it is good to know which are the less important to compromise on. The responsibility for prioritizing and consolidating are the *stakeholders'* (or their representatives').

Pt.n.Rd. Examine authenticity. *Requirements* control all *black-box* restrictions and behavior, implying that they control the complete *product*. Thus, this is the decision gate *task* which *governs* how *stakeholders* review and control all *requirements*. At such a decision gate meeting, among others, *stakeholders* and *line management* must approve every *requirement* (outsourcing *requirement* capturing work to external *providers* should not be considered).

Pt.n. Tailored white-box technical process schedule. This is a *tailored process schedule* for this *black-/white-box nesting level*, adapted for the specific *product* and company need. This *tailored schedule* must be provided to all allocated *roles* on this *nesting level*. *Tailoring processes* used to be a *task* for the *quality management* department organization, but these *activities* are not fully described by this book. However, some examples of *tailored schedules* are used for the examples of this book (see below ch. 6.13, "EXAMPLE: Tailored process overall schedules," p. 160).

Pt.n.Re. Plan development roles and providers. This is the *task* which *governs* how to allocate *roles* on this *black-/white-box nesting level*, being specified by somebody with good overview of available competency, for example the *project leader* for the *product development project*. Start to evaluate own development capacity, and whether available competence is enough for all development work up to *verification* of the *black-/white-box*. Apart from specifying *requirements* and allocations, remaining *phases* on this *nesting level* may be successfully outsourced to specialized subcontractors. If *terms* for *roles* are absent, it is a good opportunity to specify them now, especially for third-party subcontractors.

Masters referring to this schedule

Pt.n.Sa. Prioritized requirements. This is the *master* which *governs* the complete *requirements* on this *nesting level* (see **Pt.n.Sa.**, pp. 130, 132, 133, 137, 152).

Pt.n.Sb. Black-/white-box development staffing. This is the *master* which *governs* the staffing allocation documentation, which constitutes a list of distinct *roles* together with which companies may provide them (including in-house-provided *roles*). This allocated staff will be used in the accounting of all *product requisites*. Allocation of *roles* from *providers* to run this *phase* is done as early as possible.

Tasks referring to this schedule

Xb.a. Develop turnkey items. This is the *task* which *governs* how to *demand turn-key items.* The best way to inform *suppliers* about what is needed is to share proper *requirements* with them. Note that the third-party *turnkey supplier* is not decided within this *phase*, and so far only *white-box solution alternatives* are asked for (also see **Xb.a.,** pp. 100, 114, 116, 133, 135, 137).

6.12.2 Pt.n.P. Predetermine solutions with sourcing options

Pt:2. Cpdm generic advanced technical schedule

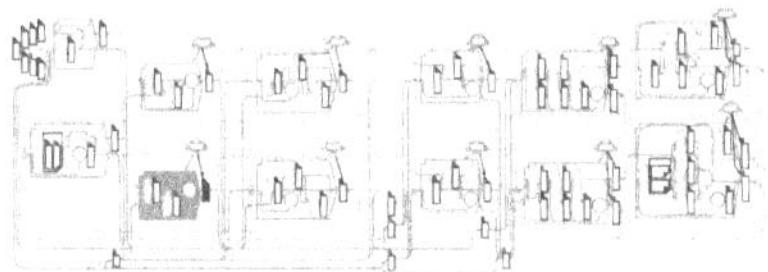

This chapter explains a *schedule*:

• The *schedule* is part of **Pt:2.n. Develop outermost black-/white-box and each inward embedded black-/white-box (left half),** pages 106 and 107 (see left thumbnail).

• The *schedule* **Pt.n.P. Predetermine solutions with sourcing options** is explained below (highlighted in left thumbnail).

This *phase* tries to identify *white-box solution alternatives* to the newly performed *requirement refinement,* and then to identify possible sourcing options for these *white-box solution alternatives* (see Fig. 6-24 below).

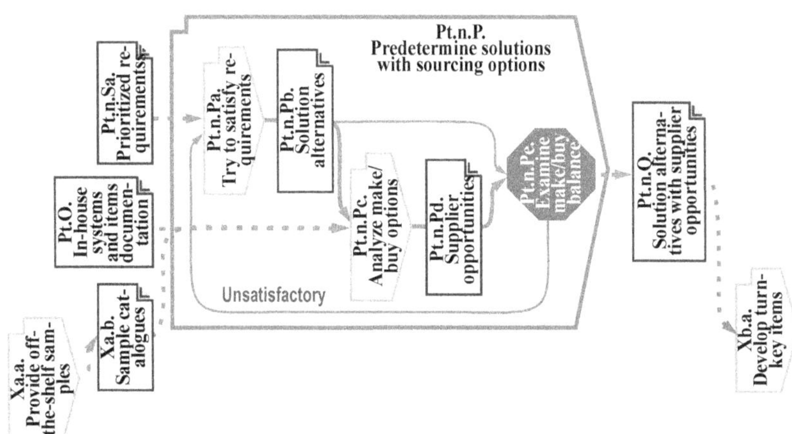

FIGURE 6-24 Predetermine solutions with sourcing options schedule

Masters and tasks referred by this schedule

Pt.n.Sa. Prioritized requirements. This is the *master* which *governs* the newly prior-itized *requirements,* which are the main *result* when trying to find *white-box solu-tion alternatives* (see **Pt.n.Sa.,** pp. 130, 132, 133, 137, 152).

Xa.a. Provide off-the-shelf samples. This is the *task* which *governs off-the-shelf suppliers*. These *suppliers* often have a longtime relation with developing companies, and get updated continuously with information on future needs. What ends up in *sample catalogues* is quite often results indirectly from this relation. If it is important to not let crucial *samples* show up and get offered through public *sample catalogues*, contract instead *turnkey suppliers* (also see **Xa.a.**, pp. 100, 113, 134).

Pt.O. In-house systems and items documentation. This is the *master* which *governs* how in-house *source systems* and *white-box design items* are documented and how to search and find them, a bit like browsing an in-house *sample catalogue*. Reusing known and reliable in-house and *supplier design items* as *white-box solution alternatives* should always be a top priority.

Xa.b. Sample catalogues. This is the *master* which *governs* how to *demand sample catalogues*, which now play the role of enabling and catalyzing the invention of *white-box solution alternatives*. Especially in mechanics and electronics, much effort is expended to find *reusable items* to populate the *product* to be developed. *Sample catalogues*, data sheets, and offerings from *suppliers* are really good sources to survey the market for what is available off the shelf. At this time, preliminary information is often necessary, such as technical specifications, prices by different volumes, and delivery times. Later in the *process* of selecting suitable *samples*, direct contact with the *supplier* is most often needed. Bigger companies may have an internal list of preferred *suppliers* and their approved *samples* (also see **Xa.b.**, pp. 100, 119, 134, 137, 141).

Schedule content

Pt.n.Pa. Try to satisfy requirements. This is the *task* which *governs* how the *black-box requirements* are carefully evaluated in order to find satisfactory solutions. This is one of the trickiest parts of the whole development, and solid experience particularly from similar technology and industry is very helpful. One by one, it may be simple to satisfy each *requirement*, but one single solution may interfere or break a solution from another *requirement*. Preferably, a solution should be found that satisfies many *requirements* collectively in a cost-efficient way.

Pt.n.Pb. Solution alternatives. This is the *master* which *governs* the appearance of the *white-box solution alternatives*. Each *requirement* must have at least one solution alternative. *Requirements* may be grouped, and together be satisfied by *white-box solution alternatives*. *White-box solution alternatives* may be accompanied by one or more assessments (such as price, availability, delivery time, etc.), to facilitate future selections.

Pt.n.Pc. Analyze make/buy options. This is the *task* which *governs* how *white-box solution alternatives* are analyzed in order to find out how to realize them. Many *white-box solution alternatives* may be developed in-house, but usually there are many alternatives to that. For example, most *samples* matching *white-box solution alternatives* can (at least with some compromises) be bought off the shelf. *Sample catalogues* are scanned, *supplier* sales people are contacted, *samples* are collected and evaluated, all to find how to best realize and match *white-box solution alternatives*, and which *suppliers* may be preferred. It may be hard to find *samples* from *suppliers* matching all *white-box solution alternatives* exactly, and some of them need to be developed in-house to glue remaining third-party *samples* together.

Pt.n.Pd. Supplier opportunities. This is the *master* which *governs* documentation of preferred *suppliers*. A list or database should show at least one possible *supplier* for each solution alternative. Candidate *suppliers* can preferably be separately documented, and referenced from the *white-box solution alternatives*.

Pt.n.Pe. Examine make/buy balance. This is a decision gate *task* which *governs* approval of *samples* and responsibilities of *suppliers*. Even more important is to check and approve that no unnecessary in-house development has been conveniently chosen, instead of making the effort to find cheap and efficient *samples*. If approved, the *samples* must be evaluated by technology experts for their capability to match *white-box solution alternatives*. Also, readiness to consider third-party solutions is important to examine, since many developers are egocentric and believe that in-house development is always the best, regardless of cost and time considerations. All these evaluations must be prepared before the decision gate meeting, to be approved and decided on at the decision gate meeting.

Masters referring to this schedule

Pt.n.Q. Solution alternatives with supplier opportunities. This is the overall *master* which *governs* the total **Pt.n.P.** *results*. Proposed *white-box solution alternatives* are documented in a structured way, together with possible *suppliers* able to deliver *samples* matching the solutions.

Tasks referring to masters referring to this schedule

Xb.a. Develop turnkey items. This is the *task* which *governs* how to *demand turnkey items*. A *turnkey supplier* probably has the best competence to develop the solutions needed, but may not understand how the *sample* will be used by the *product* being developed. The best way to use the turnkey developers is to provide them with *requirements* and let them consider the *white-box solution alternatives* to these *requirements* themselves (also see **Xb.a.,** pp. 100, 114, 116, 133, 135, 137).

6.12.3 Pt.n.A. Satisfy requirements by decomposing outward nesting level black-box into white-box design containing possible black-boxes

Pt:2. Cpdm generic advanced technical schedule

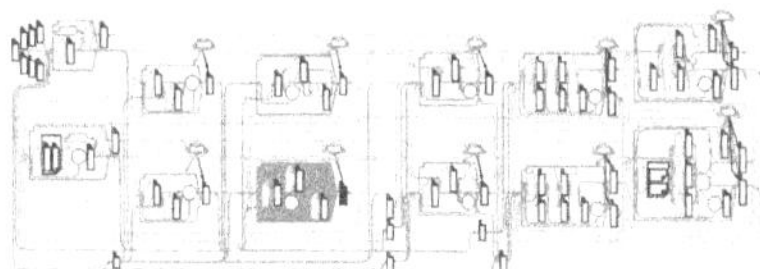

This chapter explains a *schedule*:

• The *schedule* is part of **Pt:2.n. Develop outermost black-/white-box and each inward embedded black-/white-box (left half),** pages 106 and 107 (see left thumbnail).

• The *schedule* **Pt.n.A. Satisfy requirements by decomposing outward nesting level black-box into white-box design containing possible black-boxes** is explained below (highlighted in left thumbnail).

The previous *phase* described the *white-box solution alternatives* and possible *suppliers* to match these solutions with *samples*. It is now about time to open the *black-box* into a *white-box* and settle which *white-box solution alternatives* to go for, and to document them by *architectures* of several kinds (see Fig. 6-25 below). Of course, it is very important that the *white-box* on this *nesting level* still looks the same and has exactly the same *interfaces*, as when it was an unopened *black-box* one *nesting level outward*.

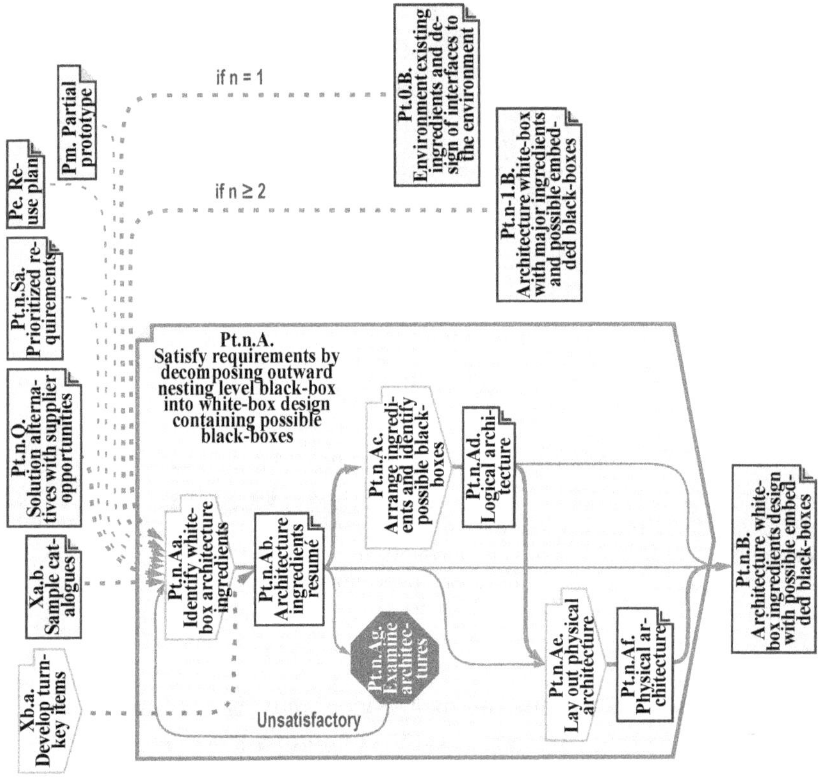

FIGURE 6-25 Satisfy requirements by decomposing outward nesting level black-box into white-box design containing possible black-boxes schedule

Masters referred to in this schedule

Pe. Reuse plan. This *master* is not explicitly described in this book but instead see description in chapter 4.4.1, page 42.

Pm. Partial prototype. This *master* is not explicitly described in this book but instead see description in chapter 4.6.1, page 45.

If n = 1 the intended master is Pt.0.B. Environment existing ingredients and design of interfaces to the environment. This is the *master* which *governs* the *environment architecture*, which is the starting point for the *architecture design* work on this *nesting level.* The *environment architecture* shows potential *users*, existing *elements* and *interfaces*, and new *interfaces* to the *product's* outermost *black-box*, which now should be opened and designed by satisfying its newly refined and prioritized *requirements.*

If n ≥ 2 the intended master is Pt.n-1.B. Architecture white-box with major ingredients and possible embedded black-boxes. This is the *master* which *governs* the *architectures* from the *outward nesting level*, which is the starting point for the *architecture design* work on this *nesting level.* The *outward architectures* show *interfaces* and *elements*, including *embedded black-boxes*, which now should be opened and designed by satisfying their newly refined *requirements* (see the *master* Pt.n.B., which is the below *master* referring to this *schedule* **Pt.n.A.**).

Pt.n.Sa. Prioritized requirements. This is the *master* which *governs* the prioritized *requirements* on this *nesting level.* When making *architectures*, the *white-box solution alternatives* are the main enablers, but it is always a good help to also check back to the *requirements* behind the *white-box solution alternatives.*

Pt.n.Q. Solution alternatives with supplier opportunities. This is the *master* which *governs* the developed *result* in the previous *phase.*

Xa.b. Sample catalogues. This is the *master* which *governs* how to *demand supplier sample catalogues.* These are very useful for architects to select *samples*, especially if they are crucial for a certain solution alternative (also see **Xa.b.**, pp. 100, 119, 134, 137, 141).

Tasks referred by this schedule

Xb.a. Develop turnkey items. This is the *task* which *governs* how to *demand* selected *turnkey items.* They now need to deliver an *architecture* of their *turnkey items*, to let the *architecture designers* fit them into the total *architecture* (also see **Xb.a.**, pp. 100, 114, 116, 133, 135, 137).

Schedule content

Pt.n.Aa. Identify white-box architecture ingredients. This is the *task* which *governs* a tricky *activity*, which now is to open the *outward black-box* to become a *white-box*, and populate it with ingredients from preferred *solution alternatives.* Some turnkey development is already started, and *architectures* from these *suppliers* must also fit into the total *architecture.* Furthermore, a minimum of efficient *samples* ought to cover as many as possible, in order to be cost-efficient. Sometimes there are *quality requirements* for chosen *samples*, but more often there are price limits. It may also happen that a *sample* is found that is a better solution than any proposed solution alternative. Some *elements* are uncomplicated and noncrucial, and *sample* selection can wait until the finalize-design *phase.*

Pt.n.Ab. Architecture ingredients resumé. This is the *master* which *governs* the appearance of the *architecture resumé*, which is a list of needed *elements* and *interfaces*, each of them assigned a unique ID and name. This list contains all possible ingredients, some of which may be crucial *samples* and/or from crucial *suppliers*, but most are probably generic and can later be populated with *reusable items.* *Architecture* drawings must always refer to the *architecture resumé*, and may never invent any separate *elements* or *interfaces* not in the *architecture resumé* (apart from *abstract elements*).

Pt.n.Ac. Arrange ingredients and identify possible black-boxes. This is the *task* which *governs* the mandatory development of the *logical architectures*. Now as many as possible of the *requirements* will finally be satisfied by invented *elements* and *interfaces*. To make this drawing, use trial and error to twist and turn around the *elements* connected by *interfaces*. Many of them already are obvious and part of the *architecture resumé*, but they might be added, changed, or replaced in the *architecture resumé* during the *architecture design* work. If any *element* seems to hold overly *intrinsic complexity*, this *element* may be judged to be a *black-box* which needs to be specified and developed one *nesting level inwards*.

Pt.n.Ad. Logical architecture. This is the *master* which *governs* how to illustrate how all *elements* are arranged and how they are interconnected by *interfaces*. *Architecture ingredient* symbols to use can be found in Table 6-7, page 96 above. To be holistic, break the silly rule of a maximum of seven things per page, and fit in as many as possible on one page by making it large enough. (An exception to this rule is examples of this book, which must be squeezed into small and nonextendable pages).

Pt.n.Ae. Lay out physical architecture. This is the *task* which *governs* how to develop *physical architectures*. To see is to believe. Everything that isn't behavior, but has appearance, can be illustrated by a *physical architecture*. *Stakeholders*, and even possible potential *customers*, typically want to contribute and influence by reviewing these drawings.

Pt.n.Af. Physical architecture. This is the *master* which *governs* physical drawings or illustrations of physical appearance. If the *product* is merely a piece of fashion, it can even be a full physical 3D mock-up. Other examples of *physical architectures* are computer 3D simulations of mechanical constructions like houses, machines, and so forth, which can be printed as 2D-perspective drawings. For electronics, the example can be *component* diagrams or PCB placing of *components*. As for *programs*, memory maps, *source file* structures, and screen shots, MMI might be helpful.

Pt.n.Ag. Examine architectures. This is the decision gate *task* which *governs* how to scrutinize the *architecture resumé* first when all *architectures* have been prepared. At that time the *architecture design* work has challenged and tested out which ingredients are really needed to satisfy the *requirements*, forcing the *architecture resumé* to be optimal and consistent. In this way the *architecture resumé* is the proof that right *elements* and *interfaces* have been decided on, and the right *samples* have been found, and this is the reason to let developers review the *architecture resumé*, and decision makers to approve it. If doubtful *elements*, *interfaces*, or *samples* are found, decision makers may lock this decision gate and order an *architecture* redesign, bringing changes to the *architecture resumé*. After this decision gate, final negotiations can start with *suppliers* who will deliver *samples* for approved *elements* and *interfaces*.

Masters referring to this schedule

Pt.n.B. Architecture white-box ingredients design with possible embedded black-boxes. This is the *master* which *governs* the total package of documented *white-box architectures*. If the *white-box* contains *embedded black-boxes*, each of them must become *target* for *requirements* and developed on the next *inward nesting level*. What is visible in the *white-box* on this *nesting level*, and not visible in any *black-boxes*, is finalized in the next *phase* on this *nesting level*.

6.12.4 Pt.n.F. Finalize design of white-box with product requisites

Pt:2. Cpdm generic advanced technical schedule

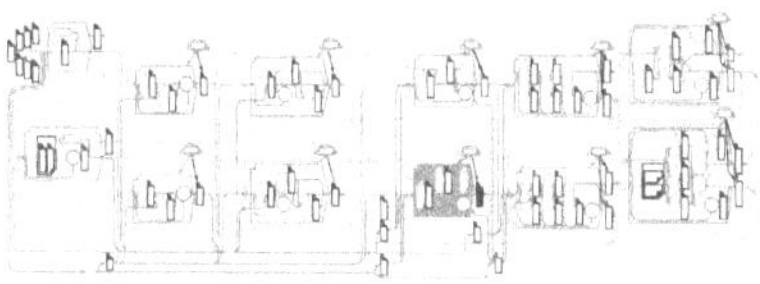

This chapter explains a *schedule*:

• The *schedule* is part of **Pt:2.n. Develop outermost black-/white-box and each inward embedded black-/white-box (left half),** pages 106 above and 107 (see left thumbnail).

• The *schedule* **Pt.n.F. Finalize design of white-box with product requisites** is explained below (highlighted in left thumbnail).

When the *architecture* is completed, it is straightforward, albeit a bit laborious, to finalize the design (see Fig. 6-26 below). Observe that finalizing design is done inside-out, beginning with the *innermost white-boxes* having no *embedded black-boxes*, ending with the *environment*, contrary to the previous three *phases*, which were developed outside-in, from the *environment* to the *innermost white-boxes* (see Fig. 6-32, p. 161).

Pm. Partial prototype. This *master* is not explicitly described in this book but instead see description in chapter 4.6.1, page 45.

Pt.n.Sb. Black-/white-box development staffing. This is the *master* which *governs* staffing on this *nesting level*, which is useful when estimating development costs in the *product requisites*.

Pt.n.Q. Solution alternatives with supplier opportunities. This is the *master* which *governs* the appearance of *solution alternatives* and *supplier opportunities*. *Supplier opportunities* are needed to derive *product requisite*.

Pt.n.B. Architecture white-box ingredients design with possible embedded black-boxes. This is the *master* which *governs embedded white-boxes*, which now will be finalized (see **Pt.n.B. Architecture white-box ingredients design with possible embedded black-boxes,** p. 139).

Pt.E. In-house legacy source system. This is the *master* which *governs* the needed appearance of in-house legacy *source systems* of earlier *product versions*, or even of other *products*, if to be *recycled* for the *product* under development. If a development company has been active for a long time, there are always a lot of old developed *products* to *demand* when developing a new *product*. The cleanest way is to *reuse* an entire *black-box* from a previous *product*, or if it is highly structured and documented, demarcated parts of them might fit as *elements* and *interfaces* without too much rework. The least attractive but sometimes beneficial way, is to completely *recycle* the legacy *source system* (also see **Pt.E.,** pp. 140, 119).

Observe that this book only considers new *product development* from scratch, and does not repeatedly extend it with more and more functionality. This demands *configuration management*, which is something different and yet much trickier, and a topic candidate for the next book *"Cpdm* technical overhead."

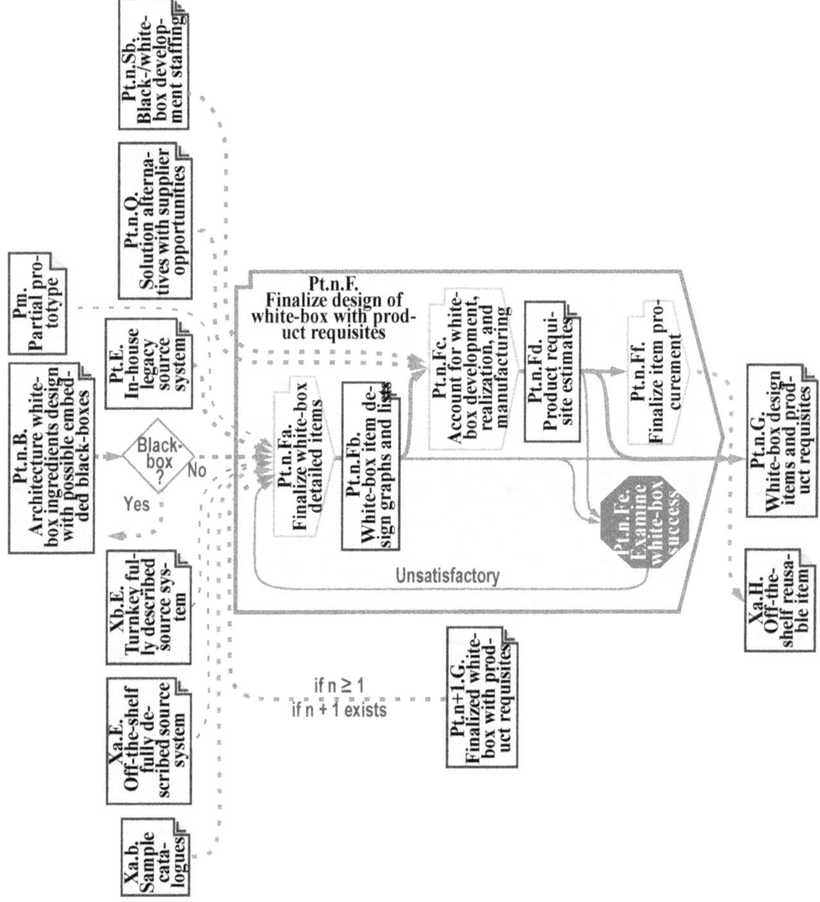

FIGURE 6-26 Finalize design of white-box with product requisites schedule

Masters referred to in this schedule

Xa.E. Off-the-shelf fully described source system. This is the *master* which *governs* the needed appearance of *source systems* when buying from *off-the-shelf suppliers*. During finalization, a purchased *source system* is integrated forever, and will in the future be treated in the same way as any in-house systems would be. If the *source* *system* is acquired off the shelf, it is probably very generic and needs more or less reconstruction in the finalization design (also see **Xa.E.**, pp. 140, 119).

Xb.E. Turnkey fully described source system. This is the *master* which *governs* the needed appearance of *source systems* when buying from *turnkey suppliers*. A *turnkey item* fits better than an off-the-shelf *source system* and also fits better than a legacy *source system*, because it has been ordered custom-made for a particular purpose, and needs almost no adaptations (also see **Xb.E.**, pp. 140, 141).

Xa.b. Sample catalogues. This is the *master* which *governs* how to *demand sample catalogues*. This late in development *sample catalogues* must be complemented with detailed data sheets, and negotiated prices for different *design items* volumes. Big development companies often have an internally published *sample catalogue* of preferred *suppliers* and approved *samples*, as well as an internal storage of these preferred *samples*. Another aspect is delivery times and reliability for *product prototypes*, and in manufacturing, some *suppliers* are quick to deliver but expensive, other are cheap but slow in delivery of large volumes (also see **Xa.b.**, pp. 100, 119, 134, 137, 141).

If $n \geq 1$ and if $n + 1$ exists, the intended master is Pt.n+1.G. Finalized white-box with product requisites. This is the *master* which *governs* the already finalized *white-box* one *nesting level inwards*. *Interfaces* to all finalized *inward embedded white-boxes* are now much more developed and easier to understand and are much more reliably designed, and it becomes much safer to connect them to *elements* in the current *white-box*, thus *enfold* the detailed design *outward*. Also, a lot of *design item properties* emerging from the *embedded white-box* may now be accumulated to the current *white-box*, for example, costs, current consumption, electricity supply, and so forth (this is the *master* which also *governs* the output from this *schedule* on this *nesting level* (see **Pt.n.G. White-box design items and product requisites** below).

Schedule content

Pt.n.Fa. Finalize white-box detailed items. This is the *task* which *governs* how to finalize the design. Everything apart from newly defined *black-boxes* needs to be detailed, whether it is a mechanical construction, an electronic circuit, or a piece of a *program*. During the finalization, the *architecture* may be shown to be premature (it isn't easy to develop a correct *architecture* in one go), and must be changed or extended in this *phase*. It should also be noted that this *task* is among the most labor-intensive of development work, and is often outsourced in countries with lower labor salaries. Also, be aware that there is still nothing but documentation produced so far in the development work at this *nesting level*.

Pt.n.Fb. White-box item design graphs and lists. This is the *master* which *governs* the appearance of the design now being finalized. From now on, *white-box design items* become the designation for *samples* chosen for use in *realizations*. *Samples* matching *elements* and *interfaces* now must be transformed to exact *white-box design items*, known for sure to be available on the market or occur in *sample catalogues* or data sheets. Sometimes one *element* or *interface* corresponds exactly to one known *white-box design item*, sometimes it takes several *white-box design items* to realize them. Some crucial *elements* and *interfaces* may have been matched with *samples* a long time before, whereas other noncrucial *elements* and *interfaces* are now being matched directly with *supplier design items* for the first time. Preferably, documentation will be a table package or database of various formats, depending on the nature of the *elements* and *interfaces*.

Pt.n.Fc. Account for white-box development, realization, and manufacturing. This is the *task* which *governs* how to account for all development costs, some of them already known and the others to be estimated. All *items* are collected, bringing cost to the developed *product*, such as material, equipment and manufacturing work, *product prototypes* and development. If a database is used, it is rather simple to organize the *traceability* between *architecture ingredients* and *white-box design items*, otherwise it is a heavy manual administration (like in this book).

Pt.n.Fd. Product requisite estimates. This is the *master* which *governs* the appearance of *product requisites*. The most important *product requisites* for business and profitability are manufacturing costs, where all material, assembly work, and equipment must be specified. These costs are rather similar to *product prototype* builds, which also need material, building work, and equipment. Development cost is often sloppily calculated as a huge lump sum, but in *Cpdm* this cost must be distributed over all *roles* of each development *phase*. For *programs*, where development costs are often lumped together even more, cost in *Cpdm* must also be distributed to *program elements* in the *architecture*, to allow a huge scale-up with preserved cost control.

Pt.n.Fe. Examine white-box success. This is the decision gate *task* which *governs* the final description work before *realization* of *product prototypes* and real manufacturing, which make these descriptions crucial for *quality* and economy. Before realizing anything in *product prototypes* and manufacturing, a review of *white-box design items* must be made by manufacturing engineers, and *product requisites* must be reviewed by sourcing experts and *managers* responsible for economy. This is the decision gate to approve the *white-box design item* and *product requisites*, thus giving *product prototype* building the green light.

Pt.n.Ff. Finalize item procurement. This is the *task* which *governs* how all *white-box design items* finally might be ordered from *suppliers*. In the predetermine solutions & suppliers *phase*, or even earlier, *samples* now turning to *white-box design items* have already been evaluated. But some *white-box design items* may have more recently been identified and must immediately be procured. Observe that *complex white-box design items* recently identified can be hard to quickly find off the shelf or may require some delivery times. Always be careful to elaborate *complexity* before the **Pt.A. Design architectures** *phase*, when there is still time to consider simplifications, reconstructions, extended in-house development, or ordering from *turnkey suppliers*.

Masters referring to this schedule

Pt.n.G. White-box design items and product requisites. This is the *master* which *governs* all finalized documentation. This includes technical documentation of *white-box design items* with *design item properties* and *product requisites* with cost accounting for manufacturing, *product prototype* builds, and development. This *result* is the basis for *product prototype* building and later manufacturing of the *product*. It is also the basis for economic estimates of margins and profitability.

Xa.H. Off-the-shelf reusable item. This is the *master* which *governs* handling and appearance of *off-the-shelf items* bought from *off-the-shelf suppliers*. These *off-the-shelf items* are ordered and received just in time for *white-box integration* and *realization* (also see **Xa.H.**, pp. 144 122, 143, 144).

6.12.5 Pt.n.I. Procure items, linkable programs, and integrate/await embedded black-boxes in order to realize white-box

Pt:2. Cpdm generic advanced technical schedule

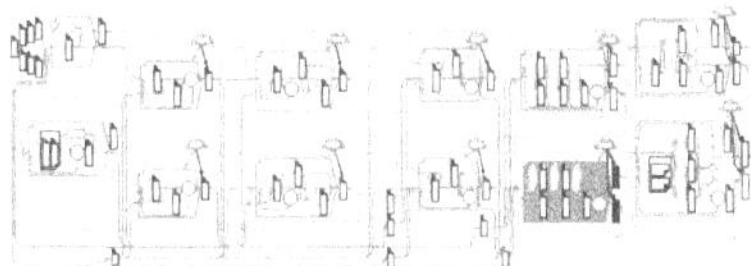

in left thumbnail).

This chapter explains a *schedule*:

• The *schedule* is part of **Pt:2.n. Develop outermost black-/white-box and each inward embedded black-/white-box (left half),** pages 106 and 107 (see left thumbnail).

• The *schedule* **Pt.n.I. Procure items, linkable programs, and integrate/await embedded black-boxes in order to realize white-box** is explained below (highlighted in left thumbnail).

According to the *product requisites* described in a previous *schedule, supplier design items* are ordered and now being received, *in-house design items* are gathered, and so are *linkable programs* (see Fig. 6-27 below).

Note that *integration* can be realized either inside-out, beginning with realizing the *innermost white-boxes* and ending with realizing the *environment interfaces*, or the reverse, outside-in *integration*, beginning with realizing the *environment interfaces* and ending with the *innermost white-boxes*, or even a mix of inside-out and outside-in *integration*. The inside-out *integration* is called *incorporative integration* and the outside-in *integration* is called *invasive integration*.

If *embedded white-boxes* are integrated as fakes, the *white-box* on this *nesting level* can be partially verified. *Invasive integration* with *fake black-/white-boxes* is further described in chapter 11.1.4, "Outside-in invasive integration," page 595.

Masters referred to in this schedule

Pt.n.G. White-box design items and product requisites. This is the *master* which *governs product requisites*, which are now received to build the *product prototype* (see **Pt.n.G. White-box design items and product requisites,** p. 142).

Pt.H. In-house legacy reusable items. This is the *master* which *governs* the needed appearance of *in-house design items* already completed. Sometimes developing companies have in-house-developed *products* from the past that can be used in the same way as a *white-box design item* would be used from an external *supplier*. Such a *component design item* is *reused* as is, without any reshaping or adaptation **Pt.H.**, pp. 144, 122). If changes are needed, a full in-house *source system* is instead *recycled* (see **Pt.E.,** pp. 140, 119).

Xa.H. Off-the-shelf reusable item. This is the *master* which *governs* handling and appearance of *off-the-shelf items* bought from *off-the-shelf suppliers.* Now these *off-the-shelf items* are received from third-party shelves. Mechanical *off-the-shelf items* may sometimes be called just parts, the electronic industry calls them *components,* and in *programs* they are called *program libraries* or subroutines. Characteristic for *off-the-shelf items* like these, is that they are not easily changed or adapted, and have to be perfectly selected to fit into the *realization* (also see **Xa.H.,** pp. 144 122, 143, 144).

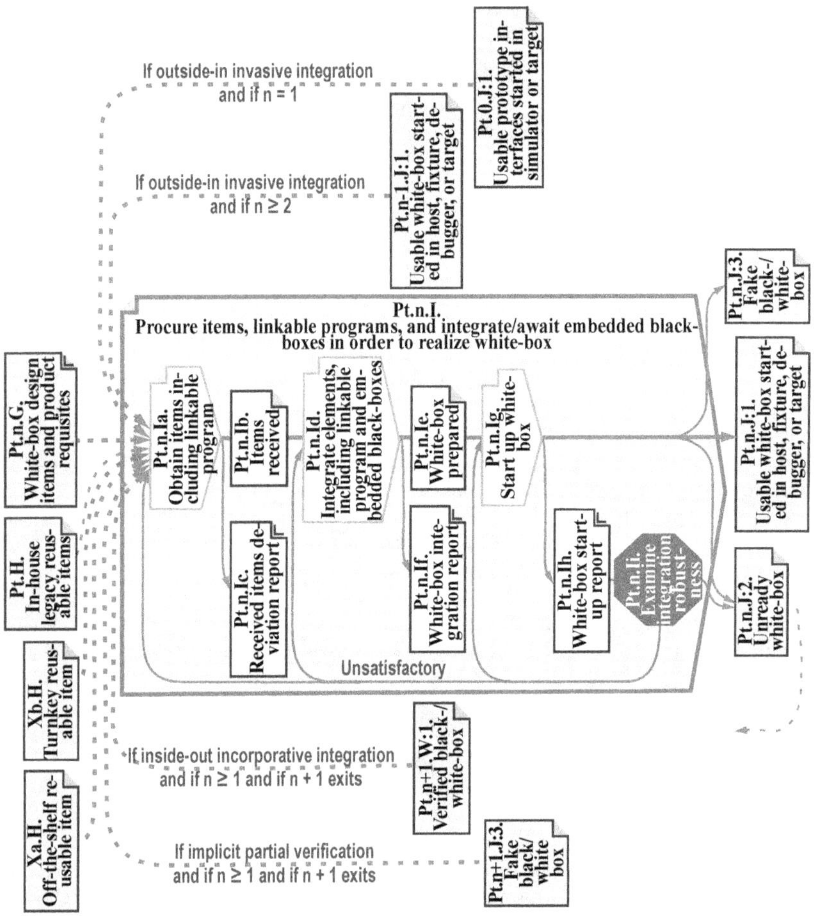

FIGURE 6-27 Procure items, linkable programs, and integrate/await embedded black-boxes in order to realize white-box schedule

Xb.H. Turnkey reusable item. This is the *master* which *governs handling and appearance of turnkey items* bought from *turnkey suppliers.* These *turnkey items* are ordered from own *requirements* and have been negotiated to fit into the total *architecture*, implying that these *turnkey items* should instantly fit perfectly into the *realization* (also see **Xb.H.**, pp. 144, 123).

If outside-in invasive integration and if n = 1, the intended master is Pt.0.J:1. Usable prototype interfaces started in simulator or target. This is the *master* which *governs* judging start-up of usable *environment interfaces*, either for *partial verification* of the outermost *fake black-/white-box*, or for *implicit verification* after the *outermost white-box* is fully completed without any *inward* fakes at all.

If outside-in invasive integration and if n ≥ 2, the intended master is Pt.n-1.J:1. Usable white-box started in host, fixture, debugger, or target. This is the *master* which *governs* judging start-up of a usable *black-/white-box*, either for *partial verification* if containing any embedded *fake black-/white-boxes*, or for *implicit verification* if containing only fully invaded embedded *black-/white-boxes* without any *inward* fakes at all.

If outside-in fake invasive integration and if n ≥ 1 and if n + 1 exists, the intended master is Pt.1.J:3. Fake black-/white-box. This is the *master* which *governs fake black-/white-boxes*. The purpose is to quickly prepare embedded fake *black-/white-boxes* in order to expose the *white-box* for *partial verification* (this is the same *master* as one of the three *masters* that *govern* the output from this *schedule* (also see **Pt.n.J:3. Fake black-/white-box** below).

If inside-out incorporative integration and if n ≥ 1 and if n + 1 exists, the intended master is Pt.n+1.W:1. Verified black-/white-box. This is the *master* which *governs* the verified *black-/white-boxes* one *nesting level inwards*, which now are ready to be integrated.

Schedule content

Pt.n.Ia. Obtain items including linkable program. This is the *task* which *governs* how to obtain *white-box design items* to be used for realizing *white-boxes*. *Samples* have been evaluated since these were proposed, and have been ordered and tried since the *architecture resumé* was stable, and have recently been ordered as *white-box design items*. Big companies usually have a storage of often-used approved *white-box design items*, and before placing an external order, a check can be made of internal availability. In this respect, *linkable program* is also regarded as a *white-box design item*. (Just regard assembling or compiling of finalized *programs* as being an automated procedure to obtain *linkable program*.)

Pt.n.Ib. Items received. This is the *master* which *governs* how to handle received *white-box design items*. All *white-box design items* received should be expected to arrive with a documentation of the *white-box design item*. Acquired *white-box design items* from *suppliers* often have a delivery note attached, which describes what has been sent and can as such be used for the deviation report shown below. It is also highly recommendable that internal delivery is arranged in a formal way, with documentation and delivery notes. Since a *white-box* often contains a lot of *white-box design items*, and a *product prototype* may consist of a lot of *white-boxes*, the *white-box design items* must be properly marked with an internal *white-box design item* number and sorted into stores and shelf bins so as not to get mixed together.

Pt.n.Ic. Received items deviation report. This is the *master* which *governs* the appearance of the deviation report. When *white-box design items* arrive, a check should be made to ensure that what is received is what was ordered. Any deviations between ordered and received should be reported, for example if the *white-box design item* has sold out, been replaced by a similar *white-box design item*, or if price or any other *design item properties* have changed.

Pt.n.Id. Integrate elements, including linkable program and embedded black-boxes. This is the *task* which *governs* the *realization* (build) of *white-boxes*. For mechanics and electronics, this is one of the most expensive and time-consuming *activities*. In the case of *programs*, building is one of the most mechanized *phases*, using a linker to join all *program* parts to produce a *machine program*, and a loader to install the *machine program* into the *product prototypes' program* memory.

Pt.n.Ie. White-box prepared. This is the *master* which *governs* the appearance of the realized *white-box*. This *master* may advantageously be *tailored* to a specific *product*, and extended with more specific internal regulations on how the completed *white-box* must appear, with name and unique identification and so forth.

Pt.n.If. White-box integration report. This is the *master* which *governs* how to document the progressing *realization*. During the *white-box realization* (build), it is recommended to keep some kind of diary of key events to serve as the basis for this *integration* report. When building a *white-box*, the most interesting aspect is how well *interfaces* of *elements* fit with each other, and the report should focus on problems with *interface* matching.

Pt.n.Ig. Start up white-box. This is the *task* which *governs* how to start up the realized *white-box*. The *white-box* should be started in a selected execution *environment*, such as in the *host development systems*, in any kind of debugger, in any available *fixture*, in a development kit or board, or in the *target environment*. When the *white-box* starts up and begins to function as expected, this is the final acknowledgement that the *white-box* is ready for *verification*.

Pt.n.Ih. White-box start-up report. This is the *master* which *governs* the appearance of the *start-up report*, which becomes the basis for the future readiness decision gate. The report should focus on how stable the *white-box* appears and executes after it is started. There is no point in sending the *white-box* for *verification* if several engineers must be present to keep it intact all the time.

Pt.n.Ii. Examine integration robustness. This is the decision gate *task* which *governs* how to scrutinize how ready the *black-/white-box* is for *verification*. If the *white-box* has a problem starting without a lot of care from development engineers or if it stops working whenever being approached, it is not ready for *verification*. In this case there are problems with the design, and the *white-box* must be sent back for redesign.

Masters referring to this schedule

Pt.n.J:1. Usable white-box started in host, fixture, debugger, or target. This is the *master* which *governs* the appearance of an approved *white-box*, sufficient to be verified in *host*, debugger, *fixture*, and/or *target*.

Pt.n.J:2. Unready white-box. This is the *master* which *governs* the appearance of a *white-box* that is judged by the decision gate to be useless for *verification*.

Pt.n.J:3. Fake black-/white-box. This is the *master* which *governs* a *fake black-/white-box*. It contains only temporary rudimentary *white-box design items*, in order to be temporarily integrated to a *white-box*, which may be exposed to *partial verification*.

6.12.6 Expanded Pt.n.I. Procure items, linkable programs, and integrate/await embedded black-boxes in order to realize white-box

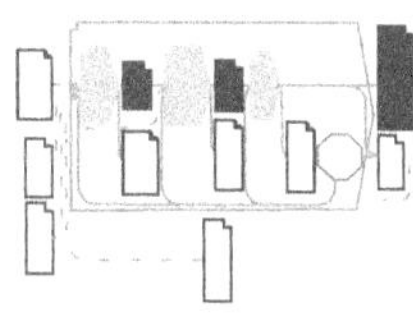

This chapter expands and explains in detail a *schedule*:

- **Pt.n.I. Procure items, linkable programs, and integrate/ await embedded black-boxes in order to realize white-box,** page 144 (see left thumbnail).
- The expanded *tasks* and *masters* explained in this chapter are highlighted in left thumbnail.

Integration of *white-boxes* can be really *complex* if they contain technology of very different kinds, for example, mechanical and electronic integrating, integrating passive electronics with microcontrollers or processors, integrating these with *programs*, or integrating many pieces of *programs*. Some hierarchy patterns are the only ones possible—it is mechanics that keeps together electronics, and it is electronics that holds the *programs*—nothing else.

Even integrating *programs* with other *programs* running on the same microcontroller or *process* can be tricky, especially if this *white-box* should be *ported* to another *target environment*.

Since *integration* might be really *complex* and nasty in details, the *integration schedule* is extended, but of course, fully compliant with the nonextended (see Fig. 6-28 below).This extension only explains inside-out *integration* of *white-boxes*. How to perform outside-in *invasive integration* is shown in the above Figure 6-27, Procure items, linkable programs, and integrate/await embedded black-boxes in order to realize white-box schedule, page 144.

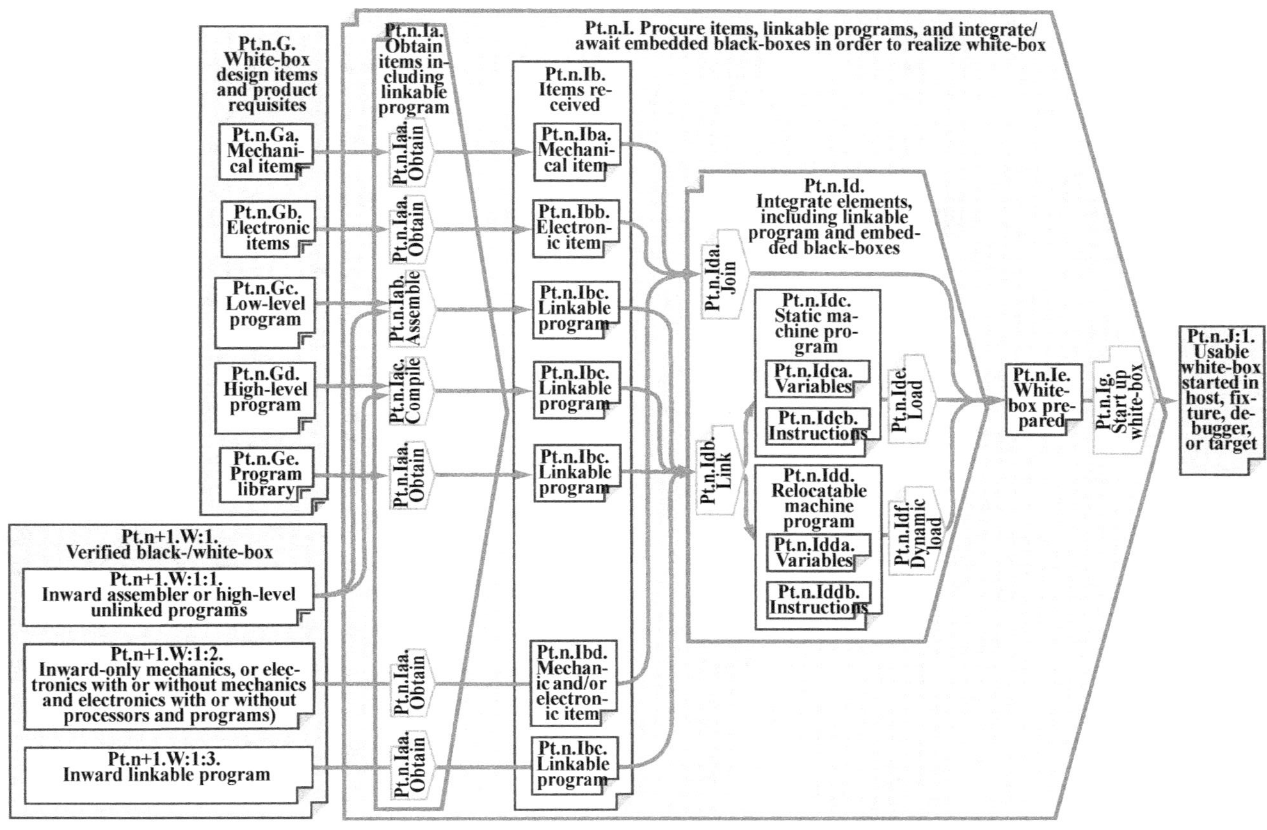

FIGURE 6-28 Procure items, linkable programs, and integrate/await embedded black-boxes in order to realize white-box expanded schedule

Masters referred to in this schedule

Pt.n.G. White-box design items and product requisites. This is the *master* which *governs* the *white-box design items* and *product requisite* (see **Pt.n.G. White-box design items and product requisites,** p. 142).

Pt.n.Ga. Mechanical items. This is the *master* which *governs* how mechanical *white-box design items* need to be designed to be integrated.

Pt.n.Gb. Electronic items. This is the *master* which *governs* how electronic *white-box design items* need to be designed to be integrated.

Pt.n.Gc. Low-level program. This is the *master* which *governs* how *low-level program white-box design items* (e.g. *assembler programs*) need to be designed to be integrated.

Pt.n.Gd. High-level program. This is the *master* which *governs* how *high-level program white-box design items* (e.g. *structured programs* and *object-oriented programs*) need to be designed (e.g. according to style guides) to be integrated.

Pt.n.Ge. Program library. This is the *master* which *governs* how *program libraries white-box design items* need to be designed to be integrated. Typically, *program libraries* are provided together with compilers as *development system libraries*. Sometimes *program libraries* are provided as stand-alones as special-purpose *programs*. *Program libraries* are also sometimes delivered as *source systems*, totally integrable in the in-house *source system*.

Pt.n+1.W:1. Verified black-/white-box. This is the *master* which *governs* all kinds of verified *black-/white-boxes*, to be integrated from the *inward nesting level*.

Pt.n+1.W:1:1. Inward assembler or high-level unlinked programs. This is the *master* which *governs* *black-box programs* to be integrated into a microcontroller or processor chip set of any kind.

Pt.n+1.W:1:2. Inward-only mechanics, or electronics with or without mechanics and electronics with or without processors and programs). This is the *master* which *governs* hierarchies of *black-box* possibilities. The closest *inward black-box* might be mechanical or electronic *black-boxes*. The mechanical *black-box* may hold solely mechanical *black-boxes* or can hold *black-boxes* with printed circuit boards with electronics. Electronic *black-boxes* can hold solely nonprocessor electronic *black-boxes* or *black-boxes* with processor electronics with *programs*.

Pt.n+1.W:1:3. Inward linkable program. This is the *master* which *governs* *linkable programs*. Unfortunately, *programs* cannot be hierarchically organized as mechanics and electronics, since a linker can't produce *linkable programs* to be used for nested linking from an *outward nesting level*. However, *programs* are as rich in structure as mechanics and electronics, and it is strongly recommended that large *programs* should be *partitioned* into many adjacent *black-boxes*, which after being compiled can be linked in one single operation.

Schedule content

Pt.n.Ia. Obtain items including linkable program. This is the *task* which *governs* how to obtain *white-box design items*.

Pt.n.Iaa. Obtain. This is the *task* which *governs* how to obtain mechanical, electronic, and *program libraries*.

Pt.n.Iab. Assemble. This is the *task* which *governs* how to obtain *linkable program* from *assembler programs* by use of an assembler tool.

Pt.n.Iac. Compile. This is the *task* which *governs* how to obtain *linkable program* from *high-level programs* by use of a compiler tool.

Pt.n.Ib. Items received. This is the *master* which *governs* the appearance of *white-box design items* received (see **Pt.n.Ib. Items received,** p. 145).

Pt.n.Iba. Mechanical item. This is the *master* which *governs* the appearance of mechanical *white-box design items* received.

Pt.n.Ibb. Electronic item. This is the *master* which *governs* the appearance of electronic *components* received.

Pt.n.Ibc. Linkable program. This is the *master* which *governs* the appearance of *linkable programs*.

Pt.n.Id. Integrate elements, including linkable program and embedded black-boxes. This is the *task*, now being expanded to a *schedule*, which *governs* how to realize (build) the *white-box* on this *nesting level,* maybe complemented by integrating *inward embedded black-boxes* (see **Pt.n.Id. Integrate elements, including linkable program and embedded black-boxes,** p. 146).

Pt.n.Ida. Join. This is the *task* which *governs* how to build together mechanical and electronic *white-box design items*. For electronics, often a printed circuit board is used to keep *components* together.

Pt.n.Idb. Link. This is the *task* which *governs* how to execute a linker to build *machine programs* from all *linkable programs*. The linker must be properly selected to generate *machine programs* for the intended controller or processor, so also for *programs* being *ported*.

Pt.n.Idc. Static machine program. This is the *master* for the *machine programs* to be loaded in fixed predestined memory locations, to be executed by controllers or processors.

Pt.n.Idca. Variables. This is the *master* which *governs* statically allocated *machine variables*.

Pt.n.Idcb. Instructions. This is the *master* which *governs* statically allocated *machine instructions*.

Pt.n.Idd. Relocatable machine program. This is the *master* which *governs machine programs* for computers which have capacity enough to hold operating systems, that dynamically load *program* applications to flexible machine locations for execution while the computer is being operated.

Pt.n.Idda. Variables. This is the *master* which *governs* relocatable *machine variables.*

Pt.n.Iddb. Instructions. This is the *master* which *governs* relocatable *machine programs.*

Pt.n.Ide. Load. This is the *master* which *governs* how small microcontrollers are loaded and integrated with *machine instructions,* which stay there during the whole lifetime of the microcontroller. In these systems, *machine variables* are specifically directed to random-access memory and *machine instructions* must be permanently loaded into read-only memory.

Pt.n.Idf. Dynamic load. This is the *master* which *governs* how large computer systems, through an operating system, can load their *programs* from secondary memory to any working memory location, to be executed there.

Pt.n.Ie. White-box prepared. This is the *master* which *governs* the expected appearance of the integrated and realized *white-box* (see **Pt.n.Ie. White-box prepared,** p. 146).

Pt.n.Ig. Start up white-box. This is the *task* which *governs* how to start up a white-box (see **Pt.n.Ig. Start up white-box,** p. 146).

Pt.n.J:1. Usable white-box started in host, fixture, debugger, or target. This is the *master* which *governs* an approved *white-box* for *verification* (see **Pt.n.J:1. Usable white-box started in host, fixture, debugger, or target,** p. 146).

6.12.7 Pt.n.V. Verify black-/white-box

Pt:2. Cpdm generic advanced technical schedule

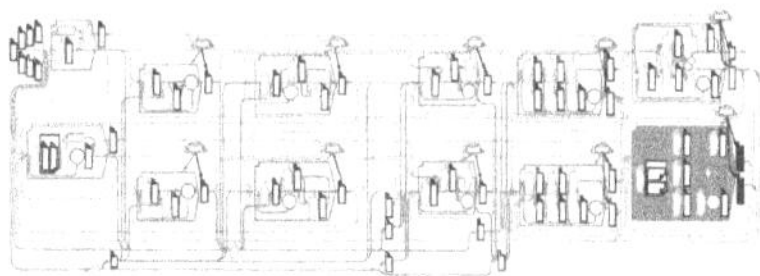

This chapter explains a *schedule:*

• The *schedule* is part of **Pt:2.n. Develop outermost black-/white-box and each inward embedded black-/white-box (left half),** pages 106 and 107 (see left thumbnail).

• The *schedule* **Pt.n.V. Verify black-/white-box** is explained below (highlighted in left thumbnail).

The last thing to do on this *nesting level* is to verify all prepared *white-boxes* (see Fig. 6-29 below).

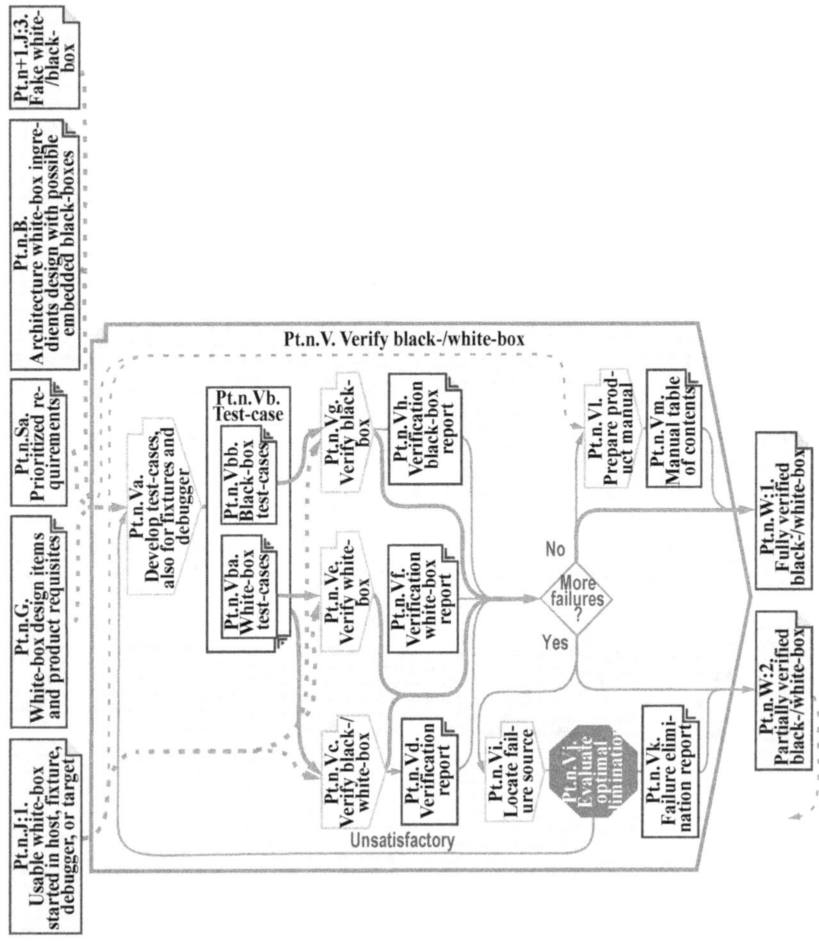

FIGURE 6-29 Verify black-/white-box schedule

Masters referred to in this schedule

Pt.n.J:1. Usable white-box started in host, fixture, debugger, or target. This is the *master* which *governs* that the *white-box* now functions well enough to be verified in *host*, debugger, *fixture*, and/or *target* (see **Pt.n.J:1. Usable white-box started in host, fixture, debugger, or target,** p. 146).

Pt.n.Sa. Prioritized requirements. This is the *master* which *governs* the most important enabler for developing *test-cases.* If they are *restriction requirements*, they will cause *white-box test-cases* for **Pt.n.Ve. Verify white-box**; if they are *behavior requirements*, they will cause *black-box test-cases* for **Pt.n.Vg. Verify black-box**; and if they are of unspecified kind, they will cause *test-cases* for a compound **Pt.n.Vc. Verify black-/white-box**.

Pt.n.B. Architecture white-box ingredients design with possible embedded black-boxes. This is the *master* which *governs architecture design*, which will cause *white-box test-cases* (see **Pt.n.B. Architecture white-box ingredients design with possible embedded black-boxes,** p. 139).

Pt.n.G. White-box design items and product requisites. This is the *master* which *governs* detail design, which will cause *white-box test-cases* (see **Pt.n.G. White-box design items and product requisites,** p. 142).

Schedule packages:

Pt.n.Va. Develop test-cases, also for fixtures and debugger. This is the *task* which *governs* how to develop *test-cases*. From specified *requirements* on this *nesting level, test-cases* are now developed, both for *white-box* and *black-box verification*. If *requirements* do not distinguish between *restriction requirements* and *behavior requirements*, the *test-cases* may not do so either, and *white-box* and *black-box verification* are done from compound *test-cases* executed at the same time.

Pt.n.Vb. Test-case. This is the *master* which *governs test-cases*. Each *requirement* must refer to at least one *test-case*, and each *test-case* must refer to at least one *requirement* that should be verified. Each *test-case* contains a description of how to arrange the *verification*, and contains an expected outcome and a pass-fail criterion. *White-box test-cases* and *black-box test-cases* are very similar, but the *verification activity* using the various *test-cases* is rather different.

Pt.n.Vba. White-box test-cases. This is the *master* which *governs white-box test-cases*, which are derived from *restriction requirements*, and used explicitly for *white-box verification*.

Pt.n.Vbb. Black-box test-cases. This is the *master* which *governs black-box test-cases*, which are derived from *behavior requirements*, and used explicitly for *black-box verification*.

Pt.n.Vc. Verify black-/white-box. This is the *task* which *governs* how to perform a *black-/white-box verification* with compound *test-cases*. (If *restriction requirement* and *behavior requirements* are kept apart and generate separate *test-cases*, the *verification* can be more dedicated and efficient. If *test-cases* do not distinguish between *black-box* and *white-box*, *white-box* and *black-box verification* are done at the same time.)

Pt.n.Vd. Verification report. This is the *master* which *governs* the *verification report* containing a mix of executing *black-box test-cases* and *white-box test-cases*. During *verification*, outcomes from executing each *test-case* are documented. If the *test-case* execution meets specified criteria, no more information needs to be reported, but if the *test-case* fails to meet specified criteria, it is important to document what happened instead of the expected outcome, in order to find and eliminate the failure by subsequent *activities*.

Pt.n.Ve. Verify white-box. This is the *task* which *governs* how to perform a white-box *verification*. This is similar to verifying *black-/white-box* explained above, with the difference that only the *white-box* has been verified by using *white-box test-cases*.

Pt.n.Vf. Verification white-box report. This is the *master* which *governs* the *white-box verification report*. This is similar to the compound *verification report* explained above, but with the difference that only the *white-box* has been verified.

Pt.n.Vg. Verify black-box. This is the *task* which *governs* how to perform a black-box *verification*. This is similar to the *black-/white-box verification* explained above, with the difference that only the *black-box* has been verified by using *black-box test-cases*.

Pt.n.Vh. Verification black-box report. This is the *master* which *governs* the black-box *verification report*. This is similar to the *verification report* explained above, but with the difference that only the *black-box* has been verified.

Pt.n.Vi. Locate failure source. This is the *task* which *governs* how to locate failures. *Failure detection* is often the easier part of a *verification*. *Failure elimination* is more difficult, and first a *failure localization* must be made, and that might be nasty detective work. The failure location and reasons why it happened are important facts for the *failure elimination report*, which is prepared in the upcoming decision gate.

Pt.n.Vj. Evaluate optimal elimination. This is the business gate *task* which *governs* how to eliminate a failure detected, and determining if the *verification* is sufficient to be approved. Thus, the gate examines the following:

- Failures might be of very different severity for the *stakeholders* and might be very different in cost to correct. The gate meeting addresses each failure and how it may be eliminated, and decisions can be made on priority, severity, or cost of elimination.

- If the verified *white-box* is found to be very failure-prone and unlikely to meet high expectations, maybe a redesign must be decided on, including a new *white-box* to be prepared. Even if *failure detection* results in few failures, it is important to examine the cause. Maybe some restrictions and behavior of the *black-/white-box* to be verified have too few *test-cases* and have not been sufficiently tested; maybe the existing *test-cases* are testing only uncomplicated restrictions and behavior of the *black-/white-box*, or maybe the pass-fail criteria are too weak. If so, more and better *test-cases* must be ordered, or even better *requirements* must be specified.

Pt.n.Vk. Failure elimination report. This is the *master* which *governs* the *failure elimination report*, which is the glue for all *failure eliminations* being made when sending around the *black-/white-box* to all organizations responsible for the locations of the failures. The primary failure is eliminated by the organization which once gave rise to it, including correcting all other documentation to align with the primary *failure elimination*. All corrective actions are documented in the *failure elimination report*. The *black-/white-box* is sent further on to organizations having possible spin-off secondary failures to eliminate. The *failure elimination report* is then forwarded with the *black-/white-box* for elimination of the next failure in turn. When all *failure eliminations* are made, including ensuring that developed documentation is consistent internally and with the failure-free *black-/white-box*, it is returned for *reverification*. *Failure detections*, *failure localizations*, and *failure eliminations* are tricky (for more details see chapter 12.1, p. 660).

Pt.n.Vl. Prepare product manual. This is the *task* which *governs* how to prepare the *product* manual. As soon as a *white-box* has been verified failure-free, no time should be lost to finish manual work. It is not seldom seen that manual work appears on the *critical path* meaning that other organizations are impatiently waiting for the manual to get completed.

Pt.n.Vm. Manual table of contents. This is the *master* which *governs* the appearance of the table of contents of the manual. Note that the *target* for manuals is mainly potential *users* of the *product*, even if manuals also may be prepared for installation personnel, personnel to *reuse* the *product*, recycling organizations, and so forth. It is mandatory to at least prepare the table of contents for all *user* aspects of the *product*, and the final manual can be finished when all *product prototypes* have been verified.

Masters referred to in this schedule

Pt.n.W:1. Fully verified black-/white-box. This is the *master* which *governs* the appearance of a completely verified *black-/white-box*. When all *test-cases* have passed *verification* and the decision gate has judged the *verification* sufficient, this *white-box* is ready for *integration* in its surrounding *white-box* or into the *product environment*.

Pt.n.W:2. Partially verified black-/white-box. This is the *master* which *governs* the appearance of a partially verified but not yet completely verified *black-/white-box*. As long as *failure detection* results in new failures, the *white-box* is sent back to the organization that has developed the location of the failure.

6.12.8 Detailed Pt.n.Vg. Verify black-box manually or through automation

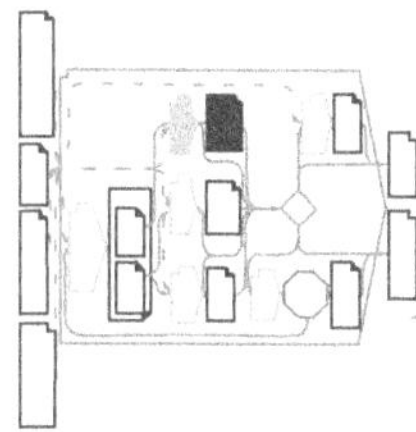

This chapter expands and explains in detail the *task* and *master*.

- **Pt.n.Vg. Verify black-box** and **Pt.n.Vh. Verification black-box report** in *schedule* **Pt.n.V. Verify black-/white-box,** page 152 (see left thumbnail).
- The expanded *task* and *master* explained in this chapter are highlighted in left thumbnail.

After a *failure elimination*, the *black-box* must be reverified to ensure that the failure is eliminated. However, when eliminating a failure, it is often the fact that secondary failures get introduced, which will probably not be detected if only the *test-case* that identified the primary failure is used in the *reverification*.

Consequently, after any correction or other changes of the *black-box*, it is not enough to verify with only some few *test-cases*, but to always verify with the full set of *test-cases*, which is called *regression verification*.

Of course, always verifying with the full set of *test-cases* after any *failure elimination* requires a lot of resources. To make a regression *verification* really efficient, it must be made automatic by using a *test-case* executor *fixture*. Note that the *black-*

box or *product prototype* also must be designed and extended in such a way that both input *interfaces* for injections of *stimuli* and output *interfaces* for judging correct *responses* must be easily connectable to the executor *fixture*.

Note that this chapter is an expansion of the *task* **Pt.n.Vg. Verify black-box** to become the *schedule* containing more detailed *tasks* and *masters*.

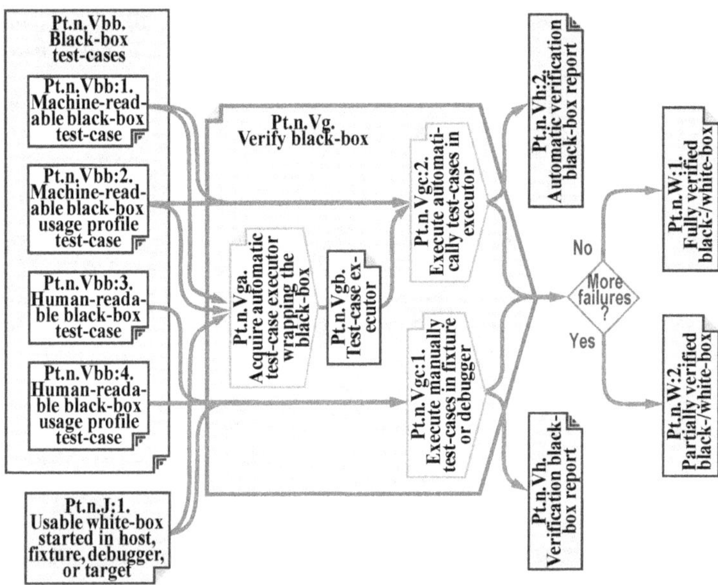

FIGURE 6-30 Verify black-box task expanded to a detailed schedule

Masters referred to in this schedule

Pt.n.Vbb. Black-box test-cases. This is the *master* which *governs* different kinds of *black-box test-cases*, which can be used to stimulate a *black-box*, and to check if its *response* is according to its expected *behavior requirements*.

Pt.n.Vbb:1. Machine-readable black-box test-case. This is the *master* which *governs* *black-box* text cases, written formally enough for a machine to read it.

Pt.n.Vbb:2. Machine-readable black-box usage profile test-case. This is the *master* which *governs* the *usage profiles*, written formally enough for a machine to read it.

Pt.n.Vbb:3. Human-readable black-box test-case. This is the *master* which *governs* *black-box test-cases* written for a human verifier.

Pt.n.Vbb:4. Human-readable black-box usage profile test-case. This is the *master* which *governs usage profiles* written for a human verifier.

Pt.n.J:1. Usable white-box started in host, fixture, debugger, or target. This is the *master* which *governs* the *black-box* to verify (see **Pt.n.J:1. Usable white-box started in host, fixture, debugger, or target,** p. 146).

Schedule content

Pt.n.Vg. Verify black-box. This is the *task*, extended to a *schedule*, which *governs* how to verify a *black-box* manually or automatically.

Pt.n.Vga. Acquire automatic test-case executor wrapping the black-box. This is the *task* which *governs* how to run the executor, which can be developed in-house or ordered from a *turnkey supplier.*

Pt.n.Vgb. Test-case executor. This is the *master* which *governs* the appearance of the automatic verifying executor tool. *Test-cases* must be retrieved by this tool, and *test-case stimuli* must be automatically injected into the *black-box* via its *interfaces.* *Responses* must be automatically received from the *black-box* via its *interfaces*, in order to be automatically compared with what is expected by the *test-case*, and the pass-fail results from the *verification* must be reported automatically. For more explanations, see Figure 12-27, Simplified architecture of a test-case executor, page 743.

Pt.n.Vgc:1. Execute manually test-cases in fixture or debugger. This is the *task* which *governs* how to manually verify the *black-box* by using human-readable *test-cases.*

Pt.n.Vgc:2. Execute automatically test-cases in executor. This is the *task* which *governs* how to automatically verify a *black-box* by a *test-case* executor using machine-readable *test-cases.*

Masters referring to this schedule

Pt.n.W:1. Fully verified black-/white-box. This is the *master* which *governs* the verified *black-box* having no residual fails.

Pt.n.W:2. Partially verified black-/white-box. This is the *master* which *governs* the verified *black-box* still having residual fails (see **Pt.n.W:2. Partially verified black-/white-box,** p. 155).

Pt.n.Vh. Verification black-box report. This is the *master* which *governs* the *verification black-box* report.

Pt.n.Vh:2. Automatic verification black-box report. This is the *master* which *governs* the automatic *verification black-box* report.

6.12.9 Detailed Pt.n.Va:2. Develop test-cases, also for fixtures and debugger, by usage profile modeling

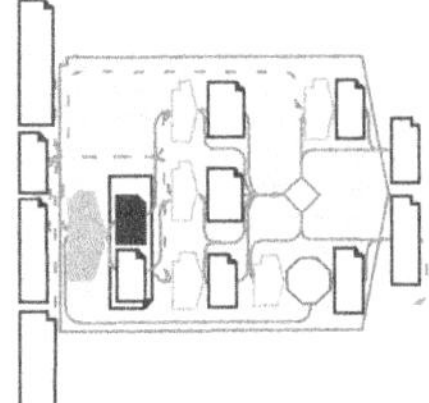

This chapter expands and explains in detail the *task* and *master*.

- **Pt.n.Va. Develop test-cases, also for fixtures and debugger** and **Pt.n.Vb. Test-case** in *schedule* **Pt.n.V. Verify black-/white-box,** page 152 (see left thumbnail).
- The expanded *task* and *master* explained in this chapter are highlighted in left thumbnail.

This *verification* method is rather mathematical, but if it is applied in a practical way and using common sense, it is very elegant and efficient (see Fig. 6-31 below). The best way to understand it is to eagerly study the example beginning at chapter 12.20.5, page 736.

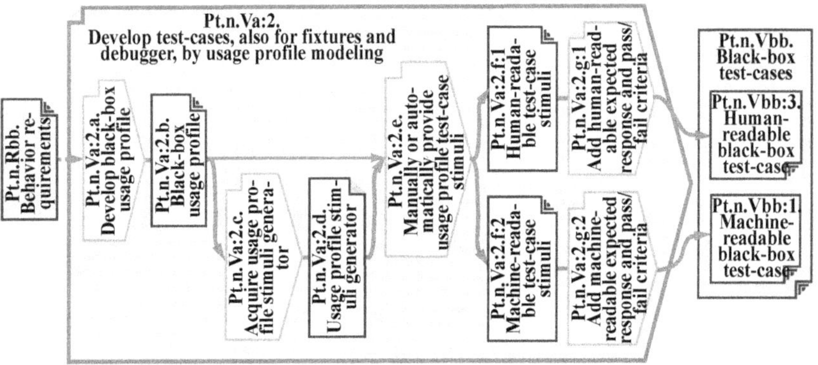

FIGURE 6-31 Develop test-cases, also for fixtures and debugger, by usage profile modeling detailed schedule

Masters referred to in this schedule

Pt.n.Rbb. Behavior requirements. This is the *master* which *governs behavior requirements*, which enables *usage profile verification* of a *black-box* and is ideally specified as *formal scenarios*.

Schedule content

Pt.n.Va:2.a. Develop black-box usage profile. This is the *task* which *governs* how to develop a *black-box usage profile*, which is developed from the *wait-state-machine*. If the behavior is specified by *formal scenarios* (see ch. 7.19, "Scenario behavior requirements," p. 230), these *formal scenarios* can easily be changed to a *wait-state-machine* (see ch. 7.20, "Wait-state-machine," p. 232). Building a *usage profile* from a *wait-state-machine* implies that all transitions departing from all its *wait-states* will be assigned a probability value, reflecting the chance that this transition happens when using the *black-box*, and the sum of all transitions from each *wait-state* will, of course, be 100%. The way the *black-box* is used can vary; for example, a machine is a very rigid *user*, and a human can be a more or less advanced *user*. To cover differences in usage, many different *usage profiles* can be developed.

Pt.n.Va:2.b. Black-box usage profile. This is the *master* which *governs* the appearance of a *black-box usage profile*. There are several ways to document the *usage profile*, and to note probabilities directly on a *wait-state-machine* transition is good for human interpretation. This probability notation can easily be transferred to a table (see Table 12-61, p. 737). When generating *test-case stimuli* automatically, the probability values must be noted either in a data description file or in a *variable* area within the *stimuli* generator *program* (see A.b.Probabilities table, p. 738).

Pt.n.Va:2.c. Acquire usage profile stimuli generator. This is the *task* which *governs* how to develop a *usage profile stimuli* generator. It is rather easy to develop a *program* for this purpose in any language available. (Just use *Cpdm* as the *process* for this development :-)

Pt.n.Va:2.d. Usage profile stimuli generator. This is the *master* which *governs* the construction of a usage *stimuli* generator. As an example of a *program* to generate *stimuli* from a *usage profile* model, see Figure 12-26, page 741.

Pt.n.Va:2.e. Manually or automatically provide usage profile test-case stimuli. This is the *task* which *governs* how to generate *test-case stimuli*, which now will be automatically generated or manually obtained. The probabilities in the *wait-state-machine* now must be transferred to a stream of *stimuli*, where each *stimulus* occurs as frequently as its probability value indicates. For example, say that two different *stimuli* can initiate a transition from a certain *wait-state*, and the probability for one of these *stimuli* is much higher than for the other, there will be many more *test-cases* containing the higher-probability *stimulus* than there are *test-cases* containing the low-probability *stimulus*. Or in conclusion, if there is a certain way to use the *black-box* that is more common, there will be more *test-cases* generated according to this usage. Or the reverse, if there are some obscure and rare ways to use the *black-box*, these will not be tested very much. If the obscure way to use the *black-box* is also important to verify (safety against crashes), define a separate *usage profile* for such a way.

Pt.n.Va:2.f:1 Human-readable test-case stimuli. This is the *master* which *governs* the appearance of *test-case stimuli* which are readable by a human verifier.

Pt.n.Va:2.f:2 Machine-readable test-case stimuli. This is the *master* which *governs* the appearance of machine-readable *test-case stimuli*, formal enough to be read by a computer. For an example of generated *stimuli*, see LIST 12-4, p. 741, LIST 12-5, p. 742, and LIST 12-6, p. 742.

Pt.n.Va:2.g:1 Add human-readable expected response and pass/fail criteria. This is the *task* which *governs* how to add human-readable *response* and pass-fail criteria to *test-case stimuli*. When the *test-cases* have been generated containing *stimuli*, expected *results* and pass/fail criteria should always be assigned to those *stimuli* that would cause a *response*. Adding the expected *response* with pass-fail criteria is often the most tedious manual work, because it is hard to find any automated procedure that is intelligent enough to judge this.

Pt.n.Va:2.g:2 Add machine-readable expected response and pass/fail criteria. This is the *task* which *governs* how to add machine-readable *response* and pass-fail criteria to *test-cases stimuli*. This is similar to the *task* above, with the difference that the expected *response* and pass-fail criteria must be formally noted, to be easy for a computer to read.

Masters referring to this schedule

Pt.n.Vbb:3. Human-readable black-box test-case. This is the *master* which *governs* human-readable *black-box test-cases* for *usage profile verification*, used to perform a manual *black-box verification*.

Pt.n.Vbb:1. Machine-readable black-box test-case. This is the *master* which *governs* machine-readable *black-box test-cases* for *usage profile verification*. These *test-cases* are now formalized, to allow computers to read them, and the *verification* can be performed automatically by a *test-case* executor (see previous ch. 6.12.8, p. 155). For an example of a *test-case* sequence (see List 12-7, p. 743).

6.13 EXAMPLE: Tailored process overall schedules

To make it easier to follow used *processes* by the numerous examples in this book, some *tailored* overall *process schedules* have been prepared in advance. All these *tailored schedules* are derived from the **Pt:2. Cpdm generic advanced technical schedule,** pages 104 and page 105. Note that only the overall *schedule* is *tailored*, which refers to generic detailed *schedules* described in chapter 6.11 and 6.12 above. It is no problem to also *tailor* these *schedules* if they need to be used in any particular way.

6.13.1 EXAMPLE: Pt:2:1. Tailored advanced technical overall V-schedule n = 3, using incorporative integration

Technical development of overall *processes* is often illustrated as a V (see Fig. 6-32 below). The V-*schedule* below is *tailored* from the *Cpdm* generic *process schedule* **Pt:2. Cpdm generic advanced technical schedule,** pages 104 and page 105, by "grabbing the top left and top right corners and stretching them apart." This *V-model* assumes that the **Pt.I. Realize & integrate white-box** *phase* uses *incorporative integration* inside-out.

Since the number of *nesting levels* in the *tailored schedule* now is fixed to three, *nesting levels* are more comprehensibly illustrated below each other, and not nested inside each other as the generic overall *schedule*.

The main advantage of this *tailored* overall *schedule* is that time can be imagined to strictly elapse (no arrows going left) along an imaginary x-axis from left to right. Thus, right direction is to trace continuously *downstream* along development, and the opposite is to trace *upstream* contrary to development flow. The *nesting level* is more obvious along an imaginary y-axis running from *environment* to *innermost black-/white-boxes*. This way of drawing a *process schedule* can be very useful to indicate time durations when *schedules demand masters*.

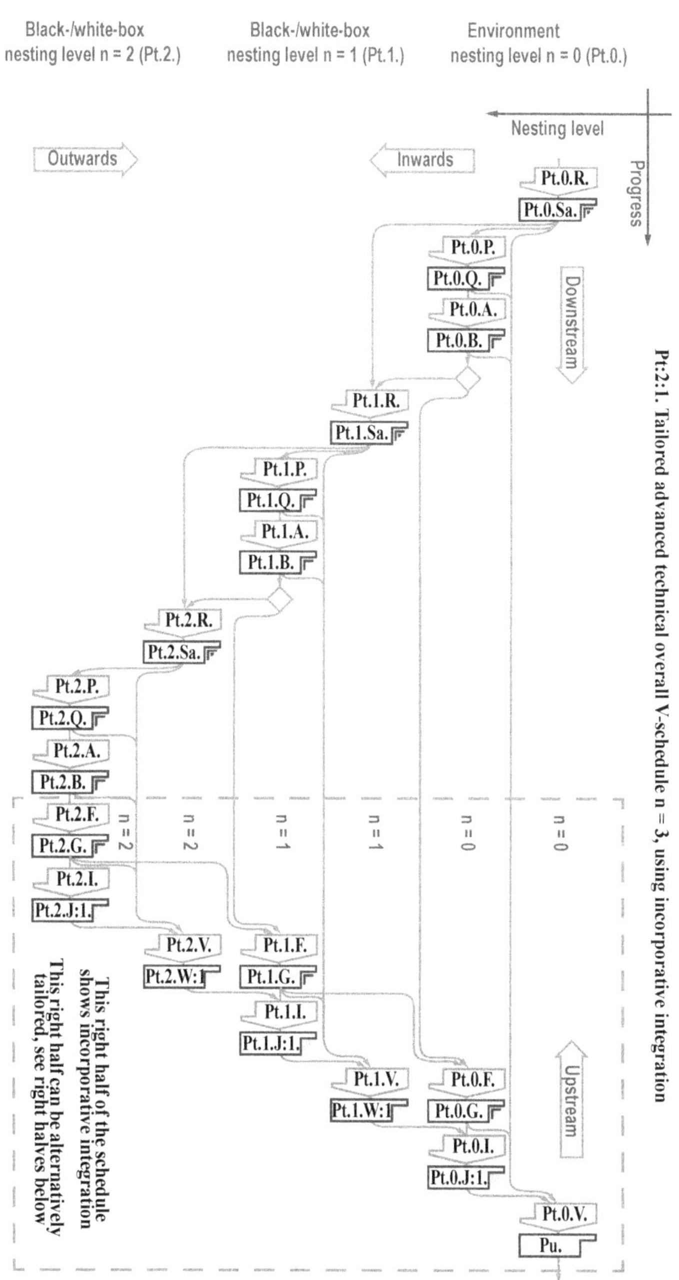

FIGURE 6-32 Pt:2:1. Tailored advanced technical overall V-schedule n = 3, using incorporative integration

The *V-model* can be symbolized with arrows instead of showing all detailed *schedules* and *masters* (see Fig. 6-33 below).

Pt:1. Cpdm generic elementary technical schedule (arrow view)

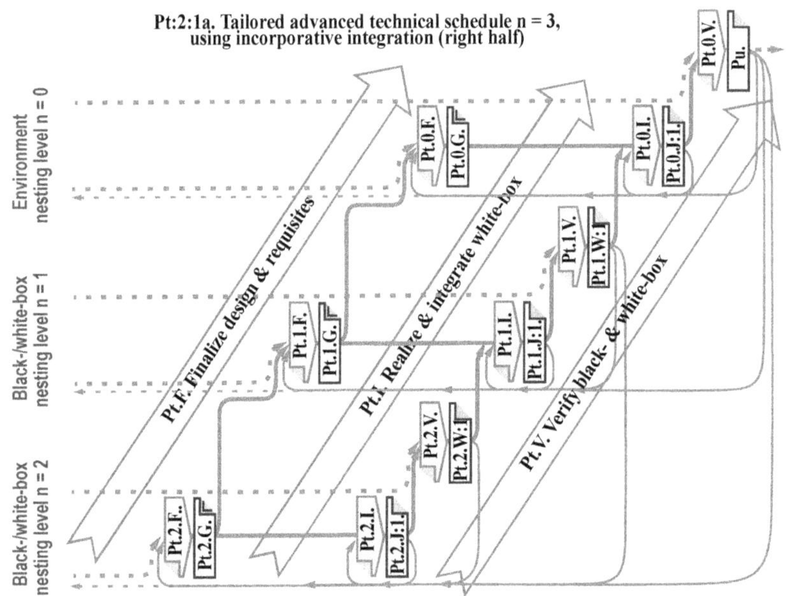

FIGURE 6-33 Tailored advanced technical overall V-schedule (symbolic view)

By using such arrows, the *incorporative integration* can be more comprehensively illustrated (see Fig. 6-34 below), especially when comparing it to other *integration* alternatives.

Pt:2:1a. Tailored advanced technical schedule n = 3, using incorporative integration (right half)

FIGURE 6-34 Inside-out incorporative integration

When developing *products* that are tricky to realize (for example very cost reduced *products*), there might be a need to *tailor* an overall *schedule* for this purpose (see Fig. 6-35 next page).

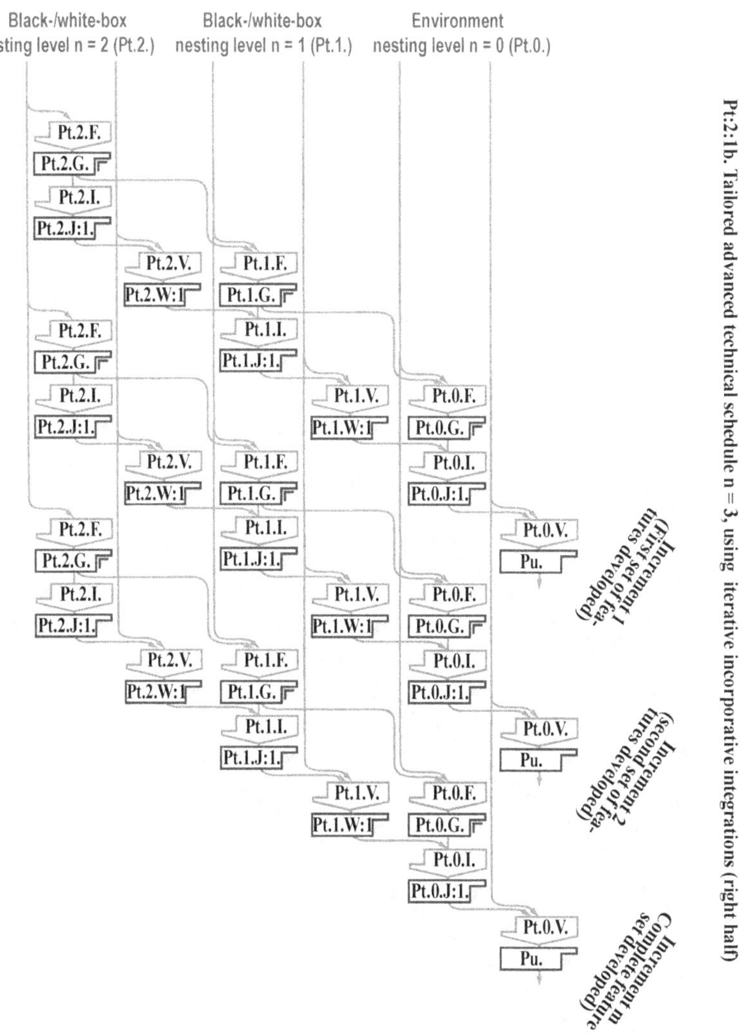

FIGURE 6-35 Iterative development schedule for stepwise emerging prototypes

6.13.2 EXAMPLE: Pt:2:2. Tailored advanced technical schedule n = 4

To suit the house example, the generic *Cpdm schedule* **Pt:2. Cpdm generic advanced technical schedule,** pages 104 and page 105, has been *tailored* to a 4-nesting-level *schedule* (see Fig. 6-36 and Fig. 6-37 below).

The left half of this *schedule* is further *tailored* to better fit *invasive integration* (see Fig. 6-37, p. 165). However, these *schedules* are fully compatible with each other.

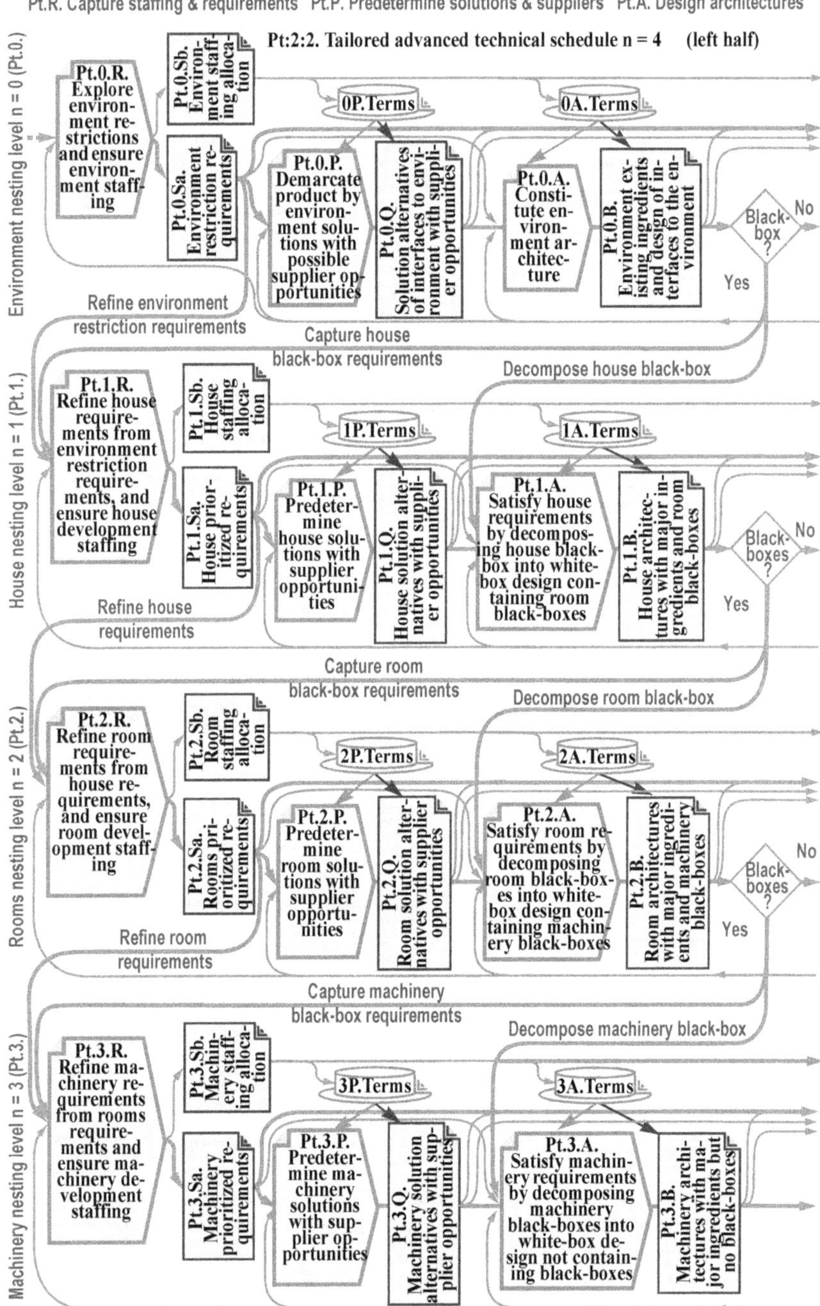

FIGURE 6-36 Tailored advanced technical schedule n = 4 (left part)

Pt.F. Finalize design & requisites   Pt.I. Realize & integrate white-box   Pt.V. Verify black- & white-box

Pt:2:2. Tailored advanced technical schedule n = 4 (right half)

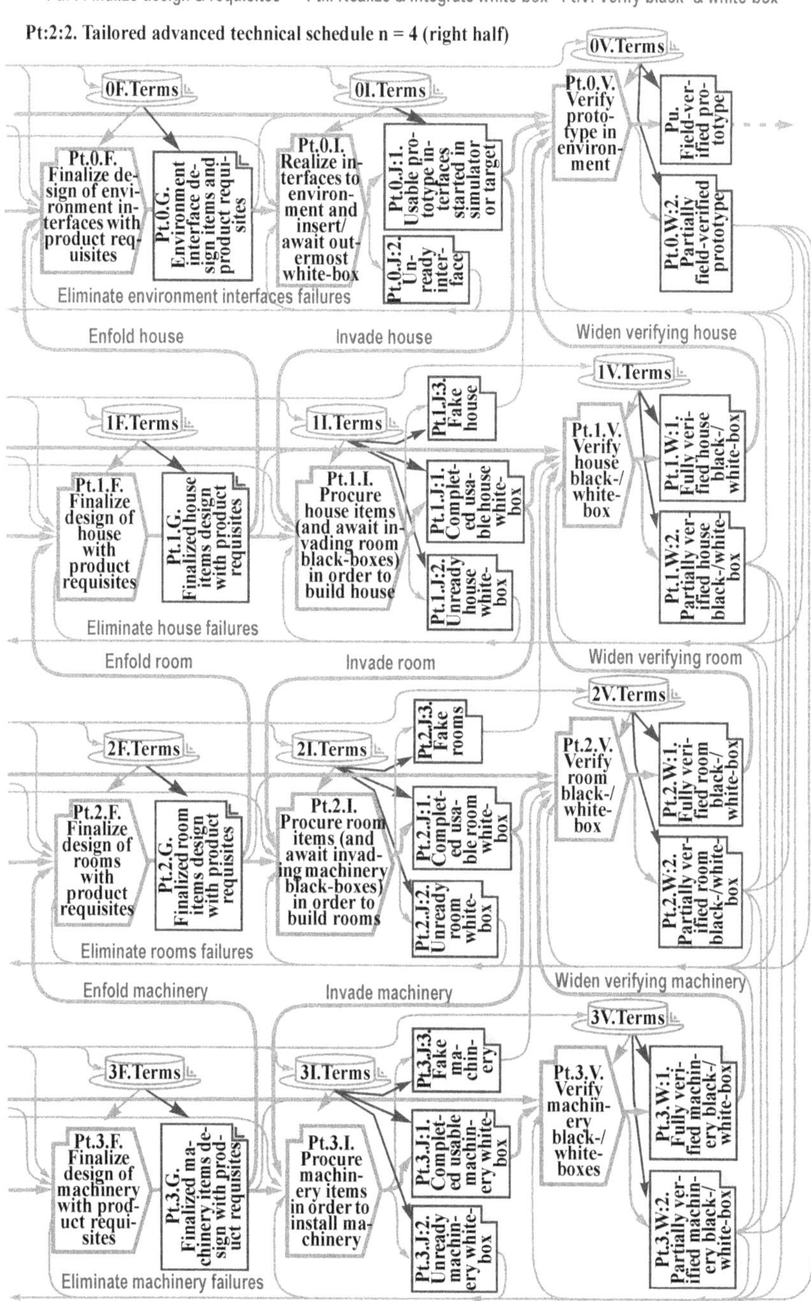

FIGURE 6-37 Tailored advanced technical schedule n = 4 (right part)

Since the number of *schedule nesting levels* is now fixed to 4, and each *nesting level* corresponds to easily recognizable magnifications, such as house environment, house, rooms, and machinery, these *nesting levels* now get more comprehensible names, and are better illustrated below each other in the *tailored schedule* (not nested inside each other as the **Pt:2.n.**, p. 104). The *environment nesting level* is not *tailored* at all, since it is rather easy to understand in the context of a house.

Note that there may be different numbers of *nesting levels* in different parts of the house example. Where the *complexity* and amounts of details are large, all 4 *nesting levels* from this *tailored schedule* might be used (for example, house *environment*, house, kitchen, machinery), but where *complexity* is low (for example in bedroom without *complex* machinery), the *innermost black-box* may be sufficient to stop at *nesting level* Pt.2.

A house is big and heavy, making it impossible to integrate it inside-out, which would be to start to integrate interiors into rooms, then integrate rooms into a house, and then integrate the house into the *environment*. Because of this, the overall *process* is *tailored* for outside-in *invasive integration*, which instead means to start to realize the *environment interfaces*, then realize the house in the *environment*, then the rooms inside the house, and then the interiors inside the rooms.

Also an overall *schedule* (right part) is *tailored* for *partial verification* with use of a *fake black-/white-box* (see Fig. 6-38, p. 167 below). For more explanations about *integration* in the house example, also see chapter 11.2, page 601.

6.13.3 EXAMPLE: Pt:2:2b. Tailored advanced technical schedule n = 4 (right half)

To better illustrate how to use *invasive integration, partial verification* with fakes and *implicit verification* when building a house, the right half of above overall *schedule* **Pt:2:2.** has been *tailored* in the same way as the overall *schedule* **Pt:2:1.**, page 161 (see below right half overall *schedule* **Pt:2:2b.**).

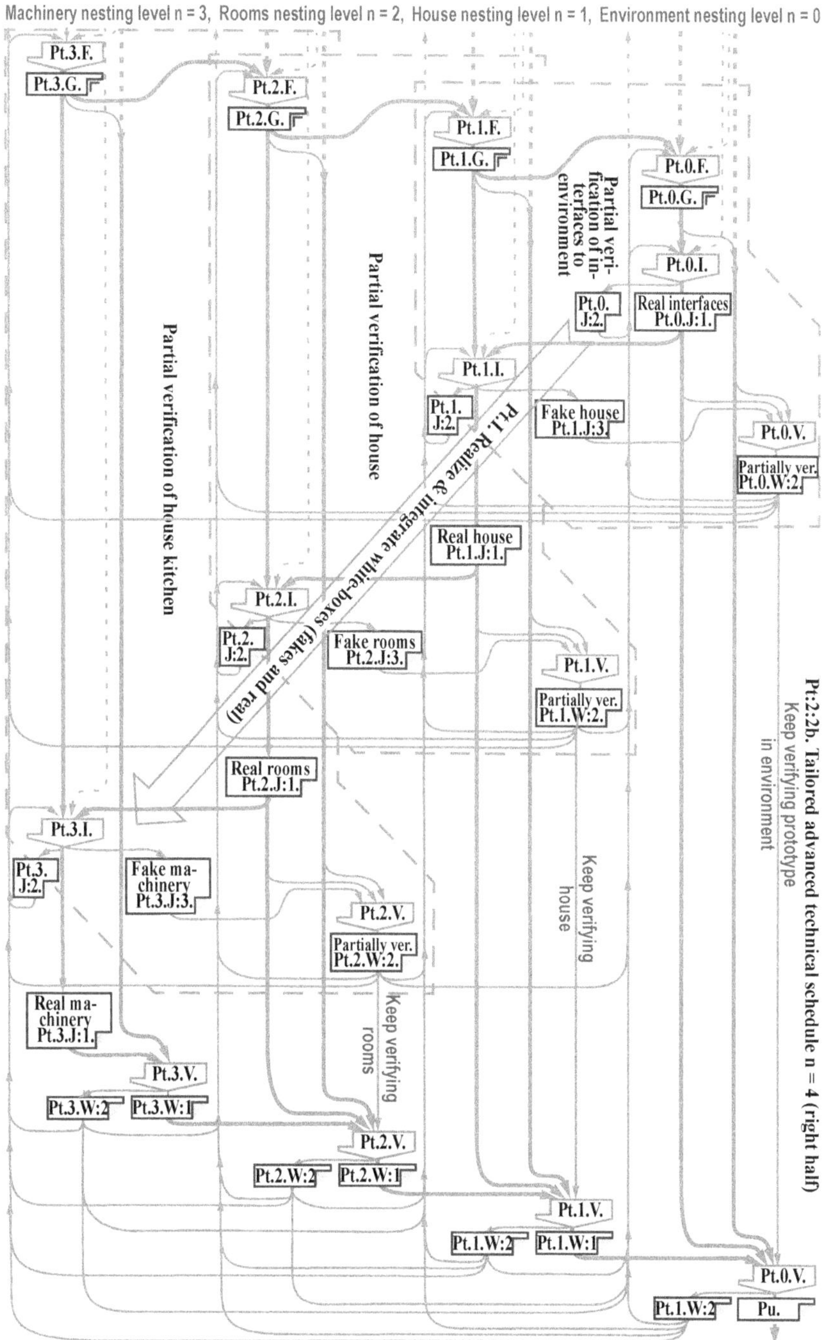

FIGURE 6-38 Outside-in invasive integration with fakes, house schedule n = 4

When performing *schedule* **Pt.0.V.** the first time for *partial verification*, only the *environment interfaces* are realized, but failures might be detected on other *nesting levels*, since for instance the already completed **Pt.2.F.** and **Pt.1.F.** may contain primary failures detected in *schedule* **Pt.0.V.**

Invasive integration is not too *complex* in itself (see the house example), but combined with faking embedded *black-/white-boxes* to prepare for *partial verification*, and after that performing *implicit verification*, certainly does result in some *complexity*. But once again, this is quite useful, so just dig into it to master its *intrinsic complexity*.

Some explanations may be in place for the above *tailored* right half *schedule*:

Pt.0.I. Realize interfaces to environment and insert/await outermost white-box.
This is the *schedule governing* how to realize the *interfaces* to the *environment* as usual.

Pt.1.I. Procure house items (and await invading room black-boxes) in order to build house, Pt.2.I. Procure room items (and await invading machinery black-boxes) in order to build rooms, and Pt.3.I. Procure machinery items in order to install machinery. These are the *schedules governing realizations* of both the fake and real *results* during *partial verification* with embedded *fake black-/white-boxes*. Before the *partial verification* is performed, these *schedules* govern preparation of the fake *realizations*, and after the *partial verification* the fake is removed, and this *schedule governs* preparation of the real *realizations*. The extent and *complexity* of a fake *realization* is by definition moderate and not crucial, thus *formalism* may be reduced here.

Pt.0.I. Realize interfaces to environment and insert/await outermost white-box.
This is the *schedule governing* how to realize the *interfaces* to the *environment* as usual.

Pt.3.V. Verify machinery black-/white-boxes. This is the *schedule governing* how to verify the *interfaces* to the *environment* as usual.

Pt.2.V. Verify room black-/white-box, Pt.1.V. Verify house black-/white-box, and Pt.0.V. Verify prototype in environment. These are the *schedules governing* how to verify the same *black-/white-box* in two steps, first performing *partial verification* with embedded *fake black-/white-boxes*, and when this is completed performing *implicit verification* with the real integrated *black-/white-boxes*. The extent of a *partial verification* is by definition moderate, thus *formalism* may be reduced here.

More about *invasive integration* will be explained in chapter 6.13.3, page 166, and chapter 11.1.4, page 595.

6.13.4 EXAMPLE: Pt:2:3. Tailored advanced technical schedule n = 2

To suit the three examples Multiplication toy, Calc_logic, and Phonebook, the *Cpdm* generic *schedule* **Pt:2. Cpdm generic advanced technical schedule**, pages 104 and page 105, has been *tailored* to a minimal *schedule* n = 2 (see Fig. 6-39, p. 169).

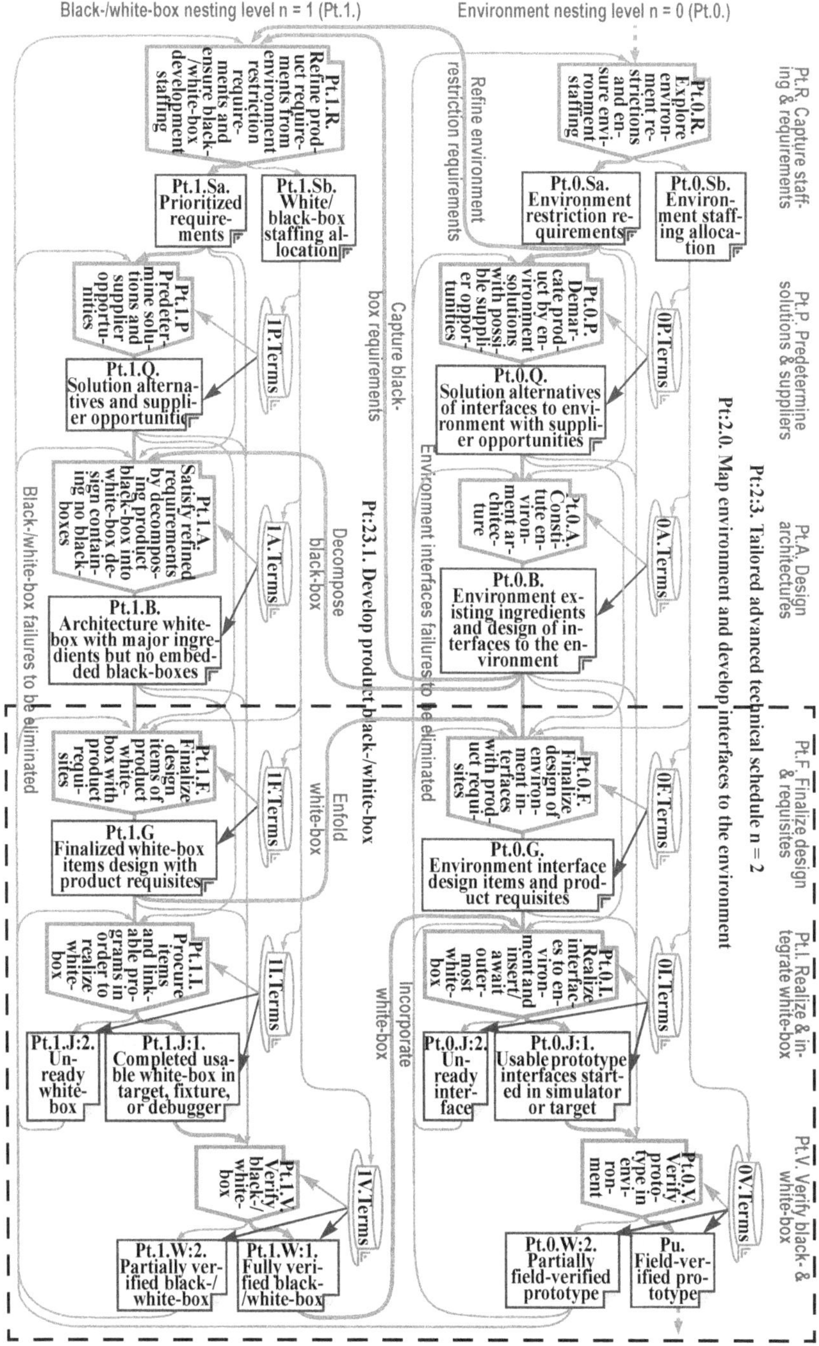

FIGURE 6-39 Most simplified advanced technical schedule n = 2

Incorporative integration is applied in this *tailored* overall *schedule*, since all these examples are small enough to allow inside-out *integration*.

6.13.5 EXAMPLE: Pt:2:3b. Tailored advanced technical schedule n = 2, porting to 3 targets (right half)

The calc_logic example assumes that the developed *product* shall be *ported* to many *targets* for *reuse* in different *products*. For this purpose an overall *schedule* (right part) is *tailored* to fit the dashed square in **Pt:2:3. Tailored advanced technical schedule n = 2,** page 169 to support *porting* (see Fig. 6-40 below). From **Pt.1.F.** the three *schedules* **Pt.0.F.** receive *porting results* **Pt.1.G** for the different *targets*, such as recompiled and reassembled *programs*.

Pt:2:3b. Tailored advanced technical schedule n = 2, porting to 3 targets (right half)

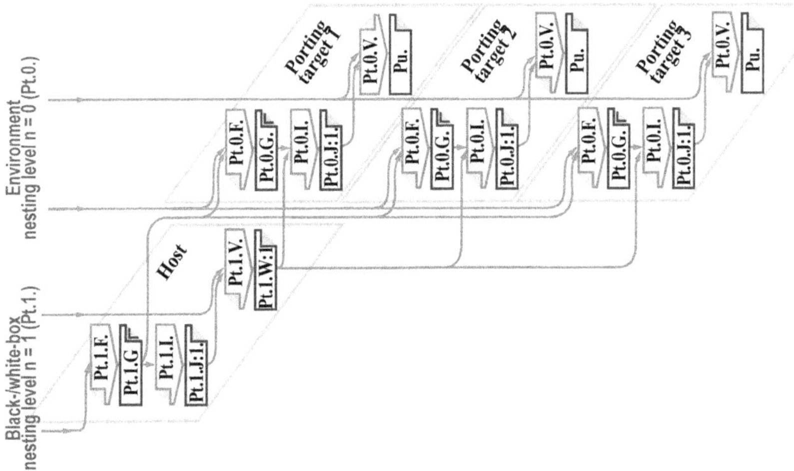

FIGURE 6-40 The advanced technical schedule n = 2 with portability

Porting in above *schedule* is performed to the *outermost white-box*, but can also be done on a deeper *nesting level*, for example, when a *program* is developed on the *host* for a deeply embedded *target black-/white-box*, which is supposed to be *ported* to an *outward* electronics *black-/white-box*, which may be the *host*, a *development board*, or the *target* electronics.

6.13.6 EXAMPLE. Pt:2:4. Tailored advanced technical short form n=3

Generic *schedule* **Pt:2.** is below *tailored* to a one-page *schedule* n = 3 short form.

1. Explore *environment restriction requirements*, such as authority regulations and market obstacles.

2. Analyze in/out and make/buy options on *environment* E. to determine demarcation of *black-box* **A.**

3. Constitute *environment* with *outermost black-box* **A.** and *interface* **-E-1.**

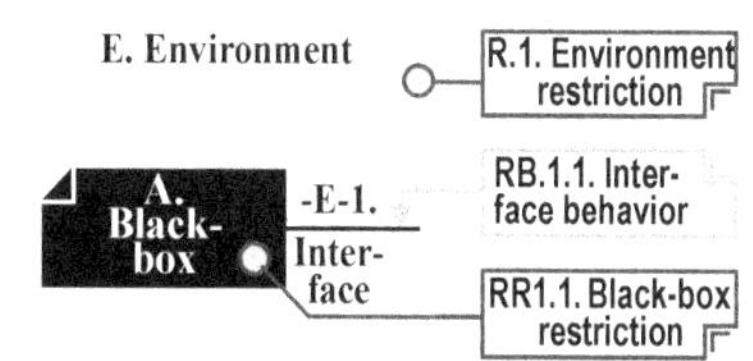

4. Refine *environment restriction requirements* and capture more *black-box* **A.** and *interface* **-E-1.** *requirements.*

5. Open *black-box* **A.** to become *white-box* **A.**

6. Find *solution alternatives* and *supplier opportunities* for **A.**

7. Design *black-box* **A.** (assumed to be one *element* and one *black-box* below).

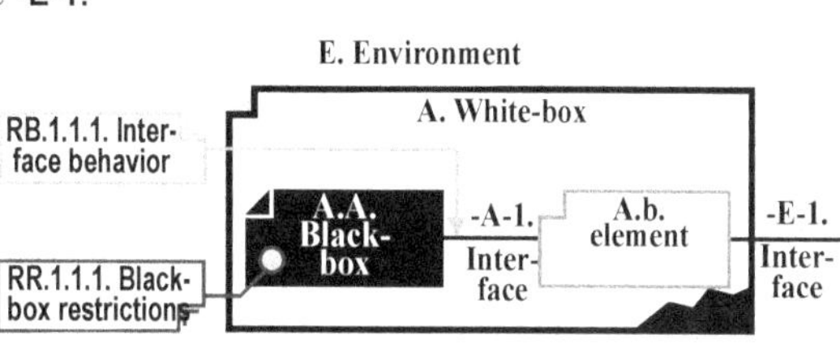

8. Refine **-E-1.** *requirements* and capture more *requirements* for *black-box* **A.A.**

9. Open *black-box* **A.A.** to a *white-box.*

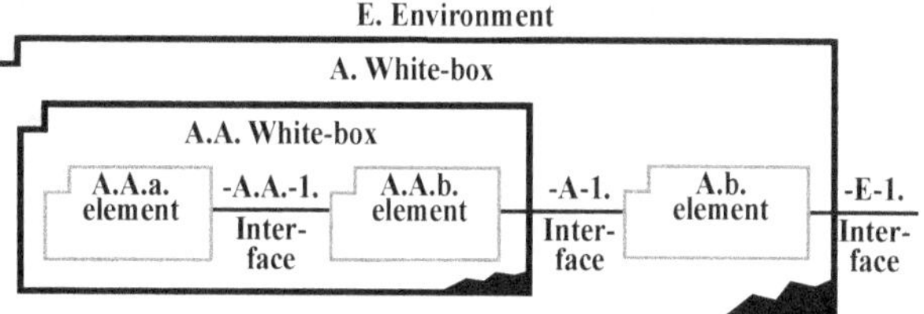

10. Find *solution alternatives* and *supplier opportunities* for **A.A.**

11. Design *white-box* **A.A.** *elements* (assumed to be two *elements* in figure at right).

12. Finalize *white-box* **A.A.** *elements* **A.A.a.** and **A.A.b.**, and *interface* - **A.A-1.** and **-A-1.** Complete their *product requisites.*

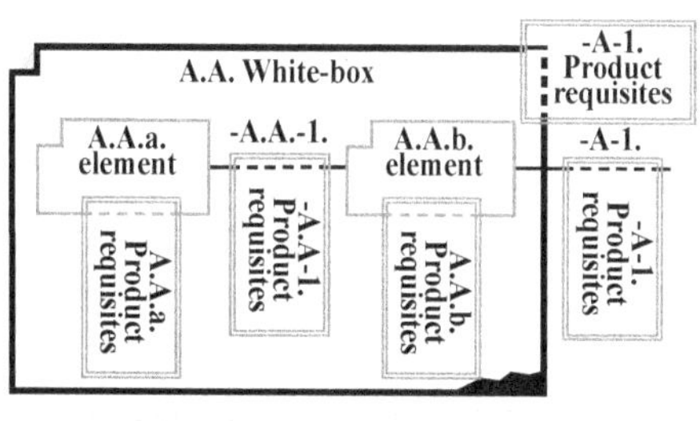

13. Procure *white-box design items* and *linkable program* to realize **A.A.** and verify it.

14. Finalize *white-box* **A.** and *interface* **-E-1.** and complete their *product requisite.*

15. Procure *white-box design items* and *linkable program* to realize *product prototype* **A.** and verify it.

16. Finalize *interfaces* to *environment.* Field-test *product prototype* **A.**

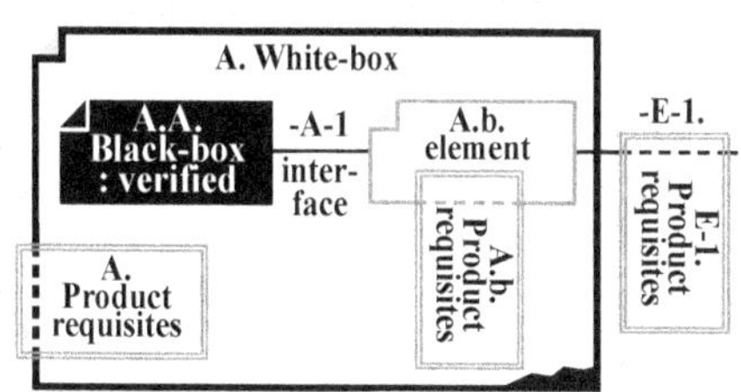

6.14 Schedules seen as programs

6.14.1 A schedule versus a program

A *schedule governs* the execution of *activities* with *results*. In a human organization the *schedule* is executed by employees, not very different from a *program* being executed by a computer processor.

Such a basic concept as executable *instructions* can be more elaborate. Comparisons between a *schedule* and a *program* are made in Table 6-8 below.

A process schedule is executed by people, in the same way as a processor executes a program.

TABLE 6-8 Comparison between schedule and program

Aspect	Schedule	Program
Execution governed by	Tasks	Instructions
Substantiation governed by	Masters	Variables
Arena	Organization	Computer
Executor	Humans	Processors
Concurrence	Individual	Thread
Obedience	As decided by the individual [a.]	Strictly according to the instructions [a.]
Efficiency	Human capability	Processor capability

a. Individuals have the possibility to choose to work after their own will or to be governed by the task. A CPU has no own will (not yet), and will always strictly execute the instructions.

6.15 Some other process models

Out in the world there are a lot of *product development* models, and a lot of methods are suggested. Some of these have influenced the *complex product development* model *Cpdm* and are briefly explained below.

6.15.1 Clean room engineering

The *clean-room engineering process* for *program* development recalls how meticulous the *process* needs to be when producing silicon chips. Why not also when developing *programs*? Clean room is built on strict mathematics and there are precise procedures going from *requirements* to finalized *programs* [5], [6], and [2].

The clean-room approach is a very far-reaching concept, but too theoretical to be directly applied on the floor.

Many really nice concepts have certainly been built into the *clean-room engineering* model, but there are also some severe obstacles. The worst is maybe the poor support for coming up with the textual *requirement* specification, which must be complete and allow only *stimuli history*. It may be concluded that the effort and difficulties are transferred from designing the *product* to developing the *requirement*.

The *Cpdm formal scenarios* are graphic representations of the *clean-room engineering requirement* specification. In *Cpdm wait-states* are also allowed as part of *requirements* to facilitate specifying *requirements* even more. Further, the statistical usage testing originates from clean room, which in fact uses *wait-states*, which in clean-room engineering are referred to as "usage states."

6.15.2 Unified Modeling Language, UML

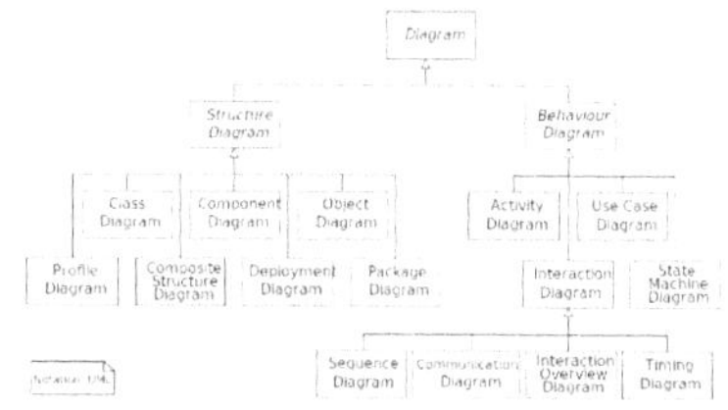

UML is now an ISO/IEC diagram-oriented standard for modeling *products*, and promotes visual and graphic documentation [8] and [9], which can be used by any kind of *processes*. The supported diagrams are illustrated by Figure 6-41 to the left.

FIGURE 6-41 UML diagrams

UML is a very serious approach and covers many good diagrams needed by *complex product development*.

There is also heavy criticism about UML, for example, that diagrams are not very well integrated with each other. Each diagram is inherently *complex*, and the poor *consistency* between them creates much frustration and added *complexity* [3].

UML is a great initiative to promote documentation, but too fragmented to be really useful

Cpdm uses many of the diagram types, but has changed and adapted them in order to perfectly fit together.

6.15.3 Rational Universal Process, RUP

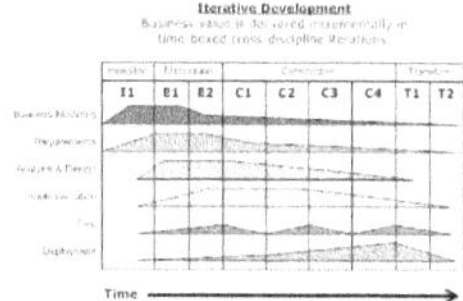

The benefit of this model is the holistic approach and *incremental development*, and most *process phases* are at hand for *complex* development of *products*. This model is known by its graphic description of the *phases* and disciplines (see Fig. 6-42, left). The development of this model was started by [4], and taken over by IBM [7].

FIGURE 6-42 RUP phases and disciplines

Incremental development is a very nice way to *partition* a *complex* large development work into smaller increments in the way that all increments deliver a useful partial *product*. However, if not keeping good control of the increments, it might lead to a *product management* battle to press all *requirements* into the first increments in an uncontrolled way. *Incremental development* will be included in *Cpdm*, in the next book *"Cpdm* technical overhead" about management and overhead.

RUP is a great initiative to promote processes, but not stringent enough to be a powerful tool.

6.15.4 Agile concepts

Agile development is a reaction against rigid models like the one mentioned above, and their heavy documentation. To break up such stagnant models, the agile lightweight concept suggests, for example, delivery after very short increments, and promotes simplicity and openness to change. There are now 12 principles in the Agile manifesto, each of them clearly supportive, but very ambitious when practiced all together. [1]

The agile concept is a sound reaction against rigid processes, but seldom scales up to the degree where complexity becomes the real problem.

Agile development may seem contrary to *Cpdm*—in some respects it is, and in some it isn't. The agile model highlights trusted senior individuals, in deep contact with *customers* and business people, and letting *program* details be the major documentation, which is, of course, beneficial on a team level.

But the agile concept gives poor support when scaled up to *complex products*. If something is very viable on a team level, it is not necessarily the solution for large *complex products*. *Cpdm* claims, for example, that large *products* (also containing large *programs*) need to be intelligently *partitioned*, and *architectures* must illustrate this before detailed design starts. Otherwise things get messy as usual. A realistic use of *Cpdm* with agile whistleblowers is a good combination.

Capture staffing & requirements

If you specify your product right, you get it good,
if you specify the right product, you get it sold.

Dear Santa,

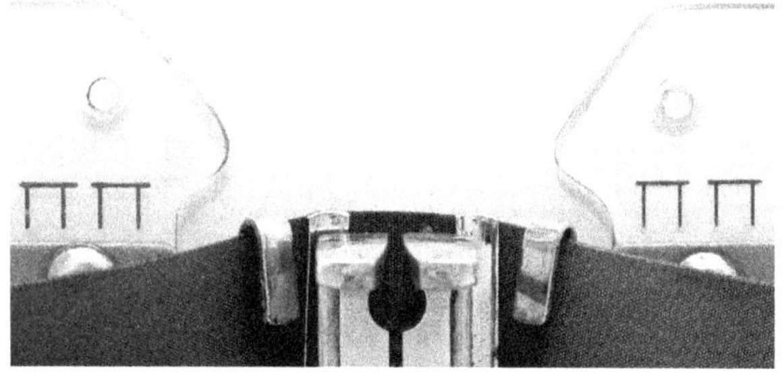

Getting a bit desperate?

7.1 About requirements

7.1.1 Purpose

Requirements are specified by stakeholders, describing behavior and limitations of the product to be developed.

Requirements are identified and prepared before a *product* is developed, to technically describe what is expected from it or wished for, based on all kinds of previous development *results*. *Requirements* should be seen as belonging to *stakeholders*, which can be authorities, *marketers*, *line management*, *customers* or *users* (or representatives of any of these) in a *product* planned to be developed.

Since *product development* might be *complex*, there is an imminent danger that wishes from the *stakeholders* get lost somewhere in the development organization, and other parties take over decision making, leading to a developed *product* that will end up as something not matching the *stakeholders' requirements*.

Requirements are said to be boring to write, hopeless to understand, and impossible to update, but they are the ultimate way to secure proper *verification* of a developed *product prototype* that guarantees that *stakeholders* (in particular the end *users*) eventually get what they want.

There are many methods and ways to capture staffing & requirements, but there are few standard ways. The choice depends on many things, for example:

- What competency is at hand in the developing company
- How *complex* the developed *product* is expected to be
- What *quality* is expected from the developed *product*
- How much control must be accorded to *suppliers*
- To what degree the *product* must be user-friendly

In this chapter presenting **Pt.R. Capture staffing & requirements**, the most common methods of specification are referenced. But as always, it is not the method itself that is most important, but its adequate and intelligent application. For example, *requirement refinement* must be traceable and *requirements* must be connected to the *architectures*.

As said, *requirements* are specified by *stakeholders* scattered all around a company, and it is very important that the *requirements* not be distorted on their way to being used by engineers for design of the *product* (see Fig. 7-1 below). It may get even worse if strong engineer *gurus* replace specified *requirements* with their own biased desire to develop a technique they enjoy or to promote their careers.

Requirements specified by stakeholders must be transferred to engineering departments without bias or distortion.

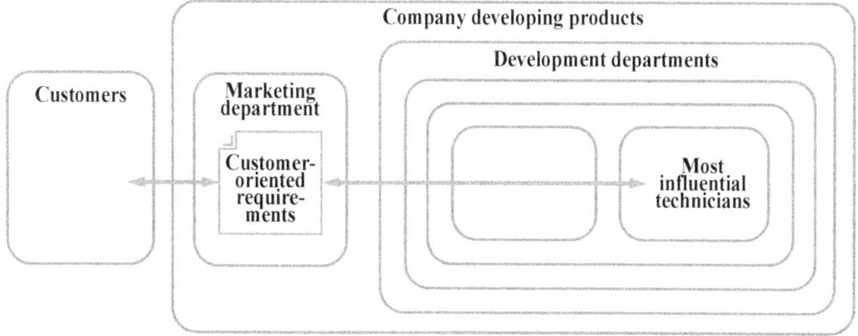

FIGURE 7-1 Undistorted connection between customers and most influential technicians

The *stakeholders* who define and capture the *requirements* are by definition those who decide how the *product* will look and function. Thus, specification of *require-ments* is too important to be outsourced to subcontractors or third-party develop-ers. However, specialists in such fields as man-machine *interfaces* and specification methodology are often included in specification teams.

7.2 About explore restrictions

When developing a new *product*, many circumstances always exist in the *environ-ment* that influence the *product* to be developed, and which might restrict it. These restrictions must be explored and understood, to ensure that the *product* will work as planned in its *environment*. Restrictions might be existing technical *interfaces*, or laws and regulations from authorities, or other fixed conditions set by the *environment*.

In *Cpdm* such *requirements* are seen as *environment restriction requirements*, rather than *requirements*, because they are not freely set by the developing com-pany and thus they might be hard to change. For this reason it is not worthwhile to prioritize *environment restriction requirements*.

Restrictions must always be carefully refined, since they might affect details deep inwards in an architecture hierarchy.

Environment restriction requirements might not only touch upon the outermost *black-/white-box nesting level*, but they might stipulate design rules or material restrictions located deeply *inward* in the *product structure* to be devel-oped. It is best to capture all *requirements* as early as possi-ble, and keep track of them, and later *refine requirements* to the appropriate *nesting level* where they are applicable.

An *environment* is what currently exists, and what will become the surroundings of the *product* to be developed. All imaginable restrictions that could influence the *product* to be developed or influence the coming *interfaces* between the *product* and the *environment* are captured and specified in *environment restriction require-*

ments. For the moment, do not analyze the *environment restriction requirement* too much, because that will soon be done when trying to find solutions for the *interfaces* between the *product* and its *environment,* and later when refining the *environment restriction requirements* to become *requirements* on the *product outermost black-box.*

7.2.1 Cpdm definitions

In *Cpdm, environment restriction requirement* is defined according to Table 7-1 below.

TABLE 7-1 Cpdm definition

Aspect	Environment restriction requirement
Definition	Limitations in the existing environment that will impact the product to be developed
Synonyms	Existing environment considerations, laws, regulations, patents
Symbol	**E. Environment** **TABLE 7-2** Environment restriction requirement E.R.n. Restriction requirement

For a more thorough explanation of *black-boxes,* see chapter 9.1, "About architecture," page 334.

7.2.2 Textual requirements

Chapter 7.4, page 180 to chapter 7.10, page 198 below will elaborate only on basic textual *requirements,* since these are the simplest and easiest to begin with.

Some guidelines for basic textual *requirements* are:

- *Requirements* shall not be arranged in large, continuous text blocks like a novel. If so, they will be hopeless to find, reference, and rearrange.
- Each *requirement* shall be short, precise, and self-contained, and each separate *requirement* shall be assigned a unique identifier to be easily traceable.
- It is preferable to include verifying people as *stakeholders,* to ensure that all *requirements* will be easily verifiable.

Capturing *requirements* and allocating development support are among the few *activities* impossible (or stupid) to leave to external *providers* or anyone else externally. *Requirements* control what makes the "soul" of a *product.*

7.3 About environment staffing allocation

Development companies seldom have all competencies in place to develop everything themselves. Consequently, to decide on and allocate development support is yet another *activity* that must precede all *activities* on the *environment nesting level*. The reason for doing this urgently is that both in-house and external development support may be needed very soon to prepare *white-box solution alternatives* in the next *phase*, **Pt.0.Pa. Try to satisfy restrictions.**

7.3.1 Cpdm definition

In *Cpdm, development staffing* is defined according to Table 7-3 below.

TABLE 7-3 Cpdm definition

Aspect	Development staffing
Definition	From whom to get support for surveying the existing environment, developing the product interfaces to the environment, and developing each inward embedded white-box
Synonyms	Staffing plan, and contractors and in-house workforce
Table	**TABLE 7-4** Development allocation staffing | Role item	Provider	Terms for role item[a]	Cpdm schedule or task | |---|---|---|---| |			| a. Terms for roles are typically development overhead not illustrated in this book

7.3.2 Environment staffing allocation process schedule

The *result development staffing* is *demanded* by all subsequent *phases* and *activities* on the *environment nesting level* (see Fig. 7-2 below). Capturing staffing can also be done prior to capture *requirements* (see **Pt:2. Cpdm generic advanced technical schedule,** p. 104) or immediately in the beginning of capturing *requirements* (see **Pt.R. Capture staffing & requirements,** p. 101). Allocators are supposed to be an in-house competency, and these developers should always be available in-house at the developing company.

The total *environment* development *activities* are not particularly abundant or time consuming, but might be tricky to perform, such as demarcating the *environment* from the *product* to be developed, understanding and describing the *environment* by *architectures*, constructing and realizing the *interfaces* to it, and finally verifying the prototype in its *target environment.*

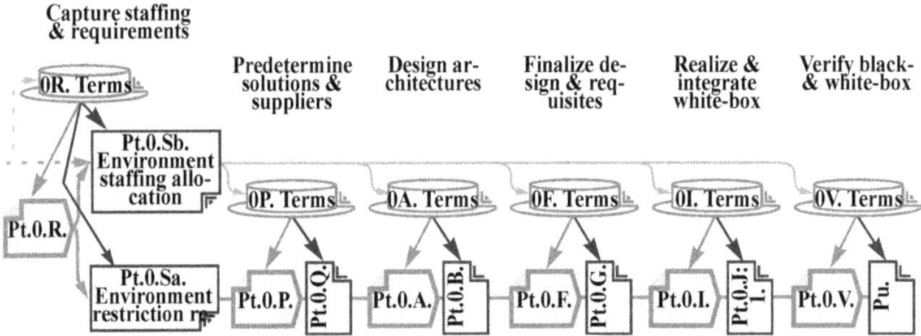

FIGURE 7-2 Allocating environment developers in capture staffing & requirements phase

The *result* **Pt.0.Sb. Environment staffing allocation** is a list of allocated developers. Developers are allocated for their particular competencies, which are described in *terms* documents, marked with a hat symbol in *schedules*. Each term describes a particular kind of competency together with appropriate responsibilities and authorities needed to perform a certain development *activity* or to prepare a certain developed *result*. It might be competency of in-house developers of any kind, of hired expert consultants, or of outsourced development companies. For in-house competency, some parts of the *terms* might be found in the employee contract. If external developers are used, the *terms* are often referred in a formal contract.

Since *terms* are part of *line management*, they belong to an administrative super-structure above the technical *value chain* not elaborated on in detail in this book. The next book *"Cpdm* technical overhead" will take care of this.

7.4 EXAMPLE House environment: Explore environment restrictions and ensure environment staffing

7.4.1 EXAMPLE House: About this example

To many people it is rather well known how to build a house, and everyone can relate to how to use a house. That makes it a good example of capturing house *requirements* and designing house *architectures*. The house example is also chosen to clearly show how *requirements* and *architectures* are interconnected, and how to organize a large number of details into a *black-/white-box hierarchy* (also see ch. 3.2.4, p. 34).

In this example it is imagined that a family, here called the house proprietors, has saved some money in order to build a house in a beautiful place. They have found a building plot for the house and started to develop it.

7.4.2 EXAMPLE House environment: Process schedule to use

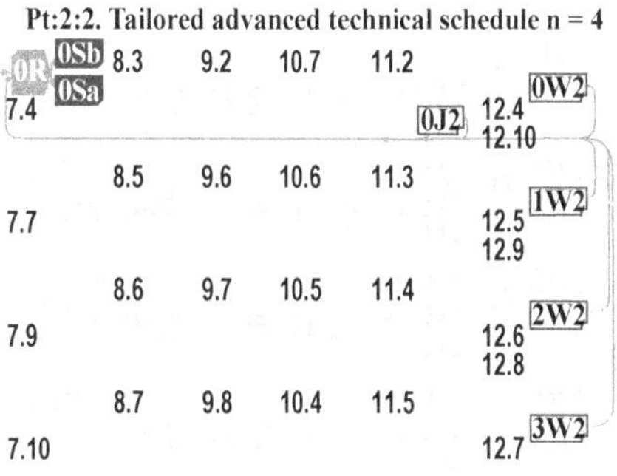

Pt:2:2. Tailored advanced technical schedule n = 4

OR OSb / 7.4 OSa	8.3	9.2	10.7	11.2
				0J2 12.4 0W2 / 12.10
7.7	8.5	9.6	10.6	11.3
				12.5 1W2 / 12.9
7.9	8.6	9.7	10.5	11.4
				12.6 2W2 / 12.8
7.10	8.7	9.8	10.4	11.5
				12.7 3W2

For the house example, a special technical overall *schedule* is *tailored* (illustrated in Fig. 7-3, left):

- **Pt:2:2. Tailored advanced technical schedule n = 4,** page 164

In this *tailored schedule* the generic *schedule* to use is (filled-in symbols):

- **Pt.0.R. Explore environment restrictions and ensure environment staffing,** page 110.

FIGURE 7-3 Position of schedule and master to use (filled in) in overall schedule. Masters to demanded results are highlighted. Chapter numbers of the house example are also indicated.

Below *results* may be demanded (their *masters* are not indicated in Fig. 7-3 above, but in the *Cpdm* overall *schedule* **P. Cpdm generic development schedule,** p. 100):

- Pk Pk. Concept brief
- Pe Pe. Reuse plan
- Pm Pm. Partial prototype
- Pg Pg. Business case
- Po Po. User profile
- Pi Pi. Product proposition

For possible *failure localization* and *failure elimination,* below *results* may be demanded (their *masters* are indicated *upstream* from the filled-in *schedule* in the figure above or in greater detail in Fig. 11-5, p. 602 and Fig. 11-6, p. 603):

- 0J2 Table 11-9. House environment start-up report 1, page 606
- 0W2 List 12-1. House environment interfaces with fake house failure elimination report 1, page 674
- 0W2 Table 12-27. House environment verification report 1, page 701
  Table 12-29. House environment reverification report 2, page 703
  Table 12-28. House environment failure elimination report 1, page 702
- 1W2 List 12-2. House with fake rooms failure elimination report 1, page 676
- 1W2 Table 12-22. House verification report 1, page 697
  Table 12-24. House verification report 2, page 698
  Table 12-23. House failure elimination report 1, page 698
- 2W2 List 12-3. House rooms with fake machinery failure elimination report 1, page 678
- 2W2 Table 12-14. House kitchen verification report 1, page 688

Table 12-17. House kitchen verification report 2, page 691

Table 12-15. House kitchen failure elimination report 1, page 689

3W2 Table 12-9. House kitchen machinery verification report 1, page 682

Table 12-11. House kitchen machinery verification report 2, page 683

Table 12-10. House kitchen machinery failure elimination report 1, page 683

7.4.3 EXAMPLE House environment: Explore restrictions

Table 7-5 below shows some *environment restriction requirements* encountered, which need to be fulfilled, in particular when a building permit is required before starting to build the house.

TABLE 7-5 House environment restriction requirement	
House restriction requirement	
E. R.	Environment restrictions
E. R.1.	Swedish environmental code, 7th chapter
E. R.2.	Malmö urban planning and local construction ordinance
E. R.3.	Swedish national building regulations
E. R.4.	Swedish Financial Supervisory Authority
Pf.1.	If public utilities are available, business cases show this to be cheaper than having the proprietor furnish them.
E. R.5.	All public utilities available shall be used and comply with building standard 2011:45

Note that one restriction originates from the **Pi. Product proposition** documentation prepared earlier.

Also note that these restrictions affect not only the *environment*, but also the house and even small details of the house, for example, fireplaces, electricity installation, gas equipment, and so forth.

Since laws and regulations are stipulated, and are not very likely to change, there is no point in copying all restriction details into one's own *requirement* database. It is better to refer to the originals in some way, by hyperlinks or other referencing facilities.

7.4.4 EXAMPLE House environment: Plan development roles and providers

The *task* is now to plan for development support needed for the *environment nesting level*. Building a house may involve authorities and regulations of many kinds, as well as artistic *architects* and *designers*, which the house proprietor most often needs help for and which must be allocated to the team (see Table 7-6 below.)

TABLE 7-6 House environment development allocation staffing

House environment role item	Provider	Cpdm schedule or task
i10. **House user**	**PR.10. Proprietor**	Pt.0.Rb. Plan development roles and providers.
i11. **City architect**	**PR.11. Robber HB**	Pt.0.Ra. Explore restrictions.
i12. **City architect** • For public utilities • Reviewed by proprietor	**PR.11. Robber HB**	Pt.0.Pa. Try to satisfy restrictions.
i13. **House user** • For driveway • Reviewed by Robber HB	**PR.10. Proprietor**	
i14. **House user** • Assistance from Robber HB	**PR.10. Proprietor**	Pt.0.Pc. Analyze in/out and make/buy options.
i15. **Chief architect** • Robber HB to comply with local regulations and requirements • Reviewed by proprietor	**PR.11. Robber HB**	Pt.0.A. Constitute environment architecture. Pt.0.F. Finalize design of environment interfaces with product requisites.
i16. **Electrician**	**PR.12. Elert AB**	
i17. **Excavator with driver and plumber** i18. **Plumber** i19. **Excavator with driver and electrician** i20. **Excavator with driver and authorized gas plumber** i21. **Dozer with 6 landscapers (including paving)** • Not yet selected • The house builder may use preferred subcontractors. House builder and preferred subcontractors are responsible to comply with official building regulations. • Elimination of detected failures	**PR.13. House builder**	Pt.0.I. Realize interfaces to environment and insert/await outermost white-box (build the house connections to the environment).
i22. **Electrician** • For electricity • Elimination of detected failures	**PR.12. Elert AB**	
i23. **Quality inspector** • Each supplier and subcontractor approves independent quality inspector. • Detects and documents deviations from requirements and design. • Attests invoices for payment. when detected failures are eliminated.	**PR.14. Q-lnc**	Pt.0.V. Verify prototype in environment.
i24. **Inspector** • Approves building permit and approves the house for use.	**PR.15. Commune**	Pt.0.A. Constitute environment architecture. Pr. Validate prototype.

7.4.5 EXAMPLE House: Proceed reading

When restrictions are captured and staffing is specified, the overall *schedule* **Pt:2:2.** continues with subsequent *schedules,* demanding these specified restrictions (again see above Fig. 7-3 or the more detailed Fig. 6-36, p. 164 and Fig. 6-37, p. 165). Since there are several *schedules* demanding these restrictions, the house example will now fork into many different chapters of this book.

To follow this example, these are the alternatives:

- **0P** Proceed to chapter 8.3, "EXAMPLE House environment: Demarcate product by environment solutions with possible supplier opportunities," page 290.
- **0A** Proceed to chapter 9.2, "EXAMPLE House environment: Constitute environment architecture," page 339.
- **0F** Proceed to chapter 10.7, "EXAMPLE House environment: Finalize design of environment interfaces with product requisites," page 477.
- **0V** Proceed to chapter 12.10, "EXAMPLE House environment: Fully verify prototype in environment," page 699.
- **1R** Proceed to chapter 7.7, "EXAMPLE House: Refine house requirements from environment restriction requirements, and ensure house development staffing," page 187.
- To leave the house example and to dig deeper into the chapter 7, "Capture staffing & requirements" topic, proceed to next chapter 7.5 below.

7.5 About black-box requirements

7.5.1 Cpdm definition

In *Cpdm, black-box requirement* is defined according to Table 7-7 below.

TABLE 7-7 Cpdm definition

Aspect	Black-box requirement
Definition	Stakeholder-documented and prioritized expectations from a black-box
Synonyms	Requirement, demand, claim, wish, desire, what-question
Symbol	

As can be seen from the definition above, general *requirements* may be a mix of *behavior requirements* on the *black-box interfaces*, and *restriction requirements* that will later control design of the inside of the *black-box*.

The *boundary interface* around the *black-box* can be treated as belonging to the *black-box*, or to the *black-box environment*, or the *boundary interface* can even be split to an inside belonging to the *black-box* and an outside belonging to the *environment*. In any case, the *boundary interface* can be specified with *restriction requirements*, either belonging to the *black-box*, the *black-box environment*, or both. For more information about *boundary interfaces*, see chapter 9.4.2, "Architecture preparation procedure," page 347. For a more thorough explanation of *black-boxes*, see chapter 9.1, "About architecture."

7.5.2 Refining traceable requirements from restrictions

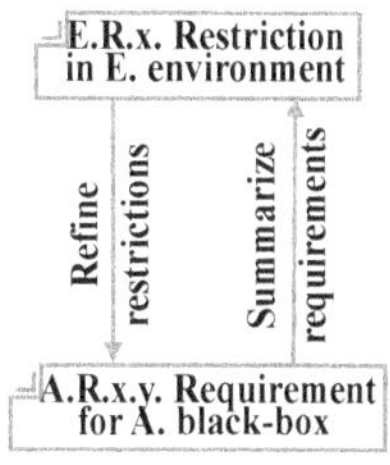

When moving *inwards* in a big, *complex product*, more and more *requirements* will gradually appear. It is important that *requirements* concerning the *environment* are clearly linked to the original *environment restriction requirements*. This *process* of detailing restrictions is called *requirement refinement*, the opposite is called *requirement summarization*, and both must be *managed information* in order to be controlled (see Fig. 7-4, left). For a refine and summarize scale-up see Figure 7-7, page 192.

FIGURE 7-4 Refining and summarizing

A computer database is needed sooner or later for *traceability* between all *requirements*. This also gives a good basic order for grouping, sorting, and keeping track of them.

7.6 About development staffing

It is often very important to get support for *environment* development, and that is even more important for any *white-box* to be developed. These can contain different technologies, some of them pertaining to mechanics, others to electronics, some to *programs*, and some *white-boxes* might be a mix of the three. To concentrate on developing *white-boxes* where in-house capability is available, to hire support, or to outsource other *black-boxes*, or to let in consultant specialists to strengthen the in-house teams may be strategically correct.

7.6.1 Staffing in the process schedule

As for the *development staffing* (see ch. 7.3, p. 179 above), all kinds of *roles* for development of the particular *white-box* are now staffed (see Fig. 7-2 below). Staffing of *specifiers* for this *phase* may have been done prior to start-up of this *phase* or just in the beginning of this *phase*. Since it is a bit difficult to get everything perfectly right from the beginning, staffing may need to be further updated, for example, while *solution alternatives* get settled, requiring staffing with extended technology competency.

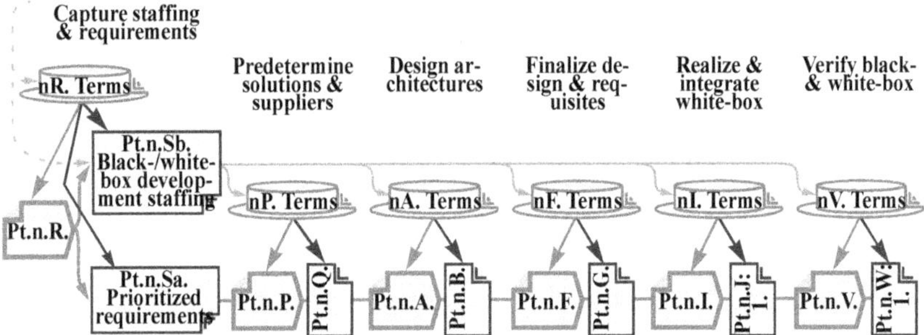

FIGURE 7-5 Allocating white-box developers in capture staffing & requirements phase

The *result* **Pt.n.Sb. Black-/white-box development staffing** is a list of *role* descriptions, which in *Cpdm* are called *terms*, represented by hat symbols. Each term describes a particular kind of competency needed to perform a certain development work. It might be competency of in-house developers of any kind, of hired expert consultants, or of outsourced development companies. For in-house competency, some parts of the *terms* might be found in the employee contract. If external developers are used, the *terms* are often referenced in a formal contract.

Specifiers and allocators are probably in-house competencies, since *requirements* describe the soul of a *product*, which is too important to leave to anybody external to the company. Allocators are staffing up, among others with third-party developers, and should therefore not in themselves be a third party.

In *complex products*, there might be many *nesting levels*, each of them containing many *white-boxes* to be developed. Each *black-box* probably contains different technology, and needed developer competencies for allocation will be very different accordingly.

Since *terms* are part of *line management*, they belong to a superstructure above the technical foundation, and are not in detail elaborated on in this book. The next book "*Cpdm* technical overhead" will take care of that.

7.7 EXAMPLE House: Refine house requirements from environment restriction requirements, and ensure house development staffing

7.7.1 EXAMPLE House: Process schedule to use

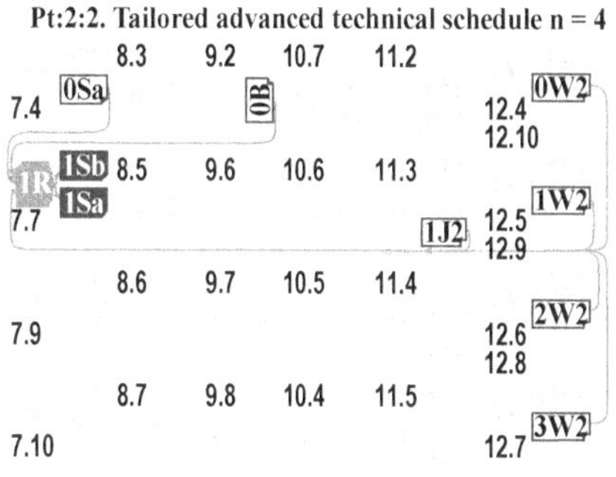

For the house example, a special technical overall *schedule* is *tailored* (illustrated in Fig. 7-6, left):

- **Pt:2:2. Tailored advanced technical schedule n = 4,** page 164

In this *tailored schedule* the generic *schedule* to use is (filled-in symbols):

- **Pt.n.R. Refine requirements from outward nesting level requirements and ensure black-/white-box development staffing,** page 130.

FIGURE 7-6 Position of schedule and master to use (filled in) in overall schedule

Below *results* may be demanded (their *masters* are indicated *upstream* from the filled-in *schedule* in the figure above):

0Sa Environment restriction requirements
Table 7-5. House environment restriction requirement, page 182

0B Environment existing ingredients and design of interfaces to the environment
Figure 9-5, House environment logical architecture, page 341

For possible *failure localization* and *failure elimination*, below *results* may be demanded (their *masters* are indicated *upstream* from the filled-in *schedule* in the figure above):

1J2 Table 11-14. House start-up report 1, page 612

0W2 List 12-1. House environment interfaces with fake house failure elimination report 1, page 674

0W2 Table 12-27. House environment verification report 1, page 701
Table 12-29. House environment reverification report 2, page 703
Table 12-28. House environment failure elimination report 1, page 702

1W2 List 12-2. House with fake rooms failure elimination report 1, page 676

1W2 Table 12-22. House verification report 1, page 697
Table 12-24. House verification report 2, page 698
Table 12-23. House failure elimination report 1, page 698

2W2 List 12-3. House rooms with fake machinery failure elimination report 1, page 678

2W2 Table 12-14. House kitchen verification report 1, page 688

Table 12-17. House kitchen verification report 2, page 691

Table 12-15. House kitchen failure elimination report 1, page 689

3W2 Table 12-9. House kitchen machinery verification report 1, page 682

Table 12-11. House kitchen machinery verification report 2, page 683

Table 12-10. House kitchen machinery failure elimination report 1, page 683

7.7.2 EXAMPLE House: Refine and capture requirements

It is a rewarding experience for one's own understanding to make a formal *requirement* specification when planning to build a house. Too often people only think about how houses are designed (their *architecture* and layout) instead of how they might be used (their behavior and functionality). For an ego showpiece, the *architecture* certainly is ultimate, but for convenient living, the house behavior is, in fact, more important.

The house *environment restriction requirements* in Table 7-5. House environment restriction requirement, page 182, will now be refined to become house *black-box requirements* (see Table 7-9 below). The *outward environment restriction requirements* R.1. to R.5. remain in the new table, and under each of them new refined *requirements* may be specified. Also, *requirements* emerging from earlier *phases* can be the basis for *requirement refinement*, such as Pn.1. and Pa.1. At the end, new desired *requirement* group not now emerging from anywhere, may be captured, such as R.6. to R10. The refined *requirements* now get two numbers in their identity, like R.1.1. and R.1.2. and so forth. The first number refers to the *environment restriction requirement* or new *requirement* group, and the second number is a running number starting at 1.

Such a house example is, of course, too large for this book, so only a limited selection is squeezed into the tables. The phrasing is also kept short to fit into a line, and should in reality be better explained.

Some of the *requirements* seem very imprecise, but at this *nesting level* this is not too problematic. Even if the *stakeholders* are vague, nobody knows better than they what they wish. These *requirements* will be further refined and detailed later on, to more *inward* concrete *requirements*. It is also possible at any time to go back to change, extend, or remove already specified *requirements* as understanding grows. Just remember that it costs significantly more to make changes to design later than to go back and improve the *requirements* immediately.

In this house example, *requirements* are not explicitly separated into restrictions and behavior, which will be further explained later on.

Note that for the moment nobody knows about rooms or any other *partition* of the house *black-box*. How to use and restrict the house must first be specified, and later on some room *partitions* may be proposed as *white-box solution alternatives*. If there are heavy requests on room *partitioning*, or any room is mandatory for the house, of course this may be put in as a restriction on the house *black-box*, for example, the need for a separate kitchen. But try to avoid this, because other attractive *white-box solution alternatives* for *partitioning* the house may emerge when opening the house *black-box* and trying to satisfy the house *requirements* in the next *process phase*, for example, to merge the kitchen with the dining room.

TABLE 7-9 House black-box compound requirements (small selection)

House compound requirement		Priority
E. R.1.	Swedish environmental code, 7th chapter	
A. R.1.1.	During cooking and dining, the lake shall be as visible as possible.	High
A. R.1.2.	Façade and roofing material appearance shall be chosen by proprietor.	High
E. R.2.	Malmö urban planning and local construction ordinance	
A. R.2.1.	The house style shall fit in with neighbors' houses.	Low
A. R.2.2.	The house shall be accessible for disabled persons.	Mid
A. R.2.3.	A building permit is mandatory before building starts.	Crucial
E. R.3.	Swedish national building regulations	
A. R.3.1.	The house shall follow proven methods, regulations, and standards.	High
A. R.3.2.	The house shall have an economical energy consumption.	High
E. R.4.	Swedish Financial Supervisory Authority	
A. R.4.1.	The house shall be attractive to allow maximum bank mortgage.	Crucial
E. R.5.	All public utilities available shall be used and comply with building standard 2011:45	
A. R.5.1.	Provide the house with a utility room for all public utilities.	Crucial
A. Pn.1.	Modern people in wealthy city districts cook and party.	
A. R.6.	Gourmet cooking and dining requirements	
A. R.6.1.	The atmosphere during dining shall be relaxed.	Mid
A. R.6.2.	The dining space shall support up to 8 guests.	High
A. R.6.3.	Serving support shall exist for one course at a time.	Low
A. R.6.4.	Storing, cooking, and dining support shall allow a 5-course French menu for 8 persons.	Mid
A. R.6.5.	Dishwashing support needed, shall hold complete dinner.	Crucial
A. R.6.6.	Cooking shall support warming, boiling, frying, wokking, and baking.	Crucial
A. R.6.7.	Smell from cooking shall be eliminated.	High
A. R.6.8.	Waste and garbage from cooking shall be odorlessly disposed of.	Mid
A. R.7.	Reunion and party requirements	
A. R.7.1.	Good space shall exist for 24 party guests, half standing, half sitting.	Mid
A. R.7.2.	Serving snack meals shall support 12 party guests.	Low
A. R.7.3.	Storing capacity shall exist for 24 party guests' snack meals.	Low
A. R.7.4.	Toileting capacity shall exist for 2 guests simultaneously.	High
A. R.7.5.	Hanger capacity shall exist for 24 coats and such.	Mid
A. R.8.	Sleeping accommodation requirements	
A. R.8.1.	(No requirements included here due to book space limitation)	
A. R.9.	Health and body caring requirements	
A. R.9.1.	The house shall be safe for children.	High
A. Pa.1.	Every house shall be prepared for optic fiber communication	
A. R.10.	Hobbies and electronic communication requirements	
A. R.10.1.	(No requirements included here due to book space limitation)	

7.7.3 EXAMPLE House: Prioritize requirements

The *stakeholders* prioritize their *requirements* in Table 7-9 above, by marking in the column called priority.

7.7.4 EXAMPLE House: Examine authenticity

A **Pt.n.Rd. Examine authenticity** gate meeting is held for the *stakeholders* to improve *requirement* authenticity and resolve possible prioritization conflicts. If there are many *stakeholders* with different opinions about the house, this can be a very time-consuming and exhausting battle. In this case, the house proprietor is the chairman and can easily resolve all conflicts.

7.7.5 EXAMPLE House: Plan development roles and providers

On the house *nesting levels* a great deal of development support is needed, and the allocation list is rather extensive (see Table 7-10 below).

TABLE 7-10 House development allocation staffing

House role item	Provider	Cpdm schedule or task
i25. **House user**	**PR.10. Proprietor**	Pt.1.R. Refine house requirements from environment restriction requirements, and ensure house development staffing.
i26. **Architect** • Reviewed by the house proprietor	**PR.11. Robber HB**	Pt.1.P. Predetermine house solutions with supplier opportunities.
i27. **Architect** • Reviewed by proprietor • Responsibility for official regulation and requirement compliance • Submitted for authority permit • All documentation to be paid and owned by the proprietor • When building permit is approved, ask for offers from 5 candidate house builders. • proprietor selects one of the candidates to be the house builder.	**PR.11. Robber HB**	Pt.1.A. Satisfy house requirements by decomposing house black-box into white-box design containing room black-boxes.
i28. **Artistic architect** • To be included in the Robber HB team	**PR.16. Homey Inc**	
i29. **House layout specialist** • To be included in the Robber HB team	**PR.16. Homey Inc**	
i30. **Electrician**	**PR.12. Elert AB**	

TABLE 7-10 House development allocation staffing

House role item	Provider	Cpdm schedule or task
i31. Calculation engineer • If nothing else stated below	PR.11. Robber HB	Pt.1.F. Finalize design of house with product requisites.
i32. Electrician • For electricity	PR.12. Elert AB	
i33. Construction laborer i34. Excavator with driver and 1 Laborer i35. Pipe-fitter i36. Authorized gas plumber i37. Founder i38. Tinsmith i39. Plumber i40. Bricklayer i41. Façade bricklayer i42. Chimney mason i43. Carpenter i44. Cabinet maker i45. Roofer i46. Woodworker • If nothing else stated below • The house builder may use own preferred subcontractors. • The building company and sub-contractors are responsible to comply with official regulations of procured building material. • Authorized personnel for gas installation • Eliminate detected failures.	PR.13. House builder	Pt.1.I. Procure house items (and await invading room black-boxes) in order to build house (build the house roof, walls, and floor).
i47. Electrician • For electricity • Eliminate detected failures.	PR.12. Elert AB	
i48. Quality inspector • Each supplier and subcontractor approves independent quality inspector. • Detects and documents deviations from requirements and design. • Attests invoices for payment when detected failures are eliminated.	PR.10. Proprietor	Pt.1.V. Verify house black-/white-box.

For large *products*, many technologies used to be gathered. *Designers* of the outermost house *black-box* are not necessarily best suited to design all deeper *inward nesting levels*. In this house example, the allocation of people typically changes a lot on each *nesting level.*

7.7.6 EXAMPLE House: Proceed reading

When the outermost house *black-box requirements* are completed the overall *schedule* forks as usual (see again Fig. 7-6).

To follow this example, these are the alternatives:

[1P] Proceed to chapter 8.5, "EXAMPLE House: Predetermine solutions with sourcing options," page 300.

[1A] Proceed to chapter 9.6, "EXAMPLE House: Satisfy house requirements by decomposing house black-box into white-box design containing room black-boxes," page 357.

[1F] Proceed to chapter 10.6, "EXAMPLE House: Finalize design of house with product requisites," page 465.

[1V] Proceed to chapter 12.9, "EXAMPLE House: Fully verify house black-/white-box," page 693.

[2R] Proceed to chapter 7.9, "EXAMPLE House rooms: Refine room requirements from house requirements, and ensure room development staffing," page 193.

- To leave the house example and to dig deeper into the chapter 7, "Capture staffing & requirements" topic, proceed to next chapter 7.8 below.

7.8 Requirements refinement hierarchy

7.8.1 Traceability to architecture hierarchy

To keep track of *requirements* and *architectures* of *complex products*, it is necessary to *partition* them in a clever way. *Cpdm* proposes to arrange a hierarchical structure of all *requirements and architectures*, beginning with the *environment restriction requirements* (see Fig. 7-7 below).

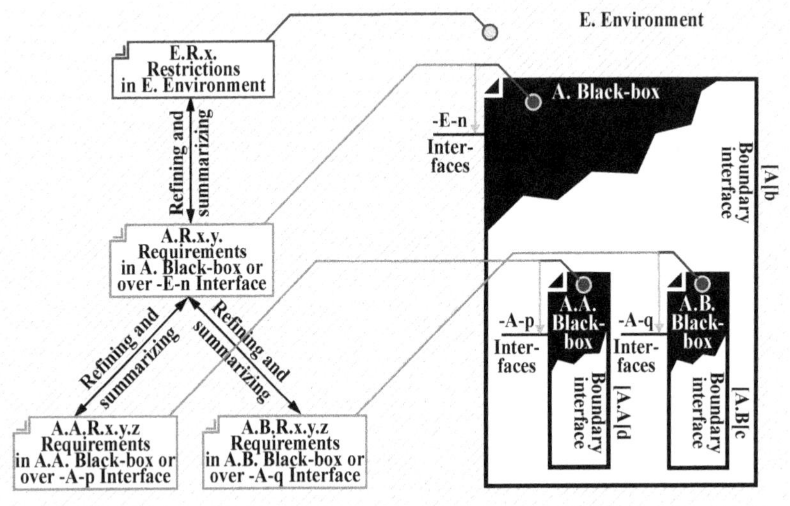

FIGURE 7-7 Requirement hierarchy traceability to a black-box hierarchy

For an example of *traceability* between *requirements* within their hierarchy, also see Figure 7-10, page 208 below.

Each *requirement* must be connected to the *black-box* it specifies. Each *nesting level* of *requirement* thus specifies each *nesting level* of *black-boxes*.

The *boundary interface* can be specified with the *restriction requirements*, either belonging to the *black-box*, or the *black-box* surroundings or both. For more information about *boundary interfaces*, see chapter 9.4.2, "Architecture preparation procedure," page 347. For a more thorough explanation of *black-boxes*, see chapter 9.1, "About architecture," page 334.

Requirement and *architecture* hierarchies with arbitrary connections between them can be really *complex*, but generally necessary. It is highly recommended to simplify this *formalism* to a manageable number of *nesting levels*, and to use databases to keep track of *traceability*. It is better to have some incomplete *formalism* than having no *formalism* at all.

7.9 EXAMPLE House rooms: Refine room requirements from house requirements, and ensure room development staffing

7.9.1 EXAMPLE House kitchen: Process schedule to use

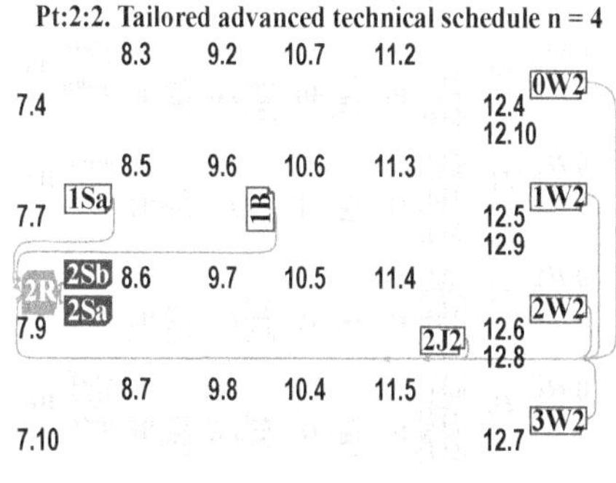

For the house example, a special technical overall *schedule* is *tailored* (illustrated in Fig. 7-8, left):

• **Pt:2:2. Tailored advanced technical schedule n = 4,** page 164

In this *tailored schedule* the generic *schedule* to use is (filled-in symbols):

• **Pt.n.R. Refine requirements from outward nesting level requirements and ensure black-/white-box development staffing,** page 130.

FIGURE 7-8 Position of schedule and master to use (filled in) in overall schedule

Below *results* may be demanded (their *masters* are indicated *upstream* from the filled-in *schedule* in the figure above):

1Sa House prioritized requirements

Table 7-9. House black-box compound requirements (small selection), page 189

1B House architectures with major ingredients and room black-boxes

Figure 9-22, House logical architecture, page 363

For possible *failure localization* and *failure elimination*, below *results* may be demanded (their *masters* are indicated *upstream* from the filled-in *schedule* in the figure above):

`2J2` Table 11-18. House kitchen start-up report 1, page 617

`0W2` List 12-1. House environment interfaces with fake house failure elimination report 1, page 674

`0W2` Table 12-27. House environment verification report 1, page 701
Table 12-29. House environment reverification report 2, page 703
Table 12-28. House environment failure elimination report 1, page 702

`1W2` List 12-2. House with fake rooms failure elimination report 1, page 676

`1W2` Table 12-22. House verification report 1, page 697
Table 12-24. House verification report 2, page 698
Table 12-23. House failure elimination report 1, page 698

`2W2` List 12-3. House rooms with fake machinery failure elimination report 1, page 678

`2W2` Table 12-14. House kitchen verification report 1, page 688
Table 12-17. House kitchen verification report 2, page 691
Table 12-15. House kitchen failure elimination report 1, page 689

`3W2` Table 12-9. House kitchen machinery verification report 1, page 682
Table 12-11. House kitchen machinery verification report 2, page 683
Table 12-10. House kitchen machinery failure elimination report 1, page 683

7.9.2 EXAMPLE House kitchen: Refine and capture requirements

Now refine the *requirements* for Table 7-9. House black-box compound requirements (small selection), page 189, to become room *requirements* for each of the rooms identified in the house *architecture* (see Fig. 9-22 House logical architecture, p. 363). Since space in this book is limited, only the house *requirements* that might be refined for the A.aC. Kitchen *black-box* are specified (see Table 7-11 below). There should, of course, be similar tables for all room *black-boxes* of the house *architecture*.

Note that not all house *requirements* are refined, since these may have impact only on the house, and may not have impact on the kitchen, or even may not have impact on any room at all.

TABLE 7-11 House kitchen black-box compound requirements (small selection)

House kitchen compound requirement		Priority
A. R.1.1.	During cooking and dining, the lake shall be as visible as possible.	High
A.aC. R.1.1.1.	The lake shall be visible from kitchen food preparation area.	High
A. R.2.2.	The house shall be accessible for disabled persons..	Mid
A.aC. R.2.2.1.	Kitchen may be inaccessible by disabled guests.	Mid
A. R.3.1.	The house shall follow proven methods, regulations, and standards.	High
A.aC. R.3.1.1.	Kitchen shall follow established norms, regulations, and standards.	Crucial
A.aC. R.3.1.2.	Kitchen interiors may not have any damage, scratches, or stains.	High
A.aC. R.3.1.3.	Kitchen interior appearance must be chosen by proprietor.	Crucial
A. R.4.1.	The house shall be attractive to allow maximum bank mortgage.	Crucial
A.aC. R.4.1.1.	Kitchen including interiors shall cost maximally 20 000 € .	Crucial
A.aC. R.4.1.2.	Purchase only ordinary interiors from few suppliers to keep prices down.	High
A. R.6.3.	Serving support shall exist for one course at a time.	Low
A.aC. R.6.3.1.	Cheap equipment keeps one course warm during eating.	Mid
A. R.6.4.	Storing, cooking, and dining support shall allow a 5-course French menu for 8 persons.	Mid
A.aC. R.6.4.1.	Include large food preparation area with hot and cold water.	High
A.aC. R.6.4.2.	Storage shall hold 12 pans of 3 liters near cooking area.	High
A.aC. R.6.4.3.	Storage shall hold 6 plates and 6 bowls near baking area.	High
A.aC. R.6.4.4.	Knife storage shall be child-safe and near preparation area.	Mid
A.aC. R.6.4.5.	Storage shall hold all kinds of cutlery and kitchen utensils which shall be easily accessible from preparation area.	High
A.aC. R.6.4.6.	Storage shall hold 8 bags (of 30 liters each) of groceries, 4 of which are chilled, and 1 frozen.	Crucial
A. R.6.5.	Dishwashing support needed, shall hold complete dinner.	Crucial
A.aC. R.6.5.1.	Dishwashing shall clean 5-course tableware (5 x 8 plates, cutlery, glasses, and cooking utensils) in same run.	Mid
A.aC. R.6.5.2.	Gentle dishwashing shall support fragile tableware.	Crucial
A.aC. R.6.5.3.	Child-safe storage shall hold chemicals for dishwashing etc.	Low
A. R.6.6.	Cooking shall support warming, boiling, frying, wokking, and baking.	Crucial
A.aC. R.6.6.1.	Food thawing shall be near preparation area.	High
A.aC. R.6.6.2.	Warming of all kinds of food shall be near preparation area.	High
A.aC. R.6.6.3.	Food boiling shall be near preparation area.	Crucial
A.aC. R.6.6.4.	Food frying shall be near preparation area.	Crucial
A.aC. R.6.6.5.	Food wokking shall be near preparation area.	Mid
A.aC. R.6.6.6.	Food baking shall be near preparation area.	High
A. R.6.7.	Smell from cooking shall be eliminated.	High
A.aC. R.6.7.1.	Odor from cooking may not spread to other rooms.	High
A. R.6.8.	Waste and garbage from cooking shall be odorlessly disposed of.	Mid
A.aC. R.6.8.1.	Garbage odor may not be recognizable in kitchen.	Mid
A. R.7.3.	Storing capacity shall exist for 24 party guests' snack meals.	Low
A.aC. R.7.3.1.	Storage shall hold 24 snack meals (2 grocery bags).	Low
A. R.9.1.	The house shall be safe for children.	High
A.aC. R.9.1.1.	Kitchen shall hold family medicines in child-safe way.	Low

Some house *requirements* (like A.R.2.2. The house shall be accessible for disabled persons.) in the above table get refined also to other rooms than the kitchen. When *requirements* are being refined they are often split to many *black-boxes*. This is no problem, just sort them by room and use them separately when designing and verifying each room. It is possible to further use these refined room *requirements* when in turn refining them *inwards* to *embedded black-boxes* in the room *black-box*.

New kitchen *requirements* may now be desired, but it is not possible to find any house *black-box requirement* to refine them from. Such problems may be caused by incomplete house *requirements*, but whenever this happens, just go back and supplement the house *black-box* with needed *requirements*.

Sometimes *requirements* need to be added and no adequate *outward nesting level requirements* can be found to refine from. In these cases it is better to not capture anything ridiculous to refer to on *outward nesting levels*, just for the sake of *formalism*.

7.9.3 EXAMPLE House kitchen: Prioritize requirements

Prioritization was important on a *product outermost black-box*, and so also on *inward nesting levels*. This *nesting level* is more concrete and it often shows that there are conflicts between different *stakeholders*. The prioritization column in Table 7-11, page 195 was filled in by the *stakeholders*.

7.9.4 EXAMPLE House kitchen: Examine authenticity

A gate meeting was held to balance and settle all opinions from the *stakeholders*.

The wife of the house proprietor family wants the kitchen to be prioritized, whereas the husband prefers a better-equipped garage. Everybody in the proprietor family must like to live in the house, so it is very important to solve such conflicts before they begin to get expensive.

The decision was that the kitchen should not be unnecessarily expensive because of fancy design. Standard interiors were sufficient for cooking even the most delicious dishes. On the other hand, it was also decided that the garage should be equipped after the house was completed, when it was easier to see how much money could be spent on that.

7.9.5 EXAMPLE House rooms: Plan development roles and providers

At room *nesting level* there is also a great need for development support (see allocations in Table 7-12 below). Note that only design of the kitchen is shown in this book, and there may be other rooms with their machinery.

TABLE 7-12 House kitchen development allocation staffing

House rooms role item	Provider	Cpdm schedule or task
i49. **House user**	**PR.10. Proprietor**	Pt.2.R. Refine room requirements from house requirements, and ensure room development staffing.
i50. **Interior designer** • For rooms if not otherwise stated below • Reviewed by the house proprietor	**PR.11. Robber HB**	Pt.2.P. Predetermine room solutions with supplier opportunities.
i51. **Kitchen designer** • For kitchen • Reviewed by the house proprietor	**PR.17. K-Studio**	
i52. **Interior senior designer** **i53.** **Interior junior designer** • General room architecture design • Approved by house proprietor before house builder starts procurement • Responsible for not make any design that violates official regulations	**PR.11. Robber HB**	Pt.2.A. Satisfy room requirements by decomposing room black-boxes into white-box design containing machinery black-boxes. Pt.2.G. Finalized room items design with product requisites.
i54. **Electrician** • For electricity design • Responsible to follow electricity regulations	**PR.12. Elert AB**	
i55. **Kitchen designer** • For kitchen interior design	**PR.17. K-Studio**	
i56. **Interior senior worker** **i57.** **Interior junior worker** • For house builder's subcontractors • For rooms inside boundary finish • Responsibility by the house builder to comply with official regulations of procured building material • House builder shall ask the proprietor before any procurement orders are done, to decide all unspecified visible solutions such as colors, surfaces, fittings, etc. • Eliminate detected failures.	**PR.13. House builder**	Pt.2.I. Procure room items (and await invading machinery black-boxes) in order to build rooms (i.e. coat inner room surfaces and install permanent interiors).
i58. **Electrician** • For electricity • Responsible to follow electricity regulations • Eliminate detected failures	**PR.12. Elert AB**	
i59. **Kitchen installer** • For kitchen interiors • Eliminate detected failures	**PR.17. K-Studio**	
i60. **Quality inspector** **i61.** **Payment endorsement** • Each supplier and subcontractor approves independent quality inspector. • Detects and documents deviations from requirements and design. • Attests invoices for payment when detected failures are eliminated.	**PR.10. Proprietor**	Pt.2.V. Verify room black-/white-box.

7.9.6 EXAMPLE House: Proceed reading

When the room *black-box requirements* are completed, the overall *schedule* forks as usual (see again Fig. 7-8).

To follow this example, these are the alternatives:

3R Proceed to next chapter 7.10, "EXAMPLE House rooms machinery: Refine machinery requirements from rooms requirements and ensure machinery development staffing" below.

2P Proceed to chapter 8.6, "EXAMPLE House rooms: Predetermine solutions with sourcing options," page 305.

2A Proceed to chapter 9.7, "EXAMPLE House rooms: Satisfy room requirements by decomposing room black-boxes into white-box design containing machinery black-boxes," page 365.

2F Proceed to chapter 10.5, "EXAMPLE House rooms: Finalize design of rooms with product requisites," page 452.

2V Proceed to chapter 12.8, "EXAMPLE House kitchen: Fully verify room black-/white-box," page 684.

3R Proceed to next chapter 7.10, "EXAMPLE House rooms machinery: Refine machinery requirements from rooms requirements and ensure machinery development staffing" below.

7.10 EXAMPLE House rooms machinery: Refine machinery requirements from rooms requirements and ensure machinery development staffing

7.10.1 EXAMPLE House kitchen machinery: Process schedule to use

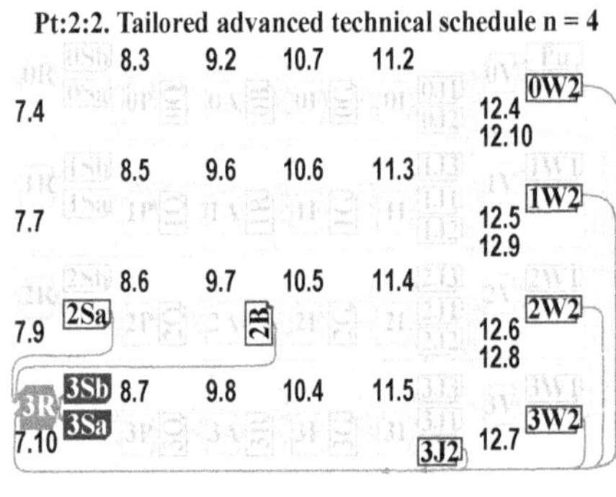

For the house example, a special technical overall *schedule* is *tailored* (illustrated in Fig. 7-9, left):

- **Pt:2:2. Tailored advanced technical schedule n = 4,** page 164

In this *tailored schedule* the generic *schedule* to use is (filled-in symbols):

- **Pt.n.R. Refine requirements from outward nesting level requirements and ensure black-/white-box development staffing,** page 130.

FIGURE 7-9 Position of schedule and master to use (filled in) in overall schedule

Below *results* may be demanded (their *masters* are indicated *upstream* from the filled-in *schedule* in the figure above):

2Sa Rooms prioritized requirements
   Table 7-11. House kitchen black-box compound requirements (small selection), page 195

2B Room architectures with major ingredients and machinery black-boxes
   Figure 9-26, House kitchen logical architecture, page 369

For possible *failure localization* and *failure elimination*, below *results* may be demanded (their *masters* are indicated *upstream* from the filled-in *schedule* in the figure above):

3J2 Table 11-22. House kitchen machinery start-up report 1, page 622

0W2 Table 12-27. House environment verification report 1, page 701
   Table 12-29. House environment reverification report 2, page 703
   Table 12-28. House environment failure elimination report 1, page 702

1W2 List 12-2. House with fake rooms failure elimination report 1, page 676

1W2 Table 12-22. House verification report 1, page 697
   Table 12-24. House verification report 2, page 698
   Table 12-23. House failure elimination report 1, page 698

2W2 List 12-3. House rooms with fake machinery failure elimination report 1, page 678

2W2 Table 12-14. House kitchen verification report 1, page 688
   Table 12-17. House kitchen verification report 2, page 691
   Table 12-15. House kitchen failure elimination report 1, page 689

3W2 Table 12-9. House kitchen machinery verification report 1, page 682
   Table 12-11. House kitchen machinery verification report 2, page 683
   Table 12-10. House kitchen machinery failure elimination report 1, page 683

7.10.2 EXAMPLE House kitchen machinery: Refine and capture requirements

The kitchen *requirements* in Table 7-11, page 195, are refined by distributing them over the machinery *black-boxes* in Figure 9-26, House kitchen logical architecture, page 369. The refined *requirements* are specified in a new table (see Table 7-13 below).

Interiors without machinery have been designed by architects earlier, based on the kitchen *requirements* in Table 7-11, page 195. Now *requirements* to machinery *black-boxes* need to be refined and captured. Also note that to limit this example to fit into this book, only the machinery in the kitchen is specified.

TABLE 7-13 House kitchen machinery black-box compound requirements

House kitchen machinery compound requirement		Priority
	A.aC.aA. Cold machinery	
A.aC. R.6.4.6.	Storage shall hold 8 bags (of 30 liters each) of groceries, 4 of which are chilled, and 1 frozen.	Crucial
A.aC.aA. R.6.4.6.1.	**Fridge shall hold 4 bags (of 30 liters each) of chilled groceries.**	**Crucial**
A.aC.aA. R.6.4.6.2.	**Freezer shall hold 1 bag of 30 liters frozen food.**	**High**
A.aC.aA. R.6.4.6.3.	**Fridge and freezer shall be maintenance-free.**	**Low**
	A.aC.aB. Warm machinery	
A.aC. R.6.3.1.	Cheap equipment keeps one course warm during eating.	Mid
A.aC.aB. R.6.3.1.1.	**Equipment shall keep cooked food warm for 2 hours.**	**Low**
A.aC. R.6.6.1.	Food thawing shall be near preparation area.	High
A.aC.aB. R.6.6.1.1	**Thawing shall work on simple food in plastic containers.**	**Crucial**
A.aC.aB. R.6.6.1.2.	**When thawing, no parts of the food may get cooked.**	**Mid**
A.aC.aB. R.6.6.1.3.	**Thawing speed must be adjustable.**	**High**
A.aC. R.6.6.2.	Warming of all kinds of food shall be near preparation area.	High
A.aC.aB. R.6.6.2.1	**Ready-to-serve food shall be warmed on heatproof plates.**	**Crucial**
	A.aC.aC. Bake machinery	
A.aC. R.6.6.6.	Food baking shall be near preparation area.	High
A.aC.aC. R.6.6.6.1	**Advanced baking shall be possible.**	**High**
A.aC.aC. R.6.6.6.2.	**Warming, grilling, and crusting shall be possible.**	**Mid**
	A.aC.aD. Heat machinery	
A.aC. R.6.6.2.	Warming of all kinds of food shall be near preparation area.	High
A.aC.aD. R.6.6.2.1.	**Fast warming when cooking shall work on food in metal pots.**	**Crucial**
A.aC. R.6.6.3.	Food boiling shall be near preparation area.	Crucial
A.aC.aD. R.6.6.3.1	**Boiling food shall work with all kinds of pans.**	**High**
A.aC.aD. R.6.6.3.2.	**Heat supply shall be adjustable and easy to control.**	**High**
A.aC.aD. R.6.6.3.3.	**Power shall be down-adjustable for small pans to simmer.**	**Mid**
A.aC. R.6.6.4.	Food frying shall be near preparation area.	Crucial
A.aC.aD. R.6.6.4.1	**Frying shall work on wide- and flat-bottom pans**	**High**
A.aC.aD. R.6.6.4.2.	**Power shall be down-adjustable, e.g., for thickening a sauce.**	**Low**
A.aC. R.6.6.5.	Food wokking shall be near preparation area.	Mid
A.aC.aD. R.6.6.5.1	**Wokking shall be possible in cupped-bottom pans.**	**Crucial**
A.aC.aD. R.6.6.5.2.	**Heat shall bring 3 dl oil to almost instant boil.**	**High**
	A.aC.aE. Ventilation machinery	
A.aC. R.6.7.1.	Odor from cooking may not spread to other rooms.	High
A.aC.aE. R.6.7.1.1	**Air evacuation from kitchen shall be stronger than air inflow.**	**Mid**
	A.aC.aF. Wet machinery	
A.aC. R.6.4.1.	Include large food preparation area with hot and cold water.	High
A.aC.aF. R.6.4.1.1.	**Cold- and warm-water faucet shall be easily accessible.**	**Crucial**
A.aC.aF. R.6.4.1.2.	**Water temperature should be efficient but not harmful.**	**High**
A.aC. R.6.5.2.	Gentle dishwashing shall support fragile tableware.	Crucial
A.aC.aF. R.6.5.3.1	**Dishwashing by hand shall be possible.**	**Crucial**
	A.aC.aG. Dish machinery	
A.aC. R.6.5.1.	Dishwashing shall clean 5-course tableware (5 x 8 plates, cutlery, glasses, and cooking utensils) in same run.	Mid
A.aC.aG. R.6.5.1.1	**One dishwasher for dirty utensils, one for cleaning tableware.**	**Mid**
A.aC.aG. R.6.5.1.2.	**Glasses shall be loadable in a rack.**	**High**

It should be admitted that developing house kitchen machinery on a separate *black-box nesting level* is a bit of overkill, but it effectively shows the idea of the *Cpdm* hierarchical concept to facilitate large *products* to scale up. If you find it funny with lack of *complexity*, imagine instead a nuclear plant machinery.

One may wonder why the *requirements* are so user-oriented and thereby hard to make precise. Wouldn't it be better to capture technical properties to the machinery, like power in watts, current in amperes, size in centimeters, weight in kilograms and so on?

The question is what is most important to the *users* of the house. Often, the way of using a *product* is the most important, and *end user stakeholders* are not familiar with expressing *user* benefit in watts, amperes, centimeters, and kilos. But don't worry, at the end of the finalized design *phase, designers* must boil everything down to watts, amperes, centimeters, and kilograms. However, these are design decisions emerging from attempts to satisfy the above *stakeholder requirements.*

7.10.3 EXAMPLE House kitchen machinery: Prioritize requirements

As can be seen from the *requirements* in Table 7-13, page 200, most are behavioral, since they were compiled in a user-friendly way by the wife of the house-buying family. They were also prioritized mostly by the wife, with support of the husband, having good understanding of machine costs.

7.10.4 EXAMPLE House kitchen machinery: Examine authenticity

Since the wife preferred standard kitchen machinery, the husband was very satisfied with his wife's cooking, and he had no garage to balance the kitchen *requirements* against, there were no battles and the *requirements* were decided as they were.

7.10.5 EXAMPLE House kitchen machinery: Plan development roles and providers

As with house and room *nesting levels*, there is still some need for development support (see Table 7-14 below).

TABLE 7-14 House kitchen machinery development allocation staffing

House kitchen machinery role item	Provider	Cpdm schedule or task
i62. **House user**	**PR.10. Pro-prietor**	Pt.3.R. Refine machinery requirements from rooms requirements and ensure machinery development staffing.
i63. **Machinery expert** • Reviewed by the proprietor	**PR.17. K-Studio**	Pt.3.P. Predetermine machinery solutions with supplier opportunities.
i64. **Machinery expert**	**PR.17. K-Studio**	Pt.3.A. Satisfy machinery requirements by decomposing machinery black-boxes into white-box design not containing black-boxes. Pt.3.F. Finalize design of machinery with product requisites.
i65. **Kitchen constructor** • Mounting back housing, punch for ventilation, etc.	**PR.17. K-Studio**	
i66. **Plumber** • House builder or house builder's subcontractor • Including detected failures elimination	**PR.17. K-Studio**	
i67. **Tin-smith** • House builder or house builder's subcontractor • Including detected failures elimination	**PR.17. K-Studio**	Pt.3.I. Procure machinery items in order to install machinery.
i68. **Gas installer** • House builder or house builder's subcontractor • Authority-approved gas installer • Including detected failures elimination	**PR.17. K-Studio**	
i69. **Electrician** • Including detected failures elimination	**PR.12. Elert AB**	
i70. **Inspector** • Verification of total machinery fitting and functionality	**PR.17. K-Studio**	Pt.3.V. Verify machinery black-/white-boxes.

7.10.6 EXAMPLE House: Proceed reading

When the machinery cabinets *compound requirements* are completed the technical *schedule* forks as usual (see again Fig. 7-9).

To follow this example, these are the alternatives:

- **3P** Proceed to chapter 8.7, "EXAMPLE House rooms machinery: Predetermine solutions with sourcing options," page 309.

- **3A** Proceed to chapter 9.8, "EXAMPLE House rooms machinery: Satisfy machinery requirements by decomposing machinery black-boxes into white-box design not containing black-boxes," page 371.

- **3F** Proceed to chapter 10.4, "EXAMPLE: House kitchen machinery: Finalize design of machinery with product requisites," page 445.

- **3V** Proceed to chapter 12.7, "EXAMPLE: House rooms machinery: Fully verify machinery black-/white-boxes," page 678.

- To leave the house example and to dig deeper into the **Pt.R. Capture staffing & requirements** topic, proceed to chapter 7.11 below.

7.11 Restriction requirements

Requirements can have a very different appearance expressed in different degrees of *formalism*, ranging from an idea residing only in a person's mind to a full-blown database integrated into other information systems.

Black-box requirements can be divided into *restriction requirement* and *behavior requirement*. Many times they can be treated the same way, as in the entire previous house example, but when verifying a prototype taking advantage of the two different types of *requirements*, the difference must be recognized (see ch. 12.11, "Verification depends on type of requirements," p. 704). So it is time to better understand this difference.

There are two types of compound requirements, called restriction and behavior requirements.

Restriction requirements override the freedom to design a *product*, but instead prescribe a static design that must be provided when opening a specified *black-box*. *Restriction* are more tangible and concrete, which makes them easier to grasp than *behavior requirements*, even if there might be a large number of them when capturing *restriction*, for example, modern machinery constructions.

Behavior requirements describe how a *black-box* is supposed to respond when a *user* operates it, or in other words, how the *interfaces* of a *black-box* behave over time. Modern *programs* are often very *complex* in behavior, which makes the *behavior requirements* large and even more difficult to deal with.

7.11.1 Cpdm definition

It can now be useful to make a stricter definition of *restriction* (see Table 7-15 below).

TABLE 7-15 Cpdm definition

Aspect	Restriction requirement
Definition	Stakeholders' documented expectations of design inside a black-box
Synonyms	Design restriction, property, attribute, characteristic, nonfunctionality, appearance
Examples	Metallic red, diesel engine, 430 gram, 380 kbit/s, 200 € retail price
Symbol	A. Black-box

TABLE 7-16 Restriction requirement

A.RR.n. Restriction requirement	Priority

The *boundary interface* of the *black-box* can also be specified with the *restriction requirements*, either belonging to the *black-box*, or the *black-box environment* or both. For more information about *boundary interfaces*, see chapter 9.4.2, "Architecture preparation procedure," page 347. For a more thorough explanation of *black-boxes*, see chapter 9.1, "About architecture," page 334.

7.11.2 Black-box restrictions predetermining architecture ingredients

Restrictions deliberately limit the freedom of the *designers* when they open the *black-box* to get it designed. Observe that too much restriction on the freedom of design gradually increases conflicts with other *requirements*, and in a worst-case a *product* that is well designed according to restrictions, may poorly satisfy large parts of the *requirements*, or poorly satisfy the *behavior requirements*. For example, demand for a high degree of reusability may end up in *components* not fitting anywhere. Another example is a house with mind-boggling *architecture* that is impossible to live in.

Capturing restrictions within a black-box too much can be seen as a desire to control the capacity of the black-box, and prevent forthcoming designers the opportunity to do their job.

Also note that *restriction requirements* may be used to limit all kinds of *architecture ingredients* within a *black-box*. For a design ingredient developed for *reuse* it is preferable to design it as a *black-box* in order to take control over its *interface* (so it will fit for *reuse* as intended).

Sometimes *restriction requirements* are said to restrict the *ability*, for example:

Cost (per unit). This is often a severe *restriction requirement* on many tangible *elements*. Most often, end consumers see the price per unit as one of the most important *requirements* of a *product*.

Procurability. These restrictions have the purpose of making it easier to buy standardized *elements* off the shelf on the open market. Buying standard *products* that sufficiently satisfy the need is often much cheaper than developing them in-house. It is quite common that the purchase of *components* within a *black-box* from selected *suppliers* is naturally predetermined, rather than having them developed in-house. Ingredients developed in-house, surrounding procured *components*, must be adapted (meshed and glued) very well to the range of *components* purchased from *suppliers*.

Producibility. This is an aspect very often ignored by development departments. Designs can be very intelligent and well made, but impossible to efficiently manufacture in a larger series. An involved *stakeholder* from the manufacturing department might mitigate this risk.

Reusability. If any *element* is planned to be *reused* by another *product* after it is initially developed, another set of restrictions may appear to fit future *products*. It is highly beneficial to try to find coming restrictions or a set of flexible restrictions as early as possible. Otherwise stated, these *requirements* must capture a degree of *variability of reusable components* to make the *component* more diverse and easy to adapt to when *reused*.

Development testability. The testability begins with checking that *requirements* can be verified. Next is identifying how *elements* are best designed so as to make *verification* easier, especially if prototypes should be automatically verified. Note that to make an ordinary *verification*, in *Cpdm* the *element* needs to be a *black-box* with its own specifications.

Manufacturing testability. Also, when the *product* is manufactured, it needs to be verified before delivery to wholesalers and retailers. This often implies that the prototype is also designed for this purpose. If not, the design of manufacturing *fixtures* becomes solely responsible for making this possible, which might be problematic and expensive.

Quality. It is most often considered to be difficult to capture the *target* grade of *quality*. If specified, it remains to be followed up and verified, which can also be difficult. However, lack of *quality* is very apparent to *customers* buying *products*, so there is no way around capturing it if good profitability and long-term market success are expected. Also observe the opposite, that zealous developers might occasionally produce a design of too high *quality*, not required by the *stakeholders*. The next book "*Cpdm* technical overhead" will address *quality* in more detail.

Reliability. The most efficient way to ensure reliability is to avoid *complexity* at first, of course, by eliminating *artificial complexity*, but also by trying to restrict *intrinsic complexity* through simple solutions.

It is not only the outermost *nesting levels elements* that might be restricted. In fact, all kinds of *inward architecture ingredients* can be restricted for any *ability* of the above. As always, the sum of all *requirements* might be conflicting and even describe something that is unobtainable, implying compromises to satisfy the *requirements* before the *white-box* design is set, or a firm sharpening up and enforcement of a tough prioritization.

7.12 Behavior requirements

7.12.1 Cpdm definition

It can also be useful to define *behavior requirement* more precisely (see Table 7-17 below). For *connection interface* definition, see Table 9-10, page 346.

TABLE 7-17 Cpdm definition

Aspect	Behavior requirement
Definition	Stakeholder-expected behavior on interfaces of a black-box
Synonyms	Interaction, function, service, procedure, method, function, functionality
Symbol example	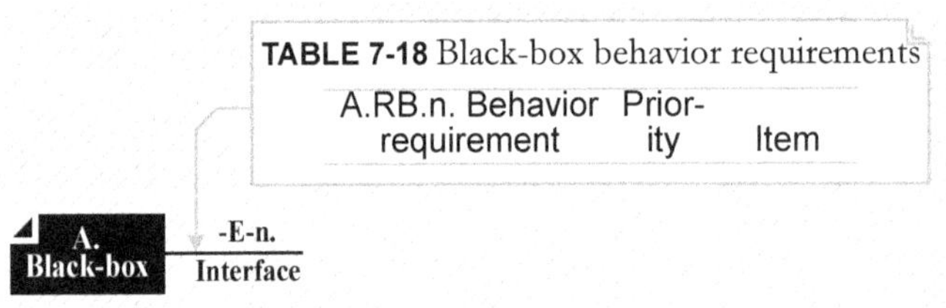 *(symbol diagram)*

The symbol example contains:

TABLE 7-18 Black-box behavior requirements

A.RB.n. Behavior requirement	Priority	Item

Items are optional, and used for detailed accounting of prerequisites (see Table 7-55, p. 266 and Table 10-108, p. 582).

It is now possible to dig deeper into capturing *behavior requirements* and thus define *interaction requirement, stimulus,* and *response* (see Table 7-17 below.)

TABLE 7-19 Cpdm definitions

Aspect	Interaction requirement	Stimulus	Response
Definition	Behavior requirement describing cooperation over a connection interface to a black-box or between black-boxes	Input to a black-box via connection interfaces	Output from a black-box via connection interfaces
Synonyms	Interaction, function, service, procedure, method, function, functionality	Impact, influence, injection, insertion	Reaction, answer, reply, comeback,
Symbol example	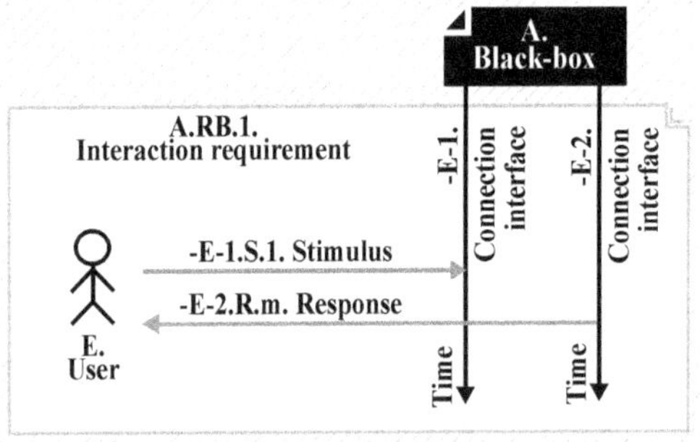 *(symbol diagram)*		

Restriction requirements are not at all unimportant, but *behavior requirements* are getting more and more important, since *programs* within *products* grow all the time, resulting in an awful lot of *product* behavior.

An additional problem is that behavior is volatile compared with static hardware, and a lack of understanding between mechanics and *program* engineers may complicate things.

Behavior requirements are about volatile events on interfaces, hence the whole remaining chapter is needed to explain them.

Altogether, *behavior requirements* are huge, hard to capture, and a challenge to verify. For that reason, the remaining part of the **Pt.R. Capture staffing & requirements** chapter will essentially explain how to capture *behavior requirements*. To illustrate *complex* behavior, *Cpdm* will propose a lot of types of *behavior requirements* to penetrate and master *complexity*.

7.13 Requirements refinement and relation to architecture

To dig deeper into restriction and *behavior requirements*, the relation to *architecture* need to be explained by a generic example.

7.13.1 EXAMPLE: Traceability within requirements hierarchy

To keep track of *architectures* and *requirements* to *complex products*, the following must be organized:

It is easy to start capturing requirements, but rather soon they must be better organized to become useful.

- *Traceability* between *requirements* within their refinement hierarchy
- *Traceability* between *architectures* within their decomposition hierarchy
- *Traceability* between *requirements* hierarchy and *architectures* hierarchy

For the example of *traceability* between *requirements* within their hierarchy, see Figure 7-10 below. Note that for space reasons, *restriction requirement* is abbreviated to restriction and *behavior requirement* is abbreviated to behavior only.

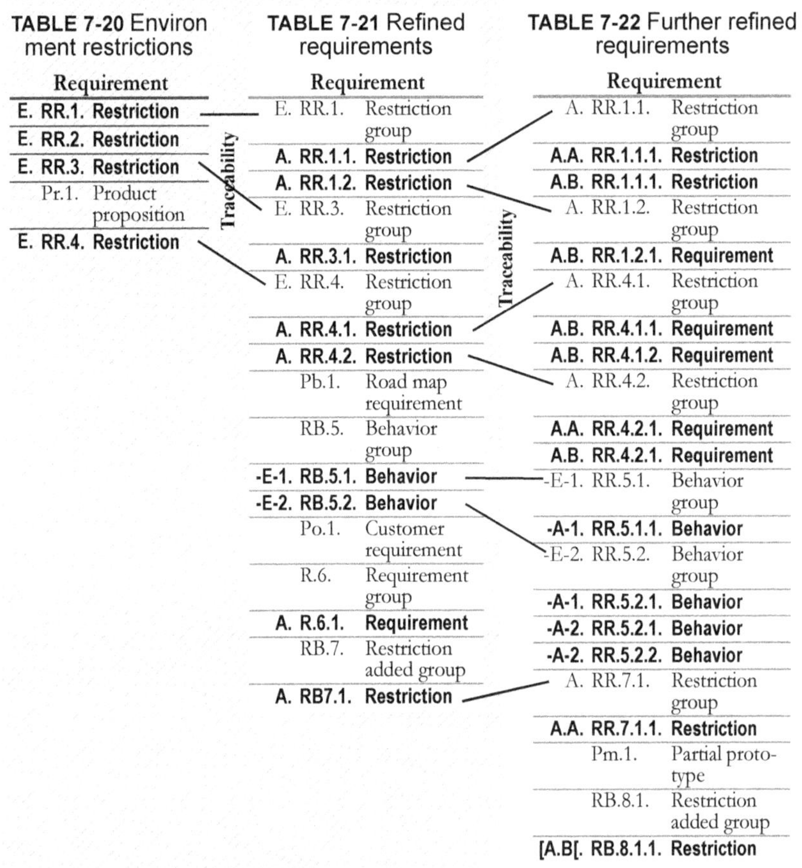

TABLE 7-20 Environment restrictions

Requirement
E. RR.1. Restriction
E. RR.2. Restriction
E. RR.3. Restriction
Pr.1. Product proposition
E. RR.4. Restriction

TABLE 7-21 Refined requirements

Requirement	
E. RR.1.	Restriction group
A. RR.1.1.	Restriction
A. RR.1.2.	Restriction
E. RR.3.	Restriction group
A. RR.3.1.	Restriction
E. RR.4.	Restriction group
A. RR.4.1.	Restriction
A. RR.4.2.	Restriction
Pb.1.	Road map requirement
RB.5.	Behavior group
-E-1. RB.5.1.	Behavior
-E-2. RB.5.2.	Behavior
Po.1.	Customer requirement
R.6.	Requirement group
A. R.6.1.	Requirement
RB.7.	Restriction added group
A. RB7.1.	Restriction

TABLE 7-22 Further refined requirements

Requirement	
A. RR.1.1.	Restriction group
A.A. RR.1.1.1.	Restriction
A.B. RR.1.1.1.	Restriction
A. RR.1.2.	Restriction group
A.B. RR.1.2.1.	Requirement
A. RR.4.1.	Restriction group
A.B. RR.4.1.1.	Requirement
A.B. RR.4.1.2.	Requirement
A. RR.4.2.	Restriction group
A.A. RR.4.2.1.	Requirement
A.B. RR.4.2.1.	Requirement
-E-1. RR.5.1.	Behavior group
-A-1. RR.5.1.1.	Behavior
-E-2. RR.5.2.	Behavior group
-A-1. RR.5.2.1.	Behavior
-A-2. RR.5.2.1.	Behavior
-A-2. RR.5.2.2.	Behavior
A. RR.7.1.	Restriction group
A.A. RR.7.1.1.	Restriction
Pm.1.	Partial prototype
RB.8.1.	Restriction added group
[A.B[. RB.8.1.1.	Restriction

FIGURE 7-10 Requirements refinement hierarchy

For the structure of *black-boxes*, see Figure 7-11, page 210. Since *requirements* will now be connected to the *architecture*, it is beneficial to also have a look in the chapter 9, "Design architectures," from page 333, and more particularly, for example, in chapter 9.5.5, page 356.

Much of Figure 7-10 above is rather self-explanatory and logical but some comments are in order:

- Table 7-20 Environment restrictions is assumed to be *environment restriction requirements*; Table 7-21 Refined requirements is assumed to be *requirements* for an *outermost black-box* **A.**; and Table 7-22 Further refined requirements are *requirements* for two *embedded black-boxes* **A.A.** and **A.B.** in **A.**

- Refining *requirements* means that *environment restriction requirements* may give rise to *black-box requirements* also for the *product outermost black-*

box **A**, and these *requirements* in turn may give rise to more detailed *black-box requirements* in *black-box* **A.A.** and **A.B.** The relationship between refined *requirements* is controlled by *traceability*, and in *Cpdm traceability* is tracked by the ID numbering of the *requirements*.

- The *environment restriction requirement* **E.RR.2.** in Table 7-20 is assumed to *target* only the *environment*, and is not further refined when capturing *black-box* **A.** *requirements* (see Table 7-21). Consequently, there is no restriction group 2 in this table. The same thing occurs for *black-box requirement* **A.R.6.1**, which is not further refined and gives rise to a gap between *requirements* **-A-1.RR.5.2.2** and **RR.7.1**. Let this gap stay, to indicate that *requirement* **A.R.6.1.** has not been refined. Don't add a new group with number 6. That would be confusing.

- Table 7-20, may also contain *requirements* from earlier *phases* of *product* development, like Pr.1. if such *requirements* in some way affect how the *product* will interact with the *environment*.

- After desired *environment restriction requirements* in Table 7-20 Environment restrictions are refined to Table 7-21 Refined requirements, they can be further extended with *requirements* from earlier *phases*, such as *road map* **Pb.1.** from **Pa. Study strategy** and *user profile* **Po.1.** from **Pn. Research market** to form new groups such as **RB.5.** and **RB.6.**

- Table 7-21 can also be extended with brand-new *requirements*, such as in group **RB.7.** However, before adding new desired *requirements* not traceable from any previous *requirements*, investigate whether maybe a previous *requirement* has been missed. It is no problem to go *outwards* in the *requirement* hierarchy and add a *requirement*, from which newly desired *requirements* can then be refined. If some design is made before updating the *requirement*, don't forget to update all succeeding design also.

- The order of *restrictions* and *behavior requirements* may get mixed, as can be seen from Table 7-21 and Table 7-22.

- Some *requirements* may be refined to many *black-boxes*, like *black-box requirement* **A.RR1.1.** is refined to both *requirement* **A.A.RR.1.1.1.** and *requirement* **A.B.RR1.1.1.** In this case, the *requirement* identity numbers might be the same and only the referred *black-box* identity differs. However, to be crystal clear, the *requirement* identity numbers may also be kept unique if desired.

- Even deep in the *requirement* hierarchy, *requirements* may occur from previous *phases*. For example **Pm.1.** Partial prototypes from **Pl. Prototype details** may impact details deep in the *product*.

7.13.2 Requirements traceability to architecture hierarchy

Requirements must be traceable to the *architecture* they *target* (see Figure 7-11).

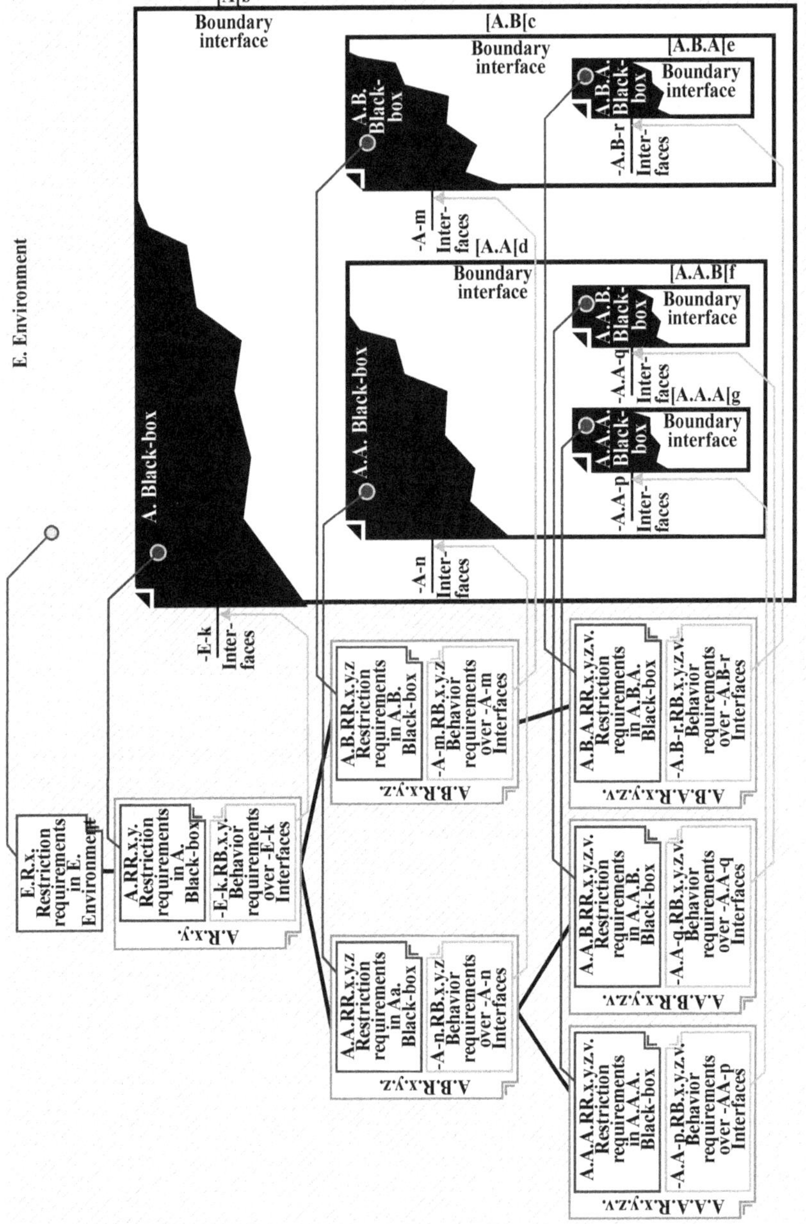

FIGURE 7-11 Requirement hierarchy traceability to the black-box hierarchy

The *boundary interface* can be specified with the *restriction requirements*, either belonging to the *black-box*, or the *black-box* surroundings, or both. For more information about *boundary interfaces*, see chapter 9.4.2, "Architecture preparation procedure," page 347. For a more thorough explanation of *black-boxes*, see chapter 9.1, "About architecture," page 334.

Figure 7-10 and Figure 7-11 together explain how the *requirement* hierarchy and *architecture* hierarchy are traceable to each other. When the *product* to develop begins to get *complex*, a database must preferably be utilized. It is not realistic that every *specifier* and *architect* can recall such *traceability* from memory.

7.14 EXAMPLE Multiplication toy environment: Explore environment restrictions and ensure environment staffing

7.14.1 EXAMPLE Multiplication toy: About this example

This vast example starts here, and like the house example, it will span the subsequent six chapters describing each *phase*, each describing a core *phase* of *Cpdm*. This example is intended to show *complexity* not primarily caused by a large *product* size with many *nesting levels* containing a lot of details, as the house example did, but is meant to show other kinds of *complexity*, such as electronics and microcontrollers (also see ch. 3.2.5, p. 35).

This example shows specification and design of one common *product* (chapters marked with C in Fig. 7-12) containing one interchangeable *element* developed and realized by two different technologies:

- hard-wired electronics (chapters marked H in Fig. 7-12), resulting only in a prototype to be the basis for possible customized ASIC chip development, to introduce how *Cpdm* masters asynchronous digital electronic networks
- microcontroller electronics fully developed for manufacturing (chapters marked M in Fig. 7-12), to introduce how *Cpdm* masters electronics with *programs*.

The behavior of this *product* is not very *complex*. The *requirements* are more focused on child safety and easy exchangeability between the two technologies.

7.14.2 EXAMPLE Multiplication toy environment: Process schedule to use

The *product* size in this example is kept modest, and a technical *process schedule* with two *nesting levels* suffices for its development. Since the technical *process schedule* is more elementary here and already explained in the house example, *process schedule* guidance will be less in this case.

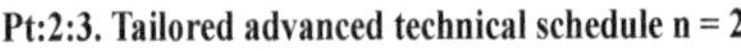

Pt:2:3. Tailored advanced technical schedule n = 2

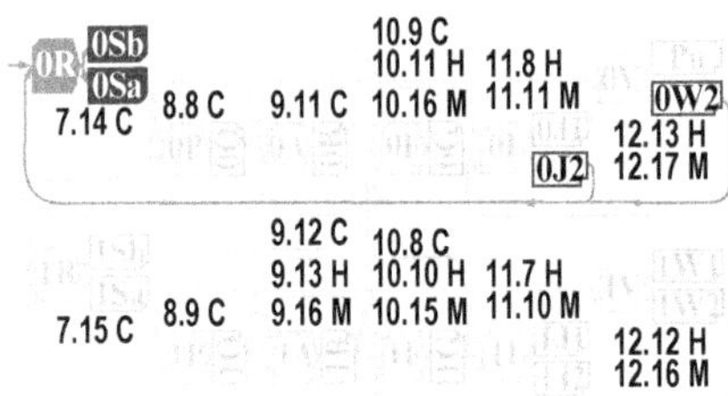

For the multiplication toy example, a special technical overall *schedule* is *tailored* (illustrated in Fig. 7-12, left):

- **Pt:2:3. Tailored advanced technical schedule n = 2,** page 169

In this *tailored schedule* the generic *schedule* to use is (filled-in symbols):

- **Pt.0.R. Explore environment restrictions and ensure environment staffing,** page 110.

FIGURE 7-12 Position of used schedule and master (filled in) in overall schedule. Chapter numbers of the multiplication toy example are also referenced.

Below *results* may be demanded (their *masters* are not indicated in Fig. 7-3 above, but in the *Cpdm* overall *schedule* **P. Cpdm generic development schedule,** p. 100):

Pk   Pk. Concept brief

Pe   Pe. Reuse plan

Pm   Pm. Partial prototype

Pg   Pg. Business case

Po   Po. User profile

Pi   Pi. Product proposition

For possible *failure localization* and *failure elimination*, below *results* may be demanded (their *masters* are indicated *upstream* from the filled-in *schedule* in the figure above or in greater detail in Fig. 11-5, p. 602 and Fig. 11-6, p. 603):

0J2   Table 11-28. Multiplication toy environment (hard-wired multiplier variant), start-up report 1, page 630

Table 11-36. Multiplication toy environment (microcontroller multiplier variant), start-up report 1, page 640

0W2   Table 12-53. Multiplication toy environment (microcontroller multiplier variant) failure elimination report 1, page 728

7.14.3 EXAMPLE Multiplication toy environment: Explore restrictions

As usual when starting technical development, the *environment restriction requirements* are first captured. In this example, there is the overall design restriction that the multiplication toy shall be a *reusable item* for many hosting cuddly toys, which means that the multiplication toy must be developed to fit many *environments*.

Otherwise, the *product* to develop is quite simple and the found appropriate restrictions are quite few (see Table 7-23 below).

TABLE 7-23 Multiplication toy environment restriction requirement
Multiplication toy restriction requirement
E. RR.1. The multiplication toy shall be childproof according to European Community regulations.
E. RR.2. The multiplication toy attachment to the hosting toy may not leak padding from the inside of the hosting toy.
E. RR.3. The multiplication toy shall be easily reusable in a safe way, by all kinds of different hosting toys.
E. RR.4. No legal responsibility is taken for incorrect mounting of the multiplication toy into the hosting toy by companies buying the multiplication toy.

Note that the *environment restriction requirements* must be concretized until they are testable, for example, the RR.3. above. This *requirement* will be used for *environment verification*, but possibly also as the base for *requirement refinement*, inwards the *product outermost black-box*.

7.14.4 EXAMPLE Multiplication toy environment: Plan development roles and providers

After the *environment restriction requirements* are specified, needed developers need to be allocated (see Table 7-24 below).

TABLE 7-24 Multiplication toy environment development allocation staffing

Multiplication toy environment role item	Provider	Cpdm schedule or task
i10. Childproofing specialist • To handle environment restrictions	PR.10.In-house project	Pt.0.Ra. Explore restrictions.
i11. Chief engineer • To allocate developers (this table)	PR.10. In-house project	Pt.0.Rb. Plan development roles and providers.
i12. General developer • To propose solutions for connection interfaces to the environment	PR.10. In-house project	Pt.0.Pa. Try to satisfy restrictions.
i13. Product manager • For technical issues	PR.10. In-house project	Pt.0.Pc. Analyze in/out and make/buy options.
i14. Manufacturing specialist • To ensure efficient manufacturing	PR.11.Mechtech	
i15. Mechanical designer • To design mechanical interfaces to environment	PR.12.Mechano	Pt.0.Aa. Identify environment architecture ingredients. Pt.0.Ac. Arrange ingredients and identify black-box. Pt.0.Ae. Lay out physical architecture. Pt.0.Fa. Finalize environment interfaces detailed items.
i16. Child MMI expert • To design the Child-Machine Interface.	PR.10. In-house project	
i17. Electric power engineer • To design power supply	PR.10. In-house project	

TABLE 7-24 Multiplication toy environment development allocation staffing

Multiplication toy environment role item	Provider	Cpdm schedule or task
i18. Product manager	PR.10. In-house project	Pt.0.Fc. Account for environment interfaces development, realization, and manufacturing.
i19. Senior mechanical designer • To evaluate fitness of the interface	PR.12. Mechano	Pt.0.I. Realize interfaces to environment and insert/await outermost white-box.
i20. MMI verifier • To verify human interfaces to environment	PR.10. In-house project	Pt.0.V. Verify prototype in environment.
i21. Hosting toy engineer • To evaluate the connection to the hosting toy	PR.12. Mechano	

7.14.5 EXAMPLE Multiplication toy: Proceed reading

To follow this example, these are the alternatives:

- Proceed to chapter 8.8, "EXAMPLE Multiplication toy environment: Demarcate product by environment solutions with possible supplier opportunities," page 313.
- Proceed to chapter 9.11, "EXAMPLE Multiplication toy environment: Constitute environment architecture," page 382.
- Proceed to chapter 10.9, "EXAMPLE Specifics for multiplication toy environment commonalities: Finalize design of environment interfaces with product requisites," page 493.
- Proceed to chapter 12.13, "EXAMPLE Multiplication toy environment (hardwired multiplier variant): Verify prototype in environment," page 712.
- Proceed to chapter 12.17, "EXAMPLE Multiplication toy environment (microcontroller multiplier variant): Verify prototype in environment," page 725.
- Proceed to chapter 7.15, "EXAMPLE Multiplication toy: Refine product requirements from environment restriction requirements and ensure black-/white-box development staffing" below

7.15 EXAMPLE Multiplication toy: Refine product requirements from environment restriction requirements and ensure black-/white-box development staffing

7.15.1 EXAMPLE Multiplication toy: Process schedule to use

The *environment* of the planned multiplication toy has been specified in the preceding chapter, and *environment restriction requirements* have been documented. It is now time to refine these restrictions and *capture requirements* to the multiplication toy *black-box*.

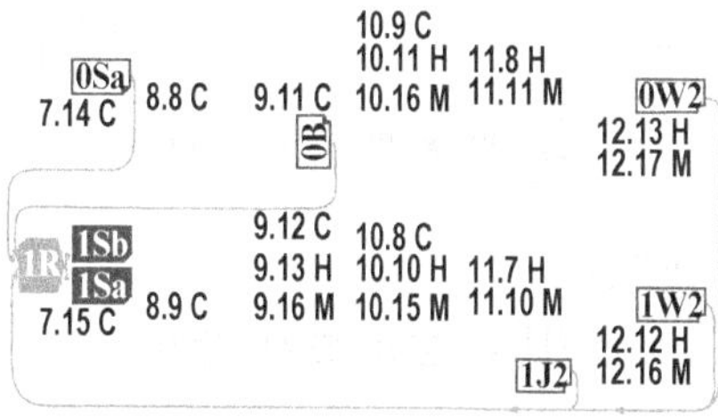

For the multiplication toy example, a special technical overall *schedule* is *tailored* (illustrated in Fig. 7-13, left):

- **Pt:2:3. Tailored advanced technical schedule n = 2,** page 169

In this *tailored schedule* the generic *schedule* to use is (filled-in symbols):

- **Pt.n.R. Refine requirements from outward nesting level requirements and ensure black-/white-box development staffing,** page 130.

FIGURE 7-13 Position of schedule and master to use (filled in) in overall schedule

Below *results* may be demanded (their *masters* are indicated *upstreams* from the filled-in *schedule* in the figure above):

0Sa Environment restriction requirements
Table 7-23. Multiplication toy environment restriction requirement, page 213.

0B Environment existing ingredients and design of interfaces to the environment
Figure 9-35, Multiplication toy environment logical architecture, page 384.

For possible *failure localization* and *failure elimination*, below *results* may be demanded (their *masters* are indicated *upstream* from the filled-in *schedule* in the figure above):

1J2 Table 11-25. Multiplication toy with hard-wired multiplier start-up report 1, page 627
Table 11-33. Multiplication toy with microcontroller multiplier start-up report 2, page 636

1W2 Table 12-38. Multiplication toy with hard-wired multiplier failure elimination report 1, page 711
Table 12-47. Multiplication toy with microcontroller multiplier failure elimination report 1, page 723

[0W2] Table 12-53. Multiplication toy environment (microcontroller multiplier variant) failure elimination report 1, page 728

7.15.2 EXAMPLE Multiplication toy: Refine and capture requirements

The restrictions are refined in a *modularized requirement* table (see Table 7-25 below).

Note that the]A]2. Casing *boundary interfaces* surrounding the A. Multiplication toy *black-box* are not considered part of the E. Multiplication toy environment, and are now instead assumed to belong to the A. Multiplication toy. See chapter 9.4.2, page 347 for more information about where a *boundary interface* belongs.

When working with *requirements*, the structure between *environment* and outermost *black-box* might not be correct in the first go. Continue to capture, change, and move back and forth, and finally everything will fall into place.

Also note that some new *requirements* are added on this *nesting level*, which are not *requirement refinements* from the *environment restriction requirements* (see RR.5. to RR.7. in Table 7-25 below and RB.8. in Table 7-26 next).

TABLE 7-25 Multiplication toy restriction requirement		
Multiplication toy restriction requirement		**Priority**
E. RR.1.	The multiplication toy shall be childproof according to European Community regulations.	
A. RR.1.1.	**Everything accessible from the outside of the multiplication toy shall be impossible for a child to remove.**	Crucial
A. RR.1.2.	**Everything accessible from the outside of the multiplication toy shall be nontoxic for a licking child.**	Crucial
Pcb.1.	The multiplication toy shall be realized through different technologies in order to obtain the most competitive product for a particular manufacturing volume.	
RR.5.	Cost reduction requirements	
A. RR.5.1.	**The multiplication calculation shall be allocated to a separate element that can be redesigned with minimal rework.**	High
A. RR.5.2.	**The casing of the hosting toy shall be exactly the same for all products independent of technology used.**	High
RR.6.	Power supply requirements	
A. RR.6.1.	**The toy shall have an internal power supply, not possible for a child to operate.**	Crucial
Pj.1.	Products like these get huge competition and get very price sensitive.	
RR.7.	Price requirements	
A. RR.7.1.	**If the toy is produced, it shall be made in such volume that the price per unit is less than € 5.**	Crucial

TABLE 7-26 Multiplication toy interaction requirement

Multiplication toy behavior requirement		Priority
RB.8.	Multiplication toy interaction requirement (see Fig. 7-14, p. 217).	
A. RB.8.1.	User amusement shall be to multiply two operand values into one product value.	Crucial
A. RB.8.2.	The toy shall have one input device for each multiplicand value. Multiplicand values shall be at least between 0 and 10. The multiplicands must be easy for a child to settle.	0 - 10 crucial > 10 high
A. RB.8.3.	The toy shall have at least a 3-digit output device for the product value. The product must be easy to read for a child.	Crucial
A. RB.8.4.	The toy calculates and shows the product value practically instantly after either multiplicand is settled.	High

To be clear, it is a good plan to also capture *stimuli* and *responses* on *black-box interfaces* (see Table 7-27 below).

TABLE 7-27 Multiplication toy interface requirements

Multiplication toy behavior requirement on interface -aE-1.	
-aE-1a.	Multiplicand input interface
-aE-1a. S.1a.	Set (operand1)
-aE-1b.	Multiplicand input interface
-aE-1b. S.1b.	Set (operand2)
-aE-1c.	Product output interface
-aE-1c. R.2.	Show (product)

Even if this example is simple, this book is intended to show development principles. Further on, more and more *complex behavior requirements* will gradually appear. The (simple) *behavior requirement* A.RB.8. to be realized in the multiplication toy can be illustrated as in Figure 7-14.

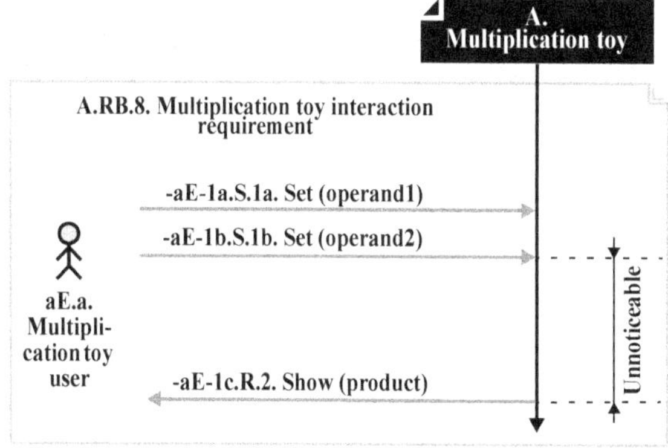

FIGURE 7-14 A. Multiplication toy interaction with its user

7.15.3 EXAMPLE Multiplication toy: Prioritize requirements

The *requirements* were reviewed by the *project sponsor* of the multiplication toy. The prioritization is documented in the same table as the *requirements* (see Table 7-25 above).

7.15.4 EXAMPLE Multiplication toy: Examine authenticity

A gate meeting was held with the *stakeholders*. An intense discussion was immediately started about the number of significant digits to set and to show.

In elementary schools most multiplication tables cover operands from 0 to 10, which would require three significant digits to show the product of 10 times 10. To show three digits instead of two would significantly raise the price and size of the toy. On the other hand, showing the multiplication table without the operand 10 would give an incomplete impression, especially since 10 is the base in our counting system.

If three digits are shown, it is possible to show rather high product numbers, such as 25 times 25, but the cost will increase significantly to find a device for input of such a high number of operands.

Glancing ahead to internal digital design, even if no such *restriction requirement* is specified, digits are assumed to be coded in a hexadecimal system. A rather good cost-effective compromise is to allow for example input of 15 times 15, resulting in the product 225. This was the final decision by the *stakeholders*.

7.15.5 EXAMPLE Multiplication toy: Plan development roles and providers

It should be noted that it is not very easy to get the *development staffing* list right from the beginning (see Table 7-28 below). Many decisions are yet to come, leading to new solutions and technologies, which need allocation of a not yet clearly identified kind. But as usual, competency is hard to get, and to allocate as early as possible is key. The allocation list can also be updated without major problems before the finalized *product requisites* are launched.

TABLE 7-28 Multiplication toy development allocation staffing

Multiplication toy role item	Provider	Cpdm schedule or task
i22. **Specifier** • To refine environment restrictions	**PR.10. In-house project**	Pt.n.Ra. Refine and capture requirements.
i23. **Chief engineer** • To allocate developers (this list)	**PR.10. In-house project**	Pt.n.Re. Plan development roles and providers.
i24. **Digital expert** • To find solution alternatives	**PR.10. In-house project**	Pt.n.Pa. Try to satisfy requirements.

TABLE 7-28 Multiplication toy development allocation staffing

Multiplication toy role item	Provider	Cpdm schedule or task
i25. **Manufacturing engineer** • To negotiate with volume suppliers	PR.11. Mechtech	Pt.n.Pc. Analyze make/buy options.
i26. **Purchase engineer** • To keep contact with suppliers • To ensure child-proof sample materials and design	PR.10. In-house project	
i27. **Electronics architect** • Among others, to gather common elements and isolate the technology differences to the fewest possible elements.	PR.10. In-house project	Pt.1.A. Satisfy refined requirements by decomposing product black-box into white-box design containing no black-boxes.
i28. **Logic designer** • For the hard-wired electronics	PR.10. In-house project	Pt.1.F. Finalize design items of product white-box with product requisites.
i29. **ASIC designer** • To ensure porting electronics to ASICs	PR.13.ASICtoGo AB	
i30. **Microcontroller designer** • For the microcontroller electronics	PR.14.Chipp	
i31. **Assembler programmer** • For the microcontroller program	PR.15.MicroHard	
i32. **Manufacturing preparer** • To estimate manufacturing costs	PR.11. Mechtech	
i33. **Automated verification specialist**	PR.11. Mechtech	
i34. **Accounting engineer** • For the hard-wired electronics	PR.10. In-house project	
i35. **Prototype mounting engineer** • To build the circuit boards	PR.10. In-house project	Pt.1.I. Procure items and linkable programs in order to realize white-box.
i36. **Programmer engineer** • To load and start the program	PR.15. MicroHard	
i37. **Electronics verifier** • To verify the black-/white-boxes	PR.10. In-house project	Pt.1.V. Verify black-/white-box.
i38. **Program verifier** • To verify the assembler program	PR.15. MicroHard	

TABLE 7-28 Multiplication toy development allocation staffing

Multiplication toy role item	Provider	Cpdm schedule or task
i39. Low-price-country worker to manually assemble PCB	PR.11. Mechtech	
i40. Process worker with PCB automated equipping conveyor belt	PR.11. Mechtech	
i41. Worker half-manually flashing microcontroller	PR.11. Mechtech	
i42. Automatic flashing on conveyor belt after PCB is assembled	PR.11. Mechtech	M. Manufacturing product (not covered by this book)
i43. Low-price-country worker to manually assemble toy mechanics	PR.11. Mechtech	
i44. Process worker with automated assembly conveyor belt	PR.11. Mechtech	
i45. Low-price-country worker manually verifying toy	PR.11. Mechtech	
i46. Process worker with automated verification conveyor belt	PR.11. Mechtech	

7.15.6 EXAMPLE Multiplication toy: Proceed reading

To follow this example, these are the alternatives:

- Proceed to chapter 8.9, "EXAMPLE Multiplication toy: Predetermine solutions with sourcing options," page 316.
- Proceed to chapter 9.12, "EXAMPLE Multiplication toy commonalities: Satisfy refined requirements by decomposing product black-box into white-box design containing no black-boxes," page 386.
- Proceed to chapter 9.13, "EXAMPLE Hard-wired multiplier: Further satisfy refined requirements by decomposing product black-box into white-box design containing no black-boxes," page 391.
- Proceed to chapter 9.16, "EXAMPLE Microcontroller multiplier: Further satisfy refined requirements by decomposing product black-box into white-box design containing no black-boxes," page 397.
- Proceed to chapter 10.8, "EXAMPLE Specifics for multiplication toy commonalities: Finalize design items of product white-box with product requisites," page 482.
- Proceed to chapter 12.12, "EXAMPLE Multiplication toy with hard-wired multiplier: Verify black-/white-box," page 706.
- Proceed to chapter 12.16, "EXAMPLE Multiplication toy with microcontroller multiplier: Verify black-/white-box," page 717.
- To leave the multiplication toy example and to dig deeper into the chapter 7, "Capture staffing & requirements" topic, proceed to chapter 7.16 below.

7.16 Use-case behavior requirements

7.16.1 Cpdm definition

A common way to capture *interaction requirements* is by *use-cases*. Depending on how successful the interaction has become for the *user*, they can be divided into *normal use-cases* and *exception use-cases*. Table 7-30 define these types of *use-cases*.

TABLE 7-29 Cpdm definitions

Aspect	Use-case	Normal use-case	Exception use-case
Description	A sequence of stimuli and responses interacting on interfaces to a black-box	A favorable sequence of stimuli and responses interacting on interfaces to a black-box	Unfavorable sequence of stimulus and responses interacting on interfaces to a black-box
Synonyms	Function, behavior, interaction	Success, breakthrough, completion	Warning, deviation, misuse
Symbol examples	Textual or graphical, see below definitions		

7.16.2 Cpdm definition

Basic use-cases is a good first attempt to formalize *behavior requirements* expressed in text. In *Cpdm, basic use-case* is defined according to Table 7-30 below.

TABLE 7-30 Cpdm definition

Aspect	Basic use-case
Description	Stakeholder-expected behavior on interfaces of a black-box, specified by one interaction or a sequence of interactions
Synonyms	Procedure, function, case, event, stimulus-response sequence, method, test-case
Symbol examples	**TABLE 7-31** A.RB. Basic use-case **A.RB.UC. Behavior requirement** RB.WS.10. Use-case precondition (optional) IA.11. **Interaction 11 name** / Interaction 11 description IA.21. **Interaction 21 name** / Interaction 21 description IA.31. **Interaction 31 name** / Interaction 31 description RB.WS.40. Use-case postcondition (optional)

Basic use-cases can be adapted to any company, and are rather easy to capture and understand.	*Basic use-cases* are a good step-up in structure, *formalism*, and control, compared to plain text, because *basic use-cases* organize *requirements* in a better way than plain text. Tables with texts in cells are also supported by many ordinary text editors. *Basic use-cases* are rather easy to fill in and understand. *Precondition* and *postcondition* are optional, but offer a possibility to attach many separately described *formal use-cases* into one or many *formal scenarios* (see ch. 7.19.1, p. 230).

7.16.3 EXAMPLE: Rudimentary dual-operator calculator use-case

The first example is a very simple *basic use-case*, which specifies the simplest imaginable pocket calculator applying the four rules of arithmetic between two operands (see Table 7-32 below).

TABLE 7-32 A.RB. Rudimentary calculator basic use-case

	A.RB.UC. Rudimentary calculator behavior requirement
A.RB.IA.11.	**Input leading operand** The user enters the leading operand by consecutively pressing digits 0-9. Show input on display.
A.RB.IA.21.	**Input dual operator** The user enters one of the dual operators +, -, * or /.
A.RB.IA.31.	**Input trailing operand** The user enters the trailing operand by consecutively pressing digits 0-9. Show input on display.
A.RB.IA.41.	**Order conclusion** The user enters the equal sign =. Calculate value. Show product on display.

The few who have developed complete pocket calculator *programs* know that such a simple device is nastier and more difficult to master than first imagined.

The main difficulty is the *artificial complexity* of most calculators, which depends on the intent to squeeze in as much mathematical computation as possible into one single device.

The challenge then becomes to make the computation comprehensible and intuitive, so the calculator computation can be sufficiently trusted by ordinary people not very familiar with mathematics and electronic devices.

7.16.4 EXAMPLE: Simple dual-operator calculator basic use-case with exceptions

Basic use-cases can contain wait-states but risk being too messy to understand

With this example, exception handling is also specified (see Table 7-33). The *normal use-case* forks into many possible *exception use-cases* in the same table, thus creating a network of different paths. This is a convenient way to still use ordinary text editor table tools.

Basic use-cases can be expanded even more than shown here, but the risk is great that the square table concept limits the ability to express *complexity*. Instead, consider changing to graphic concepts introduced in the next chapter.

TABLE 7-33 A.RB.S. Simple dual-operator calculator basic use-case

A.RB.UN. Normal behavior requirement		A.RB.UX exception behavior requirement				
	ID, name and interaction		ID, name and interaction		ID, name and interaction		ID, name and interaction
A.RB.IA.11.	Input lead operand The user enters the lead operand by consecutively pressing digits 0-9, backspace, and decimal point.	A.RB.IA.12.	Too many digits entered Notify user by blinking. Reject digit.	A.RB.IA.13.	Second decimal point entered Don't notify user. Reject point.	A.RB.IA.14.	Backspace when 1 or fewer digits Don't notify user.
A.RB.IA.21.	Input dual operator The user breaks digit input with one of the dual operators +, -, * or /.						
A.RB.IA.31.	Input trail operand The user enters the trail operand by consecutively pressing digits 0-9, backspace and decimal point.	A.RB.IA.32.	Too many digits entered Notify user by blinking. Reject digit.	A.RB.IA.33.	Second decimal point entered Don't notify user. Reject point.	A.RB.IA.34.	Backspace when 1 or fewer digits Don't notify user.
A.RB.IA.41.	Input equal sign Conclude with =. Calculate value and show it. Proceed to input new lead operand.	A.RB.IA.42.	Divide by zero Show divide by zero instead of result. Expect new lead operand.	A.RB.IA.43.	Overflow after + or * Show value as floating-point instead.	A.RB.IA.44.	Lost significance after small / big Show result 0. Don't notify user.

An even more complete specification of this calculator will follow (see ch. 7.23.2, p. 243 below), but first there is a need to raise the degree of *formalism*.

7.17 Requirements formalism

7.17.1 Cpdm definition

Interaction requirements, *use-units* and *basic use-cases* usually become snippets, which occur in ever growing numbers. When they still remain separate from each other, it becomes difficult to understand how they connect or how they fork out. The junctions between snippets may be *black-box wait-states* (or just *wait-state*), which are defined according to Table 7-34 below.

TABLE 7-34 Cpdm definition

Aspect	Black-box wait-state
Definition	A black-box being in stand-by mode, awaiting random stimuli on any of its interfaces or from internal time-out
Synonyms	Usage-state, black-box-state, stand-by-state
Examples	Idle, busy, await action, stand by for use
Symbol	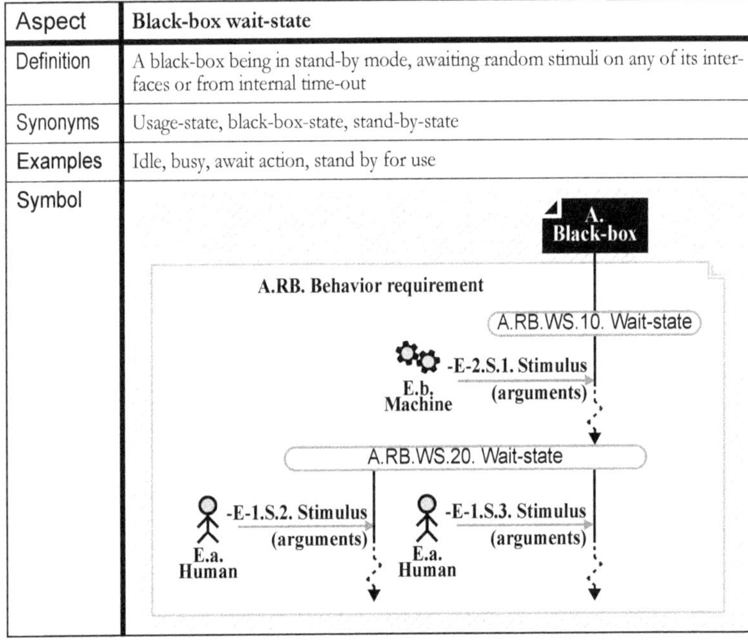

A *black-box wait-state* must be observable outside the *black-box*. When the *black-box* is in this *wait-state*, the *black-box* doesn't do anything else than wait for the next *stimulus* from outside the *black-box*, in order to continue the desired *black-box* behavior.

A black-box is standing by in a wait-state, until a new stimulus arrives on its interfaces.

Observe that when opening a *black-box* to a *white-box*, a lot more *states* become visible than just the *black-box wait-states*. *Programs* may contain thousands of *variables*, making their total *state* combinations almost infinite. When the *black-box* is opened to be designed into a *white-box*, one particular *variable* may well be designed to represent and remember the *black-box wait-states*.

Stimuli arrive on two *interfaces* to the *black-box*, but these are not drawn separately in Table 7-34 above. The *stimuli* already contain the *interface* identity, but the *interfaces* can be drawn separately from the *black-box*, if needed for clarity.

7.17.2 Cpdm definition

Now it is possible to define *interaction requirement* a bit more formally. In *Cpdm*, *use-unit* is defined according to Table 7-35 below.

TABLE 7-35 Cpdm definition

Aspect	Use-unit
Definition	One stimulus arriving on an interface to a black-box being in a certain wait-state, causing zero, one, or more responses to leave the black-box before it enters its next wait-state
Synonyms	Transition, interaction, interchange, exchange, changeover, shift, ping
Symbol	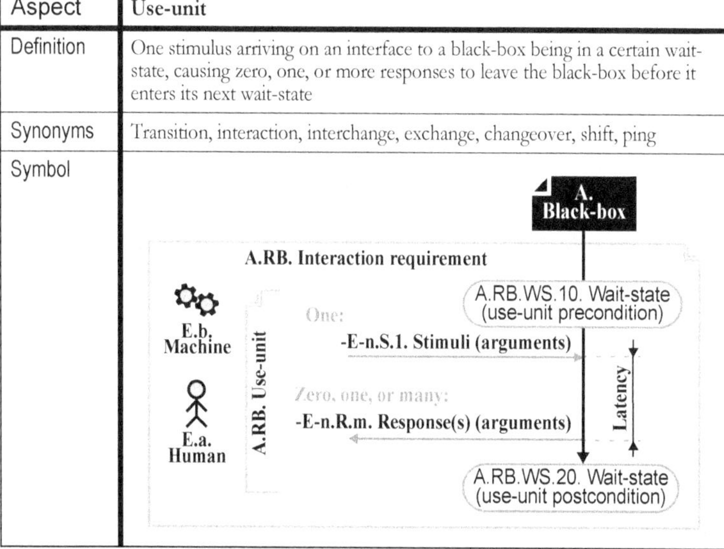

A *user* is not necessarily a human as opposed to a man-machine *interface* (MMI); it might be another *product* connected over a machine-machine *interface* to this *product*. Observe that the *stimulus* and succeeding *responses* within a *use-unit* can even occur over different *interfaces*.

A special aspect of the *black-box interface* behavior might easily be specified, namely, the latency caused by the *black-box* execution time from incoming *stimulus* to outgoing *response* (Table 7-35 above).

7.17.3 Cpdm definition

One more thing is needed to complete the *formalism* of *behavior requirements*. In *Cpdm*, *stimuli history* is defined according to Table 7-36 below.

TABLE 7-36 Cpdm definition

Aspect	Stimuli history
Definition	Referable history of all stimuli with their arguments received on interfaces of a black-box since it started up
Synonyms	Stimuli-response history, event chain, sequence of input
Symbol example	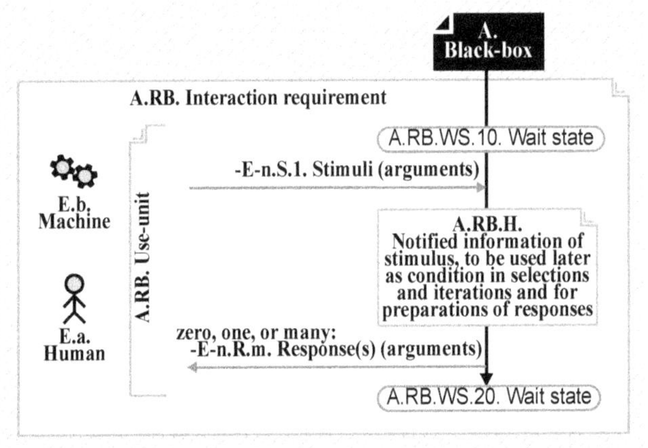

7.17.4 About stimuli history

To calculate any responses, the "only" thing necessary to know is the stimuli history.

When calculating *responses* and their arguments before sending them from the *black-box*, any *stimuli* from the history might be used for this preparation. There is obviously a need to keep track of and retrieve *stimuli* to the *black-box* that have occurred in the past.

Note that *stimuli history* may only refer to a *stimulus* that has been received over a *connection interface* to the *black-box* somewhere in the past, and not to *elements* or *interfaces* hidden somewhere internally in the *black-box*. For example, asking the *black-box* to remember a *stimulus* that arrived in the past is design-free, compared to reading a certain *variable* that refers to a design *element* with an implied and unexplained history.

To assign *variables* to remember the *stimuli history* would be to break into the *black-box* and design it, which is not a good idea when capturing *requirements* (that would cause them to become *designers*). But *stimuli history* is a smart way not to design memory *elements* to remember the history, but just to note the history itself, and this is called to express and document the *stimuli history*.

It must, in fact, be recognized that there is some design influence from using *black-box wait-states*, which may bias specification of behavior, but this is insignificant in the whole. Each *black-box wait-state* can be translated to a *stimuli history*, which is absolutely design-free, but such a *stimuli history* gets long and tricky, and does not outweigh the comprehensiveness of *black-box wait-states*.

The use of a *black-box wait-state* to represent a lot of *stimuli history* significantly simplifies the *complexity* of *behavior requirements*. *Wait-states* are not really necessary and the *clean-room engineering* specifies *requirements* entirely with *stimuli history*. For example, the *wait-state* A.RB.WS.10. Wait-for-lead-opd-1st-digit in Figure 7-28, page 247, has the *stimuli history* "the most recent *stimulus* has been eq-sign, or calc_logic has received no *stimulus* since start-up." For orthodox stimuli history usage, see Clean Room references, such as [5], [6] and [2].

7.17.5 Restriction requirements instead of stimuli history

Not all information is brought to the *black-box* as *stimuli*; some is chosen to be hard-wired into the *black-boxes*. This is a restriction to the *black-box*, and can be taken care of as a *stimuli history* that is forever present in the *black-box* (see Fig. 7-15, right).

FIGURE 7-15 Last resort from stimuli history

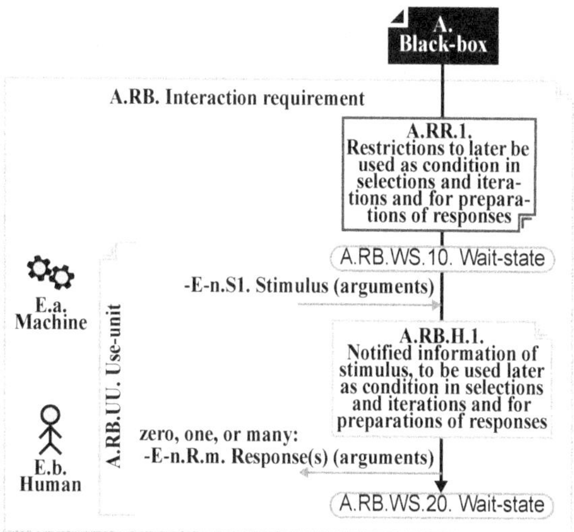

7.18 Formal use-case requirements

Formal use-cases are similar to basic use-cases, but can describe more complex behavior.

Formal use-cases are rather similar to *basic use-cases*, but constitute a further step-up in *formalism* (Table 7-37 below). Now each single *stimulus* and *response* to the *black-box* is explicitly illustrated and *black-box wait-states* are mandatory. To make the specification easier to understand, *stimuli history*, collected from *stimuli*, can also optionally be shown, in order to later explain *complex responses*.

A *use-unit* (where the number of *responses* may be zero, one, or many) is the smallest possible *use-case*, and can be seen as a *use-case* atom.

TABLE 7-37 Cpdm definition

Aspect	Formal use-case
Definition	A sequence of use-units interacting on interfaces to a black-box, starting from and ending in an explicit wait-state, with all responses in between explicitly calculated from specified stimuli histories
Synonyms	Procedure, case, event, stimulus-response sequence, method, test-case
Symbols example (table)	*See TABLE 7-38 below*
Symbols example (graphic)	*See graphic below*

TABLE 7-38 A.RB. Interaction requirement

A. Black-box

A.RB.UC Formal use-case

A.RB.WS.10. Wait-state (use-case precondition)

		Use-unit name and description	-E-1.S-1. Stimulus (arguments)		
E.a. Human	A.RB.UU.11.	Use-unit name and description	-E-1.S-1. Stimulus (arguments)		

A.RB.WS.20. Wait-state

E.b. Machine	A.RB.UU.21.	Use-unit name and description	-E-2.S-2. Stimulus (arguments)	-E-2.R3. Response (arguments)	A.RB.H.1. History

A.RB.WS.30. Wait-state

E.a. Human	A.RB.UU.31.	Use-unit name and description	-E-1.S-4. Stimulus (arguments)	-E-1.R5. Response -E-1.R6. Response	A.RB.H.2. History

RB.WS.40. Wait-state (use-case postcondition)

Symbols example (graphic)

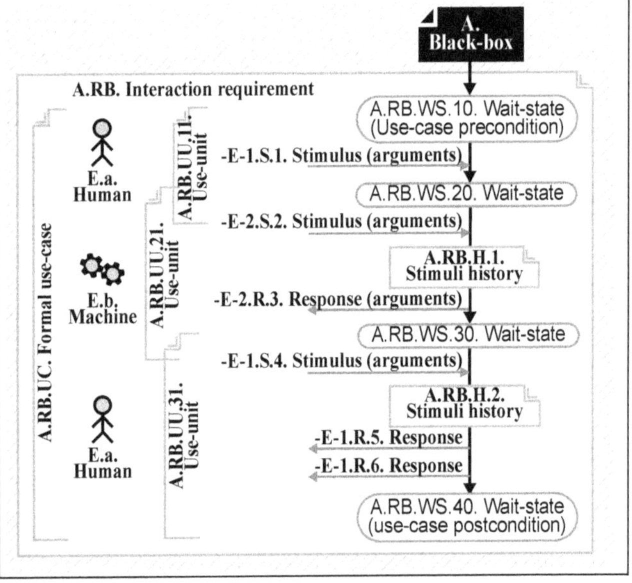

When capturing *behavior requirements*, it is powerful to arrange *black-box stimuli* and *responses* into *use-units*. Furthermore, one or many *use-units* may be linked into a sequence, which makes up a *use-case*. A *formal use-case* can also fork back and forth to *exception use-cases*.

Behavior of a *black-box* always begins in such a way that the *black-box* is in a *stand-by wait-state* and a *stimulus* is expected to arrive from a *user*. This is referred to as the *black-box* being stationed in a *precondition* (see Table 7-37 above). After a *stimulus*, the *black-box* might be preparing a *response* to the *user* (based on older *stimuli* and resulting *responses*). After possible *responses*, the *black-box* enters the next *wait-state* and again waits for a *stimulus* from the *user*.

After some sequences of *use-units*, the *black-box* enters a *postcondition* and the *use-case* has ended. Typically, this *postcondition* becomes the *precondition* for a new *use-case* to follow. Alternating pre- and postconditions are named *black-box wait-states* and will be more elaborated on later.

7.18.1 EXAMPLE: Bistro visit normal use-case

Now examples will illustrate how usage of a house can be specified in *formal use-case requirement* charts.

The house that was previously specified in this chapter, is imagined to be bought by a restaurant owner and will be used as a bistro.

The first *use-case requirement* is the A.RB.UN. Bistro visit normal use-case (see Fig. 7-16).

Note that this *formal use-case* is only one unbranched sequence of *use-units*, and in reality, behavior is more *complex* than that. The example below introduces how to cope with more *complexity*.

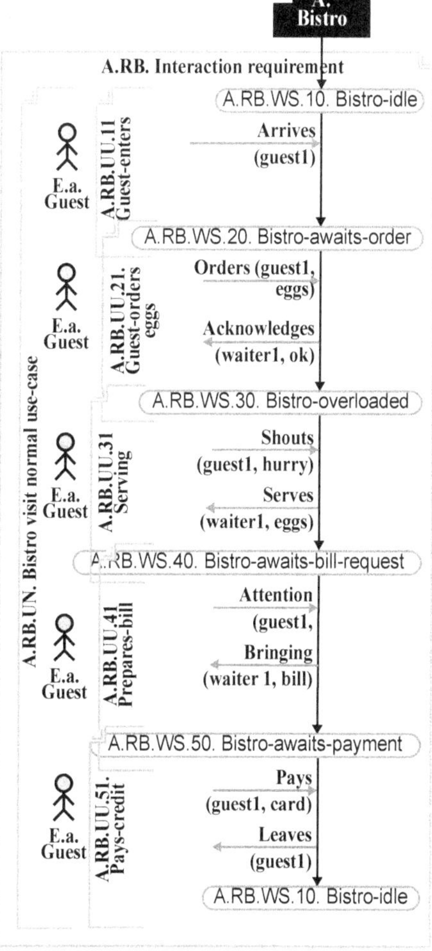

FIGURE 7-16 Bistro visit normal use-case

7.19 Scenario behavior requirements

7.19.1 Cpdm definition

In *Cpdm, formal scenario* is defined according to Table 7-39 below.

TABLE 7-39 Cpdm definition

Aspect	Formal scenario
Definition	A network of formal use-cases, assembled by attaching their wait-states
Synonyms	Course of events, event network, event scenario, episode network, episode scenario, behavior scenario, functional scenario
Symbol example	This definition is also applicable to basic use-cases (see ch. 7.16.4, "EXAMPLE: Simple dual-operator calculator basic use-case with exceptions," p. 223).
Symbol example	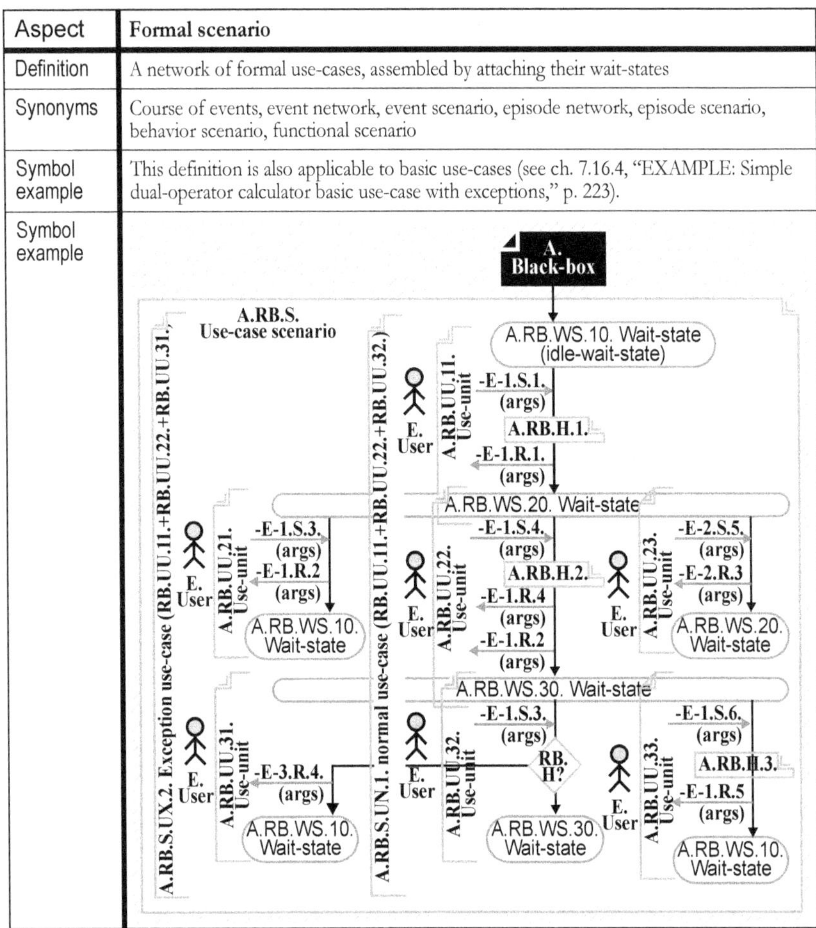

Note that returning to a previous *wait-state* does not require an explicit upward arrow. It is sufficient to just illustrate the same end *wait-state* by referring to the same exact *wait-state* identity.

Many *use-units* and *formal use-cases* can be added to get a *formal scenario*. Whenever many different *stimuli* arrive when the *black-box* is in a *wait-state*, the *formal scenarios* will fork to one of these *use-units*. The most common paths through the *scenario* form the most normal *use-cases* with specific identities and naming.

Most *formal scenarios* are specified with some kind of *idle wait-state* or *stand-by wait-state* as a starting point for the *scenario* to spread out from, and finally return to (see the A.RB.WS.10. Wait-state in Table 7-39 above). It is very common for *users* to learn to recognize the *idle wait-state*, and find it natural to start interacting with a *product* from there, and await to return to it sooner or later.

7.19.2 EXAMPLE: Visit bistro formal scenario

More *use-units* can now be added to the A.RB.UN. Bistro visit normal use-case in Figure 7-16 above (see the A.RB.S. Visit bistro formal scenario in Fig. 7-17 below).

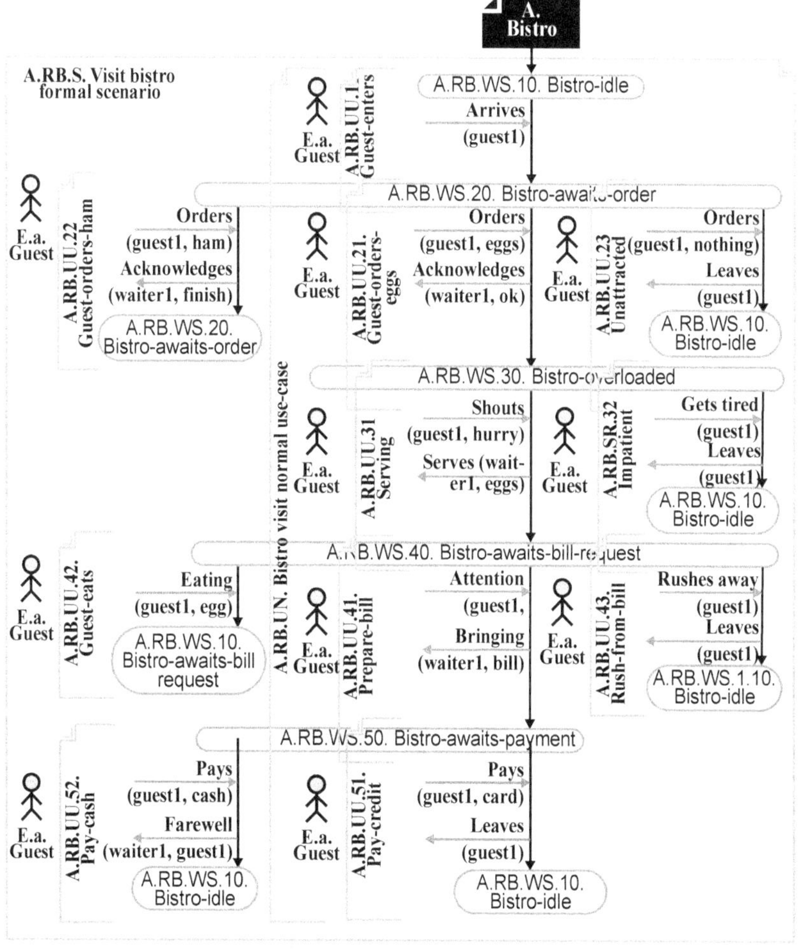

FIGURE 7-17 Visit bistro formal scenario

As can be seen in Figure 7-17 above, the original A.RB.UN. Bistro visit normal use-case from Figure 7-16 remains intact. This *use-case* has been extended with both *use-units* being parts of *normal use-cases*, as well as *use-units* being part of *exception use-cases*. In a *formal scenario* there are a lot of paths along arriving *stimuli*; some are *exception use-case* paths, while other *use-case* paths are broken by exceptions.

7.20 Wait-state-machine

7.20.1 Cpdm definition

In *Cpdm*, *wait-state-machine* is defined according to Table 7-40 below.

TABLE 7-40 Cpdm definition

Aspect	Wait-state-machine
Definition	Modeling black-box behavior by stimuli-initiated transitions between wait-states
Synonyms	Automat, transition system
Symbols	A. Black-box · A.RB.SM. Wait-state-machine · A.RB.UU. Use-unit · A.RB.WS.10 Wait-state · A.RB.UU.4 Use-unit4 · A.RB.UU. Use-unit2 · A.RB.UU.3 Use-unit3 · A.RB.WS.30 Wait-state · A.RB.WS.20 Wait-state · A.RB.UU. Use-unit

A *wait-state-machine* is very similar to a *formal scenario*. The graphic notation for a *wait-state-machine* is somewhat different and simplified; for example, *stimuli* and *responses* used to be omitted. Sometimes *stimuli* are used to mark out the transitions instead of *use-units*. In *wait-state-machine* descriptions, each *wait-state* occurs only once, contrary to *formal scenarios description*, where the very same *wait-state* can be duplicated.

7.20.2 EXAMPLE: Breakfast wait-state-machine requirements

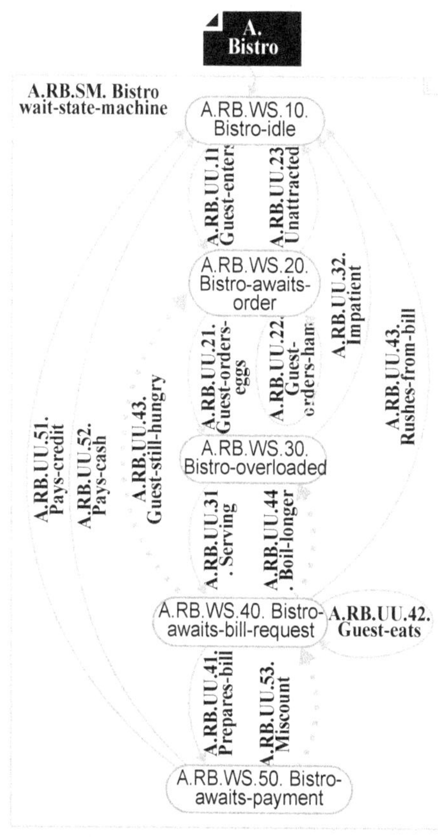

FIGURE 7-18 Bistro wait-state-machine

A *formal scenario* such as above A.RB.S. Visit bistro formal scenario in Figure 7-17 is very simple to transform into A.RB.SM. Bistro wait-state-machine (see Fig. 7-18, left).

Making this transformation can be quite rewarding, because some new *use-units* not imagined before can be easily detected, such as A.RB.SR.32 Impatient, when the guest gets no attention while waiting for the food.

A *wait-state-machine* is often said to be too detailed and mathematical to be applied during stressful development. In many simple and uncomplicated parts of the *requirement* database, *wait-state-machine* may be overkill, but in some *complex* parts they are unbeatable for understanding. Also, the close relationship to *formal scenario* lowers the barrier to capture them. Note that *white-box variable states* should never populate a *wait-state-machine*, but only the *black-box wait-states*.

7.20.3 Number of black-box wait-states

The trick is to minimize the number of wait-states, but not to the degree of introducing if-branches.

The trick to minimize *black-box wait-state complexity* and facilitate *verification* (especially *usage profile verification*), is to find the minimum number of *black-box wait-states*, but not so few that if-branches show up when capturing the transitions. For example, if there are three different *stimuli* that are very similar, a parameter can instead be used in a generic *stimulus*, like *stimuli* with a digit parameter instead of many separate digit *stimuli*.

To illustrate optimizing the number of *black-box wait-states*, the rudimentary calculator specified in Table 7-32, page 222, will be shown as single-*wait-state-*, dual-

wait-state-, quadruple-*wait-state-*, and overkill-*wait-state-machines*. Note that the *use-cases* below omit illustrating any *responses*.

7.20.4 EXAMPLE: Single-wait-state rudimentary calculator

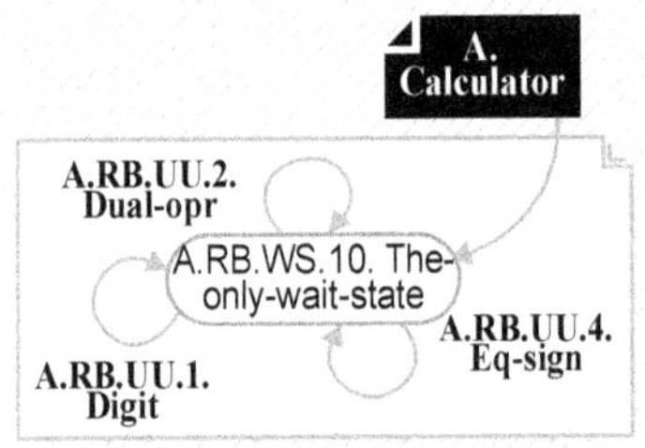

The *wait-state-machine* in Figure 7-19 is simple, but contains no substantial information. Since only one *black-box wait-state* is represented, this could be seen as a no-state-machine.

FIGURE 7-19 Rudimentary calculator single-wait-state-machine

The *formal scenario* accordingly contains only one single *black-box wait-state* (see Fig. 7-20). A need now arises to introduce many if-branches, making the *formal scenarios* hard to read and comprehend, despite the fact that it is a clear picture.

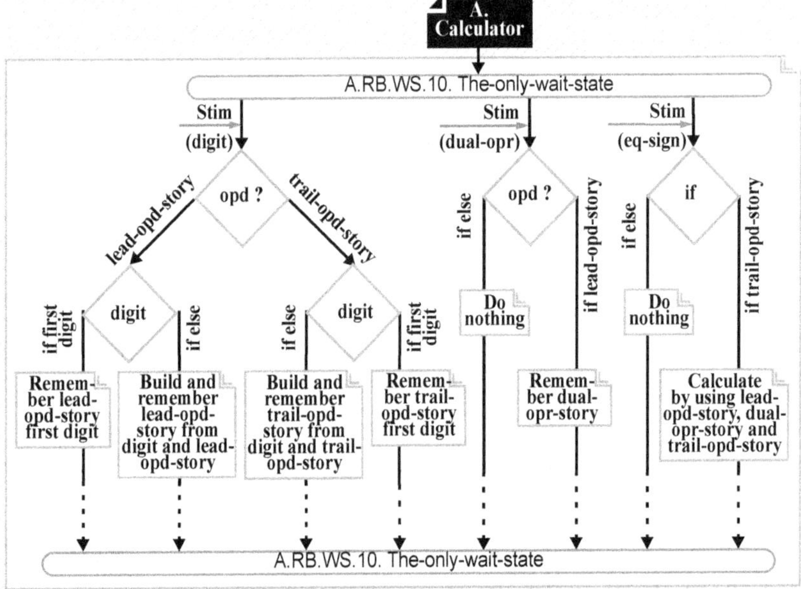

FIGURE 7-20 Rudimentary calculator single-wait-state scenario

It is not completely explained in the above figure how to know whether digit *stimuli* are building lead-opd-story or trail-opd-story, which require some extra *stimuli history* to be collected.

7.20.5 EXAMPLE: Dual-wait-state rudimentary calculator

The next attempt to improve on the previous example is to use two *black-box wait-states* instead of only one.

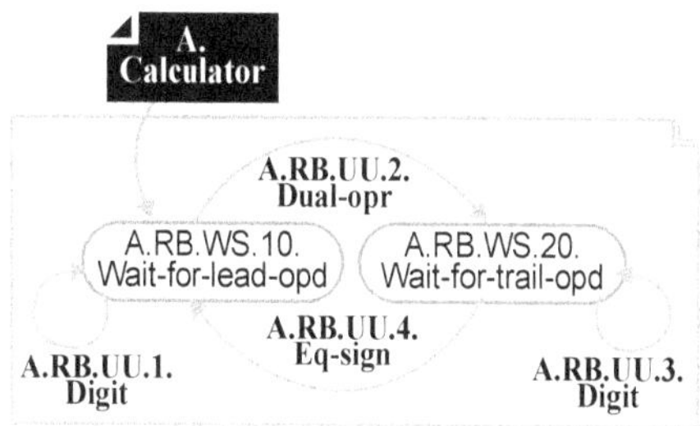

It is not surprising for a dual-operator pocket calculator to also imagine two *black-box wait-states* when using it. One *black-box wait-state* occurs while waiting for the leading operand and the other while waiting for the trailing operand (see Fig. 7-21, left).

FIGURE 7-21 Rudimentary calculator dual-wait-state-machine

The *formal scenario* now is extended to two *black-box wait-states*, but still rather hard to read and comprehend, with some if-branches for testing *stimuli histories*.

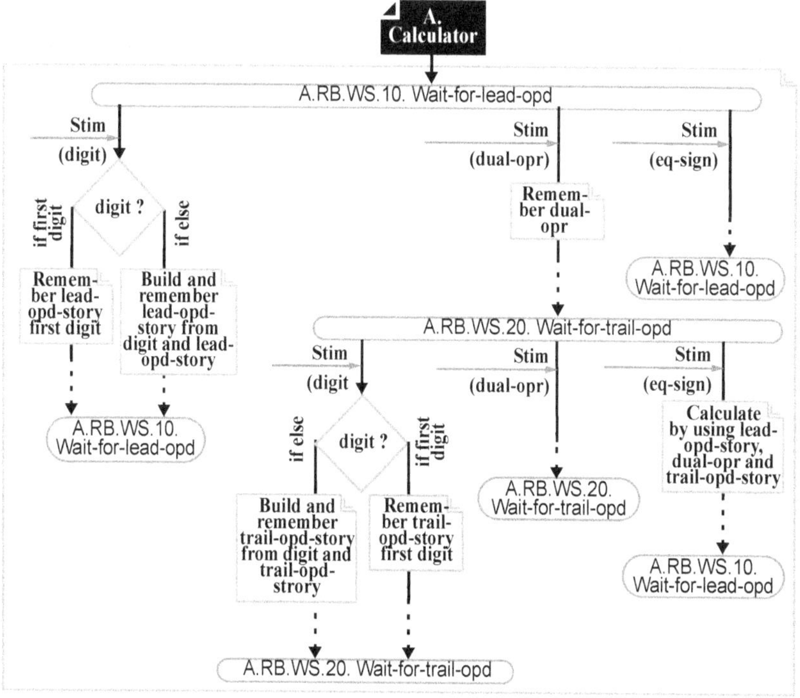

FIGURE 7-22 Rudimentary calculator dual-wait-state-scenario

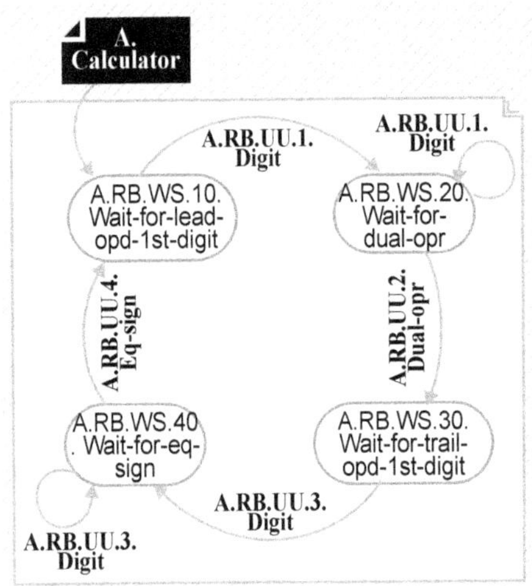

7.20.6 EXAMPLE: Quadruple-wait-state rudimentary calculator

It is also very plausible to identify four *black-box wait-states* when using a pocket calculator (see Fig. 7-23, left). For a corresponding *formal scenario* (yet a bit extended), see Figure 7-28, page 247. By using this *wait-state* model, all if-branches are eliminated, compared to Figure 7-20 above. So far, this seem to be the most comprehensive alternative.

FIGURE 7-23 Rudimentary calculator quadruple-wait-state-machine

7.20.7 EXAMPLE: Overkill-wait-state rudimentary calculator

For example, it is imaginable to use one *black-box wait-state* for each individual digit that is entered (see Fig. 7-24 below). Now the *wait-states* are obviously too many, and the illustration gets a lot of *artificial complexity*.

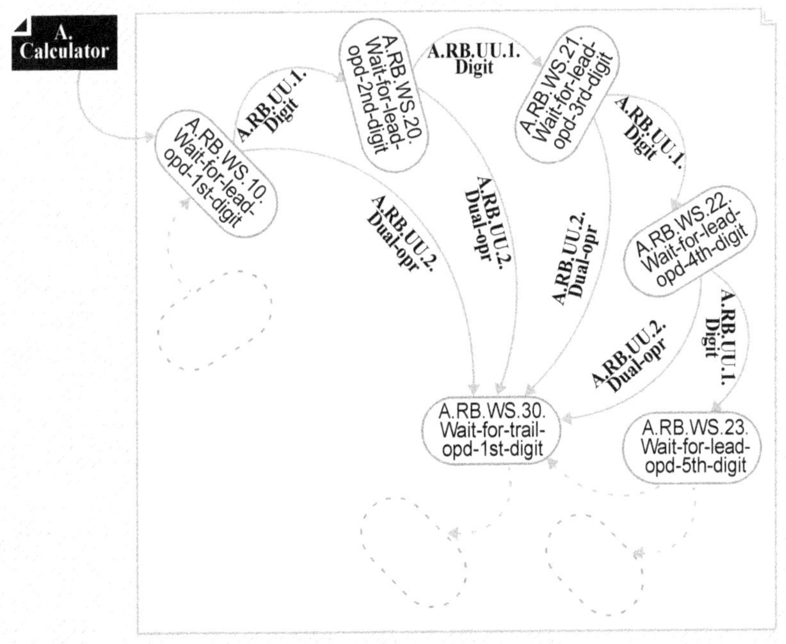

FIGURE 7-24 Rudimentary calculator overkill-wait-state-machine

7.21 EXAMPLE Calc_logic environment (the pocket calculator firmware): Explore environment restrictions and ensure environment staffing

7.21.1 Calc_logic: About this example

In this example a *program* subroutine is planned to be developed, performing mathematical calculations for calculator *program* surroundings.

Development of this calculator can be imagined to have started long before the technical development that now starts (see previous *activities* in ch. 6.6, p. 99). From these earlier *activities*, it is already assumed that two kinds of *products* shall be simultaneously developed:

- A calc_logic subroutine used by a real-time surrounding *program* in a pocket calculator (chapter numbers marked with P in Fig. 7-25 below). The technical development example of this *program* starts with chapter 7.21 below.
- A calc_logic subroutine used by an application *program* in Windows operating system (chapter numbers marked with W in Fig. 7-25 below). The technical development example of this *program* starts in chapter 7.22, page 240.

Since there will be two different surroundings to the calc_logic subroutine, one set of restrictions must be captured for each *environment*. These two restriction sets can then be merged when refining *requirements* for the *outermost black-box*.

The previous house and multiplication toy examples were provided with explicit references to *results* likely to be *demanded*. In remaining examples they are not provided, but they are rather easy to imagine by looking at the recurrent *tailored* overall *schedule*.

7.21.2 EXAMPLE: Calc_logic environment (the pocket calculator firmware): Process schedule to use

Pt:2:3. Tailored advanced technical schedule n = 2

```
           8.10 P  9.20 P   10.19 W  11.16 W
OR  SO     8.11 W  9.23 W   10.20 P  11.17 P
                                     11.18 P
7.21 P                                        12.21 W
7.22 W                                        12.22 P
                                              12.23 P


           8.12 R  9.24 R   10.18 R  11.14 H
7.23 R                                        12.20 H
```

For the calc_logic example, a special technical overall *schedule* is *tailored* (illustrated in Fig. 7-25, left):

- **Pt:2:3. Tailored advanced technical schedule n = 2,** page 169

In this *tailored schedule* the generic *schedule* to use is (filled-in symbols):

- **Pt.0.R. Explore environment restrictions and ensure environment staffing,** page 110.

FIGURE 7-25 Position of schedule and master to use (filled in) in overall schedule. Chapter numbers of the calc_logic example are also referenced.

Although this example starts with separate *activities*, the *program* will be developed as one common *component*, to be *reused* by the Windows and pocket calculator (chapter numbers marked with R in Fig. 7-25 above). The *reusable items* will be *realized* and verified in a *host development system* (chapter numbers marked with H in Fig. 7-25 above).

7.21.3 EXAMPLE Calc_logic environment (the pocket calculator firmware): Explore restrictions

Calc_logic is developed as a *program black-box*, to be *reused* by several calculator *products*. Thus, these calculators become the *environment* to this *black-box*. Typical restrictions in the stand-alone pocket *environments* are shown in Table 7-41 below. Note that the pocket calculator surrounding the calc_logic is developed in a separate *project*.

	TABLE 7-41 Calc_logic environment (the pocket calculator firmware) environment restriction requirement
	Restriction requirement
E. RR.1.	To fit market cost of an affordable everyday calculator, the number of buttons shall be no greater than 16.
E. RR.2.	To fit market cost of an affordable calculator, the stand-alone calculator display shall hold a maximum of 8 digits.
E. RR.3.	Cheap microcontrollers rarely have on-chip or external chip support for mathematical arithmetic.
E. RR.4.	Standard compilers to microcontrollers seldom have support for extensive mathematical arithmetic, but there are some microcontroller development systems with floating-point library support for float variables ($\approx$ 7 digits).
E. RR.5.	Mathematical program libraries for microprocessors can be developed in-house or outsourced, and for some microprocessors portable libraries are available from third-party shelves.
E. RR.6.	Calc_logic for reuse developed on a host system must be portable to development boards and the target pocket calculator.

7.21.4 EXAMPLE Calc_logic environment (the pocket calculator firmware): Plan development roles and providers

Based on the *tailored process schedule* **Pt:2:3. Tailored advanced technical schedule n = 2,** page 169, staffing for development *activities* of calc_logic *interfaces* to surrounding *environment* is now estimated (see Table 7-42 below).

TABLE 7-42 Calc_logic environment (the pocket calculator firmware) development allocation staffing

Calc_logic environment (the pocket calculator firmware) role item	Provider	Cpdm schedule or task
i10. **Calculator stakeholders** • To capture environment restrictions	PR.10.In-house stakeholder group	Pt.0.Ra. Explore restrictions.
i11. **Computer mathematician** • To support stakeholders to capture restrictions	PR.11.University researcher	
i12. **"Dg3" project leader** • "Dg3" is the name of the project to develop calc_logic	PR.12.In-house project "dg3"	Pt.0.Rb. Plan development roles and providers.
i13. **External embedded designer**	PR.13.Micro Hard Inc.	Pt.0.Pa. Try to satisfy restrictions.
i14. **External programs purchaser** • To investigate available sources for compilers and mathematical libraries for pocket calculator	PR.14.Hardware & Co	Pt.0.Pc. Analyze in/out and make/buy options.
i15. **Chief architect** • To support reusable architecture between the projects	PR.15.In-house common resource for projects "de1", "dw2" and "dg3"	Pt.0.A. Constitute environment architecture.
i16. **"Dg3" project leader** • To partition the work and responsibilities between the pocket calculator project "de1" and "dg3".	PR.12. In-house project "dg3"	
i17. **"Dg3" programmer**	PR.12. In-house project "dg3"	Pt.0.Fa. Finalize environment interfaces detailed items.
i18. **"Dg3" Project engineer**	PR.12. In-house project "dg3"	Pt.0.Fc. Account for environment interfaces development, realization, and manufacturing.
i19. **Development board specialist assistance** • To support porting calc_logic to development board	PR.16.Chk&Go	Pt.0.I. Realize interfaces to environment and insert/await outermost white-box.
i20. **Microcontroller developer** • To support porting calc_logic to target stand-alone pocket calculator	PR.17.In-house project "de1"	
i21. **Development board specialist assistance** • To support verifying calc_logic in development board environment	PR.16. Chk&Go	Pt.0.V. Verify prototype in environment.
i22. **Microcontroller developer** • To support verifying calc_logic in the stand-alone pocket calculator environment	PR.17. In-house project "de1"	

It can be hard to get the estimate precise at this early *phase*, but it is important to come up with an allocation plan to secure engineers immediately.

7.21.5 EXAMPLE Calc_logic for reuse: Proceed reading

To follow this example, these are the alternatives:

- Proceed to chapter 7.22, "EXAMPLE Calc_logic environment (the Windows calculator application): Explore environment restrictions and ensure environment staffing" below
- Proceed to chapter 7.23, "EXAMPLE Calc_logic for reuse: Refine product requirements from environment restriction requirements and ensure black-/white-box development staffing," page 243.
- Proceed to chapter 8.10, "EXAMPLE Calc_logic environment (the pocket calculator firmware): Demarcate product by environment solutions with possible supplier opportunities," page 319.
- Proceed to chapter 9.20, "EXAMPLE Calc_logic environment (the pocket calculator firmware): Constitute environment architecture," page 408.
- Proceed to chapter 12.22, "EXAMPLE Calc_logic environment (the pocket calculator firmware) on development board: Verify prototype in environment," page 761.

7.22 EXAMPLE Calc_logic environment (the Windows calculator application): Explore environment restrictions and ensure environment staffing

7.22.1 EXAMPLE: Calc_logic environment (the Windows calculator application): Process schedule to use

Pt:2:3. Tailored advanced technical schedule n = 2

For the calc_logic example, a special technical overall *schedule* is *tailored* (illustrated in Fig. 7-26, left):

- **Pt:2:3. Tailored advanced technical schedule n = 2**, page 169

In this *tailored schedule* the generic *schedule* to use is (filled-in symbols):

- **Pt.0.R. Explore environment restrictions and ensure environment staffing**, page 110.

FIGURE 7-26 Position of schedule and master to use (filled in) in overall schedule

7.22.2 EXAMPLE Calc_logic environment (the Windows calculator application): Explore restrictions

As for calc_logic for the pocket calculator, it is now time to identify restrictions in the calc_logic Window *environment* (see Table 7-43 below).

TABLE 7-43 Calc_logic environment (the Windows calculator application) environment restriction requirement

	Restriction requirement
E. RR.7.	Since a window in Microsoft Windows can hold almost any number of buttons, a large number of buttons shall be possible stimuli for calc_logic.
E. RR.8.	Since a window field in Microsoft Windows can show almost any number of alphanumeric characters, the floating-point precision in calc_logic shall be as high as possible.
E. RR.9.	If a math coprocessor is available in the computer, it may be used to heighten calculation speed. However, the calculating program shall also function in computers without a math coprocessor.
E. RR.10.	There are a lot of compilers for Windows, with a lot of different kinds of mathematical calculation support. Standard Windows C development systems with library supports have float ($\approx$ 7 digits), double-float ($\approx$ 15 digits) and long double-float ($\approx$ 31 digits) precision. However, long double-float variables most often are reduced to the precision of double variables. As Windows 64-bits version gets more popular, in the near future many compilers will probably offer no undiminished support for long double-floating-point ($\approx$ 31 digits) precision.
E. RR.11.	Many kinds of supplier mathematical libraries are available or portable to Windows.
E. RR.12.	Calc_logic for reuse developed on a host system must be portable to the target Windows disk and input/output system.

7.22.3 EXAMPLE Calc_logic environment (the Windows calculator application): Plan development roles and providers

Performing the *environment* work for calc_logic in Windows is not very different from performing the work for calc_logic in the stand-alone pocket calculator (see Table 7-44 below).

TABLE 7-44 Calc_logic environment (the Windows calculator application) development allocation staffing

Calc_logic environment (the Windows calculator application) role item	Provider	Cpdm schedule or task
i23. **Calculator stakeholders** • To capture Calc_logic environment (the Windows calculator application) restrictions	**PR.10. In-house stakeholder group**	Pt.0.Ra. Explore restrictions.
i24. **Computer mathematician** • To support stakeholders to capture restrictions	**PR.11. University researcher**	

TABLE 7-44 Calc_logic environment (the Windows calculator application) development allocation staffing

Calc_logic environment (the Windows calculator application) role item	Provider	Cpdm schedule or task
i25. **"Dg3" project leader** • "Dg3" is the name of the project to develop calc_logic	**PR.12. In-house project "dg3"**	Pt.0.Rb. Plan development roles and providers.
i26. **"Dg3" project designer**	**PR.12. In-house project "dg3"**	Pt.0.Pa. Try to satisfy restrictions.
i27. **External programs purchaser** • To investigate available sources for compilers and mathematical libraries for Windows calculator	**PR.14. Hardware & Co**	Pt.0.Pc. Analyze in/out and make/buy options.
i28. **Common chief architect** • To support reusable architecture between the projects	**PR.15. In-house common resource for projects "de1", "dw2" and "dg3"**	Pt.0.A. Constitute environment architecture.
i29. **"Dg3" project leader** • To partition the work and responsibilities between the Windows application project "dw2" and "dg3".	**PR.12. In-house project "dg3"**	
i30. **"Dg3" Windows programmer**	**PR.12. In-house project "dg3"**	Pt.0.Fa. Finalize environment interfaces detailed items.
i31. **"Dg3" project engineer**	**PR.12. In-house project "dg3"**	Pt.0.Fc. Account for environment interfaces development, realization, and manufacturing.
i32. **Developer assistance** • To support porting calc_logic to target Windows environment	**PR.18.In-house project "dw2"**	Pt.0.I. Realize interfaces to environment and insert/await outermost white-box.
i33. **Developer assistance** • To support verifying calc_logic in the Windows environment	**PR.18. In-house project "dw2"**	Pt.0.V. Verify prototype in environment.

7.22.4 EXAMPLE Calc_logic for reuse: Proceed reading

To follow this example, these are the alternatives:

• Proceed to chapter 7.21, "EXAMPLE Calc_logic environment (the pocket calculator firmware): Explore environment restrictions and ensure environment staffing," page 237.

• Proceed to chapter 7.23, "EXAMPLE Calc_logic for reuse: Refine product requirements from environment restriction requirements and ensure black-/white-box development staffing" below.

• Proceed to chapter 8.11, "EXAMPLE Calc_logic environment (the Windows calculator application): Demarcate product by environment solutions with possible supplier opportunities," page 323.

- Proceed to chapter 9.23, "EXAMPLE Calc_logic environment (the Windows calculator application): Constitute environment architecture," page 415.
- Proceed to chapter 12.21, "EXAMPLE Calc_logic environment (the Windows calculator application): Verify prototype in target environment," page 758.

7.23 EXAMPLE Calc_logic for reuse: Refine product requirements from environment restriction requirements and ensure black-/white-box development staffing

7.23.1 EXAMPLE: Calc_logic for reuse: Process schedule to use

For the calc_logic example, a special technical overall *schedule* is *tailored* (illustrated in Fig. 7-27, left):

- **Pt:2:3. Tailored advanced technical schedule n = 2,** page 169

In this *tailored schedule* the generic *schedule* to use is (filled-in symbols):

- **Pt.n.R. Refine requirements from outward nesting level requirements and ensure black-/white-box development staffing,** page 130.

Pt:2:3. Tailored advanced technical schedule n = 2

```
          8.10 P   9.20 P   10.19 W   11.16 W
          8.11 W   9.23 W   10.20 P   11.17 P
                                      11.18 P
7.21 P                                12.21 W
7.22 W                                12.22 P
                                      12.23 P

1R IS     8.12 R   9.24 R   10.18 R   11.14 H
7.23 R                                12.20 H
```

FIGURE 7-27 Position of schedule and master to use (filled in) in overall schedule

7.23.2 EXAMPLE Calc_logic for reuse: Refine and capture requirements

It is now time to merge the pocket calculator *environment restriction requirements* and the Windows calculator *environment restriction requirements*, and refine these into the common calc_logic *requirements*. When all restrictions are analyzed and these *requirements* are captured, it is also time to capture if anything else is desired from the calc_logic for reuse.

Both sets of calculator *environment restriction requirements* are now refined to get the *requirements* for the outermost calc_logic *black-box* (see Table 7-45 below).

TABLE 7-45 Calc_logic for reuse restriction requirement		
Restriction requirement		**Priority**
E. RR.1.	To fit market cost of an affordable everyday calculator, the number of buttons shall be no greater than 16.	
E. RR.7.	Since a window in Microsoft Windows can hold almost any number of buttons, a large number of buttons shall be possible stimuli for calc_-logic.	
A. RR.1.1.	**The calc_logic element shall be able to process 16-button stimuli at a minimum and be capable of processing many more future button stimuli.**	**Critical**
E. RR.2.	To fit market cost of an affordable calculator, the stand-alone calculator display shall hold a maximum of 8 digits.	
E. RR.8.	Since a window field in Microsoft Windows can show almost any number of alphanumeric characters, the floating-point precision in calc_logic shall be as high as possible.	
A. RR.2.1.	**The calc_logic element shall be able to process 8-digit responses, and be capable of processing a much larger future number of digits.**	**Critical**
E. RR.3.	Cheap microcontrollers rarely have on-chip or external chip support for mathematical arithmetic.	
E. RR.9.	If a math coprocessor is available in the computer, it may be used to heighten calculation speed. However, the calculating program shall also function in computers without a math coprocessor.	
A. RR.3.1.	**To reduce complexity from the unpredictable existence of coprocessor, math coprocessors shall not be used at all.**	**Medium**
E. RR.4.	Standard compilers to microcontrollers seldom have support for extensive mathematical arithmetic, but there are some microcontroller development systems with floating-point library support for float variables ($\approx$ 7 digits).	
E. RR.10.	There are a lot of compilers for Windows, with a lot of different kinds of mathematical calculation support. Standard Windows C development systems with library supports have float ($\approx$ 7 digits), double-float ($\approx$ 15 digits) and long double-float ($\approx$ 31 digits) precision. However, long double-float variables most often are reduced to the precision of double variables. As Windows 64-bits version gets more popular, in the near future many compilers will probably offer no undiminished support for long double-floating-point ($\approx$ 31 digits) precision.	
A. RR.4.1.	**A maximum calculation precision of $\approx$ 15 digits during host development of the calc_logic program is sufficient.**	**High**
E. RR.5.	Mathematical program libraries for microprocessors can be developed in-house or outsourced, and for some microprocessors portable libraries are available from third-party shelves.	
E. RR.6.	Calc_logic for reuse developed on a host system must be portable to development boards and the target pocket calculator.	
E. RR.11.	Many kinds of supplier mathematical libraries are available or portable to Windows.	
E. RR.12.	Calc_logic for reuse developed on a host system must be portable to the target Windows disk and input/output system.	
Pcb.1.	Calc_logic shall always be reused, and never exist in many versions.	
A. RR.5.1.	**Calc_logic for reuse shall be portable to the Windows calculator application, to development boards, and to the pocket stand-alone calculator.**	**Critical**

The calc_logic restrictions are now complemented with *behavior requirements* (see Table 7-45 below).

TABLE 7-46 Calc_logic for reuse black-box behavior requirements

	Behavior requirement use-case	Priority
RB.6.	Calc_logic for reuse behavior requirement no 6	
A. RB.6.S1.	**Calc_logic for reuse, behavior requirement 6, scenario 1** (see Fig. 7-28, p. 247)	
A. RB.6.S1.U1.	**Lead-operand-input normal use-case** sequence of RB.UU.11. + n * RB.UU.21.	Critical
A. RB.6.S1.U2.	**Trail-operand-input normal use-case** sequence of RB.UU.31. + n * RB.UU.41.	Critical
A. RB.6.S1.U3.	**Conclusion normal use-case** sequence of RB.UU.11. + n * RB.UU.21. + RB.UU.22. + RB.UU.31. + n * RB.UU.41. + RB.UU.42.	Critical
A. RB.6.S1.U4.	**Repeating normal use-case** sequence of RB.UU.11. + n * RB.UU.21. + RB.UU.22. + RB.UU.31. + n * RB.UU.41. + RB.UU.43. + ...	High
A. RB.6.S1.U5.	**Repeating-after-conclusion normal use-case** sequence of RB.UU.12. + RB.UU.31. + ...	Critical

The graphic *formal scenario* for the calculator subroutine can be seen in Figure 7-28, page 247, below. Observe that despite the presence of many structures in this graphic describing behavior, there is yet nothing referring to design inside the *black-box*.

Note that the *user* of calc_logic is the surrounding *program*, which performs *subroutine calls* or *structured function calls* to calc_logic. The *user* of the calculator is located *outwards* in the *environment* to calc_logic (see Fig. 9-54, p. 410 and Fig. 9-58, p. 417). Thus, the *stimuli* to and *responses* from calc_logic are named "stim" and "resp" even if they derive from the *user* pressing a calculator button and reading the calculator display.

For *consistency*, it might be a good idea to also prepare a specification for all constituents of the *formal use-case* (see Table 7-47 below). The constituents can be used by all *behavior requirements*, and are not bound to *requirement* number RB.6.

TABLE 7-47 Calc_logic for reuse black-box behavior requirements

Behavior requirement wait-states, use-units, and histories

A. RB. calc_logic for reuse behavior requirement	
A. RB.WS.10.	Wait-for-lead-opd-1st-digit
A. RB.WS.20.	Wait-for-further-digit-or-dual-opr
A. RB.WS.30.	Wait-for-trail-opd-1st-digit
A. RB.WS.40.	Wait-for-further-digit-or-eq-sign
A. RB.UU.01.	Start up
A. RB.UU.11.	Input-1st-digit-of-lead-opd
A. RB.UU.12.	Prepare-continuation-after-conclusion
A. RB.UU.21.	Input-rest-of-lead-opd
A. RB.UU.22.	Input-dual-opr
A. RB.UU.31.	Input-1st-digit-of-trail-opd
A. RB.UU.41.	Input-rest-of-trail-opd
A. RB.UU.42.	Do-conclusion
A. RB.UU.43.	Do-conclusion-with-continuation
A. RB.H.0.	lead-opd-story = 0
A. RB.H.1.	Lead-opd-story = digit
A. RB.H.2.	Expand lead-opd-story rightwards with digit
A. RB.H.3.	Prev-dual-opr-story = dual-opr
A. RB.H.4.	Trail-opd-story = digit
A. RB.H.5.	Expand trail-opd-story rightwards with digit
A. RB.H.6.	Manipulate lead-opd-story by trail-opd-story according to prev-dual-opr-story

Stimuli and *responses* are also good to capture (see Table 7-48 below). Note that *stimuli* are bound to the *black-box connection interfaces* (like -E-1.).

TABLE 7-48 Calc_logic for reuse black-box behavior requirements

Behavior requirement at interface -E-1.

-E-1. Common calc_logic interface	
-E-1. S.1.	Stim (digit)
-E-1. S.2.	Stim (dual-opr)
-E-1. S.3.	Stim (eq-sign)
-E-1. R.4.	Resp (lead-opd-story)
-E-1. R.5.	Resp (trail-opd-story)

Now all specified constituents can be assembled to the full *formal scenarios* 1 of *behavior requirement* 6 (see Fig. 7-28 below). Again note that it is not the *user* who directly uses calc_logic by sending *stimuli* and receiving *responses*, but it is the *program* surrounding calc_logic, which are respectively E.ab. Button interrupt thread (Table 9-30, p. 409). and E.adb. Case WM_COMMAND (see Table 9-32, p. 416).

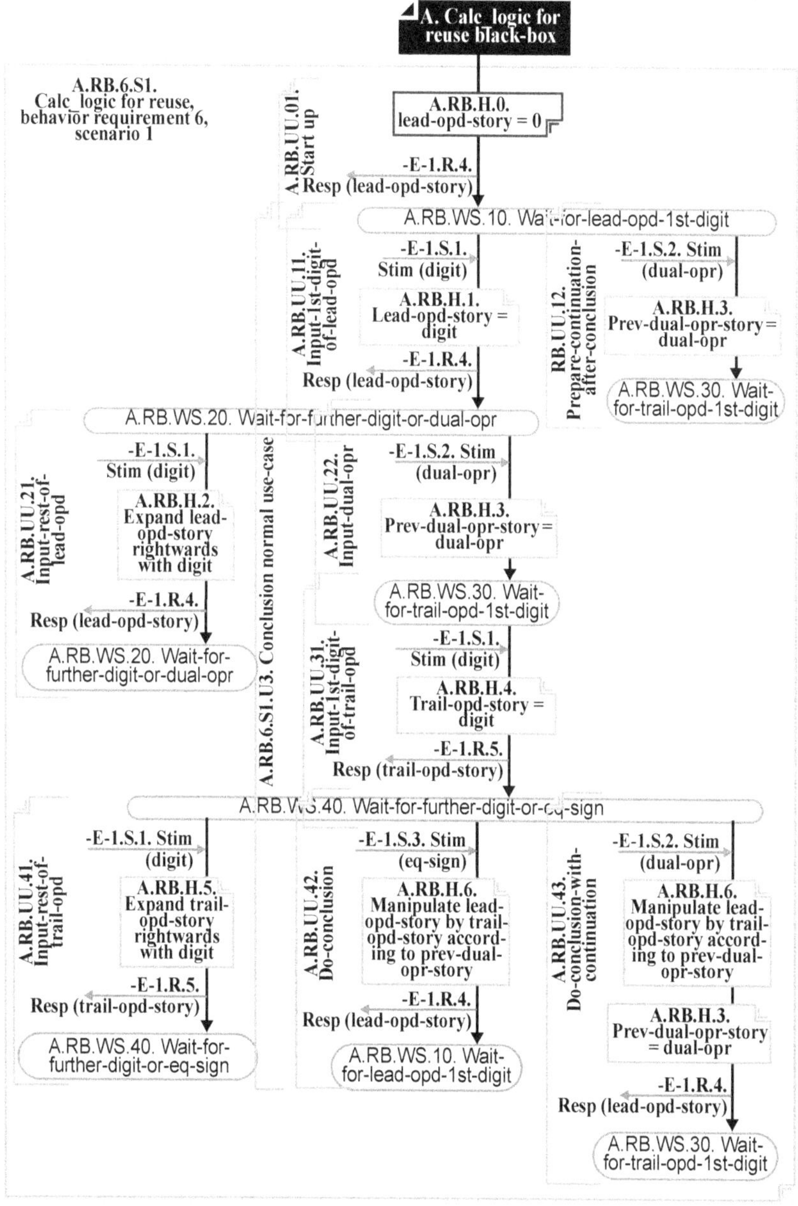

FIGURE 7-28 Calc_logic for reuse, behavior requirement 6, scenario 1

7.23.3 EXTRA EXAMPLE Calc_logic for reuse: Test-use-cases to check requirements

Another advantage of *formal scenarios* is that they are very easy to simulate in the mind; just make a *test-use-case* like Table 7-49 below. More of this will follow in chapter "Verify black- & white-box," page 659.

TABLE 7-49 Calculator test-use-cases

Stim	Lead-opd-story	Prev-dual-opr-story	Trail-opd-story	Resp	Use-unit	Use case
Start of A.	A.RB.H.0. = 0			-E-1.R.4. Resp (0)	A.RB.UU.01. Start up	A.RB.6.S1.U4. Repeating normal use-case
A.RB.WS.10. Wait-for-lead-opd-1st-digit						
-E-1.S.1. Stim (2)	A.RB.H.1. = 2			-E-1.R.4. Resp (2)	A.RB.UU.11. Input-1st-digit-of-lead-opd	
A.RB.WS.20. Wait-for-further-digit-or-dual-opr						
-E-1.S.1. Stim (4)	A.RB.H.2. = 10 * 2 + 4			-E-1.R.4. Resp (24)	A.RB.UU.21. Input-rest-of-lead-opd	
A.RB.WS.20. Wait-for-further-digit-or-dual-opr						
-E-1.S.2. Stim (-)	24	A.RB.H.3. = -			A.RB.UU.22. Input-dual-opr	
A.RB.WS.30. Wait-for-trail-opd-1st-digit						
-E-1.S.1. Stim (6)	24	-	A.RB.H.4. = 6	-E-1.R.5. Resp (6)	A.RB.UU.31. Input-1st-digit-of-trail-opd	
A.RB.WS.40. Wait-for-further-digit-or-eq-sign						
-E-1.S.2. Stim (+)	A.RB.H.6. = 24 - 6	A.RB.H.3. = + 6		-E-1.R.4. Resp (18)	A.RB.UU.43. Do-conclusion-with-continuation	
A.RB.WS.30. Wait-for-trail-opd-1st-digit						
-E-1.S.1. Stim (4)	18	+	A.RB.H.4. = 4	-E-1.R.5. Resp (4)	A.RB.UU.31. Input-1st-digit-of-trail-opd	A.RB.6.S1.U3. Conclusion normal use-case
A.RB.WS.40. Wait-for-further-digit-or-eq-sign						
-E-1.S.3. Stim (=)	A.RB.H.6. = 18 + 4	+	4	-E-1.R.4. Resp (22)	A.RB.UU.42. Do-conclusion	
A.RB.WS.10. Wait-for-lead-opd-1st-digit						
-E-1.S.2. Stim (/)	22	A.RB.H.3. = / 4			A.RB.UU.12. Prepare-continuation-after-conclusion	A.RB.6.S1.U5. Repeating-after-conclusion normal use-case
A.RB.WS.30. Wait-for-trail-opd-1st-digit						
-E-1.S.1. Stim (2)	22	/	A.RB.H.4. = 2	-E-1.R.5. Resp (2)	A.RB.UU.31. Input-1st-digit-of-trail-opd	
A.RB.WS.40. Wait-for-further-digit-or-eq-sign						
-E-1.S.3. Stim (=)	A.RB.H.6. = 22 / 2	/	2	-E-1.R.4. Resp (11)	A.RB.UU.42. Do-conclusion	
A.RB.WS.10. Wait-for-lead-opd-1st-digit						

7.23.4 EXAMPLE Calc_logic for reuse: Prioritize requirements

An extra column for prioritization was inserted in Table 7-45, p. 244 and Table 7-46, p. 245. *Stakeholders* filled in degree of prioritization for all newly refined *requirements*, and no major problems was encountered with this.

7.23.5 EXAMPLE Calc_logic for reuse: Examine authenticity

All members of the decision gate were asked to review the *requirements*, and all reported this was a true pleasure. It became obvious that a simple calculator logic could become quite *complex*, and in great need of advanced *requirement* methodology to finally become crystal clear.

The restrictions for the calc_logic *environment* are also valid for the Windows calculator application developed by in-house project "dw2". The gate decision was that the same chief *architect* would be the in-house common resource for projects "de1", "dw2" and "dg3", to ensure that calc_logic would become compatible.

The meeting participants decided to introduce this *requirement* methodology all over the company, and to buy and distribute books on *Cpdm* :-).

7.23.6 EXAMPLE Calc_logic for reuse: Plan development roles and providers

The in-house project "dg3" *project leader* prepared the needed staff allocation. Most desired in-house competencies were free to allocate in the desired period, but a few must be allocated from external *providers*.

TABLE 7-50 Calc_logic for reuse development allocation staffing

	Calc_logic for reuse role item	Provider	Cpdm schedule or task
i34.	"Dg3" project specifier	PR.10. In-house stakeholder group	Pt.n.Ra. Refine and capture requirements.
i35.	"Dg3" project leader	PR.12. In-house project "dg3"	Pt.n.Re. Plan development roles and providers.
i36.	Chief architect • To support reusable architecture between the projects	PR.15. In-house common re-source for pro-jects "de1", "dw2" and "dg3"	Pt.n.Pa. Try to satisfy require-ments. Pt.1.A. Satisfy refined require-ments by decomposing product black-box into white-box design containing no black-boxes.
i37.	"Dg3" project programmer • To prepare the development sys-tem • To write the C program for calc_-logic	PR.12. In-house project "dg3"	Pt.n.Fa. Finalize white-box detailed items.
i38.	"Dg3" project engineer • To prepare all product requisites	PR.12. In-house project "dg3"	Pt.n.Fc. Account for white-box development, realization, and manufacturing.

TABLE 7-50 Calc_logic for reuse development allocation staffing

	Calc_logic for reuse role item	Provider	Cpdm schedule or task
i39.	**"Dg3" project programmer** • To compile, link, and start calc_logic in the Windows host development system	**PR.12. In-house project "dg3"**	Pt.1.I. Procure items and linkable programs in order to realize white-box.
i40. i41.	**Junior programmer** • To prepare fixture for automated verification **Senior programmer** • To prepare efficient verification methods	**PR.16. Chk&Go**	Pt.1.V. Verify black-/white-box.
i42.	**"Dg3" project verifier** • To prepare automatic stimuli test-cases and perform the calc_logic verification in Windows host debugger	**PR.12. In-house project "dg3"**	

7.23.7 EXAMPLE Calc_logic for reuse: Proceed reading

To follow this example, these are the alternatives:

- Proceed to chapter 8.12, "EXAMPLE Calc_logic for reuse: Predetermine solutions with sourcing options," page 326.
- Proceed to chapter 9.24, "EXAMPLE Calc_logic for reuse: Satisfy refined requirements by decomposing product black-box into white-box design containing no black-boxes," page 419.
- Proceed to chapter 12.20, "EXAMPLE Hosted calc_logic for reuse: Verify black-/white-box," page 733.
- Proceed to chapter 7.24 below to leave the calc_logic example and to dig deeper into the capture staffing & requirements topic.

7.24 EXAMPLE Formal and complete requirements for calculator

It is quite possible to capture a full and complete calculator in one *requirement* graph of four parts (see Fig. 7-29 to Fig. 7-32 below).

The model for this specification is the Microsoft 7 Calculator, which is included as an application within the Windows 7 operating system. Note that only numeric *responses* in its standard *view* are specified. The *formalism* with identifiers in this example is also lowered a bit because of limited space in the graphs below.

It takes a day or two of intense work to capture such *requirements* as below. It can be hard to find the right *stimuli histories*, but the nice thing is that when finished, this specification is 100% clear and complete. When testing the finished calculator *product* and when writing the *user's* manual, there will be no lack of clarity.

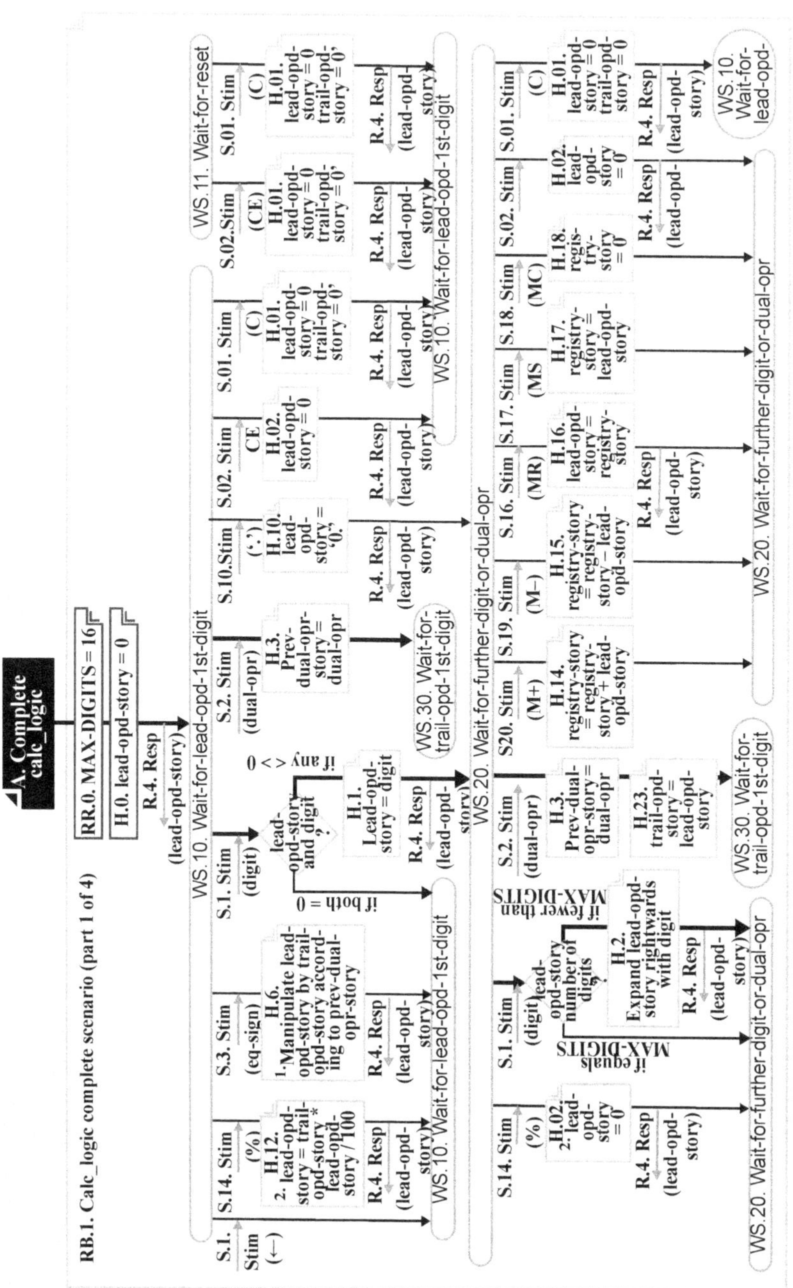

FIGURE 7-29 Calculator complete scenario requirements (part 1 of 4).

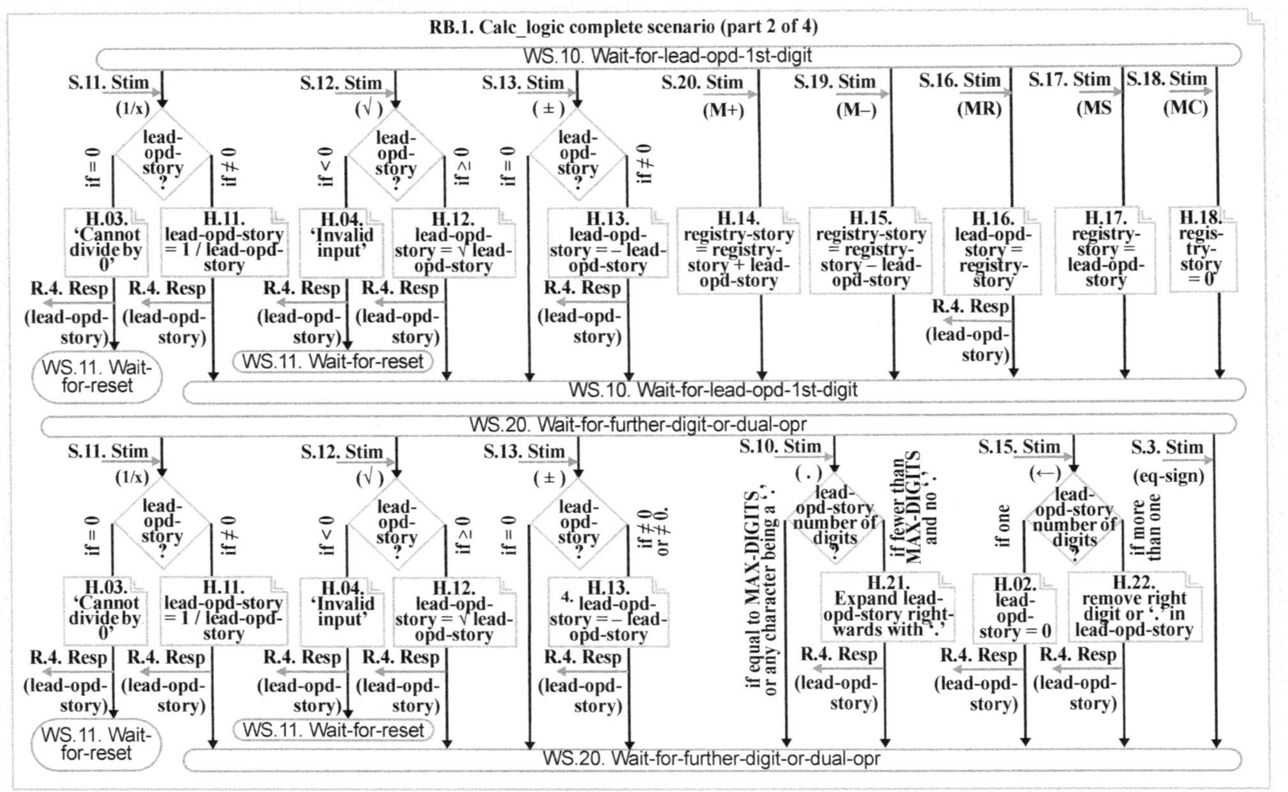

FIGURE 7-30 Calculator complete scenario requirements (part 2 of 4).

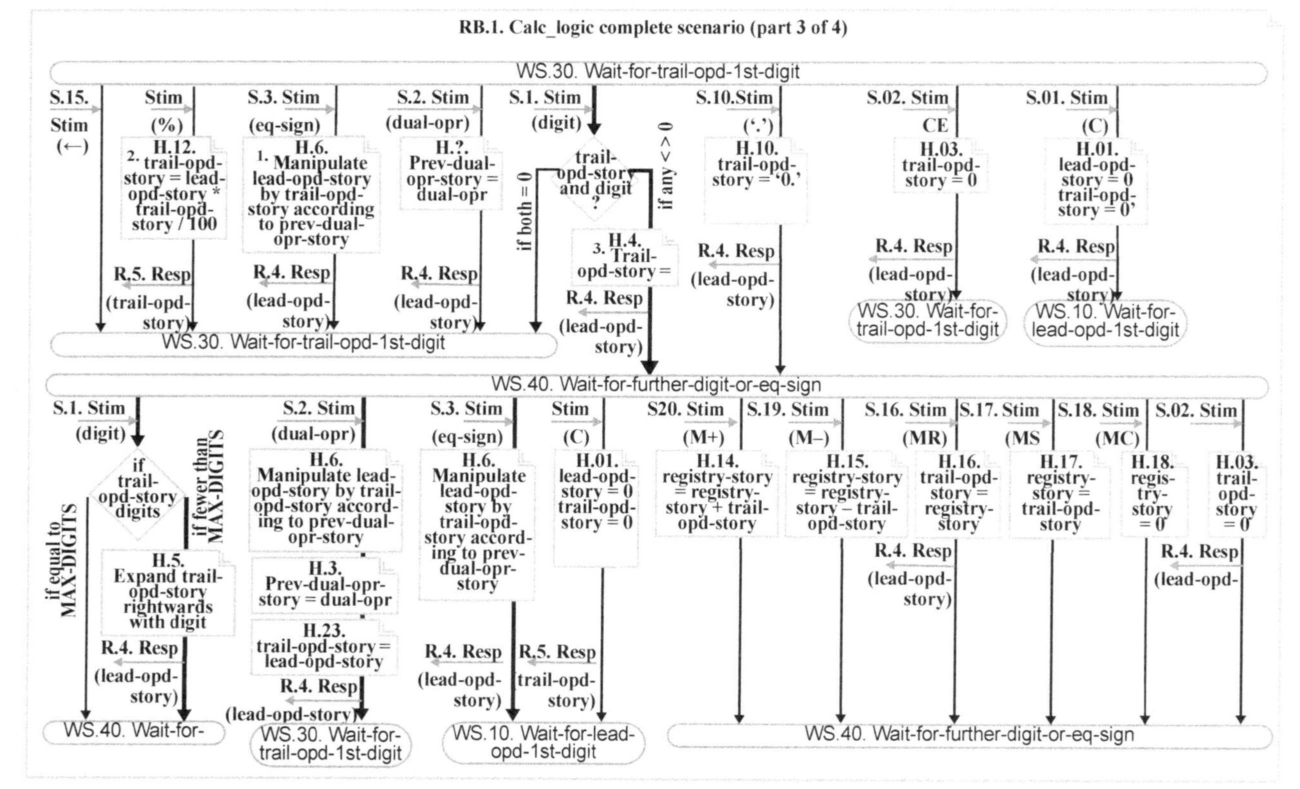

FIGURE 7-31 Calculator complete scenario requirements (part 3 of 4).

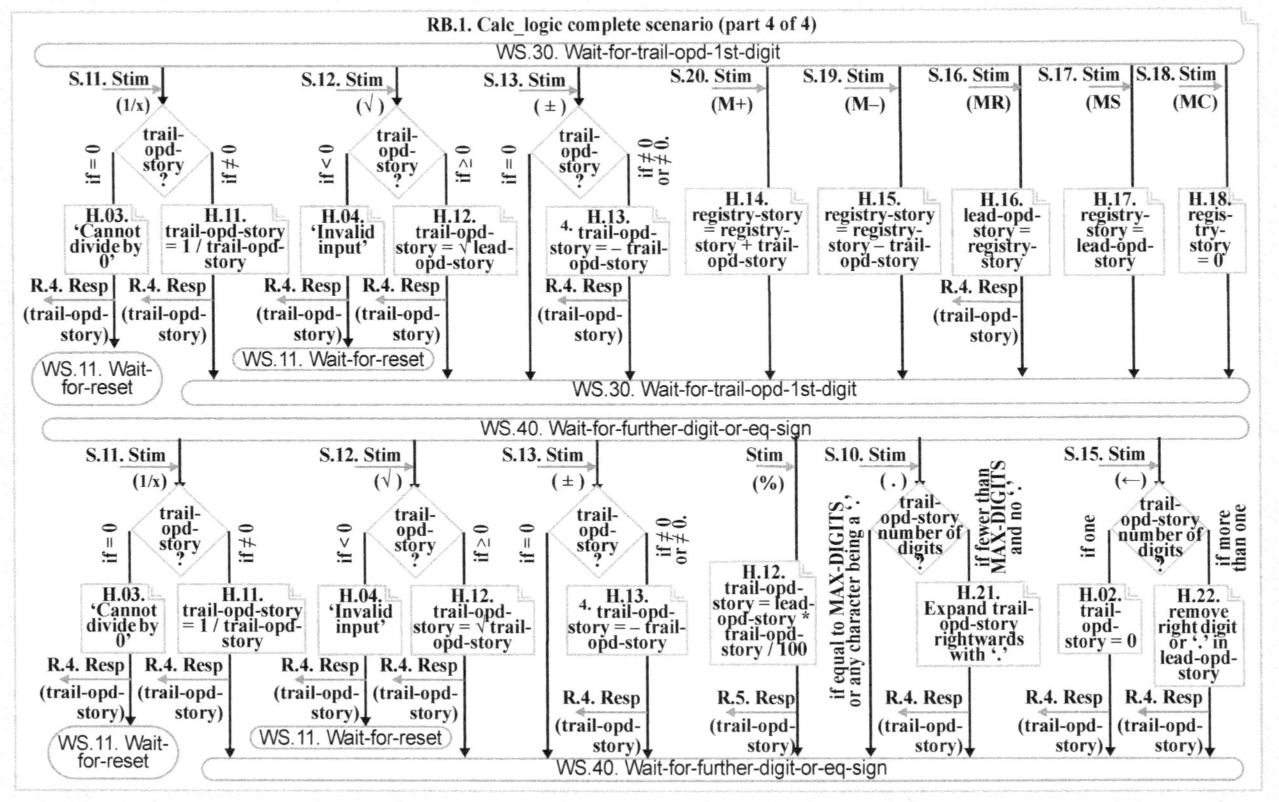

FIGURE 7-32 Calculator complete scenario requirements (part 4 of 4).

A disclaimer by me may be in order here. Possibly some minor failures might have found their way into this example. The trick to get everything right is most often to have it reviewed by highly meticulous persons (readers of this book :-). Since this is a book about development, not enough time has been spent on reviewing the examples to make them technically perfect (even if possible).

During capturing the *requirements*, some strange behaviors have been discovered in this calculator. It is very typical of Microsoft to offer too many possibilities to a *user*, which makes the calculator hazardous to use. With this kind of *formal scenarios* it has been very easy to prohibit or reject *stimuli* causing unnecessary and confusing behavior:

1. Repeats operation of last dual operator and operand, difficult to understand.
2. Incomprehensible in the context (no behavior to these *stimuli* would have been better).
3. By entering first trail digit, possible single operations on lead operand get lost.
4. If H.10. has been passed from WS.10. the '±' appends a leftward minus sign to 0.

7.25 Uncorrelated stimuli on different interfaces

Much *complexity* usually occurs when a *product* has *interfaces* with overflow of *stimuli* on one *interface*, or uncorrelated *stimuli* on many *interfaces*, which the *product* must be able to handle. The best way to illustrate this is a couple of examples.

7.25.1 EXAMPLE: Overflowing stimuli solved by parallelism

The previous bistro *scenarios* had only one single *customer* to take care of. In a profitable bistro, there should be a crowd of guests, and a big team of waiters to take care of them. Then one waiter can no longer follow each guest through the bistro visit, and wait for the guest to decide on the menu, eat the food, and retrieve money for payment.

One way to handle overbearing *stimuli* and not to risk missing a single one, is to allow any waiter in the team to make any service possible for the guests (see Fig. 7-33 below). This results in a very wide *scenario* with few *wait-states*, without leading into long branches with a lot of built-in waiting time.

Another observation that can be easily made, is that there are two distinct *subscenarios*, each of them being separate and cohesive, one for waiters and one for chefs. On a few occasions these two *scenarios* interact with each other (dashed arrows in Fig. 7-33 below). This indicates that the design solution most probably needs at least two separate *threads* of the type *scenario-thread-machines* with asynchronous *interfaces* in between. A possible design solution for these *requirements* can be found in Figure 9-50, page 404.

Since these *use-cases* are mostly to show principles, they are a bit less formal than the previously shown examples.

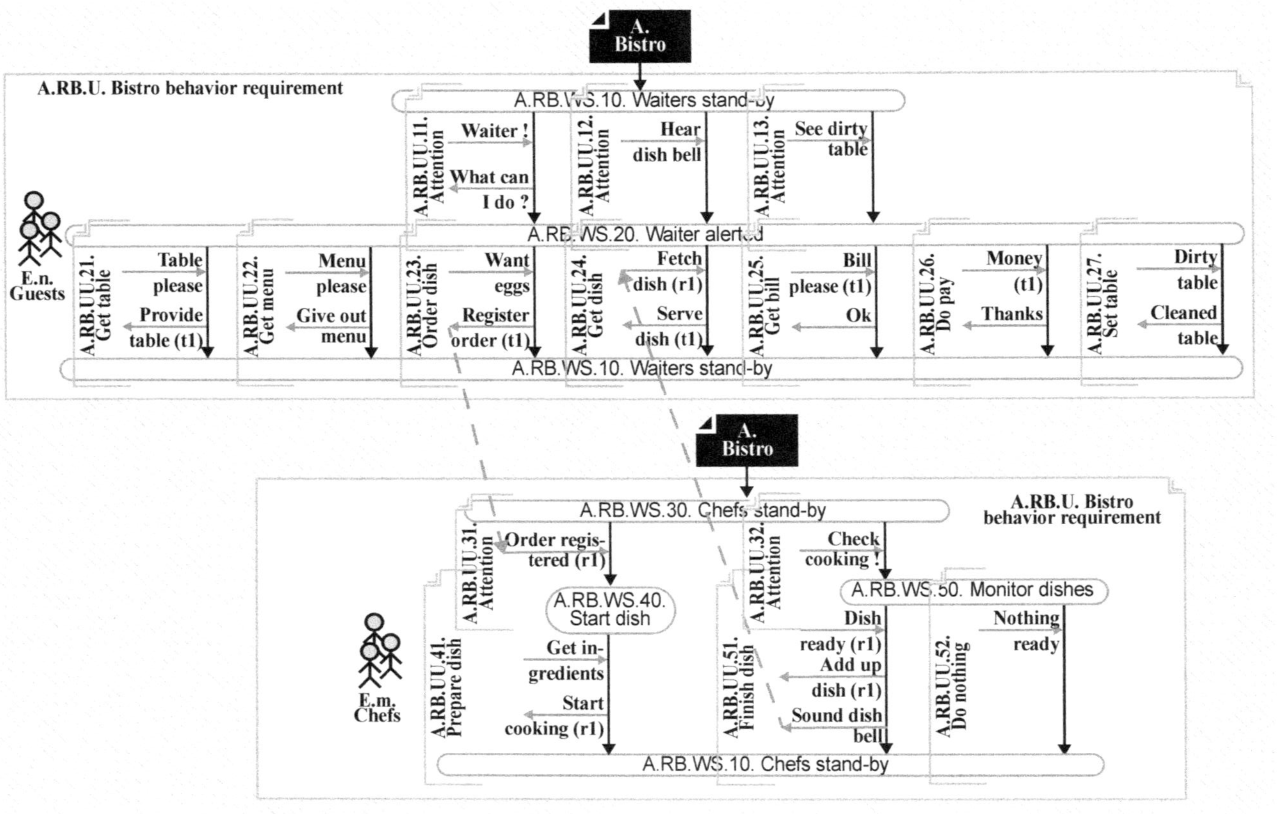

FIGURE 7-33 Instantiate waiter and chef to avoid waiting queues

7.25.2 EXAMPLE: Uncorrelated stimuli solved by busy handling

A telephone switch of the same age as the telephone in the picture on the right is a large *product* with a lot of mechanics, electronics, and *programs*.

Nevertheless, by using a *formal scenarios* diagram as in Figure 7-34, page 258 below, it is really simple to explain the principles of this old-fashioned switch (whose principles are not very different from those of a switch for modern cell phones). The *formal scenario* specifies the functionality for each subscriber, thus the switch (including the subscriber telephone device itself) executes as many instances as there are subscribers.

Specified *use-units* can describe both a subscriber initiating a call, and a subscriber being called. The beauty in using the graphic *formal scenarios* is that they capture caller *use-units* in one instance, creating *responses* which become *stimuli* in other called *use-units* of another instance, which both are described by the same *formal scenario* (so to say, the *formal scenario* speaks for itself).

Possible connections between the caller and called subscriber are:

- If there is no subscriber with the number dialed so far, the dialing may continue.
- If a called number exists and responds idle, the caller and called subscribers are connected.
- If a called subscriber exists but the called subscriber is busy with another call, there is no chance to serve the caller. The caller thus receives a busy tone and has to hang up and call again a bit later.

Since this *formal scenario* is mostly to show principles, it is a bit less formally specified than the previously shown examples.

For a possible principle solution of these *requirements* see Figure 9-51, page 405.

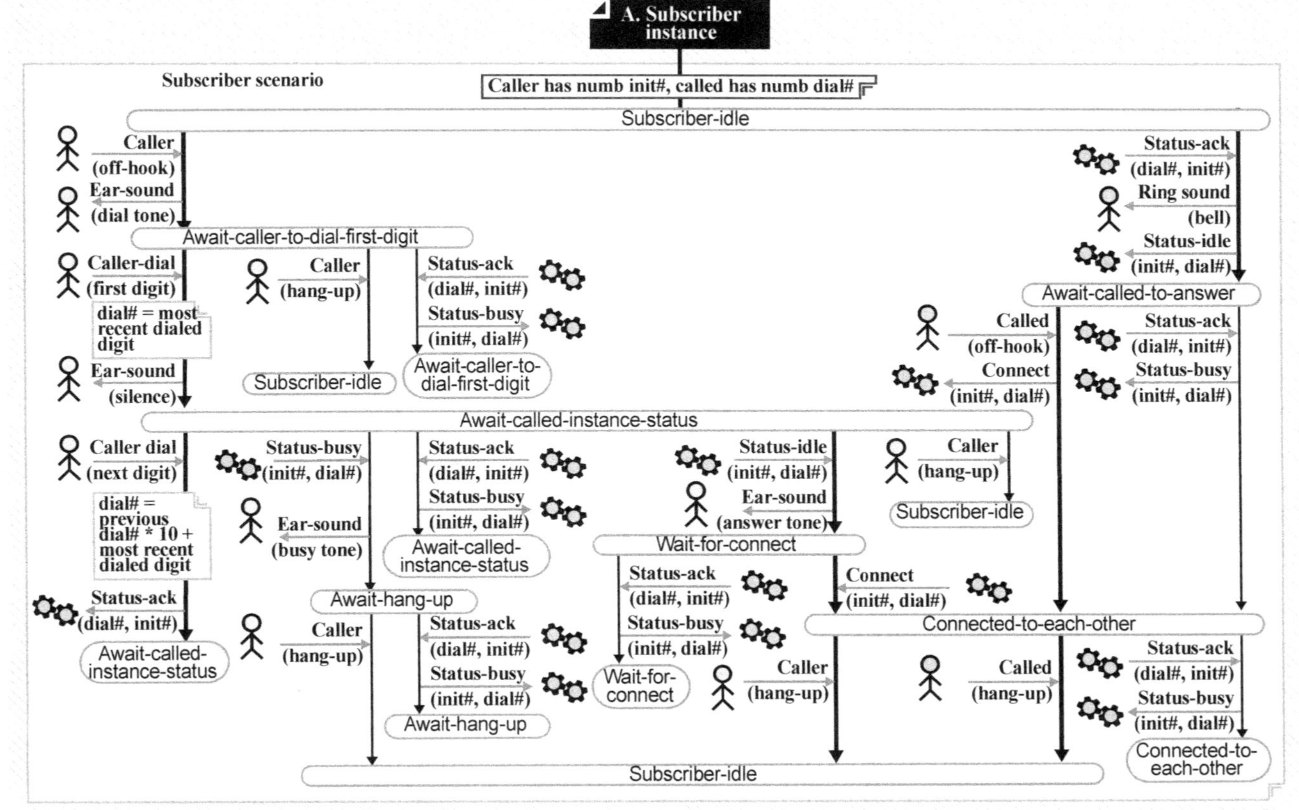

FIGURE 7-34 Old-fashioned telephone switch scenario

7.25.3 EXAMPLE Test-use-case to check telephone switch requirements

A good check of *requirements* is to simulate a *test-use-case* (see Table 7-51 below).

TABLE 7-51 A.RB. Subscriber test-use-case

<table>
<tr><th colspan="3">Caller has number init# = 67</th><th colspan="3">Called has number dial# = 52</th></tr>
<tr><th>Stimulus</th><th>Response</th><th>Use-unit name (not shown in scenario)</th><th>Stimulus</th><th>Response</th><th>Use-unit name (not shown in scenario)</th></tr>
<tr><td colspan="3">Subscriber-idle</td><td colspan="3"></td></tr>
<tr><td>Caller (off-hook)</td><td>Ear-sound (dial tone)</td><td>Caller-starts-call</td><td></td><td></td><td></td></tr>
<tr><td colspan="3">Await-caller-to-dial-first-digit</td><td colspan="3"></td></tr>
<tr><td>Caller-dial (5)</td><td>Ear-sound (silence)</td><td>Caller-dials-first-digit</td><td></td><td></td><td></td></tr>
<tr><td colspan="3">Await-called-instance-status</td><td colspan="3"></td></tr>
<tr><td>Caller dial (2)</td><td>Status-ack (52, 67)</td><td>Caller-dials-rest-of-digits</td><td></td><td></td><td></td></tr>
<tr><td colspan="3">Await-called-instance-status</td><td colspan="3">Subscriber-idle</td></tr>
<tr><td></td><td></td><td></td><td>Status-ack (52, 67)</td><td>Ring sound (bell) Status-idle (67, 52)</td><td>Called-make-status-check</td></tr>
<tr><td colspan="3"></td><td colspan="3">Await-called-to-answer</td></tr>
<tr><td>Status-idle (67, 52)</td><td>Ear-sound (answer tone)</td><td>Dialed-instance-make-itself-known</td><td>Called (off-hook)</td><td>Connect (67, 52)</td><td>Called-subscriber-answer</td></tr>
<tr><td colspan="3">Wait-for-connect</td><td colspan="3"></td></tr>
<tr><td>Connect (67, 52)</td><td></td><td>Connect-to-found-instance</td><td></td><td></td><td></td></tr>
<tr><td colspan="6">Connected-to-each-other</td></tr>
<tr><td>Caller (hang-up)</td><td></td><td>Caller-finish-call</td><td>Called (hang-up)</td><td></td><td>Called-finish-call</td></tr>
<tr><td colspan="3">Subscriber-idle</td><td colspan="3">Subscriber-idle</td></tr>
</table>

7.26 Capturing data structures

In chapter 7.28.2, "EXAMPLE Phonebook: Refine and capture requirements," page 264, it is hinted that the more obvious the data structure is, compared to the behavior of a *program*, the more effort should be put into identifying *elements* as being design *restriction requirements* compared to identifying *behavior requirements*.

In database programs, it might be better to capture the complex data structure than the simple behavior.

It may seem that object orientation in this case overturns the *Cpdm* model of this book. One might ask, wouldn't it be even better to start with designing *architecture elements* (e.g. *objects*), and then capture their *behavior requirements* when the *elements* are created and known?

Generally, it is easier to deal with design restrictions than with behavior, because it is much easier to discuss and illustrate tangible *artifacts* than volatile events. But the overall reason for behavior is more important than structure, if the behavior of a *program* in the end is what is visible to *customers* and determines *customer* satisfaction (despite the fact that many *customers* superficially argue that how things look is more important than what they do). *Objects* are most often hidden from the *customers*, and are sufficient if *architecture* is solid and efficient.

7.26.1 EXAMPLE: Capturing a wine cellar database

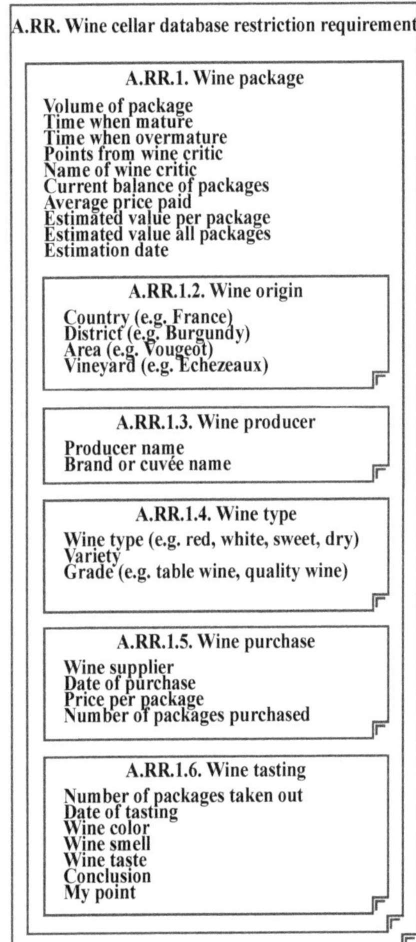

FIGURE 7-35 Restriction requirements for a wine cellar database

In a big cellar of collected wine bottles, there might be a need to keep order in what is stored (at least the insurance company might so request). At first sight it might sound easy to develop a database *program* for this purpose, but it really isn't so simple. The problem is that a package of wine can be so differently labeled, depending on its origins. A lot of wine traditions have grown for hundreds of years, resulting in a large number of wine types, areas of origin, legal regulations, and in addition to this, a lot of insignificant information. The problem is to fit this variation into one standard database, without complicating it until it gets out of hand, or simplifying it until the needed information doesn't get registered.

Before dealing with the wine cellar database it might be very beneficial to conduct a test by registering ten to twenty wines from all over the world, in order to find out which restrictions are necessary and which are not. This simple investigation preferably ends up capturing needed *restriction requirements* (see Fig. 7-35, left). Note that so far, these are *restriction requirements* and not design of a *white-box*.

7.27 EXAMPLE Phonebook environment: Explore environment restrictions and ensure environment staffing

7.27.1 EXAMPLE Phonebook: About this example

This example illustrates *product development* of an *object-oriented program*. A simple phonebook prototype will be founded for future large-scale expansion. Consequently, the design will be heavily *modularized* to be future-proof, also allowing *product requisites* to be powerful for accounting and follow-up.

7.27.2 EXAMPLE Phonebook environment: Process schedule to use

Pt:2:3. Tailored advanced technical schedule n = 2

	8.13	9.26	10.25	11.20
7.27				12.27 W

	8.14	9.27	10.24	11.19
7.28				12.26 H

For the phonebook example, a special technical overall *schedule* is *tailored* (illustrated in Fig. 7-36, left):

- **Pt:2:3. Tailored advanced technical schedule n = 2,** page 169

In this *tailored schedule* the generic *schedule* to use is (filled-in symbols):

- **Pt.0.R. Explore environment restrictions and ensure environment staffing,** page 110.

FIGURE 7-36 Position of schedule and master to use (filled in) in overall schedule. Chapter numbers of the phonebook example are also referenced.

7.27.3 EXAMPLE Phonebook environment: Explore restrictions

It is straightforward to capture the *environment restriction requirements* (see Table 7-52 below).

TABLE 7-52 Phonebook environment restriction requirement

Restriction requirement	
E. RR.1.	The phonebook program shall run in any standard Windows environment, and need not be portable elsewhere.
E. RR.2.	For the first evaluation, textual interaction with the user is sufficient.
E. RR.3.	Keeping all phonebook data in an ordinary Windows disk file is sufficient.
Pcb.1.	This is only the first simple version of a strategic product. The portfolio will include a better and better phonebook until it is one of the best on the market.
E. RR.4.	The program shall scale up easily in future development and be adapted for Windows graphic user interaction.
E. RR.5.	The program shall scale up easily in future development and be adaptable to database storage tools, such as SQL.

7.27.4 EXAMPLE Phonebook environment: Plan development roles and providers

For many of the development *activities* on the *environment interfaces*, in-house developers had sufficient competence (see Table 7-53 below). Windows *programming* needed to be strengthened with some external help.

TABLE 7-53 Phonebook environment development allocation staffing

Phonebook environment role item		Provider	Cpdm schedule or task
i10.	Product manager • To capture restrictions and prepare environment restrictions	PR.10. In-house	
i11.	Senior Windows graphic MMI designer • To propose a simple user interaction mechanism, that easily transforms to graphics	PR.11. Zide Kft	Pt.0.Ra. Explore restrictions.
i12.	Development department management	PR.10. In-house	Pt.0.Rb. Plan development roles and providers.
i13.	Windows programmer consultant • To evaluate and propose Windows development system for this application	PR.12. DeZ-ine Kft	Pt.0.Pb. Interfaces from environment solution alternatives
i14.	Development department management	PR.10. In-house	Pt.0.Pd. Environment demarcated by interface solutions with supplier opportunities.

TABLE 7-53 Phonebook environment development allocation staffing

Phonebook environment role item	Provider	Cpdm schedule or task
i15. **Windows programmer consultant** • To evaluate and propose how to establish a future-save database structure	PR.12. DeZine Kft	Pt.0.A. Constitute environment architecture.
i16. **C++ programmer**	PR.10. In-house	Pt.0.Fa. Finalize environment interfaces detailed items.
i17. **Development department management**	PR.10. In-house	Pt.0.Fc. Account for environment interfaces development, realization, and manufacturing.
i18. **C++ development system analyst**	PR.10. In-house	Pt.0.I. Realize interfaces to environment and insert/await outermost white-box.
i19. **C++ programmer**	PR.10. In-house	Pt.0.V. Verify prototype in environment.

7.27.5 EXAMPLE Phonebook: Proceed reading

To follow this example, these are the alternatives:

- Proceed to chapter 8.13, "EXAMPLE Phonebook environment: Demarcate product by environment solutions with possible supplier opportunities," page 328.

- Proceed to chapter 9.26, "EXAMPLE Phonebook environment: Constitute environment architecture," page 425.

- Proceed to chapter 12.27, "EXAMPLE Phonebook environment: Verify prototype in environment," page 780.

- Proceed to chapter 7.28, "EXAMPLE Phonebook: Refine product requirements from environment restriction requirements and ensure black-/white-box development staffing" below.

7.28 EXAMPLE Phonebook: Refine product requirements from environment restriction requirements and ensure black-/white-box development staffing

7.28.1 EXAMPLE Phonebook: Process schedule to use

For the phone example, a special technical overall *schedule* is *tailored* (illustrated in Fig. 7-6, left):

- **Pt:2:3. Tailored advanced technical schedule n = 2**, page 169

In this *tailored schedule* the generic *schedule* to use is (filled-in symbols):

- **Pt.n.R. Refine requirements from outward nesting level requirements and ensure black-/white-box development staffing,** page 130.

FIGURE 7-37 Position of schedule and master to use (filled in) in overall schedule

7.28.2 EXAMPLE Phonebook: Refine and capture requirements

When dealing with *assembler programs* (see ch. 10.14, "Assembler programs," p. 516) and *structured programs* (see ch. 10.17, "Structured programs," p. 529), the behavior of a *program* is often the natural start of the development. However, when approaching *object-oriented programs* with better support for structures (ch. 10.22, "Object-oriented programs," p. 554), it might be quite natural to also capture restrictions of the *objects* before capturing behavior, or to alternate between these two.

As can be seen from the *requirements* in Table 7-54 below, there is a restriction to use an *object-oriented program* style of programming. There are also restrictions on the contacts and phone numbers structure. Consequently, there is an obvious chance that contacts and phone numbers will end up as *objects* in the coming *architecture design*.

When developing *programs*, where data structures are large and not more difficult to capture than the behavior of the *program*, put more effort into capturing the data structure as restrictions and spend even more time to get the *architecture* right. If it is important to override *designers'* freedom, the entire data disk structure could have been specified in the same way as Figure 7-35, page 260. In this example this was not done, and consequently the *designers* could invent the data structure in Figure 9-67, page 433.

TABLE 7-54 Phonebook restriction requirement		
Restriction requirement		**Priority**
E. RR.1.	The phonebook program shall run in any standard Windows environment, and need not be portable elsewhere.	
A. RR.1.1.	**The phonebook program shall be developed for a standard Windows computer.**	Critical
E. RR.2.	For the first evaluation, textual interaction with the user is sufficient.	
A. RR.2.1.	**First phonebook program shall use simple text-oriented console interaction.**	High
E. RR.3.	Keeping all phonebook data in an ordinary Windows disk file is sufficient.	
A. RR.3.1.	**The program shall load phonebook data at start-up and save them on program termination.**	High
E. RR.4.	The program shall scale up easily in future development and be adapted for Windows graphic user interaction.	
A. RR.4.1.	**Choose an object-oriented program development system, design a strict modularization, and keep user interaction together with used objects.**	Critical
E. RR.5.	The program shall scale up easily in future development and be adaptable to database storage tools, such as SQL.	
A. RR.5.1.	**The first phonebook program version shall use only compiler built-in fixed-size variable storage. The phonebook file shall hold an array of contacts with contact data, one of which is an array of phone numbers with phone number data.**	Critical
A. RR.5.2.	**MAX_CONT_PER_BOOK = 128**	Medium
A. RR.5.3.	**MAX_NUMB_PER_PERS = 16**	
A. RR.5.4.	**MAX_PERS_LEN = 128**	
A. RR.5.5.	**MAX_NUMB_LEN = 32**	

The behavior of the phonebook *program* is rather straightforward. It is largely about how to operate the phonebook database, for example, to register a contact, and to register a phone number in a contact, and so forth. To maintain the database it is also necessary to erase contacts and phone numbers.

How to operate the phonebook database is specified by *normal use-cases* and *exception use-cases* (see Table 7-55 below). This table is developed at the same time as the graphic *formal scenario* is specified in this table.

The *item* column in Table 7-55 will be used later when accounting for *product requisite* (see Table 10-108, p. 582).

TABLE 7-55 Phonebook scenario			
Behavior requirement use-case		**Priority**	**Item**
RB.6.	Phonebook behavioral requirement no 6 defined in this table below		
A. RB.6.S.1.	**Initiate-phonebook scenario** defined in Figure 7-38, page 269	Critical	i1. Start scenario.
A. RB.6.S.1.U.1.	**Program-loading-file-and-start-up normal use-case**		
A. RB.6.S.1.X.2.	**No-file-found exception use-case**		
A. RB.6.S.1.X.3.	**Found-file-empty exception use-case**		
A. RB.6.S.1.X.4.	**Found-file-unexpected-end exception use-case**		
A. RB.6.S.2.	**Append-contact-to-phonebook scenario** defined in Figure 7-39, page 270	High	i2. Append scenario.
A. RB.6.S.2.U.1.	**Append-contact-to-phonebook normal use-case** (RB.UU.11. + RB.UU.21. + RB.UU.31. + RB.UU.41. + RB.UU.32.)		
A. RB.6.S.2.X.2.	**Append-contact-to-full-phonebook exception use-case** (RB.UU.15.)		
A. RB.6.S.2.X.3.	**Append-phone-number-to-full-phone-number-list exception use-case** (RB.UU.11. + RB.UU.21. + RB.UU.33.)		
A. RB.6.S.3.	**List-phonebook-contacts scenario** defined in Figure 7-40, page 271.	High	i3. List scenario.
A. RB.6.S.3.U.1.	**List-phonebook-contacts normal use-case** (RB.UU.12.)		
A. RB.6.S.3.U.2.	**List-empty-phonebook normal use-case** (RB.UU.14.)		
A. RB.6.S.4.	**Delete-phonebook-contact scenario** defined in Figure 7-41, page 271.	High	i4. Delete scenario.
A. RB.6.S.4.U.1.	**Delete-phonebook-contact normal use-case** (RB.UU.13. + RB.UU.51.)		
A. RB.6.S.4.X.2.	**Delete-contact-in-empty-phonebook exception use-case** (RB.UU.14.)		
A. RB.6.S.4.X.3.	**Delete-contact-not-existing exception use-case** (RB.UU.13. + RB.UU.52.)		
A. RB.6.S.5.	**Terminate-phonebook scenario** defined in Figure 7-42, page 272.	Critical	i5. Terminate scenario.
A. RB.6.S.5.U.1.	**Program-saving-file-and-termination normal use-case**		
A. RB.6.S.5.X.2.	**File-not-created exception use-case**		
A. RB.6.S.5.X.3.	**File-not-written exception use-case**		
A. RB.6.S.5.X.4.	**File-incompletely-written exception use-case**		

In order to be future-proof, be sure to keep *formalism* and *consistency*. All *formal scenario* constituents need to be specified with unique identities and comprehensible names (see Table 7-56 below). Table 7-56 is developed at the same time as

Table 7-57 above is being completed.

TABLE 7-56 Phonebook scenario

Behavior requirement for wait-states, use-units, and histories

A. RB. Phonebook behavioral requirements

A. RB.WS.00.	Wait-for-file-handler
A. RB.WS.01.	Wait-for-1st-field
A. RB.WS.02.	Wait-for-remaining-fields
A. RB.WS.10.	Wait-for-phonebook-function-choice
A. RB.WS.20.	Wait-for-register-phone-contact
A. RB.WS.30.	Wait-for-register-more-phone-numbers-y/n-answer
A. RB.WS.40.	Wait-for-phone-number
A. RB.WS.50.	Wait-for-contact-number-to-delete
A. RB.WS.60.	Wait-for-file-handler
A. RB.WS.61.	Wait-for-write-status
A. RB.WS.62.	Wait-for-write-status
A. RB.UU.11.	Phonebook-add-contact use-unit
A. RB.UU.12.	Phonebook-list use-unit
A. RB.UU.13.	Contact-delete use-unit
A. RB.UU.14.	Phonebook-empty use-unit
A. RB.UU.15.	Phonebook-filled use-unit
A. RB.UU.21.	Person-register use-unit
A. RB.UU.31.	Number-further use-unit
A. RB.UU.32.	Number-no-more use-unit
A. RB.UU.33.	N-numbers-filled use-unit
A. RB.UU.41.	Number-register use-unit
A. RB.UU.51.	Contact-deleted use-unit
A. RB.UU.52.	Contact-not-existed use-unit
A. RB.H.1.	Contact-count-story = zero
A. RB.H.2.	Contact-count-story = number of read intact contacts
A. RB.H.3.	Contact-count-story = number of found contacts
A. RB.H.4.	Contact-count-story = Contact-count-story + 1
A. RB.H.5.	Contact-number-count-story = zero
A. RB.H.6.	Contact-number-count-story = Contact-number-count-story + 1
A. RB.H.7.	Contact-count-story = Contact-count-story - 1

There are a lot of *stimuli* and *responses* on the *connection interfaces* used to operate the phonebook database. In order to be future-proof, also capture these with unique identities and comprehensible names (see Table 7-57 below). This table is developed at the same time as completing Table 7-55, page 266, and Table 7-56, page 267 above.

TABLE 7-57 Phonebook scenario

Behavior requirement on interfaces

-E-2:1. Console interface	
-E-2:1. S.1.	Select-function (append-contact)
-E-2:1. S.2.	Select-function (list-contacts)
-E-2:1. S.3.	Select-function (delete-contact)
-E-2:1. S.4.	Select-function (quit-phonebook)
-E-2:1. S.5.	Input (contact-name)
-E-2:1. S.6.	Input ('y')
-E-2:1. S.7.	Input ('n')
-E-2:1. S.8.	Input (phone-number)
-E-2:1. S.9.	Select (contact-number)
-E-2:1. R.10.	Show (phonebook-functions)
-E-2:1. R.11.	Show (incomplete-file-found)
-E-2:1. R.12.	Show (empty-file-found)
-E-2:1. R.13.	Show (no-file-found)
-E-2:1. R.14.	Ask-for (contact-name)
-E-2:1. R.15.	Ask-for (more-phone-numbers)
-E-2:1. R.16.	Ask-for (phone-number)
-E-2:1. R.17.	Show (phone-numbers-full)
-E-2:1. R.18.	Show (phonebook-functions)
-E-2:1. R.19.	Show (contacts-full)
-E-2:1. R.20.	Show (all-contacts)
-E-2:1. R.21.	Show (no-contact)
-E-2:1. R.22.	Ask (contact-number-to-delete)
-E-2:1. R.23.	Show (contact-not-existing)
-E-2:1. R.24.	Show (no-file-created)
-E-2:1. R.25.	Show (empty-file-created)
-E-2:1. R.26.	Show (file-incompletely-written)
-E-3:1. File interface	
-E-3:1. R.27.	File (handler)
-E-3:1. S.28.	File (1st-field)
-E-3:1. S.29.	File (remaining-fields)
-E-3:1. S.30.	File (write-status)
-E-3:1. R.31.	File (open-to-read)
-E-3:1. R.32.	File (read-1st-field)
-E-3:1. R.33.	File (read-remaining-fields)
-E-3:1. R.34.	File (open-to-write)
-E-3:1. R.35.	File (write-1st-field)
-E-3:1. R.36.	File (write-remaining-fields)

7.28.3 EXAMPLE Phonebook: Refine and capture requirements for initiate-phonebook scenario

The *formal scenario* of starting the phonebook *program* (see Fig. 7-38 below).

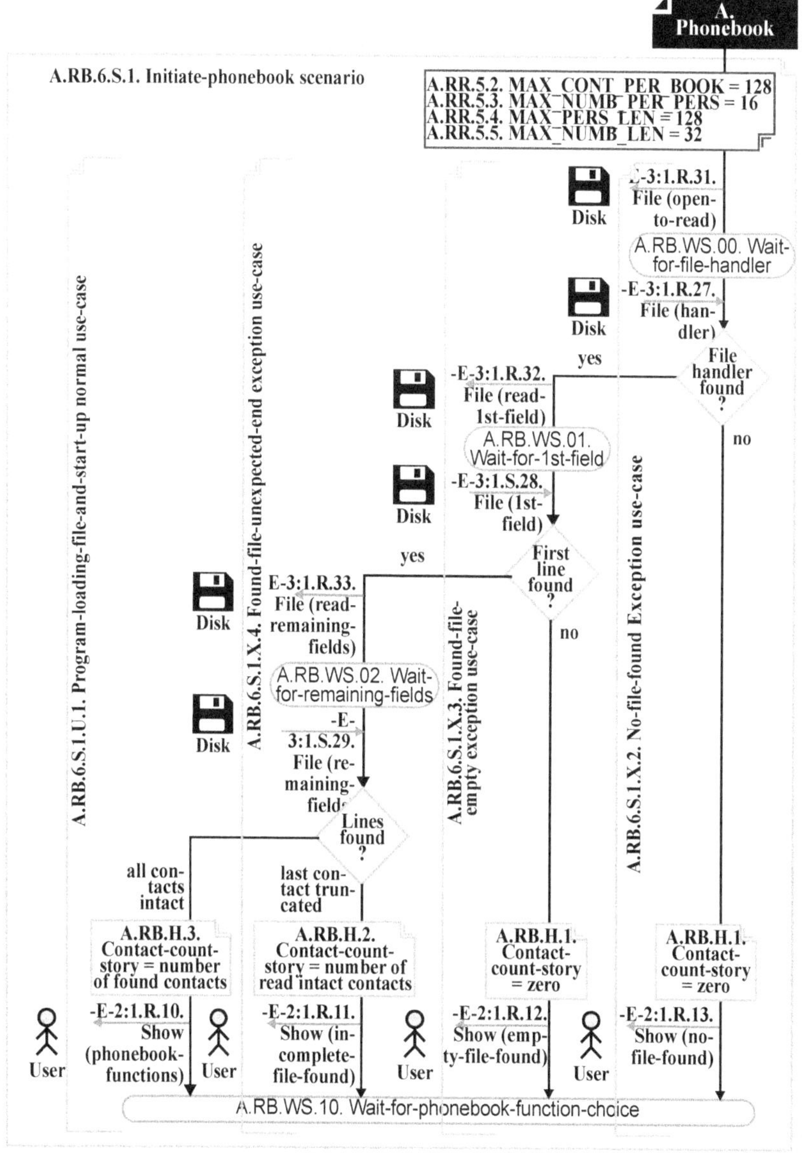

FIGURE 7-38 Initiate-phonebook scenario black-box requirements

7.28.4 EXAMPLE Phonebook: Refine and capture requirements for append-contact-to-phonebook scenario

The *formal scenario* of appending a contact to the phonebook (see Fig. 7-39 below).

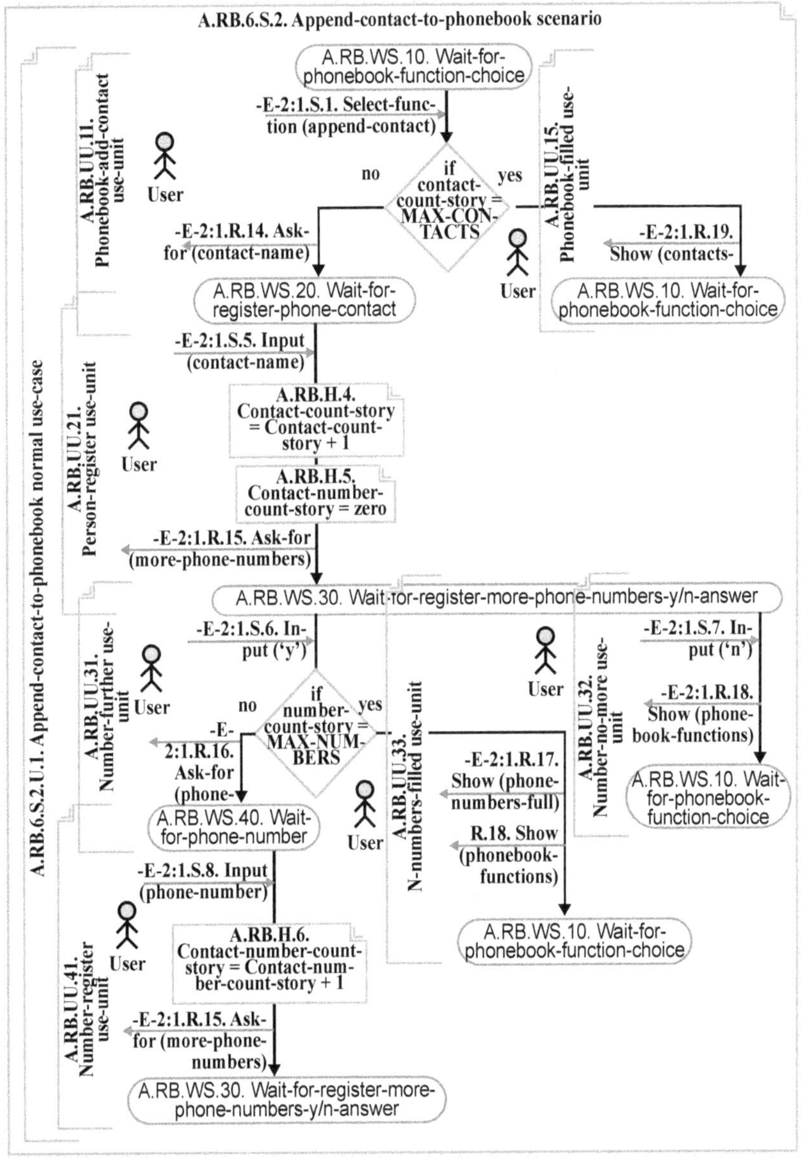

FIGURE 7-39 Append-contact-to-phonebook scenario black-box requirements

7.28.5 EXAMPLE Phonebook: Refine and capture requirements for list-phonebook-contacts scenario

The *formal scenario* of listing the phonebook (see Fig. 7-39 below).

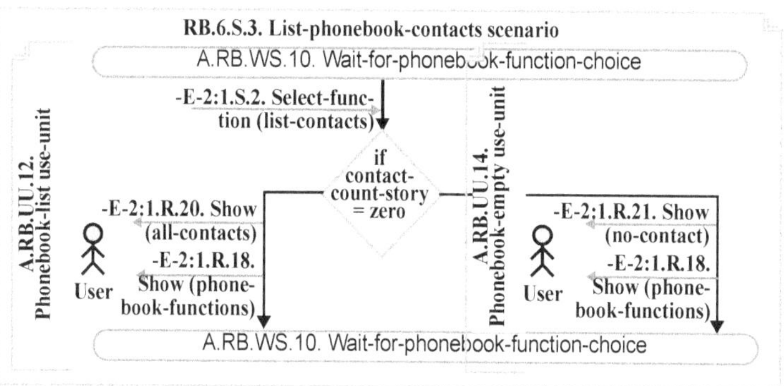

FIGURE 7-40 List-phonebook-contacts scenario black-box requirements

7.28.6 EXAMPLE Phonebook: Refine and capture requirements for delete-phonebook-contact scenario

The *formal scenario* of deleting a contact in the phonebook (see Fig. 7-39 below).

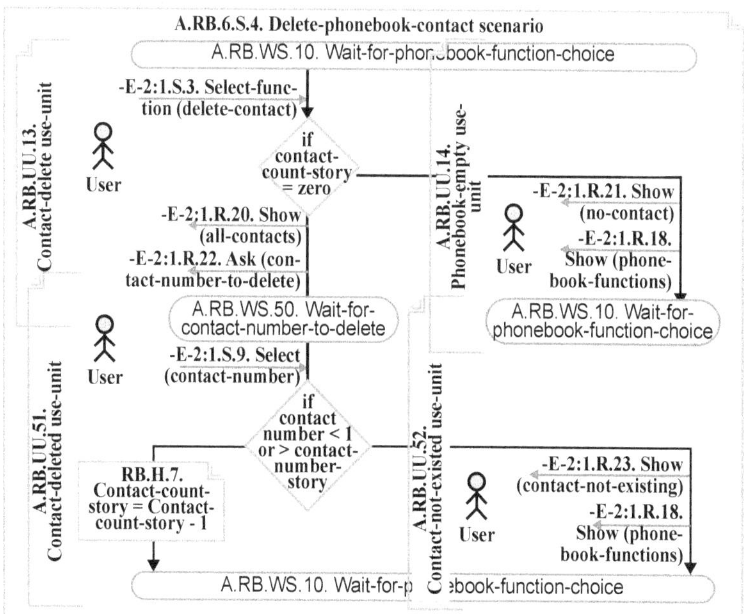

FIGURE 7-41 Delete-phonebook-contact scenario black-box requirements

7.28.7 EXAMPLE Phonebook: Refine and capture requirements for terminate-phonebook scenario

The *formal scenario* of terminating the phonebook *program* (see Fig. 7-38 below).

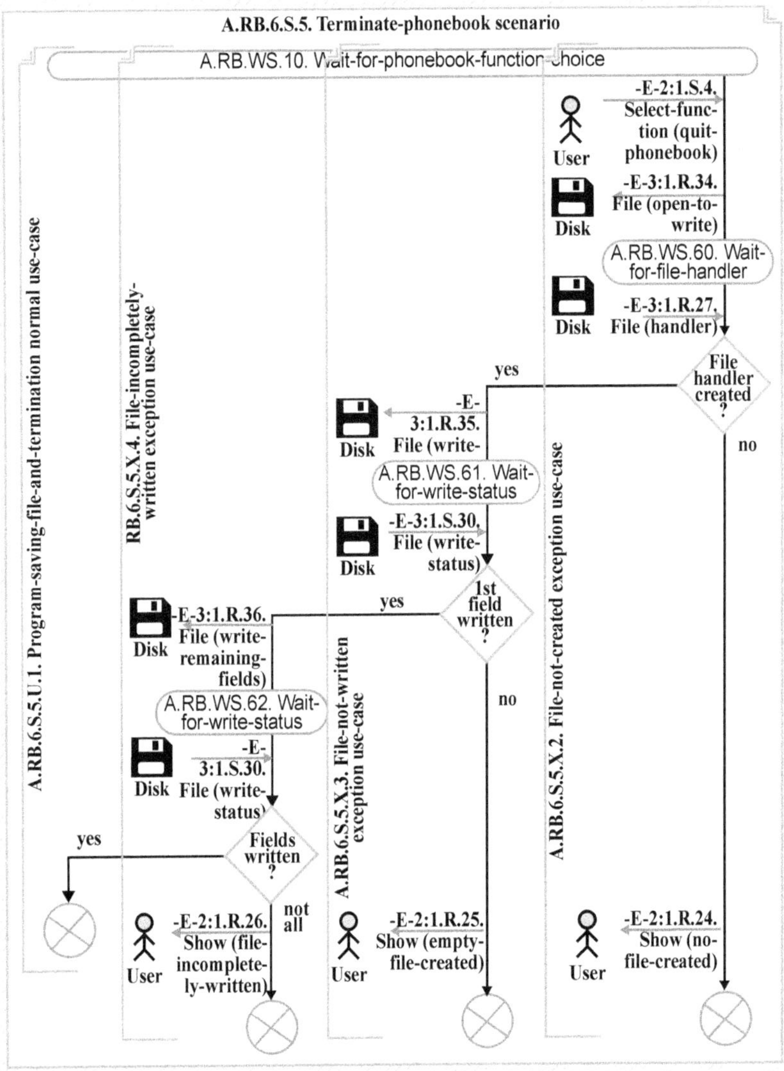

FIGURE 7-42 Terminate-phonebook scenario black-box requirements

7.28.8 EXAMPLE Phonebook: Prioritize requirements

The prioritization was made in the *restriction requirements* (see the separate column of Table 7-54, p. 265), and in each *formal use-case* of the *behavior requirements* (see the separate column of Table 7-55, p. 266).

7.28.9 EXAMPLE Phonebook: Examine authenticity

Since this was the first development of the phonebook, the capacity and appearance were not the most important. It was more important to find a future-proof structure and scalability to be the foundation for further development. Also very important was a high modularization obtained by *object-oriented program* restrictions.

At the gate meeting, the above considerations were found to be well accounted for.

7.28.10 EXAMPLE Phonebook: Plan development roles and providers

Considerable competency is available in-house for *program* development, including *object-oriented program* design (see Table 7-58 below).

TABLE 7-58 Phonebook development allocation staffing

	Phonebook role item	Provider	Cpdm schedule or task
i20.	**Specifier** • To edit requirements	**PR.10. In-house**	Pt.n.Ra. Refine and capture requirements.
i21.	**Product manager** • To review and prioritize requirements	**PR.10. In-house**	
i22.	**Development department management**	**PR.10. In-house**	Pt.n.Re. Plan development roles and providers.
i23.	**Chief architect** • To propose solution alternatives	**PR.10. In-house**	Pt.n.Pa. Try to satisfy requirements. Pt.n.Pc. Analyze make/buy options.
i24.	**Windows designer** • To evaluate and propose how to modularize the program	**PR.10. In-house**	Pt.n.Aa. Identify white-box architecture ingredients. Pt.n.Ac. Arrange ingredients and identify possible black-boxes. Pt.n.Ae. Lay out physical architecture.
i25.	**Architect programmer** • To plan, coordinate, and write interface headers	**PR.10. In-house**	
i26.	**Object-oriented programmer** • To plan, coordinate, and write classes	**PR.10. In-house**	Pt.n.Fa. Finalize white-box detailed items.
i27.	**General C++ programmer** • To plan, coordinate, and write functions, variables, and restriction headers	**PR.10. In-house**	

TABLE 7-58 Phonebook development allocation staffing

	Phonebook role item	Provider	Cpdm schedule or task
i28.	Development department management	PR.10. In-house	Pt.n.Fc. Account for white-box development, realization, and manufacturing.
i29.	Configuration manager	PR.10. In-house	Pt.n.Ia. Obtain items including linkable program. Pt.n.Id. Integrate elements, including linkable program and embedded black-boxes. Pt.n.Ig. Start up white-box.
i30.	Certified verifier	PR.10. In-house	Pt.n.V. Verify black-/white-box.
i31.	Verifier	PR.10. In-house	

7.28.11 EXAMPLE Phonebook: Proceed reading

To follow this example, these are the alternatives:

- Proceed to chapter 8.14, "EXAMPLE Phonebook: Predetermine solutions with sourcing options," page 330.
- Proceed to chapter 9.27, "EXAMPLE Phonebook: Satisfy refined requirements by decomposing product black-box into white-box design containing no black-boxes," page 429.
- Proceed to chapter 12.26, "EXAMPLE Phonebook: Verify black-/white-box," page 770.
- Proceed to next chapter 7.29 below to leave the phonebook example and to dig deeper into the capture staffing & requirements topic.

7.29 Huge networks of scenarios

7.29.1 About networks of scenario

One advantage with *formal scenarios* presented in this book is that they can be specified complete. On the other hand, the reverse is also true: they can be specified very skeleton-like.

This flexibility makes it possible to start to capture an overall *formal scenario* to a *complex black-box*, and develop this step by step. Only *complex* parts of the *formal scenario* may need to be further specified as detailed *formal scenarios*, and the *formal scenario* development can be finished when it seems that the *complex* parts are fully mastered, even if the *requirements* are not quite complete.

7.29.2 EXAMPLE: Scenarios of operating a PC

For example, capture the behavior of Microsoft Windows step by step, and begin with Figure 7-43 below.

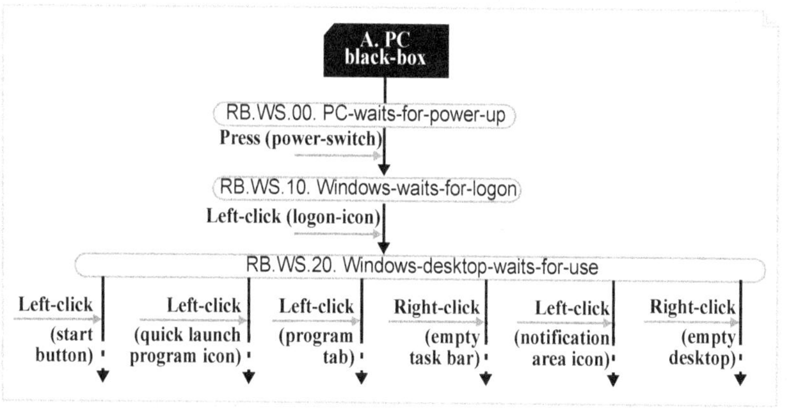

FIGURE 7-43 Overall capturing stimuli to windows

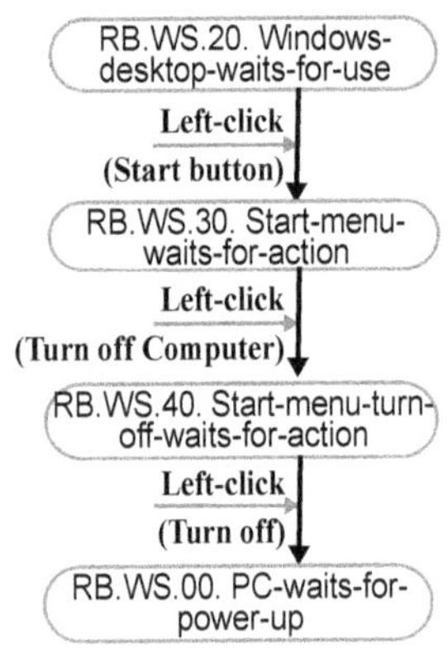

Maybe the start and shutdown are most important to describe, which continue with Figure 7-44 at left.

Formal scenarios have many advantages like:

• The *formal scenario* can be half finished, but already offers a lot of understanding on how to proceed.

• The *formal scenarios* can be made on separate sheets and be connected to each other by *black-box wait-state* naming.

• The *formal scenarios* can be extended in any direction at any time.

• The graphical notation makes you point on the *requirement* with the finger like navigating on a map.

FIGURE 7-44 Capturing PC turn off stimuli

One of the few disadvantages of graphical *formal scenarios* is that they demand a graphics editor, which may be expensive and has some learning barriers.

7.30 Road maps

Road maps are long-term and strategic *requirements*, specified far in advance, compared to usual *requirements*, specified as close to design as possible.

The purpose is to spread a feeling for how *products* will evolve, in order to gain some years to prepare the organization with new capabilities. It might be new technology that must be shifted in, lower *product* price, more advanced *products*, opening of new markets and segments, or something similar.

Road maps are long-term requirements for starting early to prepare for future demand for new capabilities.

It is important to not only follow logical technical development and standardization, but also to be prepared for *paradigm shifts*. Apple developed a new paradigm for the man-machine *interface* of a mobile phone, which entirely overturned that industry. Apple's new *interface* was not a lightning bolt from a clear sky, and despite many chances to update their *road maps* over several years, competitors obviously had big problems doing so, which ended up with Google offering the Android concept, to get incapable mobile companies going again.

7.31 Frequently asked questions about requirements

7.31.1 Why use black-boxes when capturing requirements ?

There is a wide variety of reasons why the *black-box* approach is advantageous.

- Opening *black-boxes* to *white-boxes* is a strong and easy-to-understand concept which clearly tells that design is impossible while it is black and that design may start when it turns white.

- It forces *behavior requirements* to be refined and specified, because *interfaces* and behavior are all that is visible.

- *Black-box requirements* prevent referring of solutions, because *black-box* internal ingredients are not referable.

- It is a way to *partition* a *product* into manageable and understandable pieces, by creating many adjacent to each other, or a hierarchy inside each other.

- A *black-box* can be used as a clear demarcation of responsibilities. The black-box surroundings and the content can be developed or realized by different organizations or can be easily outsourced.

- When considering an *element* to be a candidate for *reuse*, it is useful to design it as a *black-box*, in order to ensure that it will fit everywhere as decided.

- When manufacturing is in full swing, it is much easier to change a well-specified *black-box* than to change a spaghetti *module*, e.g., to make a cost reduction in the *black-box* design, and be certain not to affect the *black-box* surroundings.

- If high *quality* is an essential *requirement*, there is no way around structure and *formalism*. An abundance of *black-boxes* is not the way to save development cost, but if done intelligently it surely heightens *quality*.

- Despite many laudable attempts to improve *structured programs* and *object-oriented programs*, there is so far poor support for overall structure. *Black-box* hierarchies can be used instead of vain nesting of *objects*.

Many of the above arguments will be elaborated on later in this book.

7.31.2 EXAMPLE: Why not open the black-box early?

All during specification of house *requirements*, there is some risk of forcing the opening of the *black-box* before the *requirements* are settled. It is much easier for most humans to think in tangible design terms than in volatile behavior terms. Take, for example, the R.6. Gourmet cooking and dining requirements in Table 7-9, page 189. Most engineers would immediately have interpreted "dining *requirements*" as "a table with chairs."

It is often referred to as a big problem, that engineers peek at black-boxes to begin designing them long before it is known what they should do.

Gourmet cooking and dining *requirements*, of course, materialize to a convenient way of sitting and dining at a table, but also a lot of other things, such as:

- quiet surroundings, maybe with some background music

- a possibility to dim the light, maybe a possibility of candles
- a short path to the cooking area, without door thresholds
- no smell from the cooking area to trouble dining guests
- nearby clean, easily accessible toilet facilities
- a capability for keeping the last dish warm within the dining area

Whoever has visited a star-*quality* French restaurant knows all about this.

Reducing "dining *requirements*" to "a table with chairs" will eliminate nearly all the above bullets. So be sure to keep the *black-box* closed until specification is finished.

7.31.3 Isn't it too expensive to refine and capture requirements?

Capturing requirements is not free, but compared to total development cost it is often insignificant.

It is costly to capture *requirements*, that is true. Often this money (and calendar time) is initially saved, to get going with design. But when the time comes for *verification* and developing of the manual, problems accumulate. The extra amount of work spent on test cases and manuals often costs more than capturing *requirements* from the start.

When extending the *product* later or introducing *incremental development*, a big problem in the absence of documented *requirements* is keeping track of what is currently included in the *product*, and how new features would interact with those already existing. Beware of these problems.

7.31.4 Who are requirement stakeholders?

The most important *stakeholders* are the end *users* of the *product*. In the end they pay for it, and simply want to get value for their money, or next time they will try a competitor.

Often, retailers and wholesalers see themselves as the most important *stakeholders* because they understand their end *users* and can easily *capture requirements*. But far too often retailers and wholesalers are conservative with their offered *products*, and plan their sales short-term, reluctant to introduce risky new concepts and troublesome (for them) changes.

Requirement authenticity depends only on selection of the right stakeholders having the right competency.

In-house there are also quite a few *stakeholders*. Often, larger companies have departments for *product management* that have the responsibility of keeping an ear to the ground. But keep in mind that these employees might compete with each other, leading to cannibalism between their *products* in the common *product portfolio*, or find it easier to listen to wholesalers and retailers than to end *users*, and so forth. *Product management marketers* must be as tightly trained as *product* developers.

It is also useful to have some engineering representatives, for example, verifiers to check *requirements* to be verifiable and manufacturing people to check *requirements* that will facilitate rather than hinder smooth manufacturing.

The worst *stakeholders* of all imaginable are maybe the developers themselves. They are not evil people by any means, but their minds are preoccupied with many things unfavorable to the end *users*. Engineers are technically educated, and employed to develop technique, and cannot be expected to put themselves in the place of a technically hopeless *end user*. Don't let technicians capture *requirements*, especially not during development.

7.31.5 How a durable prioritization is made

Prioritization of requirements should be a battle, but not to the degree of destroying the stakeholders.

Not everything wished for from a *product* is equally important. Sooner or later *requirements* conflict with each other; for example, the *product* gets too expensive, is delayed, or some other problem comes up as a result of too many or too ambitious *requirements*.

With many *stakeholders* and a big *product portfolio* for a very competitive market, it can be a battle to capture and prioritize *requirements*. Each *product management marketer* with some pride will have the most competitive *product*, to satisfy the largest number of *requirements*. Development capability is always limited, and can seldom match an overflow of *requirements*.

Prioritization is easy to define and understand, but if there are many *stakeholders* producing an overflow of *requirements* (and there often are), it gets difficult to perform in reality. Many *stakeholder* individuals cannot adequately separate what is best for the company from what is best for them personally. This often leads to deadlocked negotiations and suboptimizations, which leads to damages for the developing company. My next book *"Cpdm* technical overhead" on development overhead will further address these problems.

7.31.6 Why outside-in, and not inside-out requirement refinement?

Cpdm promotes *requirement refinement* outside-in, that is, starting with *environment restriction requirements* and refining these *inwards* until capturing the *requirements of innermost black-boxes* not containing further *black-boxes*. The reverse refinement order is also imaginable—capturing the *innermost black-boxes requirements* first, and then adapt *outward black-boxes requirements* by using already developed *white-boxes*.

Outside-in refinement of requirements can feel awkward, but what are the alternatives?

There are many arguments for outside-in *requirement refinement*. Most fundamental is that the only given when starting specification of *requirements* is the *environment* the *product* will be part of. Starting inside-out will probably end up in a *product* not fitting in the existing *environment* at all.

Accepting outside-in *requirement refinement* doesn't means that nothing can be specified before starting any development at all. Applying a *black-/white-box hierarchy* is the solution, because it allows the deeper *inward requirements* to be specified later, after the more general *requirements* are specified and design is started.

The risk in capturing deep *inward requirements* after *outward nesting levels* design has started, is that some new deeper *inward nesting levels* insights may pop up that overturn some *outward requirements*. It always gets costly to recapture wide *outward requirements* influencing almost every deeper *inward requirement*. This risk can be mitigated by *prototyping* of details (see ch. 4.6.1, p. 45).

## CHAPTER 8

Predetermine solutions & suppliers

Success is often a middle ground between the extremes
of buying everything from suppliers
and developing everything in-house.

Solution alternatives off the shelf?

8.1 About predetermine solutions & suppliers

8.1.1 The two purposes of predetermine solutions & suppliers

The purpose of **Pt.P. Predetermine solutions & suppliers** is to bridge between **Pt.R. Capture staffing & requirements** and **Pt.A. Design architectures**. The purpose is twofold:

1. To propose a sufficient number of *white-box solution alternatives* for each of all captured *requirements*. From those alternatives it is much easier to later select *elements* and other *architecture ingredients* that fit best. But maybe the most important is that from the totality of these alternatives it gets much easier to optimize the total *architecture* containing all selected *white-box solution alternatives*.

2. To consider how the *white-box solution alternatives* will be procured. The choice might often be between solutions developed in-house and solutions bought from *suppliers*. The more general the *elements* appear, the more probable it becomes that they will be available and purchased from *off-the-shelf suppliers*. All major solutions planned to be sourced from *suppliers* must later be included in the *architectures* to ensure that they fit with each other and with solutions developed in-house. Consequently, *supplier samples* must be well described and promptly made available to the architects.

Pt.P. Predetermine solutions & suppliers refers to design with the more vague word *solution alternatives*, which is a wider concept that includes technology, *suppliers*, contractual and legal issues, development tools, and so forth. When entering the subsequent **Pt.A. Design architectures** *phase*, the design gets much more technically precise, with well-documented *architecture ingredients*. Until then it is important to keep the solution as open as possible, and not until *architecture* time narrow down and decide on technical *partition* of the *product architecture ingredients*.

8.1.2 The schoolbook world

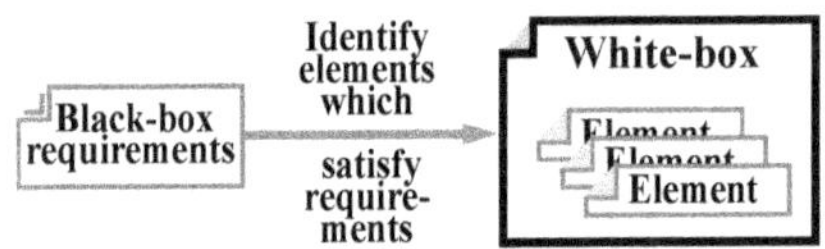

In the schoolbook world, there is a short and simple leap between *requirements* and design (see Fig. 8-1, left.)

FIGURE 8-1 Schoolbook relation between requirements and design

In the real world however, to design an *architecture* from *requirements* is far from straightforward. Design is said to be heuristic, which is to apply a trial-and-error approach, which in turn is to propose a plausible design and check back how well it fulfills the *requirements*.

To translate requirements to architecture design by intuition and trial and error is the only way, but constitutes a challenge.

Sometimes the trial is lucky and a good design appears quickly, but sometimes proposal after proposal gets rejected. Often *designers* have experience with similar solutions, which helps to achieve a good design more quickly.

Furthermore, design is sometimes included in *requirements* (in *Cpdm* such *requirements* are referred as *environment restriction requirements*); thus, many design *elements* cannot be freely selected. Restrictions are specified for a number of practical reasons, for example, stipulated *reuse* of already available (in-house) *elements*.

Special design may sometimes be of high value for "artistic" *products*, but for most *products* it is indifferent. In many companies, design of *programs* is believed to be artistic and handcrafted, but this is often a way for *gurus* to make themselves indispensable. The layout and design of most *programs* cannot be observed from the outside by *customers*, and as long as the *program* is failure-free enough and has the desired behavior, *customers* will be happy with any kind of internal design.

8.1.3 The real world

In the real world, unfortunately, many developers still jump from *requirement* to *architecture* in one cosmic schoolbook leap. But the trick to take *intellectual control* is to introduce **Pt.P. Predetermine solutions & suppliers** in the middle between *requirements* and *architectures*, to take the leap step by step (see Fig. 8-2 below).

The trick is to move stepwise from requirements to architecture design.

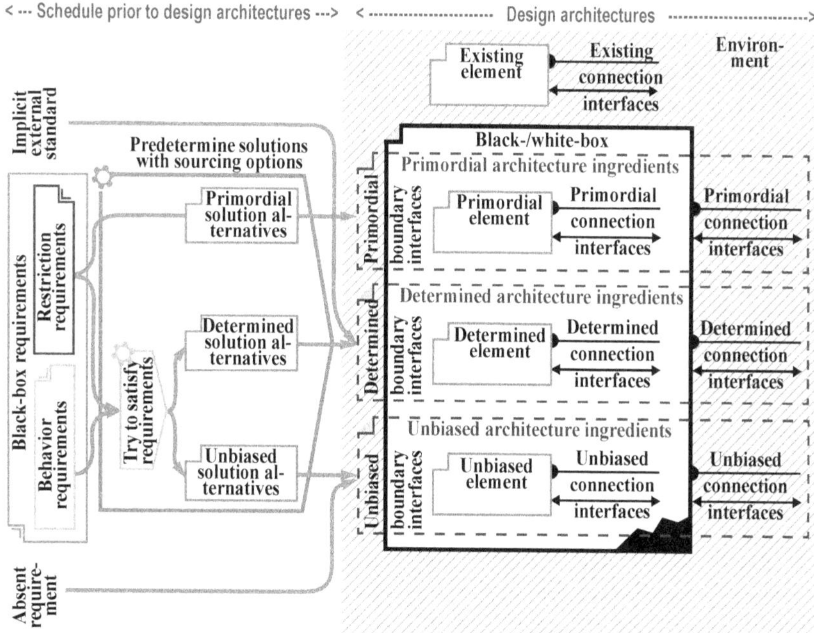

FIGURE 8-2 Stepwise transition from requirements to architecture

Some explanations to Figure 8-2 above:

Restriction requirements have the purpose of capturing limitations on *white-box* design, such as preferred *elements*, or their size, price, *quality*, and so forth. These *requirements* most often imply that a special type of solution simply must be chosen when executing **Pt.P. Predetermine solutions & suppliers**. If a restriction with lower priority must be compromised, it may end up as a *determined architecture ingredient*.

Behavior requirements give total freedom for any proposed solutions and design, and can be chosen as *unbiased architecture ingredients*. But there may be overall restrictions that indirectly influence behavior *elements*, for example, memory or bandwidth, and such designs get a bit more restricted as *determined architecture ingredients*.

Black-box requirements are a mix of *restriction requirements* and *behavior requirements*. With such *requirements* the design is usually not totally free, resulting in *determined architecture ingredients*.

Absent requirements occur when *requirements* have perhaps been missed, or more often when design is consciously transferred to engineers. Some design has to be made despite the fact that there are no underlying *requirements* at all. Beware of the extreme—if far too few *requirements* are captured, all design decisions are transferred to engineers. Absent *requirements* also violate *traceability*, because from resulting *architecture ingredients* there will be no *white-box solution alternatives* for *upstream* reference.

Implicit external standards may cause some *requirements* to be omitted, because there might be explicit or implicit standards that *designers* are used to following, for example, how ground, wall, and roof are generally restricted in a house.

Try to satisfy requirements. This is the *activity* of the trial-and-error identification of solutions. Often *sample catalogues* are helpful to catalyze invention of solutions to satisfy *requirements*. Since there are most often many possible solutions to one *requirement*, all plausible solutions must be presented for final analysis when preparing the *architecture*. Different solutions might fit together better or worse, and one benefit with many *white-box solution alternatives* is the possibility to optimize the *architecture* by picking the best-fitted set of solutions.

Solution alternatives. Documentation is in order when the trial-and-error identification of solutions is not going any further, and the found solutions seem good enough. As usual, each alternative should refer to the source *requirement(s)*, and have a unique ID, which becomes necessary for later reference.

Primordial architecture ingredient. Ingredients are already decided more or less because referenced *restriction requirements* have stipulated it.

Unbiased architecture ingredient. Ingredients are chosen totally freely, because referenced *requirements* capture only behavior, or because there is no *requirement* at all to reference.

Determined architecture ingredient. Ingredients are to some extent stipulated by restrictions, but yet chosen rather freely.

Existing architecture ingredients. Ingredients (including *users*) already exist in the *environment* and should be connected by *interfaces* to the *product* to be developed.

Note that the **Pt.P. Predetermine solutions & suppliers** *result* may not contain or add *requirements*. If *requirement* shortcomings or inaccuracies are identified, go back to the capture staffing & requirements *phase* and eliminate failures or add what is missing in the *requirement* documentation.

8.1.4 Upstream traceability from solutions to requirements

Requirement restrictions are in a way already solutions, which in turn are translated by **Pt.P. Predetermine solutions & suppliers** in a very straightforward and comprehensible way. Consequently, it is rather easy to trace from *primordial architecture ingredients* back to solutions and further back to *restriction requirements*. At *white-box verification* based on *restriction requirements*, these ingredients get rather easy to spot.

Tracing back from *unbiased architecture ingredients* to *white-box solution alternatives* is not as straightforward as from *primordial architecture ingredients*, just because *unbiased architecture ingredients* are unbiased, and by definition are freely invented by developers. Of course, these *unbiased architecture ingredients* must be traceable back to the *behavior requirements* they intend to satisfy, but there is often no simple relation between single *behavior requirements* and single *unbiased architecture ingredients*.

In fact, it is quite often a big mistake to choose one design ingredient to satisfy exactly one *behavior requirement*, leading to an *architecture* containing one new *element* developed for every additional *behavior requirement* (the *element* typically named by a verb instead of a noun), which sooner or later needs *refactoring*. This syndrome is sometimes called *functional decomposition*.

To conclude, there is no simple relation between *behavior requirements* and *architecture ingredients*. However, the situation gets better when *test-cases* are prepared based on *behavior requirements*. *Black-box* behaviors are verified on their *interfaces*, and to prepare *test-cases* is to describe *stimuli* to and expected *responses* from the *black-box*. In this way, the *traceability* from *test-cases* to *behavior requirements* ensures that the design can be judged correctly, regardless of how the design looks, provided it satisfies possible *restriction requirements*.

Another challenge in the *intrinsic complexity* relation between *behavior requirements* and *architecture ingredients* is during *verification*, when an incorrect *response* has been observed from an injected *stimulus*, causing a lot of difficulty to locate the *architecture ingredient(s)* responsible for the incorrect behavior (see ch. 12.3 "Troubleshooting," p. 665).

A question to always consider is how many *white-box solution alternatives* should be proposed for each of a group of *behavior requirements* (*formal use-cases* or *for-*

mal scenarios), and what could be left to the architects to invent. Case by case, it must be decided if the solution engineers or the architects are best suited to propose the best design, or if maybe the development should go back and forth between them.

8.1.5 Involving suppliers

Another thing not often mentioned in schoolbooks is that sourcing must be started as soon as the *requirements* are finished. It often works the other way around; when looking into sourcing offerings and *sample catalogues*, some ideas may pop up on how to satisfy the *requirements* (see Fig. 8-3 below). The figure below shows the principle of how to involve *suppliers*, and will be explained further in coming chapters.

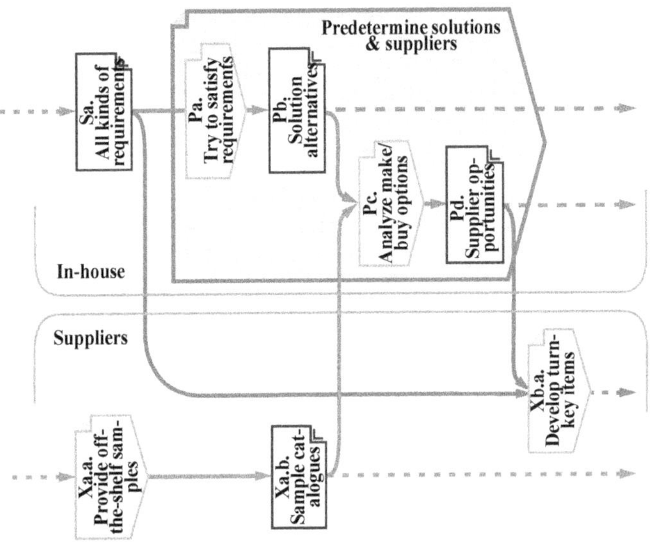

FIGURE 8-3 Interaction with suppliers starts as soon as requirements are finished.

There might be a debate whether the *white-box solution alternatives* should first be figured out, and after that suitable *suppliers* identified, or if it should be the other way around, that first *supplier* solutions should be studied, and then the *solution alternatives* chosen according to what is available. There is no simple sequence to work, and both alternatives have benefits and drawbacks; in practice, developers should go back and forth between *solution alternatives* and *supplier* offerings.

The *Cpdm process schedule* illustrates that *white-box solution alternatives* are developed first, and then an attempt is made to realize these alternatives with the best offers available from *suppliers*. However, the *process* illustration should not be rigidly interpreted, but instead be used with common sense.

8.2 Demarcate product by environment solutions with possible supplier opportunities (environment nesting level = 0)

8.2.1 Specialties on environment nesting level n = 0

On *environment nesting level* = 0, the **Pt.P. Predetermine solutions & suppliers** is a bit different, compared to *black-/white-box nesting levels* n ≥ 1. On *nesting level* 0 there are only the *interfaces* between the *environment* and the *product outermost black-box*. Accordingly, on this *environment nesting level* = 0, *white-box solution alternatives* and *white-box supplier opportunities* concern only connections between the *product* and its *environment.* (This is true, even if one single *sample* is found to be good enough to match the entire *product outermost black-box*, and the only remaining thing is to connect the *sample* to the existing *environment.*)

8.2.2 Cpdm definitions

In *Cpdm, environment interfaces solution alternatives* and *selected solutions supplier opportunities* are defined according to Table 8-1 below.

TABLE 8-1 Cpdm definitions

As-pect	Environment interfaces solution alternatives	Selected solutions supplier opportunities
Defini-tion	Various interface solution proposals how to best possible satisfy one or several specified environment restrictions	Determination of which environment solutions are part of the product, and from where (including in-house) to get appropriate samples to match these solutions
Syno-nyms	Impact analysis, implementation proposal, solution alternatives, preliminary design	Boundary of development, system limitation border
Table	**TABLE 8-2** Solution alternatives of interfaces to environment | SL.n. Interface solution alternative	As-sess-ment	Environ-ment re-striction
---|---|---|	**TABLE 8-3** Solutions excluded | SX.n. Excluded from the product to be developed	Re-sponsi-ble	Environment solution alternative
---|---|---| **TABLE 8-4** Solutions included with supplier opportunities | SU.n. Included in the product with supplier opportunity	Environment solution alternative
---	---	

8.2.3 Sourcing affects all environment schedules

In this **Pt.P. Predetermine solutions & suppliers** *phase*, the sourcing of this specific *product* begins and evaluation of *samples* starts and continues through the entire *phase* (see Fig. 8-4 below).

Crucial *samples* for this *product* may already have been developed long ago by nearby *suppliers* or in-house.

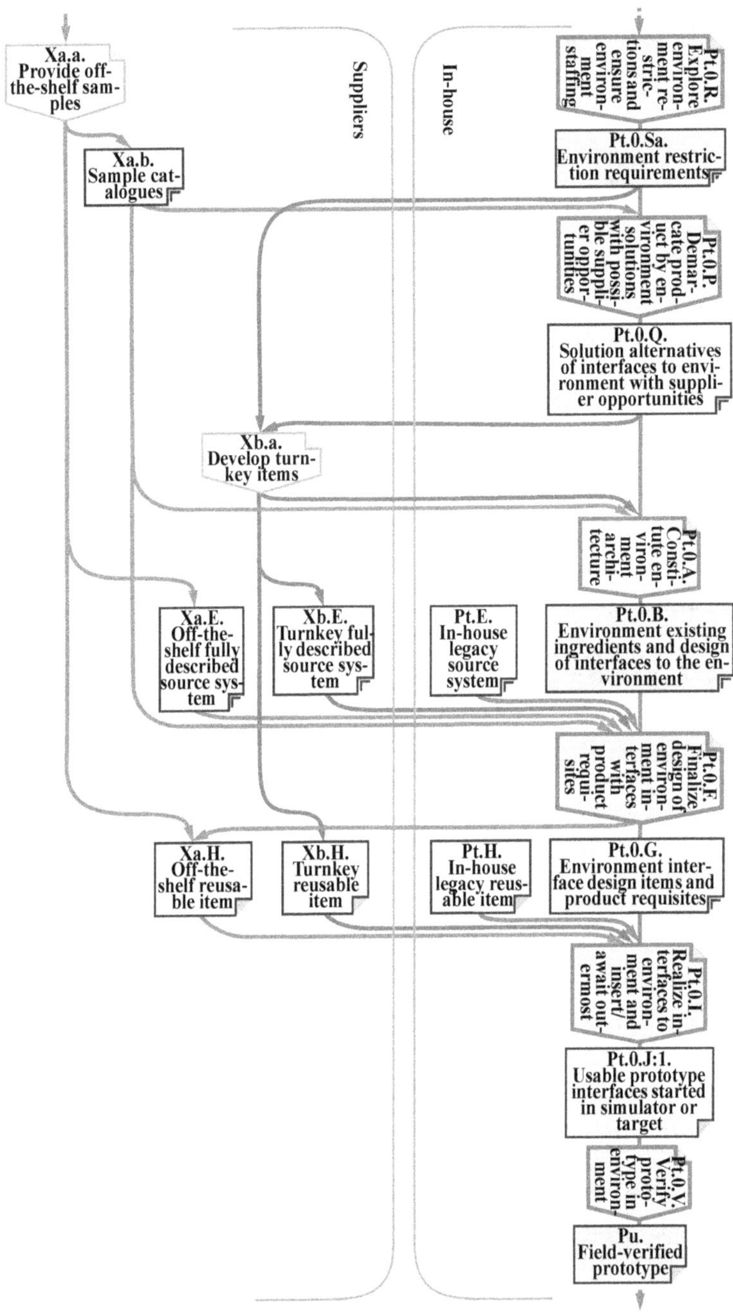

FIGURE 8-4 Process schedule for sourcing product interfaces to environment

The *environment nesting level* n = 0 has the following sourcing implications to handle:

Pt.0.Sa. Environment restriction requirements. *Environment restriction requirements* are the starting point for technical development and are possibly used by *turnkey suppliers* for the *environment interfaces*.

Pt.0.P. Demarcate product by environment solutions with possible supplier opportunities. Sometimes the *product* fits perfectly into the current *environment*, sometimes it requires some changes in the *environment*, and not seldom the *product* affects or replaces a lot of existing parts of the *environment*. By proposing *solution alternatives* for how the *product* can *interface* with the *environment*, the *product* to be developed is demarcated from its *environment*. *Sample catalogues* and *supplier* salespersons are used to identify *samples* that can match the *solution alternatives*.

Pt.0.Q. Solution alternatives of interfaces to environment with supplier opportunities. The *environment interfaces solution alternatives* are the selected *results* from executing the above *schedule*. *Turnkey suppliers* can be asked to propose their own *environment interfaces solution alternatives* for their offerings of *environment interfaces*.

Pt.0.A. Constitute environment architecture. It is now very important to document which *solution alternatives* are chosen. This depends a lot on what is found in *suppliers' sample catalogues* and which *suppliers* can develop *turnkey items*. All chosen *samples*, together with all ingredients developed in-house, must be included in the *architecture design* work, because it must become obvious that everything fits together in the very same *architectures*.

Pt.0.B. Environment existing ingredients and design of interfaces to the environment. *Results* of the execution of this *schedule* is the *environment architecture resumé, environment logical architecture*, and *environment physical architecture* containing existing and planned ingredients to be developed in-house and/or purchased from *suppliers*.

Pt.0.F. Finalize design of environment interfaces with product requisites. All *source systems* to the *product interfaces*, such as in-house and *supplier source systems* are here received and finally designed together with all design finalized in-house. Finally all orders are checked and sent to *suppliers* of *off-the-shelf items* for their delivery confirmation.

Pt.0.G. Environment interface design items and product requisites. This documented design now describes everything needed to realize the prototype *environment interfaces*, such as *environment interfaces design item* and *product requisites*. These constitute the base documentation for *interface realization*, which also will be the base for manufacturing.

Pt.0.I. Realize interfaces to environment and insert/await outermost white-box. All *off-the-shelf items* and *turnkey items* are received and integrated with all *in-house design items*, to become the *environment interfaces*. If normal inside-out *incorporative integration* is used, the prototype *white-box* is first developed and verified, and then connected by its *environment interfaces* and started for field *verification*. If outside-in *invasive integration* and inside-out *implicit verification* is used, the *interfaces* are realized and used from the start, before realizing the *product white-box*.

8.3 EXAMPLE House environment: Demarcate product by environment solutions with possible supplier opportunities

8.3.1 EXAMPLE House: About this example

Technical solutions for connections between a house and its *environment* often involve a discussion of what is available in the surrounding *environment*. If there is no public water available to the house, this is a restriction, and the solution will be to find water in another way.

Regarding the boundary solutions between the house and its *environment* (façade, ground, and roof), this will be specified as belonging to the house, and will be handled at the next *nesting level* 1.

8.3.2 EXAMPLE House environment: Process schedule to use

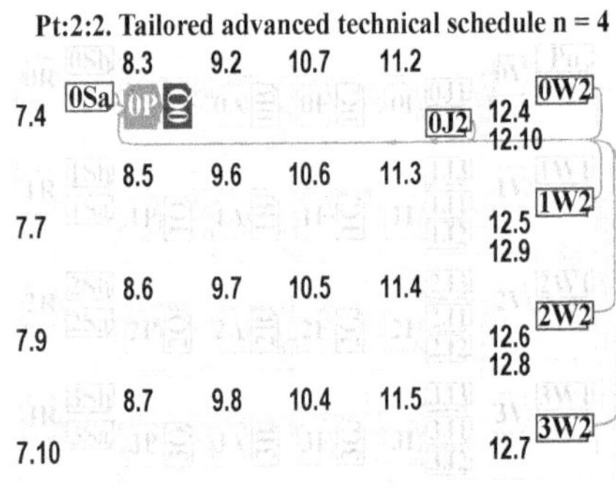

For the house example, a special technical overall *schedule* is *tailored* (illustrated in Fig. 8-5, left):

- **Pt:2:2. Tailored advanced technical schedule n = 4,** page 164

In this *tailored schedule* the generic *schedule* to use is (filled-in symbols):

- **Pt.0.P. Demarcate product by environment solutions with possible supplier opportunities,** page 112.

FIGURE 8-5 Position of schedule and master to use (filled in) in overall schedule

Below *results* may be demanded (their *masters* are indicated *upstream* from the filled-in *schedule* in the figure above):

0Sa Environment restriction requirements

Table 7-5, House environment restriction requirement, page 182

Xb Suppliers sample catalogues

For possible *failure localization* and *failure elimination*, below *results* may be demanded (their *masters* are indicated *upstream* from the filled-in *schedule* in the figure above or in greater detail in Fig. 11-5, p. 602 and Fig. 11-6, p. 603):

[0J2] Table 11-9. House environment start-up report 1, page 606

[0W2] List 12-1. House environment interfaces with fake house failure elimination report 1, page 674

[0W2] Table 12-27. House environment verification report 1, page 701

Table 12-29. House environment reverification report 2, page 703

Table 12-28. House environment failure elimination report 1, page 702

[1W2] List 12-2. House with fake rooms failure elimination report 1, page 676

[1W2] Table 12-22. House verification report 1, page 697

Table 12-24. House verification report 2, page 698

Table 12-23. House failure elimination report 1, page 698

[2W2] List 12-3. House rooms with fake machinery failure elimination report 1, page 678

[2W2] Table 12-14. House kitchen verification report 1, page 688

Table 12-17. House kitchen verification report 2, page 691

Table 12-15. House kitchen failure elimination report 1, page 689

[3W2] Table 12-9. House kitchen machinery verification report 1, page 682

Table 12-11. House kitchen machinery verification report 2, page 683

Table 12-10. House kitchen machinery failure elimination report 1, page 683

8.3.3 EXAMPLE House environment: Try to satisfy restrictions

How to satisfy the prepared house restrictions with *white-box solution alternatives* will now be analyzed, and *suppliers* for *samples* to the solutions will be considered. Penetrate the restrictions found in Table 7-5. House environment restriction requirement, page 182, and develop *environment solution alternatives* (see Table 8-5 below).

There might be more *white-box solution alternatives* found than those triggered by the specified restrictions. If so, go back to the capture staffing & requirements *phase* and add restrictions that are missing behind these newfound *white-box solution alternatives*. Always try to keep all development *results* updated and consistent with each other.

When there is more than one alternative, and alternatives are exchangeable, these are denoted with a, b, c, and so forth. Bulleted lines are comments.

TABLE 8-5 House environment solution alternatives of interfaces to environment

Interface solution alternative	Assessment	Environment restriction
E.SL1.a. Keep distance to shoreline as close as possible to 150 m.	No cost	E.R.1. Swedish environmental code, 7th chapter
E.SL1.b. Build on a higher building plot to keep lake view.	Elevate the plot ≈ € 6 000 / m	
E.SL1.c. Build higher house to keep lake view.	Elevate house ≈ € 9 000 / m	
• Building plot		
E.SL2.a. Extend the building plot by 2 000 m², to ensure a building area of 100 m².	≈ € 15 000	E.R.2. Malmö urban planning and local construction ordinance
E.SL2.b. Do not extend the building plot but build partly underground.	Underground ≈ € 1 000 / m²	
E.SL2.c. Do not extend the building plot but build smaller.	No cost	
• Building size		
E.SL3. When connections to the environment are constructed, existing neighboring environment may not be damaged.	Insurance cost ≈ € 1 800	R.3. Swedish national building regulations
E.SL4. Ramp from driveway to entry for disabled persons	≈ € 800	
E.SL5. The building plot must be located in an attractive area.	Plot extra cost ≈ € 100 / m²	R.4. Swedish Financial Supervisory Authority
E.SL6.a. Ø 32 mm gas pipe from street	≈ € 7 000	E.R.5. All public utilities available shall be used and comply with building standard 2011:45.
• Gas is available in street.	≈ € 500	
E.SL6.b. 3 m³ or 7 m³ rental tank close to house		
• Gas tank free, included in gas deliveries		
E.SL7.a. 16 Amperes from overhead lines	≈ € 1 000	
E.SL7.b. 32 Amperes from underground cable	≈ € 3 000	
• Electricity		
E.SL8.a. Drilled well with pump and purifier	≈ € 11 000	
E.SL8.b. Ø 32 mm water pipe from street	≈ € 4 000	
E.SL8.c. Pipe from lake with pump and purifier	≈ € 8 000	
• Water		
E.SL9.a. Waste-water treatment to lake recipient	≈ € 6 300	
E.SL9.b. Ø 125 mm pipe to street sewage	≈ € 2 000	
• Sewage		
E.SL10.a. Basalt-covered 2.4 m wide road	≈ € 15 000	
E.SL10.b. Four basalt-covered parking places, 2 m road extension at street, with 0.5 m wide basalt-covered footpath to house	≈ € 2 000	
• Driveway		
Minimum cost sum	≈ € 11 300	
Maximum cost sum	almost unlimited	

To facilitate which alternative to choose, there is also an assessment column, containing an approximate cost estimate for each alternative. Some of the costs are approximations, and expensive to estimate, for example, the cost per meter of height to elevate the building plot or the house. Costs are especially important for custom-built houses, because every new house is most often unique and has a different *environment*. The table might have more assessment columns to support decisions, for example, development time, or house *quality*. Decision makers then get more information when making up their minds.

It is already obvious that some decisions can be hard or frustrating to make at this early point in the technical *process*. For example, many of the *environment* connection dimensions depend on what equipment there will be in the house. But the great benefit of following a *process* is that all kinds of house questions get on the table. It is often better to be aware and make a qualified guess that later can be changed, than to forget this issue until it is quite expensive to fix (see ch. 6.3.11 "Applying a process schedule," p. 94).

Also note that the house restrictions are often massive, especially in urban areas. This house example is only a small selection, to show development principles.

8.3.4 EXAMPLE House environment: Analyze in/out and make/buy options

Now it must be finally decided which *white-box solution alternatives* are included in the house development, and which are handled separately (see Table 8-6 below). As can be seen from the table, costly solutions and management solutions are not included in the house development, but handled by the proprietor separately.

TABLE 8-6 House environment solutions excluded

Excluded from the product to be developed	Responsible	Environment solution alternative
SX.1 Buy beachside building plot larger than 2 000 m² with possible house ground at 150 m from the lake.	Proprietor	Building plot E.SL1.a., E.SL1.b., E.SL1.c.. E.SL2.a.., E.SL2.b., E.SL2.c.
SX.2. Avoid damage to neighbors' environment.	Proprietor	E.SL3.
SX.3. Secure adequate lake view.	Proprietor	E.SL5.
SX.4. Obtain water and electricity supply from environment while the house is being built.	Proprietor	Preliminary electricity and water E.SL7.a., E.SL7.b., E.SL8.a., E.SL8.b., E.SL8.c.
SX.5. Cooperate with municipal authorities.	Proprietor	Electricity and water E.SL7.a., E.SL7.b., E.SL8.a., E.SL8.b., E.SL8.c., E.SL9.a., E.SL9.b.

Connections between the existing *environment* and the house to be developed should now be considered and chosen. *Suppliers* from whom to order *samples* and *components* for these connections can also be predetermined (see Table 8-7 below).

TABLE 8-7 House environment solutions included with supplier opportunities

Included in the product with supplier opportunity	Environment solution alternative
SU.10. House builder's suppliers • Use ordinary building material, from selected house builder suppliers.	Electricity, water and sewage E.SL7.a., E.SL7.b., E.SL8.a., E.SL8.b., E.SL8.c.
SU.11. Electricity supplier • To approve connection of electricity	E.SL7.a., E.SL7.b.
SU.12. Water supplier • To approve connection of water and sewage	E.SL8.a., E.SL8.b., E.SL8.c., E.SL9.a., E.SL9.b.

TABLE 8-7 House environment
solutions included with supplier opportunities

Included in the product with supplier opportunity	Environment solution alternative
SU.13. Gas supplier • Use only authority-approved equipment. Use tank provided by gas supplier.	Gas pipe and tank E.SL6.a.., E.SL6.b.
SU.14. Ext Bas AB • Use stone sample H34.	Driveway and ramp E.SL4., E.SL10.a., E.SL10.b.
SU.15. Local supplier • Use natural material harmonizing with the environment from local suppliers.	Driveway E.SL10.a., E.SL10.b.

8.3.5 EXAMPLE House environment: Examine product scope

The above make-buy analysis and demarcations are reviewed by the proprietor and building *stakeholders* and settled as above.

A contract is prepared and signed between the proprietor and the Robber HB architects company. A quality inspector from Q-Inc is contracted and approved by Robber HB.

8.3.6 EXAMPLE House: Proceed reading

To follow this example, these are the alternatives (again see Fig. 8-5, p. 290):

0A Proceed to chapter 9.2, "EXAMPLE House environment: Constitute environment architecture," page 339.

0F Proceed to chapter 10.7, "EXAMPLE House environment: Finalize design of environment interfaces with product requisites," page 477.

0V Proceed to chapter 12.4, "EXAMPLE House environment: Partially verify interfaces to environment with house fake," page 673.

0V Proceed to chapter 12.10, "EXAMPLE House environment: Fully verify prototype in environment," page 699.

• To leave the house example and to dig deeper into the chapter 8, "Predetermine solutions & suppliers" topic, proceed to chapter 8.4 below.

8.4 Predetermine solutions with sourcing options (nesting levels n ≥ 1)

8.4.1 Purpose

Requirements will provide solution alternatives, for both interfaces to, and elements within all black-boxes.

In the above chapters about *environment restriction requirements*, there were no other *requirements* than just restrictions. On all *nesting levels inwards* from these, refined *restriction requirements* are still applied to the *black-box(es)*, but there are also a lot of refined and freely added *requirements* capturing *black-box* behavior.

These restriction and *behavior requirements* are all applied to *black-boxes*, which implies that not only *white-box solution alternatives* for *black-box interfaces* are considered, but also *white-box solution alternatives* for filling the whole *black-box* with *architecture ingredients* when it is opened to a *white-box*.

The **Pt.n.Pa. Try to satisfy requirements** *task* has to propose solutions as well as possible to satisfy the specified set of *requirements*. It is especially important that the alternatives analysis tries to include solutions that can already be obtained from *suppliers*. Furthermore, if a *requirement* is a design restriction, such design solutions must be included as proposed solutions.

Sometimes there are many *architectures* and *elements* that might satisfy a given set of *requirements*, and it is important that all reasonable solution proposals be documented from the solutions analysis. Most often the total *requirements* are too hard to all realize, and no solution proposal is capable of satisfying them all. In this case, *requirements* must be sacrificed according to their prioritization, until a reasonable solution is possible. Such compromises must also be accepted by the **Pt.n.Pe. Examine make/buy balance** gate meeting.

When *white-box solution alternatives* are proposed, every solution is scrutinized for the following judgments:

- Instead of in-house development of a solution, can better solutions be found by *suppliers?*
- Instead of in-house *architecture design*, can more available or more competent developers be found externally?
- Instead of in-house *realization* and *verification*, can more available or more competent developers be found externally?

In extreme cases, in-house development will only result in adding "glue" *architecture ingredients* between all externally found solutions. But normally, the development will result in both smaller and bigger parts being brought in from *suppliers* as well as smaller and bigger parts developed on one's own.

8.4.2 Cpdm definitions

In *Cpdm*, *white-box solution alternatives* and *white-box supplier opportunities* are defined according to Table 8-8 below.

TABLE 8-8 Cpdm definitions

Aspect	White-box solution alternatives	White-box supplier opportunities
Defini-tion	Various solution proposals for how to best possible satisfy one or several specified requirements	Proposals of which suppliers (including in-house) can possibly provide appropriate samples to match specified solution alternatives
Syno-nyms	Impact analysis, implementation proposal, solution alternatives, preliminary design	Approved suppliers and components.
Table	TABLE 8-9 White-box solution alternatives | SL.n. Solution alternative	Assess-ment	Require-ment	Pri-ority
---|---|---|---|	TABLE 8-10 White-box supplier opportunities | SU.n. Supplier opportunity	Solution alternative
---	---	

8.4.3 Different kinds of predetermined solutions

As can be seen in the *process schedule* in Figure 8-6 above, there are a lot of possible predetermined solutions, which later lead to different *architecture ingredients*.

TABLE 8-11 Explanation and exemplification of predetermined solutions.

		Source system item	Reusable item
		A source system is delivered before finalizing the design, with complete documentation that makes it possible to add to in-house development. For example, source systems contain mechanical strength calculations, electronic circuitry drawings, and programs.	A reusable sample solution, often called component, is offered to be easily included at integration time, but contains documentation only for this purpose. If the item fits in poorly, it is hard to change and must most often be exchanged for something else fitting better.
Turnkey item	A turnkey solution is developed according to own available requirements. Such solutions are custom-made and fit well, but might suffer from "early teething" problems, because there is feedback from only one user.	**Xb.E. Turnkey fully described source system** **Pt.E. In-house legacy source system** Both solutions fit very well from the start of development and allow the best freedom to change whatever does not fit. The penalty is the high cost not shared by anybody when developing a unique source system.	**Xb.H. Turnkey reusable item** These solutions are generally expensive. However, it is possible to order well-fitting turnkey component solutions at reasonable cost, if the supplier also expects to deliver training, service, and extensions of this solution in the future. Or if the supplier can transform it to an own off-the-shelf item.
Off-the-shelf item	An off-the-shelf solution is developed under hard competition to fit many different customers. Thus it is always a compromised and optimized solution, and a cheap and well-proven volume product.	**Xa.E. Off-the-shelf fully described source system** Sometimes a source system might be available off the shelf, well proven and easy to change. A new trend is to give away excellent program source systems for free, in order to capitalize on an expected boom in this community growth.	**Xa.H. Off-the-shelf reusable item** This is a very common way to procure a cheap solution. For example, house building blocks, and electronic components. Programs are often sold as licensed applications, which might be expensive per unit if volume grows high.

8.4.4 EXAMPLE: Analyze make/buy options for some well-known solutions

To make Table 8-11 above more understandable, it can be interesting to look at some existing and well-known solutions.

Xb.E. Turnkey fully described source system examples: Special customized constructions delivered with full documentation, like a detached house with finished landscaping, complete road construction, digital artwork with source images, *program source systems* and so forth.

Xb.H. Turnkey reusable item examples: Designed labels and customized market material, house alarm system, specially designed *programs* (delivered as linkable files), and so on.

Xa.E. Off-the-shelf fully described source system examples: Free download of Linux operating system, Android mobile phone, and Magento e-commerce. Another way to obtain a *source system* might be to buy the company that develops it, for example, Microsoft buying Skype, Nokia buying Scalado.

Xa.H. Off-the-shelf reusable item examples: House building blocks such as windows, doors, bricks, kitchen machinery. Electronic *components* such as resistors, capacitors, inductors and so forth. Also immaterial *design items* such as Microsoft Windows and Office, Adobe Photoshop and others.

Pt.E. In-house legacy source system examples: Quite seldom *products* are developed from scratch; most are extensions of previous *versions*. *Configuration management*, not extensively discussed in this book, must be in place to keep control of extending a *complex product*.

8.4.5 Sourcing affects all black-/white-box schedules

Everything on the current black-box nesting level is affected by sourcing items from suppliers.

The sourcing from *suppliers* affects all *phases* within a *black-/white-box nesting level* (see Fig. 8-6 below). The sourcing *process* for *white-boxes* is rather similar to sourcing for the *environment interfaces* (see again ch. 8.2.3, p. 287).

Crucial *samples*, maybe not yet on the market, and considered to be a big competitive advantage, may have already been developed by trusted *suppliers* or in-house.

Each and every *white-box* has the following procurement to handle:

Pt.n.Sa. Prioritized requirements. *Black-box requirements* are the starting point for in-house development and *turnkey suppliers* of the *white-boxes*.

Pt.n.P. Predetermine solutions with sourcing options. *Requirements* are satisfied with proposed *white-box solution alternatives,* which then get matched to *suppliers' sample catalogues* or through their salespersons. The other way around is also possible, which is to use *suppliers' sample catalogues* to catalyze own proposals of *white-box solution alternatives.*

Pt.n.Q. Solution alternatives with supplier opportunities. The *results* from executing the above *schedule* are proposed *white-box solution alternatives* and *white-box supplier opportunities. Suppliers* providing *turnkey items* can be asked to propose their own *white-box solution alternatives* for the *white-box.*

Pt.n.A. Satisfy requirements by decomposing outward nesting level black-box into white-box design containing possible black-boxes. It is now important to document which *white-box solution alternatives* to choose for the *white-box.* This depends a lot on what is found in *suppliers' sample catalogues* or otherwise must be developed by *turnkey suppliers.* All chosen *samples,* together with all ingredients developed in-house, must be included in the *architecture design* work, because they must all fit well together in the very same *architectures.*

Pt.n.B. Architecture white-box ingredients design with possible embedded black-boxes. *Results* when executing the above *schedule* are *architecture ingredients, logical architecture,* and *physical architecture* containing planned ingredients to be developed in-house and/or purchased from *suppliers.*

Pt.n.F. Finalize design of white-box with product requisites. All *source systems* useful for the *white-box,* such as in-house and *supplier source systems* are received here and finally designed together with all finalized in-house design. Finally, all orders are checked and sent to *suppliers* of *off-the-shelf items* for their delivery confirmation.

Pt.n.G. White-box design items and product requisites. This documented design now describes everything needed to realize the *product* and *embedded white-boxes,* such as *white-box design items* and *product requisites.* These constitute the base documentation for the *white-box realizations,* and will also be the base for manufacturing.

Pt.n.I. Procure items, linkable programs, and integrate/await embedded black-boxes in order to realize white-box. All *off-the-shelf items* and *turnkey items* are received and integrated with all *in-house design items,* to become the *white-box.* If normal inside-out *incorporative integration* is used, all verified *embedded white-boxes* and received *white-box design items* are integrated to the *white-box,* which is started for *verification.* If outside-in *invasive integration* and inside-out *implicit verification* are used, the *white-box* is filled with received *white-box design items* and *embedded white-boxes* are awaited to be developed and finally verified.

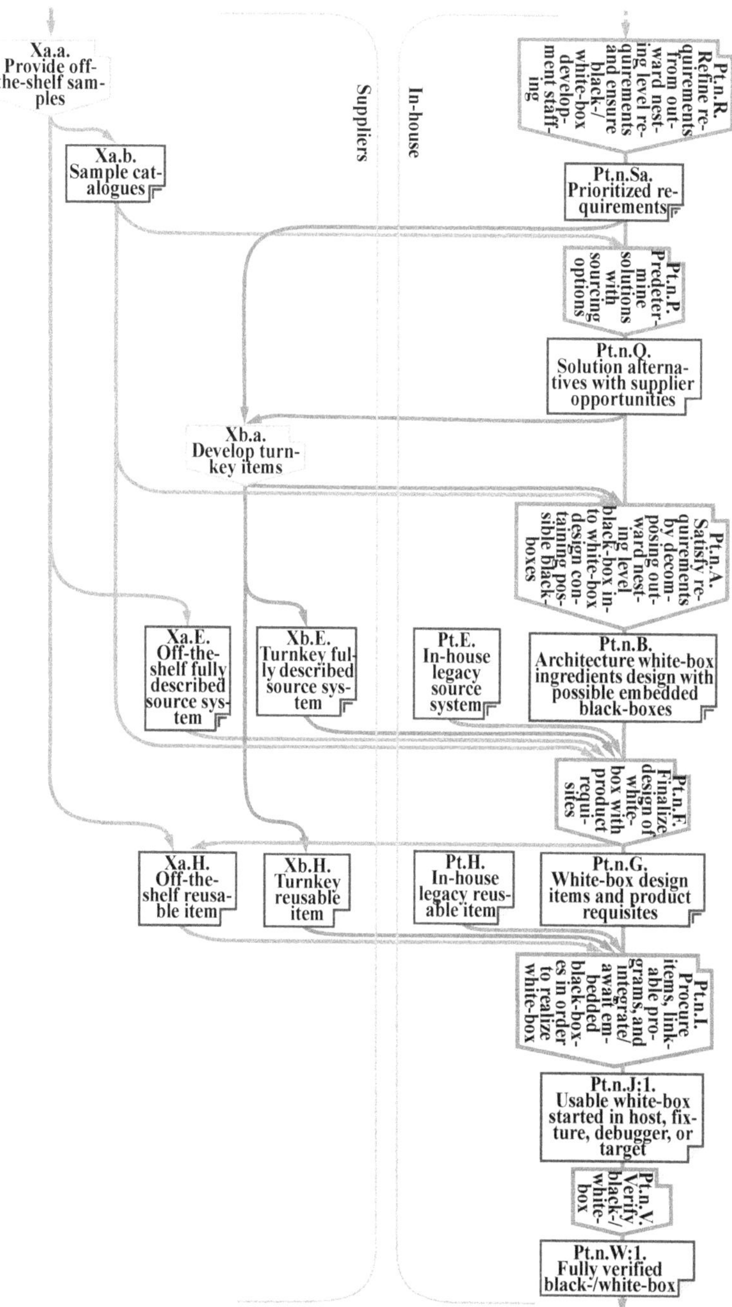

FIGURE 8-6 Sourcing of product and embedded white-boxes

8.5 EXAMPLE House: Predetermine solutions with sourcing options

8.5.1 EXAMPLE House: Process schedule to use

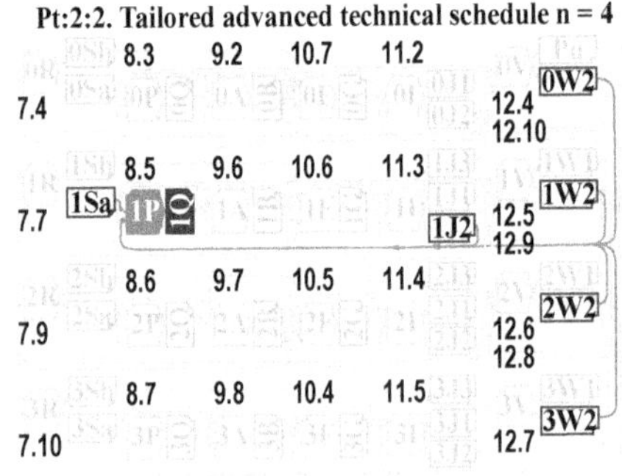

For the house example, a special technical overall *schedule* is *tailored* (illustrated in Fig. 8-7, left):

- **Pt:2:2. Tailored advanced technical schedule n = 4**, page 164

In this *tailored schedule* the generic *schedule* to use is (filled-in symbols):

- **Pt.n.P. Predetermine solutions with sourcing options**, page 133.

FIGURE 8-7 Position of schedule and master to use (filled in) in overall schedule

Below *results* may be demanded (their *masters* are indicated *upstream* from the filled-in *schedule* in the figure above):

[1Sa] House prioritized requirements
Table 7-9, House black-box compound requirements (small selection), page 189

[Xb] Suppliers sample catalogues

For possible *failure localization* and *failure elimination*, below *results* may be demanded (their *masters* are indicated *upstream* from the filled-in *schedule* in the figure above):

[1J2] Table 11-14. House start-up report 1, page 612

[0W2] List 12-1. House environment interfaces with fake house failure elimination report 1, page 674

[0W2] Table 12-27. House environment verification report 1, page 701
Table 12-29. House environment reverification report 2, page 703
Table 12-28. House environment failure elimination report 1, page 702

[1W2] List 12-2. House with fake rooms failure elimination report 1, page 676

[1W2] Table 12-22. House verification report 1, page 697
Table 12-24. House verification report 2, page 698
Table 12-23. House failure elimination report 1, page 698

[2W2] List 12-3. House rooms with fake machinery failure elimination report 1, page 678

[2W2] Table 12-14. House kitchen verification report 1, page 688
Table 12-17. House kitchen verification report 2, page 691

Table 12-15. House kitchen failure elimination report 1, page 689

3W2 Table 12-9. House kitchen machinery verification report 1, page 682

Table 12-11. House kitchen machinery verification report 2, page 683

Table 12-10. House kitchen machinery failure elimination report 1, page 683

8.5.2 EXAMPLE House: Try to satisfy requirements

The prioritized *requirements* identified (see Table 7-9, p. 189) are now analyzed, and *white-box solution alternatives* are proposed (see Table 8-12 below).

Note that this example has only a limited selection of *requirements* and *white-box solution alternatives* presented in this book. In reality, many more *requirements* than illustrated would certainly have increased the number of solutions. In reality, there would also be many more *white-box solution alternatives* than squeezed into the tables in this book.

The price per m² in the assessment is a standard estimate of € 1 000 / m², which includes material and building work for the house boundary framework and inner wall framework. To this add costs for specific house material for windows, doors, heating, water, gas, and so forth. Development cost is not included in these costs. Room boundaries, interiors, and machinery, with additional costs, will be estimated in coming development.

Another disclaimer may be in place here. In this chapter and many others a lot of costs are presented, with totals, subtotals, and accumulated totals. In reality, such cost lists should be *managed information* in a calculator spreadsheet for convenient sorting and correct summation, and such functions. In the editor used for writing this book there were no calculator facilities, hence, there may be some inaccuracies in the sums. And again, this is a book about how to master *complex* development, and not about correct solutions.

TABLE 8-12 House white-box solution alternatives

Solution alternative	Assess-ment	Requirement	Pri-ority
A.SL1. Kitchen window around 1 m²	≈ € 700	A.R.1.1. During cooking and dining, the lake shall be as visible as possible.	High
A.SL2. Dining room double outer glass door	≈ € 1 400		
A.SL3. Add 1 m² to dining room door space.	≈ € 1 000		
A.SL4. Basalt stone on walls and terracotta on roof	Included in m² cost	A.R.1.2. Façade and roofing material appearance shall be chosen by proprietor.	High
A.SL5. All windows and doors custom-made			
A.SL6. Use façade brickwork and roofing tiles.	Included in m² cost	A.R.2.1. The house style shall fit in with neighbors' houses.	Low
A.SL7. Big toilet with handles, add 2 m² to toilet room.	≈ € 2 000	A.R.2.2. The house shall be accessible for disabled persons.	Mid
A.SL8. Use wide doors without door steps.	No extra cost		
i27. Architect responsibility (see p. 190). • (No solution alternatives applicable).	-	A.R.2.3. A building permit is mandatory before building starts.	Crucial

TABLE 8-12 House white-box solution alternatives

Solution alternative	Assessment	Requirement	Priority
A.SL9. The house walls shall be of stone and bricks, and the house vaults of cast concrete.	Included in m² cost	A.R.3.1. The house shall follow proven methods, regulations, and standards.	High
A.SL10. Applied insulation solution shall achieve less than 90 kWh/m²a energy consumption.	Included in m² cost	A.R.3.2. The house shall have an economical energy consumption.	High
A.SL11. Use well-accepted house style and finish, to be judged valuable if the house is to be sold.	No extra cost	A.R.4.1. The house shall be attractive to allow maximum bank mortgage.	Crucial
A.SL12. Make simple utility room for all house connection interfaces in secondary space under the staircase.	Included in m² cost	A.R.5.1. Provide the house with a utility room for all public utilities.	Crucial
A.SL13.a. Separate dining room from other rooms with doors only.	≈ € 1 000	A.R.6.1. The atmosphere during dining shall be relaxed.	Mid
A.SL13.b. Separate dining room from other rooms with doors and from kitchen with an arch.	≈ € 1 000		
A.SL14. Use 12 m² of house for dining room.	≈ € 12 000	A.R.6.2. The dining space shall support up to 8 guests.	High
A.SL15. Use 2 m² of house for serving space.	≈ € 2 000	A.R.6.3. Serving support shall exist for one course at a time.	Low
A.SL16.a. Use 3 m² of house for kitchen with one chef.	≈ € 3 000		
A.SL16.b. Use 5 m² of house for kitchen with two chefs.	≈ € 5 000		
A.SL17. Use 4 m² of house for kitchen food preparation space.	≈ € 4 000		
A.SL18. Use 3 m² of house for kitchen grocery storage.	≈ € 3 000	A.R.6.4. Storing, cooking, and dining support shall allow a 5-course French menu for 8 persons.	Mid
A.SL19.a. Use 2 m² of house for kitchen tableware storage.	≈ € 2 000		
A.SL19.b. Use 2 m² of house for dining room tableware storage.	≈ € 2 000	A.R.6.5. Dishwashing support needed, shall hold complete dinner.	Crucial
A.SL20. Use 2 - 3 m² of house for kitchen dishwashing.	≈ € 2 000 - € 3 000		
A.SL21. Use 2 m² of house for kitchen cooking utensils.	≈ € 2 000	A.R.6.6. Cooking shall support warming, boiling, frying, wokking, and baking.	Crucial
A.SL22. Use 2 m² of house for kitchen cooking equipment.	≈ € 2 000		
A.SL23. Supply kitchen with cold and warm water.	≈ € 2 000		
A.SL24. Supply kitchen with 3-phase electricity.	≈ € 2 000		
A.SL25. Supply kitchen with floor heating.	≈ € 4 000		
A.SL26. Supply kitchen with stove gas.	≈ € 1 000		
A.SL27. Supply kitchen with sewage.	≈ € 2 000		
A.SL28. Kitchen needs forced suction 12 m³ / hour.	≈ € 700	A.R.6.7. Smell from cooking shall be eliminated.	High
A.SL29. Use 1 m² of house for kitchen disposal space.	≈ € 1 000	A.R.6.8. Waste and garbage from cooking shall be odorlessly disposed of.	Mid
A.SL30. House needs a kitchen disposal ventilation 2 m³ / day.	≈ € 200		

TABLE 8-12 House white-box solution alternatives

Solution alternative	Assess-ment	Requirement	Pri-ority
A.SL31.a. Use 20 m² of house for dining room big parties.	≈ € 20 000	A.R.7.1. Good space shall exist for 24 party guests, half standing, half sitting.	Mid
A.SL31.b. Use 12 m² of house for dining room small parties.	≈ € 12 000		
A.SL32. House needs dining room with forced ventilation 2 m³/hour.	≈ € 2 000		
A.SL33. Use another 2 m² of house for kitchen party meals.	≈ € 2 000	A.R.7.2. Serving snack meals shall support 12 party guests.	Low
A.SL34. Use 1 m² of house for kitchen party grocery storage.	≈ € 1 000	A.R.7.3. Storing capacity shall exist for 24 party guests' snack meals.	Low
A.SL35.a. Use 2 x 4 m² of house for two bathrooms.	≈ € 8 000	A.R.7.4. Toileting capacity shall exist for 2 guests simultaneously.	High
A.SL35.b. Use 5 m² of house for one bathroom.	≈ € 5 000		
A.SL36. Use 3 m² of house for entrance room.	≈ € 3 000	A.R.7.5. Hanger capacity shall exist for 24 coats and such.	Mid
A.SL37. Utility room shall be lockable.	≈ € 10	A.R.9.1. The house shall be safe for children.	High
Realization minimum cost sum	≈ € 76 010		
Realization maximum cost sum	≈ € 90 010		

Note that *requirements* with an identifier higher than R.9.1. are not presented in the table above, due to lack of space in this book.

The house construction, apart from proprietor's wishes, is strictly regulated by authorities and standardization, captured by *restriction requirements* such as R.3.1. The house shall follow proven methods, regulations, and standards. In reality, this *requirement* will certainly affect more solutions than its single appearance in the above downsized house example.

8.5.3 EXAMPLE House: Analyze make/buy options

A make/buy analysis is performed (see Table 8-13 below):

TABLE 8-13 House white-box supplier opportunities

Supplier opportunity	Solution alternative
SU16. House builder's suppliers • Officially approved supplier must be used for gas.	Utility room for water, heating, gas and sewage A.SL12. Piping for water, heating, gas and sewage A.SL23., A.SL25., A.SL26., A.SL27. Ventilation A.SL28., A.SL30., A.SL32..
SU17. AllBau • The house proprietor has a price discount to take advantage of at AllBau for the house builder selected later. • The AllBau building material catalogue with data and dimensions will be provided for the house builder by Robber HB.	A.SL3., A.SL7.,A.SL9., A.SL11., A.SL14., A.SL15., A.SL16.a., A.SL16.b., A.SL17., A.SL18., A.SL19.a., A.SL19.b., A.SL20., A.SL21., A.SL22., A.SL29., A.SL31.a., A.SL31.b., A.SL33., A.SL34., A.SL35.a., A.SL35.b., A.SL36.
SU18. All-in-Stone	Façade and roof material A.SL4., A.SL6.

TABLE 8-13 House white-box supplier opportunities

Supplier opportunity	Solution alternative
SU19. Pa&Pp • Turnkey custom-made by high quality supplier Pall & Papp • Robber HB provide dimensions to architects and to Pall & Papp.	Doors and windows A.SL1., A.SL2., A.SL5., A.SL8., A.SL13.a., A.SL13.b., A.SL37.
SU20. Elert's suppliers	Electricity A.SL12., A.SL24.
SU21. IsoBau's suppliers • Robber HB is already allocated to design the house to meet insulation requirements, and can contract additional specialist if needed. • Robber HB will also investigate if extraordinary insulation solutions are needed, not available from selected supplier. • Robber HB will notify the design decision gate if additional suppliers are needed.	Economical energy house A.SL10.
SU22. Local suppliers	Unspecified or common available building material A.SL11.

8.5.4 EXAMPLE House: Examine make/buy balance

Robber HB presented a well-known artistic landscape *architect*, and she was chosen to take part in the house *architecture design* work. Also two room layout architects were presented by Robber HB, and they were both included in the house interior *architect* work.

The documentation for the house *development staffing* allocation, house *solution alternatives*, and house *supplier opportunities* was sent to the house proprietor and Robber HB for review. After some minor corrections, they were approved by the house proprietor

The management also appointed Robber HB to monitor *supplier* progress and delivery of selected *turnkey items*.

8.5.5 EXAMPLE House: Proceed reading

To follow this example, these are the alternatives:

1A Proceed to chapter 9.6, "EXAMPLE House: Satisfy house requirements by decomposing house black-box into white-box design containing room black-boxes," page 357.

1F Proceed to chapter 10.6, "EXAMPLE House: Finalize design of house with product requisites," page 465.

1V Proceed to chapter 12.5, "EXAMPLE House: Partially verify black-/white-box with embedded fakes," page 675.

1V Proceed to chapter 12.9, "EXAMPLE House: Fully verify house black-/white-box," page 693.

• To stay within the house example, but not following the overall *schedule*, proceed to chapter 8.6. below.

8.6 EXAMPLE House rooms: Predetermine solutions with sourcing options

8.6.1 About the house kitchen

The house example is too large for all designed rooms, room *interfaces*, room walls, ceilings, and floors to be illustrated in this book. The kitchen is chosen because this is one of the more *complex* parts of a house, particularly when it comes to satisfying a demanding chef.

8.6.2 House kitchen: Process schedule to use

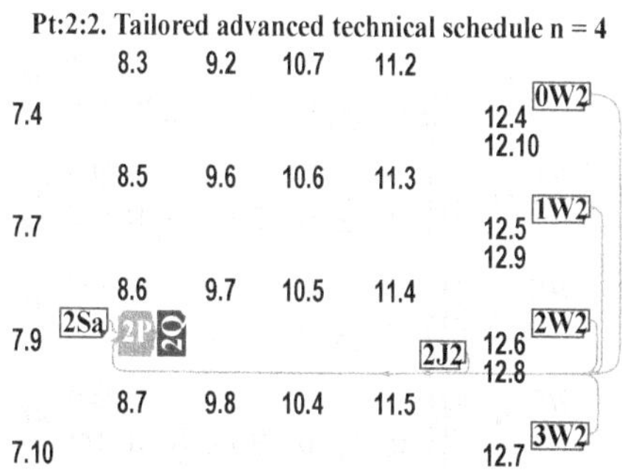

For the house example, a special technical overall *schedule* is *tailored* (illustrated in Fig. 8-8, left):

• **Pt:2:2. Tailored advanced technical schedule n = 4**, page 164

In this *tailored schedule* the generic *schedule* to use is (filled-in symbols):

• **Pt.n.P. Predetermine solutions with sourcing options**, page 133.

FIGURE 8-8 Position of schedule and master to use (filled in) in overall schedule

Below *results* may be demanded (their *masters* are indicated *upstream* from the filled-in *schedule* in the figure above):

2Sa Rooms prioritized requirements
Table 7-11, House kitchen black-box compound requirements (small selection), page 195

Xb Suppliers sample catalogues

For possible *failure localization* and *failure elimination*, below *results* may be demanded (their *masters* are indicated *upstream* from the filled-in *schedule* in the figure above):

2J2 Table 11-18. House kitchen start-up report 1, page 617

0W2 List 12-1. House environment interfaces with fake house failure elimination report 1, page 674

0W2 Table 12-27. House environment verification report 1, page 701
Table 12-29. House environment reverification report 2, page 703
Table 12-28. House environment failure elimination report 1, page 702

1W2 List 12-2. House with fake rooms failure elimination report 1, page 676

1W2 Table 12-22. House verification report 1, page 697

Table 12-24. House verification report 2, page 698

Table 12-23. House failure elimination report 1, page 698

[2W] List 12-3. House rooms with fake machinery failure elimination report 1, page 678

[2W] Table 12-14. House kitchen verification report 1, page 688

Table 12-17. House kitchen verification report 2, page 691

Table 12-15. House kitchen failure elimination report 1, page 689

[3W] Table 12-9. House kitchen machinery verification report 1, page 682

Table 12-11. House kitchen machinery verification report 2, page 683

Table 12-10. House kitchen machinery failure elimination report 1, page 683

8.6.3 EXAMPLE House kitchen: Try to satisfy requirements

The K-Studio came up with good *white-box solution alternatives* (see Table 8-14 below). The assessment column indicates price for material and work for the interiors. Many *requirements* capture kitchen machinery, which has to be connected to water, sewage, gas, ventilation, and electricity, which will add costs to the kitchen.

Requirements are also omitted for kitchen boundary appearance (but included in the *architecture*, Fig. 9-26, p. 369). Thus, there are too few *white-box solution alternatives* for this, and true kitchen cost will be substantially higher than indicated in the assessment sum.

The assessment is mainly *realization* of material costs. To this are added some *realization* work costs, which are included in the assessment.

TABLE 8-14 House kitchen white-box solution alternatives

Solution alternative	Assess-ment	Requirement	Prior-ity
AaC.SL.1. Place preparation area in front of window.	**No extra cost**	A.aC.R.1.1.1.The lake shall be visible from kitchen food preparation area.	High
• No particular solutions needed		A.aC.R.2.2.1.Kitchen may be inaccessible by disabled guests.	Mid
AaC.SL.2a. Walls of plastered bricks **AaC.SL.2b. Walls of boards on joists**	≈ € 2 000 ≈ € 1 000	A.aC.R.3.1.1.Kitchen shall follow established norms, regulations, and standards.	Crucial
AaC.SL.3. Have proprietor check quality before paying invoices.	**Heighten quality**	A.aC.R.3.1.2.Kitchen interiors may not have any damage, scratches, or stains.	High
AaC.SL.4. Have proprietor choose supplier and material before ordering.	**For attrac-tiveness**	A.aC.R.3.1.3.Kitchen interior appearance must be chosen by proprietor.	Crucial
AaC.SL.5. Use only ordinary interiors for kitchen.	**Saves cost**	A.aC.R.4.1.1.Kitchen including interiors shall cost maximally 20 000 € .	Crucial
AaC.SL.6. Choose only few sample suppliers to match proposed solutions.	**Saves cost**	A.aC.R.4.1.2.Purchase only ordinary interiors from few suppliers to keep prices down.	High
AaC.SL.7a. Estimate totally 6 m² stone preparation area. **AaC.SL.7b. Estimate totally 5 m² laminated preparation area.**	≈ € 1480 ≈ € 300	A.aC.R.6.4.1.Include large food preparation area with hot and cold water.	High

TABLE 8-14 House kitchen white-box solution alternatives

Solution alternative	Assessment	Requirement	Priority
AaC.SL.8a. Use two revolving-shelf cabinets near counter. AaC.SL.8b. Use two fixed-shelf cabinets near counter.	≈ € 940 good access ≈ € 620 low capacity	A.aC.R.6.4.2.Storage shall hold 12 pans of 3 liters near cooking area.	High
AaC.SL.9. One wide drawer for baking sheets AaC.SL.10. One cupboard for bowls	≈ € 130 ≈ € 130	A.aC.R.6.4.3.Storage shall hold 6 plates and 6 bowls near baking area.	High
AaC.SL.11a. Use magnet knife hanger at side of preparation area. AaC.SL.11b. Use small knife drawer under countertop.	≈ € 12 ≈ € 140	A.aC.R.6.4.4.Knife storage shall be child-safe and near preparation area.	Mid
AaC.SL.12. Use drawer section 60 x 60 cm beneath counter. AaC.SL.13. Use drawer section 40 x 60 cm beneath counter.	≈ € 430 ≈ € 300	A.aC.R.6.4.5.Storage shall hold all kinds of cutlery and kitchen utensils which shall be easily accessible from preparation area.	High
AaC.SL.14. Cold section 60 x 60 x 200 cm for chilled and frozen food AaC.SL.15. Full-height cupboard 60 x 40 x 200 cm for room-temperature provisions AaC.SL.16. Half-height cupboard 60 x 40 x 100 cm for room-temperature provisions	≈ € 600 ≈ € 640 ≈ € 300	A.aC.R.6.4.6.Storage shall hold 8 bags (of 30 liters each) of groceries, 4 of which are chilled, and 1 frozen. A.aC.R.7.3.1.Storage shall hold 24 snack meals (2 grocery bags).	Crucial Low
AaC.SL.17. Machinery dishwashing cabinet section 120 cm wide under preparation area	≈ € 780	A.aC.R.6.5.1.Dishwashing shall clean 5-course tableware (5 x 8 plates, cutlery, glasses, and cooking utensils) in same run.	Mid
AaC.SL.18. Partition 1 m² manual dishwashing section of left preparation area for this.	≈ € 100	A.aC.R.6.5.2.Gentle dishwashing shall support fragile tableware.	Crucial
AaC.SL.19. Chemicals cupboard, child-safe height, separate from medicines	≈ € 80	A.aC.R.6.5.3.Child-safe storage shall hold chemicals for dishwashing etc.	Low
AaC.SL.20. Cabinet section for warming equipment	≈ € 330	A.aC.R.6.3.1.Cheap equipment keeps one course warm during eating.	Mid
		A.aC.R.6.6.1.Food thawing shall be near preparation area.	High
		A.aC.R.6.6.2.Warming of all kinds of food shall be near preparation area.	High
AaC.SL.21. Partition 3 m² heating area in center of preparation area.	≈ € 100	A.aC.R.6.6.3.Food boiling shall be near preparation area.	Crucial
		A.aC.R.6.6.4.Food frying shall be near preparation area.	Crucial
		A.aC.R.6.6.5.Food wokking shall be near preparation area.	Mid
AaC.SL.22. Section 50 x 90 cm under preparation area.	≈ € 260	A.aC.R.6.6.6.Food baking shall be near preparation area.	High
AaC.SL.23. Evacuation equipment 60 cm over heating area	≈ € 160	A.aC.R.6.7.1.Odor from cooking may not spread to other rooms.	High
AaC.SL.24. Airtight cupboard 60 x 60 x 60 cm under sink	≈ € 440	A.aC.R.6.8.1.Garbage odor may not be recognizable in kitchen.	Mid
AaC.SL.25. Small cupboard 200 cm above floor	≈ € 140	A.aC.R.9.1.1.Kitchen shall hold family medicines in child-safe way.	Low
Realization minimum cost sum Realization maximum cost sum	≈ € 6 852 ≈ € 9 480		

Assessments of needed kitchen area, and the house cost for this, is already done in Table 8-12, page 301.

8.6.4 EXAMPLE House kitchen: Analyze make/buy options

The following make/buy analysis is performed (see Table 8-15 below).

TABLE 8-15 House kitchen white-box supplier opportunities

Supplier opportunity	Solution alternative
SU.23 House builder with subcontractors • To build the inside of kitchen walls, no longer visible when finished	AaC.SL.2a.. AaC.SL.2b.
SU.24 K-Studio • Robber HB and K-Studio share dimensioned kitchen architectures. • Cabinets can be adapted (e.g. holes for interface connections) on spot when later mounted in the kitchen.	Cabinets, cupboards, and drawers AaC.SL.5., AaC.SL.6., AaC.SL.8a., AaC.SL.8b., AaC.SL.9., AaC.SL.10., AaC.SL.11a., AaC.SL.11b., AaC.SL.12., AaC.SL.13., AaC.SL.14., AaC.SL.15., AaC.SL.16., AaC.SL.17., AaC.SL.19., AaC.SL.20., AaC.SL.22., AaC.SL.23., AaC.SL.24., AaC.SL.25.
SU.25 K-Studio's suppliers • To be turnkey custom-made. K-Studio to provide dimensions	Counter AaC.SL.1., AaC.SL.6., AaC.SL.7a., AaC.SL.7b., AaC.SL.18., AaC.SL.21.
• No supplier needed (see i48. Quality inspector, p. 191)	Progress of work AaC.SL.3.
SU.26 Proprietor selected supplier • No supplier needed (see i10. House user, p. 183)	Room appearance AaC.SL.4.
SU.27 Local supplier	Screw, plug etc.

8.6.5 EXAMPLE House kitchen: Examine make/buy balance

Several kitchen *suppliers* were considered good candidates, the K-Studio was selected as the kitchen *designer* company, and a contract was signed with them. A contract was also made with the electricity candidate Elert AB.

8.6.6 EXAMPLE House: Proceed reading

To follow this example, these are the alternatives:

2A Proceed to chapter 9.7, "EXAMPLE House rooms: Satisfy room requirements by decomposing room black-boxes into white-box design containing machinery black-boxes," page 365.

2F Proceed to chapter 10.5, "EXAMPLE House rooms: Finalize design of rooms with product requisites," page 452.

2V Proceed to chapter 12.6, "EXAMPLE House kitchen: Partially verify black-/white-box with embedded fakes," page 677.

2V Proceed to chapter 12.8, "EXAMPLE House kitchen: Fully verify room black-/white-box," page 684.

• To stay within the house example, but not following the overall *schedule*, proceed to chapter 8.7. below.

8.7 EXAMPLE House rooms machinery: Predetermine solutions with sourcing options

8.7.1 About house kitchen machinery

In this example this *nesting level* is chosen to be the innermost *nesting level* and thus there are no further *embedded black-boxes*.

It is always possible to define *elements* as *complex black-boxes*, and it happens not so seldom with techies who can't stop developing at the *nesting level* where purchased *components* are cheaper than those developed in-house.

8.7.2 EXAMPLE House rooms machinery: Process schedule to use

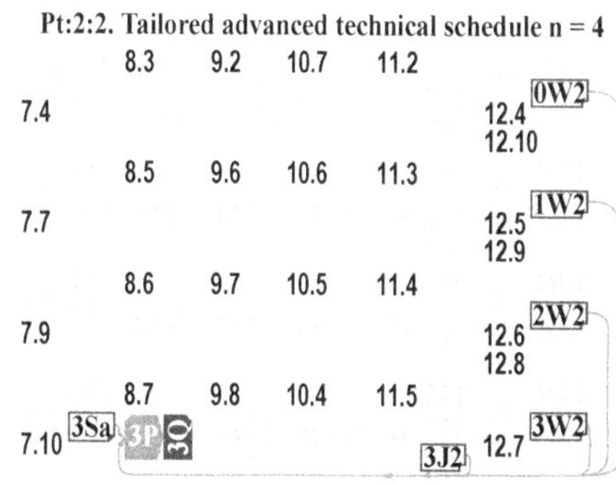

For the house example, a special technical overall *schedule* is *tailored* (illustrated in Fig. 8-9, left):

- **Pt:2:2. Tailored advanced technical schedule n = 4,** page 164

In this *tailored schedule* the generic *schedule* to use is (filled-in symbols):

- **Pt.n.P. Predetermine solutions with sourcing options,** page 133.

FIGURE 8-9 Position of schedule and master to use (filled in) in overall schedule

Below *results* may be demanded (their *masters* are indicated *upstream* from the filled-in *schedule* in the figure above):

3Sa Machinery prioritized requirements

Table 7-13, House kitchen machinery black-box compound requirements, page 200

Xb Suppliers sample catalogues

For possible *failure localization* and *failure elimination*, below *results* may be demanded (their *masters* are indicated *upstream* from the filled-in *schedule* in the figure above):

3J2 Table 11-22. House kitchen machinery start-up report 1, page 622

0W2 Table 12-27. House environment verification report 1, page 701

Table 12-29. House environment reverification report 2, page 703

Table 12-28. House environment failure elimination report 1, page 702

1W2 List 12-2. House with fake rooms failure elimination report 1, page 676

[1W] Table 12-22. House verification report 1, page 697

Table 12-24. House verification report 2, page 698

Table 12-23. House failure elimination report 1, page 698

[2W] List 12-3. House rooms with fake machinery failure elimination report 1, page 678

[2W] Table 12-14. House kitchen verification report 1, page 688

Table 12-17. House kitchen verification report 2, page 691

Table 12-15. House kitchen failure elimination report 1, page 689

[3W] Table 12-9. House kitchen machinery verification report 1, page 682

Table 12-11. House kitchen machinery verification report 2, page 683

Table 12-10. House kitchen machinery failure elimination report 1, page 683

8.7.3 EXAMPLE House kitchen machinery: Try to satisfy requirements

The house proprietor has a good overview of the kitchen machinery market and contributes to *white-box solution alternatives* (see Table 8-16 below).

The assessment column shows total prices for machinery material, not including kitchen interiors to store the machinery. Some work is required to install the machinery, which is not estimated in the assessment.

TABLE 8-16 House kitchen machinery white-box solution alternatives

Solution alternative	Assess-ment	Requirement	Prior-ity
A.aC.aA. Cold machinery			
AaC.aA.SL.1a. Separate fridge (150 liters) and separate freezer (50 liters) • Fridge with electronic thermostat adjustable between +4 °C to +18 °C • Freezer with thermostat adjustable to around - 20 °C	≈ € 600	A.aC.aA.R.6.4.6.1.Fridge shall hold 4 bags (of 30 liters each) of chilled groceries.	Crucial
	≈ € 700	A.aC.aA.R.6.4.6.2.Freezer shall hold 1 bag of 30 liters frozen food.	High
AaC.aA.SL.1.b. Combined fridge (150 liters) and freezer (50 liters) • Fridge with electronic thermostat adjustable between +4 °C and +18 °C • Freezer with thermostat adjustable to around - 20 °C	≈ € 800	A.aC.aA.R.6.4.6.1.Fridge shall hold 4 bags (of 30 liters each) of chilled groceries.	Crucial
		A.aC.aA.R.6.4.6.2.Freezer shall hold 1 bag of 30 liters frozen food.	High
AaC.aA.SL.2. Fridge and freezer with automatic defrosting	+ ≈ € 100	A.aC.aA.R.6.4.6.3.Fridge and freezer shall be maintenance-free.	Low
A.aC.aB. Warm machinery			

TABLE 8-16 House kitchen machinery white-box solution alternatives

Solution alternative	Assessment	Requirement	Priority
AaC.aB.SL.1a. Heating cabinet 600 W • Adjustable between +20 °C and +80 °C	≈ € 300 very special	A.aC.aB.R.6.3.1.1.Equipment shall keep cooked food warm for 2 hours.	Low
AaC.aB.SL.1b. Warming oven 600 W • Adjustable between +50 °C and +120 °C	≈ € 350 more useful		
AaC.aB.SL.2. Microwave oven 2 000 W • With smooth power regulator and rotating food	≈ € 400	A.aC.aB.R.6.6.1.1Thawing shall work on simple food in plastic containers.	Crucial
		A.aC.aB.R.6.6.1.2.When thawing, no parts of the food may get cooked.	Mid
		A.aC.aB.R.6.6.1.3.Thawing speed must be adjustable.	High
		A.aC.aB.R.6.6.2.1Ready-to-serve food shall be warmed on heatproof plates.	Crucial
A.aC.aC. Bake machinery			
AaC.aC.SL.1. Electrical oven 2 000 W • Adjustable between. +20 °C and +300 °C, with fan and grill	≈ € 380	A.aC.aC.R.6.6.6.1Advanced baking shall be possible.	High
		A.aC.aC.R.6.6.6.2.Warming, grilling, and crusting shall be possible.	Mid
A.aC.aD. Heat machinery			
AaC.aD.SL.1. Electric cooktop 6 000 W • With several power-regulated hotplates, at least one 2 200 W	≈ € 260	A.aC.aD.R.6.6.2.1.Fast warming when cooking shall work on food in metal pots.	Crucial
		A.aC.aD.R.6.6.3.1Boiling food shall work with all kinds of pans.	High
		A.aC.aD.R.6.6.3.2.Heat supply shall be adjustable and easy to control.	High
		A.aC.aD.R.6.6.3.3.Power shall be down-adjustable for small pans to simmer.	Mid
		A.aC.aD.R.6.6.4.1Frying shall work on wide- and flat-bottom pans.	High
		A.aC.aD.R.6.6.4.2.Power shall be down-adjustable, e.g., for thickening a sauce.	Low
AaC.aD.SL.2a. Gas hub 4.8 kW **AaC.aD.SL.2b. Gas hub 2.3 kW**	≈ € 280 ≈ € 130	A.aC.aD.R.6.6.5.1Wokking shall be possible in cupped-bottom pans.	Crucial
		A.aC.aD.R.6.6.5.2.Heat shall bring 3 dl oil to almost instant boil.	High
A.aC.aE. Ventilation machinery			
AaC.aE.SL.1a. Wall vent 240 W **AaC.aE.SL.1b. Hood 180 W**	≈ € 170 ≈ € 110	A.aC.aE.R.6.7.1.1Air evacuation from kitchen shall be stronger than air inflow.	Mid
A.aC.aF. Wet machinery			
AaC.aF.SL.1a. Separate cold- & warm-water faucets **AaC.aF.SL.1b. Combined cold- & warm-water faucets**	≈ 2 x € 80 ≈ € 80 one hand	A.aC.aF.R.6.4.1.1.Cold- and warm-water faucet shall be easily accessible.	Crucial
AaC.aF.SL.2. Warm-water distribution from adjustable mixer blending cold and hot water	≈ € 100	A.aC.aB.R.6.4.1.2.Water temperature should be efficient but not harmful.	High
AaC.aF.SL.3. Stainless steel sink 2 bowls prepared for faucets	≈ € 300	A.aC.aF.R.6.5.3.1Dishwashing by hand shall be possible.	Crucial
A.aC.aG. Dish machinery			

TABLE 8-16 House kitchen machinery white-box solution alternatives

Solution alternative	Assess-ment	Requirement	Prior-ity
AaC.aG.SL.1. Simple dishwasher (40 liters)	≈ € 400	A.aC.aG.R.6.5.1.1One dishwasher for dirty utensils, one for cleaning tableware.	Mid
AaC.aG.SL.2. Advanced dish-washer (40 liters)	≈ € 600		
AaC.aG.SL.3. Glass racks in dish-washers	≈ € 20	A.aC.aG.R.6.5.1.2.Glasses shall be load-able in a rack.	High
Material minimum cost sum	≈ € 4 480		
Material maximum cost sum	≈ € 5 020		

8.7.4 EXAMPLE House kitchen machinery: Analyze make/ buy options

The below sourcing analysis is performed (see Table 8-17 below):

TABLE 8-17 House kitchen machinery white-box supplier opportunities

Supplier opportunity	Solution alternative
SU.28 K-Studio. • Kitchen machinery shall be standard reusable compo-nents. • Machinery shall fit well into cover cabinets or onto countertops, and all their interfaces shall match kitch-en connections. Thus comprehensive catalogues and installation instructions must promptly be made avail-able from selected supplier for architects and design-ers.	Cold machinery AaC.aA.SL.1.b., AaC.aA.SL.2.
	Warm machinery AaC.aB.SL.1a., AaC.aB.SL.1b., AaC.aB.SL.2.
	Bake machinery AaC.aC.SL.1.
	Evacuation machinery AaC.aE.SL.1a., AaC.aE.SL.1b.
	Dishwasher machinery AaC.aG.SL.1., AaC.aG.SL.2., AaC.aG.SL.3.
SU.29 House builder's supplier • An authority-approved hub shall be proposed by a certified supplier.	Heat machinery AaC.aD.SL.1., AaC.aD.SL.2a., AaC.aD.SL.2b.
SU.30 Wat Son	Wet machinery AaC.aF.SL.1a., AaC.aF.SL.1b., AaC.aF.SL.2., AaC.aF.SL.3.

8.7.5 EXAMPLE House kitchen machinery: Examine make/ buy balance

No issues were brought up.

8.7.6 EXAMPLE House: Proceed reading

To follow this example, these are the alternatives:

3A Proceed to chapter 9.8, "EXAMPLE House rooms machinery: Satisfy machinery requirements by decomposing machinery black-boxes into white-box design not containing black-boxes," page 371.

3F Proceed to chapter 10.4, "EXAMPLE: House kitchen machinery: Finalize design of machinery with product requisites," page 445.

3V Proceed to chapter 12.7, "EXAMPLE: House rooms machinery: Fully verify machinery black-/white-boxes," page 678.

- To leave the house example and to enter the multiplication toy example, proceed to below chapter 8.8.

8.8 EXAMPLE Multiplication toy environment: Demarcate product by environment solutions with possible supplier opportunities

8.8.1 EXAMPLE Multiplication toy: About this example

Finding safe solutions for a toy intended for children now gets tricky, as does keeping solutions very concrete and down-scaled to press the price per unit.

8.8.2 EXAMPLE Multiplication toy environment: Process schedule to use

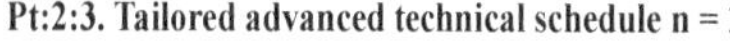
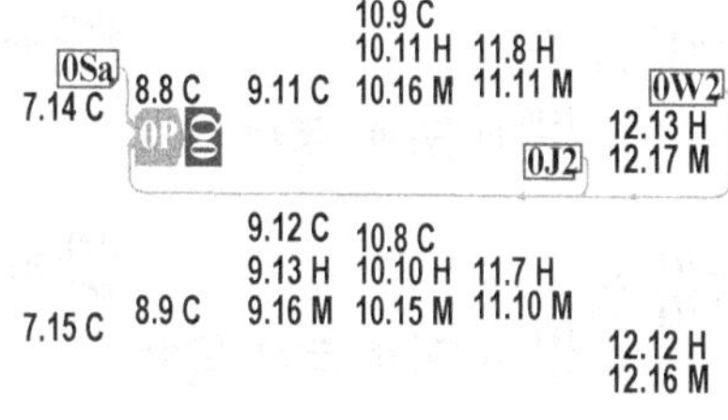

For the multiplication toy example, a special technical overall *schedule* is *tailored* (illustrated in Fig. 8-10, left):

- **Pt:2:3. Tailored advanced technical schedule n = 2,** page 169

In this *tailored schedule* the generic *schedule* to use is (filled-in symbols):

- **Pt.0.P. Demarcate product by environment solutions with possible supplier opportunities,** page 112.

FIGURE 8-10 Position of schedule and master to use (filled in) in overall schedule

Below *results* may be demanded (their *masters* are indicated *upstreams* from the filled-in *schedule* in the figure above):

0Sa Environment restriction requirements
   Table 7-23. Multiplication toy environment restriction requirement, page 213

Xb Suppliers sample catalogues

For possible *failure localization* and *failure elimination*, below *results* may be demanded (their *masters* are indicated *upstreams* from the filled-in *schedule* in the figure above):

0J2 Table 11-28. Multiplication toy environment (hard-wired multiplier variant), start-up report 1, page 630
   Table 11-36. Multiplication toy environment (microcontroller multiplier variant), start-up report 1, page 640

OW2 Table 12-53. Multiplication toy environment (microcontroller multiplier variant) failure elimination report 1, page 728

8.8.3 EXAMPLE Multiplication toy environment: Try to satisfy restrictions

Specified restrictions (see Table 7-23, p. 213) are now analyzed to get some ideas how to satisfy them by proposing *environment interfaces solution alternatives* (see Table 8-18 below).

It is very important by now to understand which *interface* solutions should be proposed on the current *environment nesting level,* and which wi ll be skipped for later development. The rule of thumb is that only solutions connecting the *product* to the *environment* should be taken care of now, and boundaries between *product* and *environment* can be taken care of later on. It seems that only E.RR.3. will lead to a connection solution and needs to be taken care of right now.

TABLE 8-18 Multiplication toy environment
solution alternatives of interfaces to environment

Interface solution alternative	Assess-ment	Environment restriction
E.SL1. The toy may not be sharp, toxic, or demountable to be harmful for its environment or users in the environment.	Not more expensive	E.RR.1. The multiplication toy shall be childproof according to European Community regulations.
E.SL2.a. Bolts or similar for fastening the toy to the hosting toy E.SL2.a. Gasket to tighten between the multiplication toy and the hosting toy E.SL2.b. Glue to fasten and tighten between the multiplication toy and the hosting toy.	$\approx$ €0.1 - €0.5 per toy $\approx$ €0.1 - €0.5 per toy	E.RR.2. The multiplication toy attachment to the hosting toy may not leak padding from the inside of the hosting toy.
E.SL3. Holes or guide pins for fastening bonds, located asymmetric to each other, which fit together in only one way E.SL4. Soft hosting toys are potentially too floppy, and in need of some kind of strengthening support when attaching the multiplication toy to it.	Not expensive	E.RR.3. The multiplication toy shall be easily reusable in a safe way, by all kinds of different hosting toys.
E.SL5. Disclaimer to be signed by hosting toy company E.SL6.a. Insurance against product failures E.SL6.b. Hosting toy company requested to take product failure insurance	Exclude from product development	E.RR.4. No legal responsibility is taken for incorrect mounting of the multiplication toy into the hosting toy by companies buying the multiplication toy.

8.8.4 EXAMPLE Multiplication toy environment: Analyze in/out and make/buy options

Now the demarcation must be made of what will and will not be included (see Table 8-19 and Table 8-20 below).

TABLE 8-19 Multiplication toy environment solutions excluded

Excluded from the product to be developed	Responsible	Environment solution alternative
SX.1 Disclaimer paperwork and insurance against third-party claimer	Product management department	E.SL5. Disclaimer to be signed by hosting toy company E.SL6.a. Insurance against product failures E.SL6.b. Hosting toy company requested to take product failure insurance

TABLE 8-20 Multiplication toy environment solutions included with supplier opportunities

Included in the product with supplier opportunity	Environment solution alternative
SU.10. Gasket-Seal	E.SL2.a.
SU.11. Specialist glue supplier • For selection of gluing method	E.SL2.b.
SU.12. Gross Hardware • To supply with environment interface parts SU.13. In-house prototype factory • Prototype material SU.14. In Steal Inc • Series manufacturing steal material	E.SL1., E.SL2.a., E.SL3., E.SL4.

8.8.5 EXAMPLE Multiplication toy environment: Examine product scope

The *product* should be offered as a straightforward built-in toy for cuddly toy producers, and no problems with the *product* scope have been identified by the gate members.

However, there was some anxiety concerning the safety of the toy. Selecting safe technology for the *environment interfaces* still seems a bit premature, and it was decided to look more carefully at appropriate material and *suppliers* when the *white-box solution alternatives* are prepared later on.

8.8.6 EXAMPLE Multiplication toy: Proceed reading

To follow this example, these are the alternatives:

- Proceed to chapter 9.11, "EXAMPLE Multiplication toy environment: Constitute environment architecture," page 382.
- Proceed to chapter 10.9, "EXAMPLE Specifics for multiplication toy environment commonalities: Finalize design of environment interfaces with product requisites," page 493.
- Proceed to chapter 12.13, "EXAMPLE Multiplication toy environment (hardwired multiplier variant): Verify prototype in environment," page 712.

- Proceed to chapter 12.17, "EXAMPLE Multiplication toy environment (microcontroller multiplier variant): Verify prototype in environment," page 725.
- To stay within the multiplication toy example, but not following the overall *schedule*, proceed to chapter 8.9 below.

8.9 EXAMPLE Multiplication toy: Predetermine solutions with sourcing options

8.9.1 Multiplication toy: Process schedule to use

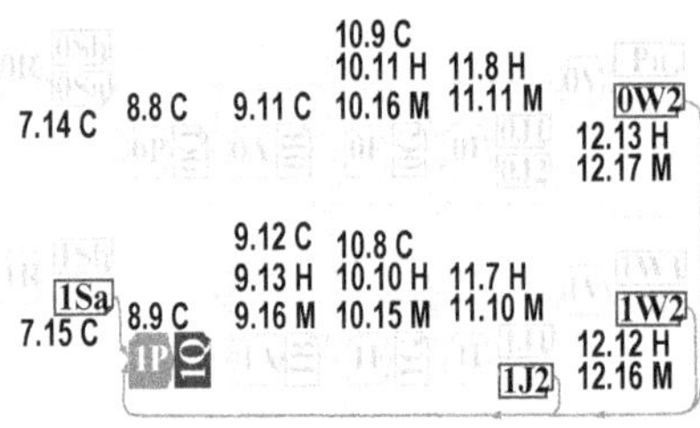

For the multiplication toy example, a special technical overall *schedule* is *tailored* (illustrated in Fig. 8-11, left):

- **Pt:2:3. Tailored advanced technical schedule n = 2,** page 169

In this *tailored schedule* the generic *schedule* to use is (filled-in symbols):

- **Pt.n.P. Predetermine solutions with sourcing options,** page 133.

FIGURE 8-11 Position of schedule and master to use (filled in) in overall schedule

Below *results* may be demanded (their *masters* are indicated *upstream* from the filled-in *schedule* in the figure above):

1Sa Prioritized requirements
Table 7-25. Multiplication toy restriction requirement, page 216
Table 7-26. Multiplication toy interaction requirement, page 217
Table 7-27. Multiplication toy interface requirements, page 217

Xb Suppliers sample catalogues

For possible *failure localization* and *failure elimination*, below *results* may be demanded (their *masters* are indicated *upstream* from the filled-in *schedule* in the figure above):

1J2 Table 11-25. Multiplication toy with hard-wired multiplier start-up report 1, page 627
Table 11-33. Multiplication toy with microcontroller multiplier start-up report 2, page 636

1W2 Table 12-38. Multiplication toy with hard-wired multiplier failure elimination report 1, page 711
Table 12-47. Multiplication toy with microcontroller multiplier failure elimination report 1, page 723

[OW2] Table 12-53. Multiplication toy environment (microcontroller multiplier variant) failure elimination report 1, page 728

8.9.2 EXAMPLE Multiplication toy: Try to satisfy requirements

The *requirements* from Table 7-25, page 216 are analyzed and appropriate solutions are proposed (see Table 8-21 below).

TABLE 8-21 Multiplication toy white-box solution alternatives

Solution alternative	Assessment	Requirement	Priority
A.SL1. Use clamping or friction screws for fastening parts on the front panel.	Not more expensive	A.RR.1.1.Everything accessible from the outside of the multiplication toy shall be impossible for a child to remove.	Crucial
A.SL2. Use only guaranteed nontoxic painting and surface material for the casing.	Not more expensive	A.RR.1.2.Everything accessible from the outside of the multiplication toy shall be nontoxic for a licking child.	Crucial
A.SL3.a. Gather all multiplication intelligence to a separate element and manage its interface. A.SL3.b. Gather all multiplication intelligence to a separate black-box with own requirements and verification.	The multiplication intelligence is not too complex to master. An embedded black-box will increase the development time and cost, without big benefits.	A.RR.5.1.The multiplication calculation shall be allocated to a separate element that can be redesigned with minimal rework.	High
A.SL4.a. One technology shall be a hard-wired gate network, to ensure a correct prototype, before transferring this gate network to a customized ASIC. A.SL4.b. One technology shall contain a cheap single-chip microcontroller with a minimal program intended for series manufacturing.	These two technologies should both be developed.		
A.SL5. To switch between the two product variants, only the circuit boards should need to be exchanged.	No extra cost	A.RR.5.2.The casing of the hosting toy shall be exactly the same for all products independent of technology used.	High
A.SL6.a. Battery A.SL6.b. Solar panel	≈ € 0.1 ≈ € 3	A.RR.6.1.The toy shall have an internal power supply, not possible for a child to operate.	Crucial
A.SL7.a. One customized ASIC variant for 25 000 products at € 2 each A.SL7.b. One microcontroller variant for 100 000 products at € 2 each	≈ € 5 / product ≈ € 5 / product	A.RR.7.1.If the toy is produced, it shall be made in such volume that the price per unit is less than € 5.	Crucial
A.SL8. Design for efficient manufacturing test.	To reach ≈ € 5 / pcs		
A.SL9. The multiplier element shall calculate the product from two operands.	No extra cost	A.RB.8.1. User amusement shall be to multiply two operand values into one product value.	Crucial

TABLE 8-21 Multiplication toy white-box solution alternatives

Solution alternative	Assessment	Requirement	Priority
A.SL10.a. Cheap robust rotating switch A.SL10.b. Cheap key row system A.SL10.c. Numeric pad	≈ 2 x € 0,5 at 1 000 pcs ≈ 2 x € 1 at 1 000 pcs ≈ € 2 at 100 000 pcs	A.RB.8.2. The toy shall have one input device for each multiplicand value. Multiplicand values shall be at least between 0 and 10. The multiplicands must be easy for a child to settle.	0 - 10 crucial > 10 high
A.SL11.a. LED panel A.SL11.b. LCD panel	≈ € 1 at 10 000 pcs ≈ € 1 at 100 000 pcs	A.RB.8.3. The toy shall have at least a 3-digit output device for the product value. The product must be easy to read for a child.	Crucial
A.SL12. Maximum delay between operand settling and display reading is 0.1 second.	This modest speed may save some money.	A.RB.8.4. The toy calculates and shows the product value practically instantly after either multiplicand is settled.	High

With the *requirements* **RR.5.1.** the stakeholders provide the design of the multiplier to be separated, but the two *solution alternatives* **A.SL4.a.** and **A.SL4.b.** are purely technical, provided that the cost for either gets below € 5.

8.9.3 EXAMPLE Multiplication toy: Analyze make/buy options

Now the solutions have got more concrete than when the *environment interfaces* were first studied. Possible *suppliers* were chosen (see Table 8-21 below).

TABLE 8-22 Multiplication toy white-box supplier opportunities

Supplier opportunity	Solution alternative
SU.15. Tooly • Tools supplier for metal manufacturing • For series manufacturing **SU.16. In Steal Inc** • Customized metal parts manufacturer • For series manufacturing **SU.17. In-house prototype factory** • For prototypes	Boundary casing A.SL1., A.SL2.
• No sourcing impact	Element or black-box A.SL3.a., A.SL3.b.
SU.18. In-house storage • For prototypes only, replan for manufacturing **SU.19. Elmech** • For electromechanical components to prototypes and series manufacturing **SU.20. Passta** • For passive components to prototypes and series manufacturing	Electronic components A.SL4.a., A.SL6.a., A.SL6.b., A.SL7.a., A.SL7.b., A.SL9, A.SL10.a., A.SL10.b., A.SL10.c., A.SL11.a.., A.SL11.b.
SU.21. SiliC • New supplier to evaluate and secure • Ask for complete data sheets. • For prototypes and series manufacturing	Single-chip microcontroller A.SL4.b., A.SL9.

TABLE 8-22 Multiplication toy white-box supplier opportunities

Supplier opportunity	Solution alternative
• No sourcing impact	Calculation speed A.SL12.
SU.22. PeCeBe • To make PCB from electronics schedule • To suggest PCB layout for manufacturing test • For prototypes and series manufacturing	Printed Circuit Board (PCB) and manufacturing test A.SL5., A.SL8.

8.9.4 EXAMPLE Multiplication toy: Examine make/buy balance

Especially important is selection of childproof material and design for *samples* exposed to the *user* in the *environment*. A special "safety" petition was prepared for the *project leader* to be announced in his team.

It is also very important to proactively plan for *product* manufacturing already, to be able to get under the maximum price of € 5 per *product* unit. It was decided that the *project leader* should have a cost awareness meeting for purchase engineers and manufacturing engineers once a week.

8.9.5 EXAMPLE Multiplication toy: Proceed reading

To follow this example, these are the alternatives:

- Proceed to chapter 9.12, "EXAMPLE Multiplication toy commonalities: Satisfy refined requirements by decomposing product black-box into white-box design containing no black-boxes," page 386.
- Proceed to chapter 10.8, "EXAMPLE Specifics for multiplication toy commonalities: Finalize design items of product white-box with product requisites," page 482.
- Proceed to chapter 12.12, "EXAMPLE Multiplication toy with hard-wired multiplier: Verify black-/white-box," page 706.
- Proceed to chapter 12.16, "EXAMPLE Multiplication toy with microcontroller multiplier: Verify black-/white-box," page 717.
- To leave the multiplication toy example and to enter the calc_logic example proceed to chapter 8.10 below.

8.10 EXAMPLE Calc_logic environment (the pocket calculator firmware): Demarcate product by environment solutions with possible supplier opportunities

8.10.1 About this example

The two different sets of *environment restriction requirements* to the calc_logic *black-box* result in different *solution alternatives* for how to *interface* calc_logic. Since calc_logic will be made for *reuse* and is provided with merged *requirements*, the *environment interfaces* must be a superset of the two solutions.

8.10.2 EXAMPLE Calc_logic environment (the pocket calculator firmware): Process schedule to use

For the calc_logic example, a special technical overall *schedule* is *tailored* (illustrated in Fig. 8-12, left):

- **Pt:2:3. Tailored advanced technical schedule n = 2,** page 169

In this *tailored schedule* the generic *schedule* to use is (filled-in symbols):

- **Pt.0.P. Demarcate product by environment solutions with possible supplier opportunities,** page 112.

Pt:2:3. Tailored advanced technical schedule n = 2

			11.16 W
8.10 P	9.20 P	10.19 W	11.17 P
8.11 W	9.23 W	10.20 P	11.18 P
7.21 P			12.21 W
7.22 W			12.22 P
			12.23 P
	8.12 R	9.24 R	10.18 R	11.14 H
7.23 R				12.20 H

FIGURE 8-12 Position of schedule and master to use (filled in) in overall schedule

8.10.3 EXAMPLE Calc_logic environment (the pocket calculator firmware): Try to satisfy restrictions

The allocated i13. External embedded designers assisted by developers from the in-house project "dg3" analyzed the pocket calculator restrictions and came up with calc_logic solution alternatives of interfaces to environment (see Table 8-23 below).

TABLE 8-23 Calc_logic environment (the pocket calculator firmware) solution alternatives of interfaces to environment

Interface solution alternative	Assessment	Environment restriction
E.SL1. Only the 16 most common calculator stimuli are expected to enter calc_logic. E.SL2. Proposed are digits 0 - 9, decimal point, equal sign, subtraction, addition, multiplication, and division.	It is optimal to arrange the buttons in a 4 x 4 square. Integers are not sufficient for a useful calculator, floats with digital point are necessary.	E.RR.1. To fit market cost of an affordable everyday calculator, the number of buttons shall be no greater than 16.
E.SL3. There is no need to respond with more than 8 digits from calc_logic. E.SL4. Only numeric responses must appear.	Alphanumeric display is far too expensive.	E.RR.2. To fit market cost of an affordable calculator, the stand-alone calculator display shall hold a maximum of 8 digits.
E.SL5. Don't consider using any mathematical chip.	Mathematical chips are expensive.	E.RR.3. Cheap microcontrollers rarely have on-chip or external chip support for mathematical arithmetic.
E.SL6.a. Microcontroller development system with no floating-point E.SL6.b. Microcontroller development system with ≈ 7 digits float support E.SL6.c. Microcontroller development system with ≈ 15 digits float support.	For very cheap microcontrollers For some cheap microcontrollers Only for expensive microcontrollers	E.RR.4. Standard compilers to microcontrollers seldom have support for extensive mathematical arithmetic, but there are some microcontroller development systems with floating-point library support for float variables (≈ 7 digits).

TABLE 8-23 Calc_logic environment (the pocket calculator firmware)
solution alternatives of interfaces to environment

Interface solution alternative	Assessment	Environment restriction
E.SL7.a. Microcontroller compiler generated floating-point linkable program	No extra cost	E.RR.5. Mathematical program libraries for microprocessors can be developed in-house or outsourced, and for some microprocessors portable libraries are available from third-party shelves.
E.SL7.b. Microcontroller development system built-in floating-point libraries	No extra cost	
E.SL7.c. In-house-developed microcontroller floating-point source system and/or linkable library	Needs high competency & high cost	
E.SL7.d. Supplier off-the-shelf floating-point microcontroller library	Medium cost	
E.SL7.e. Supplier turnkey-developed microcontroller floating-point source system library	Huge cost	
E.SL8. Porting calc_logic to a microcontroller experimental board supporting ≈ 7 digits floating-point precision.	≈ € 2 000	E.RR.6. Calc_logic for reuse developed on a host system must be portable to development boards and the target pocket calculator.
E.SL9. Port calc_logic to the target stand-alone pocket calculator supporting ≈ 7 digits floating-point precision.	≈ € 2 000	

8.10.4 EXAMPLE Calc_logic environment (the pocket calculator firmware): Analyze in/out and make/buy options

The *solution alternatives* above were studied and the boundaries to calc_logic were settled (see Table 8-24 below and Table 8-25 next). At this moment it is very important to define what parts in particular belong to calc_logic, and what parts belong to the surrounding pocket calculator reusing the calc_logic *program*.

TABLE 8-24 Calc_logic environment (the pocket calculator firmware)
solutions excluded

Excluded from the product to be developed	Responsible	Environment solution alternative
SX.1 Mechanics, electronics with microcontroller, and firmware for the pocket calculator. • The pocket calculator surrounding to calc_logic	In-house project "de1"	Pocket calculator
SX.2 Development system for the pocket calculator firmware • Procured by in-house project "de1" for in-house project "dg3" usage		Development systems including compiler and standard built-in libraries
SX.3 Development board with development system for calculator firmware • Procured by in-house project "de1" for in-house project "dg3" usage		Development board
SX.4 Floating-point library with appropriate library interface to calc_logic • Procured by in-house project "de1" for in-house project "dg3" usage		Separate floating-point library (if not supported by compiler) E.SL7.c., E.SL7.d., or E.SL7.e.
SX.5 Mathematical chip on pocket calculator		Mathematical chip E.SL5.

Typically, *program libraries* can be included as *development system libraries*, or excluded as a separate *element* somebody else is responsible for. As seen in Table 8-24 above, the *development systems* can also be excluded.

Included in the *product* is the calc_logic *product black-box*, *ported* and field verified in defined *environments* (see Table 8-25 below).

TABLE 8-25 Calc_logic environment (the pocket calculator firmware) solutions included with supplier opportunities

Included in the product with supplier opportunity	Environment solution alternative
SU.10. In-house project "dg3" • To develop calc_logic with interfaces including field verification on development boards and in target pocket calculator	Calc_logic design E.SL1., E.SL2., E.SL3., E.SL4., E.SL6.a., E.SL6.b. E.SL6.c., E.SL7.a., E.SL7.b., E.SL8., E.SL9.

8.10.5 EXAMPLE Calc_logic environment (the pocket calculator firmware): Examine product scope

It is important that the *reusable item* calc_logic really will fit the planned calculators regarding *development system* and floating-point capacity.

The restrictions for calc_logic *environment* are also valid for the pocket calculator developed by in-house project "de1". The gate decision is that the in-house project "de1" and the in-house project "dg3" shall allocate the same chief *architect*, to ensure that the design will be compatible.

The gate also decided to appoint a group of *stakeholders* representing pocket calculators, Windows calculators, and strategic *products* for the future, to come up with a calc_logic *reuse road map* before compilers and *program libraries* are decided.

8.10.6 EXAMPLE Calc_logic for reuse: Proceed reading

To follow this example, these are the alternatives:

- Proceed to chapter 8.12, "EXAMPLE Calc_logic for reuse: Predetermine solutions with sourcing options," page 326.
- Proceed to chapter 9.20, "EXAMPLE Calc_logic environment (the pocket calculator firmware): Constitute environment architecture," page 408.
- Proceed to chapter 10.20, "EXAMPLE Calc_logic environment (the pocket calculator firmware) development board and pocket calculator: Finalize design of environment interfaces with product requisites," page 549
- Proceed to chapter 8.11 "EXAMPLE Calc_logic environment (the Windows calculator application): Demarcate product by environment solutions with possible supplier opportunities" below.

8.11 EXAMPLE Calc_logic environment (the Windows calculator application): Demarcate product by environment solutions with possible supplier opportunities

8.11.1 *EXAMPLE* Calc_logic environment (the Windows calculator application): Process schedule to use

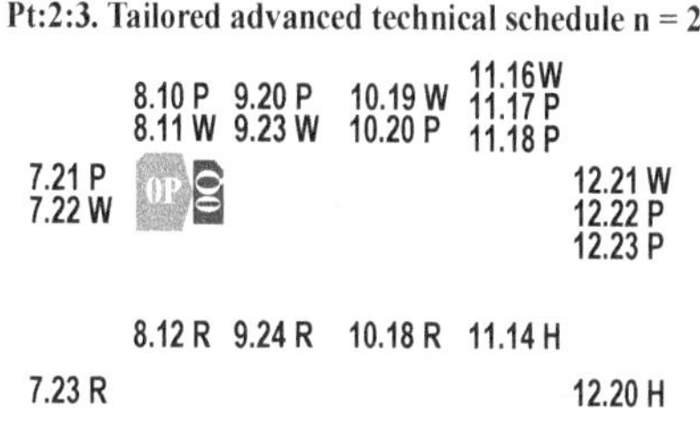

For the calc_logic example, a special technical overall *schedule* is *tailored* (illustrated in Fig. 8-13, left):

• **Pt:2:3. Tailored advanced technical schedule n = 2,** page 169

In this *tailored schedule* the generic *schedule* to use is (filled-in symbols):

• **Pt.0.P. Demarcate product by environment solutions with possible supplier opportunities,** page 112.

FIGURE 8-13 Position of schedule and master to use (filled in) in overall schedule

8.11.2 EXAMPLE Calc_logic environment (the Windows calculator application): Try to satisfy restrictions

As for the pocket calculator, the Windows restrictions were also analyzed in order to identify possible *interfaces* to calc_logic in this *environment* (see Table 8-26 below).

TABLE 8-26 Calc_logic environment (the Windows calculator application) solution alternatives of interfaces to environment

Interface solution alternative	Assessment	Environment restriction
E.SL10. Calc_logic shall expect a large number of different stimuli.	This doesn't cost extra.	E.RR.7. Since a window in Microsoft Windows can hold almost any number of buttons, a large number of buttons shall be possible stimuli for calc_logic.
E.SL11. Calc_logic must respond with a large number of digits and at least more than accuracy requires. E.SL12. Text response is also conceivable.	This doesn't cost extra.	E.RR.8. Since a window field in Microsoft Windows can show almost any number of alphanumeric characters, the floating-point precision in calc_logic shall be as high as possible.
E.SL13.a. Use but hide the usage of the mathematical coprocessor. E.SL13.b. Emulate the mathematical coprocessor with a program	If there is no co-processor, the program must contain floating-point solutions anyway.	E.RR.9. If a math coprocessor is available in the computer, it may be used to heighten calculation speed. However, the calculating program shall also function in computers without a math coprocessor.

TABLE 8-26 Calc_logic environment (the Windows calculator application) solution alternatives of interfaces to environment

Interface solution alternative	Assessment	Environment restriction
E.SL14.a. Windows development system with no floating-point	Very seldom seen	E.RR.10.There are a lot of compilers for Windows, with a lot of different kinds of mathematical calculation support. Standard Windows C development systems with library supports have float ($\approx$ 7 digits), double-float ($\approx$ 15 digits) and long double-float ($\approx$ 31 digits) precision. However, long double-float variables most often are reduced to the precision of double variables. As Windows 64-bits version gets more popular, in the near future many compilers will probably offer no undiminished support for long double-floating-point ($\approx$ 31 digits) precision.
E.SL14.b. Windows development system with $\approx$ 7 digits floating-point	Seldom seen	
E.SL14.c. Windows development system with $\approx$ 15 digits floating-point	The most common	
E.SL14.d. Windows development system with $\approx$ 31 or more digits floating-point	Probably the standard in the future	
E.SL15. Only a recompilation and change of header files must be necessary when changing float precision.	Makes it easy to port calc_logic to various environments.	
E.SL16.a. Windows compiler-generated floating-point linkable program	No extra cost	E.RR.11.Many kinds of supplier mathematical libraries are available or portable to Windows.
E.SL16.b. Windows development system built-in standard libraries	No extra cost	
E.SL16.c. In-house-developed Windows floating-point source system and/or library	Needs special competency, mid cost.	
E.SL16.d. Supplier off-the-shelf Windows floating-point library	Low cost	
E.SL16.e. Supplier turnkey-developed Windows floating-point source system library	High cost	
E.SL17. Port calc_logic to a Windows application development system supporting at $\approx$ 15 digits floating-point precision.	$\approx$ € 1 000	E.RR.12.Calc_logic for reuse developed on a host system must be portable to the target Windows disk and input/output system.
E.SL18. When available, procure compiler for Windows, supporting better double-floating-point precision.	No extra cost	

8.11.3 EXAMPLE Calc_logic environment (the Windows calculator application): Analyze in/out and make/buy options

How to demarcate calc_logic from the Windows *environment* was now studied (see Table 8-27 and Table 8-28 below).

TABLE 8-27 Calc_logic environment (the Windows calculator application) solutions excluded

Excluded from the product to be developed	Responsible	Environment solution alternative
SX.6 Calculator Windows application reusing calc_logic • The calculator surrounding to calc_logic		Windows calculator application
SX.7 Development system for developing the Windows application • Procured by in-house project "dw2" for in-house project "dg3" usage	In-house project "dw2"	Development system
SX.8 Floating-point library with appropriate interface to calc_logic • Procured by in-house project "dw2" for in-house project "dg3" usage		Separate floating-point library (if not supported by development system) E.SL16.c., E.SL16.d., E.SL16.e.

TABLE 8-28 Calc_logic environment (the Windows calculator application) solutions included with supplier opportunities

Included in the product with supplier opportunity	Environment solution alternative
SU.11. In-house project "dg3" • To develop calc_logic with interfaces including field verification in target Windows application	Calc_logic design E.SL10., E.SL11., E.SL12., E.SL13.a., E.SL13.b., E.SL14.a., E.SL14.b., E.SL14.c., E.SL14.d., E.SL15., E.SL16.a., E.SL16.b., E.SL17., E.SL18.

8.11.4 EXAMPLE Calc_logic environment (the Windows calculator application): Examine product scope

As said by the pocket calculator decision gate, it is important that the *reusable item* calc_logic really fit the planned calculators, regarding compiler and floating-point capacity.

The gate also decided to join the group of *stakeholders* representing pocket calculators, Windows calculators, and strategic *products* for the future, to come up with a calc_logic *reuse road map* before compilers and *program libraries* are decided.

8.11.5 EXAMPLE Calc_logic for reuse: Proceed reading

To follow this example, these are the alternatives:

• Proceed to chapter 9.23, "EXAMPLE Calc_logic environment (the Windows calculator application): Constitute environment architecture," page 415.

- Proceed to chapter 10.19, "EXAMPLE Calc_logic environment (the Windows calculator application): Finalize design of environment interfaces with product requisites," page 545.
- Proceed to chapter 8.12 "EXAMPLE Calc_logic for reuse: Predetermine solutions with sourcing options" below.

8.12 EXAMPLE Calc_logic for reuse: Predetermine solutions with sourcing options

8.12.1 EXAMPLE Calc_logic for reuse: Process schedule to use

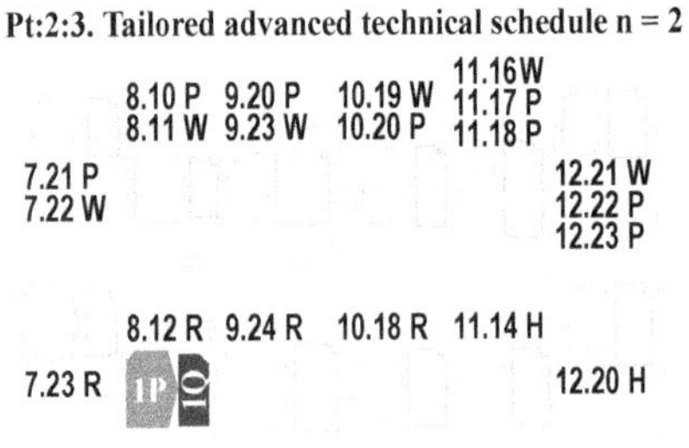

Pt:2:3. Tailored advanced technical schedule n = 2

			11.16 W
8.10 P	9.20 P	10.19 W	11.17 P
8.11 W	9.23 W	10.20 P	11.18 P

7.21 P
7.22 W

12.21 W
12.22 P
12.23 P

8.12 R   9.24 R   10.18 R   11.14 H

7.23 R   IP 10

12.20 H

For the calc_logic example, a special technical overall *schedule* is *tailored* (illustrated in Fig. 8-14, left):

- **Pt:2:3. Tailored advanced technical schedule n = 2,** page 169

In this *tailored schedule* the generic *schedule* to use is (filled-in symbols):

- **Pt.n.P. Predetermine solutions with sourcing options,** page 133.

FIGURE 8-14 Position of schedule and master to use (filled in) in overall schedule

8.12.2 EXAMPLE Calc_logic for reuse: Try to satisfy requirements

The two sets of *environment restriction requirements* have been merged and refined to one set of common *black-box requirements*, to be revisited for identification of common *white-box solution alternatives* (Table 8-29 below).

TABLE 8-29 Calc_logic for reuse white-box solution alternatives

Solution alternative	Assess-ment	Requirement	Prior-ity
A.SL1. Calc_logic must internally handle 64 buttons.	No extra cost	A.RR.1.1. The calc_logic element shall be able to process 16-button stimuli at a minimum and be capable of processing many more future button stimuli.	Critical
A.SL2. Up to 128 digits must be the internal capacity, even if not all are accurate. A.SL3. Handling must include both numeric and non-numeric characters.	No extra cost	A.RR.2.1. The calc_logic element shall be able to process 8-digit responses, and be capable of processing a much larger future number of digits.	Critical
A.SL4. No mathematical coprocessor shall ever be used by calc_logic, but instead a built-in or external floating-point library is used.	Saving cost	A.RR.3.1. To reduce complexity from the unpredictable existence of coprocessor, math coprocessors shall not be used at all.	Medium

TABLE 8-29 Calc_logic for reuse white-box solution alternatives

Solution alternative	Assessment	Requirement	Priority
A.SL5. Develop first calc_logic for reuse on a host development system supporting ≈ 15 digits precision.	≈ € 40 000 Small cost	A.RR.4.1. A maximum calculation precision of ≈ 15 digits during host development of the calc_logic program is sufficient.	High
A.SL6. If different precision or different environments demand calc_logic-specific solutions, these must be hidden behind compiler switches.		A.RR.5.1. Calc_logic for reuse shall be portable to the Windows calculator application, to development boards, and to the pocket stand-alone calculator.	Critical
A.SL7. Make implementation in C, and use scenario wait-states and use-units in program.	Cost ≈ € 20 000 (excluded compiler and math library)	A.RB.6.S1. Calc_logic for reuse, behavior requirement 6, scenario 1	Critical

8.12.3 EXAMPLE Calc_logic for reuse: Analyze make/buy options

Most of the *solution alternatives* to calc_logic for *reuse* consider the *program*, which will be developed mainly in-house. There are rather few restrictions ruling the internal design of the *program*, thus the *program* is a rather *unbiased architecture ingredient* and sourcing is fairly simple for the calc_logic for *reuse* (see Table 8-30 below).

Sourcing of *development system* and boards, and possibly sourcing of the *supplier program libraries*, are assigned outside the in-house project "dg3" to the two calculator *projects* in-house project "de1" and in-house project "dw2".

TABLE 8-30 Calc_logic for reuse white-box supplier opportunities

Supplier opportunity	Solution alternative
SU.11. In-house project "dg3" • To develop the calc_logic for reuse	A. Calc_logic for reuse A.SL1., A.SL2., A.SL3., A.SL4., A.SL5., A.SL6., A.SL7.
SU.12. in-house project "de1" and in-house project "dw2" • To lend out a host computer and development systems (compiler, linker debugger and so forth)	Host computer with development system A.SL5.

8.12.4 EXAMPLE Calc_logic for reuse: Examine make/buy balance

The gate meeting members were rather satisfied with the good planning for *reuse*. The calculator in-house project "de1" and in-house project "dw2" get the responsibility to choose *development systems*, experimental boards, and so forth, that fit their *product development* best, but with one mutual condition—both of them must concurrently be compatible with the reusable and portable calc_logic.

8.12.5 EXAMPLE Calc_logic for reuse: Proceed reading

To follow this example, these are the alternatives:

- Proceed to chapter 9.24, "EXAMPLE Calc_logic for reuse: Satisfy refined requirements by decomposing product black-box into white-box design containing no black-boxes," page 419.
- Proceed to chapter 10.18, "EXAMPLE Hosted calc_logic for reuse: Finalize design items of product white-box with product requisites," page 533.
- Proceed to chapter 8.13 below, to leave the calc_logic example and to enter the phonebook example.

8.13 EXAMPLE Phonebook environment: Demarcate product by environment solutions with possible supplier opportunities

8.13.1 EXAMPLE Phonebook environment: Process schedule to use

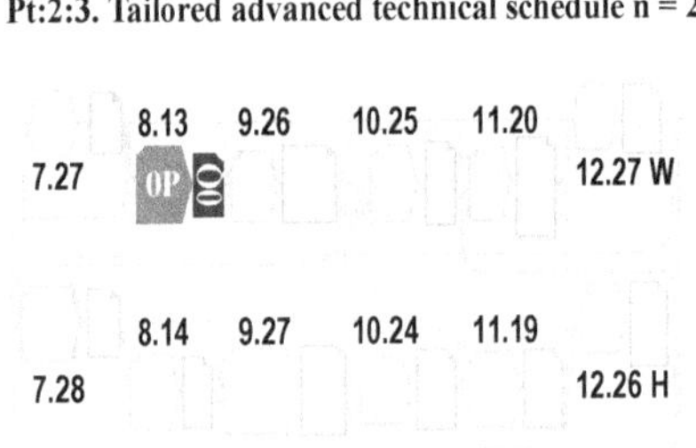

For the phonebook example, a special technical overall *schedule* is *tailored* (illustrated in Fig. 8-15, left):

- **Pt:2:3. Tailored advanced technical schedule n = 2,** page 169

In this *tailored schedule* the generic *schedule* to use is (filled-in symbols):

- **Pt.0.P. Demarcate product by environment solutions with possible supplier opportunities,** page 112.

FIGURE 8-15 Position of schedule and master to use (filled in) in overall schedule

8.13.2 EXAMPLE Phonebook environment: Try to satisfy restrictions

The phonebook *program* is expected to be a substantial attraction in the *product portfolio*. A big part of a successful *program* is nice user-oriented *interfaces*. The first development step must prepare a flexible foundation (see Table 8-31 below).

TABLE 8-31 Phonebook environment
solution alternatives of interfaces to environment

Interface solution alternative	Assessment	Environment restriction
E.SL1. Choose any standard Windows development system with built-in libraries to develop the phonebook.	Cheap or free	E.RR.1. The phonebook program shall run in any standard Windows environment, and need not be portable elsewhere.
E.SL2. For phonebook first evaluation, the command prompt window can be used.	Cheap and simple, but must be replaced	E..RR.2. For the first evaluation, textual interaction with the user is sufficient.
E.SL3. For accessibility and easy verification, the disk file shall be plain textual and line based, and readable and writable with a simple editor.	Cheap, but must be replaced with a more expensive database	E.RR.3. Keeping all phonebook data in an ordinary Windows disk file is sufficient.
E.SL4. User textual interaction shall be grouped in such a way that it is easy to redesign it for graphic windows.	Some initial extra cost	E.RR.4. The program shall scale up easily in future development and be adapted for Windows graphic user interaction.
E.SL5. The phonebook object structure must be grouped in such a way that it is easy to redesign it for an external database.	Some initial cost that quickly pays off	E.RR.5. The program shall scale up easily in future development and be adaptable to database storage tools, such as SQL.

8.13.3 EXAMPLE Phonebook environment: Analyze in/out and make/buy options

Most of the solutions for this first phonebook development step were included, with some few exceptions (see Table 8-32 below).

TABLE 8-32 Phonebook environment
solutions excluded

Excluded from the product to be developed		Responsible	Environment solution alternative
SX.1	Registering of phonebook data stored at the phonebook file is not part of the project (beyond what is needed for verifying the phonebook program).	Customer care department	Content on phonebook disk file E.SL3.

The first phonebook *environment* step involves some *suppliers* (see Table 8-33 below).

TABLE 8-33 Phonebook environment
solutions included with supplier opportunities

Included in the product with supplier opportunity	Environment solution alternative
SU.10. Development system for interfaces to environment from reliable supplier	Phonebook program environment E.SL1.
SU.11. Graphical Windows tool investigation	E.SL4.
SU.12. Database tool investigation	E.SL5.
SU.13. In-house development of interfaces • The rest to be developed in-house, no more suppliers needed	E.SL2., E.SL3., E.SL4., E.SL5.

8.13.4 EXAMPLE Phonebook environment: Examine product scope

The scope is not very big on behavior, but more important is finding future-proof structures and easy scalability. The gate members all agreed on this.

8.13.5 EXAMPLE Phonebook: Proceed reading

To follow this example, these are the alternatives:

- Proceed to chapter 9.26, "EXAMPLE Phonebook environment: Constitute environment architecture," page 425.
- Proceed to chapter 10.25, "EXAMPLE Phonebook environment: Finalize design of environment interfaces with product requisites," page 585.
- Proceed to chapter 12.27, "EXAMPLE Phonebook environment: Verify prototype in environment," page 780.
- Proceed to chapter 8.14 "EXAMPLE Phonebook: Predetermine solutions with sourcing options" below.

8.14 EXAMPLE Phonebook: Predetermine solutions with sourcing options

8.14.1 EXAMPLE Phonebook: Process schedule to use

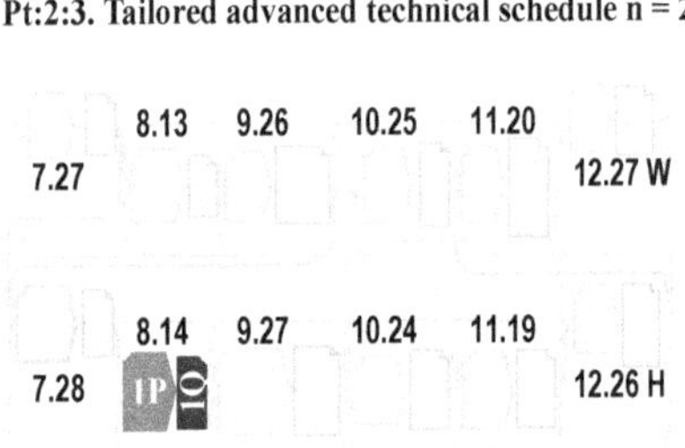

For the phonebook example, a special technical overall *schedule* is *tailored* (illustrated in Fig. 8-16, left):

- **Pt:2:3. Tailored advanced technical schedule n = 2,** page 169

In this *tailored schedule* the generic *schedule* to use is (filled-in symbols):

- **Pt.n.P. Predetermine solutions with sourcing options,** page 133.

FIGURE 8-16 Position of schedule and master to use (filled in) in overall schedule

8.14.2 EXAMPLE Phonebook: Try to satisfy requirements

The phonebook *program* is expected to be a substantial attraction in the future *product portfolio.* Such a large *product* must be gradually developed, and this is the first step. It is essential that the proposed *solution alternatives* are future-proof (see Table 8-34 below)

TABLE 8-34 Phonebook white-box solution alternatives

Solution alternative	Assessment	Requirement	Priority
A.SL1. Write program in a C++ object-oriented program language with function calls to libraries for Windows operating system.	Free download No extra cost	A.RR.1.1. The phonebook program shall be developed for a standard Windows computer.	Critical
		RR.4.1. Choose an object-oriented program development system, design a strict modularization, and keep user interaction together with used objects.	High
A.SL2. The program shall from the start be separated into self-contained items.		RR.2.1. First phonebook program shall use simple text-oriented console interaction.	High
A.SL3. Use line-oriented text streams for saving and loading phonebook on file.	+ 5% on below cost estimates	RR.3.1. The program shall load phonebook data at start-up and save them on program termination.	Critical
A.SL4. Use array of objects to store phonebook data.	Cheapest alternative	RR.5.1. The first phonebook program version shall use only compiler built-in fixed-size variable storage. The phonebook file shall hold an array of contacts with contact data, one of which is an array of phone numbers with phone number data.	Critical
A.SL5. Implement specified restrictions RR.5.2., RR.5.3., RR.5.4. and RR.5.5. A.SL6. Stimuli may exceed these limits, but the program may not crash for this reason.	+ 10% on below cost estimates	RR.5.2. MAX_CONT_PER_BOOK = 128 RR.5.3. MAX_NUMB_PER_PERS = 16 RR.5.4. MAX_PERS_LEN = 128 RR.5.5. MAX_NUMB_LEN = 32	Medium
A.SL7. All possible exception use-cases must be implemented with informative user interaction.	Included in below estimates	RB.6. Phonebook behavioral requirement no 6	Critical
A.SL8. Implement in phonebook constructor.	No extra cost	RB.6.S.1. Initiate-phonebook scenario.	Critical
A.SL9. Implement in phonebook destructor.	No extra cost	RB.6.S.5. Terminate-phonebook scenario.	Critical

8.14.3 EXAMPLE Phonebook: Analyze make/buy options

TABLE 8-35 Phonebook white-box supplier opportunities

Supplier opportunity	Solution alternative
SU.14. Windows development system from reliable supplier • Choose one of the biggest.	Development equipment A.SL1.
SU.15. In-house development • The rest to be developed in-house, no more suppliers needed	Phonebook A.SL2., A.SL3., A.SL4., A.SL5., A.SL6., A.SL7., A.SL8., A.SL9.

8.14.4 EXAMPLE Phonebook: Examine make/buy balance

The gate meeting was surprised by the fact that so few *suppliers* were needed. The concern was raised that possibly no *supplier* could be found to offer *programs* for *reuse.* The simple answer was that the phonebook design should form the foundation for many years ahead, and it would be rather foolish to limit design by some short-term benefits from *reuse.*

The gate meeting was also concerned that the rather *complex behavior requirement* RB.6. Phonebook behavioral requirement no 6 resulted in very few *whitebox solution alternatives.* Was it really intended that the engineers be free to make the design at their sole discretion? However, the conclusion from the meeting was that the restrictions were quite sufficient, and the *unbiased architecture ingredients* designed by the engineers were fully acceptable, provided that the *product* fulfilled the specified behavior.

8.14.5 EXAMPLE Phonebook: Proceed reading

To follow this example, these are the alternatives:

- Proceed to chapter 9.27, "EXAMPLE Phonebook: Satisfy refined requirements by decomposing product black-box into white-box design containing no black-boxes," page 429.
- Proceed to chapter 10.24, "EXAMPLE Phonebook: Finalize design items of product white-box with product requisites," page 561.
- Proceed to chapter 12.26, "EXAMPLE Phonebook: Verify black-/white-box," page 770.

## CHAPTER 9

Design architectures

Architecture is a heavily overused word,
not seldom concealing a skill that is not present.

Time and money for architects?

9.1 About architecture

9.1.1 Purpose

In some cases an *architecture* is easy to explain, for example, the *architecture* of a building or the layout of an apartment. But sometimes it is almost impossible to explain what it is, and the more abstract the technology is, the worse it gets. *Programs* are truly immaterial information in strict sequences of characters and words, creating assignments, loops, and selections that can be packaged and illustrated as *architectures*.

Mechanics, electronics, and programs can all be structured in a similar way by an architecture.

For many people, *managers* included, such packages are not as real as houses, rooms, cabinets, and machinery. However, coming chapters will show that the *program* comparison to mechanics and electronics is real and even obvious, and will deal with *program* packages as if they were ordinary house-building materials, being nailed, screwed, and mortared together.

9.1.2 SHORT EXAMPLE: If programs had been mechanics

If there is no strong discipline to prepare *program architectures*, *interfaces* will be invisible and unstructured, and *integration* of all *program elements* will be tedious and time consuming. *Program architecture* is very similar to mechanical *architecture*, but often poorly understood. If *programs* were mechanics, the following might happen.

An ordered and received "*program* bathtub" simply appears to be too big for the "*program* bathroom," because no proper *program architecture* was prepared containing measurements for the *program* bathroom before the *program* bathtub was ordered. A new *program* bathtub had to be procured, but nothing was found, because no *supplier* could offer a bathtub 40 cm long and 130 cm high. The developers instead simply displaced the "*program* wall" towards the "*program* kitchen," and now the first ordered *program* bathtub fits very well. Until the "*program* sink" arrives to the *program* kitchen . . .

9.1.3 Cpdm definitions

First, to get in sync with *architecture*, some definitions are needed.

In *Cpdm*, *user*, *interface*, and *element* are defined according to Table 9-1 below.

TABLE 9-1 Cpdm definitions

Aspect	User	Interface	Element
Definition	Human role interacting with an element through a man-machine interface	The means for two or more elements to influence each other, or to keep each other apart	Architecture ingredients package, which has interfaces to its surroundings
Synonyms	End-user, operator, handler, manipulator, exploiter, maneuverer	Combine, intermix, merge, unite, join, or separator, shelter, barrier, shield	Entity, section, layout, unit, item, component, module, part
Symbol	A.b. User — -A-1 Interface — A.a. Element of ingredients		

A *user* is a kind of *role*. This symbol has no hat, because this *user* is not acting according to any documented *terms*. For a more thorough explanation of *interface* types, see chapter 9.4 "Architecture interfaces."

In *Cpdm*, *environment* and *black-box* are defined according to Table 9-2 below.

TABLE 9-2 Cpdm definition

Aspect	Environment	Black-box
Description	The existing surroundings and the interfaces to the product being developed	Element which conceals all of its inward content, but shows its dynamic behavior via its interfaces. Can be opened to become a white-box
Synonyms	Surroundings, neighborhood, ambient setting, context	Unit, device, part, module, component, element, object
Symbol examples	E. Environment — User — Existing interface — Existing element — machine-machine interface (to be developed) — User — man-machine interface (to be developed) — A. Black-box (product to be developed)	
Alternative symbol (to indicate that there are hidden contents inside the black-box)	Interface — A. Black-box (hidden content partly visible) ... a ...lack-box	

As indicated in the definition in Table 9-2 above, before the *product* to be developed is introduced, there are existing ingredients in the *environment* to relate to. These ingredients should be the surroundings of the *product* to be developed, and it is most important from *environment nesting level* n = 0 to identify *interfaces* from these *environment* ingredients to the *product* being developed.

The name *black-box* refers to the fact that from the outside nothing at all can be perceived of the inside, other than what is conveyed on *interfaces* to and from it.

The restrictions and behaviors of a *black-box* are by definition specified by its *requirements*, which are later used to verify the *black-/white-box* as a stand-alone *element*. This implies that all *interfaces* to a *black-box* must be known, described, and attainable from the outside. Two *black-boxes* with the same *connection interface* and behavior, no matter what differs inside the *black-box*, can be exchanged, allowing a step from one internal technology to another, without the possibility of detection from the outside.

In *Cpdm, environment architecture resumé, environment logical architecture*, and *environment physical architecture* are defined according to Table 9-3 below.

TABLE 9-3 Cpdm definitions

Aspect	Environment architecture resumé	Environment logical architecture	Environment physical architecture
Defini-tion	List of existing relevant ingredients in the environment and newly developed interfaces to the product	Logical drawing of the environment ingredients, including the product black-box with its interfaces	Appearance illustration of the environment surrounding the black-box of the product
Syno-nyms	Approved solution, solution domain, how-question	System architecture, product interface drawing	Product illustration, system image
Symbol	**TABLE 9-4** Environment architecture ingredients resumé **Existing and product interface ingredient** / **Explanation or chosen solution alternative or sample thereof**	Existing environment / New inter-face / Existing element / Black-box / New MMI / Poten-tial user **FIGURE 9-1** Environment logical architecture	Environment / Black-box **FIGURE 9-2** Environment physical architecture

9.1.4 Gather all design into environment architecture resumé

If a solution is decided to be bought from *off-the-shelf suppliers* or from *turnkey suppliers*, it is very important to include their specific solutions in the *environment architectures*, because all used solutions must be part of the *architecture* analysis, in order to fit together when later integrating the *product outermost black-box* into the *environment*.

In Figure 9-3, the *upstream* references are also indicated by dashed lines, to make the *process* from *requirements* to *architectures* visible and traceable *upstream* and *downstream*. Thus, a chosen *element* can be derived from the *requirement* which gave rise to it, and the reverse: from *requirements* it can be understood how these lead to the *architecture*. If later on a *requirement* happens to be changed, the *architecture* can be quickly modified at any time.

As can be seen from the figure, this is a typical example of a database system keeping track of all references. This book shows the most fundamental and minimalistic references within *Cpdm*, and a more comprehensive database system should be used if the *product* is significantly *complex* or must meet a considerable *quality*.

Also see Figure 10-4, page 444, for a more complete description of these *cascading results*.

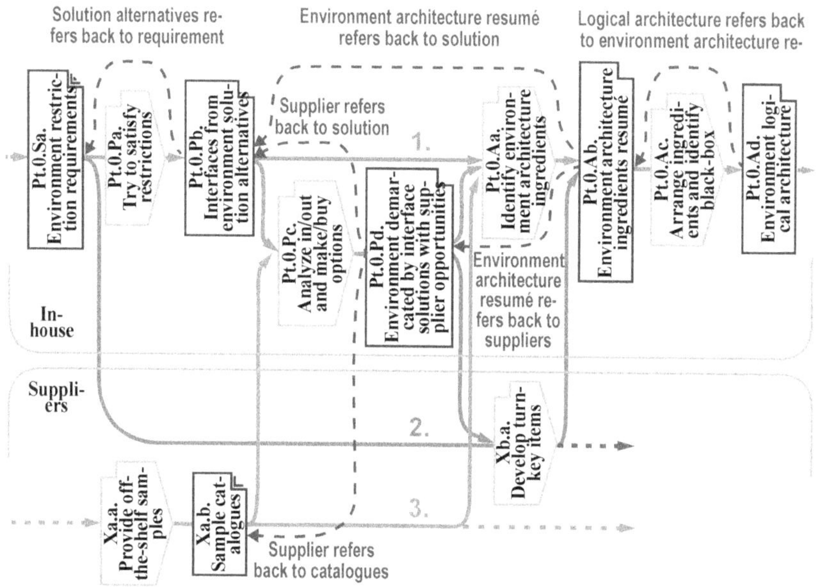

FIGURE 9-3 Schedule to trace solutions in the environment architecture resumé

Utilizing all kinds of *suppliers* to achieve the *environment interfaces* is illustrated in Figure 9-3 above:

1. In-house development (upper arrow lines)
2. Ordering turnkey (middle arrow lines)
3. Buying off the shelf (lower arrow lines)

9.1.5 References from

In an *environment architecture resumé* the purpose is to gather all *environment architecture* phenomena (see Table 9-5 below).

TABLE 9-5 Environment architecture ingredients resumé

Ingredient type	Existing and product interface ingredient	Architecture phenomena to document
Primordial	Black-box element	The black-box to be developed
	Element	Existing elements in environment
	User	Black-box users populating the environment
	Interface	Existing interfaces in environment
		Newly designed interface to the black-box according to solution alternatives derived from environment restrictions
Determined	Interface	Newly designed interface to environment according to solution alternatives not derived from explicit restrictions, such as implicit or explicit standardization
Unbiased	Interface	Newly designed interface decided freely by purpose
		Newly designed interface due to absence of solution alternatives
		Newly designed interface due to absence of requirements causing absence of solution alternatives

Generally, all *environment* design originates from restrictions. However, cases may occur when an *interface* solution to the existing *environment* seems necessary, despite the absence of explicit *environment restriction requirements* forcing this *interface*, or other *requirements* forcing interaction between the *environment* and the *product* to be developed.

Be aware that lack of *environment interfaces solution alternatives* also creates an absence of *supplier* opportunities. It is recommended that the *environment restriction requirements* be reviewed, to evaluate if the problem has been caused simply by a forgotten restriction or missed *solution alternatives*.

9.2 EXAMPLE House environment: Constitute environment architecture

9.2.1 EXAMPLE House: About this example

When it comes to *architecture*, houses are strikingly good examples to discuss. They show several *inward nesting levels* of decomposing *architectures*, in order to get more and more detailed without losing control. How *requirements* and *architecture* are interconnected is also shown.

A house example is a concrete way to begin explaining architecture.

It is obvious that a single house *nesting level* with all detailed ingredients would create a too flat and wide structure to be comprehensible. The question is whether we should go for two, three, four or still more *nesting levels* to find an appropriate house hierarchy. The house example proposes four house *nesting levels*:

1. *Environment* of the house (shown in this chapter)
2. The house (shown partly in this book)
3. Rooms (kitchen shown in this book)
4. Machinery (kitchen machinery shown in this book)

Interfaces are among the most difficult to handle with *consistency* in *product development*. Interestingly, a house is, in fact, very *complex* in this matter and has four kinds of *interfaces*:

1. *Connection interfaces* for people, such as doors and doorways which allow people to move around in the house
2. *Connection interfaces* for supplies, such as pipes, conductors, and tubes for commodities to be obtained in the house
3. Static *connection interfaces*, such as mortar keeping bricks together and screws keeping various building parts together
4. Static *boundary interfaces*, such as façades, inner walls, and different kinds of cabinets for protection and beauty value

9.2.2 EXAMPLE House environment: Process schedule to use

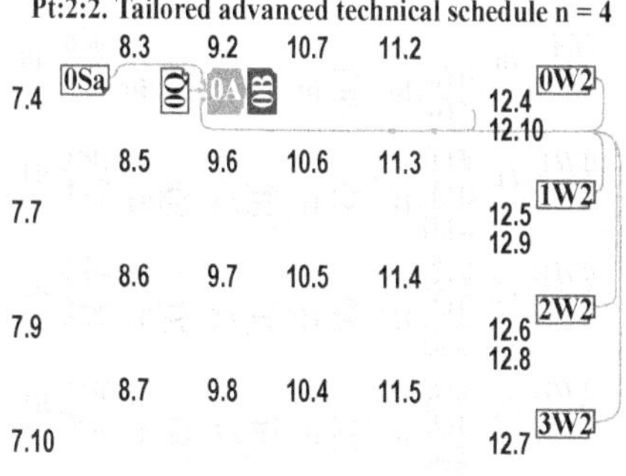

For the house example, a special technical overall *schedule* is *tailored* (illustrated in Fig. 9-4, left):

- **Pt:2:2. Tailored advanced technical schedule n = 4,** page 164

In this *tailored schedule* the generic *schedule* to use is (filled-in symbols):

- **Pt.0.A. Constitute environment architecture,** page 115.

FIGURE 9-4 Position of schedule and master to use (filled in) in overall schedule

Below *results* may be demanded (their *masters* are indicated *upstream* from the filled-in *schedule* in the figure above):

0Q Solution alternatives of interfaces to environment with supplier opportunities

Table 8-5, House environment solution alternatives of interfaces to environment, page 292

Table 8-6, House environment solutions excluded, page 293

Table 8-7, House environment solutions included with supplier opportunities, page 293

0Sa Environment restriction requirements

Table 7-5, House environment restriction requirement, page 182

Xb Suppliers' sample catalogues

For possible *failure localization* and *failure elimination*, below *results* may be demanded (their *masters* are indicated *upstream* from the filled-in *schedule* in the figure above or greater detail in Fig. 11-5, p. 602 and Fig. 11-6, p. 603):

0J2 Table 11-9, House environment start-up report 1, page 606

0W2 List 12-1. House environment interfaces with fake house failure elimination report 1, page 674

0W2 Table 12-27, House environment verification report 1, page 701

Table 12-29, House environment reverification report 2, page 703

Table 12-28, House environment failure elimination report 1, page 702

1W2 List 12-2. House with fake rooms failure elimination report 1, page 676

1W2 Table 12-22, House verification report 1, page 697

Table 12-24, House verification report 2, page 698

Table 12-23, House failure elimination report 1, page 698

2W2 List 12-3. House rooms with fake machinery failure elimination report 1, page 678

2W2 Table 12-14, House kitchen verification report 1, page 688

Table 12-17, House kitchen verification report 2, page 691

Table 12-15, House kitchen failure elimination report 1, page 689

3W2 Table 12-9, House kitchen machinery verification report 1, page 682

Table 12-11, House kitchen machinery verification report 2, page 683

Table 12-10, House kitchen machinery failure elimination report 1, page 683

9.2.3 EXAMPLE House environment: Identify environment architecture ingredients

From the existing *environment*, appropriate ingredients are now identified in Table 9-6 below. Additionally, ingredients derived from Table 7-5, page 182, can now be identified for the *interface* between the existing *environment* and the *product* to be developed.

TABLE 9-6 House environment architecture ingredients resumé

Existing and product interface ingredient		Explanation or chosen solution alternative or sample thereof
E.	House environ-ment	The existing surroundings of the house E.SL3. When connections to the environment are constructed, existing neighboring environment may not be damaged.
E.a.	House user	Proprietor, relatives, and guests coming from the environment
E.b.	Lake	Existing beautiful lake, south of the planned house
E.c.	Street	Existing residential street, east of the planned house
-E-	Interfaces to environment	Collective identity for all interfaces to the environment
-E-1.	Shoreline distance	Product interface to be developed E.SL1.a. Keep distance to shoreline as close as possible to 150 m.
-E-2.	Water	Product interface to be developed E.SL8.b. Ø 32 mm water pipe from street
-E-3.	Sewage	Product interface to be developed E.SL9.b. Ø 125 mm pipe to street sewage
-E-4.	Electricity	Product interface to be developed E.SL7.b. 32 Amperes from underground cable
-E-5.	Gas	Product interface to be developed E.SL6.b. 3 m³ or 7 m³ rental tank close to house Must go shortest way from outer wall to inside burner.
-E-6.	Driveway	Product interface to be developed E.SL10.a. Basalt-covered 2.4 m wide road Stone sample H34 from Ext Bas AB
-E-7.	Entrances	Product interface to be developed E.SL4. Ramp from driveway to entry for disabled persons
A.	House	Product to be developed E.SL2.a. Extend the building plot by 2 000 m², to ensure a building area of 100 m². E.SL5. The building plot must be located in an attractive area.

9.2.4 EXAMPLE House environment: Arrange ingredients and identify black-box

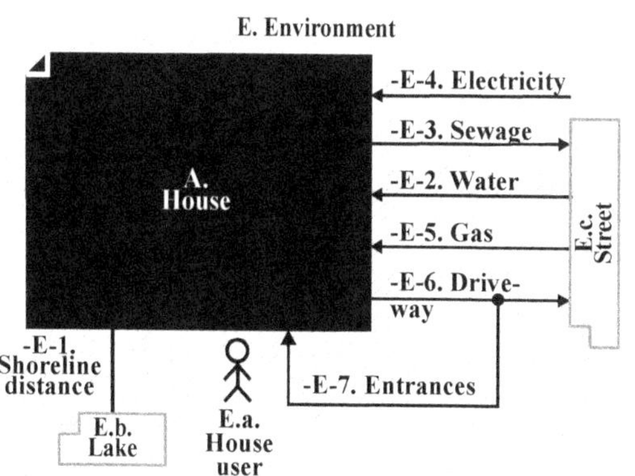

The *architecture ingredients* from the *architecture resumé* are now documented in a *logical architecture* (see Fig. 9-5 below).

FIGURE 9-5 House environment logical architecture

9.2.5 EXAMPLE House environment: Lay out physical architecture

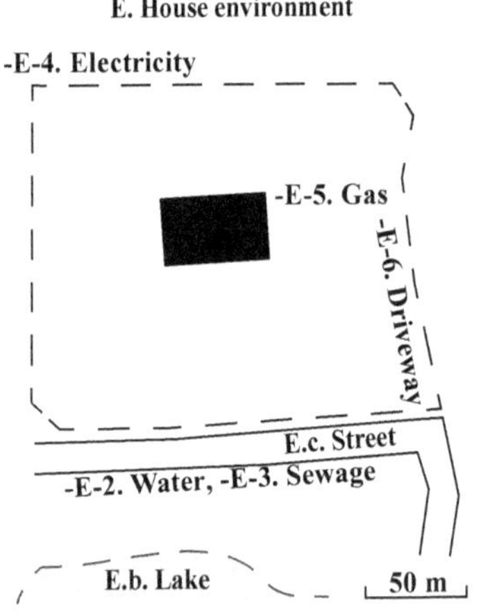

FIGURE 9-6 House environment physical architecture

The *physical architecture* layout for an *environment architecture* draws the house into a map of the *environment*. This must show the house *black-box* with its right measurements and scale in relation to the surrounding map (see Fig. 9-6).

At this *phase*, it might be hard to know the exact physical form of the 100 m² house body, so some updating is expected after the house is better known, a bit later in the *schedule*.

9.2.6 EXAMPLE House environment: Examine product context

It is crucial that identified *environment architecture resumés* are almost failure-free. These will soon be finalized and will heavily influence the *product* being developed.

To ensure that all *architecture ingredients* are right, a gate meeting was held with the *architecture* company Robber HB. The *architecture resumé* was walked through, and all *interfaces* and *elements* were discussed. Only small weaknesses were detected and eliminated.

9.2.7 EXAMPLE House: Proceed reading

To follow this example, these are the alternatives (again see Fig. 9-4, p. 339):

OF Proceed to chapter 10.7, "EXAMPLE House environment: Finalize design of environment interfaces with product requisites," page 477.

OI Proceed to chapter 11.2, "EXAMPLE House environment: Realize interfaces to environment and insert/await outermost white-box," page 601.

OV Proceed to chapter 12.4, "EXAMPLE House environment: Partially verify interfaces to environment with house fake," page 673.

OV Proceed to chapter 12.10, "EXAMPLE House environment: Fully verify prototype in environment," page 699.

IR Proceed to chapter 7.7, "EXAMPLE House: Refine house requirements from environment restriction requirements, and ensure house development staffing," page 187.

1A. Proceed to chapter 9.6, "EXAMPLE House: Satisfy house requirements by decomposing house black-box into white-box design containing room black-boxes," page 357.

- To leave the house example and to dig deeper into the chapter 9, "Design architectures" topic, proceed to next chapter 9.3 below.

9.3 Architecture white-box

9.3.1 Cpdm definition

In *Cpdm, white-box* is defined according to Table 9-7 below.

TABLE 9-7 Cpdm definition

Aspect	White-box
Definition	Element which is an opened black-box, containing disclosed architecture ingredients, some of them possibly embedded black-boxes
Synonyms	Opened black-box, specified subsystem, self-sufficient element
Symbol	-E-1 Connection interface — A. White-box (opened black-box) — interface — All. Boundary

A *black-box* that has been opened turns by definition into a *white-box*. Inside the *white-box* a number of *elements* will appear, of which zero or more might be appointed *black-boxes* (see Table 9-7 above). Also, *interfaces* to the *white-box* can now be extended and connected to newly designed ingredients inside the *white-box*, and new *interfaces* will also appear between totally internal *elements*.

There is no big difference between a *black-box*, *white-box*, and an *element*. The main idea of distinguishing *black-boxes* and *white-boxes* from *elements*, is that *black-boxes* are assigned *requirements* and the *black-/white-box* can thus be separately verified. This allows *black-/white-boxes* to be bigger and more *complex*, without loss of control.

In *Cpdm*, *architecture resumé*, *logical architecture*, and *physical architecture* are defined according to Table 9-8 below.

TABLE 9-8 Cpdm definitions

Aspect	Architecture resumé	Logical architecture	Physical architecture
Definition	List of decided ingredients, derived from solution alternatives or supplier samples thereof	Drawing illustrating how ingredients logically relate to each other	Drawing illustrating how ingredients physically appear and how they combine with each other
Synonyms	Approved solution, solution domain, how-question	Symbol architecture, product interface drawing	Element illustration, part image
Symbol	**TABLE 9-9** Architecture ingredients resumé **White-box ingredient** — Chosen solution alternative or sample thereof	FIGURE 9-7 Logical architecture	FIGURE 9-8 Physical architecture

If a *sample* is considered in the *architecture resumé*, it is also a good idea to document the *supplier*, particularly if there is only one *supplier* offering such *sample*.

9.3.2 Architecture decomposition (partition)

Partition is often used as a special technical word, for example, how to *partition* the storage space of a hard disk into smaller and more manageable storage pieces. The opposite, *aggregate*, is also a technical word, for example, to assemble a construction of Lego toy pieces, or to assemble furniture from an IKEA flat package. For definitions, see Table 6-3, page 85. In the *architecture* field, *partition* is often called *decompose*.

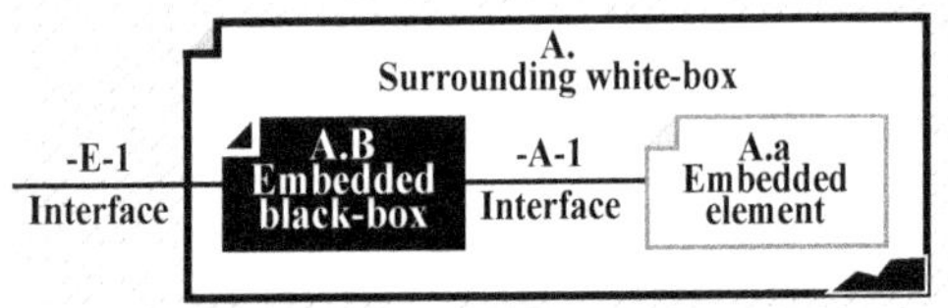

Architecture decomposition is a rather well known and understood concept (see Fig. 9-9).

FIGURE 9-9 Architecture decomposition (partitioning)

To divide a system into subsystems is straightforward and is done intuitively by most engineers.

Black-boxes can be *decomposed* into parts, which might also be *black-boxes*, which may also be *decomposed* into parts, thus creating a *black-/white-box hierarchy*, like a Russian doll. When referring to *architecture* hierarchies throughout this book, the terms in Figure 9-10 are used.

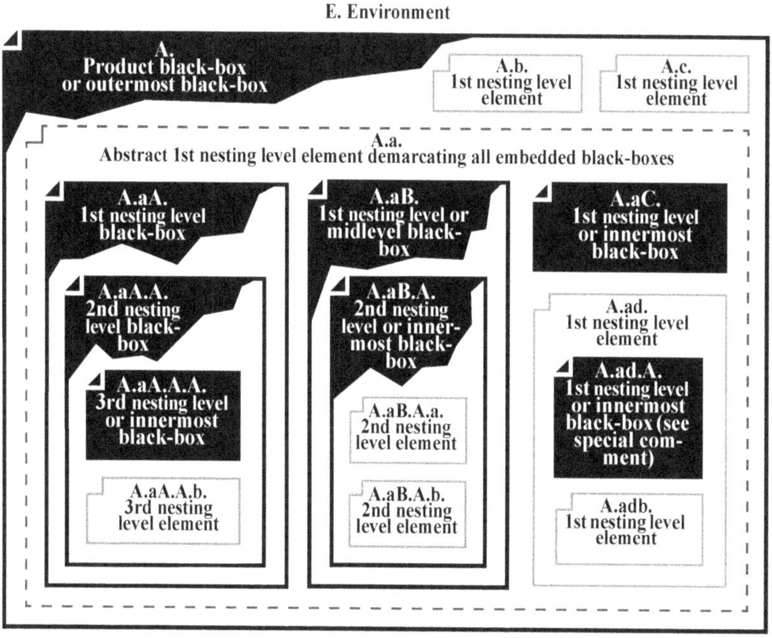

FIGURE 9-10 Black-/white-box hierarchy principles and naming

Some further explanations to Figure 9-10 above:

- Some areas of the *black-/white-box hierarchy* go deeper *inward* than others. This is fully natural, because some parts contain more *elements* and are more *complex* than other parts, which creates variable depth of hierarchies.

- *Abstract elements*, like **A.a.** in Figure 9-10, can be used to group *architecture elements* together and give that group a name. Note that such groups should by definition not hold any own *design items* other than contained *elements*. The *abstract element* identifier can be useful but is not mandatory.

- It is permissible for *black-boxes* to reside within *elements* that have never been a *black-box*, like **A.ad.** in Figure 9-10 above. However, keep in mind that **A.ad.** must be finalized before **A.ad.A.** *black-box* can be specified by *requirements*

- Identifiers of a *black-box* are separated by dots. Many *element* identifiers can be written together without separating dots, and the principle is that the identifier should be kept short, but not so short that it gets difficult to read.

The above settlement of box hierarchies is a lightweight *version* of the *clean-room engineering* concept (see references [5], [6] and [2]).

9.4 Architecture interfaces

9.4.1 Cpdm definitions

In *Cpdm, connection interface* and *boundary interface* are defined according to Table 9-10 below.

TABLE 9-10 Cpdm definitions

Aspect	Connection interface	Boundary interface
Description	The means for two or more elements to influence each other	The means for two or more elements to keep apart
Synonyms	Joint, splice, seam, connection, interconnection, interplay, channel	Border, boundary, demarcation, separator, finish, exterior, interior
Static examples	Screws, bolts, nails, clips, strings, weld, solder, mortar	Walls, shields, cabinets, housings, casings, firewalls (both buildings and programs), safes, cases, paint, plating, coating, plaster, filler, variable visibility, encapsulation, information hiding, physical memory protection
Dynamic examples	Hoses, cables, copper paths on PCB, pipes, doors, tick-boxes, api (application interface), mmi (man-machine or machine-machine interface), gui (graphic user interface)	
Symbols		

For many reasons *interfaces* are more troublesome to understand and handle than *elements* are.

First, *interfaces* connect only *elements*, and seem not to bring much of value to a system. They may be easy to understand as concepts, but can be very *complex* in reality. For example, a driveway to a house can be seen as a simple *connection interface* between the house and outside road, but can be seen as a *complex* construction with a lot of *elements*. Context and clarity must determine how to model the reality.

> **An interface is often seen to be just a simple link between elements, but is, in fact, often more complex than most elements.**

Second, *interfaces* are like *elements* in the sense that the more abstract they get, the less they are understood. *Program interfaces* are known to most developers, but are most often not actively documented and *managed information*. One reason is that in the tangible world, an *interface* like a pipe or a door costs money, but in *program interfaces*, material seems to come for free, and does not need to be maintained. Even if a big, immaterial "*program* pipe" is designed to hold together all *function* calls from one subsystem to another, some *program* developers do not hesitate to draw extra immaterial *program* pipes to sneak past with some obscure calls. Everybody knows that to lay down pipes in a road in different places and to dig even once more to sneak in an extra pipe will cost a lot of extra money and effort.

If one particular end of a dynamic *interface* acts as the source, this might be marked with an arrow in the far end. If it is unclear (at least for the moment) in what direction the transfer is going, use a straight line without arrows.

9.4.2 Architecture preparation procedure

Architecture development is by far best performed as outside-in decomposition. The existing *environment architecture* is always known, and a good starting point for moving *inwards* (see Fig. 9-11 below).

The principal sequence of development events is:

0.0. The *environment restriction requirement* **E.RR.** and *environment interfaces solution alternatives* **E.SL.** have first been captured and proposed.

0.1. Next, all existing ingredients found in the *environment* are mapped into the *environment architecture* (for example, the *element* **E.a.**, the *connection interface* **-E-2.**, and the *user* **E.b.**).

0.2. The *outermost black-box* **A.** is then illustrated to show how it is surrounded by all existing ingredients in the *environment*.

0.3. *Environment restriction requirements* (such as **E.RR.**) may lead to the *solution alternatives* (like **E.SL.**), causing design of *connection interfaces* (for example, the *connection interface* **-E-1.**, with its outer end connected to the *element* **E.a.**, and its inner end connected somewhere inside the *black-box* **A.**). Since such *connection interfaces* are first identified in the *environment*, their identity refers to the *environment* (like **-E-1.**).

0.4. *Environment restriction requirements* are then refined to become *black-box requirements* (like **E.RR.** might be refined to **A.RB.**).

0.5. When all restrictions are refined, more *requirements* can be added for the *black-box* **A.** (for example, adding more *requirements* in **A.RB.**).

0.6. After *black-box requirements* are captured, the *black-box* **A.** can be opened to become the *white-box* **A.**, which now gets ready to be populated by *architecture ingredients*.

1.0. *Black-box requirements* (among *requirements* **A.RB.**) may lead to boundary interface solutions causing architectural design of *boundary interfaces* (for example, **[A[2.**, being the boundary of the *white-box* **A.**). Since this *boundary interface* is considered to belong to the *outermost white-box*, its identity refers to this *white-box* ().

1.1. Other *black-box requirements* (among *requirements* **A.RB.**) may lead to *element solution alternatives* **A.SL.**, causing design of *elements* of different types (like **A.b.**). Note that these *elements* may in turn have *boundary interfaces* and *connection interfaces*.

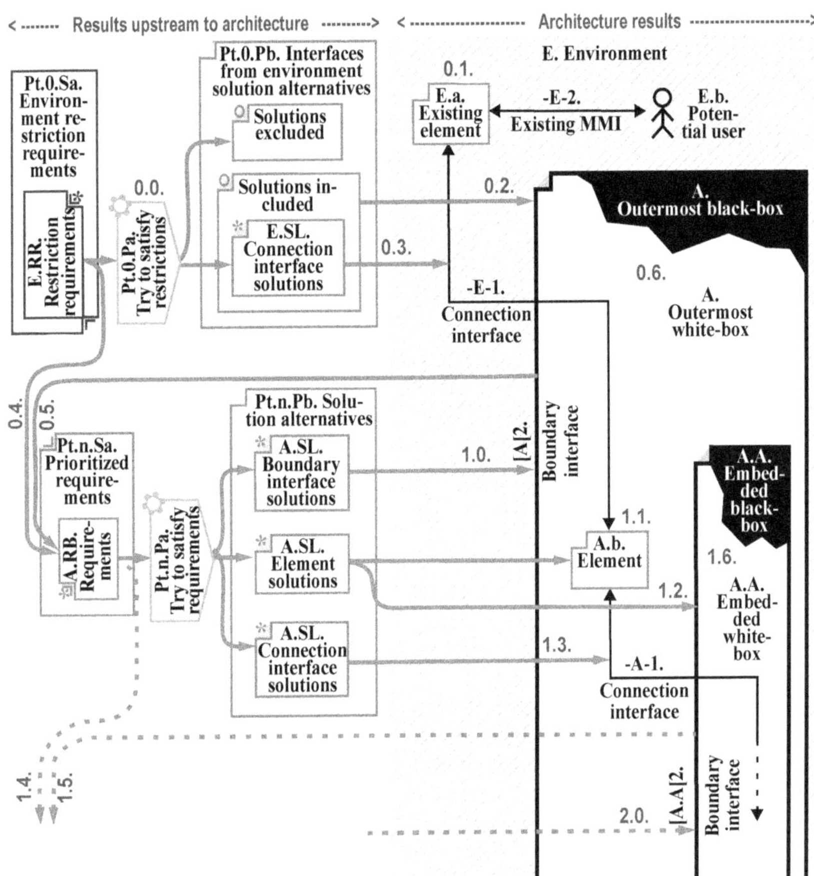

FIGURE 9-11 Architecture design of connection and boundary interfaces

1.2. Some of the designed *elements* in *white-box* **A.** may be found very *complex*, and are appointed as *embedded black-boxes* (like **A.A.**).

1.3. Still other *black-box requirements* (among *requirements* **A.RB.**) may lead to *connection interface solution alternatives* **A.SL.**, causing design of *connection interfaces* to *embedded black-boxes* (such as *connection interface* **-A-1.**, with its outer end connected to *element* **A.b.**, and its inner end connected somewhere inside *embedded black-box* **A.A.**). Since this *connection interface* was first identified in the *outermost white-box* **A.**, its identity refers to this *black-box* (e.g. **-A-1.**).

1.4. *Black-box requirements* **A.RB.** are now being refined further for possible *embedded black-boxes* (such as the **A.A.**).

1.5. More *requirements* can be captured and added as in 0.5. above.

1.6. After all *black-box requirements* are captured, the *embedded white-box* **A.A.** is opened to become the *embedded white-box* **A.A.**, which is now ready to be populated by *architecture ingredients*.

2.0. The *architecture* development procedure now repeats from 1.0. above but at one *nesting level inwards.*

Connection interfaces are discovered in outside-in order, because the *architecture* work starts with mapping the *environment* and proceeds *inwards*. Thus, it is quite natural to name *connection interfaces* according to the *element* where they are first detected.

Boundary interfaces of an *element* are often first identified when the *element* itself is detected; it becomes natural to consider the boundary to belong to that *element*, and identity then corresponds to the *element*, such as **[A[2.** However, a boundary can also be seen as divided into one outside and one inside the same *element* (as when the connection occurs both outside and inside an *element*). For example, a house outer façade can be considered to belong to the *environment*, and the house internal walls as belonging to the house. To distinguish between these, the boundary identity might become **]E[3.** for the outside of *white-box* **A.** *boundary interface*, and **[A]2.** for the inside of *white-box* **A.** In practice, this discussion of belonging has no big significance, so beware of introducing unnecessary *complexity* before splitting *boundary interfaces* into separate outside and inside *interfaces.*

9.4.3 Module coupling and strength

<table>
<tr><td>

These rules are really two old goodies still going strong.

</td><td>

Even if the transformation of *requirements* is heuristic with trial and error as the troublesome main method, there are some traditional golden principles for how *elements* should be designed. These are sometimes referred to as *module coupling* and *module strength*, the latter also called *module cohesion.*

</td></tr>
</table>

For an *element* to be well made, it must have low coupling to the surroundings, and high strength (cohesion) within itself (see Fig. 9-12 below).

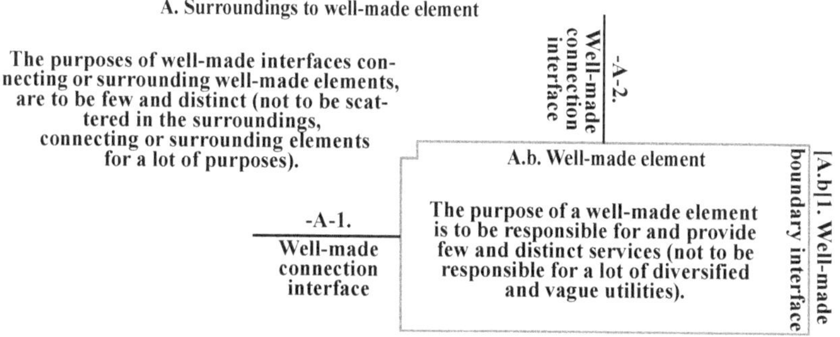

FIGURE 9-12 Interface coupling and element cohesion

In practical life, there might be conscious exceptions to these rules. For example, a small flat consisting of one room with kitchen is attractive, because it can be built and rented cheap. But one room with kitchen must be built really multipurpose, and thus not close to fulfil low cohesion as defined above.

Another example is *interfaces* to operating systems, which offer a broad range of different service calls, to be as versatile as possible. But this is in direct conflict with having low coupling to applications.

But in most cases these rules are great. They make the *architecture* clean and comprehensible, and bring a high-*quality* structure to the *product* itself.

Static *boundary interfaces* are also covered by this rule; for example, a cover to an electronic device shall preferably be fastened by few and easily accessible screws, to be cheap to produce and simple to open for maintenance.

9.4.4 Verifiability of a black-box or a white-box

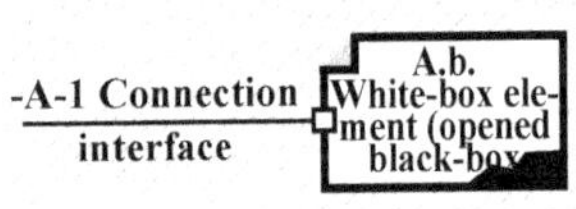

It is ultimately important for *verification* that all *interfaces* to a *black-/white-box* are disconnectable or provided with *observation posts*. In Figure 9-13, the *interface* **-A-1** is explicitly marked with small squares, symbolizing the possibility for accessibility and stand-alone *verification* of **A.b.**

FIGURE 9-13 Accessible observation post

In reality, such an access point might, for example, be a water scale on an incoming water pipe to measure the water consumption in a house, a pin on the PCB for connecting an instrument, or it might be a *program instruction* to send information to a printer.

There is no point in designing a white-box not accessible for verification.

Note that when the *interface* **-A-1** is disconnectable, the *element* **A.b.** by definition also becomes interchangeable. If not disconnectable or not with reachable posts, a stand-alone *verification* of the *black-/white-box* is almost impossible (especially if the *verification* is automatized), as is an easy exchange of the *black-box*.

Especially when applying outside-in *invasive integration* it may be easy to slip on this rule. For more information about this, see chapter 11.1.4, page 595.

9.4.5 Buffered interfaces to isolate elements

Buffers do not add more value than making an element more isolated.

To make a *white-box* or any *element* more independent of its surroundings, a buffer can also be located at the square symbol in Figure 9-13. Including buffers on *element* connectors or boundaries is a good way to create robustness between the *element* and its surroundings.

For example, there are usually buffers on all in- and out-connections to an integrated circuit, for example, the microcontroller in Figure 9-44, page 395. The buffer on the input is barely loading the sourcing, and is thereby getting independent and trouble-free for the surroundings. In the same way, a buffer on an output can drive almost any kind of load in the surroundings without affecting the function of the circuit inside.

Another buffer might be to arrange a wall-fixed bearing beam over a house window frame, thus making the frame freely selectable because it is independent of load bearing of the wall above.

Program elements for *reuse* as well as system *elements* with MMI to humans often have built-in extended control of input *stimuli*, to take care of everything that might go wrong. If an operation failure is detected inside such an *element*, specified failure codes are returned, instead of risking the *element* to crash, and instead pass back predictable return values that do not crash anything, including the surrounding *program* calling it.

9.4.6 Static interfaces

There are two kinds of static *interfaces:* the static *connection interface* and the static *boundary interface.* Static *connection interfaces* tie together *elements,* and *boundary interfaces* cover *elements* and separate them from their surroundings. Static *interfaces* don't convey any information or behavior; they are only keeping everything statically together.

A static interface can be either a connector between elements or a boundary separating them.

9.4.7 Static interface junctions

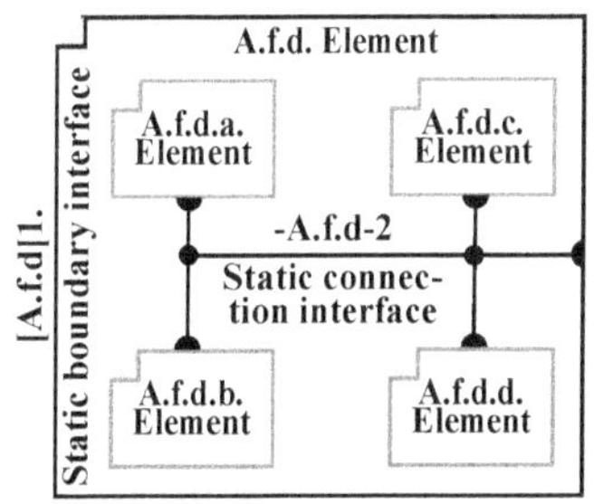

Elements in reality might be tied together with each other in such a *complex* way that it is not meaningful to detail them in the *architecture,* unless they are simplified in a *logical architecture.*

In an *architecture* drawing, such *interfaces* might be lumped together in one or many junction points, illustrated by a small filled circle (see Fig. 9-14). This figure also shows a connection to the **A.f.d-1** Static *boundary interface.*

FIGURE 9-14 Static interfaces between many elements and the boundary

9.4.8 Encapsulation by static boundary interfaces

Boundaries may be somewhat conspicuous as *architecture ingredients,* but they are really common in practice (see Table 9-11 below). It is hard to regard them as *elements,* and even harder to regard them as *connection interfaces.* To handle them as *boundary interfaces* with certain specific characteristics is an elegant and rewarding methodology.

TABLE 9-11 Examples of boundaries

Boundary example	Purpose	Inwards	Outwards
House	Weather protection	Convenient climate	Wild nature
	Theft protection	Valuables	Acquisitiveness
Shield	Radiation restriction and protection	Own radiation	Human-generated and cosmic radiation
Visibility	Variable access protection	Local variables	Global variables

9.4.9 Dynamic interfaces

Dynamic *interfaces* are penetrating boundaries to *elements* that convey interaction between its inner structures and its surrounding structures.

Dynamic interfaces always convey something between elements or users.

When looking from the *environment inwards*, the first detectable *interfaces* are those spanning from *users* or *elements* in the *environment* into the *black-box* to be developed. During *architecture* development the design progresses *inward*, and more and more *elements* with *interfaces* are designed all the time.

As described in Figure 9-10, page 345, classification of *elements* and identities is rather straightforward. It is a bit harder to classify *interfaces*, because they may span between *elements* that may be very far from each other, or reside in very different depths of the *element* hierarchy.

Only a rough part of *interface* identification may be made while moving *inwards* from the *environment* to the inner *elements*, because it is not until the inner end of the *interface* terminates that it becomes possible to fully understand it. For example, it can be hard to design and calculate the dimensions of water pipe *interfaces* to a house before it is known where every pipe terminates in faucets, toilets, and so forth. The same goes for a power supply to a *product* that consumes current of a so far unknown amount.

The final identification and design of *interfaces* are consequently developed during finalization of the *elements*, while going back from innermost *elements* and *outwards* to the *environment* (see next chapter, **Pt.F. Finalize design & requisites**).

9.4.10 Dynamic interface sourcer and absorber

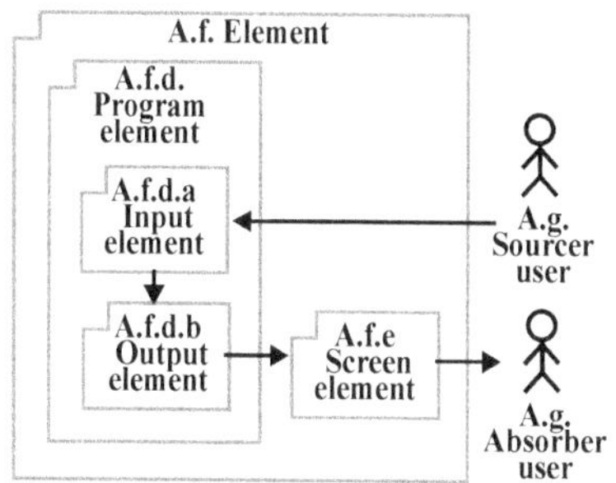

One-way *interfaces* always have sourcer and absorber *elements*. (Two-way *interfaces* can almost always be split into two one-way *interfaces*).

For example, a *user* can source a system by pressing a keyboard button, and an input *instruction* can absorb the letter typed (see Fig. 9-16). The next *program instruction* might transfer the key letter to a screen for the *user* to absorb it.

FIGURE 9-15 Sourcer and absorber

9.4.11 Dynamic interface junctions

Simple dynamic one-way service *interfaces* can be lumped together as static *interfaces* (see Fig. 9-16). This is sufficient, for example, to illustrate water distribution to many faucets in a house or direct current supply to a lot of *components* in an electronic circuit.

FIGURE 9-16 One-way interface for many elements

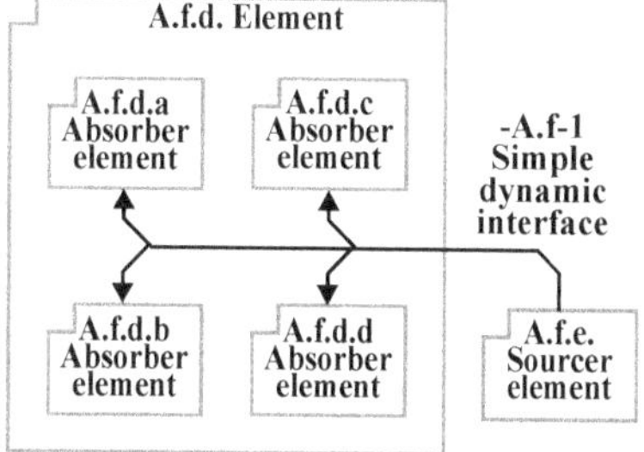

A *variant* of a simple *interface* is the buses often occurring in computers. These *interfaces* contain many conductors, which is the same as one conductor, but multiplied by a width factor. The *elements* are *complex* and intelligent, and there is no big problem in making them reversible two ways (again see Fig. 9-17).

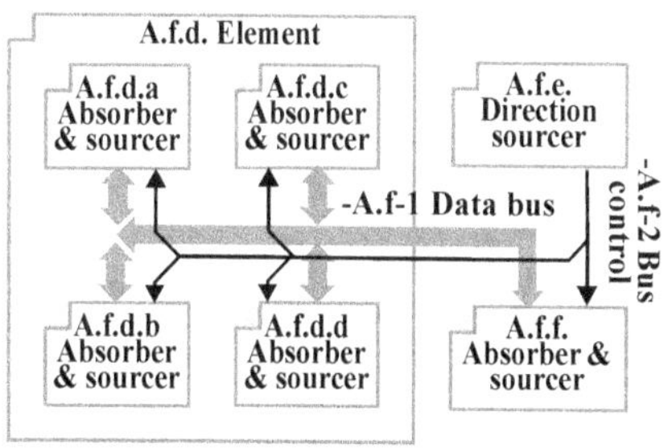

FIGURE 9-17 Data bus between many elements

If there is a mix of one- and two-way dynamic *interfaces* or a mix of different kinds of *interfaces*, it is most often not good to lump them together. Such *architectures* get difficult to read and follow, and there is no point in making *complex* drawings in this way. Keep the *interfaces* apart and give each of them an identification (see Fig. 9-18).

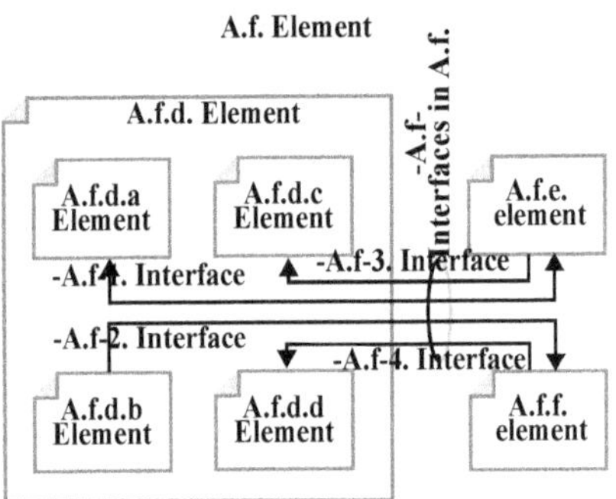

FIGURE 9-18 Complex interfaces between many elements

9.5 Transforming requirements into architecture ingredients

9.5.1 Neglecting architectural work

At this point, many engineers get the feeling that *Cpdm* is overkill, with too much academic equilibration, taking time from getting the *product* designed and finished.

When there are more than two coffee machines in a development department, the architecture must be formalized.

In my lifelong experience, it seems that one not seldom sees *product structures* resembling a shantytown or magpie's nest. This is clearly the result of absent architectural capabilities. If a *product* is small, one person or small team can mentally encompass the entire structure, but when such a *product* grows bigger and more developers are involved, the structure degenerates gradually. This can, in fact, continue until the *product* slips beyond control.

9.5.2 When to designate an element to a black-box

When identifying design *elements* from a set of *solution alternatives*, the most *complex elements* must be identified, so that it can be decided whether they should be made *black-boxes* instead, to create an *inward nesting level* of *requirements* and *architectures*.

Typical situations in which to consider designation of a *black-box* instead of an ordinary *element*:

- If the *elements* seem to be so *complex* that they require high *formalism* to understand.

- If development of such *elements* must be outsourced to a subcontractor (or to remote parts of one's own organization).

- If a new technology must be used for an *element* which is not familiar to the development company.

- If there is a need to consider and develop many interchangeable alternatives for an *element*.

- If an *element* is planned to be *reused* (not *recycled*) by another in-house *product* or to be sold to external *customers*.

- If a *product* is so large that the *complexity* is significantly reduced by encapsulation into smaller parts.

- If there is a need for taking care of *scenario-thread-machines* to many concurrent *stimuli*.

There are also situations when a *black-box* designation can be questioned:

- If an *element* is small and simple, and it is obvious that it should stay that way.

- If an *element* is a support system to other *black-boxes*. If designated to a *black-box*, the risk is that its *requirements* assume a life of their own, and the *user* of the support system might be forgotten.

9.5.3 Gather all design into environment architecture resumé

The *environment architectures* are now prepared based on available *results* from previous *phases*. *Results* to *demand* are *white-box solution alternatives* for in-house development, *sample catalogues*, real *samples* from *off-the-shelf suppliers*, or architectural proposals from *turnkey suppliers* (see Fig. 9-19 below). If a solution is proposed to be bought from an *off-the-shelf supplier* or from a *turnkey supplier*, it is very important to include their specific solutions, because all proposed solutions must be part of the *architecture* analysis, in order to fit together when later integrating the *white-box*.

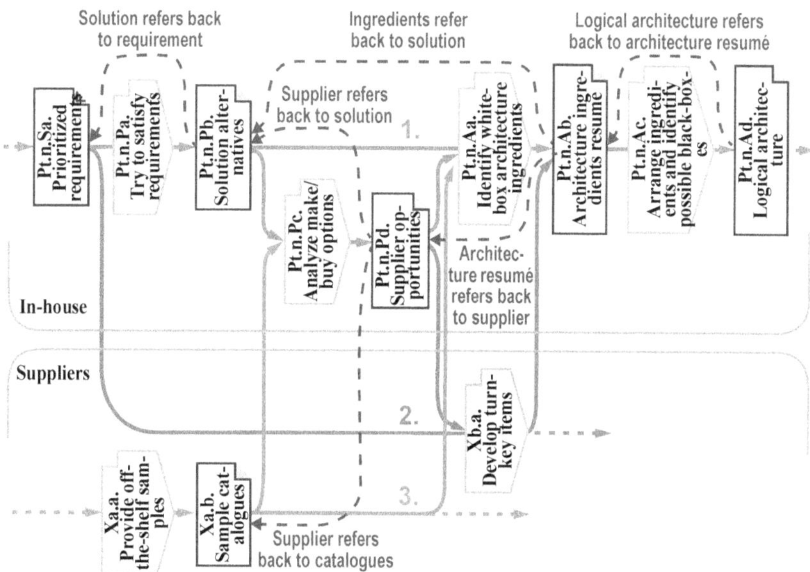

FIGURE 9-19 Process schedule to gather solutions into architecture resumé

The references from **Pt.n.Ab. Architecture ingredients resumé** to **Pt.n.Pd. Supplier opportunities** are needed only for immediate convenience and this *traceability* is later solved by the *result* **Pt.n.Fd. Product requisite estimates** coming in the next *phase* **Pt.F. Finalize design & requisites** (see Fig. 10-4, p. 444, for a more complete description of these *cascading results*).

Utilizing all kinds of *suppliers* to obtain ingredients to a *white-box* is illustrated in Figure 9-19 above:

1. In-house development (upper arrow lines)
2. Ordering turnkey (middle arrow lines)
3. Buying off the shelf (lower arrow lines)

9.5.4 Architecture phenomena for the architecture resumé

Just as for the *environment architecture resumé*, there will be an *architecture resumé* for each *white-box* of the *product* to be developed. The *white-box architecture resumé* is somewhat more focused than the *environment architecture resumé*, because there are no existing ingredients to worry about, but only ingredients to be developed or purchased (see Table 9-12 below).

TABLE 9-12 Architecture ingredients resumé

White-box ingredient	Ingredient type	Architecture phenomena to document
Black-box element Element Connection interface Boundary interface	Primordial	New design according to solution alternative derived from restriction requirement
	Determined	New design according to solution alternatives derived from satisfied requirements
		New design according to implicit or explicit solution alternatives derived from standardization
	Unbiased	New design free by purpose
		New design due to absence of solution alternatives
		New design due to absence of requirements causing absence of solution alternatives

There might be cases when *elements* or *interfaces* are technically necessary, despite the fact that there are no explicit *requirements* that force them and no *white-box solution alternatives* to reference. Be aware that lack of *white-box solution alternatives* might also create an absence of *supplier opportunities*. For *unbiased architecture ingredients* without *white-box solution alternatives*, *supplier opportunities* must be made explicit as soon as these *elements* are identified in the *architectures*.

9.5.5 EXAMPLE: Requirements transformed to architecture

An example of how to transform *requirements* to *architecture* can be seen in Figure 9-20 below. The left part illustrates the *requirements*, which may cause an assumed *architecture* illustrated in the right part.

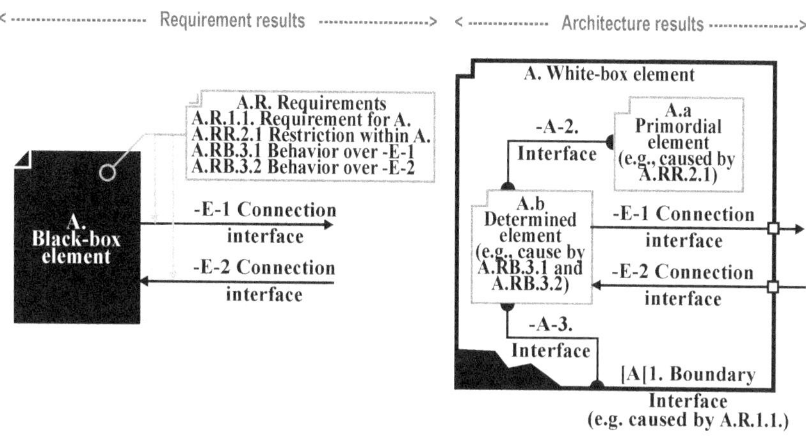

FIGURE 9-20 Designed architecture is based on requirements

Note that the *black-box* **A.** with the *connection interfaces* **-E-1** and **-E-2** was already designed in the *environment architecture* when the *black-box requirement* **A.R.** was specified.

9.6 EXAMPLE House: Satisfy house requirements by decomposing house black-box into white-box design containing room black-boxes

9.6.1 EXAMPLE House: Process schedule to use

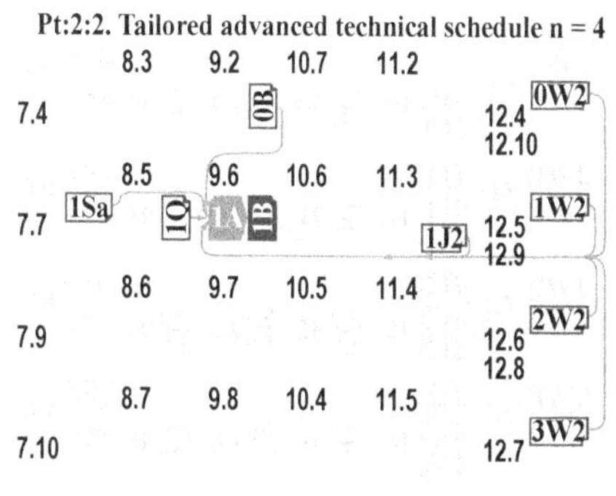

For the house example, a special technical overall *schedule* is *tailored* (illustrated in Fig. 9-21, left):

• **Pt:2:2. Tailored advanced technical schedule n = 4,** page 164

In this *tailored schedule* the generic *schedule* to use is (filled-in symbols):

• **Pt.n.A. Satisfy requirements by decomposing outward nesting level black-box into white-box design containing possible black-boxes,** page 136.

FIGURE 9-21 Position of schedule and master to use (filled in) in overall schedule

Below *results* may be demanded (their *masters* are indicated *upstream* from the filled-in *schedule* in the figure above):

0B Environment existing ingredients and design of interfaces to the environment
Table 9-6, House environment architecture ingredients resumé, page 341
Figure 9-5, House environment logical architecture, page 341
Figure 9-6, House environment physical architecture, page 342

1Q House solution alternatives with supplier opportunities
Table 8-14, House kitchen white-box solution alternatives, page 306
Table 8-15, House kitchen white-box supplier opportunities, page 308

1Sa House prioritized requirements
Table 7-9, House black-box compound requirements (small selection), page 189

Xb Suppliers' sample catalogues

For possible *failure localization* and *failure elimination*, below *results* may be demanded (their *masters* are indicated *upstream* from the filled-in *schedule* in the figure above):

1J2 Table 11-14, House start-up report 1, page 612

0W2 List 12-1. House environment interfaces with fake house failure elimination report 1, page 674

0W2 Table 12-27, House environment verification report 1, page 701
Table 12-29, House environment reverification report 2, page 703
Table 12-28, House environment failure elimination report 1, page 702

1W2 List 12-2. House with fake rooms failure elimination report 1, page 676

1W2 Table 12-22, House verification report 1, page 697
Table 12-24, House verification report 2, page 698
Table 12-23, House failure elimination report 1, page 698

2W2 List 12-3. House rooms with fake machinery failure elimination report 1, page 678

2W2 Table 12-14, House kitchen verification report 1, page 688
Table 12-17, House kitchen verification report 2, page 691
Table 12-15, House kitchen failure elimination report 1, page 689

3W2 Table 12-9, House kitchen machinery verification report 1, page 682
Table 12-11, House kitchen machinery verification report 2, page 683
Table 12-10, House kitchen machinery failure elimination report 1, page 683

From Figure 9-21 above, it is obvious that *architecture* is the spider in the overall technical *schedule*, tying together *schedules* on current *nesting level*, with *schedules* on *inward* and *outward nesting levels*.

9.6.2 EXAMPLE House: Identify white-box architecture ingredients

It is now time to select a set of solutions from Table 8-12, House white-box solution alternatives, page 301, that will create a nice design while still satisfying the *requirements* and thus fit the *stakeholders'* needs. The chosen design must be described for all involved parties of the development of the house, and all selected *elements* must be described and illustrated.

In order not to consume a disproportionate amount of book space, the design in Table 9-13 and Table 9-14 below lacks many *architecture ingredients*, but assumes that a full solution alternative analysis has been done on a complete *requirement* set.

Some comments on the decided house *architecture*:

- On purpose, to show that a house is mainly a functional *product*, there are not many restrictions in the example of house *requirements*.

- Many of the selected *interfaces* and *elements* are already determined by *white-box solution alternatives* in a previous *phase*. If many alternatives are at hand, one of them is selected, mostly because it fits together best with other *solution alternatives*.

- *Interfaces* and *elements* are not restricted, and have no determined *solution alternatives*, and are therefore unbiased and free for the architects to design.

- In the case of a house, the proprietor often wants to take part in *architecture design* work, and make design decisions. This can be a problem if late design overrides do violate design-satisfying *behavior requirements*.

- The house *boundary interface* is chosen to belong to the house, and consequently has the ID **[A[**. The façade, roof, ground, outer doors, and windows are considered to belong to the house (rather expected). According to *Cpdm*, nothing prevents that all inner *interfaces*, such as inner doors and inner walls, also belong to the house, and not to the rooms, also see below.

- However, rooms are also boundary encapsulations, which can be considered to belong to the rooms. It might be a bit confusing to keep track of whether an inner wall belongs to the house or to the room, but a good rule is that the surface of the inner wall belongs to the room, and the invisible construction of the inner wall belongs to the house (for example, the invisible construction of a wall may well bear the upstairs floor of the house, but the surface of the wall is independent of the house).

- A house is very much about *boundary interfaces*, protecting from weather and intruders, and encapsulating human conditions. In turn, rooms are like houses, also being mainly boundaries. In rooms humans have a lot of storage containers, like rooms in the house, and so forth.

- When designing the house *architecture*, the need for some more *environment interfaces* is discovered, and according to *Cpdm*, these belong to the *environment* with an ID -E-.

- For security reasons, the -E-5. Gas pipe has to fork outside the house to also feed the kitchen gas stove. The pipe for the kitchen can be named -E-5a. K-Gas to indicate it is a branch of the main -E-5. Gas pipe.

The *architecture resumé* is divided into two tables, the first for identified *elements*, and the second for identified *interfaces*.

TABLE 9-13 House rooms black-boxes architecture ingredients resumé

White-box ingredient		Chosen solution alternative or sample thereof	
A.	House	Already defined in environment architectures	
A.a.	Rooms	Primary house spaces (abstract black-box)	
A.aA.	Dining room	A.SL3. Add 1 m² to dining room door space. A.SL13.b. Separate dining room from other rooms with doors and from kitchen with an arch. from Pa&Pp. A.SL14. Use 12 m² of house for dining room.. A.SL15. Use 2 m² of house for serving space. A.SL31.a. Use 20 m² of house for dining room big parties. A.SL32. House needs dining room with forced ventilation 2 m³/hour. A.SL33. Use another 2 m² of house for kitchen party meals. [a.]Total dining room area ≈ 22 m²	A.SL11. Use well-accepted house style and finish, to be judged valuable if the house is to be sold.
A.aB.	Entrance	A.SL36. Use 3 m² of house for entrance room. Due to space limitation in the book, there are few requirements resulting in solutions for the bedroom.	
A.aC.	Kitchen	A.SL16.b. Use 5 m² of house for kitchen with two chefs. A.SL17. Use 4 m² of house for kitchen food preparation space. A.SL18. Use 3 m² of house for kitchen grocery storage. A.SL19.a. Use 2 m² of house for kitchen tableware storage. A.SL21. Use 2 m² of house for kitchen cooking utensils. A.SL22. Use 2 m² of house for kitchen cooking equipment. A.SL29. Use 1 m² of house for kitchen disposal space. A.SL30. House needs a kitchen disposal ventilation 2 m³ / day. A.SL34. Use 1 m² of house for kitchen party grocery storage. [a.]Total kitchen room area ≈ 20 m² *	
A.aD.	Toilet	A.SL7. Big toilet with handles, add 2 m² to toilet room. A.SL35.a. Use 2 x 4 m² of house for two bathrooms.	
A.aE.	Bed room	Due to space limitation in the book, there are few requirements resulting in solutions for the bedroom.	

a The total area is smaller than the sum, because some areas might overlap each other.

Note that some more interfaces belonging to the *environment* are now detected and specified.

TABLE 9-14 House interfaces architecture ingredients resumé

White-box ingredient		Chosen solution alternative or sample thereof
-E-	Interfaces to environment	Already defined in environment architectures
-E-5a.	K-Gas	A.SL26. Supply kitchen with stove gas.
-E-6.	Air outlet	A.SL28. Kitchen needs forced suction 12 m³ / hour. A.SL30. House needs a kitchen disposal ventilation 2 m³ / day.
-E-7.	Chimney	[a]Exhaust pipe from gas boiler
-E-8.	Door section	A.SL2. Dining room double outer glass door from Pa&Pp
-E-9.	K-window	A.SL1. Kitchen window around 1 m² from Pa&Pp
-E-10.	T-window	Raw glass window
-E-11.	Front door	Standard wooden outer energy-saving front entry door
-E-12.	Back door	Standard outer energy-saving back entry door
-A-	Interfaces inside the house	Common identity for all interfaces not visible outside the house
]A[1.	House boundary	House boundary. A.SL4. Basalt stone on walls and terra-cotta on roof from All-in-Stone A.SL6. Use façade brickwork and roofing tiles.
-A-2.	House inner walls	A.SL9. The house walls shall be of stone and bricks, and the house vaults of cast concrete.
-A-3.	E-door	Inner door with footstep
-A-4.	Lock door	Inner door with footstep and lock
-A-5.	Open arch	Width 2.0 m, upper radius 1.0 m
-A-6.	Stairway	With 0.8 m hand rails on both sides
-A-7.	S-door	Simple door to the staircase space
-A-8.	L-door	Simple door to the stairway loft
-A-9.	Warm water	A.SL23. Supply kitchen with cold and warm water.
-A-10.	Cold water	
-A-11a.	K-heating	A.SL25. Supply kitchen with floor heating.
-A-11b.	D-heating	[a]Heat conduction for dining room
-A-11c.	E-heating	[a]Heat conduction for entry
-A-11d.	T-heating	[a]Heat conduction for toilet
-A-11e.	B-heating	[a]Heat conduction for bedroom
-A-12a.	K-Wiring	A.SL24. Supply kitchen with 3-phase electricity.
-A-12b.	O-Wiring	[a]Electrical wiring for remaining rooms
-A-13.	Sewer	A.SL27. Supply kitchen with sewage..

Spanning solution notes on the right side of the table:

A.SL5. All windows and doors custom-made

A.SL10. Applied insulation solution shall achieve less than 90 kWh/m²a energy consumption.

A.SL11. Use well-accepted house style and finish, to be judged valuable if the house is to be sold.

A.SL5. All windows and doors custom-made

A.SL8. Use wide doors without door steps.

a   Due to limited space in this book, solutions alternatives are missing.

TABLE 9-15 House secondary space elements architecture ingredients resumé

	White-box ingredient	Chosen solution alternative or sample thereof		
A.	House	Already defined in environment architectures		
A.b.	Loft	The unavoidable A-shaped space under the roof		
A.c.	Staircase space	A.SL12. Make simple utility room for all house connection interfaces in secondary space under the staircase. A.SL37. Utility room shall be lockable.	A.SL12. Make simple utility room for all house connection interfaces in secondary space under the staircase.	A.SL11. Use well-accepted house style and finish, to be judged valuable if the house is to be sold.
A.ca.	Main distr. box	Electricity hub in staircase space		
A.cb.	Boiler	Gas scale, heater, and boiler in staircase space		
A.cc.	Scale	Water scale in staircase space		
A.cd.	Cesspit	Junction sump for all sewers in house		

As already pointed out, due to limited space in this book, the house secondary spaces have no **Pt.n.Sa. Prioritized requirements** and **Pt.n.Pb. Solution alternatives**, and consequently, there are no complete *solution alternatives* references in Table 9-15 above.

9.6.3 EXAMPLE House: Arrange ingredients and identify possible black-boxes

Based on the *architecture resumé* in Table 9-13, page 360, and Table 9-14, page 361, the logical *architecture* drawing can now be prepared and illustrated (see Fig. 9-22 below). This description includes newfound *black-boxes* as well as other *architecture ingredients*.

The house outermost *architecture* is now shown as an opened *white-box*, from being a closed *outermost black-box*. In this *white-box, inward nesting level* room *black-boxes* now appear together with doors and ports between them. Some simple room *elements* (A.c. Staircase space and A.b. Loft) have also been identified as simple *elements*.

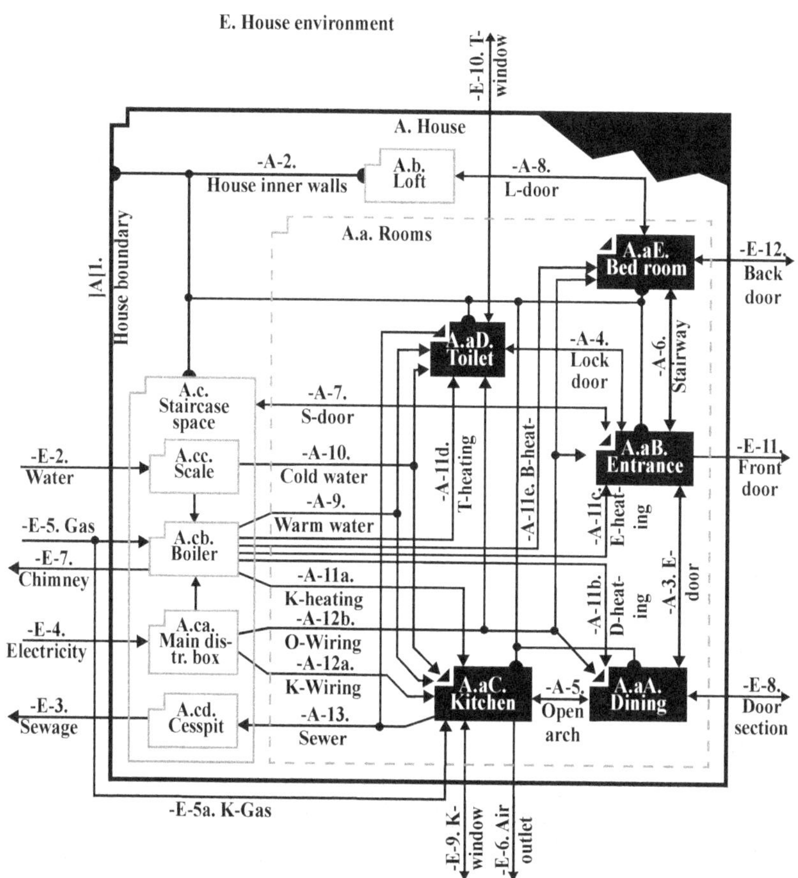

FIGURE 9-22 House logical architecture

Not all *elements* of the façade, roof, and ground are lumped together in the *boundary interface*]A[1. House boundary.

9.6.4 EXAMPLE: House: Lay out physical architecture

The developers came up with a *physical architecture* and a computer-simulated façade printing (see Fig. 9-23 below and Fig. 9-24 next).

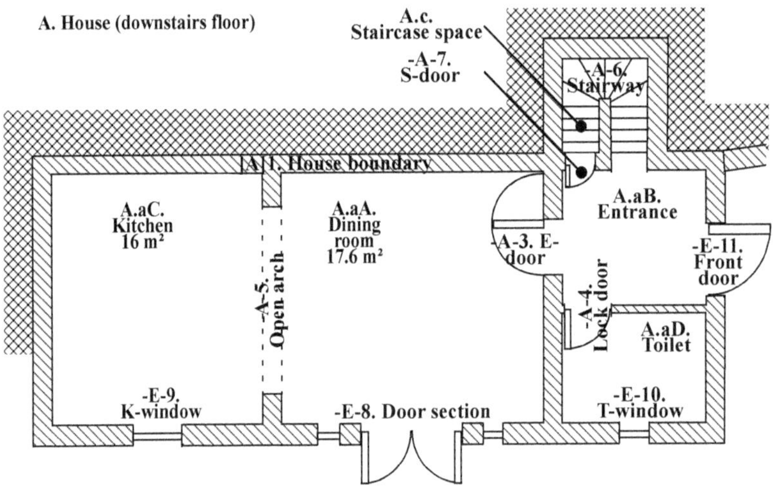

FIGURE 9-23 House physical architecture

FIGURE 9-24 House façade physical architecture

9.6.5 EXAMPLE House: Examine architectures

After this *activity* is done, almost everything is finished for submitting the building permit to the authorities.

Robber HB now presented all documentation so far to the proprietor for review. After some corrections (not shown in this book) the documentation was approved and sent to the authorities by Robber HB.

9.6.6 EXAMPLE House: Proceed reading

To follow this example, these are the alternatives:

1F   Proceed to chapter 10.6, "EXAMPLE House: Finalize design of house with product requisites," page 465.

1I   Proceed to chapter 11.3, "EXAMPLE House: Procure house items (and await invading room black-boxes) in order to build house," page 607.

1V   Proceed to chapter 12.5, "EXAMPLE House: Partially verify black-/white-box with embedded fakes," page 675.

1V   Proceed to chapter 12.9, "EXAMPLE House: Fully verify house black-/white-box," page 693.

2R   Proceed to chapter 7.9, "EXAMPLE House rooms: Refine room requirements from house requirements, and ensure room development staffing," page 193.

2A   Proceed to chapter 9.7 "EXAMPLE House rooms: Satisfy room requirements by decomposing room black-boxes into white-box design containing machinery black-boxes" below.

9.7 EXAMPLE House rooms: Satisfy room requirements by decomposing room black-boxes into white-box design containing machinery black-boxes

9.7.1 EXAMPLE House kitchen: Process schedule to use

For the house example, a special technical overall *schedule* is *tailored* (illustrated in Fig. 9-25, left):

- **Pt:2:2. Tailored advanced technical schedule n = 4,** page 164

In this *tailored schedule* the generic *schedule* to use is (filled-in symbols):

- **Pt.n.A. Satisfy requirements by decomposing outward nesting level black-box into white-box design containing possible black-boxes,** page 136.

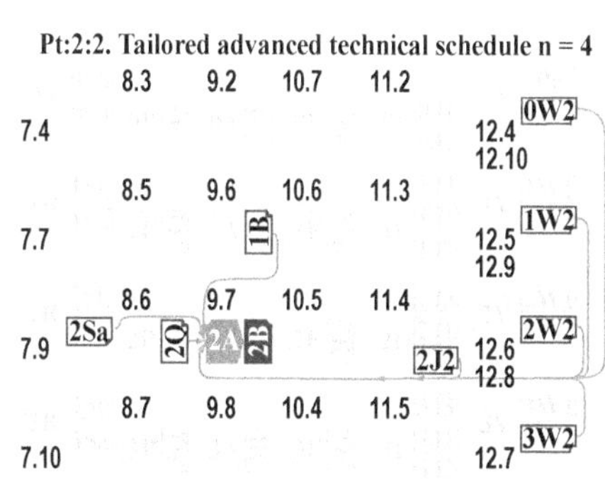

FIGURE 9-25 Position of schedule and master to use (filled in) in overall schedule

Below *results* may be demanded (their *masters* are indicated *upstream* from the filled-in *schedule* in the figure above):

1B   House architectures with major ingredients and room black-boxes

Table 9-13, House rooms black-boxes architecture ingredients resumé, page 360

Table 9-14, House interfaces architecture ingredients resumé, page 361

Table 9-15, House secondary space elements architecture ingredients resumé, page 362

Figure 9-22, House logical architecture, page 363

Figure 9-23, House physical architecture, page 364

Figure 9-24, House façade physical architecture, page 364

2Q Room solution alternatives with supplier opportunities

Table 8-14, House kitchen white-box solution alternatives, page 306

Table 8-15, House kitchen white-box supplier opportunities, page 308

2Sa Rooms prioritized requirements

Table 7-11, House kitchen black-box compound requirements (small selection), page 195

X Suppliers' sample catalogues

For possible *failure localization* and *failure elimination*, below *results* may be demanded (their *masters* are indicated *upstream* from the filled-in *schedule* in the figure above):

2J2 Table 11-18, House kitchen start-up report 1, page 617

0W2 List 12-1. House environment interfaces with fake house failure elimination report 1, page 674

0W2 Table 12-27, House environment verification report 1, page 701

Table 12-29, House environment reverification report 2, page 703

Table 12-28, House environment failure elimination report 1, page 702

1W2 List 12-2. House with fake rooms failure elimination report 1, page 676

1W2 Table 12-22, House verification report 1, page 697

Table 12-24, House verification report 2, page 698

Table 12-23, House failure elimination report 1, page 698

2W2 List 12-3. House rooms with fake machinery failure elimination report 1, page 678

2W2 Table 12-14, House kitchen verification report 1, page 688

Table 12-17, House kitchen verification report 2, page 691

Table 12-15, House kitchen failure elimination report 1, page 689

3W2 Table 12-9, House kitchen machinery verification report 1, page 682

Table 12-11, House kitchen machinery verification report 2, page 683

Table 12-10, House kitchen machinery failure elimination report 1, page 683

9.7.2 EXAMPLE House kitchen: Identify white-box architecture ingredients

The room *black-boxes* are now opened and designed. Fore space reasons, only the kitchen *black-box* is opened to a kitchen *white-box*, and this *white-box* will now be filled with ingredients, satisfying its *requirements* and matching proposed *white-box solution alternatives* (see Table 9-16 below and Table 9-17 next).

TABLE 9-16 House kitchen elements architecture ingredients resumé

White-box ingredient		Chosen solution alternative or sample thereof	
A.aC.	Kitchen	Defined in house architecture resumé	
A.aC.a.	Machinery	Abstract element cover machinery black-boxes	
A.aC.aA.	Cold machinery	AaC.SL.14. Cold section 60 x 60 x 200 cm for chilled and frozen food	
A.aC.aB.	Warm machinery	AaC.SL.20. Cabinet section for warming equipment	
A.aC.aC.	Bake machinery	AaC.SL.22. Section 50 x 90 cm under preparation area.	
A.aC.aD.	Heat machinery	AaC.SL.21. Partition 3 m² heating area in center of preparation area.	
A.aC.aE.	Ventilation machinery	AaC.SL.23. Evacuation equipment 60 cm over heating area	
A.aC.aF.	Wet machinery	AaC.SL.18. Partition 1 m² manual dishwashing section of left preparation area for this.	
A.aC.aG.	Dish machinery	AaC.SL.17. Machinery dishwashing cabinet section 120 cm wide under preparation area AaC.aG.SL.3. Glass racks in dishwashers	AaC.SL.5. Use only ordinary interiors for kitchen.
A.aC.b.	Medicine cupboard	AaC.SL.25. Small cupboard 200 cm above floor	
A.aC.c.	Grocery cupboard	AaC.SL.16. Half-height cupboard 60 x 40 x 100 cm for room-temperature provisions	
A.aC.d.	Grocery cupboard	AaC.SL.15. Full-height cupboard 60 x 40 x 200 cm for room-temperature provisions	
A.aC.e.	Trash receptacle	AaC.SL.24. Airtight cupboard 60 x 60 x 60 cm under sink	AaC.SL.6. Choose only few sample suppliers to match proposed solutions.
A.aC.f.	Tiny utensil drawers	AaC.SL.13. Use drawer section 40 x 60 cm beneath counter.	
A.aC.g.	Baking cupboard	AaC.SL.9. One wide drawer for baking sheets	
A.aC.h.	Tableware drawers	AaC.SL.12. Use drawer section 60 x 60 cm beneath counter.	
A.aC.i.	Big utensil cupboard	AaC.SL.8a. Use two revolving-shelf cabinets near counter.	
A.aC.j.	Preparation counter	AaC.SL.1. Place preparation area in front of window. AaC.SL.7a. Estimate totally 6 m² stone preparation area.	
A.aC.k.	Chemical cupboard	AaC.SL.19. Chemicals cupboard, child-safe height, separate from medicines	
A.aC.l.	Tableware cupboard	AaC.SL.10. One cupboard for bowls	
A.aC.m.	Tableware cupboard	AaC.SL.8a. Use two revolving-shelf cabinets near counter.	
A.aC.n.	Knife hanger	AaC.SL.11a. Use magnet knife hanger at side of preparation area.	
A.aC.o.	Distribution box	Kitchen electric distribution box	
A.aC.p.	Chef	Kitchen user	

TABLE 9-17 House kitchen interfaces architecture ingredients resumé

White-box ingredient	Chosen solution alternative or sample thereof	
-A.aC-1. Interior access	Chef's interface to access kitchen machinery	AaC.SL.3. Have proprietor check quality before paying invoices.
-A.aC-2. Interior attachment	Mechanical interfaces to attach all interiors to ceiling, walls, and floor	
-A.aC-3. Machinery el. feed	Electrical interface to kitchen machinery	AaC.SL.4. Have proprietor choose supplier and material before ordering.
[A.aC]4. Floor	Floor covering	
[A.aC]5. Backsplash	Water protection behind the counter	
[A.aC]6. Wall and ceiling	AaC.SL.2a. Walls of plastered bricks	

9.7.3 EXAMPLE House kitchen: Examine architectures

After a review meeting with the *stakeholders* and kitchen studio contractor, the *architecture resumé* was approved.

9.7.4 EXAMPLE House kitchen: Arrange ingredients and identify possible black-boxes

The *complexity* in cabinet design will be to fit in the kitchen machinery not yet selected. Since most machinery is standardized to fit many kinds of cabinets, the machinery decisions can be left to later, provided that standardized space is reserved for the machinery.

Consequently, in the kitchen *architecture* there are *complex* machinery *black-boxes* that need to be specified later. But there are also kitchen cabinets that are relatively simple (drawers and cupboards), needing no further specifications, which can be designed immediately (see Fig. 9-26 below).

The other rooms also contain some cabinets (for space reasons not illustrated in this book), but all of them may be designed on this *nesting level* (none of these are so *complex* that they need further *black-box* specifications to be understood).

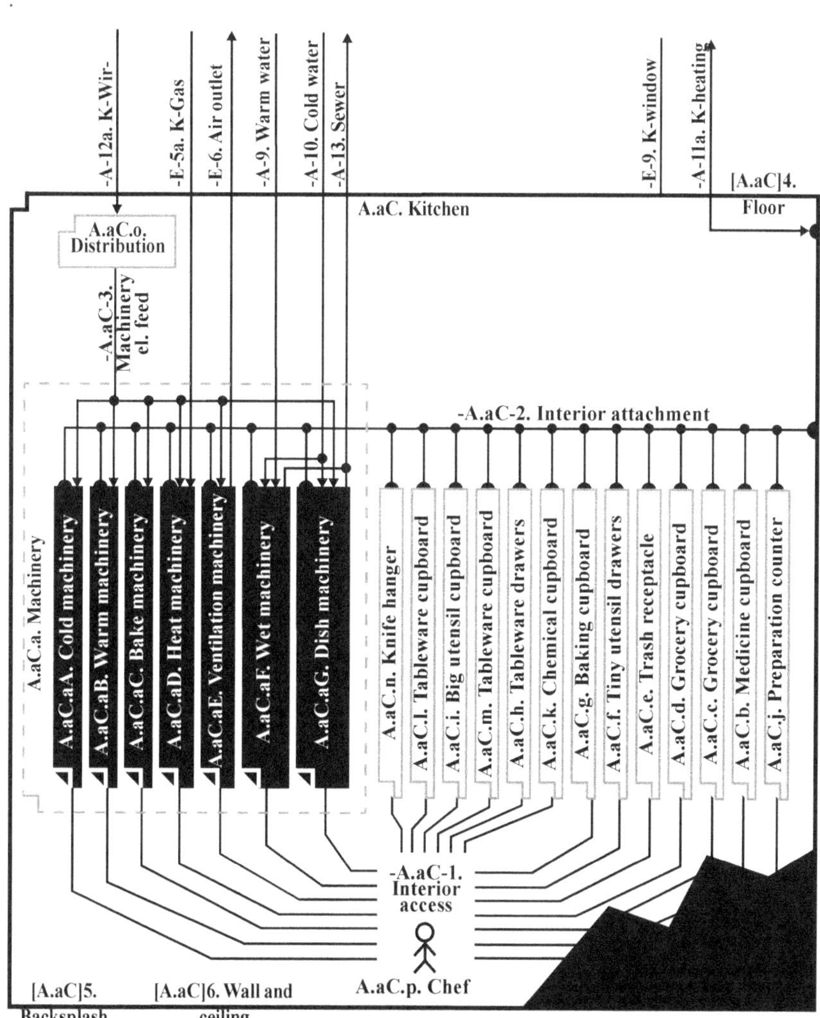

FIGURE 9-26 House kitchen logical architecture

Since efficient and professional use of the kitchen is specified to be very important, a chef *user* of the house kitchen machinery is assumed, as can be seen from Figure 9-26 above. This will be helpful for understanding cooking *test-cases* when verifying the kitchen.

It may seem a bit awkward to put a *user* within the construction being developed, but from a logical perspective there is no difference between having *users* inside or outside a system.

9.7.5 EXAMPLE House kitchen: Lay out physical architecture

The more *partitioned*, the more detailed the design gets. What was meters in the house *architecture* now is centimeters in cabinets. Otherwise, the principles do not change much from *outward nesting levels*.

It is always helpful to see and understand a *physical architecture* of any *product*. With computer-aided design the possibilities have exploded with all kinds of 3D simulations that can be done.

The kitchen with its cabinets was entered in a computer 3D simulator and a layout could almost immediately be presented (see Fig. 9-27 below).

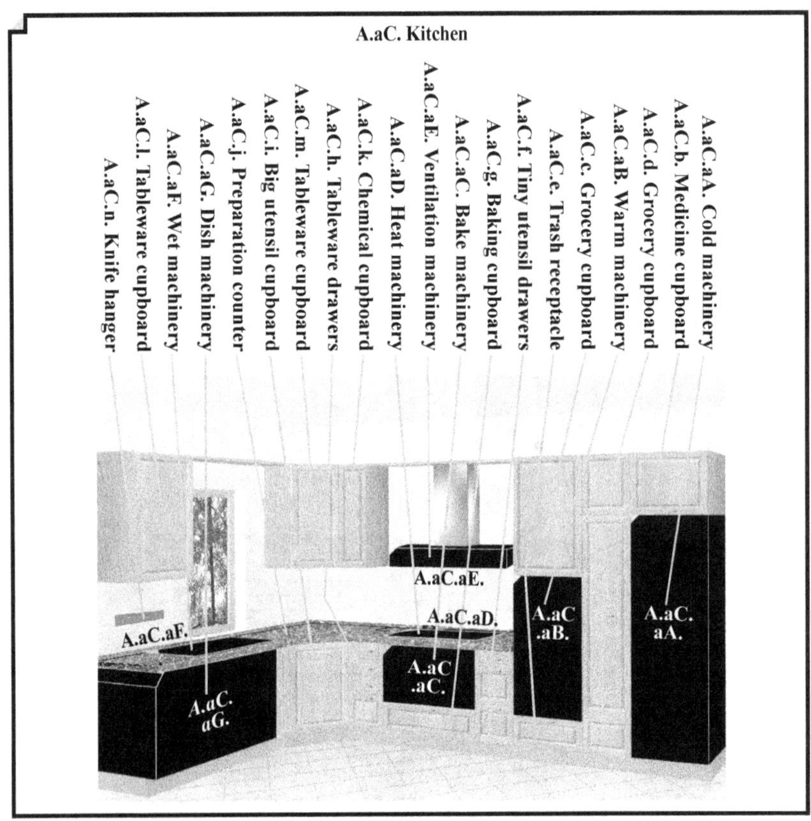

FIGURE 9-27 House kitchen physical architecture

9.7.6 EXAMPLE House kitchen: Examine architectures

After a review meeting with the kitchen studio contractor, the *architecture resumé* was approved.

9.7.7 EXAMPLE House: Proceed reading

To follow this example, these are the alternatives:

2F Proceed to chapter 10.5, "EXAMPLE House rooms: Finalize design of rooms with product requisites," page 452.

2I Proceed to chapter 11.4, "EXAMPLE House rooms: Procure room items (and await invading machinery black-boxes) in order to build rooms," page 613.

2V Proceed to chapter 12.6, "EXAMPLE House kitchen: Partially verify black-/white-box with embedded fakes," page 677.

2V Proceed to chapter 12.8, "EXAMPLE House kitchen: Fully verify room black-/white-box," page 684.

3R Proceed to chapter 7.10, "EXAMPLE House rooms machinery: Refine machinery requirements from rooms requirements and ensure machinery development staffing," page 198.

3A Proceed to chapter 9.8 "EXAMPLE House rooms machinery: Satisfy machinery requirements by decomposing machinery black-boxes into white-box design not containing black-boxes" below.

9.8 EXAMPLE House rooms machinery: Satisfy machinery requirements by decomposing machinery black-boxes into white-box design not containing black-boxes

9.8.1 EXAMPLE House kitchen machinery: Process schedule to use

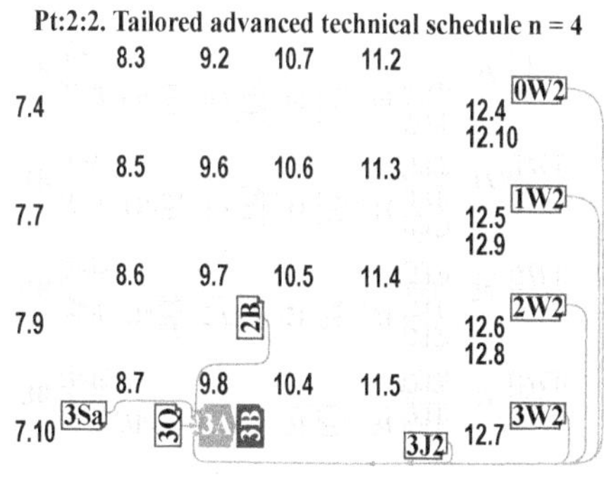

For the house example, a special technical overall *schedule* is *tailored* (illustrated in Fig. 9-28, left):

• **Pt:2:2. Tailored advanced technical schedule n = 4,** page 164

In this *tailored schedule* the generic *schedule* to use is (filled-in symbols):

• **Pt.n.A. Satisfy requirements by decomposing outward nesting level black-box into white-box design containing possible black-boxes,** page 136.

FIGURE 9-28 Position of schedule and master to use (filled in) in overall schedule

Below *results* may be demanded (their *masters* are indicated *upstream* from the filled-in *schedule* in the figure above):

2B Room architectures with major ingredients and machinery black-boxes

Table 9-16, House kitchen elements architecture ingredients resumé, page 367

Table 9-17, House kitchen interfaces architecture ingredients resumé, page 368

Figure 9-27, House kitchen physical architecture, page 370

Figure 9-26, House kitchen logical architecture, page 369

3Q Machinery solution alternatives with supplier opportunities

Table 8-12, House white-box solution alternatives, page 301

Table 8-13, House white-box supplier opportunities, page 303

3Sa Machinery prioritized requirements

Table 7-13, House kitchen machinery black-box compound requirements, page 200

Xb Suppliers' sample catalogues

For possible *failure localization* and *failure elimination*, below *results* may be demanded (their *masters* are indicated *upstream* from the filled-in *schedule* in the figure above):

3J2 Table 11-22, House kitchen machinery start-up report 1, page 622

0W2 Table 12-27, House environment verification report 1, page 701

Table 12-29, House environment reverification report 2, page 703

Table 12-28, House environment failure elimination report 1, page 702

1W2 List 12-2. House with fake rooms failure elimination report 1, page 676

1W2 Table 12-22, House verification report 1, page 697

Table 12-24, House verification report 2, page 698

Table 12-23, House failure elimination report 1, page 698

2W2 List 12-3. House rooms with fake machinery failure elimination report 1, page 678

2W2 Table 12-14, House kitchen verification report 1, page 688

Table 12-17, House kitchen verification report 2, page 691

Table 12-15, House kitchen failure elimination report 1, page 689

3W2 Table 12-9, House kitchen machinery verification report 1, page 682

Table 12-11, House kitchen machinery verification report 2, page 683

Table 12-10, House kitchen machinery failure elimination report 1, page 683

9.8.2 EXAMPLE House kitchen machinery: Identify white-box architecture ingredients

The *architecture resumé* now must be established for the house kitchen machinery. The question is which ingredients are most important and must be decided first and which are subordinated. Is it determined by the casing dimensions, electrical capacity, or the front panel appearance? The proprietor and the architects start with machinery types (see Table 9-18 below).

TABLE 9-18 House kitchen machinery elements architecture ingredients resumé

White-box ingredient	Chosen solution alternative or sample thereof
A.aC.a. Machinery	Defined in House kitchen architecture resumé
A.aC.aA.a. Combined fridge/ freezer	AaC.aA.SL.1.b. Combined fridge (150 liters) and freezer (50 liters) AaC.aA.SL.2. Fridge and freezer with automatic defrosting
A.aC.aB.a. Micro oven	AaC.aB.SL.2. Microwave oven 2 000 W
A.aC.aB.b. Warming oven	AaC.aB.SL.1b. Warming oven 600 W
A.aC.aC.a. Wide oven	AaC.aC.SL.1. Electrical oven 2 000 W
AaC.aDa. Electrical hubs	AaC.aD.SL.1. Electric cooktop 6 000 W possible to install not far above a warm oven
A.aC.aD.b. Gas burner	AaC.aD.SL.2a. Gas hub 4.8 kW
A.aC.aE.a. Wide hood	AaC.aE.SL.1b. Hood 180 W
A.aC.aF.a. Sink 2 bowls	AaC.aF.SL.3. Stainless steel sink 2 bowls prepared for faucets
A.aC.aF.b. Combined faucet	AaC.aF.SL.1b. Combined cold- & warm-water faucets
A.aC.aG.a. Dishwasher coarse	AaC.aG.SL.2. Advanced dishwasher (40 liters)
A.aC.aG.b. Dishwasher gentle	AaC.aG.SL.1. Simple dishwasher (40 liters)

After these decisions, the electrical *interfaces* are decided (see Table 9-19 below).

TABLE 9-19 House kitchen machinery connection interface architecture ingredients resumé

White-box ingredient	Chosen solution alternative or sample thereof
-A.aC-3. Machinery el. feed	Identified in kitchen architecture resumé
-A.aC-3a. El. fridge	Electricity connector for AaC.aA.SL.1.b. Combined fridge (150 liters) and freezer (50 liters)
-A.aC-3b. El. micro	Electricity connector for AaC.aB.SL.2. Microwave oven 2 000 W
-A.aC-3c. El. warm	Electricity connector for AaC.aB.SL.1b. Warming oven 600 W
-A.aC-3d. El. bake	Electricity connector for AaC.aC.SL.1. Electrical oven 2 000 W
-A.aC-3e. El. hub	Electricity 3-phase connector for AaC.aD.SL.1. Electric cooktop 6 000 W
-A.aC-3f. El. hood	Electricity connector for AaC.aE.SL.1b. Hood 180 W
-A.aC-3g. El. adv. dish	Electricity connector for AaC.aG.SL.2. Advanced dishwasher (40 liters)
-A.aC-3h. El. smp. dish	Electricity connector for AaC.aG.SL.1. Simple dishwasher (40 liters)
-A.aC-3i. El. lamp	Electricity to kitchen ceiling lamp
-A.aC-3j. El. switch	Switch for ceiling lamp

After deciding the house kitchen machinery types, the casing is considered crucial for matching all kitchen connections and fitting to all other cabinets in the kitchen (see Table 9-20 below).

TABLE 9-20 House kitchen machinery boundaries interface architecture ingredients resumé	
White-box ingredient	**Chosen solution alternative or sample thereof**
A.aC.a. Machinery	Identified in kitchen architecture resumé
[A.aC.a]. Machinery housing	Designation of all machinery housings
[A.aC.aA]1. Cold housing	Housing for AaC.aA.SL.1.b. Combined fridge (150 liters) and freezer (50 liters) with outer pipe grid at the back needing ventilation space
[A.aC.aB]1. Upper housing	Housing for AaC.aB.SL.2. Microwave oven 2 000 W, ventilation space needed for in- and outlet of cooling air
[A.aC.aB]2. Lower housing	Housing for AaC.aB.SL.1b. Warming oven 600 W
[A.aC.aC]1. Oven housing	Back housing for AaC.aC.SL.1. Electrical oven 2 000 W, needing ventilation space all around
[A.aC.aD]1. Gasket seal	Rubber seal with screw clips between A.aC.j. Preparation counter and both AaC.aD.SL.1. Electric cooktop 6 000 W and AaC.aD.SL.2a. Gas hub 4.8 kW
[A.aC.aE]1. Air tin box	Outlet pipe for air from AaC.aE.SL.1b. Hood 180 W with cleaning doors for sweeping
[A.aC.aF]1. Gasket seal	Rubber seal between A.aC.j. Preparation counter and AaC.aF.SL.3. Stainless steel sink 2 bowls prepared for faucets
[A.aC.aG]1. Left housing	Left housing for AaC.aG.SL.2. Advanced dishwasher (40 liters) Both dishwashers must have water dissipative floor covering to let leaking water run to the front and be observed.
[A.aC.aG]2. Right housing	Right housing for AaC.aG.SL.1. Simple dishwasher (40 liters)

Since the proprietor lets the functionality outweigh appearance, the front panels with such characteristics as colors, buttons, displays and options are the last to be decided (see Table 9-21 below).

TABLE 9-21 House kitchen machinery user interface architecture ingredients resumé	
White-box ingredient	**Chosen solution alternative or sample thereof**
-A.aC-1. Interior access	Identified in kitchen architecture resumé
-A.aC-1a. Cold panel	Front panel for AaC.aA.SL.1.b. Combined fridge (150 liters) and freezer (50 liters)
-A.aC-1b. Cold door	Right-hinged cover door for AaC.aA.SL.1.b. Combined fridge (150 liters) and freezer (50 liters)
-A.aC-1c. Micro panel	Hand-operated front panel for AaC.aB.SL.2. Microwave oven 2 000 W
-A.aC-1d. Micro door	Hand-operated right-hinged front door for AaC.aB.SL.2. Microwave oven 2 000 W
-A.aC-1e. Warm panel	Front panel for AaC.aB.SL.1b. Warming oven 600 W
-A.aC-1f. Warm door	Front door for AaC.aB.SL.1b. Warming oven 600 W
-A.aC-1g. Bake panel	Front panel for AaC.aC.SL.1. Electrical oven 2 000 W
-A.aC-1h. Bake door	Bottom-hinged front door for AaC.aC.SL.1. Electrical oven 2 000 W
-A.aC-1i. Hub panel	Front panel for AaC.aD.SL.1. Electric cooktop 6 000 W
-A.aC-1j. Burner panel	Controls with one-hand gas igniter for AaC.aD.SL.2a. Gas hub 4.8 kW
-A.aC-1k. Hood panel	Simple front panel for AaC.aE.SL.1b. Hood 180 W
-A.aC-1l. Handles	Two dimensions moveable handles for AaC.aF.SL.1b. Combined cold- & warm-water faucets
-A.aC-1m.Dish panel	Front panel with powerful dishwasher programs for AaC.aG.SL.2. Advanced dishwasher (40 liters)
-A.aC-1n. Dish door	Bottom-hinged cover door for AaC.aG.SL.2. Advanced dishwasher (40 liters)
-A.aC-1o. Dish panel	Front panel for AaC.aG.SL.1. Simple dishwasher (40 liters)
-A.aC-1p. Dish door	Bottom-hinged cover door for AaC.aG.SL.1. Simple dishwasher (40 liters)

The machinery brands need not be selected yet, and since there are many to select from, these questions can be addressed in the next *phase*, **Pt.F. Finalize design & requisites**. It must be said that it is quite uncommon to choose a kitchen from a functionality viewpoint; often fashion and appearance are discussed long before functionality.

9.8.3 EXAMPLE House kitchen machinery: Arrange ingredients and identify possible black-boxes

The *logical architecture* doesn't tell too much about dimensions and how the machinery fits the kitchen interior, but is very good for checking that all *interfaces* have been remembered and no details have been forgotten.

The *logical architecture* for the machinery is prepared (see Fig. 9-29 below).

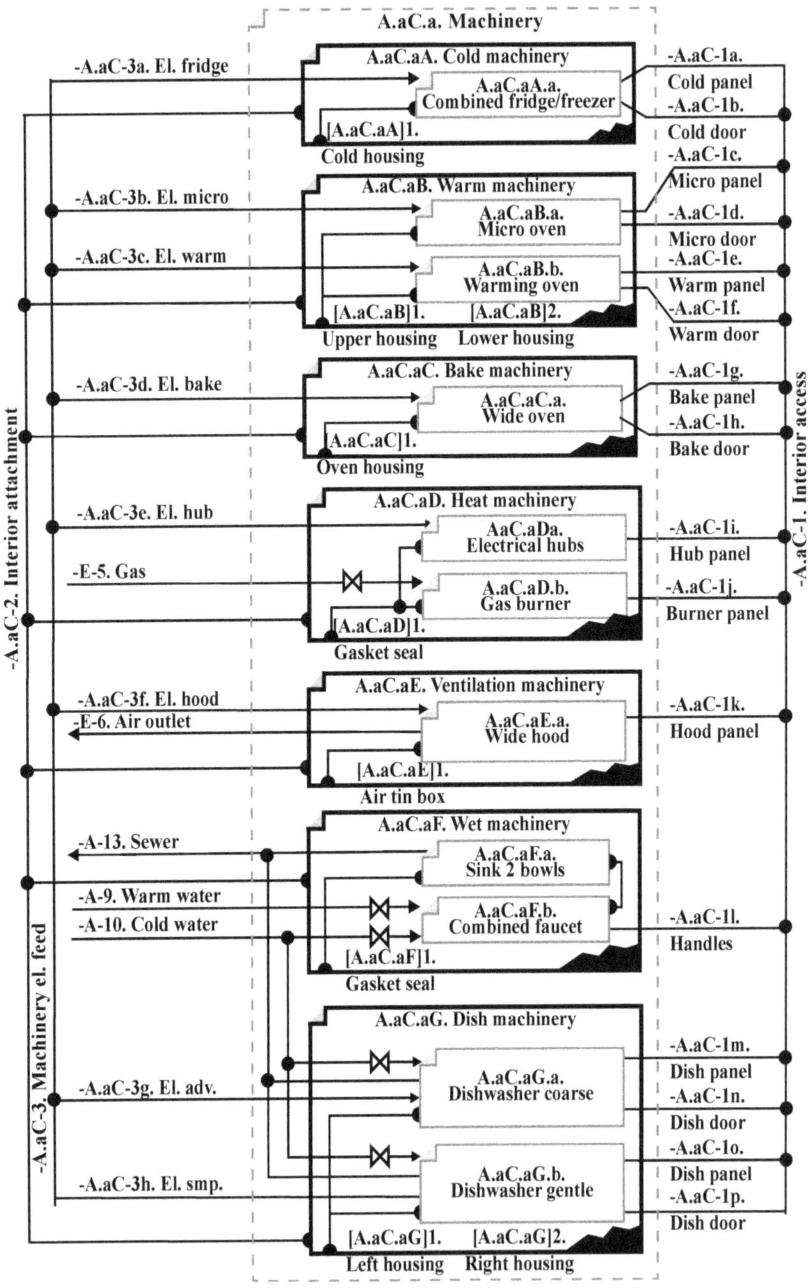

FIGURE 9-29 House kitchen machinery logical architecture

As can be seen from this *logical architecture*, there are no more *embedded black-boxes* to be specified and *decomposed* on this machinery *nesting level* in the *architecture*.

9.8.4 EXAMPLE House kitchen machinery: Lay out physical architecture

Again, it is very rewarding to see the approximate *physical architecture* with your own eyes, before the design is done. The more an end *user* can be involved in the development, the bigger the chance of satisfaction with the finished *product*.

The husband proprietor of the house-buying family used the previous 3D Figure 9-27, page 370, and just taped some typical machinery front pictures to it (see Fig. 9-30 below). He also wrote the machinery ID numbers by hand.

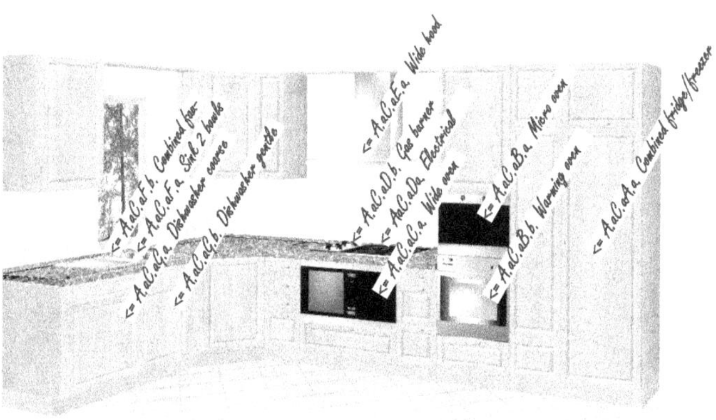

FIGURE 9-30 House kitchen machinery physical architecture

9.8.5 EXAMPLE House kitchen machinery: Examine architectures

Access *interfaces* to machinery and interiors are crucial for a chef to work efficiently, and one was invited to the gate meeting for a review. The *architectures* were reviewed and it was concluded that the *architecture resumé* was in good order and the chef judged the *physical architecture* to not be optimal but sufficient.

The review also covered the fit of the machinery with its particular casings, doors, and front panels to the kitchen interior.

9.8.6 EXAMPLE House: Proceed reading

To follow this example, these are the alternatives:

- **3F** Proceed to chapter 10.4, "EXAMPLE: House kitchen machinery: Finalize design of machinery with product requisites," page 445.

- **3I** Proceed to chapter 11.5, "EXAMPLE House kitchen machinery: Procure machinery items in order to install machinery," page 618.

- **3V** Proceed to chapter 12.7, "EXAMPLE: House rooms machinery: Fully verify machinery black-/white-boxes," page 678.

- To leave the house example and dig deeper into the chapter 9, "Design architectures" topic, proceed to chapter 9.9 below.

9.9 Managed interfaces

9.9.1 About managed interfaces

To make two elements ever compatible with each other, the interface between them must be actively managed.

Any development *result* might be defined as an artifact with *managed information*. This means that the artifact must have a unique identifier, and be under revision control. One of the most important artifacts to be managed is *requirements*. The next in importance may be *interfaces* (see Fig. 9-31). This fact is unfortunately not often identified by many developing companies.

In this book, an *interface* is marked with a knot if it is supposed to be *managed information* (see Fig. 9-31 below).

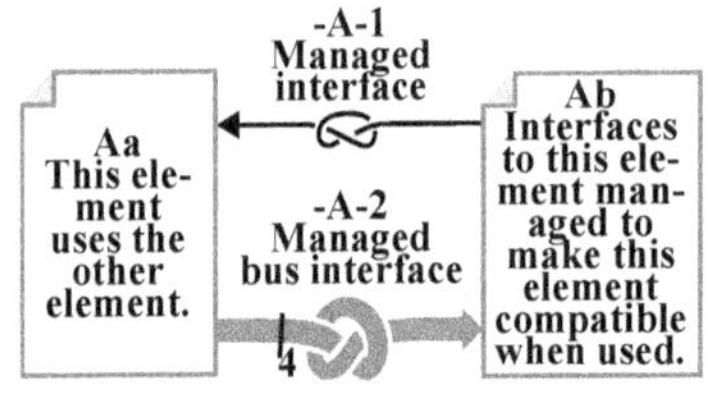

FIGURE 9-31 Managed interfaces are marked with a knot

The impact of *interfaces* being managed or failing to be, is really far-reaching. For a second book to follow this one, this topic is a hot candidate for several chapters on compatibility, *configuration management*, and planned *reuse*. The chapters below are only appetizers :-)

9.9.2 Exchangeability

One advantage with a managed interface is that connected elements can easily be exchanged.

Typically, exchange of *elements* occurs when they are broken or worn out, for example, brake pads to a car. To support the exchange, the *interface* between the *host* and the new spare part must be the same. This requires that the developers must manage the *interface* between the *user* and the exchangeable *element*. Car owners who have changed brake pads know that this has not always happened.

Exchange does not necessarily require that an *element* be broken, but may also come from a wish to improve or cheapen the *element*. Maybe new technology has reduced the price significantly, and the new *element* can replace the old also in new manufacturing.

If an *interface* between two *product* parts is not owned by the same company, it is both more important and more difficult to manage them. For example, the old electrical dimmers don't work well with new LED lamps, even if the lamp base happens to be compatible. Sometimes standardization organizations have the mission to define and manage *interfaces* applicable for commercial companies.

9.9.3 Reusability

Managing *interface*s becomes crucially important when efficient *reuse* must function well.

Reuse is maybe more common than we sometimes realize. In the construction industry, there is a lot of *reuse* of windows, doors, stone, ceramics, and so forth, and a lot of electronic *components*, such as resistors, capacitors, semiconductors, and so forth, are not developed by the *product development* company itself, but bought as *reusable items* from *suppliers*.

The perplexing fact is how small the *reused components* often are (*granularity* is small) compared to the full *product*. This fact still causes a lot of development work to *interface* them with each other and keep track of all *components* for their *reuse*, and does not release the full potential of *reuse*. However, there are also some examples of high *granularity reuse* with well-*managed information* of *interfaces* (see Fig. 9-32).

FIGURE 9-32 Obviously interfaces between sections need to be well managed.

9.9.4 Recycling

Recycling is often better than nothing, but it is far better to have architects detect and re-use design similarities early.

Recycling is the simplest form of *reuse*. When a particular *element* is designed, it may be easier to start with something similar than to start from scratch. An old *architecture* drawing or a legacy *program* can be the source, and these are remolded until they fit the new *requirements*. It is like recycling cans and bottles; they are melted down and new cans and bottles of any new form can be resurrected. However, this is not *reuse* without rework, it is recycling with significant rework.

9.10 Illustrate interactions between elements

9.10.1 Cpdm definition

In *Cpdm, message sequence chart (MSC)* is defined according to Table 9-22.

TABLE 9-22 Cpdm definition

Aspect	Message sequence chart
Definition	Chart illustrating how stimuli and responses are distributed between architectural elements, including black-/white-boxes
Synonyms	Interaction diagram, communication diagram
Symbols	

A. white-box
A.A. white-box

-E-1. Interface A.b. Thread -A-1. Interface A.A.b. Element -A.A-1. Interface A.A.A. Black-box

A.b.WS.1.
-E-1.S1(a1)
-E-1.R1(a2)
-A-1.S1(a4)
A.A.WS.1. Wait-state
A.A.A.WS.1. Wait-state
A.A.RB.1 Behavior requirement
A.A.A.RB.1. Behavior requirement
-A.A-1.S1(a6)
A.A.A.H.1.
-A.A-1.R1(a7)
A.A.A.WS.2. Wait-state
A.A.RR.1
A.b.WS.2.
-A-1.R1(a5)
A.b.RR.1
-E-1.R2(a3)
A.A.WS.2. Wait-state
A.b.WS.1.

9.10.2 EXAMPLE: Cook and serve message sequence chart

Assume that the two *use-units* UU.21. and UU.31. in Figure 7-16, page 229, penetrate the house example *elements* in Figure 9-26, page 369 (see Fig. 9-33 below).

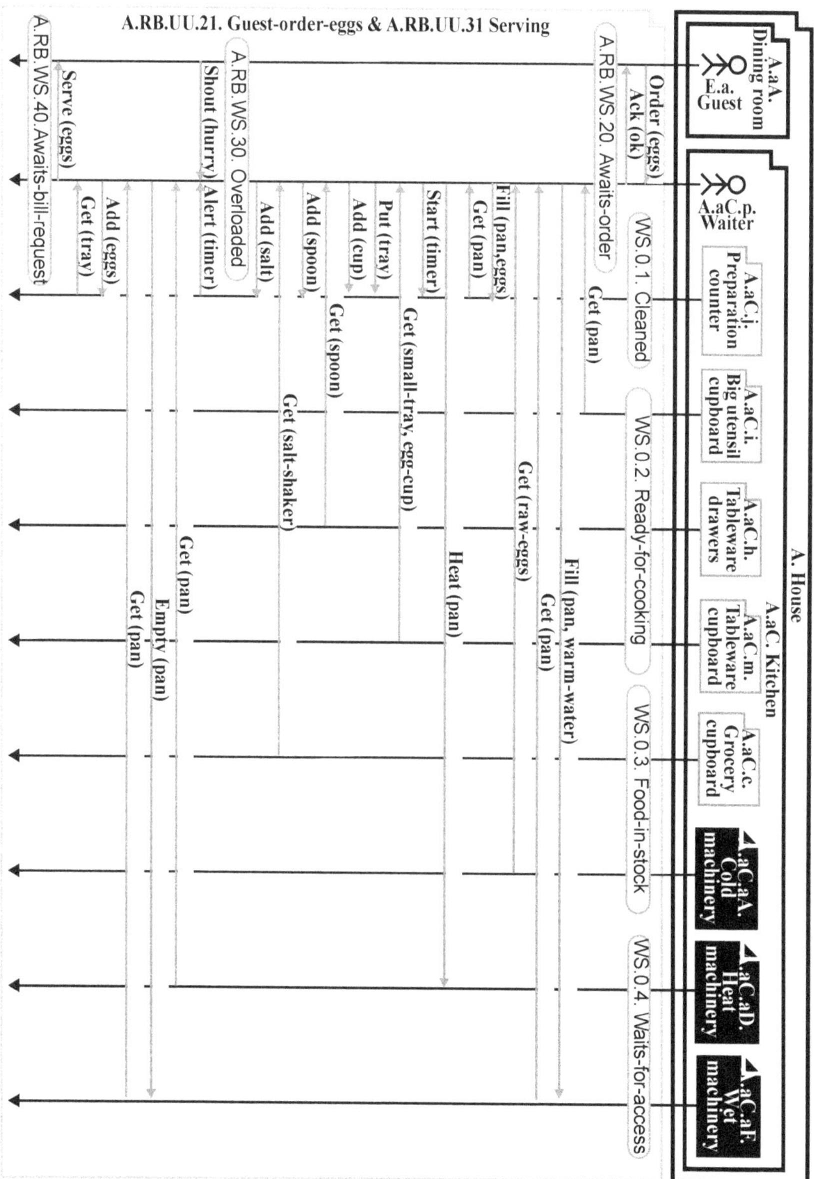

FIGURE 9-33 Two bistro use-units penetrate the house example

When starting to capture *requirements*, there is only the *product outermost black-box* to interact with. After opening and filling the *product outermost black-box* with ingredients to create the *outermost white-box*, many new *black-boxes* might be found to interact with. The strength of *message sequence charts* is that they serve well to connect behavior of *outward black-boxes* or *elements* with behavior of many *embedded black-boxes* or *elements* on *inward nesting levels*.

Note that *stimuli* and *responses* on all *interfaces* of an *element* connect to the same vertical arrow. Also note that with overuse of ingredients, the *message sequence chart* risks getting very *complex*.

9.11 EXAMPLE Multiplication toy environment: Constitute environment architecture

9.11.1 EXAMPLE Multiplication toy: About this example

The important *architecture design* work is to make way for the separation of the two technologies selected for use. The *architecture* must be split into one common *product architecture* (described by chapters marked with C in Fig. 9-34 below) and two separate *architectures*, one for each of the two interchangeable *elements* (described by chapters marked with H and M in Fig. 9-34 below).

9.11.2 EXAMPLE Multiplication toy environment: Process schedule to use

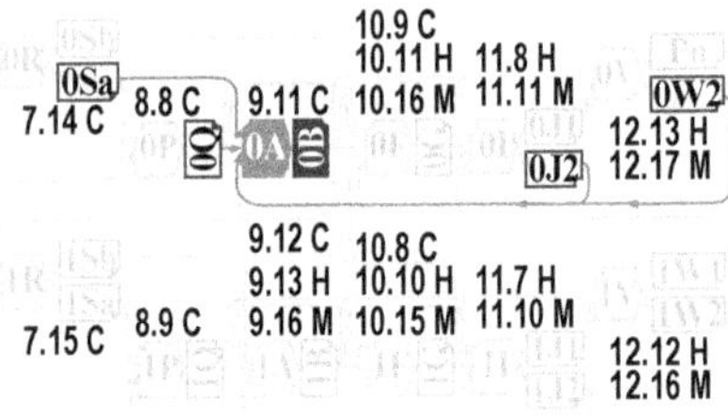

For the multiplication toy example, a special technical overall *schedule* is *tailored* (illustrated in Fig. 9-34, left):

- **Pt:2:3. Tailored advanced technical schedule n = 2,** page 169

In this *tailored schedule* the generic *schedule* to use is (filled-in symbols):

- **Pt.0.A. Constitute environment architecture,** page 115.

FIGURE 9-34 Position of schedule and master to use (filled in) in overall schedule

Below *results* may be demanded (their *masters* are indicated *upstreams* from the filled-in *schedule* in the figure above):

0Q Solution alternatives of interfaces to environment with supplier opportunities

Table 8-18, Multiplication toy environment solution alternatives of interfaces to environment, page 314

Table 8-19, Multiplication toy environment solutions excluded, page 315

Table 8-20, Multiplication toy environment solutions included with supplier opportunities, page 315

[0Sa] Environment restriction requirements

Table 7-23, Multiplication toy environment restriction requirement, page 213.

[Xb] Suppliers sample catalogues

For possible *failure localization* and *failure elimination*, below *results* may be demanded (their *masters* are indicated *upstream* from the filled-in *schedule* in the figure above):

[0J2] Table 11-28, Multiplication toy environment (hard-wired multiplier variant), start-up report 1, page 630

Table 11-36, Multiplication toy environment (microcontroller multiplier variant), start-up report 1, page 640

[0W2] Table 12-53, Multiplication toy environment (microcontroller multiplier variant) failure elimination report 1, page 728

9.11.3 EXAMPLE Multiplication toy environment: Identify environment architecture ingredients

By demanding Table 8-18, p. 314, an *environment architecture resumé* is prepared to describe which ingredients must be created to best satisfy the *requirement* (see Table 9-23 below).

TABLE 9-23 Multiplication toy
environment architecture ingredients resumé

Existing and product interface ingredient		Explanation or chosen solution alternative or sample thereof
aE.	Hosting toy and multiplication toy environment	Existing environment to the hosting toy and to the multiplication toy panel
aE.a.	Multiplication toy user	Existing child meant to use the hosting toy and multiplication toy
aE.b.	Hosting toy producer	Existing hosting toy producers, meant to mount the multiplication toy into their hosting toys E.SL5. Disclaimer to be signed by hosting toy company E.SL6.b. Hosting toy company requested to take product failure insurance
-aE-1.	Multiplying	Interface between the user and the multiplication toy E.SL1. The toy may not be sharp, toxic, or demountable to be harmful for its environment or users in the environment.
-aE-2.	Padding leakage	E.SL2.a. Gasket to tighten between the multiplication toy and the hosting toy
E.	Multiplication toy environment	Environment back casing for the multiplication toy
-E-1.	Host attachments	Product static interface to be developed E.SL2.a. Bolts or similar for fastening the toy to the hosting toy E.SL3. Holes or guide pins for fastening bonds, located asymmetric to each other, which fit together in only one way E.SL4. Soft hosting toys are potentially too floppy, and in need of some kind of strengthening support when attaching the multiplication toy to it.
]E]1.	Hosting toy boundary	Boundary interface of the hosting toy
A.	Multiplication toy	Product to be developed

9.11.4 EXAMPLE Multiplication toy environment: Arrange ingredients and identify black-box

From the above *architecture resumé*, a *logical architecture* is now arranged (see Fig. 9-35 below). Note that the *environment* now is an *element* hierarchy of two *nesting levels*, and those are numbered from the inside out. Thus, for clarity, the *outermost black-box* always gets the identification **A.** and its *environment* always gets identification **E.**

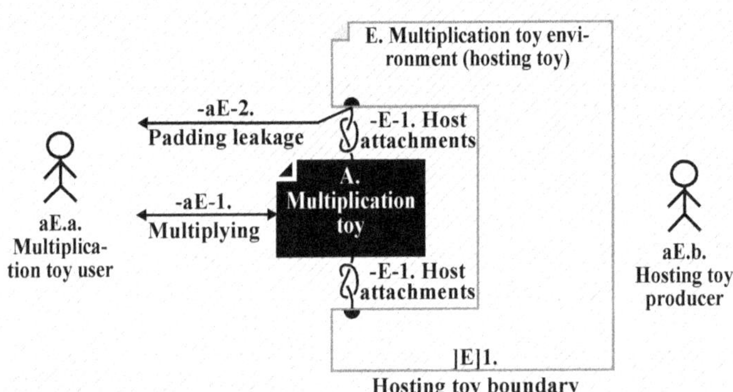

FIGURE 9-35 Multiplication toy environment logical architecture

9.11.5 EXAMPLE Multiplication toy environment: Lay out physical architecture

Some graphic material has already been developed for marketing the multiplication toy to potential hosting toy *customers* in the coming *activity* **Pp. Promote campaigns**. One picture is included as a *physical architecture* layout to illustrate the multiplication toy in a cuddly toy (see Fig. 9-35 below).

FIGURE 9-36 Multiplication toy environment physical architecture

9.11.6 EXAMPLE Multiplication toy environment: Examine product context

The *environment architecture* is rather simple. There are two *interfaces* from the multiplication toy to the *environment*, one MMI *interface* to the toy *user*, and one mechanical *interface* to the host toy. Both *interfaces* were approved by the gate members.

9.11.7 EXAMPLE Multiplication toy: Proceed reading

To follow this example, these are the alternatives:

- Proceed to chapter 10.9, "EXAMPLE Specifics for multiplication toy environment commonalities: Finalize design of environment interfaces with product requisites," page 493.
- Proceed to chapter 11.8, "EXAMPLE Multiplication toy environment (hardwired multiplier variant): Realize interfaces to environment and insert/await outermost white-box," page 628.
- Proceed to chapter 11.11, "EXAMPLE Multiplication toy environment (microcontroller multiplier variant): Realize interfaces to environment and insert/await outermost white-box," page 637.
- Proceed to chapter 12.13, "EXAMPLE Multiplication toy environment (hardwired multiplier variant): Verify prototype in environment," page 712.
- Proceed to chapter 12.17, "EXAMPLE Multiplication toy environment (microcontroller multiplier variant): Verify prototype in environment," page 725.
- Proceed to chapter 7.15, "EXAMPLE Multiplication toy: Refine product requirements from environment restriction requirements and ensure black-/white-box development staffing," page 215.
- Proceed to chapter 9.12 "EXAMPLE Multiplication toy commonalities: Satisfy refined requirements by decomposing product black-box into white-box design containing no black-boxes" below.

9.12 EXAMPLE Multiplication toy commonalities: Satisfy refined requirements by decomposing product black-box into white-box design containing no black-boxes

9.12.1 EXAMPLE Multiplication toy commonalities: Process schedule to use

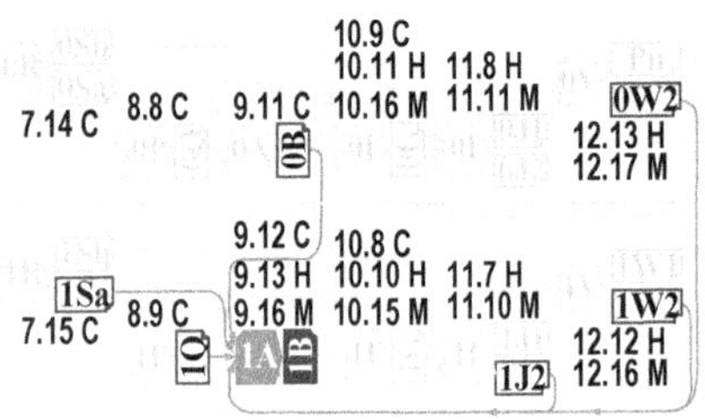

For the multiplication toy example, a special technical overall *schedule* is *tailored* (illustrated in Fig. 9-37, left):

- **Pt:2:3. Tailored advanced technical schedule n = 2,** page 169

In this *tailored schedule* the generic *schedule* to use is (filled-in symbols):

- **Pt.n.A. Satisfy requirements by decomposing outward nesting level black-box into white-box design containing possible black-boxes,** page 136.

FIGURE 9-37 Position of schedule and master to use (filled in) in overall schedule

Below *results* may be demanded (their *masters* are indicated *upstream* from the filled-in *schedule* in the figure above):

0B Environment existing ingredients and design of interfaces to the environment
Table 9-23, Multiplication toy environment architecture ingredients resumé, page 383
Figure 9-35, Multiplication toy environment logical architecture, page 384
Figure 9-36, Multiplication toy environment physical architecture, page 384

1Sa Prioritized requirements
Table 7-25, Multiplication toy restriction requirement, page 216.
Table 7-26, Multiplication toy interaction requirement, page 217
Table 7-27, Multiplication toy interface requirements, page 217

1Q Solution alternatives and supplier opportunities
Table 8-21, Multiplication toy white-box solution alternatives, page 317

Xb Suppliers' sample catalogues

For possible *failure localization* and *failure elimination*, below *results* may be demanded (their *masters* are indicated *upstream* from the filled-in *schedule* in the figure above):

1J2 Table 11-25, Multiplication toy with hard-wired multiplier start-up report 1, page 627

Table 11-33, Multiplication toy with microcontroller multiplier start-up report 2, page 636

1W2 Table 12-38, Multiplication toy with hard-wired multiplier failure elimination report 1, page 711

Table 12-47, Multiplication toy with microcontroller multiplier failure elimination report 1, page 723

0W2 Table 12-53, Multiplication toy environment (microcontroller multiplier variant) failure elimination report 1, page 728

9.12.2 EXAMPLE Multiplication toy commonalities: Identify white-box architecture ingredients

In this case, an *architecture resumé* with new *environment* ingredients is specified (see Table 9-24 below). Among others, a new *black-box user* called aE.c. User guardian has been identified. Other ingredients are also identified that were not yet part of the previous *architecture*, and the -aE-1. Multiplying has also been detailed.

TABLE 9-24 Multiplication toy common architecture ingredients resumé		
White-box ingredient		**Chosen solution alternative or sample thereof**
aE.	Hosting toy and multiplication toy environment	From environment architecture ingredients resumé
aE.c.	**User guardian**	User in possession of battery flap opening tool
-aE-1.	Multiplying	From environment architecture ingredients resumé
-aE-1a.	**Multiplicand input**	Input device for a multiplicand
-aE-1b.	**Multiplicand input**	Input device for a multiplicand
-aE-1c.	**Product output**	Output device for the product
-aE-2.	Padding leakage	From environment architecture ingredients resumé
-aE-2.	**Immovable parts**	A.SL1. Use clamping or friction screws for fastening parts on the front panel.
-aE-3.	**Toxic surface treatment**	A.SL2. Use only guaranteed nontoxic painting and surface material for the casing.
-aE-4.	**Battery exchange**	Change battery when LED does not light up anymore.

Now back to the *outermost black-box*, opening it and filling it with *architecture ingredients* from previously prepared *solution alternatives* (see Fig. 9-25 below).

When opening the *outermost black-box* and filling its *architecture resumé* with *architecture ingredients* from *solution alternatives* or data sheets of *samples*, it is not uncommon to discover new ingredients even outside the *outermost black-box* in the *environment*. However, there is no problem with that, just add them now to the *environment*, but try to put them into the *architecture resumé* before starting with the ingredients within the *black-box*.

TABLE 9-25 Multiplication toy variants architecture ingredients resumé

White-box ingredient		Chosen solution alternative or sample thereof	
A.	Multiplication toy	A.SL7.a. One customized ASIC variant for 25 000 products at € 2 each A.SL7.b. One microcontroller variant for 100 000 products at € 2 each A.SL8. Design for efficient manufacturing test.	
Aa.	Multiplier	A.SL9. The multiplier element shall calculate the product from two operands. A.SL3.a. Gather all multiplication intelligence to a separate element and manage its interface. A.SL4.a. One technology shall be a hard-wired gate network, to ensure a correct prototype, before transferring this gate network to a customized ASIC. A.SL4.b. One technology shall contain a cheap single-chip microcontroller with a minimal program intended for series manufacturing. For example, from the PIC16Fxy family	A.SL12. Maximum delay between operand settling and display reading is 0.1 second..
Ab.	Setting device	Device to generate binary code from switch	
Ab.a.	Rotor switch	A.SL10.a. Cheap robust rotating switch	
Ac.	Setting device	Device to generate binary code from switch	
Ac.a.	Rotor switch	A.SL10.a. Cheap robust rotating switch	
Ad.	Presenting device	3 decimal digit 7-LED sections	
Ad.a.	3 conv. BDC to 7-seg.	BCD to 7-segment converter	
Ad.a1.	Converter 1	BCD Converter 1	
Ad.a2.	Converter 2	BCD Converter 2	
Ad.a3.	Converter 3	BCD Converter 3	
Ad.b.	3 digits 7-seg. LED	A.SL11.a. LED panel	
Ad.b1.	Digit 1	LED digit1	
Ad.b2.	Digit 2	LED digit2	
Ad.b3.	Digit 3	LED digit3	
Ae.	Power supply	A.SL6.a. Battery	
Af.	Assembly work	Work to assemble the toy components	
-A-1.	Circuit board	A.SL5. To switch between the two product variants, only the circuit boards should need to be exchanged.	
]A]2.	Casing	Encapsulation to protect user from toy hazards	
-A-3.	Operand bus	4-bit-wide binary coded electronic interface	
-A-4.	Operand bus	4-bit-wide binary coded electronic interface	
-A-5.	Product bus	10-bit-wide BCD coded electronic interface	
-A-6.	Power feed	Power interface to all consumers	
-A-7.	Ground	Ground for all circuits	

9.12.3 EXAMPLE Multiplication toy commonalities: Arrange ingredients and identify possible black-boxes

From the previous *architecture resumé*, the *outermost black-box* can now be presented as a *outermost white-box* (see Fig. 9-38 below).

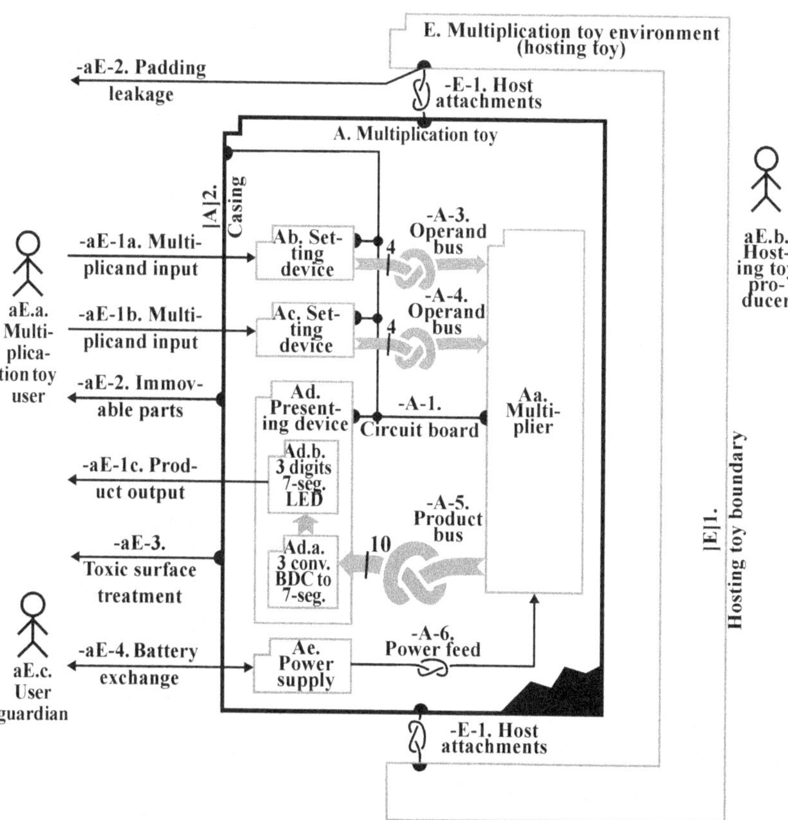

FIGURE 9-38 Multiplication toy common logical architecture

All *interfaces* to the Aa. Multiplier must be *managed information* (having knots in the *architecture* above) to ensure that both technology implementations of the Aa. Multiplier will be instantly interchangeable (for more about managed *interfaces*, see ch. 9.9, page 378).

The *interface* -E-1. Host attachments must also be *managed information*, to ensure that all potential cuddly toy producers share all documented revisions of this *interface*, to be instantly prepared to use the multiplication toy in their cuddly toy.

9.12.4 EXAMPLE Multiplication toy variants of multipliers

To make the *architectures* very clear, the *elements*, with their *interfaces* to be interchangeable *variants*, are identified and drawn in a *specialization* chart according to Figure 9-39 (for more about *specialization*, see Table 6-6, p. 89).

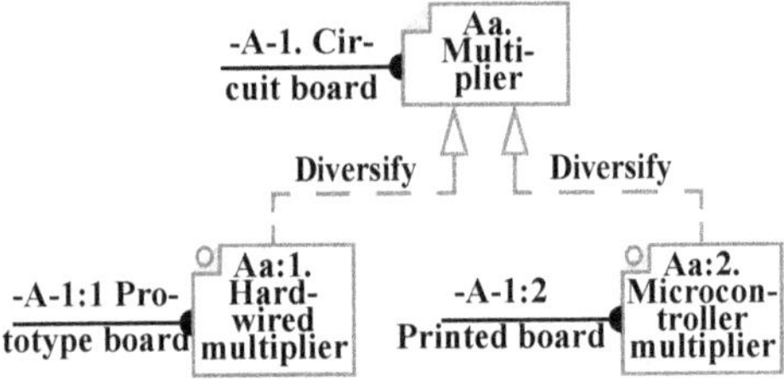

FIGURE 9-39 Diversifying the multiplier element and circuit board interface

In order to be instructive and clear, the documentation is split further on in different sections for the two *variants*.

9.12.5 EXAMPLE Multiplication toy commonalities: Physical architecture

The microcontroller multiplication toy should have a user-friendly front layout (see Fig. 9-40).

The same front layout shall also fit the multiplication toy with hard-wired multiplier prototype, but does not need to be mounted on the prototype, since this *variant* is assumed to be only an evaluation for a possible custom-designed ASIC.

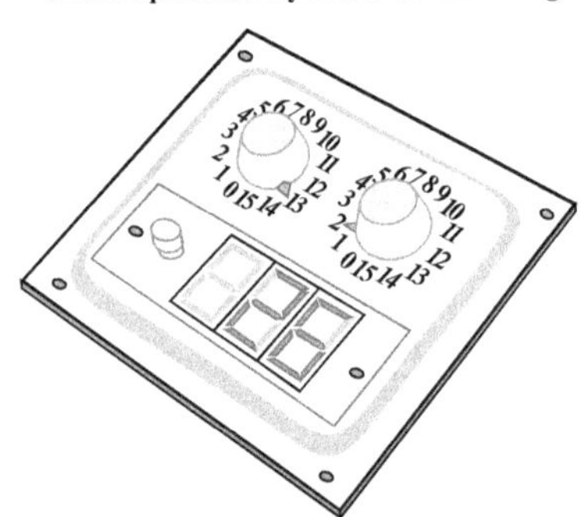

A. Multiplication toy with mounted casing

FIGURE 9-40 Multiplication toy common physical architecture

9.12.6 EXAMPLE Multiplication toy commonalities: Examine architectures

There was an intense discussion about why the Aa. Multiplier was not denoted an *embedded black-box* with refined *requirements*, which is highly promoted by *Cpdm*. But the behavior of the Aa. Multiplier is, in fact, to multiply two operands into one product, which is not *complex* enough to need an own *black-box*. But beware: not defining a separate *black-box* is not a big overall saving, because much of the development and accounting must be split into commonalities and the two specialties in any case. Skipping *formalism* gives immediate relief, but penalties will occur subsequently.

9.12.7 EXAMPLE Multiplication toy: Proceed reading

To follow this example, these are the alternatives:

- Proceed to chapter 10.8, "EXAMPLE Specifics for multiplication toy commonalities: Finalize design items of product white-box with product requisites," page 482.
- Proceed to chapter 11.7, "EXAMPLE Multiplication toy with hard-wired multiplier: Procure items and linkable programs in order to realize white-box," page 624.
- Proceed to chapter 11.10, "EXAMPLE Multiplication toy with microcontroller multiplier: Procure items and linkable programs in order to realize white-box," page 632.
- Proceed to chapter 12.12, "EXAMPLE Multiplication toy with hard-wired multiplier: Verify black-/white-box," page 706.
- Proceed to chapter 12.16, "EXAMPLE Multiplication toy with microcontroller multiplier: Verify black-/white-box," page 717.
- Proceed to chapter 9.16, "EXAMPLE Microcontroller multiplier: Further satisfy refined requirements by decomposing product black-box into white-box design containing no black-boxes," page 397.
- Proceed to chapter 9.13 "EXAMPLE Hard-wired multiplier: Further satisfy refined requirements by decomposing product black-box into white-box design containing no black-boxes" below.

9.13 EXAMPLE Hard-wired multiplier: Further satisfy refined requirements by decomposing product black-box into white-box design containing no black-boxes

9.13.1 Microcontroller multiplier: Process schedule to use

For the multiplication toy example, a special technical overall *schedule* is *tailored* (illustrated in Fig. 9-41, left):

- **Pt:2:3. Tailored advanced technical schedule n = 2,** page 169

In this *tailored schedule* the generic *schedule* to use is (filled-in symbols):

- **Pt.n.A. Satisfy requirements by decomposing outward nesting level black-box into white-box design containing possible black-boxes,** page 136.

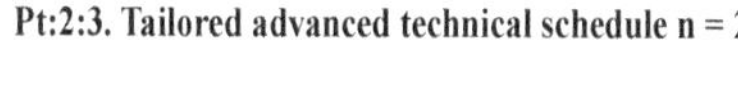
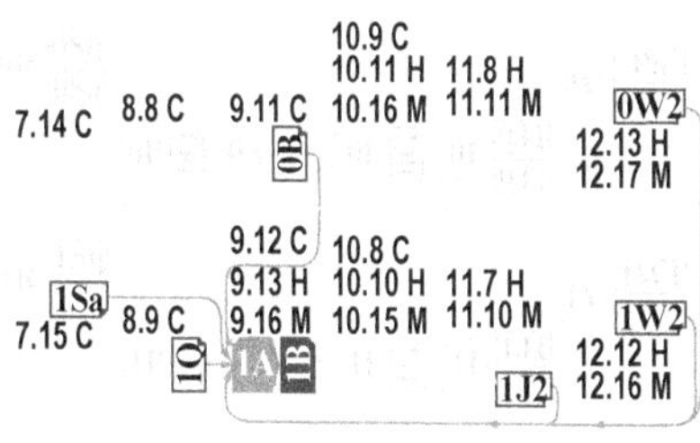

FIGURE 9-41 Position of schedule and master to use (filled in) in overall schedule

Below *results* may be demanded (their *masters* are indicated *upstream* from the filled-in *schedule* in the figure above):

1B Common Architecture white-box with major ingredients but no embedded black-boxes

Table 9-25, Multiplication toy variants architecture ingredients resumé, page 388

Figure 9-38, Multiplication toy common logical architecture, page 389

Figure 9-39, Diversifying the multiplier element and circuit board interface, page 390

Figure 9-40, Multiplication toy common physical architecture, page 390

0B Environment existing ingredients and design of interfaces to the environment

Table 9-23, Multiplication toy environment architecture ingredients resumé, page 383

Figure 9-35, Multiplication toy environment logical architecture, page 384

Figure 9-36, Multiplication toy environment physical architecture, page 384

1Sa Prioritized requirements

Table 7-25, Multiplication toy restriction requirement, page 216.

Table 7-26, Multiplication toy interaction requirement, page 217

Table 7-27, Multiplication toy interface requirements, page 217

1Q Solution alternatives and supplier opportunities

Table 8-21, Multiplication toy white-box solution alternatives, page 317

Xb Suppliers' sample catalogues

For possible *failure localization* and *failure elimination*, below *results* may be demanded (their *masters* are indicated *upstream* from the filled-in *schedule* in the figure above):

1J2 Table 11-25, Multiplication toy with hard-wired multiplier start-up report 1, page 627

Table 11-33, Multiplication toy with microcontroller multiplier start-up report 2, page 636

1W2 Table 12-38, Multiplication toy with hard-wired multiplier failure elimination report 1, page 711

Table 12-47, Multiplication toy with microcontroller multiplier failure elimination report 1, page 723

0W2 Table 12-53, Multiplication toy environment (microcontroller multiplier variant) failure elimination report 1, page 728

9.13.2 EXAMPLE Hard-wired multiplier: Identify white-box architecture ingredients

The specific hard-wired multiplier *architecture* will now be prepared, by decomposing the *element* Aa:1. Hard-wired multiplier, to get it more detailed. The content will consist of a large network of logic gates, to asynchronously perform the product calculation from two operands.

There are a lot of different solutions for making a multiplier unit. In this book, two design alternatives will be thoroughly shown. The first is a hard-wired gate network to show a classic hardware solution (see Table 9-42 below) and the second is with a microcontroller in chapter 9.16.

TABLE 9-26 Hard-wired multiplier architecture ingredients resumé

White-box ingredient		Chosen solution alternative or sample thereof
-A-1.	Circuit board	Printed circuit board forming the mechanical and electrical connection between components
-A-1:1	**Prototype board**	Circuit boards with manually wired connections
Aa.	Multiplier	Multiplier with two kinds of realizations
Aa:1.	**Hard-wired multiplier**	Multiplier prototype realized with wired logic
Aa:1.a.	**Arithmetic unit**	2 x 4 bits to 8 bits hard-wired gate network ALU
Aa:1.b.	**Bin-BCD converter**	8-bit-wide binary to 12-bit (3 digits) BCD hard-wired gate converter
-Aa:1-1.	**Bin-BCD interface**	8-bit-wide binary coded interface

9.13.3 EXAMPLE Hard-wired multiplier: Arrange ingredients and identify possible black-boxes

The multiplier is a straight binary multiplier, generating an 8-bit binary coded product from two 4-bit binary coded operands. The output from the multiplier should be BDC-coded, which requires a binary to BCD converter (see Fig. 9-42 below).

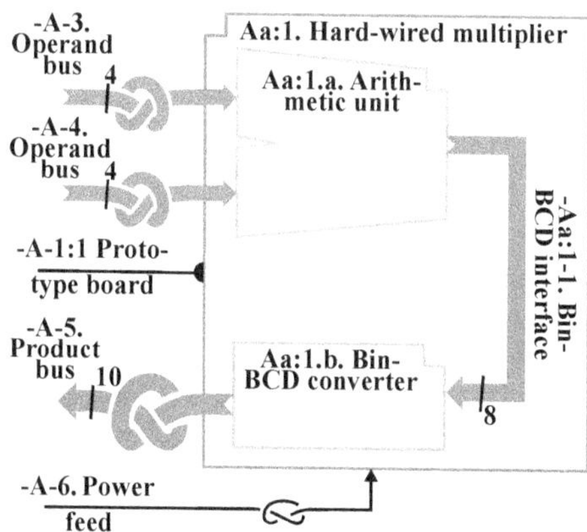

FIGURE 9-42 Hard-wired multiplier logical architecture

9.13.4 EXAMPLE Hard-wired multiplier: Lay out physical architecture

The multiplier will be designed from a lot of logic gates, which are available in dual in-line plastic packages (see Fig. 9-43 right). These will be connected to each other by wrapping wires on an experimental circuit board.

FIGURE 9-43 14-pin dual in-line packages

9.13.5 EXAMPLE Hard-wired multiplier: Examine architectures

The *white-box design items* of all logic gates will be made in the next *phase*, **Pt.F. Finalize design & requisites**. Until then the current *architecture design* work is decided to be sufficient.

9.13.6 EXAMPLE Multiplication toy: Proceed reading

To follow this example, these are the alternatives:

- Proceed to chapter 10.10, "EXAMPLE Specifics for multiplication toy with hard-wired multiplier: Finalize design items of product white-box with product requisites," page 499.
- Proceed to chapter 11.7, "EXAMPLE Multiplication toy with hard-wired multiplier: Procure items and linkable programs in order to realize white-box," page 624.
- Proceed to chapter 12.12, "EXAMPLE Multiplication toy with hard-wired multiplier: Verify black-/white-box," page 706.
- To leave the multiplication toy example and to dig deeper into the chapter 9, "Design architectures" topic, proceed to next chapter 9.14 below.

9.14 EXAMPLE: Typical microcontroller architecture

The *architectures* so far have dealt only with hardware and electronics such as houses and networks of logic gates. The revolution in miniaturizing transistor gates has led to *architectures* that are much more general and adaptable than those designed with hard-wired gate networks.

Microcontrollers are, of course, *complex* in their design, but it is enough to briefly understand how they are internally designed (see Fig. 9-44 below). To illustrate the connections between the previously designed network of gates and an *element* in a microcontroller, the multiplier part of the ALU might be built something like the multiplier for the toy (see Fig. 10-24, p. 502).

The point is to understand enough of the microcontroller hardware to:

- connect the microcontroller to surrounding hardware circuits;
- understand how to write *programs* to control the surrounding hardware.

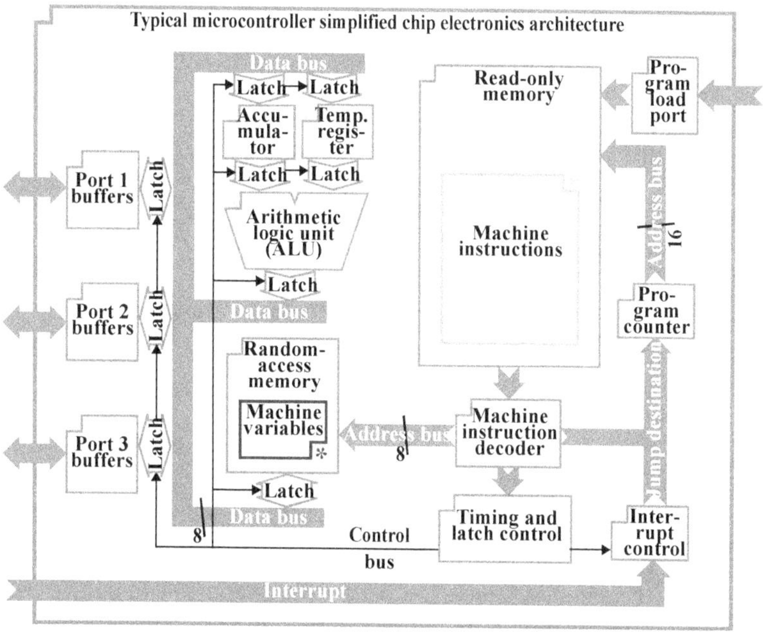

FIGURE 9-44 Typical microcontroller architecture

The *machine program* being developed is transferred into the microcontroller via a *program*-loading port. This can be arranged in a number of ways, but that is not important to understand the *program* itself. When powering up the microcontroller, the *program* counter is zero, and points to the first *machine instruction* to get decoded. The decoded *machine instructions* can retrieve and save bits, bytes, and words between the different parts of the controller. If a *machine instruction* is a branch, the new address is loaded to the *program* counter, and the execution continues at this address. In this way the *machine instructions* are executed in a permanent sequence, until the power is shut off.

9.15 Low-level programs

9.15.1 Cpdm definitions

In *Cpdm, variable, instruction, jump instruction, address label,* and *restriction* are defined according to Table 9-27 below.

TABLE 9-27 Cpdm definitions

Aspect	Low-level program architecture ingredients				
	Variable	**Instruction**	**Jump instruction**	**Address label**	**Restriction**
Definition	Storage for arrived stimuli	Executable limitless command	Command to continue execution at an address label	Entrance location among assembler instructions	Limitation of any parameter in a program
Synonyms	Data location, data segment	Low level command	Branch, skip over	Jump location, jump entry, branch target	Size limitation, number limitation
Typical structure of symbols					

For more detailed information, see chapter 10.14, "Assembler programs," page 516.

9.16 EXAMPLE Microcontroller multiplier: Further satisfy refined requirements by decomposing product black-box into white-box design containing no black-boxes

9.16.1 EXAMPLE Microcontroller multiplier: Process schedule to use

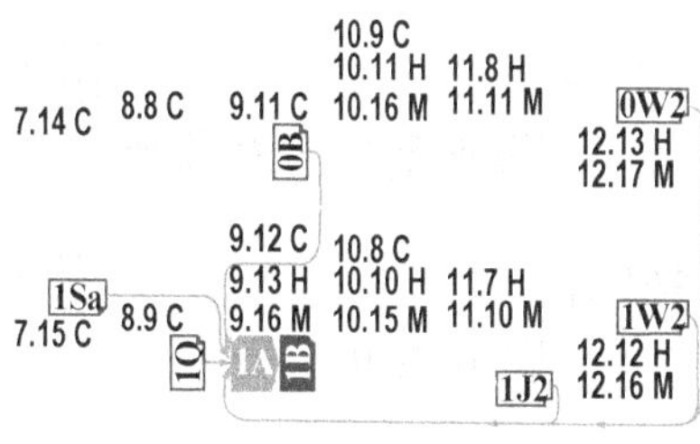

For the multiplication toy example, a special technical overall *schedule* is *tailored* (illustrated in Fig. 9-45, left):

- **Pt:2:3. Tailored advanced technical schedule n = 2,** page 169

In this *tailored schedule* the generic *schedule* to use is (filled-in symbols):

- **Pt.n.A. Satisfy requirements by decomposing outward nesting level black-box into white-box design containing possible black-boxes,** page 136.

FIGURE 9-45 Position of schedule and master to use (filled in) in overall schedule

Below *results* may be demanded (their *masters* are indicated *upstream* from the filled-in *schedule* in the figure above):

1B Common Architecture white-box with major ingredients but no embedded black-boxes

Table 9-25, Multiplication toy variants architecture ingredients resumé, page 388

Figure 9-38, Multiplication toy common logical architecture, page 389

Figure 9-39, Diversifying the multiplier element and circuit board interface, page 390

Figure 9-40, Multiplication toy common physical architecture, page 390

0B Environment existing ingredients and design of interfaces to the environment

Table 9-23, Multiplication toy environment architecture ingredients resumé, page 383

Figure 9-35, Multiplication toy environment logical architecture, page 384

Figure 9-36, Multiplication toy environment physical architecture, page 384

1Q Solution alternatives and supplier opportunities

Table 8-21, Multiplication toy white-box solution alternatives, page 317

1Sa Prioritized requirements

Table 7-25, Multiplication toy restriction requirement, page 216.

Table 7-26, Multiplication toy interaction requirement, page 217

Table 7-27, Multiplication toy interface requirements, page 217

[Xb] Suppliers' sample catalogues

For possible *failure localization* and *failure elimination*, below *results* may be demanded (their *masters* are indicated *upstream* from the filled-in *schedule* in the figure above):

[0W] Table 12-53, Multiplication toy environment (microcontroller multiplier variant) failure elimination report 1, page 728

[1J] Table 11-33, Multiplication toy with microcontroller multiplier start-up report 2, page 636

[1W] Table 12-47, Multiplication toy with microcontroller multiplier failure elimination report 1, page 723

9.16.2 EXAMPLE Microcontroller multiplier: Identify white-box architecture ingredients

Let's revisit the design of the multiplier in chapter 9.13.2, "EXAMPLE Hard-wired multiplier: Identify white-box architecture ingredients," page 392, but let's use a microcontroller instead of hard-wired logic. Then we need the ingredients in Table 9-28 below.

TABLE 9-28 Microcontroller multiplier refined architecture ingredients resumé

White-box ingredient		Chosen solution alternative or sample thereof
-A-1.	Circuit board	Printed circuit board with two kinds of layout
-A-1:2	**Printed board**	Printed circuit board with layout for wired logic
Aa.	Multiplier	Multiplier to be realized as wired gates and microprocessor
Aa:2.	**Microcontroller multiplier**	Multiplier realized with a Microchip single-chip microcontroller with enough parallel ports, e.g. [a]PIC16F57 *
-Aa:2-1.	**In-circuit programmable port**	Port for loading the program into the microcontroller
Aa:2a.	**Microcontroller**	2- x 4-bit input ports, 12-bit (3 x 4 bits) output port, low-cost, 8-bit-based microcontroller
Aa:2aa.	**Microcontroller program**	Microcontroller assembler program
Aa:2aaa.	**Variables**	The main program instruction body
Aa:2aab.	**Instructions**	Possible program instruction sub-bodies

a PIC16F57 is a trademark of Microchip Technology Inc.

9.16.3 EXAMPLE Microcontroller multiplier: Arrange ingredients and identify possible black-boxes

The *logical architecture* now resembles Figure 9-46. The operands are received and the product is delivered via ports of the microcontroller.

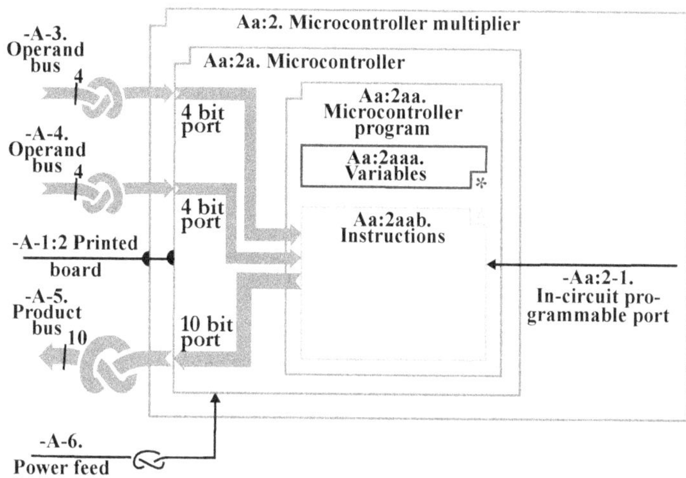

FIGURE 9-46 Microcontroller multiplier logical architecture

One may question the ports being of 4- and 12-bit width. Maybe there is no such special microcontroller available?

But it is not difficult to assume that the two 4-bit buses are two halves of an 8-bit bus, and the 12-bit bus is two 8-bit buses in parallel, using 8 bits of one and 4 bits of the other.

9.16.4 EXAMPLE Microcontroller multiplier: Lay out physical architecture

A microcontroller can be contained in various types of packages, mostly depending on how wide the ports are, thus needing more pins to make the ports connectable. Since the multiplier chosen must have a minimum of 18 ports, and must be fairly cheap, a 28 dual in-line package will be good.

FIGURE 9-47 28-pin dual in-line packages

9.16.5 EXAMPLE Microcontroller multiplier: Examine architectures

A lot of microcontrollers are offered on the market. The chosen one has a nice in-circuit programmable port and enough in-/out-ports, a decent price, and enough memory for *machine instructions* and *machine variable*s. The gate members agreed with the choice.

9.16.6 EXAMPLE Multiplication toy: Proceed reading

To follow this example, these are the alternatives:

- Proceed to chapter 10.15, "EXAMPLE Specifics for multiplication toy with microcontroller multiplier: Finalize design items of product white-box with product requisites," page 518.
- Proceed to chapter 11.10, "EXAMPLE Multiplication toy with microcontroller multiplier: Procure items and linkable programs in order to realize white-box," page 632.
- Proceed to chapter 12.16, "EXAMPLE Multiplication toy with microcontroller multiplier: Verify black-/white-box," page 717.
- To leave the multiplication toy example and to dig deeper into the chapter 9, "Design architectures" topic, proceed to chapter 9.17 below.

9.17 Multi-threads and interrupts

To take care of arriving *stimuli*, there is an important *architecture* and real-time concept called multithreading, that needs to be considered when digging into *product complexity*. Before explaining the concept, a few subchapters are needed in order to define the problem.

9.17.1 EXAMPLE: Polled multiplication toy

Simple technical solutions let the program poll input ports, to detect if any stimuli have arrived.

In the MSC diagram in Figure 9-48 below, the flow is illustrated for the multiplication toy simple *program* found in Figure 10-32, page 521. The *program* is running continuously in an infinite loop, constantly inputting the two digits and outputting the product. This way of continuously reading the inputs is called a polled system.

If preparation and calculation of the product took considerable time to perform, there would be a risk that some inputs are not polled until they have already been changed a couple of times. If so, the multiplier would have missed some input operands while the microcontroller was busy for a substantial time calculating the product. In this example that would not happen, because the preparation of the product is much faster than the *user* is to set the operands, thus no operands are missed.

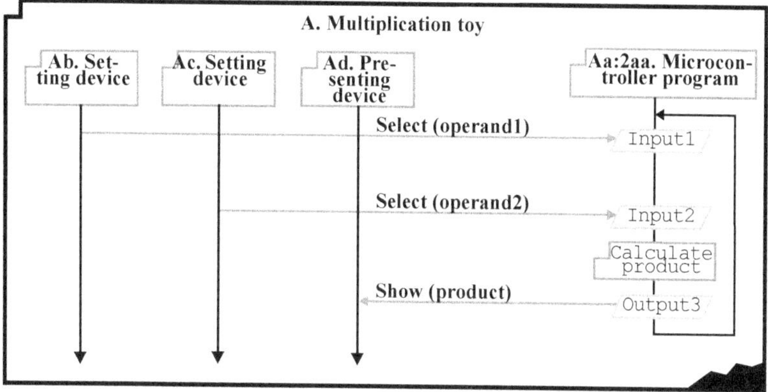

FIGURE 9-48 Unsynchronized stimuli to the polled multiplication toy

9.17.2 Wait-state explosion

When scaling up a *product* with a lot more time-consuming calculations and adding more and more inputs, each of them asynchronous with the others, there will be growing problems with sometimes missing inputs arriving too fast one after the other.

Everybody who has programmed real-time *programs* knows what happens if they are not *partitioned* into *threads* controlled by *interrupts*. All input ports must be polled all the time. When the *program* expands, there is always a risk that some iteration loop takes too much time, and an input is missed before the next input has erased it. The quick fix is to poll the input port also within the iteration, and then a new iteration is added, which also needs to poll the input and so forth. It ends with polling all input in all *program* iteration loops, which at a certain point becomes impossible, and the *program* breaks down altogether. This syndrome is sometimes called *wait-state explosion*.

The cure is to change the paradigm by *partitioning* the design into *threads* (see below).

9.17.3 Insufficient concurrence handling

Earlier in this book an example was used from the bistro world. So far only one guest was bothering the waiter, but now the bistro is getting popular.

More and more guests want the waiter's attention, and he gets overloaded with requests. When a guest has waited too long for the waiter, the guest walks away (and the problem disappears).

FIGURE 9-49 A bistro full of guests

The first thing to fix is to shift time-consuming *activities* to other persons at the bistro, maybe hire another chef and another waiter to bring out the food from the kitchen. If this is not enough, more waiters must be hired to serve the tables. Real luxury restaurants have one waiter per table, which solves the problem for good.

9.17.4 Cpdm definition

The *Cpdm* definitions of *thread, interrupt,* and *asynchronous interfaces* are rather similar to prevalent definitions (see Table 9-29 below).

TABLE 9-29 Cpdm definition

Aspect	Thread	Interrupt	Asynchronous interface
Definition	Program elements executing apparently concurrently, but in reality getting a time slot in sequence with all other threads	A signal from a port or an internal clock that causes a program to jump to a special location to perform an urgent execution before returning	A connection interface with a storage buffer function, to allow receiving at another pace and rhythm than was transmitted
Synonyms	Real-time process, process, multitask service	Break, discontinue	Ring buffer, mailbox, echo delay
Symbol example			

In case of software *programs*, *threads* can be seen as separate *programs* that execute in parallel with several others. In fact, a processor executes the total *program* sequentially, but either an *interrupt* controller or a multitasking kernel make the *threads* look like they are executing in parallel. The point is to assign only one *program user* to each *thread* that takes care of this *user's* particular *stimuli*. Even if the microcontroller or processor is very busy, the *stimuli*-receiving *thread* is let in for a short while to take care of the specific *stimulus*, and no *stimuli* are now missed.

Asynchronous interfaces have buffers that can save a signal from the sending *thread* until the receiving *thread* has the ability to receive it, for example a mailbox.

9.18 Usage of threads

9.18.1 Reentrance

Imagine there is a *program* with many *threads*, which are executing virtually in parallel, and all of them can call common sub*functions* (for example the cash register in Fig. 9-50, p. 404). It might happen that one *thread* has called the sub*function*, and while executing the sub*function*, the execution is left to another *thread*, which also happens to call the same sub*function*. This is not a big problem, but it must be ensured that such sub*functions* called from many different *threads* can handle these situations which are referred as such sub*functions* must allow reentrance (the cash register in Fig. 9-50, p. 404 must handle many calls without mixing them together).

9.18.2 EXAMPLE: Bistro real-time service

Now to propose a solution to the chapter 9.17.3, "Insufficient concurrence handling," page 402. The problem is that the bistro waiter has too much to do at lunchtime, because he has to take care of the guests one by one with a lot of waiting time for the guests to make up their minds and to finish eating.

The solution is to avoid waiting time by immediately shifting to another *thread* as soon as waiting time is risked to occur (see Fig. 9-50 below). The design trick is to *partition* the overall work for the waiter into *threads* that can be quickly done and completed, and can be terminated in favor of any other *thread* to start. Each *thread* is started by an *interrupt*, and the *thread* marked "default" takes over execution whenever no other *thread* is requested to start.

Of course, this solved the waiter problem, but moved the bottleneck to the chef preparing all the dishes. The solution was solved similarly to the waiter problem, and the chef also organized his work in *threads.*

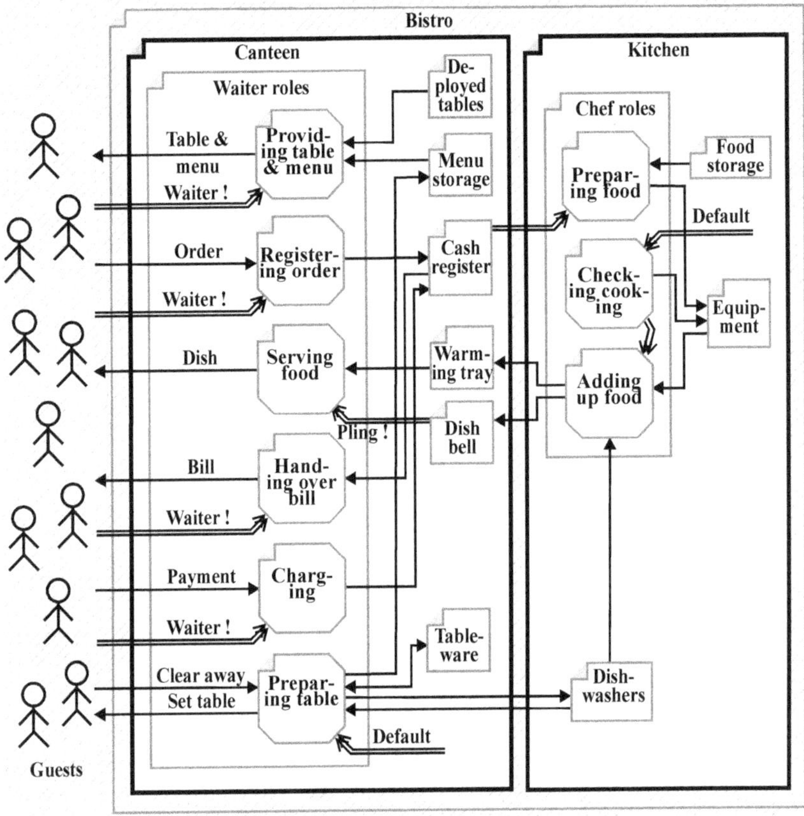

FIGURE 9-50 Multiplied resources to manage limited availability

The *asynchronous interfaces* are the various places for storage buffers:

- Menu and table storage
- Cash register with order and billing capability
- Warming tray for just-prepared dishes
- Dishwasher and storage for tableware
- Food storage for raw material
- Equipment with utensils for cooking

9.18.3 EXAMPLE: Telephone subscriber instance thread

The principle *architecture* of an old-fashioned telephone switch is illustrated in Figure 9-51 below. (The *behavior requirements* were presented in Fig. 7-34, p. 258.)

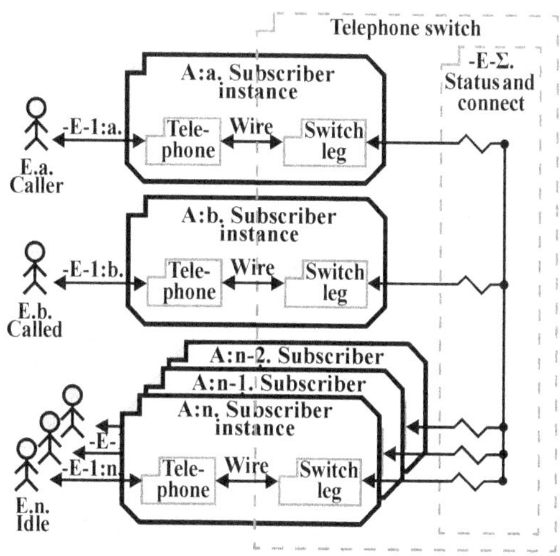

FIGURE 9-51 Subscriber idle-busy concept to manage a synchronous interface between caller and called.

The *asynchronous interface* between all subscriber instances is illustrated on a principle level, and to be fully understood certainly needs more details than are drawn in this schematic *architecture*. The *threads* are captured by only one coherent *scenario* (see Fig. 7-34, p. 258) describing predictable stimuli from only one subscriber or from the telephone switch. This kind of *thread* may be called *scenario-thread-machines*, and take care of complexity of a system with a lot of asynchronous stimuli.

9.19 EXAMPLE: Architecture ID numbering used throughout this book

In the **Pt.R. Capture staffing & requirements** *phase*, a lot was explained about *requirements* hierarchies, with proposals of how to assign each *requirement* a unique ID. The same goes for *architecture* hierarchies, and how to provide ingredients with a unique ID (see the assumed *logical architecture* in Fig. 9-52 below)

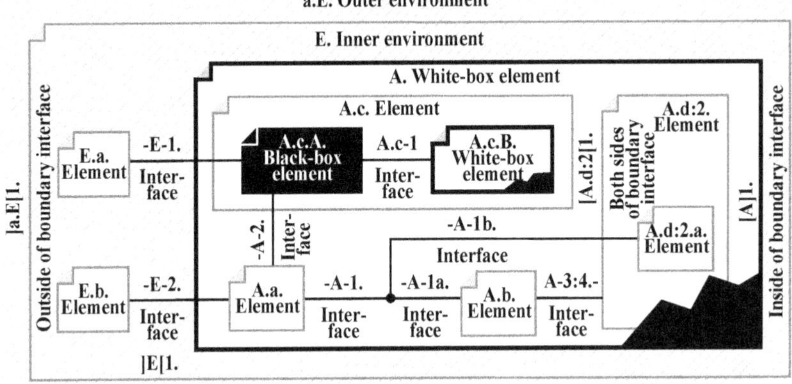

FIGURE 9-52 Architecture identification example

It may seem overkill to have unique identification numbers on all *architecture ingredients*. For small prototypes maybe it is, but as soon as they grow, the *formalism* must increase faster than *complexity* grows. To try saving money on insufficient *formalism* very soon results in lost control, with bigger costs than what was saved.

If a genius developer creates a *complex product* that fits into his or her head in its entirety, there is not so much need for formalistic documentation. But the day the genius for any reason is no longer present, nothing but the prototype itself is left for the developers taking over. If it is a *program* prototype, it might be a million lines of *program instructions* that again have to be understood.

Boundary interfaces are the trickiest to handle. The most normal case is that the *boundary interface* (both its inside and outside) belongs to the *element* limited by the boundary. For *Cpdm* completeness, the boundary may alternatively belong to the surrounding *element* of the boundary or to the *element* demarcated by the boundary.

a.E. Outer environment. This is the surroundings to the **E. Inner environment** below.

]a.E[1. Outside of boundary interface. This is the outside *boundary interface* of the **E. Inner environment** *element* belonging to the **a.E. Outer environment,**

because the outside of an *element* belongs to the surrounding *element*. If outside and inside are not specified, the *interface* may belong to either the *element* itself, or to the surrounding *element*, depending on whether the *interface* is more closely associated with the surroundings or with the *element* itself.

]E[1. Inside of boundary interface. This is the inside *boundary interface* of the **E. Inner environment** *element* (exactly the same *boundary interface* as above), but belonging to the **E. Inner environment** *element*.

E. Inner environment. This is the surrounding *element* to the **A. White-box element** below.

E.a. Element and E.b. Element. These are *elements* inside the **E. Inner environment** *element*.

-E-1. Interface and -E-2. Interface. These are *connection interfaces* with their outermost ends within the **E. Inner environment**.

A. White-box element (the product to develop). This is the *outermost white-box* identified when performing the **Pt.0.A. Constitute environment architecture**.

[A]1. Inside of boundary interface. This is the inside boundary of the **A. White-box element** belonging to this *element*.

-A-1. Interface and -A-2. Interface. These are *interfaces* with their outermost ends (in fact, both ends) within the **A. White-box element**.

-A-1a. Interface and -A-1b. Interface. These are forks of the **-A-1. Interface**.

A-3:4. - Interface. This is the fourth *variant* of an *interface* with its outermost end within the **A. White-box element**.

A.a. Element, A.b. Element and A.c. Element. These are embedded *elements* in the **A. White-box element**.

A.c.A. Black-box element and A.c.B. White-box element. These are *embedded black-/white-boxes* within the **A.c. Element**.

A.c-1 Interface. This is an *interface* with its outermost end within the **A.c. Element**.

A.d:2. Element. This is the second *variant* of an embedded *element* in the **A. White-box element**.

A.d:2.a. Element. This is an embedded *element* within the **A.d:2. Element**.

[A.d:2[1. Both sides of boundary interface. This is both outside and inside of the *boundary interface* decided to belong to the **A.d:2. Element**.

9.20 EXAMPLE Calc_logic environment (the pocket calculator firmware): Constitute environment architecture

9.20.1 Calc_logic: About this example

Now the two *environments* to calc_logic can be revealed that allow calc_logic to be demarcated. These *architectures* map a lot of the *environment,* including the end *user,* and will be extremely useful and informative for new or current developers, as well as to design the internal *architecture* of calc_logic, which is now heavily supported.

The need for reader support will also be cut down; for example, not all *results demanded* for the *activities* will be explicitly listed. A look at the overall *schedule* in Figure 9-53 below or in **Pt:2:3. Tailored advanced technical schedule n = 2,** page 169, shows *masters* which *govern results* possibly to *demand.*

9.20.2 EXAMPLE: Calc_logic environment (the pocket calculator firmware): Process schedule to use

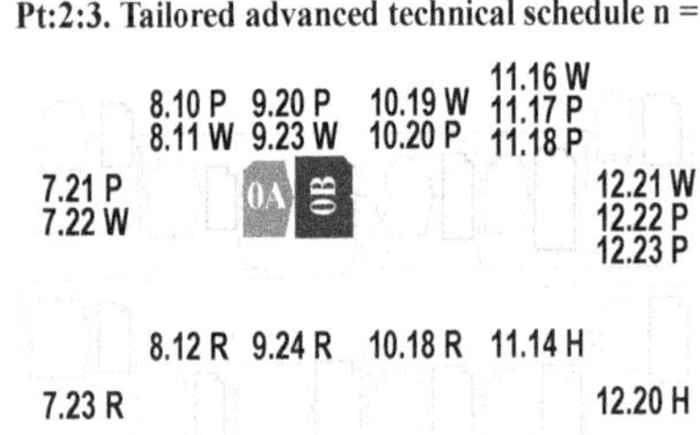

For the calc_logic example, a special technical overall *schedule* is *tailored* (illustrated in Fig. 9-53, left):

• **Pt:2:3. Tailored advanced technical schedule n = 2,** page 169

In this *tailored schedule* the generic *schedule* to use is (filled-in symbols):

• **Pt.0.A. Constitute environment architecture,** page 115.

FIGURE 9-53 Position of schedule and master to use (filled in) in overall schedule

9.20.3 EXAMPLE Calc_logic environment (the pocket calculator firmware): Identify environment architecture ingredients

A stand-alone calculator is, of course, much simpler than using a full personal computer, but on the other hand, in this example it must be developed totally from scratch. The in-house project "de1" provides a good *architecture resumé* and drawing of the pocket calculator, which holds the calc_logic deep down in surrounding structures (see Table 9-30 and Fig. 9-54, p. 410 below).

TABLE 9-30 Calc_logic environment (the pocket calculator firmware) environment architecture ingredients resumé

Existing and product interface ingredient		Explanation or chosen solution alternative or sample thereof
a.aaa.E.a.	Calculator user	The person using the calculator
-a.aaa.E-1.	Button pressing	The finger pressing the button
-a.aaa.E-2.	Display showing	The eyes reading the displays
-a.aaa.E-3.	Battery opening	The opening to use when changing batteries
a.aaa.E.	Pocket calculator environment	Everything surrounding the pocket calculator device
aaa.E.a.	Battery	Conducts electricity to all electronics.
-aaa.E-1.	Wires	Convey electricity from batteries to circuit board.
aaa.E.	Pocket calculator	The pocket calculator device
aa.E-a.	Button	The buttons on the calculator panel
aa.E-b.	8-segment LED	The LED displays in the LED window
aa.E.	Circuit board	One board holding all electronic elements
-aa.E-1.	Key port	The microcontroller port to read buttons
-aa.E-2.	Interrupt port	The microcontroller interrupt port
-aa.E-3.	LED nr port	The microcontroller port to select LED number
-aa.E-4.	LED seg port	The microcontroller port to select LED segment
a.E.a.	Timer interrupt generator	Setup on chip clock to trig interrupt for loading product values to the LED lamps
a.E.	Microcontroller	Single-chip microcontroller
E.	Calc_logic environment (the pocket calculator firmware)	The entire program in the pocket calculator
E:2.	Calc_logic environment (the development board fixture firmware)	The firmware reading the development board buttons (mounted on the experimental area), calling calc_logic, and showing the result on the displays (mounted on the experimental area)
-E-1.	Calc_logic interface to pocket calculator program	Product interface to be developed E.SL1. Only the 16 most common calculator stimuli are expected to enter calc_logic. E.SL2. Proposed are digits 0 - 9, decimal point, equal sign, subtraction, addition, multiplication, and division. E.SL3. There is no need to respond with more than 8 digits from calc_logic. E.SL4. Only numeric responses must appear. E.SL8. Porting calc_logic to a microcontroller experimental board supporting $\approx$ 7 digits floating-point precision.
E.a.	Keyboard and LED driver program	All programs surrounding "Calc_logic for reuse"
-E.a-1.	Button buffer	Storage for buttons pressed, until taken care of by the "LED interrupt thread"
E.aa.	Program initiator	Initiation of interrupt addresses etc.
E.ab.	Button interrupt thread	Real-time program thread for reading button pressings
E.ac.	LED interrupt thread	Real-time program thread for driving the LED displays
A.	Calc_logic	Product black-box to be developed

9.20.4 EXAMPLE Calc_logic environment (the pocket calculator firmware): Arrange ingredients and identify black-box

The *architecture* shows the *environment* to Calc_logic *black-box*, which is the *product* to be developed (see Fig. 9-54 below).

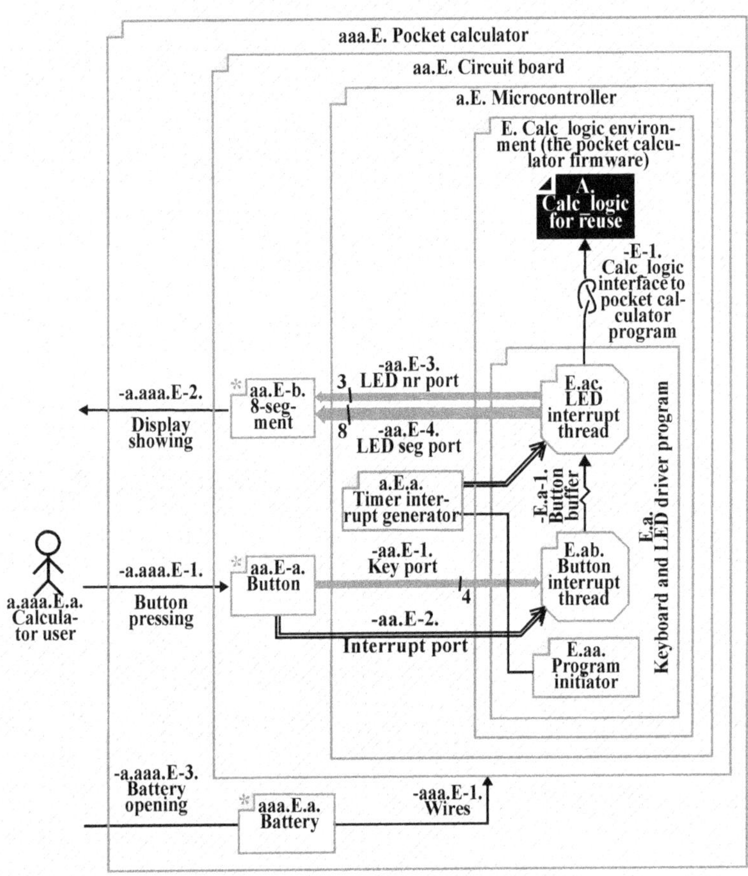

FIGURE 9-54 calc_logic environment logical architecture

Some explanations of the *environment logical architecture*:

- The keyboard buttons are connected to a microprocessor input port, but also to an *interrupt* input port. Without the *interrupt* input, the *program* has to poll the input port many times per second. With *interrupt*, the *program* reads which key was pressed only after a pressing button *interrupt* has been received.

- The LED display driver *program* has no separate outputs for every LED. That would require 7 outputs per digit times 16 digits = 14 ports each 8 bits wide, which would make the microcontroller bulky and expensive. By using port 3 to explicitly address one of the 16 LEDs in the display, and by using port 2 to address the explicit segment of the 7-segment-LEDs, all segments of all digits can be illuminated for a short time. If this is repeated often enough, the human eye doesn't discover that the LED segments are pulsating.

- The calculator logic *program* is the core of the *program*, receiving button pressures from the dispatcher, then performing all calculations, and finally delivering the output to the driver to present it at the display.

9.20.5 EXAMPLE Calc_logic environment (the pocket calculator firmware): Lay out physical architecture

The appearance of the stand-alone pocket calculator, which *reuses* the calc_logic *element*, is shown in Figure 9-55 at left. Only the most necessary buttons are present, and only 8 digits and 1 decimal point are used.

The *reusable item* calc_logic must handle both this limited man-machine *interface* and a richer Windows man-machine *interface* without significant rework.

FIGURE 9-55 calc_logic environment physical architecture

9.20.6 EXAMPLE Calc_logic environment (the pocket calculator firmware): Examine product context

Calc_logic has long ago been decided to be programmed in C with acceptable floating-point support. A discussion was held about which language to *program* the E.a. Keyboard and LED driver program in.

When *porting* a *high-level program* into assembler surroundings, that is asking for *interface* problems. However, this mix between C and assembler is rather optimal. The problems would not have been less if calc_logic had been written in the assembler, or if the E.a. Keyboard and LED driver program had been written in C. The decision was made to *program* the E.a. Keyboard and LED driver program in the assembler.

9.20.7 EXAMPLE Calc_logic for reuse: Proceed reading

To follow this example, these are the alternatives:

- Follow the calculator to chapter 7.23, "EXAMPLE Calc_logic for reuse: Refine product requirements from environment restriction requirements and ensure black-/white-box development staffing," page 243.
- Follow the calculator to chapter 9.24, "EXAMPLE Calc_logic for reuse: Satisfy refined requirements by decomposing product black-box into white-box design containing no black-boxes," page 419.
- Follow the calculator to chapter 10.20, "EXAMPLE Calc_logic environment (the pocket calculator firmware) development board and pocket calculator: Finalize design of environment interfaces with product requisites," page 549.
- Follow the calculator to chapter 11.17, "EXAMPLE Calc_logic environment (the development board fixture firmware): Realize interfaces to environment and insert/await outermost white-box," page 649.
- Follow the calculator to chapter 12.22, "EXAMPLE Calc_logic environment (the pocket calculator firmware) on development board: Verify prototype in environment," page 761.
- Proceed to chapter 9.21 below to leave the calc_logic example and to dig deeper into the **Pt.A. Design architectures** topic.

9.21 EXAMPLE Central Processing Unit (CPU)

Microcontrollers have almost everything needed on a single chip, making a complete computerized *product* cheap and simple to develop. When pushing out as many *elements* as possible in a microcontroller to become external devices, the central chip controlling the external chips can be much more specialized and powerful with wider buses. This type of computer chip is called Central Processing Unit (CPU) (see Fig. 9-56 below)

There must always be some permanent *programs* like BOOT (initial loader of the operating system) and BIOS (Basic Input/Output System) in such a chip-set, performing the initial start-up and low-level communication of the computer. When started up, the operating system can load application *programs* from a hard disk attached to the ports for peripheral devices.

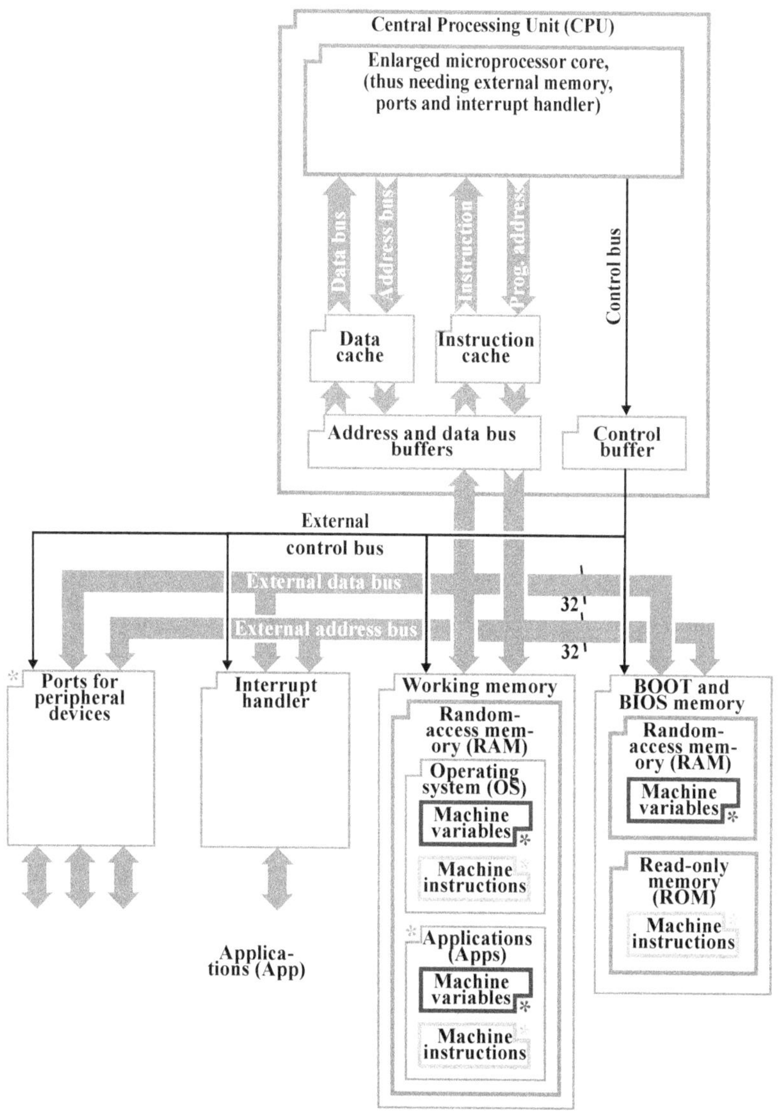

FIGURE 9-56 Chip-set with central processing unit and its external chips

9.22 High-level program architecture

9.22.1 Cpdm definition

In *Cpdm, global variable, function, local variable,* and *instruction* are defined according to Table 9-27 below. For more details, see chapter 10.17, "Structured programs," page 529.

TABLE 9-31 Cpdm definitions

Aspect	High-level program architecture ingredients			
	Global variable	Function	Local variable	Instruction
Definition	Storage of stimuli history which are accessible from any function	Encapsulation of local variables and instructions	Storage of stimuli history, which are accessible only within the function	Executable command, exclusively consisting of the types: sequence, selection, iteration, or call to functions
Synonyms	External data, global environment variable	Procedure, routine, subroutine, subprogram	Internal data, local environment variable	High-level instructions, structured instructions
Symbol				

9.23 EXAMPLE Calc_logic environment (the Windows calculator application): Constitute environment architecture

9.23.1 EXAMPLE: Calc_logic environment (the Windows calculator application): Process schedule to use

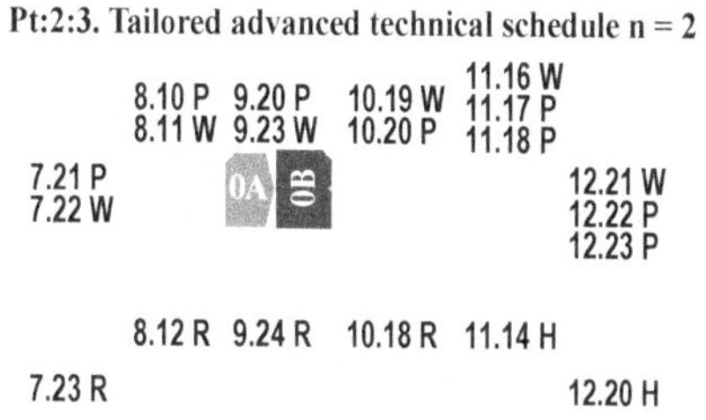

For the calc_logic example, a special technical overall *schedule* is *tailored* (illustrated in Fig. 9-57, left):

- **Pt:2:3. Tailored advanced technical schedule n = 2,** page 169

In this *tailored schedule* the generic *schedule* to use is (filled-in symbols):

- **Pt.0.A. Constitute environment architecture,** page 115.

FIGURE 9-57 Position of schedule and master to use (filled in) in overall schedule

9.23.2 EXAMPLE Calc_logic environment (the Windows calculator application): Identify environment architecture ingredients

A personal computer is a *complex* system, which covers the distance from the computer *user* through input-output *elements*, through the bios and through Windows *elements*, through the application *element* and finally ending up with the small calc_logic *element* integrated in it (see Table 9-32 and Fig. 9-58, p. 417 below).

If possible, an *environment architecture* should contain the *user* of the *product* illustrated, because it gives a much better *customer-oriented* understanding. Seeing the *user* in the *environment*, and being able to track from there all the way through the *environment architecture* up to the *black-box* to develop, gives a first-class understanding of interaction between the *user* and the *black-box*.

TABLE 9-32 Calc_logic environment (the Windows calculator application) environment architecture ingredients resumé

Existing and product interface ingredient		Explanation or chosen solution alternative or sample thereof
aa.aaa.E.	Personal computer environment	Existing environment to the personal computer
a.aaa.E.	Calculator user	Operator of the personal computer
b.aaa.E.	Screen	Screen attached to the personal computer
c.aaa.E.	Keyboard	Keyboard attached to the personal computer
d.aaa.E.	Mouse	Mouse attached to the personal computer
e.aaa.E.	Personal computer	Personal computer with Windows operating system
aaa.E.	Graphic plug-in card	Driver plug-in card for screen
baa.E.	Motherboard	Main circuit board
aa.E.	Working memory	Working memory of the personal computer
ba.E.	Main processor	Processor executing programs' working memory
ca.E.	Math coprocessor	Supporting processor to calculate floating-point
da.E.	PCI buffer	Interface to plug-in cards
ea.E.	Buffer	Buffer to external keyboard
fa.E.	Buffer	Buffer to external mouse
a.E.	Windows operating system	All programs belonging to the Windows operating system
a.E.a.	Windows disk and input/output system	Windows operating system supporting applications
a.E.aa.	User mode services	High-level operating system support
a.E.ab.	Executive mode services	Low-level operating system support
E.	Calc_logic environment (the Windows calculator application)	The calculator application program.
E.	Windows calculator application	
-E-1.	Calc_logic interface to Windows calculator program	Product interface to be developed E.SL10. Calc_logic shall expect a large number of different stimuli. E.SL11. Calc_logic must respond with a large number of digits and at least more than accuracy requires. E.SL12. Text response is also conceivable. E.SL17. Port calc_logic to a Windows application development system supporting at $\approx$ 15 digits floating-point precision.
Ea.	Application support initially generated by visual studio	When setting up a graphic Windows application project, these files are provided as a beginning framework by the developing system.
E.aa.	WinMain	Function receiving control when application starts
E.ab.	MyRegisterClass	No change needed by calc_logic
E.ac.	InitInstance	No change needed by calc_logic
E.ad.	WndProc	Communication between application and Windows
E.ada.	Case WM_CREATE	Test calculator graphics defined
E.adb.	Case WM_COMMAND	Callback when calculator buttons are clicked
E.adc.	Case WM_PAINT	No change needed by calc_logic
E.add.	Case WM_DESTROY	No change needed by calc_logic
E.ade.	About	Information if Windows About is clicked
A.	Calc_logic	Product to be developed

9.23.3 EXAMPLE Calc_logic environment (the Windows calculator application): Arrange ingredients and identify black-box

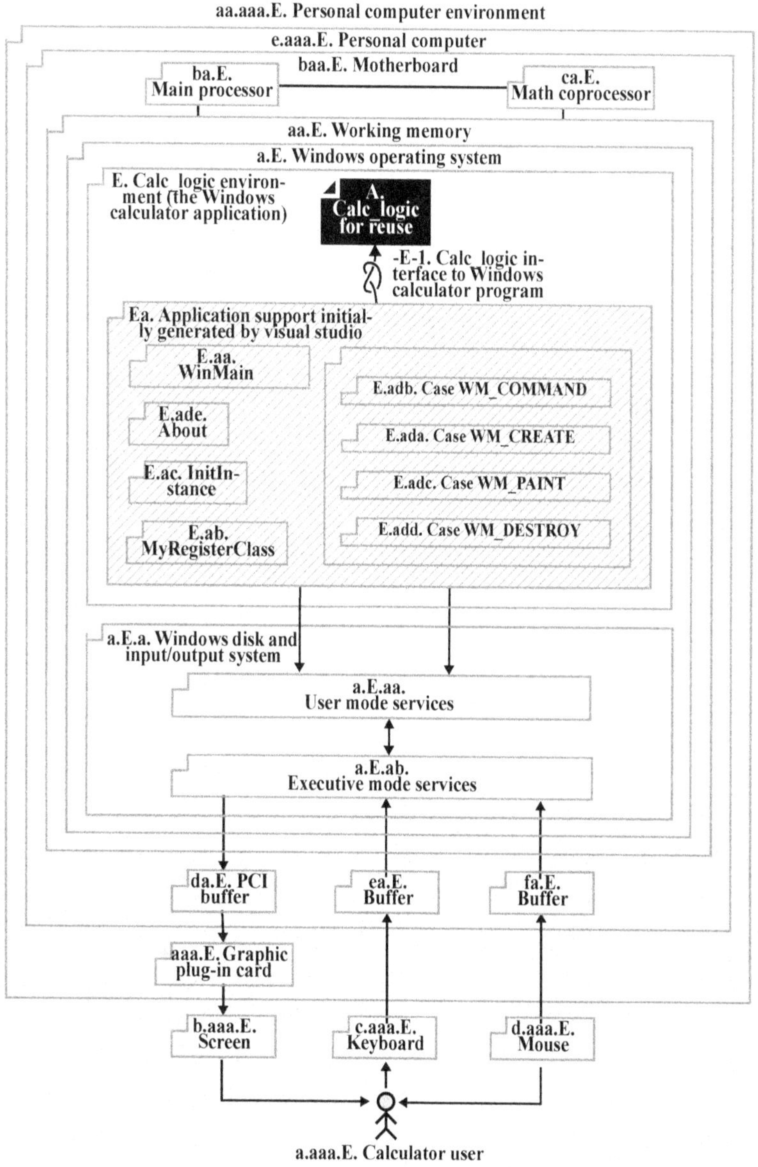

FIGURE 9-58 calc_logic environment logical architecture

9.23.4 EXAMPLE Calc_logic environment (the Windows calculator application): Lay out physical architecture

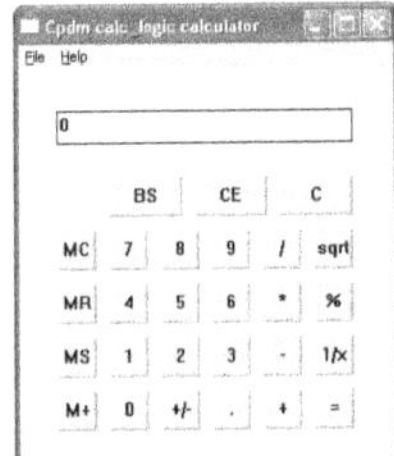

Shown in Figure 9-59 at left is the appearance of the graphic window to the Windows calculator application, which *reuses* the calc_logic *black-/white-box*. A larger set of buttons is present, and the illustrated textual output field holds a large sum of alphanumeric characters, to be taken care of by calc_logic, but not necessarily implemented in the first *version*.

FIGURE 9-59 calc_logic environment physical architecture

9.23.5 EXAMPLE Calc_logic environment (the Windows calculator application): Examine product context

The gate meeting was a bit worried about the very different *environment* of the pocket calculator and the Windows application. Engineers agreed that the *interface* to calc_logic will be a challenge for the *program* reusing it, but finally all agreed that this was the best overall *reuse* solution.

9.23.6 EXAMPLE Calc_logic for reuse: Proceed reading

To follow this example, these are the alternatives:

- Proceed to chapter 7.23, "EXAMPLE Calc_logic for reuse: Refine product requirements from environment restriction requirements and ensure black-/white-box development staffing," page 243.

- Proceed to chapter 9.24, "EXAMPLE Calc_logic for reuse: Satisfy refined requirements by decomposing product black-box into white-box design containing no black-boxes," page 419.

- Proceed to chapter 10.19, "EXAMPLE Calc_logic environment (the Windows calculator application): Finalize design of environment interfaces with product requisites," page 545.

- Proceed to chapter 11.16, "EXAMPLE Calc_logic environment (the Windows calculator application): Realize interfaces to environment and insert/await outermost white-box," page 646.

- Proceed to chapter 12.21, "EXAMPLE Calc_logic environment (the Windows calculator application): Verify prototype in target environment," page 758.

- Proceed to chapter 9.21 below to leave the calc_logic example and to dig deeper into the **Pt.A. Design architectures** topic.

9.24 EXAMPLE Calc_logic for reuse: Satisfy refined requirements by decomposing product black-box into white-box design containing no black-boxes

9.24.1 EXAMPLE: Calc_logic for reuse: Process schedule to use

Pt:2:3. Tailored advanced technical schedule n = 2

For the calc_logic example, a special technical overall *schedule* is *tailored* (illustrated in Fig. 9-60, left):

- **Pt:2:3. Tailored advanced technical schedule n = 2,** page 169

In this *tailored schedule* the generic *schedule* to use is (filled-in symbols):

- **Pt.n.A. Satisfy requirements by decomposing outward nesting level black-box into white-box design containing possible black-boxes,** page 136.

FIGURE 9-60 Position of schedule and master to use (filled in) in overall schedule

9.24.2 EXAMPLE Calc_logic for reuse: Identify white-box architecture ingredients

There are not many restrictions for designing the calc_logic *program*, thus resulting in a rather free design of *unbiased architecture ingredients*, provided it fulfills the specified behavior. However, there is one solution alternative stipulating to use *wait-states* and *use-units* already defined in the *formal scenario* in Figure 7-28, page 247.

The *architecture* contains mainly *high-level program* ingredients as already defined (see Table 9-31, p. 414).

The *architecture resumé* is prepared in Table 9-33 below. As many *program* ingredients as possible are fetched from the *wait-states* and *use-units* in Figure 7-28, page 247, which gives an excellent *traceability* from the *formal scenario requirement* to the *instruction architecture ingredients* realizing the *requirement*.

TABLE 9-33 Calc_logic for reuse architecture ingredients resumé

	White-box ingredient		Chosen solution alternative or sample thereof
A.	Calc_logic for reuse		E.SL6.b. Microcontroller development system with ≈ 7 digits float support
		The subroutine to develop for both stand-alone pocket calculator and Windows calculator	E.SL14.c. Windows development system with ≈ 15 digits floating-point
A.a.	# Includes	Inclusion of restrictions and interfaces files	A.SL1. Calc_logic must internally handle 64 buttons.
A.b.	Aliases, etc.	Declaration of aliases, compiler directives, etc.	A.SL2. Up to 128 digits must be the internal capacity, even if not all are accurate.
A.c.	Global variables	Developed global variables	
A.d.	Calc_logic main function	Entrance when calling calc_-logic	A.SL3. Handling must include both numeric and non-numeric characters.
A.da.	Local variables	Developed local variables	
	Calc_logic program instructions	Instructions of the calc_-logic function	
A.db.	Wait-for-lead-opd-1st-digit instructions	Executed when 1st lead digit awaited	
A.dba.	Input-1st-digit-of-lead-opd instructions	Executed if digit stimulus	
A.dbb.	Prepare-continuation-after-conclusion instructions	Executed if operator stimulus	
A.dc.	Wait-for-further-digit-or-dual-opr instructions	Executed when more digits or operator awaited	
A.dca.	Input-rest-of-lead-opd instructions	Executed if digit stimulus	A.SL7. Make implementation in C, and use scenario wait-states and use-units in program.
A.dcb.	Input-dual-opr instructions	Executed if operator stimulus	
A.dd.	Wait-for-trail-opd-1st-digit instructions	Executed when 1st trail digit awaited	
A.dda.	Input-1st-digit-of-trail-opd instructions	Executed if digit stimulus	
A.de.	Wait-for-further-digit-or-eq-sign instructions	Executed when more digits or equal sign awaited	
A.dea.	Input-1st-digit-of-trail-opd instructions	Executed if digit stimulus	
A.deb.	Do-conclusion instructions	Executed if equal stimulus	
A.dec.	Do-conclusion-with-continuation instructions	Executed if operator stimulus	

9.24.3 EXAMPLE Hosted calc_logic for reuse: Identify white-box architecture ingredients

TABLE 9-34 Calc_logic for reuse extended architecture ingredients resumé

White-box ingredient		Chosen solution alternative or sample thereof	
E.	Any calc_logic environment	The reusable calc_logic subroutine environment	E.SL15. Only a recompilation and change of header files must be necessary when changing float precision.
-E-1.	Common calc_logic interface	The reusable calc_logic subroutine environment interface, being the common superset of all possible reuse needs	
E.b.	Built-in standard library interface	E.SL5. Don't consider using any mathematical chip.	
E.b.	Development system built-in standard library	E.SL6.b. Microcontroller development system with ≈ 7 digits float support E.SL7.b. Microcontroller development system built-in floating-point libraries E.SL13.b. Emulate the mathematical co-processor with a program. E.SL14.c. Windows development system with ≈ 15 digits floating-point E.SL16.b. Windows development system built-in standard libraries	

9.24.4 EXAMPLE Calc_logic for reuse: Arrange ingredients and identify possible black-boxes

Now the *architecture resumé* ingredients can be arranged into a calc_logic *logical architecture* (see Fig. 9-61 below).

How to design and how not to design a *structured program* is a never ending story. Languages are not necessarily defined out of purely architectural considerations, and how to use offered language structures is in no way obvious (like playing the Tetris game by combining infinitely many odd pieces to a skyscraper). As said before, this book is hardly a design guide to smart *programs*, but a guide to organizing *product development*.

To arrange the *architectures* according to *wait-state*s and *use-units* is very rare. However, when the *program* is short and needs not be heavily *modularized*, this approach gives fantastic *intellectual control*, and hopefully some aha moments to my readers.

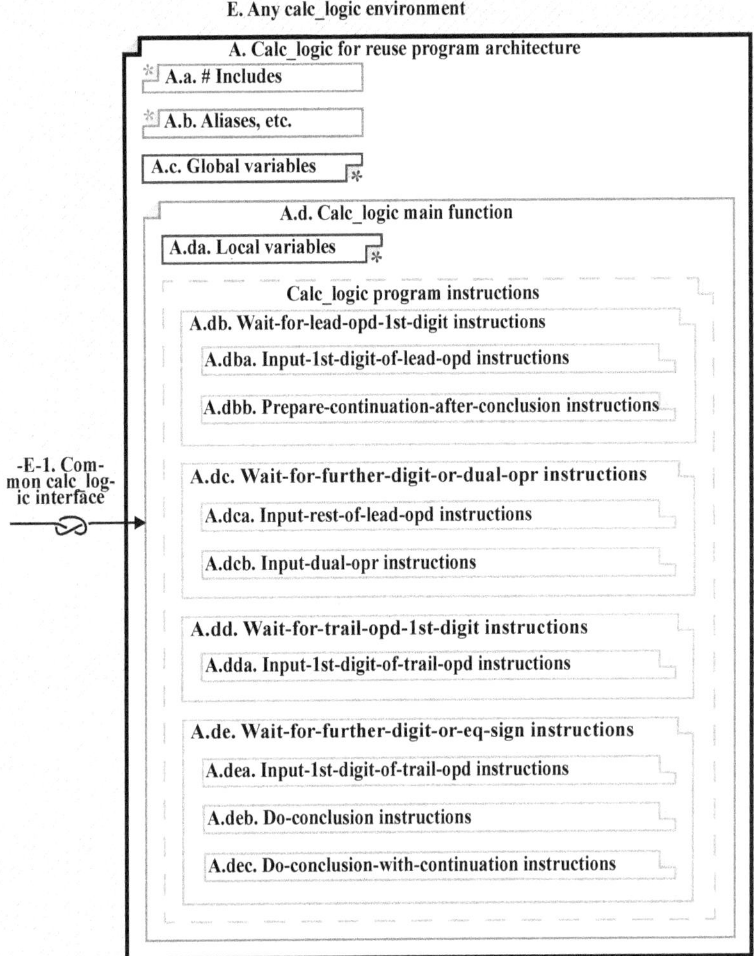

FIGURE 9-61 Calc_logic for reuse logical architecture

9.24.5 EXAMPLE Calc_logic for reuse: Examine architectures

The *architecture resumé* and the derived *logical architecture* have been scrutinized a lot by the in-house project "dg3" members. There has been a lot of discussion about whether such an awkward design (structured according to *formal scenario black-box wait-states* and *use-units*) is really beneficial and even if it will work at all.

The *quality* of a calculator and its capability to always calculate correctly is essential for the *product* to survive on the market. The calculator *requirements* are understandable for everybody and it is obvious that the *requirements* are under *intellectual control.*

The decision gate members now acquired the book on *Cpdm.* The decision was to continue keeping *intellectual control* when designing *program architectures* also, and give applying *wait-states* and *use-units* in the *program* design a try.

9.24.6 EXAMPLE Calc_logic for reuse: Proceed reading

To follow this example, these are the alternatives:

- Proceed to chapter 10.18, "EXAMPLE Hosted calc_logic for reuse: Finalize design items of product white-box with product requisites," page 533.
- Proceed to chapter 11.14, "EXAMPLE: Hosted calc_logic for reuse: Procure items and linkable programs in order to realize white-box," page 644.
- Proceed to chapter 12.20, "EXAMPLE Hosted calc_logic for reuse: Verify black-/white-box," page 733.
- Proceed to chapter 9.25 below to leave the calc_logic example and to dig deeper into the **Pt.A. Design architectures** topic.

9.25 Object-oriented program architecture

9.25.1 Cpdm definition

In *Cpdm*, *object* is defined according to Table 9-35 below. For more detailed information about *object-oriented programs*, see chapter 10.22, page 554.

TABLE 9-35 Cpdm definitions

	Object-oriented program architecture ingredients
Aspect	**Object**
Definition	Element created run-time by instantiation of a class specification. Objects may in turn contain objects, thereby enabling objects to be hierarchically organized.
Synonyms	Item, thing, body
Symbol	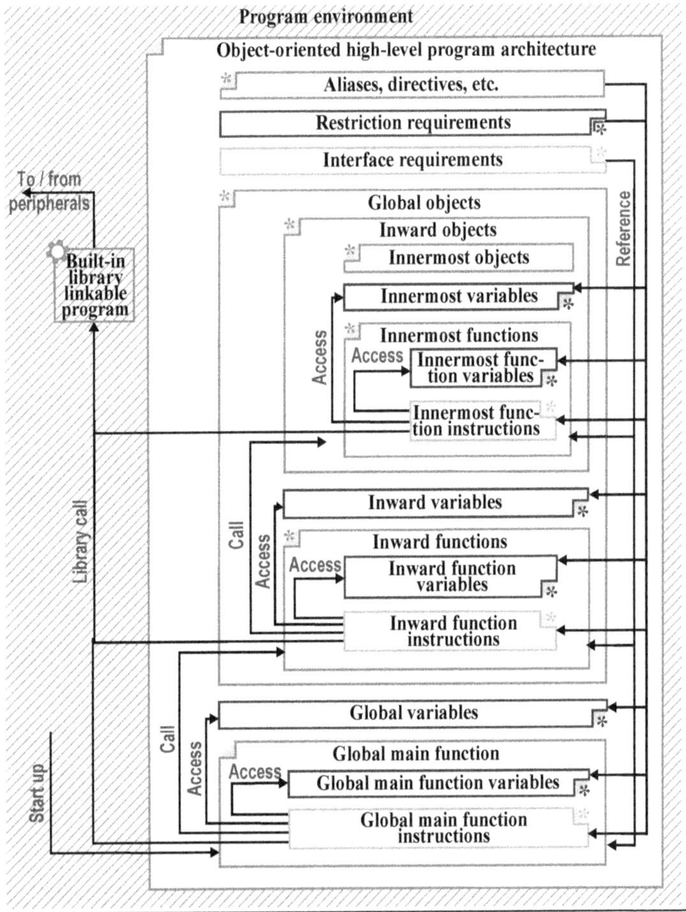

Objects used in *variables* create nested *objects*, which form hierarchical *architectures*, and may be modeled as *embedded black-boxes* in the same way as in the house example. Many programmers and *managers* imagine a *program* to be a huge flat extent without structure, impossible to *partition* and layer, illustrated as a Russian doll. Structured *programs* often support creation of hierarchical *variables*, but *object-oriented programs* support creation of hierarchical structures for the entire *program*.

When a *class* is now defined, it can be used in the same way as any language's built-in types like char, int, and float. Using a *class* like this is called to instantiate the *class* into an *object*.

More details on how to finalize *classes* and *objects* can be found in chapter 10.22, "Object-oriented programs," page 554.

9.26 EXAMPLE Phonebook environment: Constitute environment architecture

9.26.1 EXAMPLE Phonebook environment: Process schedule to use

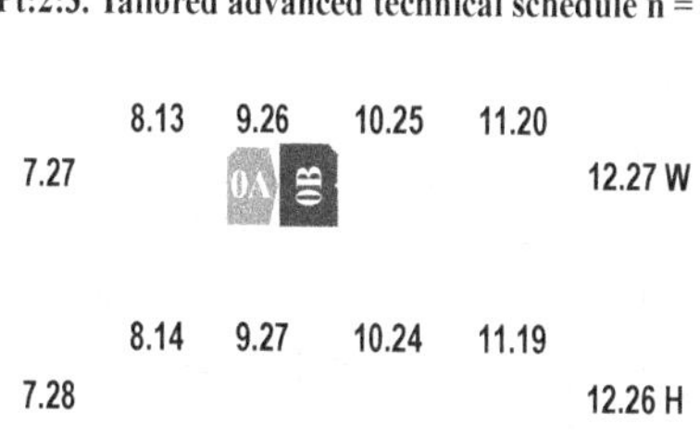

For the phonebook example, a special technical overall *schedule* is *tailored* (illustrated in Fig. 9-62, left):

• **Pt:2:3. Tailored advanced technical schedule n = 2,** page 169

In this *tailored schedule* the generic *schedule* to use is (filled-in symbols):

• **Pt.0.A. Constitute environment architecture,** page 115.

FIGURE 9-62 Position of schedule and master to use (filled in) in overall schedule

9.26.2 EXAMPLE Phonebook environment: Identify environment architecture ingredients

From the *solution alternatives*, an *environment architecture resumé* is prepared to describe ingredients that must match solutions or *samples* (see Table 9-36 below).

TABLE 9-36 Phonebook environment architecture ingredients resumé

	Existing and product interface ingredient	Explanation or chosen solution alternative or sample thereof
aaE.	Computer environment	Environment to the computer with computer peripherals and a user
aaE.a.	Phonebook user	The user of the phonebook program
-aaE-1.	User interface	The man-machine interface between the user and the peripherals
aE.a.	Motherboard	The main printed circuit board inside the computer
aE.b.	Hard disk	Built-in hard disk in the computer.
aE.ba.	Phonebook file	E.SL3. For accessibility and easy verification, the disk file shall be plain textual and line based, and readable and writable with a simple editor.
aE.c.	Peripherals	Screen, keyboard, and mouse
aE.	Computer	The computer, including mechanics
-aE-1.	Electronics/program interface	Interface between the program and the motherboard electronics
-aE-2.	Disk interface	Interface between disk electronics and motherboard electronics
-aE-3.	Wires	Cables between computer device and peripherals
E.	Phonebook environment	Console and disk program established by the development tool to interface the operating system
E.a.	Main function	The Windows computer with screen, keyboard, and mouse peripherals
E.b.	Standard C++ built-in basic library	E.SL1. Choose any standard Windows development system with built-in libraries to develop the phonebook.
E.c:1.	Standard C++ built-in file library	E.SL3. For accessibility and easy verification, the disk file shall be plain textual and line based, and readable and writable with a simple editor.
E.c:2.	Third-party database library	E.SL5. The phonebook object structure must be grouped in such a way that it is easy to redesign it for an external database.
E.d.	Windows operating system	The Windows operating system, among others, offering functions for interacting with the user via console window
E.da:1.	Console support	E.SL2. For phonebook first evaluation, the command prompt window can be used.
E.da:2.	Graphic support	E.SL4. User textual interaction shall be grouped in such a way that it is easy to redesign it for graphic windows.
-E-1.	Phonebook interface	Interface between the function main and the phonebook program
-E-2:1.	Console interface	Standard library function calls to interact with console
-E-2:2.	Future graphic interface	
-E-3:1.	File interface	Standard library function calls to interact with disk files
-E-3:2.	Future file interface	
-E-4.	Windows interface	Interface between standard library and Windows operating system
A.	Phonebook	The application program to be developed

9.26.3 EXAMPLE Phonebook environment: Arrange ingredients and identify black-box

From the above *architecture resumé*, a *logical architecture* is arranged (see Fig. 9-63 below). As usual, for clarity the *outermost black-box* has identification **A.** and its *environment* always has **E.**

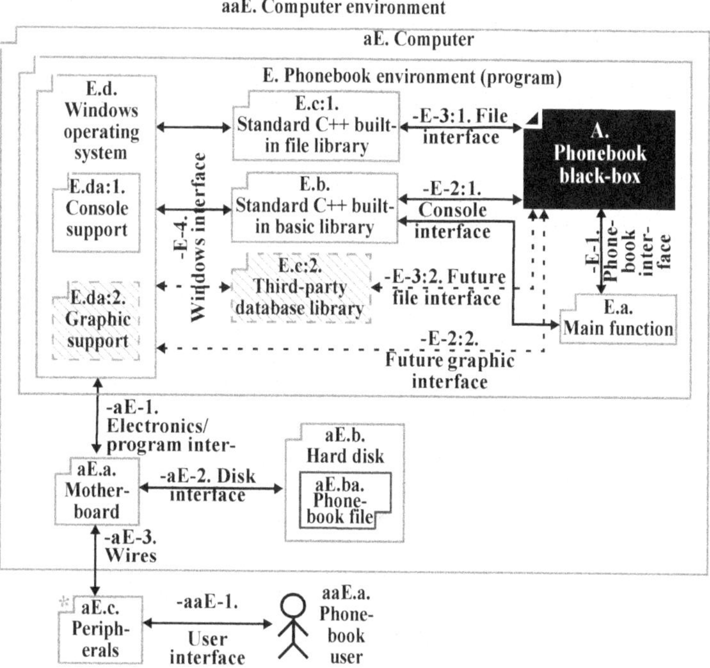

FIGURE 9-63 Phonebook environment logical architecture

9.26.4 EXAMPLE Phonebook environment: Lay out physical architecture

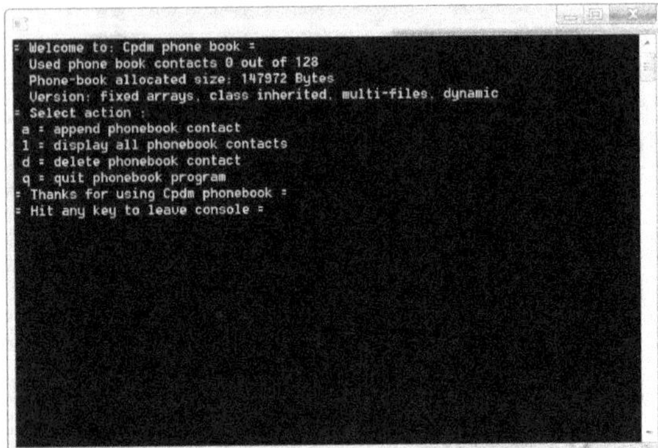

The appearance of the console is designed with simple text (see figure Fig. 9-64, left). This figure gives a flavor of the simple MMI communication between the *user* and the phonebook application.

FIGURE 9-64 Phonebook environment physical architecture

9.26.5 EXAMPLE Phonebook environment: Examine product context

The rather simple *logical architecture* was easily understood by the gate meeting and immediately approved.

The gate members realized that the *user interface* looks like a remnant from the sixties, but the purpose of this *product* is not to evaluate user-friendliness, but to settle a good *architecture* to support heavy growth of the *program* in the future.

9.26.6 EXAMPLE Phonebook: Proceed reading

To follow this example, these are the alternatives:

- Proceed to chapter 10.25, "EXAMPLE Phonebook environment: Finalize design of environment interfaces with product requisites," page 585.
- Proceed to chapter 11.20, "EXAMPLE Phonebook environment: Realize interfaces to environment and insert/await outermost white-box," page 657.
- Proceed to chapter 12.27, "EXAMPLE Phonebook environment: Verify prototype in environment," page 780.
- Proceed to chapter 9.27 "EXAMPLE Phonebook: Satisfy refined requirements by decomposing product black-box into white-box design containing no black-boxes" below.

9.27 EXAMPLE Phonebook: Satisfy refined requirements by decomposing product black-box into white-box design containing no black-boxes

9.27.1 EXAMPLE Phonebook: Process schedule to use

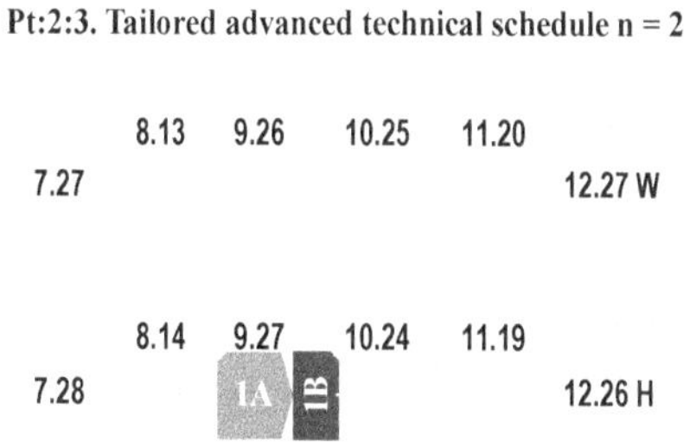

For the phonebook example, a special technical overall *schedule* is *tailored* (illustrated in Fig. 9-65, left):

- **Pt:2:3. Tailored advanced technical schedule n = 2,** page 169

In this *tailored schedule* the generic *schedule* to use is (filled-in symbols):

- **Pt.n.A. Satisfy requirements by decomposing outward nesting level black-box into white-box design containing possible black-boxes,** page 136.

FIGURE 9-65 Position of schedule and master to use (filled in) in overall schedule

9.27.2 EXAMPLE Phonebook: Identify white-box architecture ingredients

Architecture consideration was perhaps one driving force when *object-oriented programs* were defined, but unfortunately, the structures created were not very useful for analysis and design of larger *architectures*. Rather soon the object-oriented definitions were based more on adding glamor to existing *high-level programs*, than on coming up with substantial architectural support to master the massively growing *program*. Typically, when a successful object-oriented design has been made, it still becomes difficult to design accordingly with *object-oriented program* ingredients. For found phonebook ingredients, see Table 9-37 below.

TABLE 9-37 Phonebook architecture ingredients resumé

White-box ingredient		Chosen solution alternative or sample thereof	
A.	Phonebook	Defined in environment architecture resumé	A.SL1. Write program in a C++ object-oriented program language with function calls to libraries for Windows operating system..
A.a.	**Phonebook object**	A.SL4. Use array of objects to store phonebook data.. A.SL5. Implement specified restrictions RR.5.2., RR.5.3., RR.5.4. and RR.5.5. A.SL6. Stimuli may exceed these limits, but the program may not crash for this reason.	
-A-a1.	**Contact interface**	Function call to contact object	
A.xx.	**Phonebook variables and objects**	Variables and objects in phonebook object	
A.aa.	**Contact object**	Above A.SL4., A.SL5., A.SL6.	
-A.aa-1.	**Phone number interface**	Function call to phone number object	A.SL2. The program shall from the start be separated into self-contained items..
-A.aaa-1.	**Person interface**	Function call to person object	
A.xxx.x.	**Contact variables and objects**	Variables and objects in contact object	
A.aaa.	**Person object**	Above A.SL4., A.SL5., A.SL6.	
A.aaa.a.	**Person variables**	Variables and objects in person object	
A.aaa.b.	**Person functions**	Functions in person object	
A.aab.	**Phone number object**	Above A.SL4., A.SL5., A.SL6.	A.SL7. All possible exception use-cases must be implemented with informative user interaction..
A.aab.a.	**Phone number variables**	Variables and objects in phone number object	
A.aab.b.	**Phone number functions**	Functions in phone number object	
A.aac.	**Contact variables**		
A.aad.	**Contact functions**	Functions in contact object	
A.ab.	**Phonebook variables**		
A.ac.	**Phonebook functions, con- and destructors**	A.SL8. Implement in phonebook constructor. A.SL9. Implement in phonebook destructor.	

9.27.3 EXAMPLE Phonebook: Identify white-box architecture ingredients

Object-oriented programs are not the most helpful support to make *programs* more comprehensible (to build a skyscraper from Tetris pieces again comes to mind). Anyway, it is important to *modularize* large *programs* of any language into manageable pieces, to make them much more accessible for a large number of developers working simultaneously, and here object orientation is of some help.

To design the phonebook *program* into a natural structure, the *program* is divided into *objects* for persons, phone numbers, contacts, and the phonebook (see Fig. 9-

66 next).

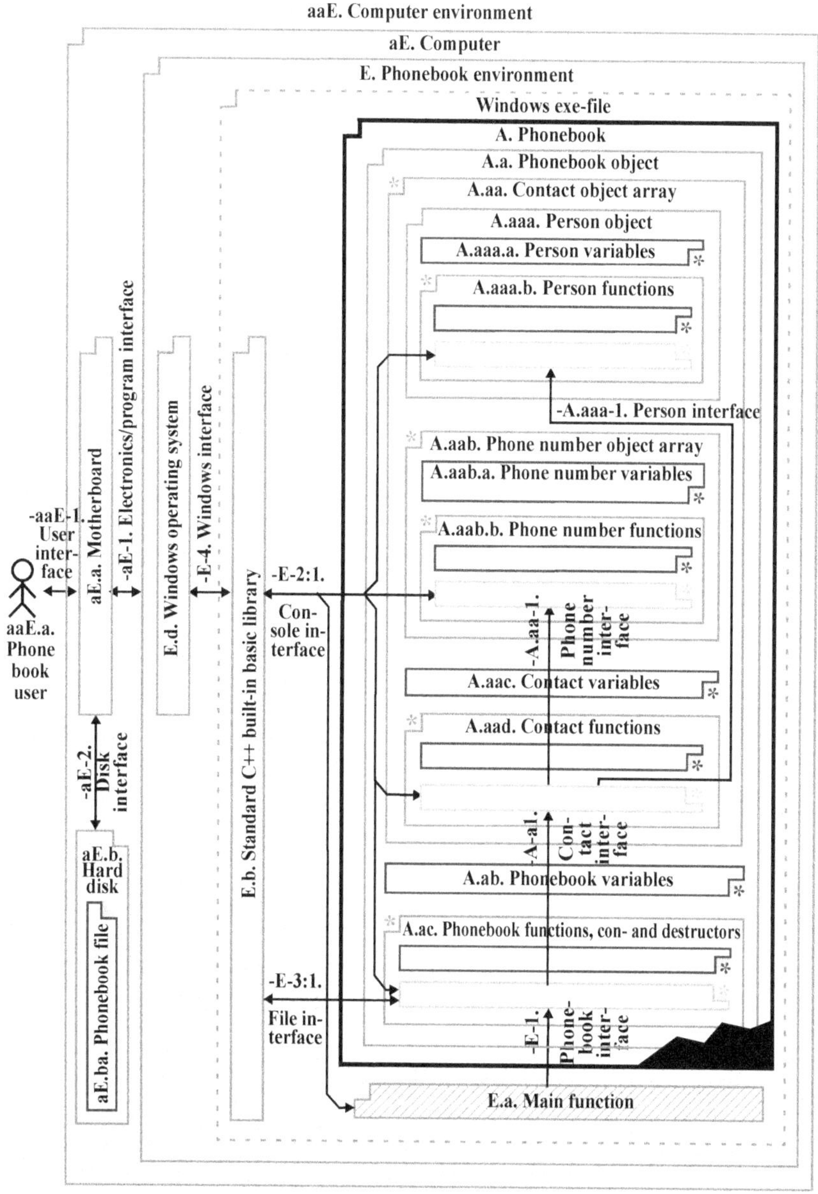

FIGURE 9-66 Phonebook logical architecture

9.27.4 EXAMPLE Phonebook file: Identify white-box architecture ingredients

Between executions of the phonebook *program*, it should be saved on disk. It is best to implement this functionality early on, because it facilitates both development and *verification* to have a lot of saved phonebooks on disk for a variety of purposes. For now, it is sufficient to save the phonebook in an easily edited text format (see Table 9-38 below).

TABLE 9-38 Phonebook file architecture ingredients resumé

White-box ingredient		Chosen solution alternative or sample thereof	
aE.ba.	Phonebook file	Defined in environment architecture resumé	
aE.ba.a.	**Title + newline**	Title of phonebook + separator	
aE.ba.b.	**Newline**	Extra separator	
aE.ba.c.	**Array of contacts**	Phonebook contacts array	
aE.ba.cd.	**Contact object**	Each contact in array	A.SL3. Use line-oriented text streams for saving and loading phonebook on file..
aE.ba.cd.a.	**Name + newline**	Contact name + separator	
aE.ba.cd.b.	**Address + newline**	Address + separator	
aE.ba.cd.c.	**Post code + newline**	Post code + separator	
aE.ba.cd.d.	**City + newline**	City + separator	
aE.ba.cd.e.	**Country + newline**	Country + separator	
aE.ba.ce.	**Array of phone numbers**	Phone number array in end of each contact	
aE.ba.ce.a.	**Phone number + newline**	Phone number + separator	
aE.ba.cf.	**Newline**	End of contact	
aE.ba.d.	**Newline**	End of phonebook	

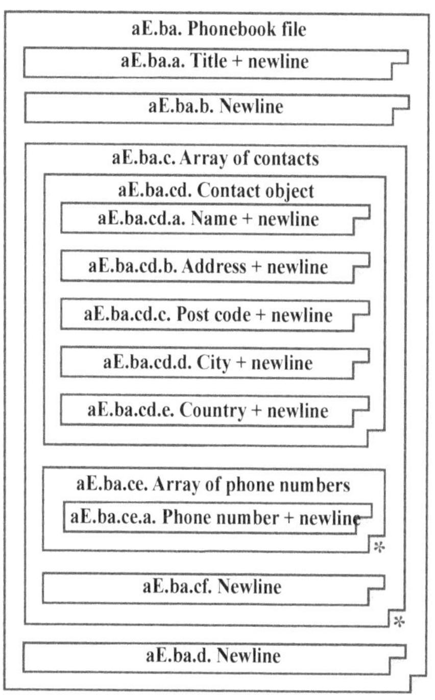

FIGURE 9-67 Phonebook file logical architecture

9.27.5 EXAMPLE Phonebook file: Arrange ingredients and identify possible black-boxes

The disk file *architecture* is hierarchical in the same way as the *objects* in the phonebook *program* (see Fig. 9-66 left).

9.27.6 EXAMPLE Phonebook: Examine architectures

A lot of discussions took place about the hierarchical structure of the *objects*. They could as well have been arranged totally flat.

But since *Cpdm* is more a book on how to develop, and not how to make smart *products*, it is more instructive to show nested *objects*. The gate meeting had to settle for this.

9.27.7 EXAMPLE Phonebook: Proceed reading

To follow this example, these are the alternatives:

- Proceed to chapter 10.24, "EXAMPLE Phonebook: Finalize design items of product white-box with product requisites," page 561.
- Proceed to chapter 11.19, "EXAMPLE Phonebook: Procure items and linkable programs in order to realize white-box," page 655.
- Proceed to chapter 12.26, "EXAMPLE Phonebook: Verify black-/white-box," page 770.

9.28 Large architecture documentation

9.28.1 Traceability between architecture drawings

The IDs of *interfaces* crossing between the outside and inside of the boundary must be specified and visualized on both sides of this boundary, especially if the outside and inside are shown by different *architecture* drawings. This is also true if the *architecture* is too large and thus must be divided into many *architecture* sheets. This is the way to keep *traceability* between *architecture* sheets.

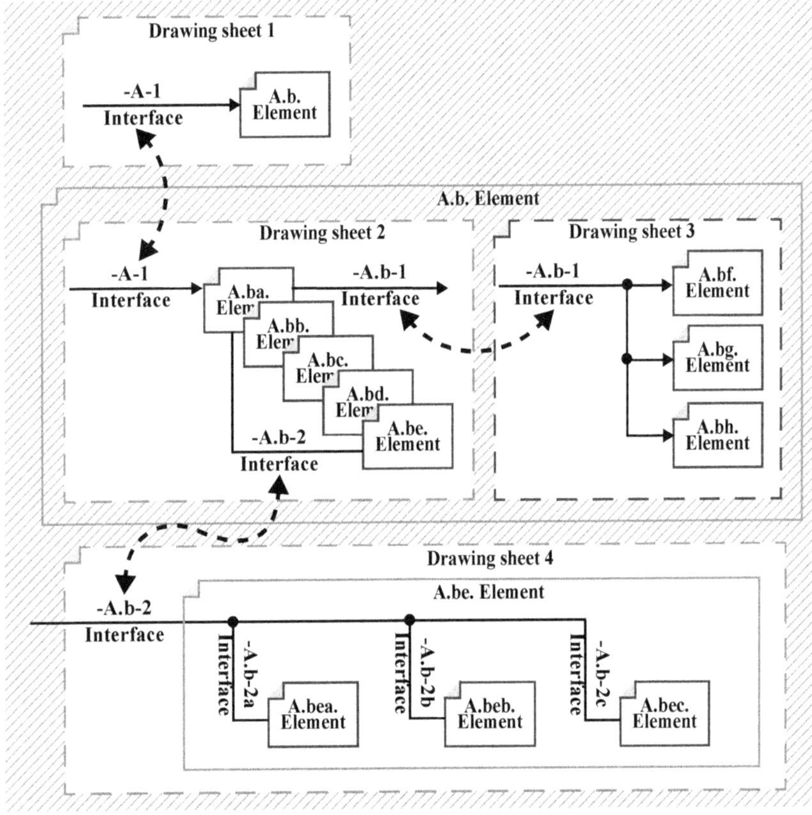

FIGURE 9-68 Traceability between splits on different sheets

## CHAPTER 10

Finalize design & requisites

Here the world unfortunately splits in two:
the tangible hardware and the immaterial software.

Programs lurking in silicon

10.1 About finalize design & requisites

10.1.1 Purpose

The previous **Pt.A. Design architectures** *phase* has identified a lot of *architecture ingredients*. Some of them are already matched by *samples* from possible *suppliers*, but many ingredients are still not detailed enough to be ordered or obtained in-house. *Programs* must be totally completed to be compiled and linked by *program* assemblers, compilers, and linkers. Mechanical parts need detailed drawings, with dimensions for prototype toolmakers and manufacturers. Electronic equipment needs detailed circuit diagrams and printed circuit board drawings.

Architecture ingredients must be transformed to *design items*, both for prototype *realization* and manufacturing. This means that some *architecture ingredients* are replaced by one or many *design items*. For example, an *element* in an *architecture* may turn out to be a resistor, capacitor, or transistor in an electronics design diagram. All identified *design items* must be described by their technical *design item properties* in enough detail to be ordered from *suppliers* or in-house.

When finalizing design, some corners of the *architectures* may be shown to not be detailed enough. In these cases, either go back to the **Pt.A. Design architectures** *phase* and extend and detail missing parts of the *architectures*, or supplement with *design item* graphs. But be aware that *design items* must not deviate from the chosen *solution alternatives*, and thereby satisfy the *requirements* less and less.

In *Cpdm*, there are many occasions where artifacts are allowed to be developed gradually to master *complexity*; the same goes for *samples* gradually turning into *design items* (see Fig. 10-1 below). Note that as usual, this chain indicates the dependency chain, and does not require that each *result* be completed before the next is started.

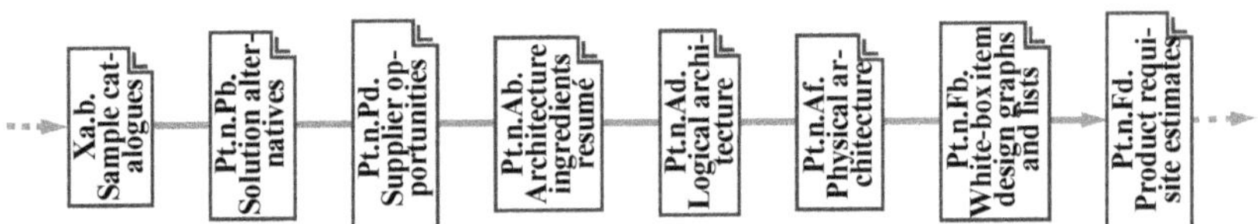

FIGURE 10-1 Sample supply chain

To find appropriate detailed *design items* that can be ordered, a lot of additional calculations may be necessary, for example, strengths of roofs and valves, voltages and current in electronic circuits, logic and timing of digital circuits, and so forth. This book does not go into such matters very much, because *Cpdm* mostly targets mastery of *complex* development and not invention of smart *products*.

When all *design items* are fully technically defined by their *design item properties*, all *design items* must be accounted for in a cost estimate. Development not only includes *design items* for prototypes and manufacturing, but also development work and development equipment *items*. At this point, some of the cost has already accrued, but many expenses have not yet occurred. However, this is the first time to get a relatively accurate total cost estimate of the *product*.

Note that finalize design & requisites is the last development *phase* to produce solely documentation. Coming *phases* will deal with *realizations* that are all tangibles (including *machine programs*). For details, see Figure 6-6, Connection between methodology, own tailored process, and manufacturing process, page 98.

10.1.2 Fight uncertainty before workload grows

When development has reached the *phase* **Pt.F. Finalize design & requisites**, the plain workload begins to grow substantially (see Fig. 10-2 below). Detailing *architectures* into compilable *programs* is by far the toughest work for programmers. For mechanics engineers, this workload is also tough, but the biggest workload will appear in the next *phase*, **Pt.I. Realize & integrate white-box**, when *realizations* should be built.

Before development work peaks, it is very important to make sure that everything thus far is carefully prepared, in order to lower the uncertainty and avoid the risk of huge amounts of rework. Time seldom allows for correction of poorly constructed *programs* after all *programs* are written, or rebuilding poorly realized prototypes after they are completed. Even if time allows, the work and cost of getting a poor *product* in shape can be astronomical.

This syndrome is often referred to as "changes are more expensive the later they occur," but paradoxically, in reality this fact does seldom result in better prepared earlier *phases*. One sign of poor *requirement* and design work is that realize & integrate white-box and verify black- & white-box occupy the time of an unreasonable number of developers.

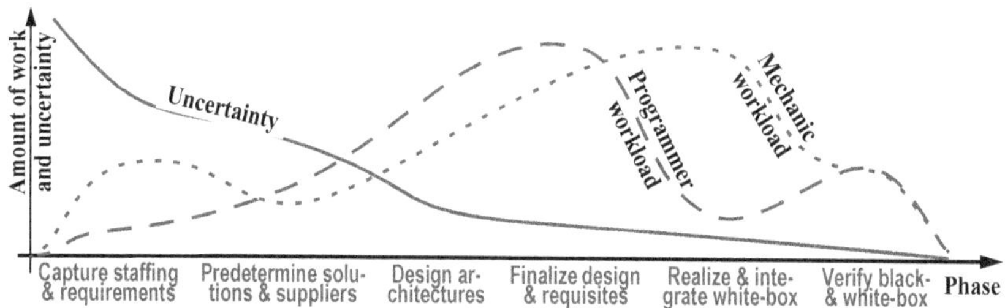

FIGURE 10-2 Amount of uncertainty and workload during the phases

10.1.3 From outside-in to inside-out process

So far the development *phases* capture staffing & requirements, predetermine solutions & suppliers, and *design architectures* have been developing the *product* outside-in. It is quite natural to begin development from the known *environment* of the *product* and then proceed *inwards* into deeper and deeper *product* details.

In other words, so far the *product* to be developed has only been *decomposed*. For many reasons it is efficient to now reverse the development from outside-in to inside-out, and start adding all developed details from *innermost black-/white-boxes* back to the *environment* (see Fig. 10-3 below).

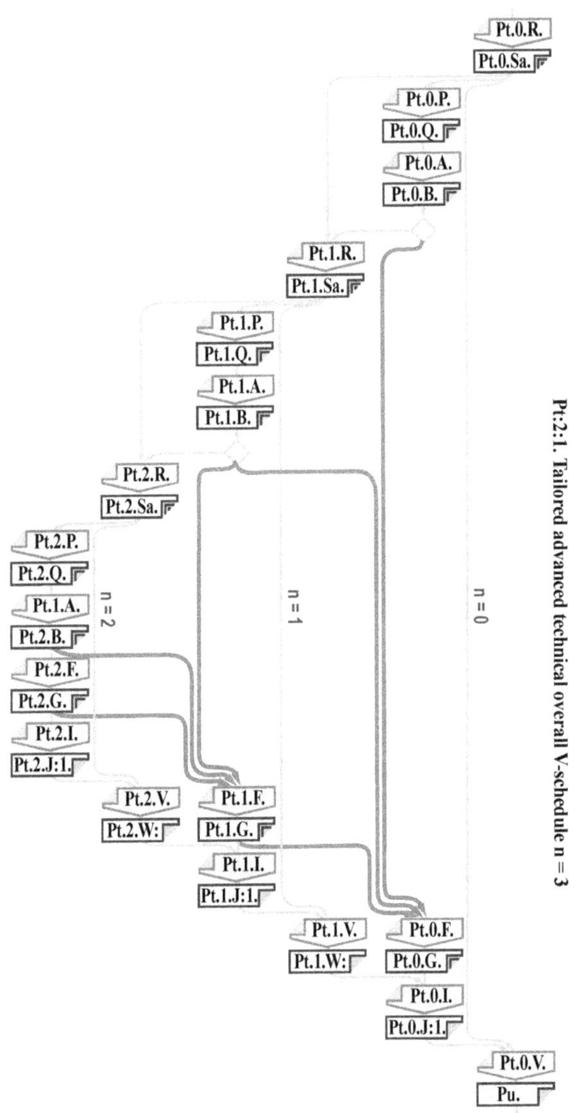

Architecture is made outside-in, to ensure that the *product* is securely attached to the existing *environment* and gradually *decomposed inwards* by populating it with needed ingredients. As development turns from outside-in to inside-out, it changes from decomposing *inwards*, to adapting the inner ingredient to *outward* surroundings.

Innermost details are first better understood and better designed to fit together, and in turn surroundings to these can be adapted and better designed, and so forth. After development has been reversed and completed it will never happen that a bathtub doesn't fit into the bathroom, or a sink into the kitchen. The same applies to electronics and *programs*.

FIGURE 10-3 First three phases outside-in and then three phases inside-out

The benefit of going inside-out during finalize design & requisites is that there will be more and more well-known inner details (again, see the highlighted references in Fig. 10-3 above). When finally designing the *environment interfaces*, all detailed *white-box design items* of all inner *nesting levels* are known for certain and those may be referred. The *architectures* that were once *decomposed* outside-in can now be completed and corrected inside-out, making *realization* and *integration* of a prototype much easier, because everything has continuously been designed to fit together.

Another benefit of *finalize design item* inside-out is that costs can be accumulated, starting with innermost details and gradually *enfold* more and more *outward* until enclosing the entire *product*. The weight, water and electricity consumption, and similar *design item properties* can be accumulated in the same way as costs.

Finalize design item from inside and *outwards* has many benefits also for *programs*. Many times *program* developers just can't get their systems to work together, because the *outermost white-box* was finalized before the design was completed of how its parts could fit together. Such mistakes are probably caused by making a design too much from *outermost white-boxes* and *inwards*. When design is finalized from *innermost black-boxes* and *outwards*, all details are discovered and can be *aggregated* in the same place as the design approaches the *outward environment*.

One perplexing fact is that the larger and more *complex elements* are, the later they can be finalized, because they have to wait for their more embedded *elements* to be finalized first. Pressure and stress on developers to finalize detailed *design items* come from waiting integrators and verifiers, not to mention the impatience of marketers eager to launch the *product*.

Large *products* and *elements* are preferably developed iteratively, which means that the *product* is developed in many steps, in such a way that each step brings value to the *stakeholders*, who can repeatedly validate the *product* during its development. How to organize this will be a good topic for the next book "*Cpdm* technical overhead".

When applying outside-in *invasive integration* (see ch. 11.1.4, "Outside-in invasive integration," p. 595), it may in the first place look like an attempt to turn finalize design & requisites to outside-in also, to get it aligned with outside-in *invasive integration* and also with **Pt.R. Capture staffing & requirements, Pt.P. Predetermine solutions & suppliers** and **Pt.A. Design architectures** *phases*, in order to save a lot of calendar time. It is true that calendar time would be saved, but the difficulty with turning finalize design & requisites outside-in is that no design *phase* will remain inside-out at all. If this is applied, everything designed on *outward nesting levels* must be accepted for all *inward nesting levels*, and no adjustments can be made on *outward nesting levels* based on findings made in *inward nesting levels*. As usual, some "devil in the details" will be found *inwards*, but there will be no *process* support to fix its consequences for *outward nesting levels*. This is such a serious issue that outside-in finalized design is not seriously considered by *Cpdm*.

10.2 About finalize design items

Now over to an area not so appreciated by engineers. Not only the *product* smartness or *quality* contributes to success, but price does also. Repeatedly, during *product development*, all costs must be watched, to ensure the *product* will be competitive. This is an extremely appropriate time for an economical analysis, since all documentation for the *product* is now finished. The last step is to define all items that form the base for the economy.

10.2.1 Cpdm definitions

In *Cpdm, item, design item, role item, in-house design item, supplier design item, turnkey item, off-the-shelf item, in-house role item,* and *provider role item* are defined according to Table 10-1 below

TABLE 10-1 Cpdm definitions

<table>
<tr><td>Aspect</td><td colspan="6">Item</td></tr>
<tr><td>Definition</td><td colspan="6">All resources needed for product development, such as material, workforce, and equipment specified with desired granularity and formal enough to be traceable</td></tr>
<tr><td>Aspect</td><td colspan="3">Design item (see Table 10-2 below)</td><td colspan="2">Role item</td><td>Any kind of items</td></tr>
<tr><td>Definition</td><td colspan="3">Material (including linkable programs) specified with desired granularity formal enough to be orderable</td><td colspan="2" rowspan="2">Accomplished and estimated work specified with desired granularity and formal enough to be traceable</td><td rowspan="2">Anything specified with desired granularity</td></tr>
<tr><td>Synonym</td><td colspan="3">Bill of material (BOM).</td></tr>
<tr><td>Aspect</td><td rowspan="3">In-house design item</td><td colspan="2">Supplier design item</td><td></td><td></td><td rowspan="3">For appropriate accounting</td></tr>
<tr><td>Definition</td><td colspan="2">Design developed by suppliers</td><td></td><td></td></tr>
<tr><td>Aspect</td><td>Turn-key item</td><td>Off-the-shelf item</td><td>In-house role item</td><td>Provider role item</td></tr>
<tr><td>Definition</td><td>Design developed by in-house workforce</td><td>Custom-made design developed to fit very well into own design</td><td>Predeveloped reusable components</td><td>Development, realization, and manufacturing in-house workforce</td><td>Development, realization, and manufacturing hired workforce</td><td>Partitioning of anything too big for desired accounting granularity (see ch. 10.23, p. 560)</td></tr>
</table>

In *Cpdm*, a *design item* may be divided into *environment interfaces design item* and *white-box design item* according to Table 10-2 below.

TABLE 10-2 Cpdm definitions

Aspect	Environment interfaces design item	White-box design item
Definition	Specified design, including technical properties, needed to realize and manufacture the product interface to environment	Specified design, including technical properties, needed to realize and manufacture the white-box
Synonyms	Detailed product interface design, detailed product connection design	Detailed system design, detailed architecture
Design graphs	**TABLE 10-3** Typically one of: _Design graph for environment interfaces:_ See table below.	**TABLE 10-4** Typically one of: _Design graph for white-box:_ See table below. For interfaces, see Table 10-3
Design descriptions	**TABLE 10-5** Interfaces to environment design items	**TABLE 10-6** White-box design items

TABLE 10-3 Typically one of:

Design type		Interface	
		Connection	Boundary
Mechanical	Static	Attachment	Casing Housing
	Dynamic	Connection	Shielding
Electronics			
Programs		Call MMI	Scope

TABLE 10-4 Typically one of:

Design type	White-box
Mechanics	Drawing
Electronics	Diagram
Programs	Flow chart Design chart File chart Class chart Program

For interfaces, see Table 10-3

TABLE 10-5 Interfaces to environment design items

Item	Design item properties	Interface ingredient

TABLE 10-6 White-box design items

Item	Design item properties	White-box ingredient

10.2.2 EXAMPLE: Interface with their design item properties

Boundary interfaces and *connection interfaces* are the trickiest to understand and because of that they are exemplified below:

EXAMPLES of mechanical static connection interface (attachment): nail, screw, mortar, glue, putty, and so on, with *design item properties* strength, dimension, and such

EXAMPLES of mechanical static boundary interface (casing or housing): ground, wall, ceiling, roof, cabin, with technical *design item properties:* material, durability, permanence, appearance, and so forth

EXAMPLES of mechanical dynamic interface (connection): water, sewage, gas, shaft, belt, and so forth, with technical *design item properties:* capacity, pressure, torque, and such

EXAMPLES of mechanical dynamic user interface (connection): lever, crank, knob, handle, etcetera, with technical and *design item properties* torque, force, material, and so forth

EXAMPLES of electronics boundary interface (shielding): lightning protection, residual current device, electromagnetic compatibility (EMC), Faraday cage, and so forth, with technical *design item properties* capacity, current, frequency, efficiency, and such

EXAMPLES of electronics interface (connection): cable, wire, plug, antenna, and so on, with technical *design item properties* voltage, current, bandwidth, impedance, directionality, and so forth

EXAMPLES of electronics user interface (connection): switch, dimmer, touch screen, etcetera, with technical *design item properties:* scale, icons, limit, and so forth

EXAMPLES of program boundary interface (scope): memory reach, memory access protection, global/local visibility, fire-wall, user permission, and so forth, with technical *design item properties:* address-width, security, authority, and such

EXAMPLES of program connection interface (call): subroutine, *interrupt*, BIOS-access, OS-access, and so on, with technical *design item properties:* parameters, name-space, reentrance, latency, feedback, etcetera

EXAMPLES of program user interface (MMI): Menus, input-fields, radio-buttons, tick-boxes, and such, with technical *design item properties:* font, icons, color, form, and so forth

10.2.3 Buying or developing detailed white-box design items

In the mechanics and electronics communities it is very common to regard detailed *white-box design items* as *components* that can be bought as affordable standard *components* and *reused* immediately. It is seldom seen that a house builder runs a brickyard instead of buying bricks, or an electronic *supplier* runs a capacitor manufacturing plant instead of buying these *components*.

> Mechanics and electronics often use supplier components, but programs are most often developed from scratch.

But in *program* development, it is rather uncommon that *reusable items* are found from any *supplier*, apart from *program libraries* offered as *development system libraries*, and very few *programs* are generalized and sold for *reuse*, so that everything has to be finalized in-house or by subcontractors. In many cases *programs* might be among the most costly items for a development company, even if much consists of heavily *recycled* own legacy *programs*. Thus, in *programs* few *elements* are planned for procurement and *reuse*, but they are most often prepared in the finalize *phase*.

But be aware, the method of incrementally extending *programs* (also for *reuse*), is not an easy game. To master this, a lot of *configuration management* must be performed to keep track of all changes and extensions; also, some strategic *architecture* redesign of the old *product* will be required in order to support the new. This book doesn't present such management overhead, but a projected second book "*Cpdm* technical overhead" will definitely take care of this.

An opposite trend is that developing companies sometimes offer *quality programs* free of charge in order to build substantial communities with other possibilities for earnings, rather than providing *programs* for *reuse*. Also, more and more free downloads are available of smaller *program elements*, which are quick to *reuse*. At the predetermine solutions & suppliers *phase* these possibilities must be evaluated carefully.

10.3 About finalize product requisites

10.3.1 Cpdm definition

In *Cpdm*, *product requisites* is defined according to Table 10-7 below.

TABLE 10-7 Cpdm definition

Aspect	Product requisite
Definition	Complete information about costs for all types of items, to allow accounting for development, realizations, and manufacturing
Synonyms	Bill of material (for the material parts), purchase list, sample compilation
Symbol	**TABLE 10-8** Product requisite estimates, one of below types: • Estimated manufacturing material requisites cost • Estimated manufacturing work requisites hourly/fixed/rental cost • Estimated manufacturing equipment requisites purchase/rental cost • Estimated realization material requisites cost • Estimated realization work requisites hourly/fixed/rental cost • Estimated realization equipment requisites purchase/rental cost • Estimated development material requisites cost • Estimated development work requisites hourly/fixed cost • Estimated development equipment requisites purchase/rental cost • Summed/accumulated requisites estimates

Volume accounting	Item	Item information

10.3.2 About accounting of product requisites

As can be seen from the above definition, there are at least 10 types of requisites, and each type of table most often requires its own set of columns to be efficient and understandable, so that the total requisite set must be divided into many separate tables. Also, division of the same requisite type into smaller tables may be preferred because of a need for better information clarity.

Many of the requisites have already accrued, but many of the requisites are still estimates. No complete *product product prototype* is yet built and verified, and manufacturing is even farther ahead in time.

Note that cost accounting for *verification* is only the cost of performing the *verification*. A substantial cost being caused by *verification* is the *failure elimination* by the developers having caused the failures. Cost accounting for *failure elimination* should accordingly be included in development cost (for example additional percentages) for each *activity* possibly causing failures.

Apart from being the base for ordering detailed *design items* to prototypes and manufacturing, the requisites gradually disclose accrued costs and improve estimates of coming costs, and profitability can be estimated better and better. One or more of the business gates **Pt.0.Fe. Examine prototype success** and **Pt.n.Fe. Examine white-box success** may well propose an immediate termination of the development if the profit analysis gets too negative.

10.3.3 Ensuring product requisites traceability

After the capture staffing & requirements *phase*, there are in total three design *phases* in *Cpdm*, which are predetermine solutions & suppliers, *design architectures* and finalize design & requisites. These three design *phases* span from *solution alternatives* to *product requisites*, which is a huge leap with many interim design *results* in between. To not lose navigation and control, it is rewarding to consider all *cascading results* in relation to each other (see Fig. 10-4 below).

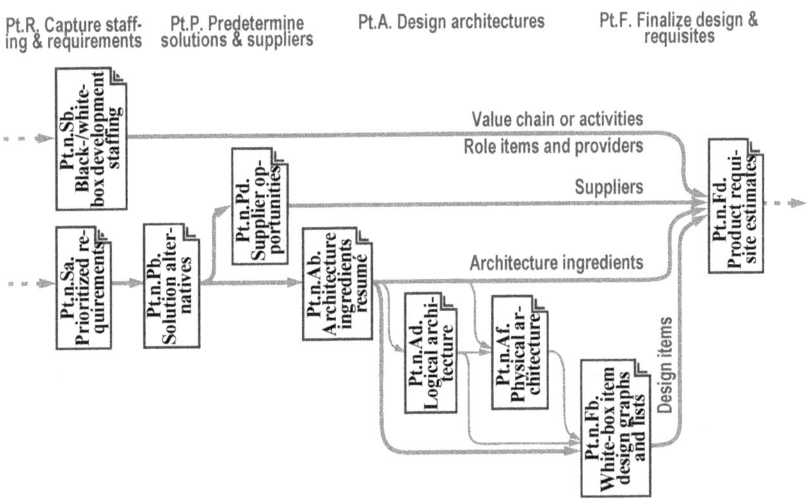

FIGURE 10-4 Results collected to prepare product requisites

In *Cpdm*, a lot of attention is given to *traceability*. Each development *result* must be *demanded* by at least one subsequent *activity*, and each *activity* must be *justified* by at least one *result*. Further on, to be able to trace in detail between *cascading results*, all *results* must be *partitioned* into separate *modules*, and each of these *modules* must have a unique identifier to be traceable. During the writing of this book, it was obvious how much better examples can be understood, even by me myself, by modularizing *results* and assigning each *module* a unique identifier.

10.4 EXAMPLE: House kitchen machinery: Finalize design of machinery with product requisites

10.4.1 EXAMPLE House: About this example

How often does it happen that machinery space doesn't fit in a kitchen, or a cupboard doesn't fit in a bathroom or bedroom, or the clearance height in a stairway is insufficient? Or that a machine needs 3-phase electricity, but only one phase is supplied, and this machine is exchanged for a similar 1-phase machine, but the main fuse on that phase isn't well enough specified. Or that a water faucet is needed, but is far from the water supply or does not have enough capacity. Or from an economical angle, a house gets much too expensive because nobody has calculated the cost of all necessary details.

The above examples require that finalize design & requisites not be made inside-out. When all *architectures* are in place, from house *environment inwards* to house room machinery, it is time to reverse the outside-in approach to get inside-out, which means to first finalize the house room machinery design. By reversing and starting with details, it becomes possible to consecutively adapt the surroundings to fit the *inward* design. For example, by deciding the exact kitchen machinery, the kitchen cabinets and interiors can be adapted to fit the machinery. And in the next step, the decided kitchen interiors can adapt and influence the kitchen room layout to best fit the interiors, and finally the house can be better adapted to fit the rooms.

The big increase in the volume of *results* with extensive details needed to complete each *phase* grows more and more evident. It may be a bit of a challenge to find and collect all *results*, but just imagine the problem if all these details were not documented at all but reside only in the heads of hundreds of developers.

10.4.2 EXAMPLE House kitchen machinery: Process schedule to use

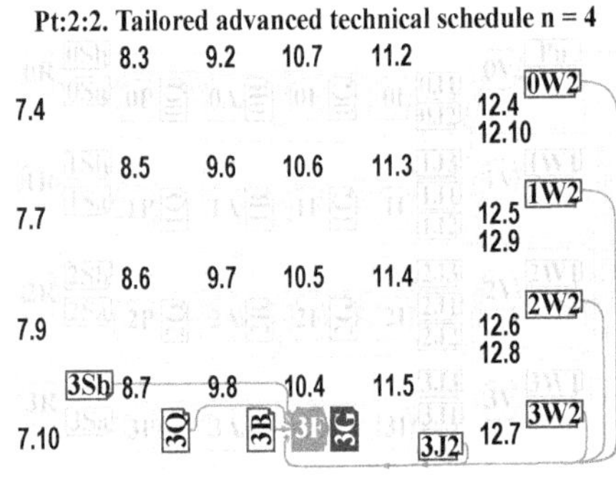

For the house example, a special technical overall *schedule* is *tailored* (illustrated in Fig. 10-5, left):

- **Pt:2:2. Tailored advanced technical schedule n = 4**, page 164

In this *tailored schedule* the generic *schedule* to use is (filled-in symbols):

- **Pt.n.F. Finalize design of white-box with product requisites**, page 140.

FIGURE 10-5 Position of schedule and master to use (filled in) in overall schedule

Below *results* may be demanded (their *masters* are indicated *upstream* from the filled-in *schedule* in the figure above):

3B Machinery architectures with major ingredients but no black-boxes

   Table 9-18, House kitchen machinery elements architecture ingredients resumé, page 373

   Table 9-21, House kitchen machinery user interface architecture ingredients resumé, page 375

   Table 9-19, House kitchen machinery connection interface architecture ingredients resumé, page 373

   Table 9-20, House kitchen machinery boundaries interface architecture ingredients resumé, page 374

   Figure 9-29, House kitchen machinery logical architecture, page 376

   Figure 9-30, House kitchen machinery physical architecture, page 377

3Q Machinery solution alternatives with supplier opportunities

   Table 8-12, House white-box solution alternatives, page 301

   Table 8-13, House white-box supplier opportunities, page 303

3Sh Machinery staffing allocation

   Table 7-14, House kitchen machinery development allocation staffing, page 202

Xb Suppliers' sample catalogues

E Picked in-house legacy source systems

XaE Ordered suppliers' off-the-shelf fully described source systems

XbE Ordered suppliers' turnkey fully described source systems

For possible *failure localization* and *failure elimination*, below *results* may be demanded (their *masters* are indicated *upstreams* from the filled-in *schedule* in the figure above):

[3J2] Table 11-22, House kitchen machinery start-up report 1, page 622

[0W2] Table 12-27, House environment verification report 1, page 701

Table 12-29, House environment reverification report 2, page 703

Table 12-28, House environment failure elimination report 1, page 702

[1W2] List 12-2. House with fake rooms failure elimination report 1, page 676

[1W2] Table 12-22, House verification report 1, page 697

Table 12-24, House verification report 2, page 698

Table 12-23, House failure elimination report 1, page 698

[2W2] List 12-3. House rooms with fake machinery failure elimination report 1, page 678

[2W2] Table 12-14, House kitchen verification report 1, page 688

Table 12-17, House kitchen verification report 2, page 691

Table 12-15, House kitchen failure elimination report 1, page 689

[3W2] Table 12-9, House kitchen machinery verification report 1, page 682

Table 12-11, House kitchen machinery verification report 2, page 683

Table 12-10, House kitchen machinery failure elimination report 1, page 683

10.4.3 EXAMPLE House kitchen machinery: Finalize white-box detailed items

Machinery is now up for *white-box design items* listing. To prepare the fake machinery, some small items are needed (see Table 10-9) below.

TABLE 10-9 House kitchen machinery fake items

Item	Color	Size	Type	White-box ingredient
i100. Marker pen	Various	Ø 2 mm	Permanent	A.aC.a. Machinery

The machinery itself does not demand a lot of design, but has many *interfaces* to the house commonalities, which must be designed carefully and properly (see Table 10-10 below).

TABLE 10-10 House kitchen machinery white-box design items

Item	Color	Mount size w x d x h cm	Max current	Other nonmechanical interface	Phases and plug	White-box ingredient
i101. Oven	Black	78 x 55 x 58	3 x 3 A	Ensure free air space behind machine.	3 x 380 V, 50 Hz Perilex plug	A.aC.aC.a. Wide oven
i102. Hub	Black	Countertop hole 40 x 50	3 x 4 A		3 x 380 V, 50 Hz Perilex plug	AaC.aDa. Electrical hubs

TABLE 10-10 House kitchen machinery white-box design items

Item	Color	Mount size w x d x h cm	Max current	Other nonmechanical interface	Phases and plug	White-box ingredient
i103. Fridge /freezer	White	58 x 56 x 110	1 x 1 A	Ensure free air space behind machine.		A.aC.aA.a. Combined fridge/ freezer
i104. Micro oven	Black	58 x 56 x 38	1 x 3 A			A.aC.aB.a. Micro oven
i105. Oven	Black	58 x 56 x 58	1 x 3 A			A.aC.aB.b. Warming oven
i106. Hood	Chrome	Stand-alone 20 x 16	1 x 1 A	Suction 12 m³/hour Outlet female air pipe Ø 12 cm		A.aC.aE.a. Wide hood
i107. Dishwasher	White	58 x 58 x 68	1 x 9 A	Cold water 3/4 inch male joint Drain by 1-inch hoses with clip		A.aC.aG.a. Dishwasher coarse
i108. Water valve				2 x Ø 3/4 inch		
i109. Dishwasher	White	58 x 58 x 68	1 x 7 A	Cold water 3/4 inch male joint Drain by 1-inch hoses with clip	230 V 50 Hz Europe AC plug with ground,	A.aC.aG.b. Dishwasher gentle
i110. Water valve				2 x Ø 3/4 inch		
i111. Sink	Chrome	Countertop hole 100 x 50		Drain 2-inch female sleeve joint		A.aC.aF.a. Sink 2 bowls
i112. Tap	Chrome	Mounted in sink, hole Ø 42 mm		Cold and warm water by 3/4 inch joint		A.aC.aF.b. Combined faucet
i113. Water valve				Ø 3/4 inch		
i114. Burner	Black	Countertop hole 30 x 50		½ inch male joint, max pressure 75 mbar		A.aC.aD.b. Gas burner
i115. Gas valve				Ø ½ inch Approved gas reducing valve		
Sum max (if distributed symmetrically)			3 x 16 A			

10.4.4 EXAMPLE House kitchen machinery: Account for white-box development, realization, and manufacturing

This example is about a house, and it is rather uncommon for more houses of this exact type to be built. All development costs are thus for one house, but with price deductions that are normal in this profession.

To support *partial verification* of how the machinery fits in the kitchen, the machinery must be prepared as fakes (see Table 10-11 and Table 10-12 below).

TABLE 10-11 House kitchen machinery fakes,
estimated realization material requisites cost

Pc	€/Pc	€	Item	Product	Type	Supplier	Element
20	2	40	i100. Marker pen	Perma	PPY-10	SU.27 Local supplier	A.aC.a. Machinery

TABLE 10-12 House kitchen machinery fakes,
estimated realization work requisites hourly/fixed/rental cost

h	€/h	Cost €	Item	Provider	Interface
2	30	60	i70. Inspector	PR.17. K-Studio	-A.aC-2. Interior attachment

Since *product requisites* are of different types, keep these in several separate tables. Begin with the material (see Table 10-13 below).

TABLE 10-13 House kitchen machinery
estimated realization material requisites cost

Pc	€/Pc	€	Item	Product	Type	Supplier	Element
1	350	350	i103. Fridge/ freezer	Wirly	W34-76	SU.28 K-Studio.	A.aC.aA.a. Combined fridge/freezer for external cover door Connector cable for -A.aC-3a. El. fridge
1	260	260	i104. Micro oven	Wirly	WMic 8	SU.28 K-Studio.	A.aC.aB.a. Micro oven -A.aC-1c. Micro panel -A.aC-1d. Micro door connector cable for -A.aC-3b. El. micro
1	220	220	i101. Oven	Wirly	W+ 4800	SU.28 K-Studio.	A.aC.aB.b. Warming oven -A.aC-1e. Warm panel -A.aC-1f. Warm door Connector cable for -A.aC-3c. El. warm
1	230	230	i105. Oven	Wirly	WZ-2000	SU.28 K-Studio.	A.aC.aC.a. Wide oven -A.aC-1g. Bake panel -A.aC-1h. Bake door Connector cable for -A.aC-3d. El. bake
1	110	110	i102. Hub	Wirly	E4 W2	SU.28 K-Studio.	AaC.aDa. Electrical hubs -A.aC-1i. Hub panel Connector cable for -A.aC-3e. El. hub 3
1	120	120	i114. Burner	Buttan	Mall-2 Ex	SU.29 House builder's supplier	A.aC.aD.b. Gas burner -A.aC-1j. Burner panel Flexible connector hose for -E-5. Gas
1	180	180	i115. Gas valve	Buttan	Reducer PR-6	SU.29 House builder's supplier	Valve for A.aC.aD.b. Gas burner
1	90	90	i106. Hood	Wirly	Vac 3411	SU.28 K-Studio.	A.aC.aE.a. Wide hood -A.aC-1k. Hood panel Connector cable for -A.aC-3f. El. hood
1	130	130	i111. Sink	Plating	K2 Fe	SU.30 Wat Son	A.aC.aF.a. Sink 2 bowls Sink trap for -A-13. Sewer

TABLE 10-13 House kitchen machinery
estimated realization material requisites cost

Pc	€/Pc	€	Item	Product	Type	Supplier	Element
1	140	140	i112. Tap	Plating	Z34-Px	SU.30 Wat Son	A.aC.aF.b. Combined faucet -A.aC-1l. Handles Flexible connector hose for -A-9. Warm waterFlexible connector hose for -A-10. Cold water
2	20	40	i113. Water valve	Plating	VV-6	SU.30 Wat Son	Valve for A.aC.aF. Wet machinery
1	480	480	i107. Dishwasher	Teller Inc	Bull-4-30	SU.28 K-Studio.	A.aC.aG.a. Dishwasher coarse for external cover door -A.aC-1m. Dish panel Connector cable for -A.aC-3g. El. adv. dish
1	520	520	i109. Dishwasher	Teller Inc	Cat-6-40	SU.28 K-Studio.	A.aC.aG.b. Dishwasher gentle for external cover door -A.aC-1o. Dish panel Connector cable for -A.aC-3h. El. smp. dish
2	19	28	i110. Water valve	Plating	VV-6	SU.30 Wat Son	A.aC.aG. Dish machinery
10	1	10	i100. Marker pen	Insignificant	Perma-10	SU.27 Local supplier	A.aC.a. Machinery fake
		2 908					

Continue with work needed to install the machinery (see Table 10-14 below). Observe that *realization* work is done mostly on *interfaces* and not on *elements* (which are already cost-accounted as material), hence references in the table *target architecture interfaces.*

TABLE 10-14 House kitchen machinery
estimated realization work requisites hourly/fixed/rental cost

h	€/h	Cost €	Item	Provider	Interface
5	30	150	i65. Kitchen constructor	PR.17. K-Studio	[A.aC.aA]1. Cold housing [A.aC.aB]1. Upper housing [A.aC.aB]2. Lower housing [A.aC.aC]1. Oven housing [A.aC.aD]1. Gasket seal [A.aC.aF]1. Gasket seal [A.aC.aG]1. Left housing [A.aC.aG]2. Right housing
10	29	290	i66. Plumber	PR.17. K-Studio	-A-9. Warm water -A-10. Cold water -A-13. Sewer
4	30	120	i67. Tin-smith	PR.17. K-Studio	-E-6. Air outlet [A.aC.aE]1. Air tin box
4	46	184	i68. Gas installer	PR.17. K-Studio	-E-5. Gas
4	35	140	i69. Electrician	PR.12. Elert AB	-A.aC-3a. El. fridge -A.aC-3b. El. micro -A.aC-3c. El. warm -A.aC-3d. El. bake -A.aC-3e. El. hub -A.aC-3f. El. hood -A.aC-3g. El. adv. dish -A.aC-3h. El. smp. dish -A.aC-3i. El. lamp -A.aC-3j. El. switch
		884			

Note that building a house demands an outside-in *realization* (see ch. 11.1.4, "Outside-in invasive integration," p. 595). Consequently, the machinery will be the innermost items to be installed in the house.

Also, all development cost should be estimated and accounted (see Table 10-15 below). Note that the development of the machinery in the staircase space is not shown in the book, but the costs are anticipated.

TABLE 10-15 House kitchen machinery
estimated development work requisites hourly/fixed cost

#	€ /h	€	Item	Provider	Schedule
		0	i62. House user	PR.10. Proprietor	Pt.3.R. Refine machinery requirements from rooms requirements and ensure machinery development staffing.
40	30	1200	i63. Machinery expert	PR.17. K-Studio	Pt.3.P. Predetermine machinery solutions with supplier opportunities.
8	30	240			Pt.3.A. Satisfy machinery requirements by decomposing machinery black-boxes into white-box design not containing black-boxes.
			i64. Machinery expert	PR.17. K-Studio	Pt.3.F. Finalize design of machinery with product requisites.
5	30	150	i70. Inspector	PR.17. K-Studio	Pt.3.V. Verify machinery black-/white-boxes.
		1 590			

Finally, all machinery costs should be summarized (see Table 10-16 below). Again, note that development of machinery for toilet and staircase space is not shown in this book, but the costs are anticipated.

TABLE 10-16 House rooms machinery
summed/accumulated requisites estimates

Item	Reference	Cost €
House kitchen machinery fakes, estimated realization material requisites cost	Table 10-11, this chapter	40
House kitchen machinery fakes, estimated realization work requisites hourly/fixed/rental cost	Table 10-12, this chapter	60
House kitchen machinery estimated realization material requisites cost	Table 10-13, this chapter	2 908
House kitchen machinery estimated realization work requisites hourly/fixed/rental cost	Table 10-14, this chapter	884
Remaining house rooms machinery estimated realization material requisites cost	Not shown in this book	1 740
Remaining house rooms machinery estimated realization work requisites hourly/fixed/rental cost		610
House kitchen machinery estimated development work requisites hourly/fixed cost	Table 10-15, this chapter	1 590
House rooms machinery summed/accumulated requisites estimates		**7 832**

10.4.5 EXAMPLE House kitchen machinery: Examine white-box success

K-Studio and Robber HB were called to the gate meeting. Robber HB reported that they were not competent to design toilet machinery, but the house proprietor was very satisfied with the kitchen. Robber HB was asked to find another subcontractor to design toilet machinery.

It was also decided that the proprietor must capture color and appearance of the machinery for the selected house builder, before the house builder procured these.

10.4.6 EXAMPLE House: Proceed reading

To follow this example, these are the alternatives (see again Fig. 10-5, p. 446):

3I Proceed to chapter 11.5, "EXAMPLE House kitchen machinery: Procure machinery items in order to install machinery," page 618.

3V Proceed to chapter 12.7, "EXAMPLE: House rooms machinery: Fully verify machinery black-/white-boxes," page 678.

2F Proceed to chapter 10.5 "EXAMPLE House rooms: Finalize design of rooms with product requisites" below.

10.5 EXAMPLE House rooms: Finalize design of rooms with product requisites

10.5.1 EXAMPLE House kitchen: Process schedule to use

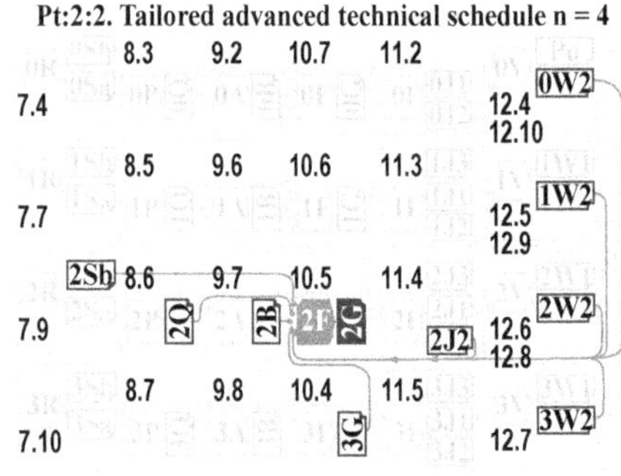

For the house example, a special technical overall *schedule* is *tailored* (illustrated in Fig. 10-6, left):

- **Pt:2:2. Tailored advanced technical schedule n = 4,** page 164

In this *tailored schedule* the generic *schedule* to use is (filled-in symbols):

- **Pt.n.F. Finalize design of white-box with product requisites,** page 140.

FIGURE 10-6 Position of schedule and master to use (filled in) in overall schedule

Below *results* may be demanded (their *masters* are indicated *upstream* from the filled-in *schedule* in the figure above):

3G Finalized machinery items design with product requisites

Table 10-9, House kitchen machinery fake items, page 447

Table 10-10, House kitchen machinery white-box design items, page 447

Table 10-12, House kitchen machinery fakes, estimated realization work requisites hourly/fixed/rental cost, page 449

Table 10-13, House kitchen machinery estimated realization material requisites cost, page 449

Table 10-14, House kitchen machinery estimated realization work requisites hourly/fixed/rental cost, page 450

2B Room architectures with major ingredients and machinery black-boxes

Table 9-16, House kitchen elements architecture ingredients resumé, page 367

Table 9-17, House kitchen interfaces architecture ingredients resumé, page 368

Figure 9-27, House kitchen physical architecture, page 370

Figure 9-26, House kitchen logical architecture, page 369

2Q Room solution alternatives with supplier opportunities

Table 8-14, House kitchen white-box solution alternatives, page 306

Table 8-15, House kitchen white-box supplier opportunities, page 308

2Sb Room staffing allocation

Table 7-12, House kitchen development allocation staffing, page 197

Xb Suppliers' sample catalogues

E Picked in-house legacy source systems

XaE Ordered suppliers' off-the-shelf fully described source systems

XbE Ordered suppliers' turnkey fully described source systems

For possible *failure localization* and *failure elimination*, below *results* may be demanded (their *masters* are indicated *upstream* from the filled-in *schedule* in the figure above):

2J2 Table 11-18, House kitchen start-up report 1, page 617

0W2 List 12-1. House environment interfaces with fake house failure elimination report 1, page 674

0W2 Table 12-27, House environment verification report 1, page 701

Table 12-29, House environment reverification report 2, page 703

Table 12-28, House environment failure elimination report 1, page 702

1W2 List 12-2. House with fake rooms failure elimination report 1, page 676

1W2 Table 12-22, House verification report 1, page 697

Table 12-24, House verification report 2, page 698

Table 12-23, House failure elimination report 1, page 698

2W2 List 12-3. House rooms with fake machinery failure elimination report 1, page 678

2W2 Table 12-14, House kitchen verification report 1, page 688

Table 12-17, House kitchen verification report 2, page 691

Table 12-15, House kitchen failure elimination report 1, page 689

3W2 Table 12-9, House kitchen machinery verification report 1, page 682

Table 12-11, House kitchen machinery verification report 2, page 683

Table 12-10, House kitchen machinery failure elimination report 1, page 683

10.5.2 EXAMPLE House kitchen: Finalize white-box detailed items

For preparing the fake rooms, some small items are needed (see Table 10-17 below).

TABLE 10-17 House kitchen fake items

Item	Color	Size	Type	White-box ingredient
i116. Chalk crayons	Various	Ø ~ 12 mm	Non-smeary	A.a. Rooms

For space reasons, many *requirements* have been left out in this example. However, kitchen appearance and behavior are well specified with *requirements*, the kitchen room well designed by *architectures*, and kitchen details will now be further finalized in this chapter. It is quite enough to focus on how to finalize kitchen design only, in order to present and understand the *process schedule*.

Since finalize design & requisites is made inside-out, it is now decided what machinery will be used in the kitchen, and let all consequences from the machinery influence the finalization of the kitchen design, which is the surroundings to the machinery.

Since the machinery is now well known, all kitchen casings and cabinets can be described by a precise *white-box design items* drawing, to ensure that everything fits together (see Fig. 10-7 below).

It also becomes more apparent how the kitchen floor area should be used, how the proportions and measurements of the kitchen walls should be decided, and even where to best locate the kitchen window. This understanding will update the kitchen drawings, but will also be useful when measurements for the house drawings are decided upon.

In order to design the house kitchen machinery, a drawing of cabinets with measurements is made by K-Studio (see Fig. 10-7 below).

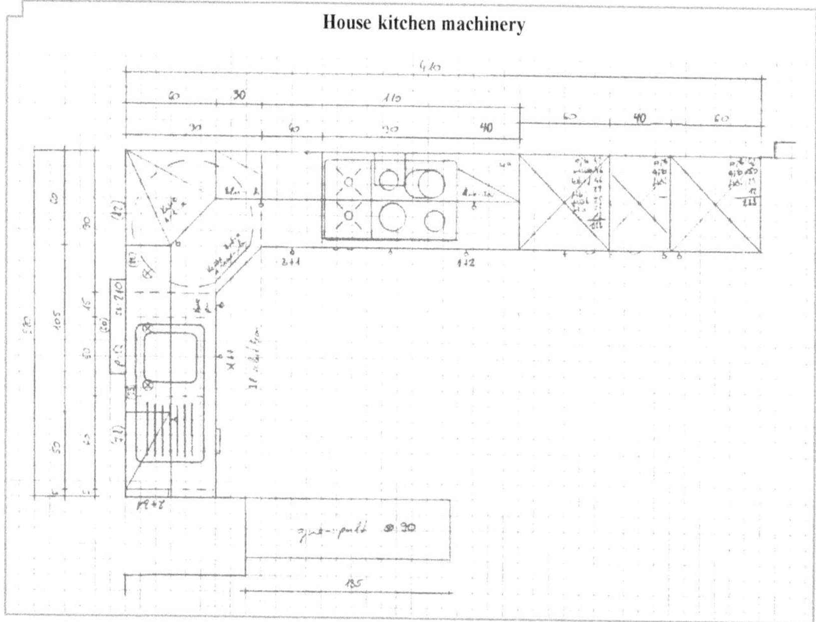

FIGURE 10-7 House kitchen interior drawing

The measurements of the cabinets are very important, since they connect to each other and to the kitchen walls and floor. Also note that parts of the decided machinery are indicated in this sketch.

When the kitchen interiors are designed, it is also rather easy to plan and draw the *connection interfaces* for the kitchen interiors (see Fig. 10-8 below).

Quite often piping design is not documented, with the consequence that later repair and maintenance of the house gets a lot more expensive and complicated, because nobody remembers how the piping was arranged. If piping is not documented in advance, at least some photos of the piping must be taken before it gets hidden.

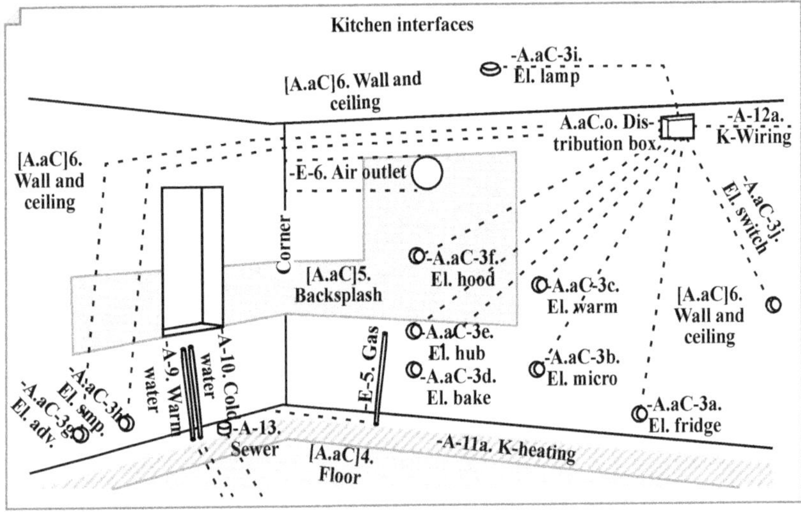

FIGURE 10-8 House kitchen connections

From above kitchen *interface* drawing in Figure 10-8, it is possible to list the *white-box design items* (see Table 10-18 and Table 10-19 below).

TABLE 10-18 House kitchen electricity white-box design items

Item i117. Junction box	i118. Socket plug	i119. El. Pipe	Pipe length meter	Item i120. Cables	Cable length meter	White-box ingredient
			2		3 x 2,5	-A.aC-3a. El. fridge
			2		3 x 2,5	-A.aC-3b. El. micro
			1		3 x 1,5	-A.aC-3c. El. warm
			2		3 x 2,5	-A.aC-3f. El. hood
			9	FK 1,5 mm² Copper wire	3 x 9,5	-A.aC-3g. El. adv. dish
			9		3 x 9,5	-A.aC-3h. El. smp. dish
			2		3 x 2,5	-A.aC-3i. El. lamp
	Plastic CEE 7/4	PVC Ø 16 mm	1		3 x 1,5	-A.aC-3j. El. switch
			2,5		5 x 3	-A.aC-3d. El. bake
Plastic Ø 84 mm	Plastic Perilex	PVC Ø 20 mm	2,5	FK 2,5 mm² Copper wire	5 x 3	-A.aC-3e. El. hub

TABLE 10-19 House kitchen electricity white-box design items

Item	Material /Finish	Inset size w x d x h cm	Outer size w x d x h cm	Openings needed	White-box ingredient
i121. Distribution box	Grey plastic	20 x 8 x 15		For connecting electricity pipes	A.aC.o. Distribution box

Based on the kitchen interior *architecture* in Figure 10-7 above, it is now possible to list the interior *white-box design items* (see Table 10-20 and Table 10-21 below).

TABLE 10-20 House kitchen interiors white-box design items
(cabinets for machinery)

Item	Material / Finish	Inset size w x d x h cm	Outer size w x d x h cm	Openings needed	White-box ingredient
i122. Housing		58 x 58 x 118	60 x 58 x 120	Back ventilation Electricity	[A.aC.aA]1. Cold housing
i123. Cover door	Oak		60 x 110		-A.aC-1b. Cold door
i124. Ventilation grid	White	58 x 8	60 x 10	Front ventilation	
i125. Housing	Grey	58 x 56 x 88	60 x 58 x 90	Back ventilation Electricity	[A.aC.aB]1. Upper housing
i126. Machine screw	Chrome	M5 x 25			
i127. Housing	Grey	58 x 56 x 88	60 x 58 x 90	Back ventilation Electricity	[A.aC.aB]2. Lower housing
i128. Machine screw	Chrome	M5 x 25			
i129. Housing	Grey	58 x 58 x 98	60 x 58 x 100	Back ventilation Electricity	[A.aC.aC]1. Oven housing
i130. Machine screw	Chrome	M6 x 25			
i131. Seal band	Heat proof rubber				[A.aC.aD]1. Gasket seal
i132. Clips	Metal				
i133. Tin box	Customized chrome		20 x 16 x 46	Air	[A.aC.aE]1. Air tin box
i134. Sheet metal screw	Stainless steel	3,5 x 16			
i135. Seal band	Rubber				[A.aC.aF]1. Gasket seal
i136. Glue	Waterproof				
i137. Housing	Grey	58 x 58 x 78	60 x 60 x 80	Cold water, Drain, Electricity	[A.aC.aG]1. Left housing [A.aC.aG]2. Right housing
i138. Machine screw	Chrome	M6 x 40			
i139. Cover doors	Oak		60 x 80		-A.aC-1n. Dish door
i140. Machine screw	Chrome	M6 x 40			-A.aC-1p. Dish door

TABLE 10-21 House kitchen interiors white-box design items
(cabinets without machinery)

Item	Material / Finish	Outer size w x h x d cm	Inset size w x h x d cm	White-box ingredient
i141. Housing	Grey	60 x 40 x 58	58 x 38 x 57	A.aC.b.Medicine cupboard
i142. Door	Oak	60 x 40		
i143. Shelf	Grey	60 x 1 x 57		
i144. Housing	Grey	60 x 210 x 48	58 x 208 x 47	A.aC.c.Grocery cupboard
i145. Sliding stand	Chrome	56 x 200 x 46		
i146. Front	Oak	60 x 210		
i147. Housing	Grey	60 x 30 x 48		A.aC.d.Grocery cupboard
i148. Door	Oak	60 x 30		
i149. Shelf	Grey	60 x 1 x 47		
i150. Housing	Grey	60 x 30 x 48	58 x 38 x 57	A.aC.e.Trash receptacle
i151. Door	Oak	60 x 30		
i152. Housing	Grey	40 x 60 x 48	38 x 58 x 57	A.aC.f.Tiny utensil drawers
i153. Drawer	Grey	38 x 10 x 47		
i154. Front	Oak	40 x 10		
i155. Drawer	Grey	38 x 25 x 47		
i156. Front	Oak	40 x 25		
i157. Housing	Grey	90 x 30 x 58		A.aC.g.Baking cupboard
i158. Door	Oak	90 x 30		
i159. Housing	Grey	40 x 80 x 48	38 x 58 x 47	A.aC.h.Tableware drawers
i160. Drawer	Grey	38 x 20 x 47		
i161. Front	Oak	40 x 20		
i162. Housing	Grey	Corner 90 x 80 x 90	88 x 58 x 88	A.aC.i.Big utensil cupboard
i163. Door	Oak	42 x 80		
i164. Revolving shelf	Grey	Ø 85 cm		
i165. Stone slab	Black granite	260 x 4 x 60	Customized hole 80 x 50	A.aC.j.Preparation counter
i166. Stone slab	Black granite	90 x 4 x 90		
i167. Stone slab	Black granite	125 x 4 x 60	Customized hole 100 x 50	
i168. Housing	Grey	30 x 80 x 28	28 x 78 x 27	A.aC.k.Chemical cupboard
i169. Door	Oak	30 x 80		
i170. Shelf	Grey	28 x 1 x 27		
i171. Housing	Grey	60 x 80 x 28	58 x 78 x 27	A.aC.l.Tableware cupboard
i172. Door	Oak	60 x 80		
i173. Shelf	Grey	58 x 1 x 27		
i174. Housing	Grey	45 x 80 x 45	28 x 78 x 28	A.aC.m.Tableware cupboard
i175. Door	Oak	21 x 80		
i176. Revolving shelf	Grey	Ø 40 cm		
i177. Magnet	Grey	60		A.aC.n.Knife hanger

Some mounting details are attached to casings, doors, and shelves, but many have to be procured separately (see Table 10-22 below).

TABLE 10-22 House kitchen interiors white-box design items
(cabinet static interfaces)

Item	Finish	Description	White-box ingredient
i178. Bolt pairs	Chrome	Each cabin is screwed together by special pairs of machine bolts M5 x 25.	
i179. Wood screw	Metal	All cabins are screwed into the wall and lower cabins into the floor socket by 4,2 x 45 mm.	
i180. Wall plug		5 x 50 plastic wall plug for screws into bricks	-A.aC-2. Interior attachment
i181. Floor socket	Oak	Floor socket under all cabinets	
i182. Outer shelf	Oak	Shelves between two gaps in upper cupboards	
i183. Door hinges	Chrome	Hinges max open angle 110° between doors and cabinet	
i184. Shelf pins	Chrome	Pins Ø 4 mm under-cupboard shelves	

The kitchen piping must be covered and the kitchen's final appearance must be designed (see Table 10-23 below).

TABLE 10-23
House kitchen boundaries white-box design items

Item	Material & size	Color	Area	Coverage	White-box ingredient
i185. Paint	Silicate	Eggshell	10 m²	0,3 liter / m²	[A.aC]6. Wall and ceiling
i186. Paint	Silicate	White	16 m²	0,3 liter / m²	
i187. Plaster	0 - 3 mm	Unfinished	25 m²	20 kg / m²	
i188. Ceramics	60 x 60 cm	Brown	5 m²	3 / m²	[A.aC]4. Floor
i189. Floating putty		Unfinished	5 m²	16 kg / m²	
i190. Adhesive		Unfinished	5 m²	3 kg/m²	
i191. Joint sealant		Dark grey	5 m²	1 kg / m²	
i192. Ceramics	30 x 60 cm	Brown	5 m²	6 / m²	[A.aC]5. Backsplash
i193. Plaster		Natural	5 m²	20 kg / m²	
i194. Adhesive		Natural	5 m²	3 kg / m²	
i195. Joint sealant		Dark grey	5 m²	1 kg / m²	

10.5.3 EXAMPLE House kitchen: Account for white-box development, realization, and manufacturing

To support *partial verification* of how the interiors fit into the kitchen, the kitchen must be marked with interior fakes (see Table 10-24 and Table 10-25 below).

TABLE 10-24 House fake rooms,
estimated realization material requisites cost

Pc	€/Pc	€	Item	Product	Type	Supplier	Element
20	1.5	30	i116. Chalk crayons	Cleany	FX	SU.27 Local supplier	A.a. Rooms

TABLE 10-25 House fake rooms,
estimated realization work requisites hourly/fixed/rental cost

h	€ /h	Cost €	Item	Provider	Interface
2	30	60	i60. Quality inspector	PR.10. Proprietor	-A-2. House inner walls

Finally, all needed kitchen building materials are compiled into a huge *supplier* purchase list (see Table 10-26 below). Note that kitchen machinery has the *product requisites* already in place, and the list below refers only to remaining kitchen interiors.

This list is far from being exhaustive; a real kitchen list would probably be much longer. The list is not very informative, but it is included to show that sometimes *complexity* depends on a huge number of simple details, which must be properly mastered.

TABLE 10-26 House kitchen estimated realization material requisites cost

pc	€/pc	€	Item	Product	Type	Supplier	Element
1	330	330	i122. Housing	IKEY	C60-120	SU.24 K-Studio	A.aC.aA. Cold machinery
1	150	150	i123. Cover door	IKEY	D60-120	SU.24 K-Studio	A.aC.aA. Cold machinery
1	35	35	i124. Ventilation grid	MedMach	W34-G	SU.24 K-Studio	A.aC.aA. Cold machinery
1	210	210	i125. Housing	IKEY	C60-80	SU.24 K-Studio	A.aC.aB. Warm machinery
8	0.04	0.32	i126. Machine screw	MedMach		SU.27 Local supplier	A.aC.aB. Warm machinery
1	230	230	i127. Housing	IKEY	C60-80	SU.24 K-Studio	A.aC.aB. Warm machinery
4	0.1	0.4	i128. Machine screw	MedMach		SU.24 K-Studio	A.aC.aB. Warm machinery
1	230	230	i129. Housing	IKEY	C60-80	SU.24 K-Studio	A.aC.aC. Bake machinery
4	0.04	0.16	i130. Machine screw	MedMach		SU.27 Local supplier	A.aC.aC. Bake machinery
2 m	5.8	11.6	i131. Seal band			SU.24 K-Studio	A.aC.aD. Heat machinery
8	0.6	4.80	i132. Clips			SU.24 K-Studio	A.aC.aD. Heat machinery
1	130	130	Tin box	Local tin-smith		SU.24 K-Studio	A.aC.aE. Ventilation machinery
4	0.04	0.16	i134. Sheet metal screw			SU.27 Local supplier	A.aC.aE. Ventilation machinery
3 m	2.6	7.8	i135. Seal band			SU.24 K-Studio	A.aC.aF. Wet machinery
1	8	8	i136. Glue			SU.24 K-Studio	A.aC.aF. Wet machinery
2	310	620	i137. Housing	IKEY	C60-80	SU.24 K-Studio	A.aC.aG. Dish machinery
8	0.04	0.32	i138. Machine screw	Wat Son		SU.24 K-Studio	A.aC.aG. Dish machinery
2	110	220	i139. Cover doors	IKEY	D60-80	SU.24 K-Studio	A.aC.aG. Dish machinery
4	0.1	0.4	i140. Machine screw	Wat Son		SU.24 K-Studio	A.aC.aG. Dish machinery
1	120	120	i141. Housing	IKEY	C60-40	SU.24 K-Studio	A.aC.b. Medicine cupboard
1	80	80	i142. Door	IKEY	C60-40	SU.24 K-Studio	A.aC.b. Medicine cupboard
2	8	16	i143. Shelf	IKEY	C60-57	SU.24 K-Studio	A.aC.b. Medicine cupboard
1	230	230	i144. Housing	IKEY	CP40-210	SU.24 K-Studio	A.aC.c. Grocery cupboard
1	490	490	i145. Sliding stand	IKEY	SL40-200 steel	SU.24 K-Studio	A.aC.c. Grocery cupboard
1	190	190	i146. Front	IKEY	C60-210	SU.24 K-Studio	A.aC.c. Grocery cupboard
1	70	70	i147. Housing	IKEY	C60-30	SU.24 K-Studio	A.aC.d. Grocery cupboard
1	70	70	i148. Door	IKEY	C60-30	SU.24 K-Studio	A.aC.d. Grocery cupboard

TABLE 10-26 House kitchen estimated realization material requisites cost

pc	€/pc	€	Item	Product	Type	Supplier	Element
2	8	16	i149. Shelf	IKEY	C60-47	SU.24 K-Studio	A.aC.d. Grocery cupboard
1	370	370	i150. Housing	IKEY	C60-30	SU.24 K-Studio	A.aC.e. Trash receptacle
1	18	18	i151. Door	IKEY	C60-30	SU.24 K-Studio	A.aC.e. Trash receptacle
1	410	410	i152. Housing	IKEY	D40-60	SU.24 K-Studio	A.aC.f. Tiny utensil drawers
1	17	17	i153. Drawer	IKEY	D40-10	SU.24 K-Studio	A.aC.f. Tiny utensil drawers
1	20	20	i154. Front	IKEY	D40-10	SU.24 K-Studio	A.aC.f. Tiny utensil drawers
2	22	44	i155. Drawer	IKEY	D40-25	SU.24 K-Studio	A.aC.f. Tiny utensil drawers
2	20	40	i156. Front	IKEY	D40-25	SU.24 K-Studio	A.aC.f. Tiny utensil drawers
1	240	240	i157. Housing	IKEY	C90-30	SU.24 K-Studio	A.aC.g. Baking cupboard
1	90	90	i158. Door	IKEY	C90-30	SU.24 K-Studio	A.aC.g. Baking cupboard
1	410	410	i159. Housing	IKEY	D40-60	SU.24 K-Studio	A.aC.h. Tableware drawers
8	20	160	i160. Drawer	IKEY	D40-10	SU.24 K-Studio	A.aC.h. Tableware drawers
8	18	144	i161. Front	IKEY	D40-20	SU.24 K-Studio	A.aC.h. Tableware drawers
1	590	590	i162. Housing	IKEY	RC90-60	SU.24 K-Studio	A.aC.i. Big utensil cupboard
1	190	190	i163. Door	IKEY	RC42-80	SU.24 K-Studio	A.aC.i. Big utensil cupboard
2	60	120	i164. Revolving shelf	IKEY	RC85	SU.24 K-Studio	A.aC.i. Big utensil cupboard
1	500	500	i165. Stone slab	Myne		SU.25 K-Studio's suppliers	A.aC.j. Preparation counter
1	400	400	i166. Stone slab	Myne		SU.25 K-Studio's suppliers	A.aC.j. Preparation counter
1	430	430	i167. Stone slab	Myne		SU.25 K-Studio's suppliers	A.aC.j. Preparation counter
1	140	140	i168. Housing	IKEY	U30-80	SU.24 K-Studio	A.aC.k. Chemical cupboard
1	30	30	i169. Door	IKEY	U30-80	SU.24 K-Studio	A.aC.k. Chemical cupboard
2	8	16	i170. Shelf	IKEY	U30-27	SU.24 K-Studio	A.aC.k. Chemical cupboard
1	210	210	i171. Housing	IKEY	U40-80	SU.24 K-Studio	A.aC.l. Tableware cupboard
1	70	70	i172. Door	IKEY	U40-80	SU.24 K-Studio	A.aC.l. Tableware cupboard
2	8	16	i173. Shelf	IKEY	U40-27	SU.24 K-Studio	A.aC.l. Tableware cupboard
1	310	310	i174. Housing	IKEY	UC45-80	SU.24 K-Studio	A.aC.m. Tableware cupboard
1	70	70	i175. Door	IKEY	UC21-80	SU.24 K-Studio	A.aC.m. Tableware cupboard
2	70	140	i176. Revolving shelf	IKEY	UC40	SU.24 K-Studio	A.aC.m. Tableware cupboard
1	29	29	i177. Magnet	IKEY	TT 60	SU.24 K-Studio	A.aC.n. Knife hanger
6 m	20	120	i181. Floor socket	IKEY		SU.24 K-Studio	-A.aC-2. Interior attachment
2 m	22	44	i182. Outer shelf	IKEY	SH-30	SU.24 K-Studio	-A.aC-2. Interior attachment
40	0.1	4	i178. Bolt pairs	IKEY		SU.24 K-Studio	-A.aC-2. Interior attachment
60	0.04	2.4	i179. Wood screw			SU.27 Local supplier	-A.aC-2. Interior attachment
30	0.02	0.60	i180. Wall plug			SU.27 Local supplier	-A.aC-2. Interior attachment
16	18	288	i183. Door hinges	IKEY		SU.24 K-Studio	-A.aC-2. Interior attachment
30	0.02	0.6	i184. Shelf pins	IKEY		SU.24 K-Studio	-A.aC-2. Interior attachment
1	34	34	i121. Distribution box			SU.27 Local supplier	A.aC.o. Distribution box
9	4	36	i117. Junction box			SU.27 Local supplier	-A.aC-3. Machinery el. feed

TABLE 10-26 House kitchen estimated realization material requisites cost

pc	€/pc	€	Item	Product	Type	Supplier	Element
7	5	35	i118. Socket plug			SU.27 Local supplier	-A.aC-3. Machinery el. feed
27	0.4	10.8	i119. El. Pipe			SU.27 Local supplier	-A.aC-3. Machinery el. feed
96 m	0.7	67.2	i120. Cables			SU.27 Local supplier	-A.aC-3. Machinery el. feed
2	9	18	i118. Socket plug			SU.27 Local supplier	-A.aC-3. Machinery el. feed
5	0.6	3	i119. El. Pipe			SU.27 Local supplier	-A.aC-3. Machinery el. feed
30 m	0.7	21	i120. Cables			SU.27 Local supplier	-A.aC-3. Machinery el. feed
33 l	4.4	145.2	i185. Paint			SU.23 House builder with subcontractors	[A.aC]6. Wall and ceiling
53 l	4.4	233.2	i186. Paint			SU.23 House builder with subcontractors	[A.aC]6. Wall and ceiling
500 kg	0,5	250	i187. Plaster			SU.23 House builder with subcontractors	[A.aC]6. Wall and ceiling & [A.aC]5. Backsplash
15	51	765	i188. Ceramics	Hazard large	Italian 4C	SU.23 House builder with subcontractors	[A.aC]4. Floor
30	34	1020	i192. Ceramics	Hazard	Italian 4C	SU.23 House builder with subcontractors	[A.aC]5. Backsplash
80 kg	4.1	328	i189. Floating putty			SU.23 House builder with subcontractors	[A.aC]4. Floor
15 kg	3.7	55.5	i190. Adhesive			SU.23 House builder with subcontractors	[A.aC]4. Floor & [A.aC]5. Backsplash
5 kg	4.3	21.5	i191. Joint sealant	Hazard	5C	SU.23 House builder with subcontractors	[A.aC]4. Floor & [A.aC]5. Backsplash
		12 198					

The cost for kitchen work is estimated and listed in Table 10-27 below.

TABLE 10-27 House kitchen
estimated realization work requisites hourly/fixed/rental cost

h	€ /h	Total €	Item	Provider	Interface
		0	i59. Kitchen installer	PR.17. K-Studio	-A.aC-1. Interior access -A.aC-2. Interior attachment
55	40	2 200	i58. Electrician	PR.12. Elert AB	-A-12a. K-Wiring -A.aC-3. Machinery el. feed -A.aC-3a. El. fridge -A.aC-3b. El. micro -A.aC-3c. El. warm -A.aC-3d. El. bake -A.aC-3e. El. hub -A.aC-3f. El. hood -A.aC-3g. El. adv. dish -A.aC-3h. El. smp. dish -A.aC-3i. El. lamp -A.aC-3j. El. switch
120	30	3 600	i57. Interior junior worker	PR.13. House builder	[A.aC]4. Floor [A.aC]5. Backsplash [A.aC]6. Wall and ceiling -E-9. K-window
30	40	1 200	i56. Interior senior worker	PR.13. House builder	-A-11a. K-heating
		7 000			

The cost for all rooms development is estimated in Table 10-28 below

TABLE 10-28 House rooms
estimated development work requisites hourly/fixed cost

h	€ /h	Total €	Item	Provider	Schedule
		0	i49. House user	PR.10. Proprietor	Pt.2.R. Refine room requirements from house requirements, and ensure room development staffing.
		0	i51. Kitchen designer	PR.17. K-Studio	Pt.2.P. Predetermine room solutions with supplier opportunities.
12	50	600	i50. Interior designer	PR.11. Robber HB	
30	40	1 200	i52. Interior senior designer	PR.11. Robber HB	Pt.2.A. Satisfy room requirements by decomposing room black-boxes into white-box design containing machinery black-boxes.
44	35	1 540	i53. Interior junior designer	PR.11. Robber HB	
20	40	800	i54. Electrician	PR.12. Elert AB	Pt.2.F. Finalize design of rooms with product requisites.
		0	i55. Kitchen designer	PR.17. K-Studio	
20	50	1 000	i60. Quality inspector	PR.10. Proprietor	Pt.2.V. Verify room black-/white-box.
		0	i61. Payment endorsement	PR.10. Proprietor	Pt.2.V. Verify room black-/white-box.
		5 140			

Finally, the accumulated costs for all rooms and their machinery are summarized and accumulated in Table 10-29 below. Note, only kitchen development has so far been presented in this book. Other rooms' costs are calculated separately and anticipated to make the house example complete.

TABLE 10-29 House rooms
summed/accumulated requisites estimates

Requisite	Reference	Cost €
House rooms machinery summed/accumulated requisites estimates	Table 10-16, chapter 10.4	7 832
House fake rooms, estimated realization material requisites cost	Table 10-24, this chapter	30
House fake rooms, estimated realization work requisites hourly/fixed/rental cost	Table 10-25, this chapter	60
House kitchen estimated realization material requisites cost	Table 10-26, this chapter	12 198
House kitchen estimated realization work requisites hourly/fixed/rental cost	Table 10-27, this chapter	7 000
House toilet estimated realization material requisites cost		3 230
House toilet estimated realization work requisites hourly/fixed/rental cost		3 560
Dining room estimated realization material requisites cost	Not shown in this book	10 560
Dining room estimated realization work requisites hourly/fixed/rental cost		3 420
Entrance estimated realization material requisites cost		7 040
Entrance estimated realization work requisites hourly/fixed/rental cost		3 760
House rooms estimated development work requisites hourly/fixed cost	Table 10-28, this chapter	5 140
House rooms summed/accumulated requisites estimates		**63 830**

The *architecture* in Figure 9-22, page 363 has divided the house into rooms and remaining secondary space, which may be accounted separately (see Table 10-30 below). Note that the secondary space contains no *embedded black-boxes*, thus everything (such as possible machinery) is included in secondary space accounting below.

TABLE 10-30 House secondary space
summed/accumulated requisites estimates

Requisite	Reference	Cost €
Staircase space estimated realization material requisites cost		5 483
Staircase space estimated realization work requisites hourly/fixed/rental cost		3 316
Loft estimated realization material requisites cost	Not shown in this book	1 770
Loft estimated realization work requisites hourly/fixed/rental cost		2 098
House secondary space estimated development work requisites hourly/fixed cost		1 227
House secondary space summed/accumulated requisites estimates		**13 894**

10.5.4 EXAMPLE House kitchen: Examine white-box success

The proprietor called together the HB Robber *architect* company and the K-Studio company to review the *white-box design items* and *product requisites*. The bottom line cost for the machinery and rooms did not deviate much from earlier estimates, and these subcontractors were asked to continue finalizing the design.

It was also decided that the proprietor must capture color and appearance of the house rooms machinery for the selected house builder, before the house builder procures these.

10.5.5 EXAMPLE: EXAMPLE House: Proceed reading

To follow this example, these are the alternatives:

2I Proceed to chapter 11.4, "EXAMPLE House rooms: Procure room items (and await invading machinery black-boxes) in order to build rooms," page 613.

2V Proceed to chapter 12.6, "EXAMPLE House kitchen: Partially verify black-/white-box with embedded fakes," page 677.

2V Proceed to chapter 12.8, "EXAMPLE House kitchen: Fully verify room black-/white-box," page 684.

1F Proceed to chapter 10.6 "EXAMPLE House: Finalize design of house with product requisites" below.

10.6 EXAMPLE House: Finalize design of house with product requisites

10.6.1 EXAMPLE House: Process schedule to use

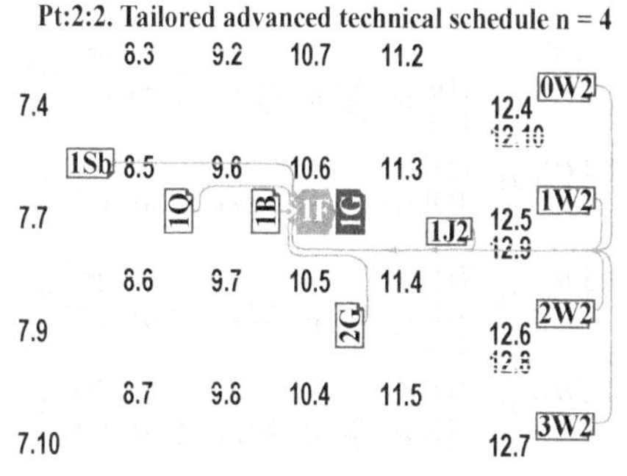

For the house example, a special technical overall *schedule* is *tailored* (illustrated in Fig. 10-9, left):

• **Pt:2:2. Tailored advanced technical schedule n = 4,** page 164

In this *tailored schedule* the generic *schedule* to use is (filled-in symbols):

• **Pt.n.F. Finalize design of white-box with product requisites,** page 140.

FIGURE 10-9 Position of schedule and master to use (filled in) in overall schedule

Below *results* may be demanded (their *masters* are indicated *upstream* from the filled-in *schedule* in the figure above):

2G Finalized room items design with product requisites

Figure 10-7, House kitchen interior drawing, page 455

Figure 10-8, House kitchen connections, page 456

Table 10-18, House kitchen electricity white-box design items, page 456

Table 10-19, House kitchen electricity white-box design items, page 457

Table 10-20, House kitchen interiors white-box design items (cabinets for machinery), page 457

Table 10-21, House kitchen interiors white-box design items (cabinets without machinery), page 458

Table 10-22, House kitchen interiors white-box design items (cabinet static interfaces), page 459

Table 10-23, House kitchen boundaries white-box design items, page 459

1B House architectures with major ingredients and room black-boxes

Table 9-13, House rooms black-boxes architecture ingredients resumé, page 360

Table 9-14, House interfaces architecture ingredients resumé, page 361

Table 9-15, House secondary space elements architecture ingredients resumé, page 362

Figure 9-22, House logical architecture, page 363

Figure 9-23, House physical architecture, page 364

Figure 9-24, House façade physical architecture, page 364

1Q House solution alternatives with supplier opportunities

Table 8-14, House kitchen white-box solution alternatives, page 306

Table 8-15, House kitchen white-box supplier opportunities, page 308

1Sb House staffing allocation

Table 7-10, House development allocation staffing, page 190

Xb Suppliers' sample catalogues

E Picked in-house legacy source systems

XaE Ordered suppliers' off-the-shelf fully described source systems

XbE Ordered suppliers' turnkey fully described source systems

For possible *failure localization* and *failure elimination*, below *results* may be demanded (their *masters* are indicated *upstream* from the filled-in *schedule* in the figure above):

1J2 Table 11-14, House start-up report 1, page 612

0W2 List 12-1. House environment interfaces with fake house failure elimination report 1, page 674

0W2 Table 12-27, House environment verification report 1, page 701

Table 12-29, House environment reverification report 2, page 703

Table 12-28, House environment failure elimination report 1, page 702

1W2 List 12-2. House with fake rooms failure elimination report 1, page 676

1W2 Table 12-22, House verification report 1, page 697

Table 12-24, House verification report 2, page 698

Table 12-23, House failure elimination report 1, page 698

2W2 List 12-3. House rooms with fake machinery failure elimination report 1, page 678

2W2 Table 12-14, House kitchen verification report 1, page 688

Table 12-17, House kitchen verification report 2, page 691

Table 12-15, House kitchen failure elimination report 1, page 689

3W2 Table 12-9, House kitchen machinery verification report 1, page 682

Table 12-11, House kitchen machinery verification report 2, page 683
Table 12-10, House kitchen machinery failure elimination report 1, page 683

10.6.2 EXAMPLE House: Finalize white-box detailed items

Since much of the *complexity* is about static material, there is no need to detail the high-level *logical architecture*. However, a *white-box design items* drawing is often needed, preferably using identifiers in the same way as *architectures* (see Fig. 10-10 below for the upper house boundary, and Fig. 10-11 below for the lower house boundary).

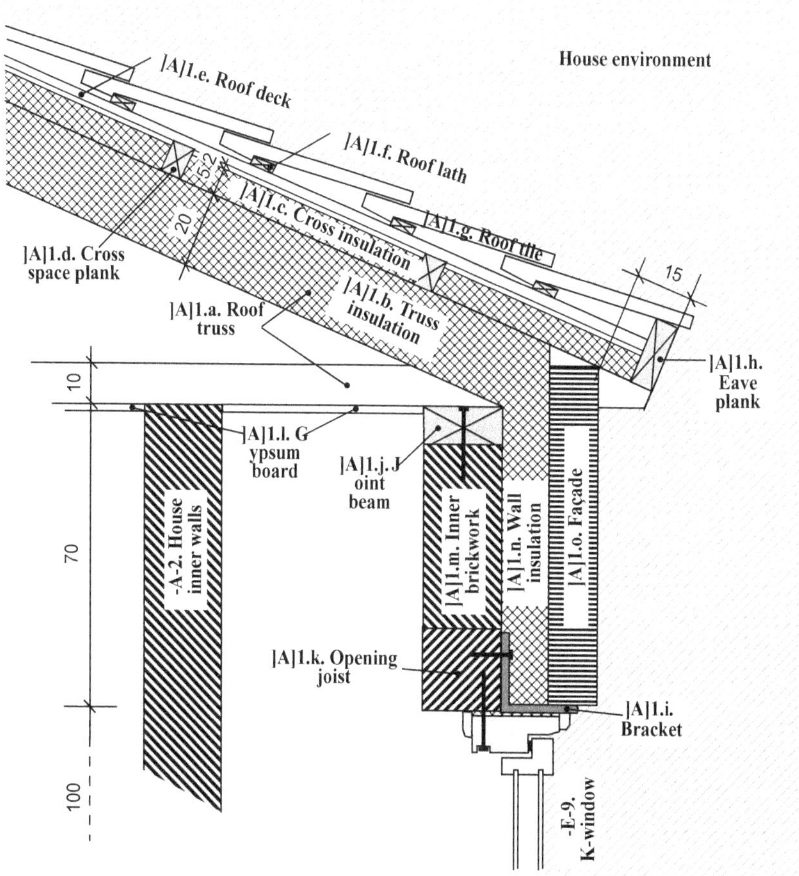

FIGURE 10-10 House boundary casing (upper part)

A general rule is to mainly describe the *complex* parts of the *product* and leave out simple parts. For example, *complexity* of a house that must be described is where roof, ceiling walls, floor, doors, and windows meet each other. There are fewer reasons to describe a central part of a wall, roof, or floor.

All rooms' inner boundaries (including those shared with the house boundary), with their machinery are now architecturally designed. The next step is to prepare *white-box design items* of the house boundary and common equipment.

The house outer boundary has many purposes:

- Protection from burglars and other intruders
- Protection from moisture
- Sufficient strength to withstand wind and bear snow on the roof.
- Insulation to keep out heat during summer, and retain heat during winter
- Appearance

House building standards together with long experience in house building often ensures a good *architecture*. A house is visible and readily understandable to *users* and owners, so despite a probably rather *unbiased architecture ingredients* design, the risk of technicians' departing from the plans is not too great.

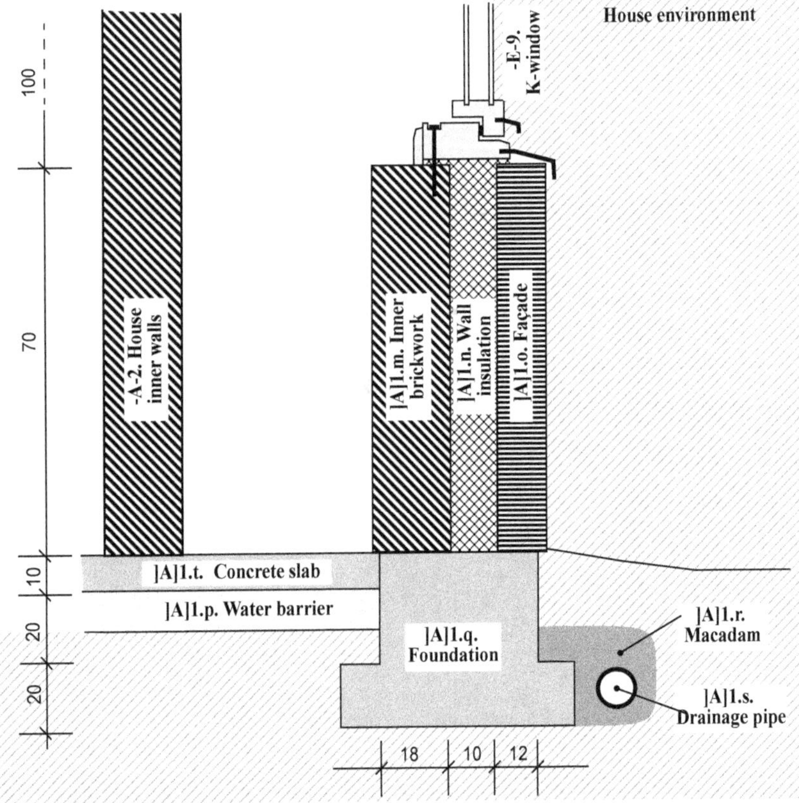

FIGURE 10-11 House boundary casing (lower part)

Also the house *connection interfaces* need to be better understood and designed, and the *physical architecture* in Figure 9-23, page 364 is updated with a *white-box design items* drawing (see Fig. 10-12 below).

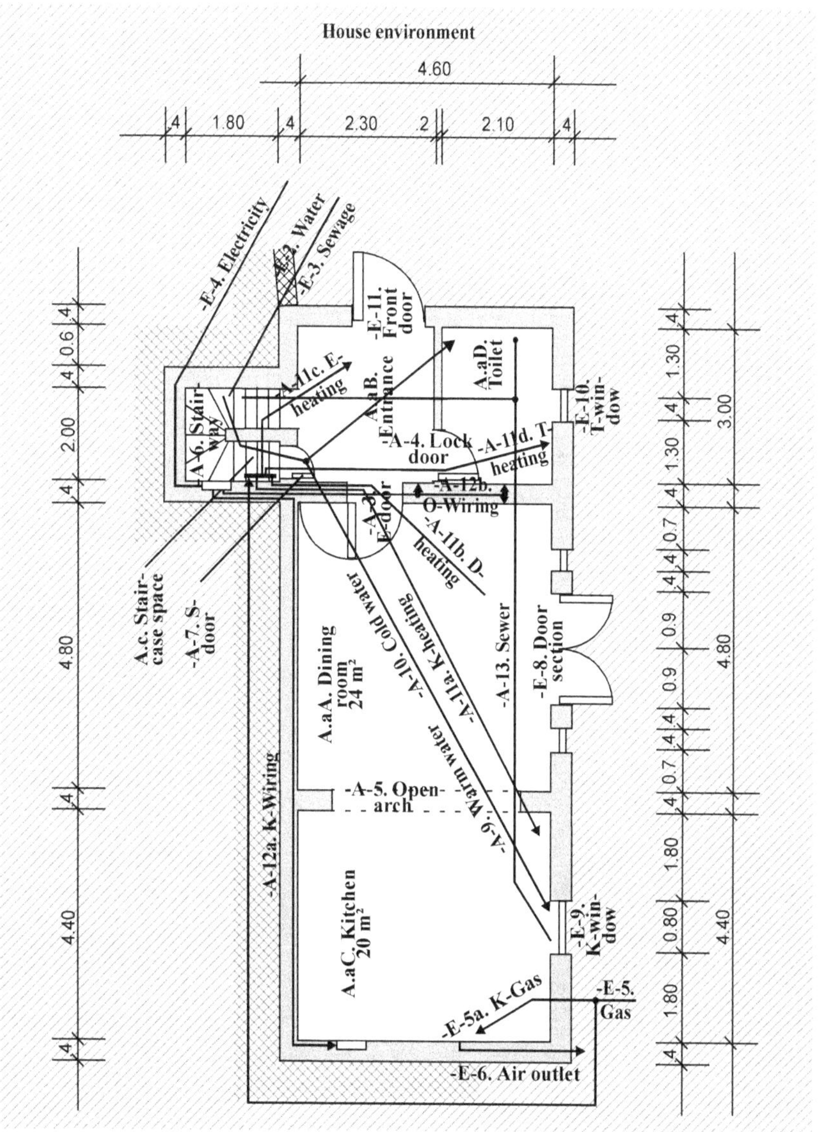

FIGURE 10-12 House connections

For preparing the fake house, some small items are needed (see Table 10-31 below.)

TABLE 10-31 House fake items

Item	Color	Size	Type	White-box ingredient
i196. Sticks	Unfinished	15 x 14 mm	Wood	A. House]A[1. House boundary
i197. Strings	White	3 mm x 100 m	Nylon	
i198. Spray paint	Various	400 ml	Permanent	
i199. Frostless faucet			Steel, outdoor type, Ø 1 inch	-E-2. Water
i200. Outdoor distribution box			Fused sockets for 3 x 20 A and 1 x 10 A	-E-4. Electricity

From above house drawings, it is easy to finalize the *white-box design items* (see Table 10-32 below).

TABLE 10-32 House boundary white-box design items

Item	Remark	Material and dimensions	White-box ingredient
i201. Roof truss	Prefabricated	Framework of spruce 5,10 wide x 2,0 m high	]A]1.a. Roof truss
i202. Fiberglass	3GS	Board 60 x 120 x 20 cm	]A]1.b. Truss insulation
i203. Fiberglass	3GS	Roll 20 m x 60 x 5 cm	]A]1.c. Cross insulation
i204. Plank		Unplaned spruce 2 x 3 inch	]A]1.d. Cross space plank
i205. Roof deck		0,5 x 5 inch, unplaned tongued spruce surplus plank	]A]1.e. Roof deck
i206. Lath		1 x 1,5 inch unplaned spruce plank	]A]1.f. Roof lath
i207. Plank		1,5 x 5 inch unplaned spruce plank	]A]1.h. Eave plank
i208. Bracket		L-shape galvanized steel 15 x 15 cm	]A]1.i. Bracket
i209. Plank		4 x 7 inch unplaned spruce plank	]A]1.j. Joint beam
i210. Joist	Prefabricated	Concrete joist 18 x 18 cm	]A]1.k. Opening joist
i211. Gypsum		Gypsum board 15 mm, 60 x 120 cm	]A]1.l. Gypsum board
i212. Brick		20/m² hollow bricks 18 x 27 cm	]A]1.m. Inner brickwork
i213. Mortar	Prefabricated	2600 kg fine concrete	
i214. Styrofoam		Styrofoam sheet 60 x 120 x 10 cm	]A]1.n. Wall insulation
i215. Gravel		Gravel 4 - 32 mm	]A]1.p. Water barrier
i216. Footer	Concrete prefabricated	Molded concreted	]A]1.q. Foundation
i217. Reinforcement bar		Iron bar Ø 15 mm	
i218. Macadam		Macadam 16 - 32 mm	]A]1.r. Macadam
i219. Pipe		Plastic perforated pipe	]A]1.s. Drainage pipe
i220.Concrete slab	Prefabricated	Fine concrete 10 cm	]A]1.t. Concrete slab
i221. Reinforcement net		Bar Ø 8 mm, 120 x 240 cm, mesh 15 x 15 cm	

From the above house connection drawing, it is also easy to finalize the *connection interface* design (see Table 10-33 below).

TABLE 10-33 House connections white-box design items

Item	Remark	Material and size	White-box ingredient
i222. Door	Prefabricated	Frame, cover oak 200 x 90	-A-3. E-door
i223. Door	Prefabricated	Frame, cover oak 200 x 80 with toilet locker	-A-4. Lock door
i224.		Plastering inside in same way as wall	-A-5. Open arch
i225. Stair	Prefabricated	Stairway left bent, 12 steps height 20 cm, depth 20 cm	-A-6. Stairway
i226. Door	Prefabricated	White painted 200 x 80	-A-7. S-door
i227. Door	Prefabricated	White painted 180 x 80	-A-8. L-door
i228. Pipe		Copper Ø 3/4 inch	-A-9. Warm water
i229. Pipe			-A-10. Cold water
i230. Hose		Polythene Ø 3/4 inch	-A-11a. K-heating
i231. Hose			-A-11b. D-heating
i232. Hose			-A-11c. E-heating
i233. Hose			-A-11d. T-heating
i234. Hose			-A-11e. B-heating
i235. 5xCable		5 x FK 2,5 mm²	-A-12a. K-Wiring
i236. 3xCable		3 x FK 1,5 mm²	-A-12b. O-Wiring
i237. Drain pipe		Plastic Ø 4 inch	-A-13. Sewer
i238. Coil pipe		Coil-pipe Ø 5 inch	-E-6. Air outlet

Partitioning the finalized design tables mostly depends on which columns are necessary to describe the *white-box design items*. Another reason to gather *interfaces* is that they give the house its outside character (see Table 10-34 below).

TABLE 10-34 House boundary and connections white-box design items

Item	Remark	Material and size	Coverage	White-box ingredient
i239. Brick		Classic red 255 x 120 x 60 mm	55 / m²	JA]1.o. Façade
i240. Mortar		Fine concrete	130kg / m²	
i241. Reinforcing		Galvanized double bars 28 mm, iron Ø 15 mm	7 m / m²	
i242. Hooks		Galvanized 20 cm	2 / m²	
i243. Tiles		Terracotta 42 x 23,5	13,3 / m²	JA]1.g. Roof tile
i244. Window	Prefabricated customized	Wood frame, primed, triple-glazed 80 x 100 cm		-E-9. K-window
i245. Window	Prefabricated customized	Wood frame, primed, triple-raw-glazed, 40 x 40 cm		-E-10. T-window
i246. Door	Prefabricated customized	Burglar-proof oak frame and cover, 200 x 90 cm		-E-11. Front door
i247. Door	Prefabricated customized	Burglar-proof hinges and lock, oak 200 x 90 + 90 cm		-E-8. Door section
i248. Window	Prefabricated customized	Oak 30 x 100 cm		

10.6.3 EXAMPLE House: Account for white-box development, realization, and manufacturing

To support *partial verification* of how the house fits the completed *interfaces* to the *environment*, the house can be marked as a fake (see Table 10-35 and Table 10-36 below).

TABLE 10-35 Fake house,
estimated realization material requisites cost

Pc	€ /Pc	€	Item	Product	Supplier	Element
40	1	40	i196. Sticks	Bank	SU17. AllBau	A. House
5	2.6	13	i197. Strings	Plasm	SU22. Local suppliers	A. House
10	4	40	i198. Spray paint	GoldPot	SU22. Local suppliers	A. House
1	53	53	i199. Frostless faucet		SU.15. Local supplier	Preliminary -E-2. Water
1	420	420	i200. Outdoor distribution box		SU.10. House builder's suppliers	Preliminary -E-4. Electricity
		566				

TABLE 10-36 Fake house,
estimated realization work requisites hourly/fixed/rental cost

h	€ /h	Cost €	Item	Provider	Interface
4	50	200	i23. Quality inspector	PR.14. Q-Inc	]A[1. House boundary
		200			

There is no plan to manufacture a series of houses, and so far only one house prototype will be developed.

In cases like a house with *integration* of a lot of smaller *components*, it takes a substantial effort to estimate all amounts of building material (see Table 10-37 below).

TABLE 10-37 House
estimated realization material requisites cost

№	Unit	€ /unit	€	Item	Product	Supplier	Interface
22	Piece	410	9 020	i201. Roof truss		SU17. AllBau	]A]1.a. Roof truss
132	Sheet	3.8	502	i202. Fiberglass		SU21. Iso-Bau's suppliers	]A]1.b. Truss insulation
8	Roll	38	304	i203. Fiberglass	Rockwool	SU21. Iso-Bau's suppliers	]A]1.c. Cross insulation
160	Meter	1.6	256	i204. Plank	Rockwool	SU22. Local suppliers	]A]1.d. Cross space plank
950	Meter	0.28	266	i205. Roof deck		SU22. Local suppliers	]A]1.e. Roof deck
260	Meter	0.13	34	i206. Lath		SU22. Local suppliers	]A]1.f. Roof lath
27	Meter	1.9	51	i207. Plank		SU22. Local suppliers	]A]1.h. Eave plank
7	Meter	8.5	60	i208. Bracket	IronWare	SU17. AllBau	]A]1.i. Bracket
27	Meter	11	297	i209. Plank		SU22. Local suppliers	]A]1.j. Joint beam
5	Meter	17.2	86	i210. Joist		SU17. AllBau	]A]1.k. Opening joist
2	Meter	17.2	34	i210. Joist		SU17. AllBau	-A-2. House inner walls

TABLE 10-37 House
estimated realization material requisites cost

Nº	Unit	€ /unit	€	Item	Product	Supplier	Interface
1600	Piece	1.2	1 920	i212. Brick	Bricket	SU17. AllBau	]A]1.m. Inner brickwork
5	m³	67	335	i213. Mortar		SU22. Local suppliers	]A]1.m. Inner brickwork
500	Piece	1.2	600	i212. Brick	Bricket	SU17. AllBau	-A-2. House inner walls
2	m³	67	134	i213. Mortar		SU22. Local suppliers	-A-2. House inner walls
50	Sheet	5.8	290	i211. Gypsum		SU22. Local suppliers	]A]1.l. Gypsum board
112	Sheet	0.7	78	i214. Styrofoam		SU21. Iso-Bau's suppliers	]A]1.n. Wall insulation
7	m³	25	175	i215. Gravel		SU22. Local suppliers	]A]1.p. Water barrier
15	m³	70	1 050	i216. Footer		SU22. Local suppliers	]A]1.q. Foundation
420	Meter	0.3	126	i217. Reinforcement bar	IronWare	SU17. AllBau	]A]1.q. Foundation
10	m³	57	570	i218. Macadam		SU22. Local suppliers	]A]1.r. Macadam
36	Meter	5	180	i219. Pipe		SU22. Local suppliers	]A]1.s. Drainage pipe
7	m³	56	392	i220. Concrete slab		SU17. AllBau	]A]1.t. Concrete slab
36	Piece	14	504	i221. Reinforcement net	IronWare	SU17. AllBau	]A]1.t. Concrete slab
4700	Piece	0.043	202	i239. Brick	Bricket	SU17. AllBau	]A]1.o. Façade
5	m³	76	380	i240. Mortar		SU22. Local suppliers	]A]1.o. Façade
600	meter	0.2	120	i241. Reinforcing	IronWare	SU17. AllBau	]A]1.o. Façade
160	Piece	0.14	22	i242. Hooks		SU17. AllBau	]A]1.o. Façade
1260	Piece	4.7	5 922	i243. Tiles	Bricket	SU17. AllBau	]A]1.g. Roof tile
1	Piece	560	460	i244. Window		SU19. Pa&Pp	-E-9. K-window
1	Piece	320	280	i245. Window		SU19. Pa&Pp	-E-10. T-window
1	Piece	872	772	i246. Door		SU19. Pa&Pp	-E-11. Front door
1	Piece	541	608	i247. Door		SU19. Pa&Pp	-E-8. Door section
2	Piece	330	560	i248. Window		SU19. Pa&Pp	-E-8. Door section
1	Piece	518	518	i222. Door		SU19. Pa&Pp	-A-3. E-door
1	Piece	530	530	i223. Door		SU19. Pa&Pp	-A-4. Lock door
1	Piece	1 387	1 187	i225. Stair		SU19. Pa&Pp	-A-6. Stairway
1	Piece	320	320	i226. Door		SU19. Pa&Pp	-A-7. S-door
1	Piece	360	360	i227. Door		SU19. Pa&Pp	-A-8. L-door
8	meter	7	56	i228. Pipe		SU16. House builder's suppliers	-A-9. Warm water
8	Meter	7	56	i229. Pipe		SU16. House builder's suppliers	-A-10. Cold water
80	Meter	7.2	576	i230. Hose		SU16. House builder's suppliers	-A-11a. K-heating
90	Meter	7.2	648	i231. Hose		SU16. House builder's suppliers	-A-11b. D-heating

TABLE 10-37 House
estimated realization material requisites cost

N⁰	Unit	€ /unit	€	Item	Product	Supplier	Interface
30	Meter	7.2	216	i232. Hose		SU16. House builder's suppliers	-A-11c. E-heating
30	Meter	7.2	216	i233. Hose		SU16. House builder's suppliers	-A-11d. T-heating
40	Meter	9.7	388	i235. 5xCable		SU20. Elert's suppliers	-A-12a. K-Wiring
48	Meter	7.1	341	i236. 3xCable		SU20. Elert's suppliers	-A-12b. O-Wiring
8	Meter	6.7	54	i237. Drain pipe		SU16. House builder's suppliers	-A-13. Sewer
3	Meter	14	42	i238. Coil pipe		SU16. House builder's suppliers	-E-6. Air outlet
			32 098				

There is a lot of work to be done to assemble all house *white-box design items* (see Table 10-38 below.)

TABLE 10-38 House
estimated realization work requisites hourly/fixed/rental cost

h	€ /h	€	Item	Provider	Interfaces
40	80	3 200	i33. Construction laborer	PR.13. House builder	-E-3. Sewage -E-2. Water
96	40	3 840	i47. Electrician	PR.12. Elert AB	-E-4. Electricity
48	150	7 200	i35. Pipe-fitter	PR.13. House builder	-A-13. Sewer -A-10. Cold water -A-9. Warm water]A]1.s. Drainage pipe
24	50	1 200	i36. Authorized gas plumber	PR.13. House builder	-E-5. Gas
140	40	5 600	i37. Founder	PR.13. House builder	]A]1.q. Foundation]A]1.t. Concrete slab
40	140	5 600	i34. Excavator with driver and 1 Laborer	PR.13. House builder	]A]1.r. Macadam]A]1.p. Water barrier
56	30	1 680	i39. Plumber	PR.13. House builder	-A-11a. K-heating -A-11b. D-heating -A-11c. E-heating -A-11d. T-heating -A-11e. B-heating -A-11d. T-heating
448	30	13 440	i40. Bricklayer	PR.13. House builder	]A]1.m. Inner brickwork,]A]1.k. Opening joist,]A]1.j. Joint beam -A-2. House inner walls
344	30	10 320	i41. Façade bricklayer	PR.13. House builder	]A]1.i. Bracket]A]1.n. Wall insulation]A]1.o. Façade
16	40	640	i38. Tinsmith	PR.13. House builder	-E-6. Air outlet
48	30	1 440	i42. Chimney mason	PR.13. House builder	-E-7. Chimney

TABLE 10-38 House
estimated realization work requisites hourly/fixed/rental cost

h	€ /h	€	Item	Provider	Interfaces
48	35	1 680	i43. Carpenter	PR.13. House builder	-E-11. Front door -E-12. Back door -E-10. T-window -E-9. K-window -E-8. Door section
124	30	3 720	i44. Cabinet maker	PR.13. House builder	-A-3. E-door -A-4. Lock door -A-5. Open arch -A-6. Stairway -A-7. S-door -A-8. L-door]A]1.l. Gypsum board
36	140	5 040	i46. Woodworker	PR.13. House builder	]A]1.a. Roof truss
252	40	10 080	i45. Roofer	PR.13. House builder	A]1.b. Truss insulation A]1.d. Cross space plank A]1.c. Cross insulation A]1.e. Roof deck A]1.f. Roof lath]A]1.h. Eave plank
		74 680			

The house development cost must also be estimated (see Table 10-39 below).

TABLE 10-39 House
estimated development work requisites hourly/fixed cost

h	€ /h	€	Item	Provider	Process schedule
		0	i25. House user	PR.10. Proprietor	Pt.1.R. Refine house requirements from environment restriction requirements, and ensure house development staffing.
30	50	1 500	i26. Architect	PR.11. Robber HB	Pt.1.P. Predetermine house solutions with supplier opportunities.
100	40	4 000	i27. Architect	PR.11. Robber HB	Pt.1.A. Satisfy house requirements by decomposing house black-box into white-box design containing room black-boxes.
20	40	800	i28. Artistic architect	PR.16. Homey Inc	Pt.n.Ae. Lay out physical architecture.
30	40	1 200	i29. House layout specialist	PR.16. Homey Inc	
8	50	400	i30. Electrician	PR.12. Elert AB	Pt.1.A. Satisfy house requirements by decomposing house black-box into white-box design containing room black-boxes.
150	30	4 500	i31. Calculation engineer	PR.11. Robber HB	Pt.1.F. Finalize design of house with product requisites.
24	40	960	i32. Electrician	PR.12. Elert AB	Pt.1.F. Finalize design of house with product requisites.
40	50	2 000	i48. Quality inspector	PR.10. Proprietor	Pt.1.V. Verify house black-/white-box.
		15 360			

Finally, the total accumulated cost should be summarized (see Table 10-40 below).

TABLE 10-40 House
summed/accumulated requisites estimates

Item	Reference	Cost €
House rooms summed/accumulated requisites estimates	Table 10-29, chapter 10.5	63 830
House secondary space summed/accumulated requisites estimates	Table 10-30, chapter 10.5	13 894
Fake house, estimated realization material requisites cost	Table 10-35, this chapter	566
Fake house, estimated realization work requisites hourly/fixed/rental cost	Table 10-36, this chapter	200
House estimated realization material requisites cost	Table 10-37, this chapter	32 098
House estimated realization work requisites hourly/fixed/rental cost	Table 10-38, this chapter	74 680
House estimated development work requisites hourly/fixed cost	Table 10-39, this chapter	15 360
House summed/accumulated requisites estimates		**200 628**

10.6.4 EXAMPLE House: Examine white-box success

The proprietor called Robber HB to a gate meeting. Now the house total cost began to appear, although the *white-box design items* of the *environment interfaces* were not yet finalized. It still looked like we were on budget.

Robber HB was asked to look around for building companies that could make their own cost estimates and make a bid on building the house.

It was also decided that the proprietor must capture color and appearance of the house *boundary interface* when required by the selected house builder, before the house builder procured these.

10.6.5 EXAMPLE House: Proceed reading

To follow this example, these are the alternatives:

[II] Proceed to chapter 11.3, "EXAMPLE House: Procure house items (and await invading room black-boxes) in order to build house," page 607.

[IV] Proceed to chapter 12.5, "EXAMPLE House: Partially verify black-/white-box with embedded fakes," page 675.

[IV] Proceed to chapter 12.9, "EXAMPLE House: Fully verify house black-/white-box," page 693.

[OF] Proceed to chapter 10.7 "EXAMPLE House environment: Finalize design of environment interfaces with product requisites" below.

10.7 EXAMPLE House environment: Finalize design of environment interfaces with product requisites

10.7.1 EXAMPLE House environment: Process schedule to use

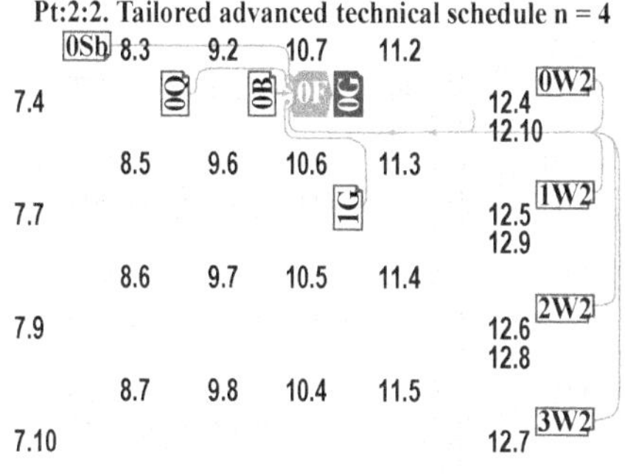

For the house example, a special technical overall *schedule* is *tailored* (illustrated in Fig. 10-13, left):

- **Pt:2:2. Tailored advanced technical schedule n = 4,** page 164

In this *tailored schedule* the generic *schedule* to use is (filled-in symbols):

- **Pt.0.F. Finalize design of environment interfaces with product requisites,** page 118.

FIGURE 10-13 Position of schedule and master to use (filled in) in overall schedule

Below *results* may be demanded (their *masters* are indicated *upstream* from the filled-in *schedule* in the figure above):

1G Finalized room items design with product requisites
Figure 10-10, House boundary casing (upper part), page 467
Figure 10-11, House boundary casing (lower part), page 468
Figure 10-12, House connections, page 469
Table 10-32, House boundary white-box design items, page 470
Table 10-33, House connections white-box design items, page 471
Table 10-34, House boundary and connections white-box design items, page 471

0B Environment existing ingredients and design of interfaces to the environment
Table 9-6, House environment architecture ingredients resumé, page 341
Figure 9-5, House environment logical architecture, page 341
Figure 9-6, House environment physical architecture, page 342

0Q Solution alternatives of interfaces to environment with supplier opportunities
Table 8-5, House environment solution alternatives of interfaces to environment, page 292
Table 8-6, House environment solutions excluded, page 293
Table 8-7, House environment solutions included with supplier opportunities, page 293

0Sb Environment staffing allocation
Table 7-6, House environment development allocation staffing, page 183

Xb) Suppliers' sample catalogues

E) Picked in-house legacy source systems

XaE) Ordered suppliers' off-the-shelf fully described source systems

XbE) Ordered suppliers' turnkey fully described source systems

For possible *failure localization* and *failure elimination*, below *results* may be demanded (their *masters* are indicated *upstreams* from the filled-in *schedule* in the figure above or greater detail in Fig. 11-5, p. 602 and Fig. 11-6, p. 603):

0J2) Table 11-9, House environment start-up report 1, page 606

0W2) List 12-1. House environment interfaces with fake house failure elimination report 1, page 674

0W2) Table 12-27, House environment verification report 1, page 701
Table 12-29, House environment reverification report 2, page 703
Table 12-28, House environment failure elimination report 1, page 702

1W2) List 12-2. House with fake rooms failure elimination report 1, page 676

1W2) Table 12-22, House verification report 1, page 697
Table 12-24, House verification report 2, page 698
Table 12-23, House failure elimination report 1, page 698

2W2) List 12-3. House rooms with fake machinery failure elimination report 1, page 678

2W2) Table 12-14, House kitchen verification report 1, page 688
Table 12-17, House kitchen verification report 2, page 691
Table 12-15, House kitchen failure elimination report 1, page 689

3W2) Table 12-9, House kitchen machinery verification report 1, page 682
Table 12-11, House kitchen machinery verification report 2, page 683
Table 12-10, House kitchen machinery failure elimination report 1, page 683

10.7.2 EXAMPLE House environment: Finalize environment interfaces detailed items

Based on the *environment architecture* in Figure 9-6, page 342, a *environment interfaces design items* drawing is made of the *environment* (see Fig. 10-14 below).

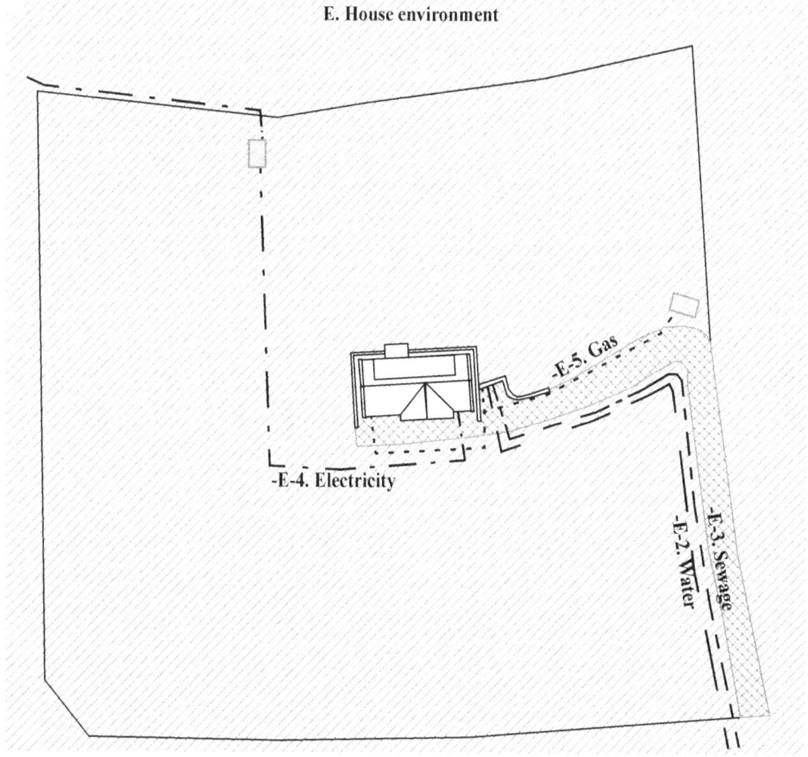

FIGURE 10-14 House environment connections

10.7.3 EXAMPLE House environment: Finalize environment interfaces detailed items

With help from the above drawing it is rather easy to estimate and list the *white-box design items* (see Table 10-41 below).

TABLE 10-41 House interfaces to environment design items

Item	Depth	Other	Material and dimensions	Interface ingredient
i249. No item		150 m	Immaterial	-E-1. Shoreline distance
i250. Water hose	Below 0,6 m		Underground cable polythene Ø 1 inch	-E-2. Water
i251. Sewage pipe	Below 0,6 m	Max 1% leaning	Plastic Ø 5 inch	-E-3. Sewage
i252. Cable	Below 0,6 m		Underground cable 3 x 2,5 + 2,5 mm²	-E-4. Electricity
i253. Gas pipe	Below 0,6 m		Galvanized steel Ø 1/2 inch	-E-5. Gas

TABLE 10-41 House interfaces to environment design items

Item	Depth	Other	Material and dimensions	Interface ingredient
i254. Paving			Basalt stones 20 x 20 cm	-E-6. Driveway
i255. Gravel			Nature 0 - 32 mm	

10.7.4 EXAMPLE House environment: Account for environment interfaces development, realization, and manufacturing

The cost of *environment interfaces design items* from the design descriptions is estimated in a material manufacturing requisite (see Table 10-41 below).

Note that only one house prototype will be built for the moment. No construction of more houses is planned as yet, and no manufacturing requisites with volume prices have been presented so far.

TABLE 10-42 House environment
estimated realization material requisites cost

Nº	Unit	€/Unit	€	Item	Type	Supplier	Elements
80	Meter	8	640	Water hose		SU.10. House builder's suppliers	-E-2. Water
80	Meter	13	1 040	Sewage pipe		SU.10. House builder's suppliers	-E-3. Sewage
60	Meter	23	1 380	Cable		SU.10. House builder's suppliers	-E-4. Electricity
30	Meter	7	210	Gas pipe		SU.13. Gas supplier	-E-5. Gas
241	m²	18	4 338	Paving	Basalt	SU.10. House builder's suppliers	-E-6. Driveway
24	m³	12	288	Gravel	Granite	SU.15. Local supplier	-E-6. Driveway
			7 896				

The *interfaces* between the house and its *environment* are cost-estimated in a work cost requisite (see Table 10-43 below).

TABLE 10-43 House environment
estimated realization work requisites hourly/fixed/rental cost

h	€/h	€	Item	Provider	IdInterfaces
32	100	3 200	i17. Excavator with driver and plumber	PR.13. House builder	-E-2. Water, -E-3. Sewage
40	30	1 200	i18. Plumber	PR.13. House builder	-E-2. Water, -E-3. Sewage
30	100	3 000	i19. Excavator with driver and electrician	PR.13. House builder	-E-4. Electricity
4	120	480	i20. Excavator with driver and authorized gas plumber	PR.13. House builder	-E-5. Gas
59	240	14 160	i21. Dozer with 6 landscapers (including paving)	PR.13. House builder	E. House environment -E-6. Driveway
		22 040			

The development cost for the *environment* is estimated (see Table 10-44 below).

TABLE 10-44 House environment
estimated development work requisites hourly/fixed cost

№	Unit	€ /unit	€	Item	Provider	Process schedule
4	Hour	50	200	i11. City architect	PR.11. Robber HB	Pt.0.Ra. Explore restrictions.
			0	i10. House user	PR.10. Proprietor	Pt.0.Rb. Plan development roles and providers.
2	Hour	50	100	i12. City architect	PR.11. Robber HB	Pt.0.Pa. Try to satisfy restrictions.
			0	i13. House user	PR.10. Proprietor	Pt.0.Pa. Try to satisfy restrictions.
			0	i14. House user	PR.10. Proprietor	Pt.0.Pc. Analyze in/out and make/buy options.
70	Hour	30	2 100	i15. Chief architect	PR.11. Robber HB	Pt.0.A. Constitute environment architecture.
40	Hour	40	1 600	i16. Electrician	PR.12. Elert AB	Pt.0.A. Constitute environment architecture.
24	Hour	40	960	i23. Quality inspector	PR.10. Proprietor	Pt.0.V. Verify prototype in environment.
1	Fixed price	800	800	i24. Inspector	PR.14. Q-Inc PR.15. Commune	Pr. Validate prototype.
			5 760			

Finally the total accumulated costs are summarized (see Table 10-45 below).

TABLE 10-45 House environment
summed/accumulated requisites estimates

Item	Reference	Cost €
House summed/accumulated requisites estimates	Table 10-40, chapter 10.6	200 628
House environment estimated realization material requisites cost	Table 10-42, this chapter	7 896
House environment estimated realization work requisites hourly/fixed/rental cost	Table 10-43, this chapter	22 040
House environment estimated development work requisites hourly/fixed cost	Table 10-44, this chapter	5 760
House environment summed/accumulated requisites estimates		**236 324**

10.7.5 EXAMPLE House environment: Examine white-box success

All design documentation is now complete for construction to start. Before starting, the documentation is presented to the house proprietor for a thorough review. The *product requisites* with total house and *interfaces* costs are approved.

It has been decided that design documentation should be spread to five house builder candidates. These contractors should be asked to review the full design documentation, including *product requisites*, but without any costs filled in.

Five candidates should be given two weeks to calculate their bids for building the house. They receive an electronic copy of the *product requisites*, which they can use to write their bids.

(Two weeks after this meeting, Mobber AB submitted the most attractive offer at a total of €165 100 for all *product requisites* (including their own margins), which is below the decided budget of €180 000 for the house. The building contract was signed with building company Mobber AB.)

10.7.6 EXAMPLE House: Proceed reading

To follow this example, these are the alternatives:

- **[OI]** Proceed to chapter 11.2, "EXAMPLE House environment: Realize interfaces to environment and insert/await outermost white-box," page 601.

- **[OV]** Proceed to chapter 12.4, "EXAMPLE House environment: Partially verify interfaces to environment with house fake," page 673.

- **[OV]** Proceed to chapter 12.10, "EXAMPLE House environment: Fully verify prototype in environment," page 699.

- Proceed to chapter 10.8 below, to leave the house example and to enter the multiplication toy example.

10.8 EXAMPLE Specifics for multiplication toy commonalities: Finalize design items of product white-box with product requisites

10.8.1 EXAMPLE Multiplication toy: About this example

Depending on the two technologies for the multiplication toy multiplier, and the two prototypes with different purposes, three extraordinary complications arise for *white-box design items* and *product requisites*.

- The two multiplication toy *variants* have the outer part in common; the inner multiplier *elements* are of two different technologies. Thus, the *white-box design items* are divided into three parts, one being commonalities for both technologies, and two separate parts for the different multiplier and PCB.

- The multiplication toy with microprocessor multiplier will be developed to a complete *product* prototype. The multiplication toy with hard-wired multiplier will be developed only to an experimental prototype, for the purpose of verifying only the mathematical correctness. This prototype will also be too big to fit the final casing, and needs no housing details and mechanical *interface* to the cuddle toy.

- However, if a multiplier ASIC is successful and it is decided to produce it, everything apart from the multiplier and PCB must be *reused* from the multiplication toy with microprocessor.

10.8.2 EXAMPLE Multiplication toy commonalities: Process schedule to use

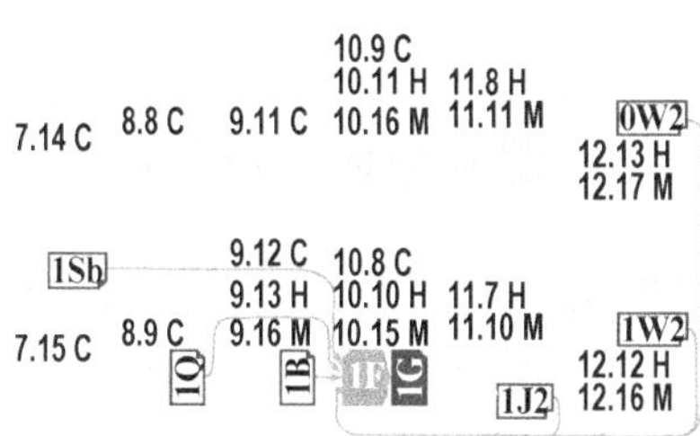

For the multiplication toy example, a special technical overall *schedule* is *tailored* (illustrated in Fig. 10-15, left):

• **Pt:2:3. Tailored advanced technical schedule n = 2,** page 169

In this *tailored schedule* the generic *schedule* to use is (filled-in symbols):

• **Pt.n.F. Finalize design of white-box with product requisites,** page 140.

FIGURE 10-15 Position of schedule and master to use (filled in) in overall schedule

Below *results* may be demanded (their *masters* are indicated *upstream* from the filled-in *schedule* in the figure above):

1B Common architecture white-box with major ingredients but no embedded black-boxes

Table 9-24, Multiplication toy common architecture ingredients resumé, page 387

Figure 9-38, Multiplication toy common logical architecture, page 389

Figure 9-39, Diversifying the multiplier element and circuit board interface, page 390

Figure 9-40, Multiplication toy common physical architecture, page 390

1B Hard-wired multiplier architecture white-box with major ingredients but no embedded black-boxes

Table 9-26, Hard-wired multiplier architecture ingredients resumé, page 393

Figure 9-42, Hard-wired multiplier logical architecture, page 393

1B Microcontroller multiplier architecture white-box with major ingredients but no embedded black-boxes

Table 9-28, Microcontroller multiplier refined architecture ingredients resumé, page 398

Figure 9-46, Microcontroller multiplier logical architecture, page 399

1Q Solution alternatives and supplier opportunities

Table 8-21, Multiplication toy white-box solution alternatives, page 317

1Sb White/black-box staffing allocation

Table 7-28, Multiplication toy development allocation staffing, page 218.

Xb Suppliers' sample catalogues

E Picked in-house legacy source systems

XaE Ordered suppliers' off-the-shelf fully described source systems

XbE Ordered suppliers' turnkey fully described source systems

For possible *failure localization* and *failure elimination*, below *results* may be demanded (their *masters* are indicated *upstream* from the filled-in *schedule* in the figure above):

1J2 Table 11-25, Multiplication toy with hard-wired multiplier start-up report 1, page 627

Table 11-33, Multiplication toy with microcontroller multiplier start-up report 2, page 636

1W2 Table 12-38, Multiplication toy with hard-wired multiplier failure elimination report 1, page 711

Table 12-47, Multiplication toy with microcontroller multiplier failure elimination report 1, page 723

0W2 Table 12-53, Multiplication toy environment (microcontroller multiplier variant) failure elimination report 1, page 728

10.8.3 EXAMPLE Multiplication toy commonalities: Finalize white-box detailed items for setting device

The *requirements* for automatic manufacturing tests result in a need for some permanent *white-box design items* of the toy construction. Some connection points must be available at the PCB for an automatic probe to connect test equipment to these points. There is a need to design these *connection interfaces* in order to make them referable in the electronics diagrams, and later to be designed on the PCB.

A design decision is made to work strictly binary coded, because this utilizes the gates most efficiently. When working with four bits, numbers from 0 to 15 can be represented, which satisfies the *requirements* with margins. A penalty is that the product number then will be 8 bits wide, representing numbers from 0 to 225, which demands a 3-decimal LED display. A robust but cheap rotating switch with 16 positions is selected (see Fig. 10-16 below).

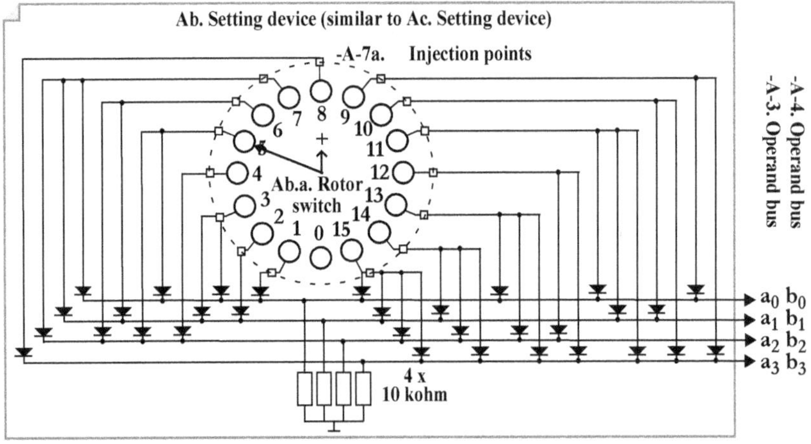

FIGURE 10-16 Setting device diagram

Note that Ab. Setting device and Ac. Setting device are identical. Also note that predetermined machine manufacturing tests require test points to inject the circuit with test input, which are marked as small squares in the figure.

10.8.4 EXAMPLE Multiplication toy commonalities: Finalize white-box detailed items for presenting device

The *product* presentation device is three 7-segment LED displays with drivers (see Fig. 10-17 below).

Also note that predetermined machine manufacturing tests again require test points at the segment drivers output to be read (see the small squares in the figure). Two test points, which can be shortened for a lamp test, are also marked.

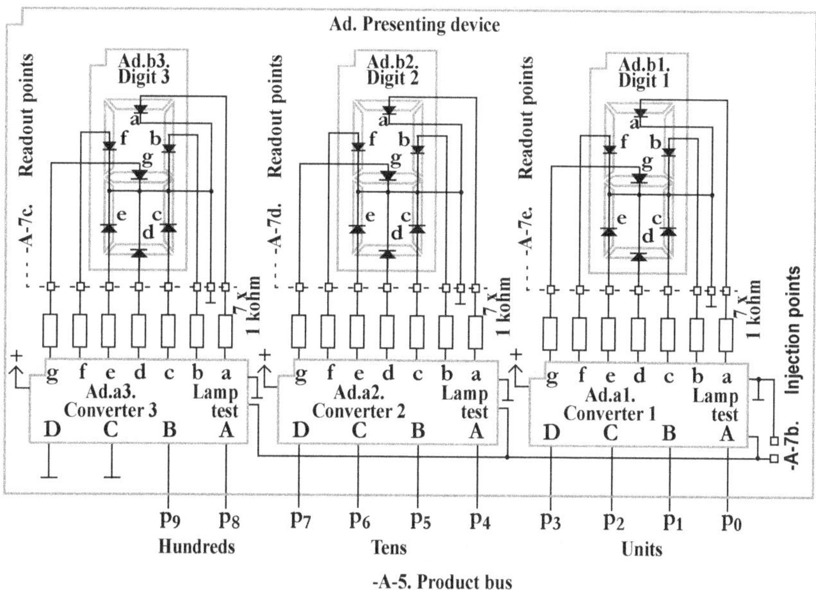

FIGURE 10-17 Presenting device diagram

10.8.5 EXAMPLE Multiplication toy commonalities: Finalize white-box detailed items for power supply

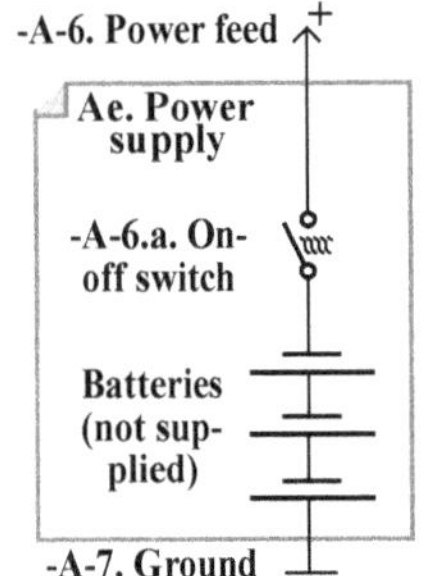

All logic gates in the multiplication toy need between 4 and 6 volts. Therefore, the power supply is three standard batteries in series, making 4.5 volts (see Fig. 10-18).

There must be some way to switch off the power, since the LED are rather power consuming. A closing button with returning spring was chosen. When pressing this the *user* can see the multiplication product as long as the button remains pressed down.

FIGURE 10-18 Power supply diagram

10.8.6 EXAMPLE Multiplication toy commonalities: Finalize white-box detailed items for casing

The multiplying toy casing mechanics are now prepared (see Fig. 10-19 below).

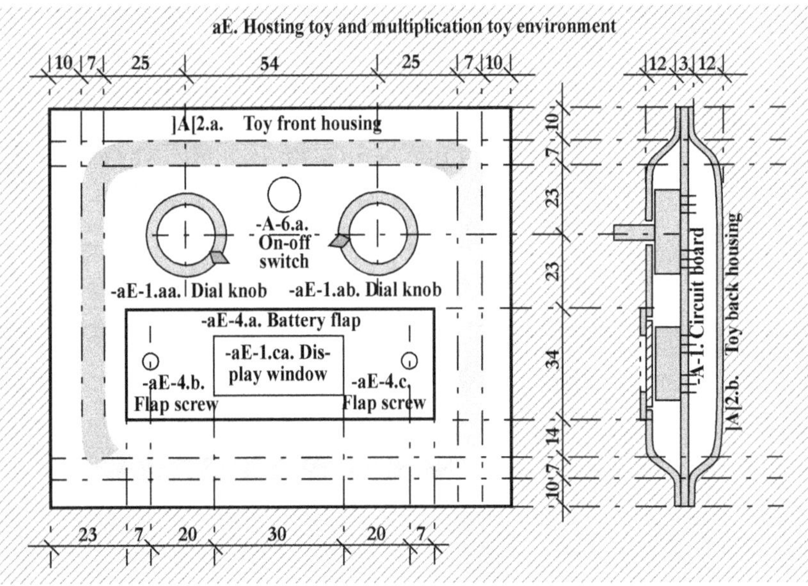

FIGURE 10-19 Multiplication toy casing

If the ASIC solution is developed, this *product* will also use these mechanics to make it fully interchangeable with the multiplication toy with microcontroller.

10.8.7 EXAMPLE Multiplication toy commonalities: Finalize white-box detailed items lists

Elements surrounding the multiplier are common to both multiplication toy *variants*, and may be listed only once (see Table 10-46 below).

TABLE 10-46 Multiplication toy common electronics, white-box design items

Item	Type	Pin-out	Size w x d x h	White-box ingredient
i100. Rotor switch	Unipolar 1 x 16 positions	16 in circle, 1 in center. Fits raster board.	20 x 9	Ab. Setting device Ac. Setting device
i101. Diode 1N4148	Diode	Axial wires	3.4 x 1.5	
i102. Resistors 4SIL10k	Resistor array	5 single in-line	13 x 3 x 8	
i103. LED-segment, no dp.	Red color, 5 mA, common cathode	10 pin in back	12 x 8 x 8	Ad. Presenting device
i104. Segment LCD driver 4511	CMOS IC with pull-up npn transistor out	16 pin dual in-line	20 x 8 x 5.3	
i105. Resistors 7SIL1k	Resistor array	8 single in-line	21 x 3 x 8	
i106. Battery bin	PCB box for 3 x AAA batteries	+ and -	44 x 32 x 10	Ae. Power supply
i107. AAA batteries	AAA type		30 x 43 x 30	

Detailed casing *white-box design items* are listed in Table 10-47 below.

TABLE 10-47 Multiplication toy with microcontroller multiplier casing, white-box design items

Item	Size mm w x h x d	Other	Material	White-box ingredient
i108. Dial knob	Conical with position tip	Fits raster board	Plastic riveted on axis (childproof)	-aE-1.aa. Dial knob -aE-1.ab. Dial knob
i109. On-off switch	M6 x 6	With nut	Ultra-slim model	-A-6.a. On-off switch
i110. Pressed front housing	138 x 128 x 1	Bulged 12 mm	Pressed painted metal plate, threaded for flap screws	]A[2.a. Toy front housing
i111. Manually slitted and folded front housing	138 x 128 x 0.6		Folded steel plate (only for evaluation)	
i112. Front housing tool			Combination tool for punching and bulging front plate	]A[2.a. Toy front housing punching and bulging manufacturing tool
i113. Back housing	138 x 128 x 1	Bulged 12 mm	Pressed metal plate	]A[2.b. Toy back housing
i114. Manually slitted and folded back housing	138 x 128 x 0.6		Folded steel plate (only for evaluation)	

TABLE 10-47 Multiplication toy with microcontroller multiplier casing, white-box design items

Item	Size mm w x h x d	Other	Material	White-box ingredient
i115. Back housing tool			Pressing tool for bulging back and front housing plate	]A[2.b. Toy back housing bulging manufacturing tool
i116. Display window	35 x 24 x 1		Transparent red plastic glued on battery flap	-aE-1.ca. Display window
i117. Cut battery flap	84 x 34 x 1		Painted metal, cut with same tool as front plate	
i118. Manually clipped battery flap	84 x 34 x 0.6		Untreated metal (only for evaluation)	-aE-4.a. Battery flap
i119. Flap screw	2 x M4 x 8	Chrome	Screws nonremovable from flap	-aE-4.b. Flap screw -aE-4.c. Flap screw

10.8.8 EXAMPLE Multiplication toy commonalities: Account for white-box development, realization, and manufacturing

Accounting will be done separately for commonalities (marked "C"), and for the two specific technologies (marked "H" and "M"), and for all three for both the *white-box* and the *environment interfaces*.

Even if the final *products* of the two technology *variants* will be quite similar, the prototypes will differ a lot, because the multiplication toy with microcontroller multiplier will be a complete *product* prototype, and the multiplication toy with hard-wired multiplier will be only an experimental prototype for correct behavior *verification*.

For accounting of electronics manufacturing commonalities, see Table 10-48 below.

TABLE 10-48 Multiplication toy white-box commonalities, estimated manufacturing material requisites cost

Pcs	1 000 series		100 000 series		Item	Product	Supplier	Elements
	€ /pc	€ /toy	€ /pc	€ /toy				
2	0.4	0.8	0.2	0.4	i100. Rotor switch	Pointer	SU.19. Elmech	Ab.Setting device
34	0.03	1.02	0.01	0.34	i101. Diode 1N4148	Any	SU.21. SiliC	Ab.Setting device
1	0.07	0.07	0.04	0.04	i102. Resistors 4SIL10k	Any	SU.20. Passta	Ab.Setting device
3	0.3	0.9	0.2	0.6	i103. LED-segment, no dp.	Texx	SU.19. Elmech	Ad.Presenting device
3	0.3	0.9	0.2	0.6	i104. Segment LCD driver 4511	Texx	SU.21. SiliC	Ad.Presenting device

TABLE 10-48 Multiplication toy white-box commonalities, estimated manufacturing material requisites cost

Pcs	1 000 series		100 000 series		Item	Product	Supplier	Elements
---	€ /pc	€ /toy	€ /pc	€ /toy				
1	0.08	0.08	0.05	0.05	i105. Resistors 7SIL1k	Any	SU.20. Passta	Ad.Presenting device
1	0.2	0.2	0.12	0.12	i106. Battery bin	Any	SU.19. Elmech	Ae.Power supply
0					i107. AAA batteries		Will not be supplied with the toy	Ae.Power supply
		3.97		2.15				

Casing material for manufacturing is shown in Table 10-49 below.

TABLE 10-49 Multiplication toy white-box commonalities, estimated manufacturing material requisites cost

Pcs	1 000 series		100 000 series		Item	Product	Supplier	Elements
---	€ /pc	€ /toy	€ /pc	€ /toy				
2	0.04	0.08	0.03	0.06	i108. Dial knob	Designer	SU.19. Elmech	-aE-1.aa.Dial knob
1	0.3	0.3	0.2	0.2	i109. On-off switch	Designer	SU.19. Elmech	-A-6.a.On-off switch
1	0.2	0.2	0.1	0.1	i110. Pressed front housing		SU.16. In Steal Inc	]A[2.a.Toy front housing
1 tool/ 50 000 pieces	5 000	5	5 000	0.1	i112. Front housing tool	In-depth press	SU.15. Tooly	]A[2.a.Toy front housing
1	0.1	0.1	0.06	0.06	i113. Back housing		SU.16. In Steal Inc	]A[2.b.Toy back housing
1 tool/ 50 000 pieces	5 000	5	5 000	0.1	i115. Back housing tool	In-depth press	SU.15. Tooly	]A[2.b.Toy back housing
1	0.02	0.02	0.015	0.015	i116. Display window	Designer	SU.19. Elmech	-aE-1.ca.Display window
1	0.025	0.025	0.02	0.02	Cut battery flap		SU.16. In Steal Inc	-aE-4.a.Battery flap
2	0.01	0.02	0.008	0.016	i119. Flap screw	Any	SU.19. Elmech	-aE-4.b.Flap screw
		10.745		0.671				

In the event that the ASIC is developed and available, commonality manufacturing work will be similar for both multiplier technologies (PCB is somewhat differently equipped), and manufacturing work can be accounted together (see Table 10-50 below).

The estimates for manufacturing work are so far rather approximate and simplified, and have been calculated as rental of equipment, not considering investments in buying own machinery. But by now it is necessary to get an idea of manufacturing costs to be able to estimate profit margins, which will be crucial for the destiny of the *product*.

TABLE 10-50 Multiplication toy white-box commonalities, estimated manufacturing work requisites hourly/fixed/rental cost

1 000 series			100 000 series			Item	Provider	Interface
h	€ /h	€ /toy	h	€ /h	€ /toy			
100	4	0.4				i39. Low-price-country worker to manually assemble PCB	PR.11. Mech-tech	-A-1:2 Printed board (same type will be used for multiplier with ASIC)
			200	80	0.16	i32. Manufacturing preparer	PR.11. Mech-tech	
			100	200	0.20	i40. Process worker with PCB automated equipping conveyor belt	PR.11. Mech-tech	
200	2	0.4				i43. Low-price-country worker to manually assemble toy mechanics	PR.11. Mech-tech	]A]2. Casing
			50	80	0.04	i32. Manufacturing preparer	PR.11. Mech-tech	
			80	100	0.08	i44. Process worker with automated assembly conveyor belt	PR.11. Mech-tech	
100	2	0.2				i45. Low-price-country worker manually verifying toy	PR.11. Mech-tech	-A-7a. Injection points -A-7c. Readout points
			50	80	0.04	i32. Manufacturing preparer	PR.11. Mech-tech	
			20	200	0.04	i46. Process worker with automated verification conveyor belt	PR.11. Mech-tech	
		1.0			0.56			

Material for the 50 prototypes are supplied mostly from a convenient but more expensive in-house storage (see Table 10-51 below).

TABLE 10-51 Multiplication toy white-box commonalities, estimated realization material requisites cost

50 prototypes			Item	Product	Supplier	Id Elements
Pcs	€ /pc	€ /toy				
2	1.2	2.4	i100. Rotor switch	Pointer	SU.19. Elmech	Ab.Setting device
34	0.3	10.2	i101. Diode 1N4148	Texx	SU.18. In-house storage	Ab.Setting device
1	0.7	0.7	i102. Resistors 4SIL10k	Resa	SU.18. In-house storage	Ab.Setting device
3	0.8	2.4	i103. LED-segment, no dp.	Texx	SU.19. Elmech	Ad.Presenting device
3	0.6	1.8	i104. Segment LCD driver 4511	Texx	SU.18. In-house storage	Ad.Presenting device
1	0.8	0.8	i105. Resistors 7SIL1k	Texx	SU.18. In-house storage	Ad.Presenting device
1	0.6	0.6	i106. Battery bin	Any	SU.19. Elmech	Ae.Power supply
3	0.3	0.9	i107. AAA batteries	Any	SU.18. In-house storage	Ae.Power supply
		19.8				

Casing material needed for both technology *product variants*, but not for the prototype with hard-wired multiplier, is shown in Table 10-52 below.

TABLE 10-52 Multiplication toy white-box commonalities, estimated realization material requisites cost

50 prototypes						
Pcs	€ /pc	€ /toy	Item	Product	Supplier	Id Elements
1	6	6	i111. Manually slitted and folded front housing	Tinsmiths	SU.17. In-house prototype factory	]A]2.a.Toy front housing
1	5.4	5.4	i114. Manually slitted and folded back housing	Tinsmiths	SU.17. In-house prototype factory	]A]2.b.Toy back housing
1	0.3	0.3	i116. Display window	Designer	SU.19. Elmech	-aE-1.ca.Display window
1	2.3	2.3	i118. Manually clipped battery flap	Tinsmiths	SU.17. In-house prototype factory	-aE-4.a.Battery flap
2	0.05	0.1	i119. Flap screw	Any	SU.18. In-house storage	-aE-4.b.Flap screw
		14.1				

Some commonality materials needed for both technology prototypes are shown in Table 10-53 below

TABLE 10-53 Multiplication toy white-box commonalities, estimated realization material requisites cost

50 prototypes						
Pcs	€ /pc	€ /toy	Item	Product	Supplier	Id Elements
2	1.1	2.2	i108. Dial knob	Any	SU.18. In-house storage	-aE-1.aa.Dial knob
1	0.9	0.9	i109. On-off switch	Any	SU.18. In-house storage	-A-6.a.On-off switch
		3.1				

Development work of *white-box* commonalities has been done during the *phases* capture staffing & requirements, predetermine solutions & suppliers, *design architectures* and finalize design & requisites (see the chapters marked "C" in *nesting level* 1 in Fig. 10-15).

The *white-box* commonalities development work is accounted in Table 10-54 below.

TABLE 10-54 Multiplication toy white-box commonalities, estimated development work requisites hourly/fixed cost

50 prototypes						
h	€ /h	€	€ /toy	Item	Provider	Schedule/task
30	40	1200	24	i22. Specifier	PR.10. In-house project	Pt.n.Ra. Refine and capture requirements.
13	60	780	15.6	i23. Chief engineer	PR.10. In-house project	Pt.n.Re. Plan development roles and providers.
16	45	720	14.4	i24. Digital expert	PR.10. In-house project	Pt.n.Pa. Try to satisfy requirements.
8	50	400	8	i25. Manufacturing engineer	PR.11. Mech-tech	Pt.n.Pc. Analyze make/buy options.

TABLE 10-54 Multiplication toy white-box commonalities, estimated development work requisites hourly/fixed cost

50 prototypes				Item	Provider	Schedule/task
h	€ /h	€	€ /toy			
6	55	330	6.6	i26. Purchase engineer	PR.10. In-house project	Pt.n.Pc. Analyze make/buy options.
24	40	960	19.2	i27. Electronics architect	PR.10. In-house project	Pt.1.A. Satisfy refined requirements by decomposing product black-box into white-box design containing no black-boxes.
19	40	760	15.2	i28. Logic designer	PR.10. In-house project	Pt.n.Fa. Finalize white-box detailed items.
15	40	600	12	i34. Accounting engineer	PR.10. In-house project	Pt.n.Fc. Account for white-box development, realization, and manufacturing (resulting in these tables).
			115			

Cost accounting for common material and development is summarized in Table 10-54 below.

TABLE 10-55 Multiplication toy white-box commonalities, summed/accumulated requisites estimates

Item	Reference	Development & 50 prototypes € /toy	1 000 series € /toy	100 000 series € /toy
Multiplication toy white-box commonalities, estimated manufacturing material requisites cost	Table 10-48, this chapter		3.97	2.15
Multiplication toy white-box commonalities, estimated manufacturing material requisites cost	Table 10-49, this chapter		10.745	0.671
Multiplication toy white-box commonalities, estimated manufacturing work requisites hourly/fixed/rental cost	Table 10-50, this chapter		1.00	0.56
Multiplication toy white-box commonalities, estimated realization material requisites cost	Table 10-51, this chapter	19.8		
Multiplication toy white-box commonalities, estimated realization material requisites cost	Table 10-53, this chapter	3.1		
Multiplication toy white-box commonalities, estimated development work requisites hourly/fixed cost	Table 10-54, this chapter	115.00		
Multiplication toy white-box commonalities, summed/accumulated requisites estimates, excluded housing design items		**137.90**		
Multiplication toy white-box commonalities, estimated realization material requisites cost	Table 10-52, this chapter	14.1		
Multiplication toy white-box commonalities, summed/accumulated requisites estimates		**152.00**	**15.715**	**3.381**

10.8.9 EXAMPLE Multiplication toy white-box commonalities: Examine white-box success

Success for the multiplication toy is heavily dependent on how low the manufacturing price can be pressed. The best price-pressers are large *component* volumes, which reduce development cost (totally contrary to the house example).

Members of the gate meeting were reminded to be very much aware of low *component* prices, and that spending some extra effort on development is fully defensible even if only small *component* price reductions can be obtained.

The gate meeting also questioned the accuracy of all manufacturing commonalities. The PCBs and the number of *components* are different, causing the assembly work to differ. This is true, but the *component* costs are separately accounted and distributed across the two technology *variants*, and only the work to assemble the PCBs is assumed to be the same. What kind of *components* are assembled is simply negligible, provided the numbers are close to equal.

10.8.10 EXAMPLE Multiplication toy: Proceed reading

To follow this example, these are the alternatives:

- Proceed to chapter 10.9, "EXAMPLE Specifics for multiplication toy environment commonalities: Finalize design of environment interfaces with product requisites," page 493.
- Proceed to chapter 10.15, "EXAMPLE Specifics for multiplication toy with microcontroller multiplier: Finalize design items of product white-box with product requisites," page 518.
- Proceed to chapter 10.10 "EXAMPLE Specifics for multiplication toy with hard-wired multiplier: Finalize design items of product white-box with product requisites" below.

10.9 EXAMPLE Specifics for multiplication toy environment commonalities: Finalize design of environment interfaces with product requisites

10.9.1 EXAMPLE EXAMPLE Multiplication toy environment interfaces commonalities: Process schedule to use

Pt:2:3. Tailored advanced technical schedule n = 2

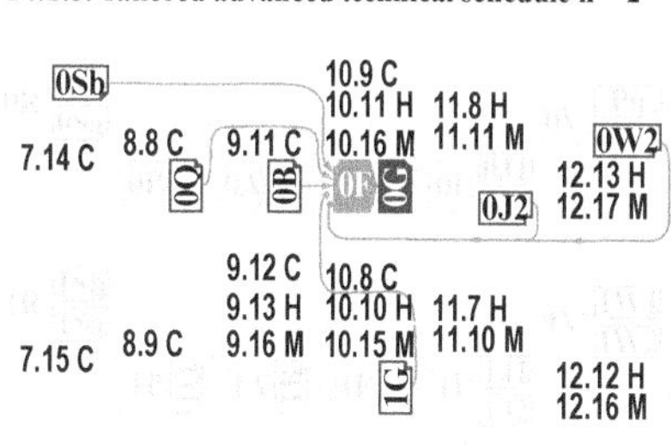

For the multiplication toy example, a special technical overall *schedule* is *tailored* (illustrated in Fig. 10-20, left):

- **Pt:2:3. Tailored advanced technical schedule n = 2,** page 169

In this *tailored schedule* the generic *schedule* to use is (filled-in symbols):

- **Pt.0.F. Finalize design of environment interfaces with product requisites,** page 118.

FIGURE 10-20 Position of schedule and master to use (filled in) in overall schedule

Below *results* may be demanded (their *masters* are indicated *upstream* from the filled-in *schedule* in the figure above):

IG Common finalized white-box items design with product requisites

Figure 10-16, Setting device diagram, page 484

Figure 10-17, Presenting device diagram, page 485

Figure 10-18, Power supply diagram, page 486

Table 10-46, Multiplication toy common electronics, white-box design items, page 487

IG Hard-wired multiplier finalized white-box items design with product requisites

Figure 10-24, Hard-wired multiplier, arithmetic unit diagram, page 502

Figure 10-25, Hard-wired multiplier, full adder multiplier diagram, page 503

Figure 10-27, Hard-wired multiplier 3-module adder diagram, page 505

Figure 10-26, Hard-wired multiplier, Bin-BCD converter diagram, page 504

Figure 10-28, Multiplication toy with hard-wired multiplier prototype board drawing, page 505

Table 10-63, Multiplication toy with hard-wired multiplier electronics white-box design items, page 506

Table 10-64, Multiplication toy with hard-wired multiplier prototype board white-box design items, page 506

IG Microcontroller multiplier finalized white-box items design with product requisites

Figure 10-31, Microcontroller multiplier diagram, page 520

Figure 10-32, Microcontroller multiplier assembler program with flow chart, page 521

Figure 10-33, Printed circuit board drawing, page 522

Figure 10-19, Multiplication toy casing, page 486

Table 10-74, Multiplication toy with microcontroller multiplier electronics white-box design items, page 522

Table 10-75, Multiplication toy with microcontroller multiplier printed board white-box design items, page 522

OB Environment existing ingredients and design of interfaces to the environment

Table 9-23, Multiplication toy environment architecture ingredients resumé, page 383

Figure 9-35, Multiplication toy environment logical architecture, page 384

Figure 9-36, Multiplication toy environment physical architecture, page 384

OSb Environment staffing allocation

Table 7-24, Multiplication toy environment development allocation staffing, page 213

Xb Suppliers' sample catalogues

E Picked in-house legacy source systems

XaE Ordered suppliers' off-the-shelf fully described source systems

XbE Ordered suppliers' turnkey fully described source systems

For possible *failure localization* and *failure elimination*, below *results* may be demanded (their *masters* are indicated *upstream* from the filled-in *schedule* in the figure above):

0J2 Table 11-28, Multiplication toy environment (hard-wired multiplier variant), start-up report 1, page 630

Table 11-36, Multiplication toy environment (microcontroller multiplier variant), start-up report 1, page 640

0W2 Table 12-53, Multiplication toy environment (microcontroller multiplier variant) failure elimination report 1, page 728

10.9.2 EXAMPLE Multiplication toy environment interfaces commonalities: Finalize environment interfaces detailed items

There is an important mechanical *interface* between the multiplication toy and the hosting cuddly toy that must have *managed information* to be cared for. It is also a matter of child safety that this *interface* cannot cause any hazards, for example, that the *interface* breaks too easily, exposing the inside of the hosting toy.

It might be difficult to find a way to mount the multiplication toy into a very soft or unstable hosting toy. For that reason, an explicit collar frame is developed that is simple to fasten to the hosting toy, after which the multiplication toy can be mounted into the collar frame.

The collar can be fastened to the hosting toy in many ways, for example:

- Cut an opening in the hosting toy fabric, fold it down through the frame, fold it *upstream*, and sew it to the fabric again.
- Make **no** openings in the hosting toy, apart from the asymmetrical holes, and just mount the collar from inside and the multiplication toy from outside.

A manual will be developed, describing the host collar and explain in detail how it can be successfully used.

The mounting of the multiplication toy into the hosting toy must be easy and foolproof. Thus, a very simple *interface* is defined (see Fig. 10-21 below).

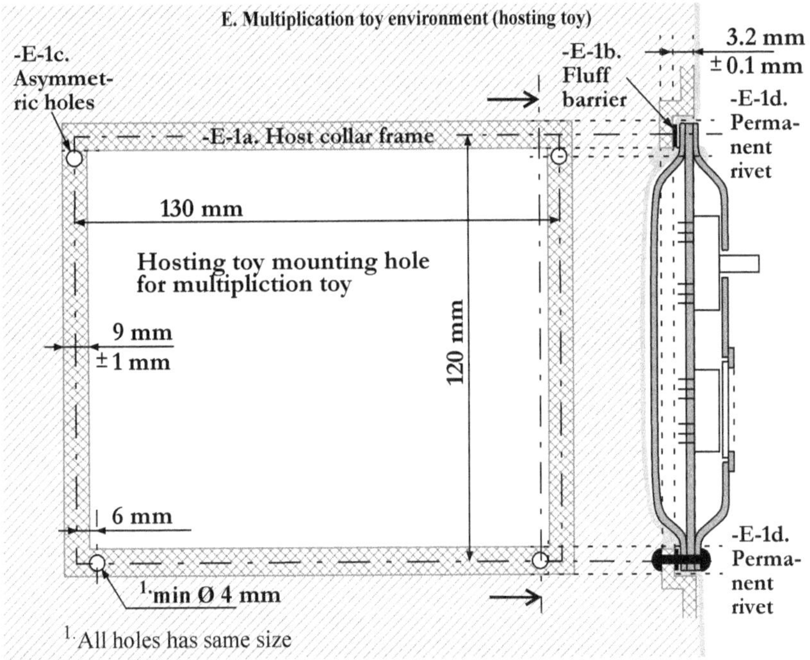

FIGURE 10-21 Multiplication toy environment attachment

10.9.3 EXAMPLE Multiplication toy environment interfaces commonalities: Finalize environment interfaces detailed items lists

Finalization of the host collar design concept is listed in Table 10-56 below.

TABLE 10-56 Multiplication toy environment interfaces commonalities, interfaces to environment design items

Item	Type	Size w x d x h	Interface ingredient
i120. Permanent rivet 2BS	Machine-mounted steel, nonremovable	Ø 4 mm, length 5 mm	-E-1d. Permanent rivet
i121. Rubber gasket	Isoprene	Band 1 mm x 6 mm. Square 120 mm x 130 mm	-E-1b. Fluff barrier
i122. Series host collar frame	Machine-folded metal plate	Outer frame measure 160 mm x 150 mm x 4 mm. Inner mounting hole measurements fit multiplication tool.	-E-1a. Host collar frame
i123. Manual host collar frame	Manually folded metal plate (only for tool evaluation)		

The multiplication toy has been planned to be connected to a laboratory power supply, but it would be nice to make it mobile for easier evaluation. The planned internal battery has limited power, and needs to be frequently exchanged. Because of this another bigger external battery in the *environment* would be appropriate (see Table 10-57 below).

TABLE 10-57 Multiplication toy environment interfaces commonalities, interfaces to environment design items

Item	Type	Size w x d x h	Interface ingredient
i124. Capacity battery with extension cable	Square-shaped with wire snap connection with charger	Not crucial	-A-6. Power feed (connection extended to environment)

10.9.4 EXAMPLE Multiplication toy environment interfaces commonalities: Account for environment interfaces development, realization, and manufacturing

For manufacturing material requisites to the *environment interfaces*, see Table 10-58 below.

TABLE 10-58 Multiplication toy environment interfaces commonalities, estimated manufacturing material requisites cost

	1 000 series		100 000 series				
Pcs	€/pc	€/toy	€/pc	€/toy	Item	Supplier	Elements
4	0.02	0.08	0.01	0.04	i120. Permanent rivet 2BS	SU.12. Gross Hardware	-E-1d. Permanent rivet
1	0.18	0.18	0.12	0.12	i121. Rubber gasket	SU.10. Gasket-Seal	-E-1b. Fluff barrier
	0.20	0.20	0.10	0.10	i122. Series host collar frame	SU.14. In Steal Inc	-E-1a. Host collar frame
1					i123. Manual host collar frame	SU.13. In-house prototype factory	-E-1a. Host collar frame
		0.46		0.26			

For prototype material requisites to the *environment interfaces*, see Table 10-58 below.

TABLE 10-59 Multiplication toy environment interfaces commonalities, estimated realization material requisites cost

	50 prototypes				
Pcs	€/pc	€/toy	Item	Supplier	Elements
4	0.2	0.8	i120. Permanent rivet 2BS	SU.12. Gross Hardware	-E-1d. Permanent rivet
1	0.4	0.4	i121. Rubber gasket	SU.10. Gasket-Seal	-E-1b. Fluff barrier
			i122. Series host collar frame	SU.14. In Steal Inc	-E-1a. Host collar frame
1	4	4	i123. Manual host collar frame	SU.13. In-house prototype factory	-E-1a. Host collar frame
		5.20			

The power battery to possibly take over power supply is accounted for in Table 10-60 below.

TABLE 10-60 Multiplication toy environment interfaces commonalities, estimated realization material requisites cost

50 prototypes			Item	Supplier	Element
Pcs	€ /pc	€ /toy			
1	50	50	i124. Capacity battery with extension cable	SU.18. In-house storage	Ae. Power supply located to environment
		50			

Development of *interfaces* commonalities through the *phases* **Pt.R. Capture staffing & requirements, Pt.P. Predetermine solutions & suppliers, Pt.A. Design architectures** and **Pt.F. Finalize design & requisites** has been kept together for both the hard-wired and microcontroller *variants* (see chapters marked "C" on *environment nesting level* = 0 in Fig. 10-20, p. 493). Consequently, accounting for these *phases* of development can be done together as *environment* commonality costs (see Table 10-61 below).

TABLE 10-61 Multiplication toy environment interfaces commonalities, estimated development work requisites hourly/fixed cost

50 prototypes				Item	Provider	Process schedule
h	€ /h	€	€ / toy			
24	50	1 200	24	i10. Childproofing specialist	PR.10. In-house project	Pt.0.Ra. Explore restrictions.
10	50	500	10	i11. Chief engineer	PR.10. In-house project	Pt.0.Rb. Plan development roles and providers.
25	36	900	18	i12. General developer	PR.10. In-house project	Pt.0.Pa. Try to satisfy restrictions.
32	40	1 280	25.6	i13. Product manager	PR.10. In-house project	Pt.0.Pc. Analyze in/out and make/buy options.
16	60	960	19.2	i14. Manufacturing specialist	PR.11. Mechtech	Pt.0.Pc. Analyze in/out and make/buy options.
30	40	1 200	24	i15. Mechanical designer	PR.12. Mechano	Pt.0.A. Constitute environment architecture. Pt.0.F. Finalize design of environment interfaces with product requisites.
16	65	1 040	20.8	i16. Child MMI expert	PR.10. In-house project	Pt.0.A. Constitute environment architecture. Pt.0.F. Finalize design of environment interfaces with product requisites.
8	65	520	10.4	i17. Electric power engineer	PR.10. In-house project	Pt.0.F. Finalize design of environment interfaces with product requisites.
		152				

Accounting for the *environment* of the multiplication toy commonalities is summarized in Table 10-68 below. The subtotal will be used for the multiplication toy with hard-wired multiplier prototype, since this will not be mounted anywhere and doesn't need a collar frame.

TABLE 10-62 Multiplication toy environment interfaces commonalities, summed/accumulated requisites estimates

Item	Reference	Development & 50 prototypes € /toy	1 000 Series € /toy	100 000 Series € /toy
Multiplication toy environment interfaces commonalities, estimated manufacturing material requisites cost	Table 10-58, this chapter		0.46	0.26
Multiplication toy environment interfaces commonalities, estimated realization material requisites cost	Table 10-60, this chapter	50.00		
Multiplication toy environment interfaces commonalities, estimated development work requisites hourly/fixed cost	Table 10-61, this chapter	152.00		
Multiplication toy environment interfaces commonalities, summed/accumulated requisites estimates, excluded host collar frame		**202.00**		
Multiplication toy environment interfaces commonalities, estimated realization material requisites cost	Table 10-59, this chapter	5.20		
Multiplication toy environment interfaces commonalities, summed/accumulated requisites estimates		**207.20**	**0.46**	**0.26**

10.9.5 EXAMPLE Multiplication toy: Proceed reading

To follow this example, these are the alternatives:

- Proceed to chapter 10.16, "EXAMPLE Multiplication toy environment (microcontroller multiplier variant): Finalize design of environment interfaces with product requisites," page 526.
- Proceed to chapter 10.11 "EXAMPLE Multiplication toy environment (hard-wired multiplier variant): Finalize design of environment interfaces with product requisites" below.

10.10 EXAMPLE Specifics for multiplication toy with hard-wired multiplier: Finalize design items of product white-box with product requisites

10.10.1 EXAMPLE Multiplication toy with hard-wired multiplier: About this example

The multiplication toy with hard-wired multiplier needs a lot of electronic gates not explicitly shown in the *architecture* in Figure 9-42, page 393. It is debatable in what *granularity* such *architectures* end and the *design item* graphs begin, but in this case there is no point in visualizing all architects' calculation logic.

Multiplying two 4-bit digital operands and converting their product to 3-decimal displays can be performed by many concepts, such as dedicated multiplication logic, one unintelligent network of gates, or a memory table look-up, and each of these can be realized by many types of logic electronics.

The purpose behind making a vast and clumsy hard-wired multiplier design at all, is to facilitate moving toward a customized ASIC design. Because of this, the *white-box design items* must be composed of rudimental logic circuits, which are available when designing the ASIC. There would certainly be a lot of simplifications if more *complex* digital circuits could be used. Also, logic networks could be simulated at low cost, without an expensive *realization* being done at all, but as said before, this book is about development models, and not about smart *products*.

10.10.2 EXAMPLE Multiplication toy white-box with hard-wired multiplier: Process schedule to use

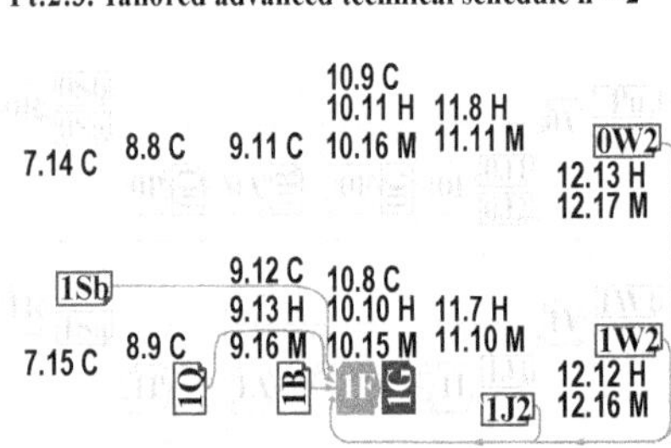

For the multiplication toy example, a special technical overall *schedule* is *tailored* (illustrated in Fig. 10-22, left):

- **Pt:2:3. Tailored advanced technical schedule n = 2**, page 169

In this *tailored schedule* the generic *schedule* to use is (filled-in symbols):

- **Pt.n.F. Finalize design of white-box with product requisites**, page 140.

FIGURE 10-22 Position of schedule and master to use (filled in) in overall schedule

Below *results* may be demanded (their *masters* are indicated *upstream* from the filled-in *schedule* in the figure above):

1G Common finalized white-box items design with product requisites

Table 10-55, Multiplication toy white-box commonalities, summed/accumulated requisites estimates, page 492

1B Common architecture white-box with major ingredients but no embedded black-boxes

Table 9-24, Multiplication toy common architecture ingredients resumé, page 387

Table 9-24, Multiplication toy common architecture ingredients resumé, page 387

Figure 9-38, Multiplication toy common logical architecture, page 389

Figure 9-39, Diversifying the multiplier element and circuit board interface, page 390

Figure 9-40, Multiplication toy common physical architecture, page 390

1B Hard-wired multiplier architecture white-box with major ingredients but no embedded black-boxes

Table 9-26, Hard-wired multiplier architecture ingredients resumé, page 393

Figure 9-42, Hard-wired multiplier logical architecture, page 393

[IQ] Solution alternatives and supplier opportunities
Table 8-21, Multiplication toy white-box solution alternatives, page 317

[1Sb] White/black-box staffing allocation
Table 7-28, Multiplication toy development allocation staffing, page 218.

[Xb] Suppliers' sample catalogues

[E] Picked in-house legacy source systems

[XaE] Ordered suppliers' off-the-shelf fully described source systems

[XbE] Ordered suppliers' turnkey fully described source systems

For possible *failure localization* and *failure elimination*, below *results* may be demanded (their *masters* are indicated *upstream* from the filled-in *schedule* in the figure above):

[1J2] Table 11-25, Multiplication toy with hard-wired multiplier start-up report 1, page 627

Table 11-33, Multiplication toy with microcontroller multiplier start-up report 2, page 636

[1W2] Table 12-38, Multiplication toy with hard-wired multiplier failure elimination report 1, page 711

Table 12-47, Multiplication toy with microcontroller multiplier failure elimination report 1, page 723

[0W2] Table 12-53, Multiplication toy environment (microcontroller multiplier variant) failure elimination report 1, page 728

10.10.3 EXAMPLE Multiplication toy with hard-wired multiplier: Finalize white-box detailed items for arithmetic unit

One way to realize a multiplication is to emulate the way we multiply by hand, as taught in elementary grades (see Fig. 10-23 below).

Decimal multiplication

```
    1 0
  * 1 2
  -----
    2 0
  + 1 0
  -----
  1 2 0
```

Same but digital multiplication

```
    1 0 1 0
  * 1 1 0 0
  ---------
    0 0 0 0
    0 0 0 0
    1 0 1 0
  + 1 0 1 0
  -----------
  0 1 1 1 1 0 0 0
```

The arithmetic unit

$$
\begin{array}{rcccccccc}
 &   &   & a_3 & a_2 & a_1 & a_0 & & \\
 & *   &   & b_3 & b_2 & b_1 & b_0 & & \\
\hline
 &   &   & r_3 & r_2 & r_1 & r_0 & & \\
 &   & s_3 & s_2 & s_1 & s_0 & & & \\
 & t_3 & t_2 & t_1 & t_0 & & & & \\
+ & u_3 & u_2 & u_1 & u_0 & & & & \\
\hline
P_7 & P_6 & P_5 & P_4 & P_3 & P_2 & P_1 & P_0 &
\end{array}
$$

FIGURE 10-23 Multiplying by hand

The *white-box design items* of the multiplier arithmetic unit can be seen in Figure 10-24 below.

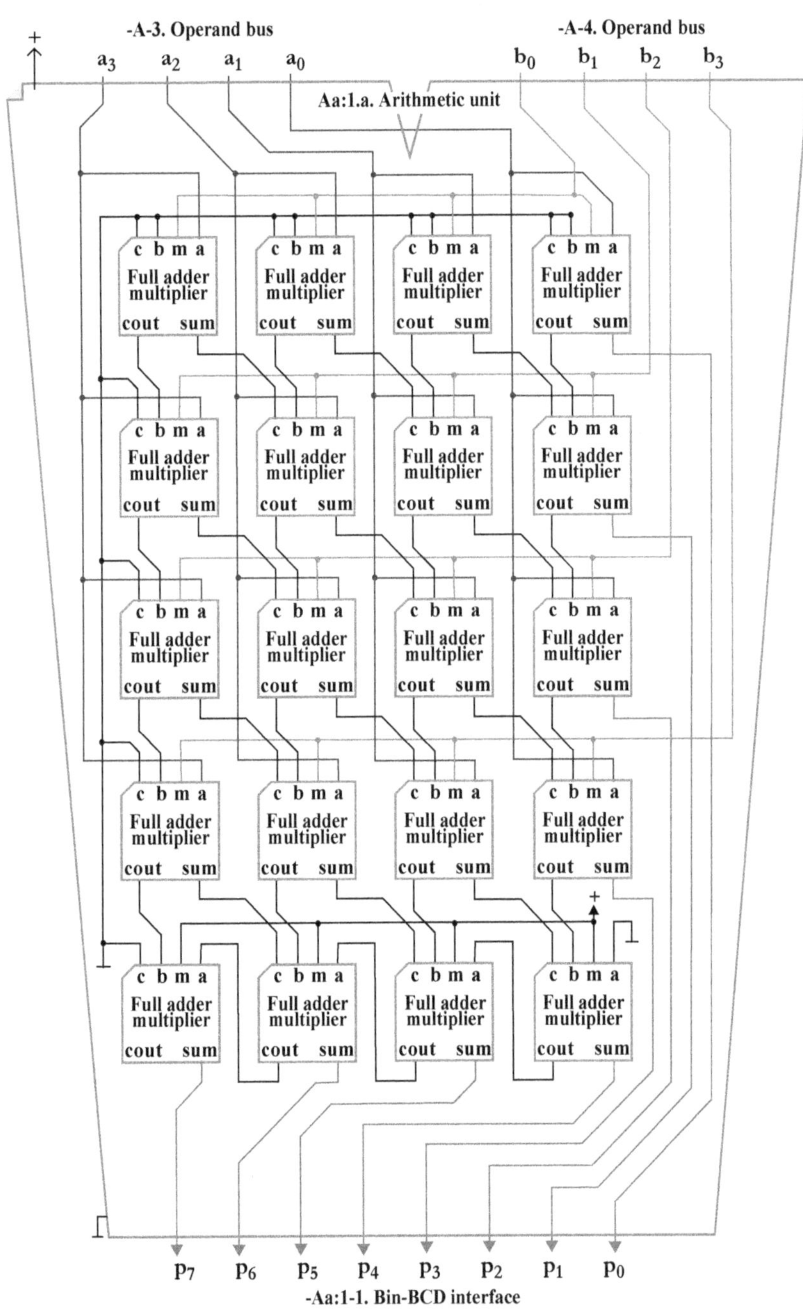

FIGURE 10-24 Hard-wired multiplier, arithmetic unit diagram

10.10.4 EXAMPLE Multiplication toy hard-wired multiplier : Finalize white-box detailed items for full adder multiplier

For the digital multiplication, there is a need for a number of *a * b bit-multipliers*, and a need of a number of *r + s + t + u bit-adders*. The multipliers and adders can be merged in the same design, as in Figure 10-25 below.

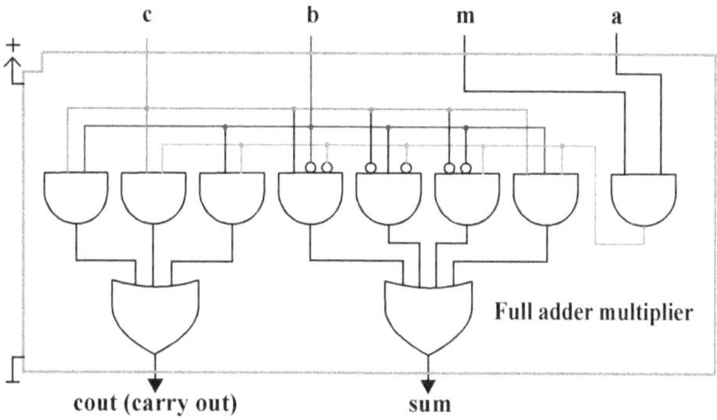

FIGURE 10-25 Hard-wired multiplier, full adder multiplier diagram

10.10.5 EXAMPLE Multiplication toy hard-wired multiplier:Finalize white-box detailed items for bin-bcd converter

Before it is possible to show the product with an LED display, it must be converted from binary to BCD format.

In the same way that the multiplication of 4 x 4 bits is realized with *full adder multipliers* to simulate a multiplication, this conversion uses a "shift-left-and-add-3-if-5-or-above" algorithm, which is designed with circuits called "3-module adder" (see Fig. 10-26 below).

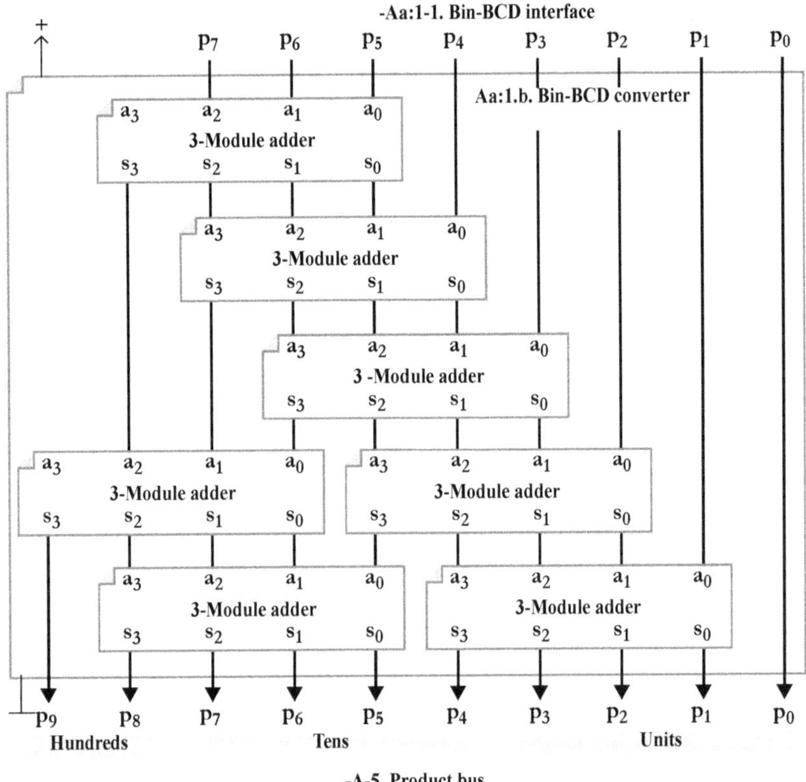

FIGURE 10-26 Hard-wired multiplier, Bin-BCD converter diagram

10.10.6 EXAMPLE Multiplication toy hard-wired multiplier: Finalize white-box detailed items for 3-module adder

The 3-module adders in above figure are designed from simple logic gates (see Fig. 10-27 below).

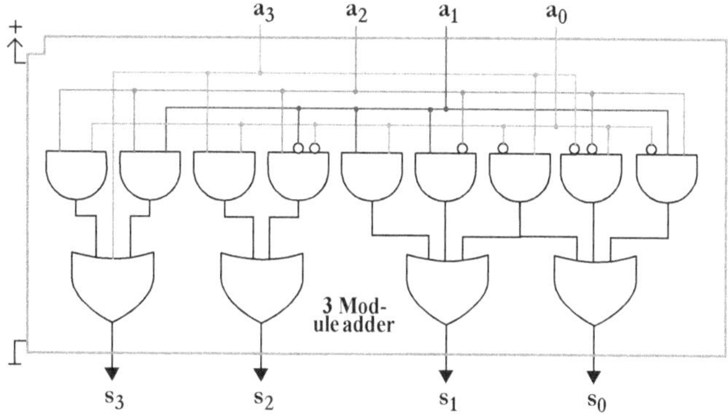

FIGURE 10-27 Hard-wired multiplier 3-module adder diagram

10.10.7 EXAMPLE Multiplication toy with hard-wired multiplier: Finalize white-box detailed items for prototype board

The prototype is built on an experimental type of board, with one motherboard for the *user interface* commonalities, and three perpendicular daughter boards for the multiplying logic networks (see Fig. 10-28). The boards are attached to each other with a detachable mother-daughter type of connector.

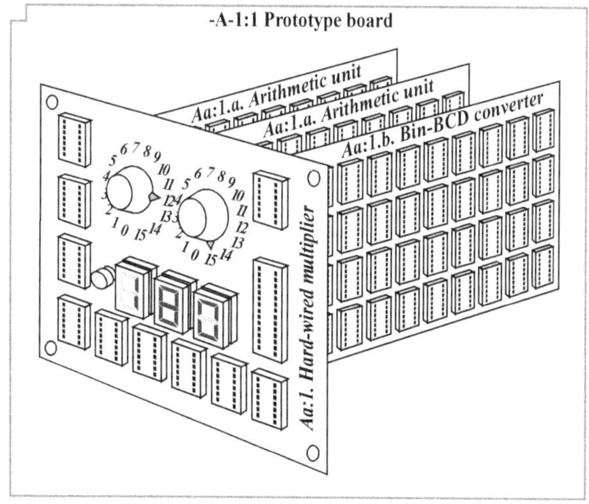

FIGURE 10-28 Multiplication toy with hard-wired multiplier prototype board drawing

10.10.8 EXAMPLE Multiplication toy with hard-wired multiplier: Finalize white-box detailed items lists

It is now time to identify all *white-box design items* illustrated in diagrams and drawings and list them with their technical *design item properties* in a *white-box design items* list (see Table 10-63 and Table 10-64 below).

TABLE 10-63 Multiplication toy with hard-wired multiplier electronics white-box design items

Item	Type	Pin-out	Size w x d x h	White-box ingredient
i125. Dual 4-input OR gate 4072	CMOS IC	14-pin dual in-line	18 x 6.4 x 4	Aa:1.a. Arithmetic unit
i126. Triple 3-input OR gate 4075	CMOS IC	14-pin dual in-line	18 x 6.4 x 4	
i127. Quad 2-input AND gate 4081	CMOS IC	14-pin dual in-line	18 x 6.4 x 4	
i128. Triple 3-input AND gate 4073	CMOS IC	14-pin dual in-line	18 x 6.4 x 4	
i129. Hex inverter 4069	CMOS IC	14-pin dual in-line	18 x 6.4 x 4	
i130. Quad 2-input OR gate 4071	CMOS IC	14-pin dual in-line	18 x 6.4 x 4	Aa:1.b. Bin-BCD converter
i131. Triple 3-input OR gate 4075	CMOS IC	14-pin dual in-line	18 x 6.4 x 4	
i132. Triple 3-input AND gate 4073	CMOS IC	14-pin dual in-line	18 x 6.4 x 4	
i133. Quad 2-input AND gate 4081	CMOS IC	14-pin dual in-line	18 x 6.4 x 4	
i134. Hex inverter 4069	CMOS IC	14-pin dual in-line	18 x 6.4 x 4	

TABLE 10-64 Multiplication toy with hard-wired multiplier prototype board white-box design items

Item	Type	Raster	Size w x d x h	White-box ingredient
i135. PCB	Experimental raster PCB	IC sockets fit raster board.	Motherboard 138 x 128 Plug-in daughters 100 x 180	-A-1:1 Prototype board
i136. PCB construction	Mother-daughter connectors	Connectors fit raster board.	96 x 240	-A-1:1 Prototype board construction

10.10.9 EXAMPLE Multiplication toy with hard-wired multiplier: Account for white-box development, realization, and manufacturing

Cost of material for the hard-wired multiplier to 50 multiplication toy prototypes is estimated in Table 10-65 below.

TABLE 10-65 Hard-wired multiplier in multiplication toy estimated realization material requisites cost

	50 prototypes					
Pcs	€ /pc	€ /toy	Item	Product	Supplier	Elements
10	0.40	4	i125. Dual 4-input OR gate 4072	Natio	SU.18. In-house storage	Aa:1.a. Arithmetic unit
7	0.25	1.75	i126. Triple 3-input OR gate 4075	Natio	SU.18. In-house storage	Aa:1.a. Arithmetic unit
20	0.35	7	i127. Quad 2-input AND gate 4081	Natio	SU.18. In-house storage	Aa:1.a. Arithmetic unit
27	0.25	6.75	i128. Triple 3-input AND gate 4073	Natio	SU.18. In-house storage	Aa:1.a. Arithmetic unit
20	0.25	5	i129. Hex inverter 4069	Natio	SU.18. In-house storage	Aa:1.a. Arithmetic unit
2	0.30	0.6	i130. Quad 2-input OR gate 4071	Natio	SU.18. In-house storage	Aa:1.b. Bin-BCD converter
7	0.25	1.75	i131. Triple 3-input OR gate 4075	Natio	SU.18. In-house storage	Aa:1.b. Bin-BCD converter
7	0.25	1.75	i132. Triple 3-input AND gate 4073	Natio	SU.18. In-house storage	Aa:1.b. Bin-BCD converter
11	0.30	3.3	i133. Quad 2-input AND gate 4081	Natio	SU.18. In-house storage	Aa:1.b. Bin-BCD converter
9	0.25	2.25	i134. Hex inverter 4069	Natio	SU.18. In-house storage	Aa:1.b. Bin-BCD converter
4	2	8	i135. PCB	PCB-era	SU.22. PeCeBe	-A-1:1 Prototype board
1	2	2	i136. PCB construction	PCB-era	SU.22. PeCeBe	-A-1:1 Prototype board
		44.15				

The cost of work to obtain *components*, to assemble and start up the 50 prototypes with hard-wired multiplier in *phase* realize & integrate white-box, is accounted in Table 10-66 below.

TABLE 10-66 Multiplication toy white-box with hard-wired multiplier estimated realization work requisites hourly/fixed/rental cost

	50 prototypes					
h	€ /h	€	€ /toy	Item	Provider	Interface
90	30	2 700	54	i35. Prototype mounting engineer	PR.10. In-house project	-A-1:1 Prototype board
			54			

Development work for the hard-wired multiplication toy includes *design architectures*, finalize design & requisites and verify black- & white-box (see chapters marked "H" on *black-/white-box nesting level* 1 in Fig. 10-22, p. 500). This accounted development work is shown in Table 10-67 below.

TABLE 10-67 Multiplication toy white-box with hard-wired multiplier estimated development work requisites hourly/fixed cost

50 prototypes				Item	Provider	Process schedule/task
h	€ /h	Tot €	€/toy			
10	60	600	12	i27. Electronics architect	PR.10. In-house project	Pt.1.A. Satisfy refined requirements by decomposing product black-box into white-box design containing no black-boxes (for Aa. Multiplier only).
10	50	500	10	i29. ASIC designer	PR.13. ASICt oGo AB	Pt.n.Fa. Finalize white-box detailed items (for Aa. Multiplier only).
20	50	1 000	20	i28. Logic designer	PR.10. In-house project	Pt.n.Fa. Finalize white-box detailed items (for Aa. Multiplier only).
30	40	1200	24	i34. Accounting engineer	PR.10. In-house project	Pt.n.Fc. Account for white-box development, realization, and manufacturing (resulting in these requisites tables).
50	36	1 800	36	i37. Electronics verifier	PR.10. In-house project	Pt.1.V. Verify black-/white-box.
			102			

Cost accounting for the multiplication toy with hard-wired multiplier is summarized in Table 10-68 below.

TABLE 10-68 Multiplication toy white-box with hard-wired multiplier summed/accumulated requisites estimates

Item	Reference	50 prototypes € /toy
Multiplication toy white-box commonalities, summed/accumulated requisites estimates, excluding housing design items	Table 10-55, chapter 10.8	137.90
Hard-wired multiplier in multiplication toy estimated realization material requisites cost	Table 10-65, this chapter	44.15
Multiplication toy white-box with hard-wired multiplier estimated realization work requisites hourly/fixed/rental cost	Table 10-66, this chapter	54.00
Multiplication toy white-box with hard-wired multiplier estimated development work requisites hourly/fixed cost	Table 10-67, this chapter	102.00
Multiplication toy white-box with hard-wired multiplier estimated development work requisites hourly/fixed cost		**338.05**

10.10.10 Multiplication toy with hard-wired multiplier: Examine white-box success

It looks like the prototypes with hard-wired multiplier will be expensive per unit. But the gate members agreed that an ASIC failure costs much more, and securing the design correctness of this prototype will be worth this cost. If this toy is a success on the market with high-volume sales, the ASIC would be an excellent way to substantially hold down its per-unit cost.

10.10.11 Multiplication toy: Proceed reading

To follow this example, these are the alternatives:

- Proceed to chapter 11.7, "EXAMPLE Multiplication toy with hard-wired multiplier: Procure items and linkable programs in order to realize white-box," page 624.

- Proceed to chapter 12.12, "EXAMPLE Multiplication toy with hard-wired multiplier: Verify black-/white-box," page 706.

- Proceed to chapter 10.11 "EXAMPLE Multiplication toy environment (hard-wired multiplier variant): Finalize design of environment interfaces with product requisites" below.

10.11 EXAMPLE Multiplication toy environment (hard-wired multiplier variant): Finalize design of environment interfaces with product requisites

10.11.1 Multiplication toy environment (hard-wired multiplier variant): Process schedule to use

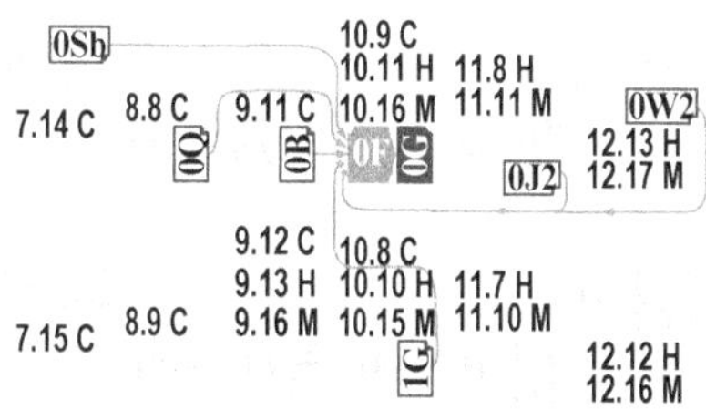

For the multiplication toy example, a special technical overall *schedule* is *tailored* (illustrated in Fig. 10-29, left):

● **Pt:2:3. Tailored advanced technical schedule n = 2,** page 169

In this *tailored schedule* the generic *schedule* to use is (filled-in symbols):

● **Pt.0.F. Finalize design of environment interfaces with product requisites,** page 118.

FIGURE 10-29 Position of schedule and master to use (filled in) in overall schedule

Below *results* may be demanded (their *masters* are indicated *upstream* from the filled-in *schedule* in the figure above):

0G Common environment interface design items and product requisites

 Table 10-62, Multiplication toy environment interfaces commonalities, summed/accumulated requisites estimates, page 499

1G Common finalized white-box items design with product requisites

 Figure 10-16, Setting device diagram, page 484

 Figure 10-17, Presenting device diagram, page 485

 Figure 10-18, Power supply diagram, page 486

 Table 10-46, Multiplication toy common electronics, white-box design items, page 487

1G Hard-wired multiplier finalized white-box items design with product requisites

 Figure 10-24, Hard-wired multiplier, arithmetic unit diagram, page 502

 Figure 10-25, Hard-wired multiplier, full adder multiplier diagram, page 503

Figure 10-27, Hard-wired multiplier 3-module adder diagram, page 505

Figure 10-26, Hard-wired multiplier, Bin-BCD converter diagram, page 504

Figure 10-28, Multiplication toy with hard-wired multiplier prototype board drawing, page 505

Table 10-63, Multiplication toy with hard-wired multiplier electronics white-box design items, page 506

Table 10-64, Multiplication toy with hard-wired multiplier prototype board white-box design items, page 506

IG Hard-wired multiplier finalized white-box items design with product requisites

Table 10-68, Multiplication toy white-box with hard-wired multiplier summed/accumulated requisites estimates, page 508

0B Environment existing ingredients and design of interfaces to the environment

Table 9-23, Multiplication toy environment architecture ingredients resumé, page 383

Figure 9-35, Multiplication toy environment logical architecture, page 384

Figure 9-36, Multiplication toy environment physical architecture, page 384

0Sb Environment staffing allocation

Table 7-24, Multiplication toy environment development allocation staffing, page 213

Xb Suppliers' sample catalogues

E Picked in-house legacy source systems

XaE Ordered suppliers' off-the-shelf fully described source systems

XbE Ordered suppliers' turnkey fully described source systems

For possible *failure localization* and *failure elimination*, below *results* may be demanded (their *masters* are indicated *upstream* from the filled-in *schedule* in the figure above):

0J2 Table 11-28, Multiplication toy environment (hard-wired multiplier variant), start-up report 1, page 630

Table 11-36, Multiplication toy environment (microcontroller multiplier variant), start-up report 1, page 640

0W2 Table 12-53, Multiplication toy environment (microcontroller multiplier variant) failure elimination report 1, page 728

The only work is to order power batteries (see Table 10-69 below).

TABLE 10-69 Multiplication toy environment (hard-wired multiplier variant) estimated realization work requisites hourly/fixed/rental cost

50 prototypes				Item	Provider	Interface
h	€ /h	Tot €	€ /toy			
70	40	2 800	39	i17. Electric power engineer	PR.10. In-house project	-A-6. Power feed -A-7. Ground
			39			

There are no specific *environment interfaces design items* to be specified on the *environment interfaces*, since the multiplication toy with hard wired multiplier doesn't have to be attached to any hosting toy to make a small field *verification*. In the *phases* finalize design & requisites and verify black- & white-box, see chapters marked "H" on *white-box nesting level 1* in Figure 10-29, page 509; there is minor work to be done on development (see Table 10-69 above).

TABLE 10-70 Multiplication toy environment (hard-wired multiplier variant), estimated development work requisites hourly/fixed cost

50 prototypes				Item	Provider	Process schedule/task
h	€ /h	Tot €	€/toy			
10	40	400	8	i34. Account-ing engineer	PR.10. In-house project	Pt.n.Fc. Account for white-box development, realization, and manufacturing (resulting in these tables).
15	30	450	9	i37. Electron-ics verifier	PR.10. In-house project	Pt.1.V. Verify black-/white-box.
			17			

Accounting for the *environment* of the multiplication toy with hard-wired multiplier is summarized in Table 10-71 below.

TABLE 10-71 Multiplication toy environment (hard-wired multiplier variant) summed/accumulated requisites estimates

Item	Reference	50 prototypes € /toy
Multiplication toy white-box with hard-wired multiplier summed/accumulated requisites estimates	Table 10-68, chapter 10.10	338.05
Multiplication toy environment interfaces commonalities, summed/accumulated requisites estimates, excluding host collar frame	Table 10-62, chapter 10.9,	202.00
Multiplication toy environment (hard-wired multiplier variant) estimated realization work requisites hourly/fixed/rental cost	Table 10-69, this chapter	39.00
Multiplication toy environment (hard-wired multiplier variant), estimated development work requisites hourly/fixed cost	Table 10-70, this chapter	17.00
Multiplication toy environment (hard-wired multiplier variant) summed/accumulated requisites estimates		**596.05**

10.11.2 EXAMPLE Multiplication toy environment (hard-wired multiplier variant): Examine prototype success

The gate meeting reminded participants that if the prototype is found satisfactory, and money is invested to make an ASIC, this *product* must get the same casing as the microcontroller *variant*, and the same type of hosting toys must fit this *product variant* as well.

10.11.3 Multiplication toy: Proceed reading

To follow this example, these are the alternatives:

- Proceed to chapter 11.8, "EXAMPLE Multiplication toy environment (hard-wired multiplier variant): Realize interfaces to environment and insert/await outermost white-box," page 628.

- Proceed to chapter 12.13, "EXAMPLE Multiplication toy environment (hard-wired multiplier variant): Verify prototype in environment," page 712.

- To leave the multiplication toy example and to dig deeper into the chapter 10, "Finalize design & requisites" topic, proceed to next chapter 10.12 below.

10.12 Enter into computer programs

10.12.1 About programs

Note that this book mainly deals with mastering *complexity*, and therefore does not dig deep into smart programming style. However, some of the mechanisms in computer languages strongly impact mastering *complexity* of *programs*. Most often, computer languages are developed to support the programmer with safety, but this is not always the case.

> This is not a book about smart programming styles, but program structures are important when mastering program complexity.

10.12.2 Differences between hardware (mechanics and electronics) and programs (software)

There are, in fact, few differences between developing *programs* and developing mechanics and electronics. However, it is much easier and takes much less time to describe the differences than the similarities.

The main differences between software and hardware are:

- *Programs* are entirely about information and are thus immaterial. To hold and keep the information, a hardware bearer must be applied, such as papers, disks, films, memory chips, and so forth. Since *programs* are immaterial, they do not feel as tangible as hardware.

- Since *programs* are immaterial, only the data carrier must be copied when multiplying them.

- Since development cost is high but copying costs are low, specific and unique *programs* become very expensive, but general and widespread *programs* get very cheap. In contrast, hardware cost per unit decreases rather insignificantly when scaling up a manufacturing series.

- *Programs* can multiply themselves during execution. By defining a *class* in a *program*, you can order as many instances of *objects* of that *class* as you wish, as long as you have free working memory.

- *Program integration* into prototypes is heavily automated by compilers and linkers. Thus, *integration* time between finalizing *programs* and verifying them takes only a few hours compared to ordering hardware *samples*, waiting for them to arrive, and finally assembling them to prototypes, which might take months.

- Since immaterial *programs* can be handled much faster than physical hardware, the time between deciding on a change and implementing that change in the prototype (*turnaround time*) is much shorter for *programs* (typically hours) than for hardware (typically months).

* Do not draw the wrong conclusion from the above. Many *managers* and other people not very familiar with *programs* mistakenly use the fast *turnaround time* as evidence that *programs* are easy and fast to produce, and any change to a prototype can be made swiftly at any time.

Other aspects are very similar, including:

* Both hardware and software have an *architecture*, that is, both can be divided into parts and pieces that have connections (*interfaces*) or boundaries between them. This book, as an example of software for human access, can similarly be divided into chapters, subchapters, paragraphs, pictures, tables, and so forth.

* Both hardware and software can be *modularized* into smaller parts (including underlying *black-boxes*) for easier *reuse*, scale-up, and understanding.

* Both software and hardware have behaviors and restrictions. Thus, they should preferably be described in the same way by *requirements*, offering the opportunity later on to be realized with either software or hardware or a mix of these.

* *Architecture* and *design items* are equally "hard"—regardless of hardware or software. If a book lacks structure or its structure is significantly inadequate, it takes a big effort to get it redesigned, and it can even be more efficient to rewrite it from scratch. Since *turnaround times* are very short for *programs*, it is often (unfortunately) assumed that heavy redesign of structure is equally short and easy.

10.13 Machine program

10.13.1 Microcontroller recap

To refresh understanding of the fundamental constituents of a microcontroller, see chapter 9.14, "EXAMPLE: Typical microcontroller architecture," page 394.

Machine programs are rarely handled manually, but it is very rewarding to understand the principle of a microcontroller.

All *elements* of the microcontroller *architecture* shown reside on one single silicon chip. To give a feeling of how it works, and to relate this to the hard-wired logic in previous chapters, a brief explanation follows below.

* Before the *machine instructions* can be executed, they are loaded into the read-only memory through a specially designed port. The *machine instructions* loaded into the memory are a sequence of *instructions*, which might be of different lengths (one to several bytes).

* Whenever the microcontroller is powered up, the *program* counter is automatically reset to zero, which causes *machine instruction* located to address 0000 to appear in the *machine instruction* decoder.

- The *machine instructions* are addressed by the *program* counter, which steps up and points to the next *machine instruction* after the previous *machine instruction* has been executed.

- The *machine instruction* under execution is decoded by the *machine instruction* decoder. If longer than one byte, it can be decoded in several sequential steps.

- Depending on the specific decoded *machine instruction*, the necessary latches are activated with correct timing. The microcontroller buses shown are connections of several parallel digital bits, in the same way as the -A-5. Product bus in Figure 10-26, page 504 or -A-4. Operand bus in Figure 10-16, page 484. A latch is a like a valve unit between two buses, and can either let the digital bits through or allow them to be blocked.

- EXAMPLE: Imagine that a *machine instruction* is decoded, which means to receive a value from outside the microcomputer, through port 3, and convey that value into the accumulator. To implement this, open Latch p3 and Latch a1, and block the rest of the data bus latches.

- The accumulator is a special register that can do a lot of operations on the value stored in it. It can also feed the arithmetic logic unit (ALU) with one of two operands. The ALU might (to some extent) be designed as the Aa:1.a. Arithmetic unit in Figure 10-24, page 502.

- The random-access memory holds bytes which can be linked to words, long words, double words, and so forth. The random-access memory has an address input to point out the byte that is intended to appear on the memory output.

- When the *interrupt* port is activated, the *program* counter is loaded with preset *interrupt* addresses, where the execution now continues. It can happen regardless of which address the microcontroller executes for the moment, and the *interrupt program* must save and restore all those registers that are used by the *interrupt program* before returning from the *interrupt*.

10.13.2 Cpdm definition

In *Cpdm*, *machine program*, *machine instruction* and *machine variable* are defined according to Table 10-72 below..

TABLE 10-72 Cpdm definitions

Aspect	Machine program	
	Machine instruction	Machine variable
Definition	Coded instruction to be decoded and executed by the controller or processor	A read-write memory location to save and retrieve stimuli history data for further calculation
Synonyms	Machine-oriented language, machine language, machine code	Register, variable location, machine storage
Architecture symbols	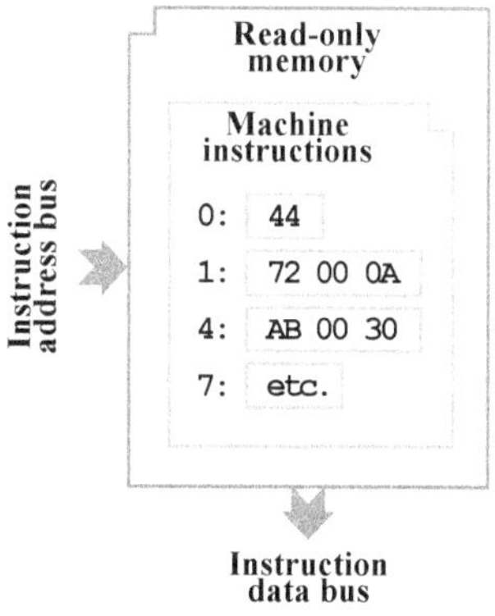 Instruction address bus → Read-only memory — Machine instructions: 0: 44; 1: 72 00 0A; 4: AB 00 30; 7: etc. → Instruction data bus	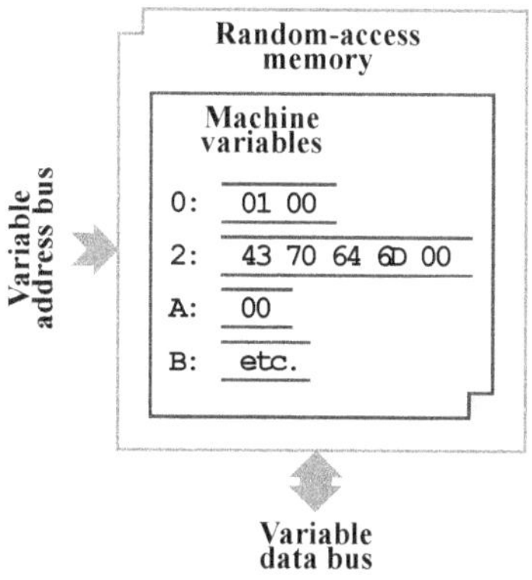 Variable address bus → Random-access memory — Machine variables: 0: 01 00; 2: 43 70 64 6D 00; A: 00; B: etc. → Variable data bus

How the memories are organized in the *architecture* of a microcontroller is illustrated in Figure 9-44, Typical microcontroller architecture, page 395.

Machine programs are very hard for humans to read, as can be seen in the three *machine instruction* examples in the definition above. Thus, *machine programs* are rarely directly written by humans, and consequently more understandable languages were introduced very early in computer history (see following chapters).

Note that there are big differences in *machine programs* between microcontroller brands. Some microcontrollers handle *variables* in predefined registers, others handle *variables* more freely in random addresses.

10.14 Assembler programs

In the beginning of the computer era, before compilers and *high-level programs* were invented, there was a lot of demand for *machine programs* to be developed. In the early days computers were programmed by switches and lamps on the front panel. A huge improvement was the introduction of assembler tools, which translated a text-edited *program* into a machine *program*.

It was a big improvement to write programs with an ordinary editor and let the assembler tool translate it to machine programs.

10.14.1 Cpdm definition

In *Cpdm*, *assembler program* is defined according to Table 10-73 below.

TABLE 10-73 Cpdm definition

Aspect	Assembler program
Definition	Programs containing symbolic mnemonics instead of machine-dependent instructions, arithmetic calculation support, user naming of variables and address labels, which consequently need assembler and linker tools for translation to machine instructions and variables
Synonyms	Low-level language program, second-generation language program, unstructured program
Examples	Each controller or processor families have their own assembler tool.
Architecture symbols	**Assembler program** **Variables of type :** Bit Byte Word Pointer **Program instructions of type :** generic instruction unconditional jump Address label : subroutine call conditional jump input or output **Can be a black-box**

10.14.2 Basic concepts of assembler programs

To make *machine instructions* more understandable for humans, the following assembler explanations can be helpful:

1. **Bring structure to different parts of the machine instruction:**
   a. The *machine instruction* type which gives the mnemonic its names
      EXAMPLE: If the *machine instruction* is found to be 44 hex (exemplified in definition above), this might mean that   INC A

the accumulator is to be incremented by one, and the mnemonic might be called INC A.

b. Separation of the data part of the *machine instruction*

EXAMPLE: If the 3-byte *machine instruction* is 72 00 0A hex (exemplified in definition above), the *machine instruction* 72 might mean to load the succeeding 2 bytes to the accumulator, which means that 000A hex is loaded to the accumulator. This mnemonic might be named as MOV A, 000A.

`MOV A, 000A`

c. Separation of the address part of the *machine instruction*

EXAMPLE: If the 3-byte *machine instruction* is AB 00 30 hex (exemplified in definition above), the *machine instruction* AB might be a *jump instruction* to the absolute address of succeeding 2-byte address, which means that 0030 is loaded to the *program* counter, thus causing the next *machine instruction* to be read at address 0030 hex. This mnemonic might be named JMP 0030.

`JMP 0030`

2. Allow trivial names for variables and address labels.

a. Definition of types of *variables*, such as bit, nibble, byte, word, and so forth

EXAMPLE: if the 2-byte *variable* is found to be 01 00 hex (exemplified in definition above) and is declared an unsigned integer, then this value equals 256 decimals.

`word variable_1`

EXAMPLE: If the 1-byte *variable* is 43 hex (string of characters exemplified in definition above) and declared a char, this means character C.

`char variable_2`

b. Address in a *jump instruction* to get calculated by the assembler tool

EXAMPLE: In a *jump instruction* in the *assembler program*, the first byte of a 3-byte *machine instruction* is AB hex (like 1.c. above). Since the destination for the jump can be written as a label, whose memory location is known by the assembler, the two succeeding bytes of the *jump instruction* are set to the absolute address of this destination by the assembler tool.

`JMP start`

3. Encapsulation:

a. There is seldom any way to hide *address labels* or *variables*. Most assemblers allow jumps to any possible *address label* and access to any possible *variables*.

4. Modularization:

a. Most microcontrollers have a possibility to call a subroutine and return to the calling *program*. Thus, a *program* can be divided into parts. However, this is a very weak mechanism, since jumping anywhere and accessing any *variable* is free anyway.

5. Structure:

a. To master *complexity* with good documentation, it is highly preferable to comment in the *program* and to draw separate flow charts. On the right side in Figure 10-32, page 521, the execution flow is illustrated, which makes it much easier to identify and follow these execution paths.

b. This kind of *program* is sometimes called *unstructured program*, since there is a possibility to jump anywhere from anywhere, and access any *variables* from anywhere.

10.15 EXAMPLE Specifics for multiplication toy with microcontroller multiplier: Finalize design items of product white-box with product requisites

10.15.1 EXAMPLE Multiplication toy with microcontroller multiplier: About this example

A microcontroller is said to be a very large scale integration (VLSI) circuit chip, built up of several thousands of transistors and integrated on one single chip. This book will not dig further into the microcontroller, but takes the advantage of using such a single-chip controller in the multiplication toy development examples, to gradually come over from mechanics and electronics to *program architecture* (with *program black-boxes*, *elements*, and *interfaces*) and further to detailed *program* design.

To avoid far too many details, in this book the *assembler programs* are written in a generic way instead of for a particular microcontroller from a certain manufacturer. Sometimes such *programs* are referred to as pseudo-language assemblers, which are easy to translate to the *target* assembler *program* for a specific microcontroller.

10.15.2 EXAMPLE Multiplication toy with microcontroller multiplier: Process schedule to use

For the multiplication toy example, a special technical overall *schedule* is *tailored* (illustrated in Fig. 10-30, left):

- **Pt:2:3. Tailored advanced technical schedule n = 2,** page 169

In this *tailored schedule* the generic *schedule* to use is (filled-in symbols):

- **Pt.n.F. Finalize design of white-box with product requisites,** page 140.

FIGURE 10-30 Position of schedule and master to use (filled in) in overall schedule

Below *results* may be demanded (their *masters* are indicated *upstream* from the filled-in *schedule* in the figure above):

1G Common finalized white-box items design with product requisites
Table 10-55, Multiplication toy white-box commonalities, summed/accumulated requisites estimates, page 492

1B Common architecture white-box with major ingredients but no embedded black-boxes

Table 9-24, Multiplication toy common architecture ingredients resumé, page 387

Table 9-24, Multiplication toy common architecture ingredients resumé, page 387

Figure 9-38, Multiplication toy common logical architecture, page 389

Figure 9-39, Diversifying the multiplier element and circuit board interface, page 390

Figure 9-40, Multiplication toy common physical architecture, page 390

⌷IB⌷ Microcontroller multiplier architecture white-box with major ingredients but no embedded black-boxes

Table 9-28, Microcontroller multiplier refined architecture ingredients resumé, page 398

Figure 9-46, Microcontroller multiplier logical architecture, page 399

⌷Xb⌷ Suppliers' sample catalogues

⌷E⌷ Picked in-house legacy source systems

⌷XaE⌷ Ordered suppliers' off-the-shelf fully described source systems

⌷XbE⌷ Ordered suppliers' turnkey fully described source systems

For possible *failure localization* and *failure elimination*, below *results* may be demanded (their *masters* are indicated *upstream* from the filled-in *schedule* in the figure above):

⌷IJ2⌷ Table 11-25, Multiplication toy with hard-wired multiplier start-up report 1, page 627

Table 11-33, Multiplication toy with microcontroller multiplier start-up report 2, page 636

⌷IW2⌷ Table 12-38, Multiplication toy with hard-wired multiplier failure elimination report 1, page 711

Table 12-47, Multiplication toy with microcontroller multiplier failure elimination report 1, page 723

⌷OW2⌷ Table 12-53, Multiplication toy environment (microcontroller multiplier variant) failure elimination report 1, page 728

10.15.3 EXAMPLE Multiplication toy with microcontroller multiplier: Finalize white-box detailed items for microcontroller

Let's imagine using such a simple microcontroller as an alternative realization of the multiplication toy multiplier. Input ports are used to receive the operands and output ports to deliver their product. The *interrupt* input port is not needed in this example.

To install the executable *program* into the microcontroller, there is often a special loading port for this purpose. The loading port is connected to the computer *host development system*, and a simple communication protocol transfers the *program* to the microcontroller, where it is permanently stored.

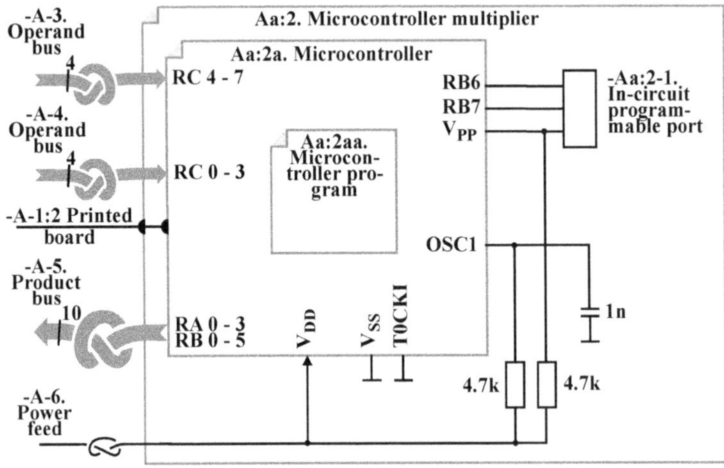

FIGURE 10-31 Microcontroller multiplier diagram

10.15.4 EXAMPLE Multiplication toy with microcontroller multiplier: Finalize white-box detailed items for program

All arithmetic and conversions that were previously made by gate networks are now performed by *program* algorithms. The same two methods are used that were used for the hard-wired logic network, but are now implemented in *programs*.

Algorithm for multiplying:

```
1101 = 13      Multiplicand
0101 = 5       Multiplyer
-----                     0  (Initial Value)
1101 = 13      0  + 13 = 13
0000  = 0      13 +  0 = 13
1101   = 52    13 + 52 = 65
0000   = 0     65 +  0 = 65
----------
01000001 = 65  Product
```

Algorithm for 8-bit binary to 3-digit BCD conversion:

1. Shift the binary number left one bit.
2. If 8 shifts have taken place, the BCD number is in the Hundreds, Tens, and Units columns.
3. If the binary value in any of the BCD columns is 5 or greater, add 3 to that value in that BCD column.
4. Go to 1.

The *program* is now handled as any *element* in an *architecture*, although it is built up with only immaterial information (see Fig. 10-32 below). In the left margin, a flow chart is drawn, illustrating the execution paths of the assembler *program*.

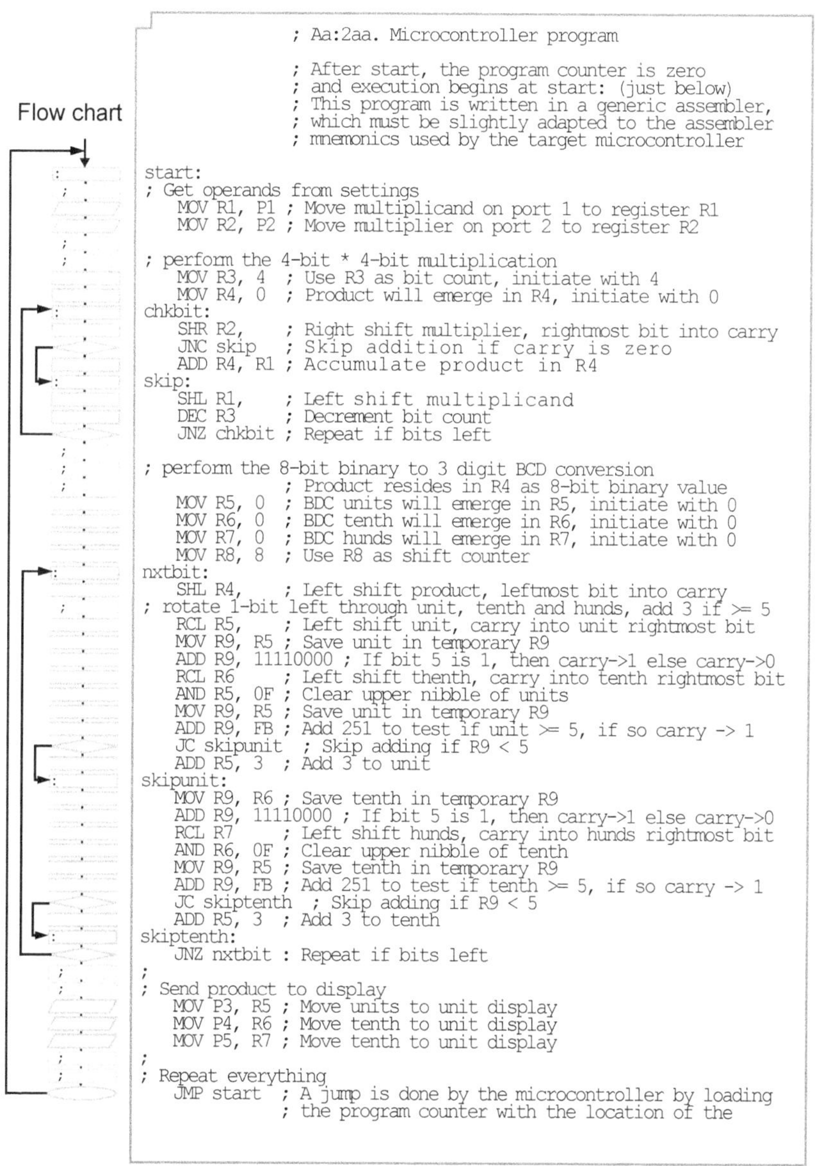


```
                        ; Aa:2aa. Microcontroller program

                        ; After start, the program counter is zero
                        ; and execution begins at start: (just below)
                        ; This program is written in a generic assembler,
                        ; which must be slightly adapted to the assembler
                        ; mnemonics used by the target microcontroller
start:
; Get operands from settings
   MOV R1, P1 ; Move multiplicand on port 1 to register R1
   MOV R2, P2 ; Move multiplier on port 2 to register R2

; perform the 4-bit * 4-bit multiplication
   MOV R3, 4  ; Use R3 as bit count, initiate with 4
   MOV R4, 0  ; Product will emerge in R4, initiate with 0
chkbit:
   SHR R2,    ; Right shift multiplier, rightmost bit into carry
   JNC skip   ; Skip addition if carry is zero
   ADD R4, R1 ; Accumulate product in R4
skip:
   SHL R1,    ; Left shift multiplicand
   DEC R3     ; Decrement bit count
   JNZ chkbit ; Repeat if bits left

; perform the 8-bit binary to 3 digit BCD conversion
            ; Product resides in R4 as 8-bit binary value
   MOV R5, 0  ; BDC units will emerge in R5, initiate with 0
   MOV R6, 0  ; BDC tenth will emerge in R6, initiate with 0
   MOV R7, 0  ; BDC hunds will emerge in R7, initiate with 0
   MOV R8, 8  ; Use R8 as shift counter
nxtbit:
   SHL R4,    ; Left shift product, leftmost bit into carry
; rotate 1-bit left through unit, tenth and hunds, add 3 if >= 5
   RCL R5,    ; Left shift unit, carry into unit rightmost bit
   MOV R9, R5 ; Save unit in temporary R9
   ADD R9, 11110000 ; If bit 5 is 1, then carry->1 else carry->0
   RCL R6     ; Left shift thenth, carry into tenth rightmost bit
   AND R5, 0F ; Clear upper nibble of units
   MOV R9, R5 ; Save unit in temporary R9
   ADD R9, FB ; Add 251 to test if unit >= 5, if so carry -> 1
   JC skipunit ; Skip adding if R9 < 5
   ADD R5, 3  ; Add 3 to unit
skipunit:
   MOV R9, R6 ; Save tenth in temporary R9
   ADD R9, 11110000 ; If bit 5 is 1, then carry->1 else carry->0
   RCL R7     ; Left shift hunds, carry into hunds rightmost bit
   AND R6, 0F ; Clear upper nibble of tenth
   MOV R9, R5 ; Save tenth in temporary R9
   ADD R9, FB ; Add 251 to test if tenth >= 5, if so carry -> 1
   JC skiptenth  ; Skip adding if R9 < 5
   ADD R5, 3  ; Add 3 to tenth
skiptenth:
   JNZ nxtbit : Repeat if bits left
;
; Send product to display
   MOV P3, R5 ; Move units to unit display
   MOV P4, R6 ; Move tenth to unit display
   MOV P5, R7 ; Move tenth to unit display
;
; Repeat everything
   JMP start  ; A jump is done by the microcontroller by loading
            ; the program counter with the location of the
```

FIGURE 10-32 Microcontroller multiplier assembler program with flow chart

10.15.5 EXAMPLE Multiplication toy with microcontroller multiplier: Finalize white-box detailed items for printed board

In the multiplication toy with microcontroller, the *components* will be soldered to a permanent PCB.

The PCB *supplier* sent a *sample* board from another, similar *product* (see Fig. 10-33).

The PC Board is the static *interface* in both the prototype and the *product*, holding all *components*, both commonality *components*, and the specific microcontroller with peripheral *components*.

FIGURE 10-33 Printed circuit board drawing

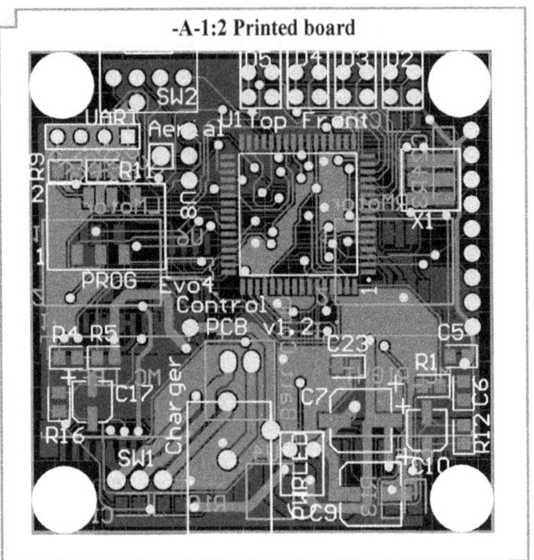

10.15.6 EXAMPLE Multiplication toy with microcontroller multiplier: Finalize white-box detailed items lists

Components for the microcontroller multiplier are listed in Table 10-74 below.

TABLE 10-74 Multiplication toy with microcontroller multiplier electronics white-box design items

Item	Type	Pin-out	Size w x d x h or program lines	White-box ingredient
i137. PIC16F57	Consumer type	28-pin SP DIL	34 x 7.8 x 5	Aa:2a. Microcontroller
i138. Resistor 4.7 k	20 %	Axial wires	7 x 3 mm	Aa:2. Microcontroller multiplier
i139. Capacitor 1 n	20 %	Axial wires	12 x 4 mm	
i140. Assembler program	Adapted to PIC16F57		63 program lines	Aa:2aa. Microcontroller program

Printed circuit board *white-box design items* are listed in Table 10-75 below.

TABLE 10-75 Multiplication toy with microcontroller multiplier printed board white-box design items

Item	Size	Type	White-box ingredient
i141. PCB	138 x 128 x 2	2-sided printing, injection points, read-out points, and in-circuit programmable port as gold-plated pads	-A-1:2 Printed board
i142. PCB preparation		Layout, clichés, drill template, etc.	-A-1:2 Printed board preparation

10.15.7 EXAMPLE Multiplication toy with microcontroller multiplier: Account for white-box development, realization, and manufacturing

Materials for manufacturing are shown in Table 10-76 below.

TABLE 10-76 Microcontroller multiplier for multiplication toy, estimated manufacturing material requisites cost

	1 000 series		100 000 series					
Pcs	€ /pc	€ /toy	€ /pc	€ /toy	Item	Product	Supplier	Element
1	0.7	0.7	0.4	0.4	i137. PIC16F57	Chip gross	SU.21. SiliC	Aa:2a. Microcontroller
2	0.1	0.2	0.075	0.15	i138. Resistor 4.7 k	Resa	SU.20. Passta	Aa:2. Microcontroller multiplier
1	0.5	0.5	0.3	0.3	i139. Capacitor 1 n	Capa	SU.20. Passta	Aa:2. Microcontroller multiplier
1	0.4	0.4	0.2	0.2	i141. PCB	PCB inc.	SU.22. PeCeBe	-A-1:2 Printed board
1	0.4	0.4	0.01	0.01	i142. PCB preparation	In-house	SU.22. PeCeBe	-A-1:2 Printed board
		2.2		1.06				

Accounting for manufacturing work to transfer the *machine instruction* to the programmable port of the microprocessor is done in Table 10-77 below.

TABLE 10-77 Microcontroller multiplier for multiplication toy, estimated manufacturing work requisites hourly/fixed/rental cost

1 000 series			100 000 series					
h	€ /h	€ /toy	h	€ /h	€ /toy	Item	Provider	Interface
200	4	0.8	400	40	0.16	i42. Automatic flashing on conveyor belt after PCB is assembled	PR.11. Mech-tech	-Aa:2-1. In-circuit programmable port

Material for the prototype is shown in Table 10-78 below.

TABLE 10-78 Microcontroller multiplier for multiplication toy, estimated realization material requisites cost

	50 prototypes					
Pcs	€ /pc	€ /toy	Item	Product	Supplier	Element
1	1.8	1.8	i137. PIC16F57	Chip gross	SU.21. SiliC	Aa:2a. Microcontroller
2	0.15	0.3	i138. Resistor 4.7 k	Resa	SU.20. Passta	Aa:2. Microcontroller multiplier
1	0.8	0.8	i139. Capacitor 1 n	Capa	SU.20. Passta	Aa:2. Microcontroller multiplier
1	1	1	i141. PCB	PCB inc.	SU.22. PeCeBe	-A-1:2 Printed board
1	4	4	i142. PCB preparation	In-house	SU.22. PeCeBe	-A-1:2 Printed board
		7.9				

Cost of prototype *realization* work is accounted in Table 10-79 below.

TABLE 10-79 Multiplication toy with microcontroller multiplier, estimated realization work requisites hourly/fixed/rental cost

50 prototypes				Item	Provider	Interface
h	€ /h	€	€ /toy			
200	25	5 000	100	i35. Prototype mounting engineer	PR.10. In-house project	-A-1:2 Printed board]A]2. Casing
80	20	1 600	32	i41. Worker half-manually flashing microcontroller	PR.11. Mech-tech	-Aa:2-1. In-circuit programmable port
			132			

Development work for the multiplication toy with microcontroller multiplier includes *design architectures*, finalize design & requisites, and verify black- & white-box (see chapters marked "M" on *black-/white-box nesting level* 1 in Fig. 10-30, p. 518). Accounted development work is shown in Table 10-80 below.

TABLE 10-80 Multiplication toy white-box with microcontroller multiplier, estimated development work requisites hourly/fixed cost

50 prototypes				Item	Provider	Process schedule
h	€ /h	€	€ /toy			
75	40	3 000	60	i30. Microcontroller designer	PR.14. Chipp	Pt.1.A. Satisfy refined requirements by decomposing product black-box into white-box design containing no black-boxes.
250	80	20 000	400	i31. Assembler programmer	PR.15. Micro-Hard	Pt.1.F. Finalize design items of product white-box with product requisites.
50	60	3 000	60	i33. Automated verification specialist	PR.11. Mech-tech	Pt.1.F. Finalize design items of product white-box with product requisites.
60	50	3 000	60	i34. Accounting engineer	PR.10. In-house project	Pt.n.Fc. Account for white-box development, realization, and manufacturing (resulting in these requisites tables).
50	40	2 000	40	i37. Electronics verifier	PR.10. In-house project	Pt.1.V. Verify black-/white-box.
125	32	4 000	80	i38. Program verifier	PR.15. Micro-Hard	Pt.1.V. Verify black-/white-box.
			700			

Accounting for the multiplication toy with microcontroller multiplier is summarized in Table 10-81 below.

TABLE 10-81 Microcontroller multiplier white-box with microcontroller multiplier, summed/accumulated requisites estimates

Item	Reference	Development and 50 prototypes € / toy	1 000 series € / toy	100 000 series € / toy
Multiplication toy white-box commonalities, summed/accumulated requisites estimates	Table 10-55, chapter 10.8	152.00	15.715	3.381
Microcontroller multiplier for multiplication toy, estimated manufacturing material requisites cost	Table 10-76, this chapter		2.20	1.06
Microcontroller multiplier for multiplication toy, estimated manufacturing work requisites hourly/fixed/rental cost	Table 10-77, this chapter		0.80	0.16

TABLE 10-81 Microcontroller multiplier white-box with microcontroller multiplier, summed/accumulated requisites estimates

Item	Reference	Development and 50 prototypes € / toy	1 000 series € / toy	100 000 series € / toy
Microcontroller multiplier for multiplication toy, estimated realization material requisites cost	Table 10-78, this chapter	7.90		
Multiplication toy with microcontroller multiplier, estimated realization work requisites hourly/fixed/rental cost	Table 10-79, this chapter	132.00		
Multiplication toy white-box with microcontroller multiplier, estimated development work requisites hourly/fixed cost	Table 10-80, this chapter	700.00		
Microcontroller multiplier white-box with microcontroller multiplier, summed/accumulated requisites estimates		**839.90**	**18.715**	**4.601**

10.15.8 EXAMPLE Multiplication toy with microcontroller multiplier: Examine white-box success

The gate meeting was surprised by the high cost of the tiny *assembler program*. A *program* gate member explained the difficulty to get unstructured *assembler programs* to work correctly. Every *program* line is a detailed assembler mnemonic, which depends on and influences every other mnemonic in the *program*. To get everything right from the beginning requires heavy *program* reviews, or if this is not done, a lot of *verification* on a detailed level in a scanty system without many development facilities.

However, when distributed across a volume of 100 000 multiplication toys, the *program* cost is only €0.15 per unit.

10.15.9 EXAMPLE Multiplication toy: Proceed reading

To follow this example, these are the alternatives:

- Proceed to chapter 11.10, "EXAMPLE Multiplication toy with microcontroller multiplier: Procure items and linkable programs in order to realize white-box," page 632.

- Proceed to chapter 12.16, "EXAMPLE Multiplication toy with microcontroller multiplier: Verify black-/white-box," page 717.

- Proceed to chapter 10.16 "EXAMPLE Multiplication toy environment (microcontroller multiplier variant): Finalize design of environment interfaces with product requisites" below.

10.16 EXAMPLE Multiplication toy environment (microcontroller multiplier variant): Finalize design of environment interfaces with product requisites

10.16.1 EXAMPLE Multiplication toy environment (microcontroller multiplier variant): Process schedule to use

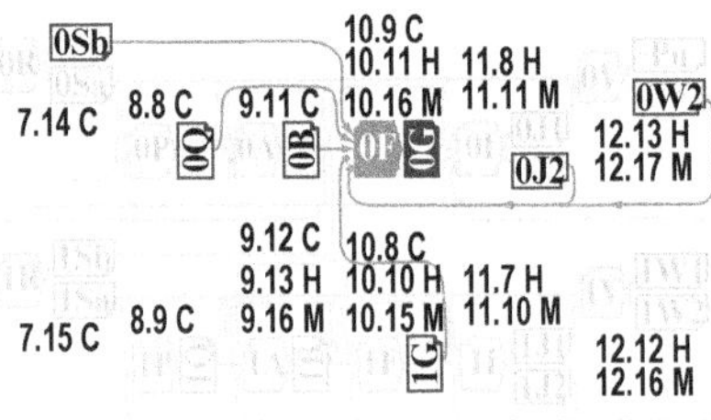

For the multiplication toy example, a special technical overall *schedule* is *tailored* (illustrated in Fig. 10-34, left):

- **Pt:2:3. Tailored advanced technical schedule n = 2,** page 169

In this *tailored schedule* the generic *schedule* to use is (filled-in symbols):

- **Pt.0.F. Finalize design of environment interfaces with product requisites,** page 118.

FIGURE 10-34 Position of schedule and master to use (filled in) in overall schedule

Below *results* may be demanded (their *masters* are indicated *upstream* from the filled-in *schedule* in the figure above):

0G Common environment interface design items and product requisites

Table 10-62, Multiplication toy environment interfaces commonalities, summed/accumulated requisites estimates, page 499

1G Common finalized white-box items design with product requisites

Figure 10-16, Setting device diagram, page 484

Figure 10-17, Presenting device diagram, page 485

Figure 10-18, Power supply diagram, page 486

Table 10-46, Multiplication toy common electronics, white-box design items, page 487

1G Microcontroller multiplier finalized white-box items design with product requisites

Figure 10-31, Microcontroller multiplier diagram, page 520

Figure 10-32, Microcontroller multiplier assembler program with flow chart, page 521

Figure 10-33, Printed circuit board drawing, page 522

Figure 10-19, Multiplication toy casing, page 486

Table 10-74, Multiplication toy with microcontroller multiplier electronics white-box design items, page 522

Table 10-75, Multiplication toy with microcontroller multiplier printed board white-box design items, page 522

1G	Microcontroller multiplier finalized white-box items design with product requisites

Table 10-81, Microcontroller multiplier white-box with microcontroller multiplier, summed/accumulated requisites estimates, page 524

0B	Environment existing ingredients and design of interfaces to the environment

Table 9-23, Multiplication toy environment architecture ingredients resumé, page 383

Figure 9-35, Multiplication toy environment logical architecture, page 384

Figure 9-36, Multiplication toy environment physical architecture, page 384

0Sb	Environment staffing allocation

Table 7-24, Multiplication toy environment development allocation staffing, page 213

Xb	Suppliers' sample catalogues

E	Picked in-house legacy source systems

XaE	Ordered suppliers' off-the-shelf fully described source systems

XbE	Ordered suppliers' turnkey fully described source systems

For possible *failure localization* and *failure elimination*, below *results* may be demanded (their *masters* are indicated *upstream* from the filled-in *schedule* in the figure above):

0J2	Table 11-28, Multiplication toy environment (hard-wired multiplier variant), start-up report 1, page 630

Table 11-36, Multiplication toy environment (microcontroller multiplier variant), start-up report 1, page 640

0W2	Table 12-53, Multiplication toy environment (microcontroller multiplier variant) failure elimination report 1, page 728

10.16.2 EXAMPLE Multiplication toy environment (microcontroller multiplier variant): Account for environment interfaces development, realization, and manufacturing

The multiplication toy *interface* to the hosting toy needs to be mounted on some selected hosting toys to be verified (see Table 10-82 below). Note that the *environment interfaces* are only delivered to the hosting toy producers, and it is their responsibility to mount them correctly.

TABLE 10-82 Multiplication toy environment (microcontroller multiplier variant) estimated realization work requisites hourly/fixed/rental cost

50 prototypes						
h	€/h	€	€/toy	Item	Provider	Interface
8	60	480	9.6	i19. Senior mechanical designer	PR.12. Mechano	-E-1.Host attachments
			9.6			

Environment interfaces specification and design development work is already accounted in the commonalities for both toy *variants*. Development work for the *environment* of multiplication toy with microcontroller multiplier includes finalize design & requisites and verify black- & white-box (see chapters marked "M" on *nesting level* 0 in Fig. 10-34, p. 526). Below is the additional accounting for the microcontroller development (see Table 10-83 below).

TABLE 10-83 Multiplication toy environment (microcontroller multiplier variant) estimated development work requisites hourly/fixed cost

50 prototypes				Item	Provider	Process schedule
h	€ /h	Tot €	€ /toy			
16	60	960	19.2	i15. Mechanical designer	PR.12. Mechano	Pt.0.Fa. Finalize environment interfaces detailed items.
24	60	1 440	28.8	i18. Product manager	PR.10. In-house project	Pt.0.Fc. Account for environment interfaces development, realization, and manufacturing.
8	50	400	8	i20. MMI verifier	PR.10. In-house project	Pt.0.V. Verify prototype in environment.
32	45	1 440	28.8	i21. Hosting toy engineer	PR.12. Mechano	Pt.0.V. Verify prototype in environment.
			84.80			

TABLE 10-84 Multiplication toy (microcontroller multiplier variant) summed/accumulated requisites estimates

Item	Reference	Development and 50 prototypes € / toy	1 000 series € / toy	100 000 series € / toy
Multiplication toy environment interfaces commonalities, summed/accumulated requisites estimates	Table 10-62, chapter 10.9	207.20	0.46	0.26
Microcontroller multiplier white-box with microcontroller multiplier, summed/accumulated requisites estimates	Table 10-81, chapter 10.15	839.90	18.715	4.601
Multiplication toy environment (microcontroller multiplier variant) estimated realization work requisites hourly/fixed/rental cost	Table 10-82, this chapter	9.60		
Multiplication toy environment (microcontroller multiplier variant) estimated development work requisites hourly/fixed cost	Table 10-83, this chapter	84.80		
Multiplication toy (microcontroller multiplier variant) summed/accumulated requisites estimates		**1 141.50**	**19.175**	**4.861**
Development cost and 50 prototypes costs distributed on manufacturing series			**20.316**	**4.872**

10.16.3 EXAMPLE Multiplication toy environment (microcontroller multiplier variant): Examine prototype success

It seems that manufacturing cost in a 100 000 series can be limited to under €5 per unit, even if the development and prototype cost loads the manufacturing. The gate decided to celebrate this with some team caps and T-shirts with the picture from Figure 9-36, page 384, with Figure 9-40, page 390, inserted.

The collar has entirely solved the problem with unstable hosting toys. Because of child safety and mounting responsibility disclaimers, it is important that the collar manual is developed together with the multiplication toy manual.

Nothing else was brought up at the gate meeting.

10.16.4 Multiplication toy: Proceed reading

To follow this example, these are the alternatives:

- Proceed to chapter 11.11, "EXAMPLE Multiplication toy environment (microcontroller multiplier variant): Realize interfaces to environment and insert/await outermost white-box," page 637.
- Proceed to chapter 12.17, "EXAMPLE Multiplication toy environment (microcontroller multiplier variant): Verify prototype in environment," page 725.
- To leave the multiplication toy example and to dig deeper into the chapter 10, "Finalize design & requisites" topic, proceed to chapter 10.17 below.

10.17 Structured programs

10.17.1 Cpdm definition

In *Cpdm*, *structured program* is defined according to Table 10-85 below.

TABLE 10-85 Cpdm definition

Aspect	Structured program
Description	Program prohibiting unstructured flow by providing only structured machine-independent instructions, exclusively being sequence, selection, iteration, and function call, needing compiler and linker tools support to be translated to machine instructions and machine variables
Synonyms	High-level program, third-generation language, procedural program
Examples	C, Pascal, PL1
Architecture symbols	Structured program Global variables Function Local variables Structured instructions, either one of : Iteration Selection Sequence Function call Can be a black-box

When structured languages were developed, this was a big step forward. Most of the structured mechanisms were very good to support programming safety without being too restrictive and rigid.

Most of the *program* descriptions in this book are based on language C, not because it might be the best, but because it is one of the most popular structured languages. Other structured languages are Pascal, PL1, C# and Ada. The language Basic is an *unstructured program*, since it accepts the "GOTO Address label" *instruction*, which heavily violates the principle of *structured programs*.

10.17.2 Basic concepts

As seen from previous chapters, there are several serious weaknesses in assembler languages. The possibility to transfer execution from anywhere to anywhere, and the possibility to change *variables* from *instructions* everywhere in the *program*, make the *program* hard to control and risky to write. When the *assembler program* spans several hundred screens (or hard copies) it is almost impossible to understand how one part of the *program* interacts with another part.

There was another big improvement in getting compiler support to avoid dangerous constructs that were supported by assemblers.

Understanding a *program* is not about reading pages of detailed *instructions*. That is as stupid as trying to understand a house by inspecting each brick the house is made of. To understand a *program*, you must be able to understand its rooms, how the doors connect them, and what their purposes are.

To understand a *program*, the language must possess useful structures. Many attempts have been made, some even dubious, to provide structured languages with proper architectural structures. The main improvements are:

1. Modularization

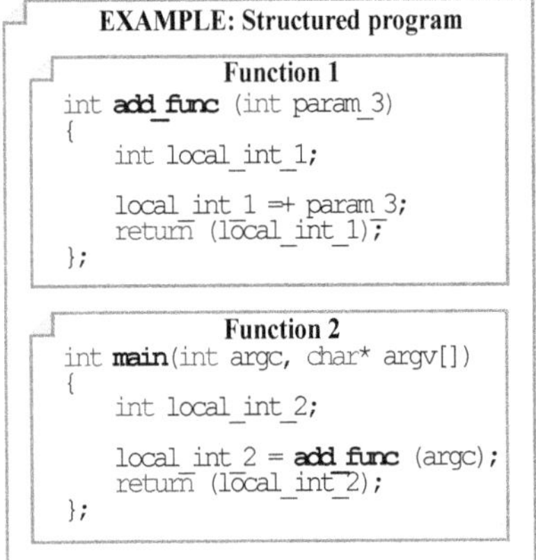

a. A *structured program* is divided into a number of *functions* (names in bold in Fig. 10-35, left).

b. A *function* can be handled as any *Cpdm* compound *element*.

c. *Functions* cannot be hierarchically nested (they make a single-level flat structure of adjacent *elements*).

d. *Functions* call each other, as in the figure the *function* 2 main calls the *function* 1 add_func.

FIGURE 10-35 EXAMPLE: Modularization of structured program.

2. Structure and encapsulation

a. *Aliases* are name assignments for expressions to better explain their meaning, and have no impact on execution of the *programv*.

b. Restriction specification originates from *restriction requirements*, for the purpose of limiting the behavior of the *program*.

c. *Function shapes* describe *interfaces* between *functions*.

d. *Variables* declared outside *functions* are global, and can then be accessed from anywhere

e. A *function* can be handled as a (small) *black-box*, provided that passed parameters, return parameters, and possibly careful usage of *global variables* are specified in the *black-box requirements*.

f. *Local variables* inside a *function* boundary cannot be accessed from outside the *function*.

g. For *instructions*, see below.

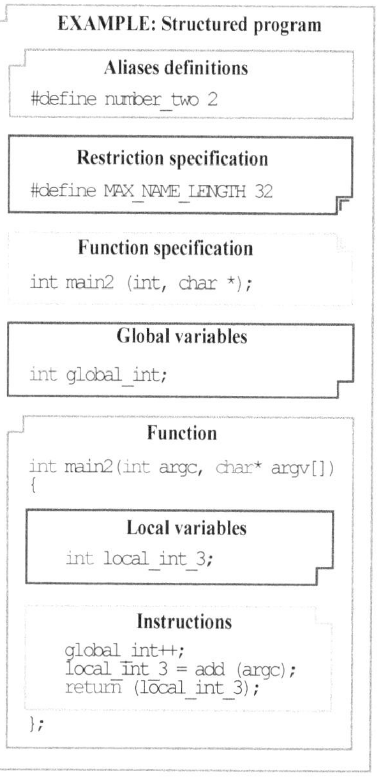

FIGURE 10-36 EXAMPLE: Encapsulation in structured program

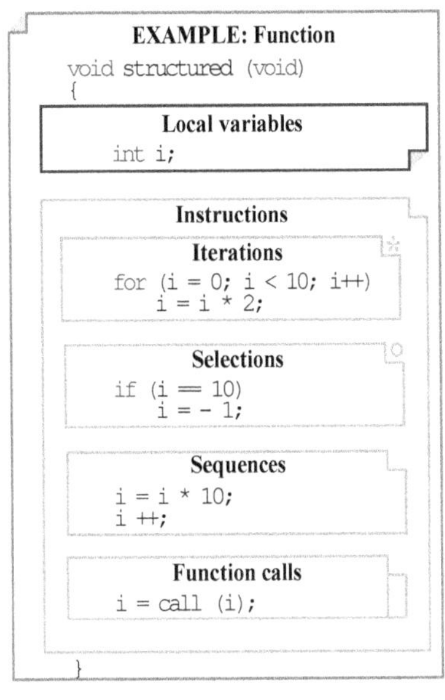

3.Function content.

a._Local variables_ are accessible only from the _instructions_ within the boundary of the same _function_.

b._Instructions_ are the executable _elements_ of a _program_. In _structured programs_, it is always possible to fit _instructions_ into one of the below _elements:_

• _Structured iteration instructions_ like while, for, and so forth.

• _Structured selection instructions_ like if-then-else, switch-case, and so forth.

• _Structured instruction sequences_ executed one _instruction_ after another

• _Structured function call_ to another _function_

c.Above _instructions_ can be freely nested with each other to any extent.

FIGURE 10-37 EXAMPLE: Parts of a structured program

10.17.3 Doubtful possibilities

There are not many obscure and risky _elements_ in structured languages. Some languages may have below doubtful possibilities.

• Conditional macros. Use only if well trained.

10.17.4 Cpdm definition

To support modularization, _programs_ can be _decomposed_ into many files. A big _program_ may consist of millions of _program_ lines (also called "Lines of Code" and abbreviated "LoC"), distributed in thousands of files, which may be nested into hierarchies. Mastering _complexity_ requires good structure and design descriptions of these.

Source files hold all the executable _instructions_ of a _program_. _Header files_ are of many types, each with its specific purpose. Unfortunately, the different types of _header files_ are not distinguished by the compiler, and are often mixed in reality, resulting in a mess of _header files_ with mixed purposes (high coupling and low cohesion).

In _Cpdm_, different _program_ file types are defined according to Table 10-86 below. _Header files_ must be included in all _source files_ referring specified _functions_ and using specified restrictions, _aliases,_ and so forth.

TABLE 10-86 Cpdm definitions

Aspect	Structured program files			
Type	Source file	Header file		
Content	Element exclusively containing program instructions, finally resulting in executable machine programs	Element exclusively containing program commands for better readability and control; never resulting in anything executable in final machine programs		
	Executable program elements complying with included function shapes	Function shape requirements	Restriction requirements	Aliases, tool directives, and similar
Content synonyms	Collection of program instructions including comments	Function prototypes, function interface, forward function declaration	#defines, directives, definitions, declarations	
Purpose	Compiled to executable machine programs	Capture how source files are obliged to use specified functions.	Design limitations like restrictions requirements	Improving readability and tool control
Architecture symbols	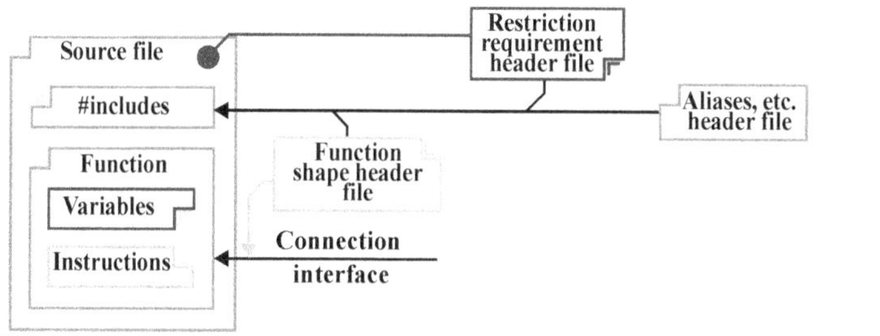 Source file / #includes / Function / Variables / Instructions — Restriction requirement header file, Aliases, etc. header file, Function shape header file, Connection interface			

10.18 EXAMPLE Hosted calc_logic for reuse: Finalize design items of product white-box with product requisites

10.18.1 EXAMPLE Calc_logic: About this example

Calc_logic is a *program* that must be developed to adapt to several *users* in different *development systems*. Calc_logic is *hosted* and developed on a Windows computer *development system*, and must first be *ported* to a Windows *target* computer, then to an external *development board*, and finally to a stand-alone pocket calculator.

10.18.2 EXAMPLE Hosted calc_logic for reuse: Process schedule to use

Pt:2:3. Tailored advanced technical schedule n = 2

Up to now, the figure at left has been used to navigate around in the **Pt:2:3. Tailored advanced technical schedule n = 2**. But from now on, the *porting* of Calc_logic requires that many *schedules* are repeated for each *target*, which can be better illustrated as shown below and in Figure 10-39, page 534, which will be used for further *process* navigation. However, the figure at left will still be used for overall navigation.

FIGURE 10-38 This overall schedule will only be used for overall navigation; for detailed navigation see below explanation and Figure 10-39, page 534.

For the calc_logic example, a special technical overall *schedule* is *tailored* (illustrated in Fig. 10-39, below):

- **Pt:2:3b. Tailored advanced technical schedule n = 2, porting to 3 targets (right half),** page 170

In this *tailored schedule* the generic *schedule* to use is (filled-in symbols below):

- **Pt.n.F. Finalize design of white-box with product requisites,** page 140.

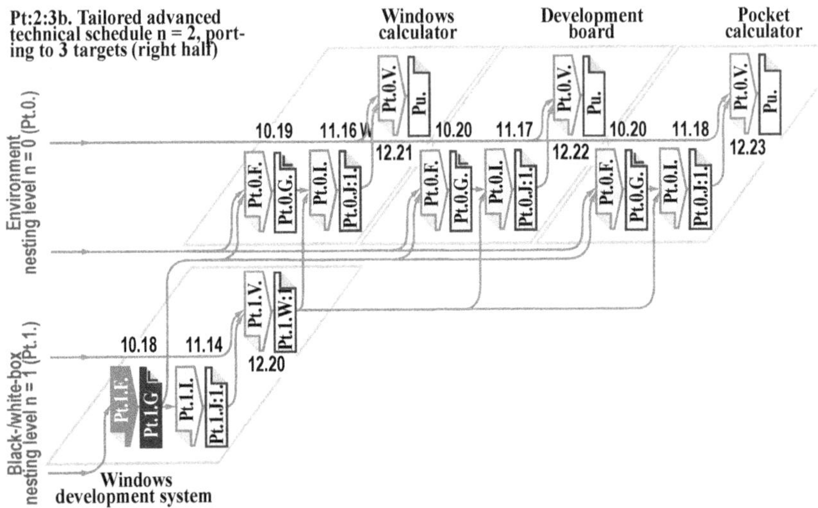

FIGURE 10-39 Position of schedule and master to use (filled in) in overall schedule

10.18.3 EXAMPLE Hosted Calc_logic for reuse: Finalize white-box detailed items

When finalizing the calc_logic programming, it is beneficial to also update and complement the *white-box design items* (see Fig. 10-40, p. 536) based on the *architecture* in Figure 9-61, page 422.

Since structured languages recognize only the *elements* sequence, iterations, selections, and *function* calls, it is rather easy to document them in charts using only structured *instruction* symbols (see again Fig. 10-40 below).

To keep track of the file structure, a file chart is prepared, describing the different types of Calc_logic files and how they relate to each other (see Fig. 10-41 next).

The detailed *program* is also prepared (see Fig. 10-42, ending on p. 538 below). The total C *program* is listed, but with graphic characters included, which lets the *architecture* be illustrated. This kind of *program* listing is not possible to compile, but it is rather illustrative, in particular for people considering *program architecture* to be a mystery.

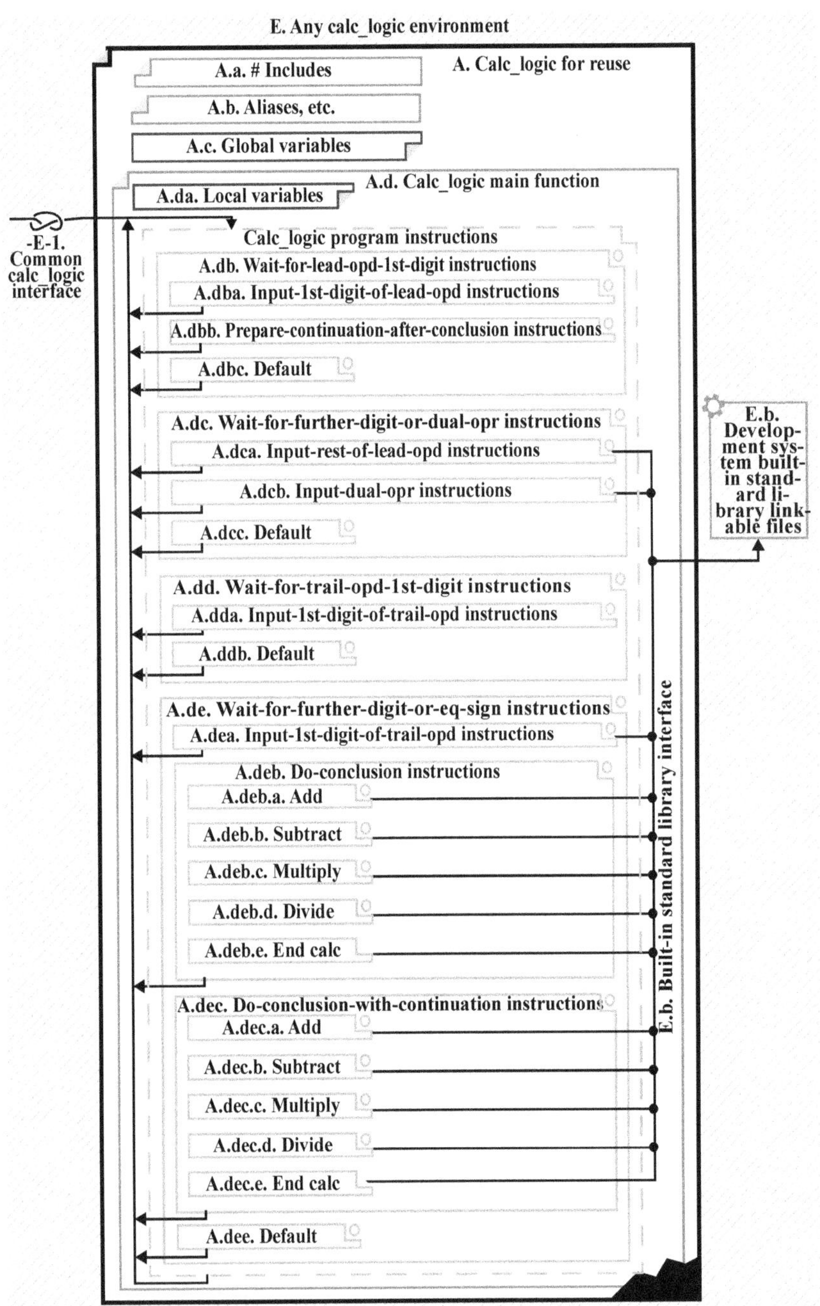

FIGURE 10-40 Calc_logic for reuse design chart

Now the physical appearance of concrete files must be illustrated. For *managers* and other people who have not written *programs*, these file *white-box design items* can be very confusing if not properly described (see Fig. 10-41 below).

Even if calc_logic might be tricky to get perfectly right, it is a quite small and transparent *program*, not in big need of modularization. However, there are a number of *white-box design items* to finalize and account for, apart from the calc_logic *program source file* itself.

Structured development languages for well-known operating *development systems* always come as *development system libraries*, connecting the developed *program* to the operating system. *Header files* capture restrictions and *connection interfaces* within the built-in *development system libraries*, and must be included in the Calc_logic *source file* utilizing the built-in *development system libraries*.

In the same way as including *development system libraries header files* in Calc_logic (like E.R.13.a. to E.R.13.c. in Fig. 10-41 below), the application that will in turn be a *user* of Calc_logic also needs to include Calc_logic *header files* (like A.RR.7. and A.RB.8. in Fig. 10-41 below). Note that these *header files* are considered to be Calc_logic *requirements*, such as *function shapes, restriction, aliases*, and so forth, provided with *requirement* symbols in Figure 10-41 below.

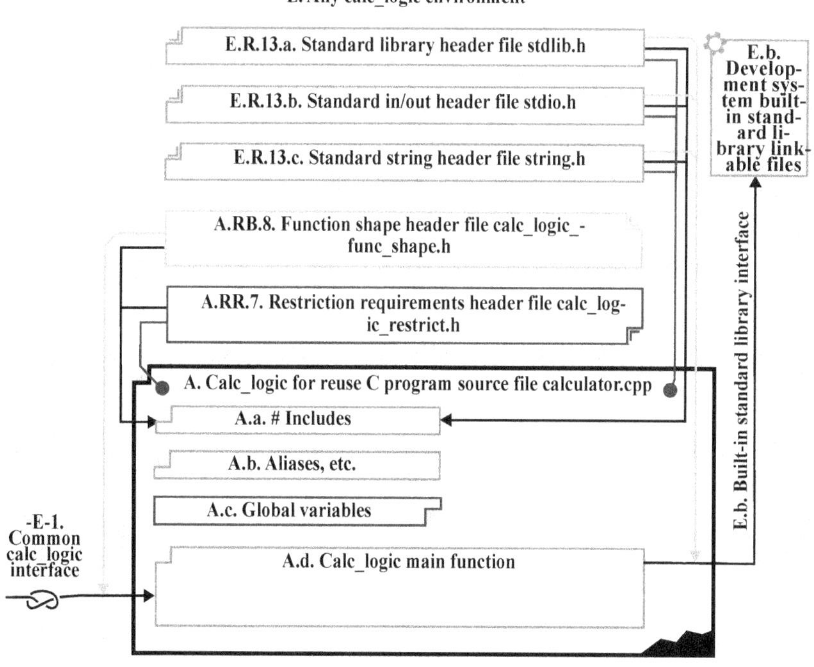

FIGURE 10-41 Calc_logic for reuse file chart

The Calc_logic C *program* is shown in Figure 10-42 below.

```
/* A. Calc_logic for reuse C program source file calculator.cpp */

                        /* A.a. # Includes */
#include "calc_logic_restrict.h"    /* E.g. Calc logic restriction header file */
#include <string.h>     /* E.R.13.c. Standard string header file string.h */
#include <stdlib.h>     /* E.R.13.a. Standard library header file stdlib.h */
#include <stdio.h>      /* E.R.13.b. Standard in/out header file stdio.h */


                        /* A.b. Aliases, etc. */
#define WAIT_FOR_LEAD_OPD_1ST_DIGIT 1
#define WAIT_FOR_FURTHER_DIGIT_OR_DUAL_OPR 2
#define WAIT_FOR_TRIAL_OPD_1ST_DIGIT 3
#define WAIT_FOR_FURTHER_DIGIT_OR_EQ_SIGN 4


                        /* A.c. Global variables */
char lead_operand[OP_MAX_LEN] = "\0";
char trial_operand[OP_MAX_LEN] = "\0";
char previous_operator = '\0';
int  wait_state = WAIT_FOR_LEAD_OPD_1ST_DIGIT;


                        /* A.d. Calc_logic main function */
char *calc_logic(char stim_char)
{
                        /* A.da. Local variables */
    char    stim_str[2];


                        /* Calc_logic program instructions */
    //  □ WP1
    switch(wait_state)
    {
                        /* A.db. Wait-for-lead-opd-1st-digit instructions */

        case WAIT_FOR_LEAD_OPD_1ST_DIGIT:
        {
            switch(stim_char)
            {
                        /* A.dba. Input-1st-digit-of-lead-opd instructions */
                case '0': case '1': case '2': case '3': case '4':
                case '5': case '6': case '7': case '8': case '9':
                case '.':
                {
                    lead_operand[0] = stim_char;
                    lead_operand[1] = '\0';
                    wait_state = WAIT_FOR_FURTHER_DIGIT_OR_DUAL_OPR;
                    //  □ WP2
                    return(lead_operand);
                }

                        /*A.dbb. Prepare-continuation-after-conclusion instructions*/

                case '+': case '-': case '*': case '/':
                {
                    previous_operator = stim_char;
                    wait_state = WAIT_FOR_TRIAL_OPD_1ST_DIGIT;
                    //  □ WP3
                    return(NULL);
                }

                        /* A.dbc. Default */
                default:
                    //  □ WP4
```

```c
                    return(NULL);

            }
        }



        case WAIT_FOR_FURTHER_DIGIT_OR_DUAL_OPR:
        {
            switch(stim_char)
            {

                case '0': case '1': case '2': case '3': case '4':
                case '5': case '6': case '7': case '8': case '9':
                case '.':
                {
                    stim_str[0] = stim_char;
                    stim_str[1] = '\0';
                    strcat(lead_operand, stim_str);
                    //   □ WP5
                    return(lead_operand);
                }


                case '+': case '-': case '/': case '*':
                {
                    previous_operator = stim_char;
                    strcpy(trial_operand, lead_operand);
                    wait_state = WAIT_FOR_TRIAL_OPD_1ST_DIGIT;
                    //   □ WP6
                    return(NULL);
                }


                default:
                //   □ WP7
                return(NULL);

            }
        }



        case WAIT_FOR_TRIAL_OPD_1ST_DIGIT:
        {
            switch(stim_char)
            {


                case '0': case '1': case '2': case '3': case '4':
                case '5': case '6': case '7': case '8': case '9':
                case '.':
                {
                    trial_operand[0] = stim_char;
                    trial_operand[1] = '\0';
                    wait_state = WAIT_FOR_FURTHER_DIGIT_OR_EQ_SIGN;
                    //   □ WP8
                    return(trial_operand);
                }


                default:
                //   □ WP9
                return(NULL);

            }
```

```c
      }

      /* A.de. Wait-for-further-digit-or-eq-sign instructions */

case WAIT_FOR_FURTHER_DIGIT_OR_EQ_SIGN:
{
    switch(stim_char)
    {
        /*A.dea. Input-1st-digit-of-trial-opd instructions*/

        case '0': case '1': case '2': case '3': case '4':
        case '5': case '6': case '7': case '8': case '9':
        case '.':
        {
            stim_str[0] = stim_char;
            stim_str[1] = '\0';
            strcat(trial_operand, stim_str);
            //   □ WP10
            return(trial_operand);
        }

            /* A.deb. Do-conclusion instructions */

        case '=':
        {
            switch(previous_operator)
            {
                    /* A.deb.a. Add */

                case '+':
                sprintf(lead_operand, format,
                strtod(lead_operand,'\0') +
                strtod(trial_operand,'\0'));
                break;

                    /* A.deb.b. Subtract */

                case '-':
                sprintf(lead_operand, format,
                strtod(lead_operand,'\0') -
                strtod(trial_operand,'\0'));
                break;

                    /* A.deb.c. Multiply */

                case '*':
                sprintf(lead_operand, format,
                strtod(lead_operand,'\0') *
                strtod(trial_operand,'\0'));
                break;

                    /* A.deb.d. Divide */

                case '/':
                // Inserted to cure Stress usage profile divide by zero
                if(strtod(trial_operand,'\0') == 0)
                {
                    wait_state = WAIT_FOR_LEAD_OPD_1ST_DIGIT;
                    return("#ZeroDivErr");
                }
                sprintf(lead_operand, format,
                strtod(lead_operand,'\0') /
                strtod(trial_operand,'\0'));
                break;

            }
                    /* A.deb.e. End calc */
            wait_state = WAIT_FOR_LEAD_OPD_1ST_DIGIT;
            //   □ WP11
            return(lead_operand);

        }
```

```c
            case '+': case '-': case '*': case '/':
            {
                switch(previous_operator)
                {

                    case '+':
                    sprintf(lead_operand, format,
                    strtod(lead_operand,'\0') +
                    strtod(trial_operand,'\0'));
                    break;


                    case '-':
                    sprintf(lead_operand, format,
                    strtod(lead_operand,'\0') -
                    strtod(trial_operand,'\0'));
                    break;


                    case '*':
                    sprintf(lead_operand, format,
                    strtod(lead_operand,'\0') *
                    strtod(trial_operand,'\0'));
                    break;


                    case '/':
                    // Inserted to cure Stress usage profile divide by zero
                    if(strtod(trial_operand,'\0') == 0
                    {
                        wait_state = WAIT_FOR_LEAD_OPD_1ST_DIGIT;
                        return("#ZeroDivErr");
                    }
                    sprintf(lead_operand, format,
                    strtod(lead_operand,'\0') /
                    strtod(trial_operand,'\0'));
                    break;

                }

                wait_state = WAIT_FOR_TRIAL_OPD_1ST_DIGIT;
                previous_operator = stim_char;
                //  □ WP12
                return(lead_operand);

            }


        default:
        //  □ WP13
        return(NULL);

        }
    }

} //  □ WP14
return(NULL);

}
```

FIGURE 10-42 Calc_logic C program

To support reusing calc_logic when compiling and linking this *program* to its *user*, for example, from the *fixture* tool **Pt.n.Vgb. Test-case executor** (see Fig. 12-27, p. 743), the adaptable specification *header files* must be submitted (see Fig. 10-43 below and Fig. 10-44 next).

A.RR.7. Restriction requirements header file calc_logic_restrict.h

```
/* calc_logic restrictions */

#define OP_MAX_LEN 256
```

FIGURE 10-43 Calc_logic restriction requirement header file used for porting

---

A.RB.8. Function shape header file calc_logic_func_shape.h

```
/* calc_logic interface */

char *calc_logic (char);
```

FIGURE 10-44 Calc_logic function shape header file used for porting

As already said, the calc_logic *program* is divided into different files (again see Fig. 10-41, p. 537), to make it easier to *port* and *reuse*. Each of these files can now be considered a *white-box design item*.

First the *source file* keeping the *program* (see Table 10-87 below).

TABLE 10-87 Hosted Calc_logic for reuse program
white-box design items

Item	Type	Purpose	Size	White-box ingredient
i100.Hosted portable calc_logic programs on CD disc	C program	Reusable program for calculators	Circa 300 program lines	A. Calc_logic for reuse C program source file calculator.cpp

The two *header files* are listed in Table 10-88 below.

TABLE 10-88 Calc_logic environment (the Windows calculator application)
interfaces to environment design items

Item	Type	Purpose	Size	Interface ingredient
i101.Adapted header files for Windows target	Function specification header	To be included when compiling environment	Circa 5 program lines	A.RB.8. Function shape header file calc_logic_func_shape.h
	Restriction specification header	To set restriction when compiling calc_logic	Circa 5 program lines	A.RR.7. Restriction requirements header file calc_logic_restrict.h

Equipment for development is specified in Table 10-89 below.

TABLE 10-89 Hosted Calc_logic for reuse development system white-box design items

Item	Product	Size / type	Environment
i102.Host computer	Well Windows PC	Computer with working memory space 8 GB, hard-disk space 400 GB and CPU speed 3 GHz	Development equipment
	MiniHard inVisual C	C development system with built-in standard libraries	Development system

10.18.4 EXAMPLE Hosted Calc_logic for reuse: Account for white-box development, realization, and manufacturing

Note that the *product requisites* in this example concern only the *reusable item* calc_logic, and do not include any requisites for the Windows application developed by the in-house project "dw2", or electronics and mechanics of the pocket calculator *product* developed by the in-house project "de1".

In this example, the used *development system* was enough to supply needed floating-point math without buying any *program libraries*, and other *program* costs are insignificant. A compact disk (CD) could hold the *programs* until they are *reused* (see Table 10-90 below).

TABLE 10-90 Hosted Calc_logic for reuse white-box, estimated realization material requisites cost

Pcs	€ /pc	€	Item	Supplier	Element
10	24	240	i100. Hosted portable calc_logic programs on CD disc	SU.11. In-house project "dg3"	A. Calc_logic for reuse C program source file calculator.cpp
		240			

Development of the Calc_logic *program white-box* is kept together, regardless of which *products* will *reuse* it. This is illustrated in Figure 10-38, page 534 with chapters marked "R" and "H". The cost for developing the *program white-box* is shown in Table 10-90 below.

TABLE 10-91 Hosted Calc_logic for reuse white-box, estimated development work requisites hourly/fixed cost

h	€ /h	€	Item	Provider	Schedule
40	50	2 000	i34. "Dg3" project specifier	PR.10. In-house stakeholder group	Pt.n.Ra. Refine and capture requirements
10	50	500	i35. "Dg3" project leader	PR.12. In-house project "dg3"	Pt.n.Re. Plan development roles and providers
40	50	2 000	i36. Chief architect	PR.15. In-house common resource for projects "de1", "dw2" and "dg3"	Pt.n.Pa. Try to satisfy requirements Pt.1.A. Satisfy refined requirements by decomposing product black-box into white-box design containing no black-boxes
156	60	9 360	i37. "Dg3" project programmer	PR.12. In-house project "dg3"	Pt.n.Fa. Finalize white-box detailed items
40	45	1 800	i38. "Dg3" project engineer	PR.12. In-house project "dg3"	Pt.n.Fc. Account for white-box development, realization, and manufacturing

TABLE 10-91 Hosted Calc_logic for reuse white-box, estimated development work requisites hourly/fixed cost

h	€ /h	€	Item	Provider	Schedule
60	40	2 400	i39. "Dg3" project programmer	PR.12. In-house project "dg3"	Pt.1.I. Procure items and linkable programs in order to realize white-box
170	70	11 900	i40. Junior programmer	PR.16. Chk&Go	Pt.1.V. Verify black-/white-box
130	60	7 800	i41. Senior programmer	PR.16. Chk&Go	Pt.1.V. Verify black-/white-box
60	40	2 400	i42. "Dg3" project verifier	PR.12. In-house project "dg3"	Pt.1.V. Verify black-/white-box
		40 160			

The *host* computer is accounted in Table 10-92 below.

TABLE 10-92 Hosted Calc_logic for reuse white-box, estimated development equipment requisites purchase/rental cost

Pcs	€ /pc	€	Item	Supplier	Element
1	3 000	3 000	i102. Host computer	SU.12. in-house project "de1" and in-house project "dw2"	Development equipment
		3 000			

The calc_logic *program white-box* is summarized in Table 10-93 below.

TABLE 10-93 Hosted Calc_logic for reuse white-box, summed/accumulated requisites estimates

Item	Reference	Cost €
Hosted Calc_logic for reuse white-box, estimated realization material requisites cost	Table 10-90, this chapter	240
Hosted Calc_logic for reuse white-box, estimated development work requisites hourly/fixed cost	Table 10-91, this chapter	40 160
Hosted Calc_logic for reuse white-box, estimated development equipment requisites purchase/rental cost	Table 10-92, this chapter	3 000
Hosted Calc_logic for reuse white-box, summed/accumulated requisites estimates		**43 400**

10.18.5 EXAMPLE Hosted Calc_logic for reuse: Examine white-box success

There was a discussion about the cost of the calc_logic *program*, especially the fact that *verification* of the prototype is estimated to equal half of the total manufacturing cost for calc_logic. The decision gate decided that the cost is justified because:

- A calculator must simply calculate correctly. Such *quality* grade demands improved *formalism* and *intellectual control.*

- The planned *verification* can be as thorough as decided, it is only to reduce or extend the generated number of *stimuli* and verify as much as desired.

- There is a big benefit in training the organization in *formalism* and improving understanding of *verification.*

- The stand-alone calculator is planned to be produced in a volume of 100 000 and the cost for the *reusable item* Calc_logic will be only €0.434 apiece. The

Windows calculator can then be offered to be downloaded free as a public relations gesture.

10.18.6 EXAMPLE Calc_logic for reuse: Proceed reading

To follow this example, these are the alternatives:

- Proceed to chapter 10.20, "EXAMPLE Calc_logic environment (the pocket calculator firmware) development board and pocket calculator: Finalize design of environment interfaces with product requisites," page 549.
- Proceed to chapter 11.14, "EXAMPLE: Hosted calc_logic for reuse: Procure items and linkable programs in order to realize white-box," page 644.
- Proceed to chapter 12.20, "EXAMPLE Hosted calc_logic for reuse: Verify black-/white-box," page 733.
- Proceed to chapter 10.19 "EXAMPLE Calc_logic environment (the Windows calculator application): Finalize design of environment interfaces with product requisites" below.

10.19 EXAMPLE Calc_logic environment (the Windows calculator application): Finalize design of environment interfaces with product requisites

10.19.1 EXAMPLE Calc_logic environment (the Windows calculator application): Process schedule to use

For the calc_logic example, a special technical overall *schedule* is *tailored* (illustrated in Fig. 10-45, below):

- **Pt:2:3b. Tailored advanced technical schedule n = 2, porting to 3 targets (right half),** page 170

In this *tailored schedule* the generic *schedule* to use is (filled-in symbols):

- **Pt.0.F. Finalize design of environment interfaces with product requisites,** page 118.

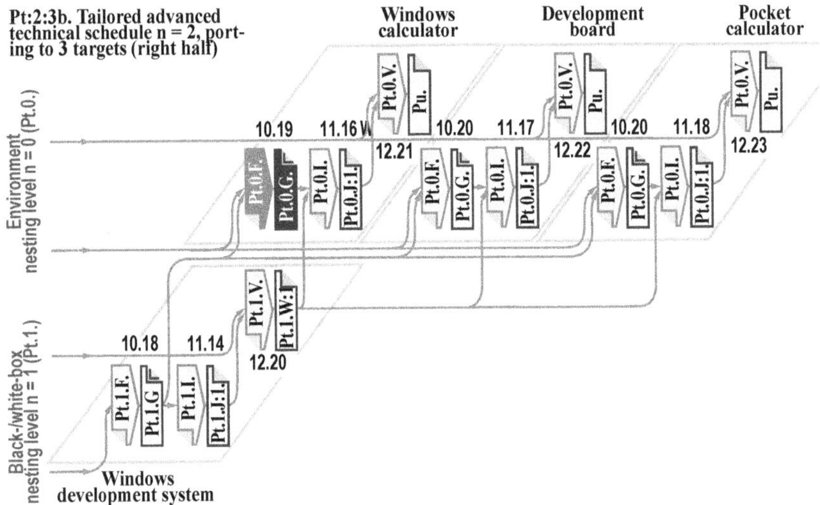

FIGURE 10-45 Position of schedule and master to use (filled in) in overall schedule

10.19.2 EXAMPLE Calc_logic environment (the Windows calculator application): Finalize environment interfaces detailed items

The *header files* are already prepared, and only need to be adapted to the Windows application *development system* (see Fig. 10-46 below and Fig. 10-47 next).

A.RR.7. Restriction requirements header file calc_logic_restrict.h

```
/* calc_logic restrictions */

#define OP_MAX_LEN 64
```

FIGURE 10-46 Calc_logic restriction requirement header file used for porting to Windows

---

A.RB.8. Function shape header file calc_logic_func_shape.h

```
/* calc_logic interface */

char    *calc_logic (char);
```

FIGURE 10-47 Calc_logic function shape header file used for porting to Windows

10.19.3 EXAMPLE Calc_logic environment (the Windows calculator application): Account for environment interfaces development, realization, and manufacturing

Adopting the *header files* and preparing CD disks are accounted in Table 10-94 below.

TABLE 10-94 Calc_logic environment (the Windows calculator application), estimated realization material requisites cost

Pcs	€ /pcs	€	Item	Supplier	Element
2	120	240	i101. Adapted header files for Windows target	SU.11. In-house project "dg3"	A.RB.8. Function shape header file calc_logic_func_shape.h A.RR.7. Restriction requirements header file calc_logic_restrict.h
	Sum	240			

Development of the Calc_logic *environment interfaces* is illustrated in Figure 10-38, page 534, in chapters marked "W". The cost of developing the *environment interfaces* is shown in Table 10-95 below.

TABLE 10-95 Calc_logic environment (the Windows calculator application), estimated development work requisites hourly/fixed cost

h	€ /h	€	Item	Provider	Schedule
12	40	480	i23. Calculator stakeholders	PR.12. In-house project "dg3"	Pt.0.Ra. Explore restrictions.
8	70	560	i24. Computer mathematician	PR.11. University researcher	Pt.0.Ra. Explore restrictions.
6	50	300	i25. "Dg3" project leader	PR.12. In-house project "dg3"	Pt.0.Rb. Plan development roles and providers.
4	40	160	i26. "Dg3" project designer	PR.12. In-house project "dg3"	Pt.0.Pa. Try to satisfy restrictions.
10	50	500	i27. External programs purchaser	PR.14. Hardware & Co	Pt.0.Pc. Analyze in/out and make/buy options.
15	40	600	i28. Common chief architect	PR.15. In-house common resource for projects "de1", "dw2" and "dg3"	Pt.0.A. Constitute environment architecture.
4	50	200	i29. "Dg3" project leader	PR.12. In-house project "dg3"	Pt.0.A. Constitute environment architecture.
2	50	100	i30. "Dg3" Windows programmer	PR.12. In-house project "dg3"	Pt.0.Fa. Finalize environment interfaces detailed items.
4	30	120	i31. "Dg3" project engineer	PR.12. In-house project "dg3"	Pt.0.Fc. Account for environment interfaces development, realization, and manufacturing.
10	40	400	i32. Developer assistance	PR.12. In-house project "dg3"	Pt.0.I. Realize interfaces to environment and insert/await outermost white-box.
8	40	320	i33. Developer assistance	PR.12. In-house project "dg3"	Pt.0.V. Verify prototype in environment.
	Sum	3740			

The accumulated cost from *host* development of the calc_logic *white-box* should be distributed to the Windows application and the stand-alone pocket calculator (this is one of the points with *reuse* :-). It was decided that the calc_logic cost should be divided fifty-fifty.

TABLE 10-96 Calc_logic environment (the Windows calculator application), summed/accumulated requisites estimates

Item	Reference	Cost €
Hosted Calc_logic for reuse white-box, summed/accumulated requisites estimates (half of 43 400)	Table 10-93, chapter 10.18	21 700
Calc_logic environment (the Windows calculator application), estimated realization material requisites cost	Table 10-94, this chapter	240
Calc_logic environment (the Windows calculator application), estimated development work requisites hourly/fixed cost	Table 10-95, this chapter	3 740
Calc_logic environment (the Windows calculator application), summed/accumulated requisites estimates		**25 680**

10.19.4 EXAMPLE Calc_logic environment (the Windows calculator application): Examine prototype success

Thanks to *reuse*, the cost for calc_logic was reduced significantly for the Windows application. Also the *quality* is considered to be higher, without adding to the total cost. The gate members approved the calc_logic development so far.

10.19.5 EXAMPLE Calc_logic for reuse: Proceed reading

To follow this example, these are the alternatives:

- Proceed to chapter 11.16, "EXAMPLE Calc_logic environment (the Windows calculator application): Realize interfaces to environment and insert/await outermost white-box," page 646.

- Proceed to chapter 12.21, "EXAMPLE Calc_logic environment (the Windows calculator application): Verify prototype in target environment," page 758.

- Proceed to chapter 10.20 "EXAMPLE Calc_logic environment (the pocket calculator firmware) development board and pocket calculator: Finalize design of environment interfaces with product requisites" below.

10.20 EXAMPLE Calc_logic environment (the pocket calculator firmware) development board and pocket calculator: Finalize design of environment interfaces with product requisites

10.20.1 Calc_logic environment (the pocket calculator firmware) development board and pocket calculator: Process schedule to use

For the calc_logic example, a special technical overall *schedule* is *tailored* (illustrated in Fig. 10-48, below):

- **Pt:2:3b. Tailored advanced technical schedule n = 2, porting to 3 targets (right half)**, page 170

In this *tailored schedule* the generic *schedule* to use is (filled-in symbols):

- **Pt.0.F. Finalize design of environment interfaces with product requisites,** page 118 (*porting* to *development board* and pocket calculator done simultaneously).

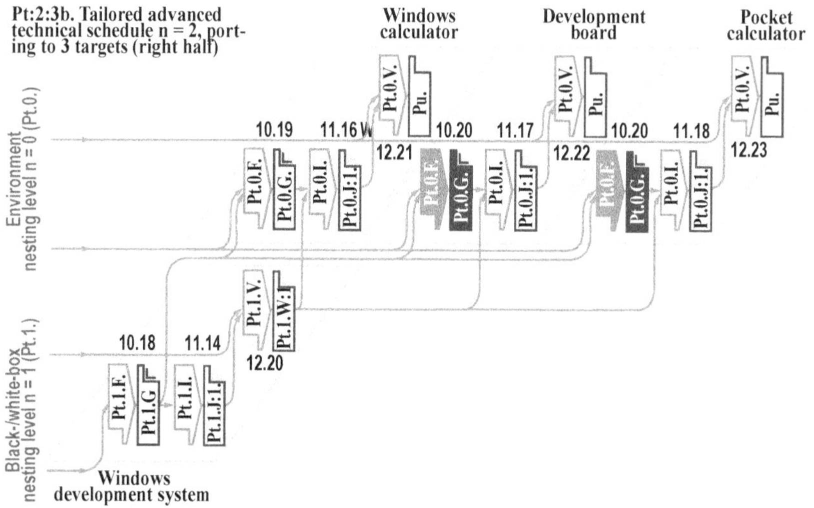

FIGURE 10-48 Position of schedule and master to use (filled in) in overall schedule

10.20.2 EXAMPLE Calc_logic environment (the pocket calculator firmware) development board and pocket calculator: Finalize environment interfaces detailed items

The *header files* for the *environment interfaces* need to be adjusted (see Fig. 10-49 below and Fig. 10-50 next). The length of operand is fixed to safe 16, because the electronic design is fixed to 8 LED characters.

```
A.RR.7. Restriction requirements header file calc_logic_restrict.h

/* calc_logic_restrictions.h for pocket calculator */

#define OP_MAX_LEN 16
```

FIGURE 10-49 Calc_logic restriction requirement header file used for porting to development board and pocket calculator firmware

```
A.RB.8. Function shape header file calc_logic_func_shape.h

/* calc_logic_interface.h for windows calculator */

char    *calc_logic (char);
```

FIGURE 10-50 Calc_logic function shape header file used for porting to development board and pocket calculator firmware

Note that there is no use of Calc_logic *function shapes* written for a C compiler, since the *development board* and pocket calculator firmware use assembler when calling calc_logic. The trick is that the *assembler program* must follow the used C compiler calling convention, when passing the *stimuli* to Calc_logic and returning the *response* from Calc_logic.

Also the *restriction requirement header file* is written for C compilers and must be translated to the assembler analogy.

TABLE 10-97 Calc_logic environment (the pocket calculator firmware) interfaces to environment design items

Item	Type	Purpose	Program lines	Interface ingredient
i103.Adapted header files for development board and pocket calculator target	Function specification header	For adaptation to microcontroller target	Circa 10 program lines	A.RB.8. Function shape header file calc_logic_func_shape.h
	Restriction specification header	To set restriction when compiling calc_logic	Circa 10 program lines	A.RR.7. Restriction requirements header file calc_logic_restrict.h

10.20.3 EXAMPLE Calc_logic environment (the pocket calculator firmware) development board and pocket calculator: Account for environment interfaces development, realization, and manufacturing

Adopting the *header files*, documenting the mechanisms of C calling conventions, and preparing CD disks with this information, are accounted in Table 10-98 below.

TABLE 10-98 Calc_logic environment (the pocket calculator firmware), estimated realization material requisites cost

Pcs	€/pc	€	Item	Supplier	Element
2	120	240	i103. Adapted header files for development board and pocket calculator target with explanations of C calling conventions	SU.10. In-house project "dg3"	A.RB.8. Function shape header file calc_logic_func_shape.h A.RR.7. Restriction requirements header file calc_logic_restrict.h
		240			

Development of the Calc_logic *environment interfaces* is illustrated in Figure 10-38, page 534 with chapters marked "P". The cost for developing *environment interfaces* is shown in Table 10-99 below

TABLE 10-99 Calc_logic environment (the pocket calculator firmware), estimated development work requisites hourly/fixed cost

h	€/h	€	Item	Provider	Schedule
20	40	800	i10. Calculator stakeholders	PR.12. In-house project "dg3"	Pt.0.Ra. Explore restrictions.
8	70	560	i11. Computer mathematician	PR.11. University researcher	Pt.0.Ra. Explore restrictions.
8	50	400	i12. "Dg3" project leader	PR.12. In-house project "dg3"	Pt.0.Rb. Plan development roles and providers.
4	50	200	i13. External embedded designer	PR.13. Micro Hard Inc.	Pt.0.Pa. Try to satisfy restrictions.
12	50	600	i14. External programs purchaser	PR.14. Hardware & Co	Pt.0.Pc. Analyze in/out and make/buy options.
4	50	200	i15. Chief architect	PR.15. In-house common resource for projects "de1", "dw2" and "dg3"	Pt.0.A. Constitute environment architecture.
16	60	960	i16. "Dg3" project leader	PR.15. In-house common resource for projects "de1", "dw2" and "dg3"	Pt.0.A. Constitute environment architecture.
4	30	120	i17. "Dg3" programmer i103. Adapted header files for development board and pocket calculator target	PR.12. In-house project "dg3"	Pt.0.Fa. Finalize environment interfaces detailed items.
2	50	100	i18. "Dg3" Project engineer	PR.12. In-house project "dg3"	Pt.0.Fc. Account for environment interfaces development, realization, and manufacturing.
17	80	1 360	i19. Development board specialist assistance	PR.16. Chk&Go	Pt.0.I. Realize interfaces to environment and insert/ await outermost white-box.

TABLE 10-99 Calc_logic environment (the pocket calculator firmware), estimated development work requisites hourly/fixed cost

h	€/h	€	Item	Provider	Schedule
10	40	400	i20. Microcontroller developer	PR.12. In-house project "dg3"	Pt.0.I. Realize interfaces to environment and insert/ await outermost white-box.
8	80	640	i21. Development board specialist assistance	PR.16. Chk&Go	Pt.0.V. Verify prototype in environment.
8	40	320	i22. Microcontroller developer	PR.12. In-house project "dg3"	Pt.0.V. Verify prototype in environment.
		6 660			

The cost for the calc_logic *white-box* development is divided fifty-fifty between the Windows application and the pocket calculator.

TABLE 10-100 Calc_logic environment (the pocket calculator firmware), summed/accumulated requisites estimates

Item	Reference	Cost €
Hosted Calc_logic for reuse white-box, summed/accumulated requisites estimates (half of 43 400)	Table 10-93, chapter 10.18	21 700
Calc_logic environment (the pocket calculator firmware), estimated realization material requisites cost	Table 10-98, this chapter	240
Calc_logic environment (the pocket calculator firmware), estimated development work requisites hourly/fixed cost	Table 10-99, this chapter	6 660
Calc_logic environment (the pocket calculator firmware), summed/accumulated requisites estimates		**28 600**

10.20.4 EXAMPLE Calc_logic environment (the pocket calculator firmware) development board and pocket calculator: Examine prototype success

The development cost for the stand-alone pocket calculator in-house project "de1" is substantial compared to the cost for calc_logic. However, to develop the calc_logic functionality in assembler has been a time-consuming nightmare, so it is very positive that Calc_logic unburdens the few stressed in-house real-time programmers. The calc_logic *component* is approved so far.

10.20.5 EXAMPLE Calc_logic for reuse: Proceed reading

To follow this example, these are the alternatives:

- Proceed to chapter 11.17, "EXAMPLE Calc_logic environment (the development board fixture firmware): Realize interfaces to environment and insert/ await outermost white-box," page 649.

- Proceed to chapter 11.18, "EXAMPLE Calc_logic environment (the pocket calculator firmware): Realize interfaces to environment and insert/await outermost white-box," page 653.

- Proceed to chapter 12.22, "EXAMPLE Calc_logic environment (the pocket calculator firmware) on development board: Verify prototype in environment," page 761.

- Proceed to chapter 12.23, "EXAMPLE Calc_logic environment (the pocket calculator firmware): Verify prototype in environment," page 767.
- Proceed to chapter 10.22 below to leave the calc_logic example and to dig deeper into the finalize design & requisites topic.

10.21 Cpdm definition

In *Cpdm*, *specialization* is defined according to Table 10-101 below.

TABLE 10-101 Cpdm definition

Aspect	Specialization
Description	To restrict an artifact in order to make it more particular with more unique characteristics (the opposite is generalization).
Synonyms	Individualize, personalize, customize
Architecture symbol	-A-1 General interface Ab. general artifact root of / kind of root of / kind of -A-1:1 Specialized interface Ab:1. Specialized artifact (inheriting from general artifact) -A-1:2 Specialized interface Ab:2. Specialized artifact (inheriting from general artifact)

The time has come to introduce the *specialization* concept in this context. It is often misunderstood and confused with the *partition* concept. In *object-oriented programs* this is called inheritance between *classes* (see Fig. 10-56, p. 558), and if wrongly used, it causes a lot of added *complexity*.

Specializing means a more cramped way to diversify into variants.

The easy way to distinguish them from each other is to interpret *decompose* as create a *part of,* and to interpret specialize as to create a *kind of.* For example, one can create a propeller which is a *part of* a ship, and I can create a steamer, which is a *kind of* ship.

The definition of specialize in Table 10-101 below is a bit simplified, but it is highly rewarding to get a grasp on this concept early on.

10.22 Object-oriented programs

The hope for further improvement of structured languages created great hype for extensions of all sorts, for example, the language C was extended to C++ several times. Unfortunately, some obscure extensions could not be rejected, offering too many risky possibilities and incomprehensible personal styles of programming.

The improvement expectations for object-oriented programs were high, but have not really been achieved.

The description below is mainly based on the language C++, not because it is the best *object-oriented program* language, but because it is one of the most popular.

The idea of creating an *object-oriented program* language was not so far-fetched. In C there has always been a possibility to create own *user* data types with the *instruction* called *typedef.* The simple object orientation enhancement of the C language, was to extend the typedef possibility to also include function types, and rename the *typedef* to *class*, in order to distinguish it from the original limited typedef.

It is debatable whether the best option really was to find the easiest way to technically extend an existing language, or if it would have been better to identify what kind of language was really needed, for example, to provide the language with architectural modeling opportunities.

Cpdm definition

In *Cpdm*, *object-oriented program* is defined according to Table 10-102 below.

TABLE 10-102 Cpdm definition

Aspect	Object-oriented program
Definition	Program containing classes instantiating objects, consisting of variables (can also be objects) and structured instruction functions
Synonyms	Few synonyms are in use.
Examples	Smalltalk, C++, C#, Java
Architecture symbols	

If only one *object* is instantiated from a *class*, this *class* can be illustrated as an "inner *requirement*" *element* of the *object*, as in Table 10-102 above. If many *objects* are instantiated from the same *class*, the same *class* shouldn't be drawn within each *object*.

When a *class* is defined, it can be used in the same way as any built-in types in a language like char, int, and float. Note that *objects* can be treated in the same way as *variables*, but have been separated by *Cpdm* for better clarity, because *objects* are compounds of *variables* and *functions*.

10.22.1 Basic object-oriented program concepts

Be careful to note that initially *object-oriented programs* may appear like relatively modest extensions, but they may easily turn very *complex* and can be hard to comprehend and control, especially if overused in large *program* systems.

Luckily, compilers of *object-oriented program* languages have been more and more competent in discovering dangerous constructs, and provided with a lot of helpful warnings to mitigate artificial risks of object-oriented programs.

Let's begin with the good object orientation extensions and look at one concept at a time:

1. Develop a *class*.

- To create an *object*, a *class* must first be developed (see Fig. 10-51, right).

- The *class* has the name my_class and contains any number of the four members (the last three similar to those in plain C):

    - nested *classes* like nested_class

    - enumerators like my_enum

    - *variables* like my_local_int and my_local_object

    - *functions* like my_func

- *Classes* can be nested into each other, like nested_class, but this is an obscure mechanism, too easily creating *artificial complexity*.

- The two most common members in *classes* are *variables* (can be *objects* like my_local_object) and *functions*.

FIGURE 10-51 Class development

```
Class
class my_class
{

        Nested class member
    class nested_class
    {
        int nested_local_int;
        int nested_func (int param)
        {
        param++;
        return (param);
        }
    };


        Enumeration members
    enum my_enum;

        Variable members
    int my_local_int;
    nested_class my_local_object;

        Function members
    int my_func (char param)
    {
        my_local_int =- param;
        return (my_local_int);
    }

};
```

2. Encapsulate *class* members:

- Encapsulation is a strong mechanism to master *complexity*.

- All member *variables* and *functions* are encapsulated by the *class*.

- *Class* members are by default not accessible from outside the *object* (see Fig. 10-52, right).

- *Class* members can be defined public, and can then be accessed from anywhere.

- Members can be handled as design *elements*.

- A *function* can be handled as a (small) *black-box* as described for *functions* in structured languages.

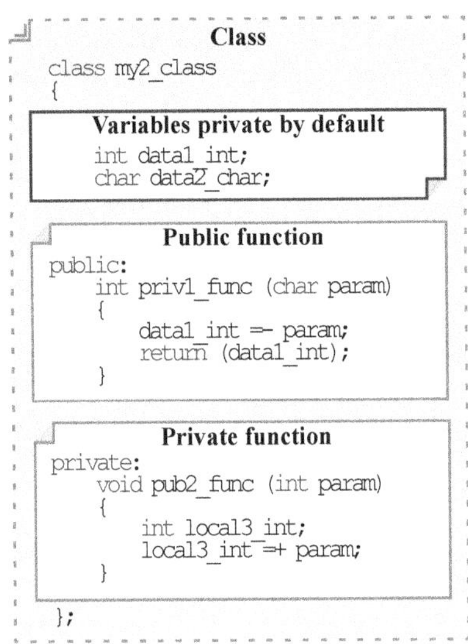

FIGURE 10-52 Encapsulation of class members

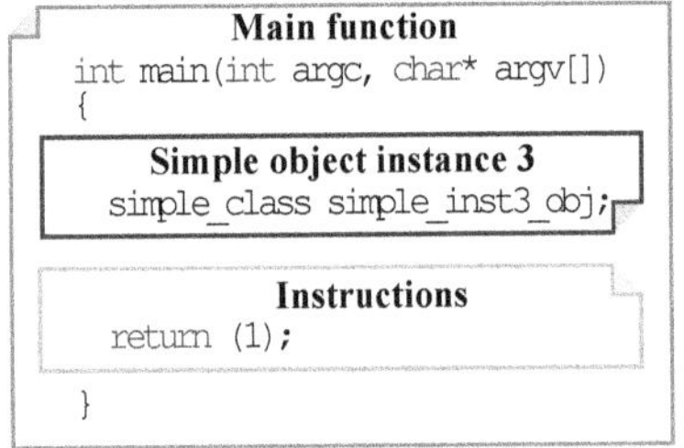

3. Instantiate a *class*:

- By capturing the *class*, an *object* instance of the *class* is created. It is the same mechanism as creating an integer *variable* by using the built-in **int** type.

- An *object* doesn't exist when a *program* starts to execute, but is created when encountered. If a hotel was built like a *program*, it would be possible to construct a room, execute the room, and destruct the room at the same pace hotel guests check in and check out.

- An *object* can be handled as an (often small) *black-box*, provided that all public *class* members (and of course all usage of *global variables*) are specified in the architecture resumé.

- In Figure 10-53 at left, three *objects* are instantiated from the *class* specification.

- The *object* simple_inst1_obj is created globally just at the end of the *class*; the *object* simple_inst2_obj is created globally; and the *object* simple_inst3_obj is created within the main *function*.

FIGURE 10-53 Instantiate an object from a class

4. Using a *class*:

* *Instructions* calls a *function* in another *object* (see Fig. 10-54, right). This is the simplest form of relation between *objects*.

* In the figure at right, it is shown how a call in the user *function* calls the *function* `func_called` in the *object* `used_obj`.

* It is also shown how the user *function* `user_func` accesses the *variable* `variable_accessed` in *object* `used_obj`.

* To access *variables* in other *objects* is risky and should be avoided when mastering *complexity*. Use a *function* to retrieve *variable* values instead.

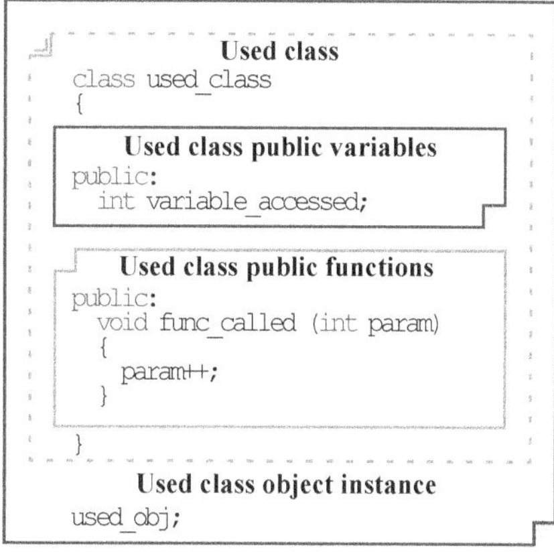

FIGURE 10-54 Function calls function in another object

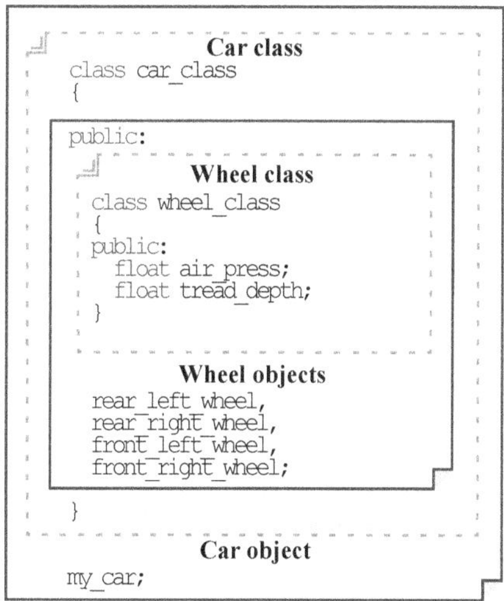

5. *Object* hierarchies ("part of" relationship)

* In Figure 10-55 left, *classes* can be organized hierarchically (contrary to inherited *classes* as in Fig. 10-56) which will create hierarchical *architecture* of instantiated *objects*. (*Objects* could just as well be placed adjacent to each other in a flat structure as most often seen, but the encapsulation will be stronger in a hierarchy).

* This relation *aggregates objects* to form a hierarchical "is part of" *object* structure.

* From the car function point of *view*, the hierarchical structure car wheels can now be used.

* This is solely an illustration of hierarchical *objects*, not an example of good programming style.

FIGURE 10-55 Hierarchical object

6. Inheritance ("kind of" relation)

- This relation means that a derived *class* inherits members from a base *class*; thus the derived *class* becomes a "kind of" the base *class*.

- In the example in Figure 10-56 at right, the *class* derived_class inherits all members from the base_class and can use all the base *class* members.

- From the user_func, all members of the *class* hierarchy become available.

- If it is important to draw inheritance in an *architecture*, then surround the base *class* with the derived *class* (with nothing in between).

- Naming convention:
  Derived *class* - Base *class*
  Sub *class* - Super *class*
  Child *class* - Parent *class*

FIGURE 10-56 Inheritance

Base class
```
class base_class
{
    int base_int;
protected:
    int base_func (int param)
    {
        base_int = param++;
        return (base_int);
    }
};
```

**Derived class
(inherit from base class)**
```
class derive_class: public base_class
{
    int derive_int;
public:
    int derive_func (char param)
    {
        derive_int = base_func (4);
        return (derive_int);
    }
}
```

Derive object
```
derive_obj;
```

User function
```
void user_func (void)
{
    derive_obj.derive_func (4);
};
```

7. Other good extensions not especially related to object orientation:

- Extended and more reliable error handling
  This improvement is less risky than the error return mechanism, but has no significant impact on better control of *complexity*.

- Constructor and destructor of *objects*
  When instantiating and destroying an *object*, a *function* can be set off for initialization and clean-up. Good but no significant impact on *complexity*.

10.22.2 Doubtful possibilities

In fact, the pressure for improving structured languages escalated to a point that unnecessary possibilities and risky extensions were incorporated just for the beauty and versatility of a computer language.

1. Some languages such as C++ have almost full backward compatibility to old structured languages. This is good and bad:

   a. It lowers the barrier between structured *programs* and *object-oriented programs*.

   b. C is structured but very close to assembler level in abstraction (good and bad) and already by plain C it is easy to write incomprehensible *programs*. Now this weakness can be even worse, ending up in an incomprehensible mix of structured *programs* and *object-oriented programs*.

   c. The improved language C# (C-sharp) is stricter in those matters.

2. Without proper training, it is easy to mix *class* inheritance with *object* aggregation. If so, the naming of *classes, objects, functions,* and *variables* might be totally misleading compared to how they are interpreted by the compiler:

   a. Thus, inheritance risk may be misused as "part of" and add unneeded *artificial complexity.*

   b. Thus, powerful aggregation hierarchies (such as nested big *black-boxes*) are not used at all, or get hidden in the mass of *program* lines.

3. Obscure and *guru*-driven extensions:

   a. Polymorphism. Use only if well trained.

   b. Operator overloading. Use only if well trained.

4. Multiple inheritance. Use only if well trained.

10.22.3 Missing in object-oriented program extensions

1. Some languages miss built-in multithreading (C++), but other languages have it (Ada, C# and Java).

2. Some languages miss garbage collection, which is a mechanism to clean and compact unused or freed memory areas.

10.22.4 Class shape header file

In the same way that *high-level programs* may contain *function shapes header files* to be included in *source files* (see Table 10-86, p. 533), *object-oriented programs* may also include these kind of *header files*, and in addition, *class shape header files* may be specified and included as separate *header files* (see Table 10-104 below).

As was the case with *function shape header files*, the *class shape header files* must be included in all *source files* referring these *classes*.

If *variables* are declared public, they are also accessible from outside the *object*, and can be seen as a kind of *connection interface* to the *object*. But since this is similar to using *global variables*, which is a poor programming style, *variables* in *object-oriented programs* are not seen as accessible *connection interfaces* in *Cpdm*.

TABLE 10-103 Cpdm definition

Aspect	Object-oriented program file	
Type	Source file	Header file
Content	Keeping elements intended to be executed	Keeping elements not intended to be executed, but to be included in source files
	Executable program elements complying with included class shapes	Compound of function, variable, and object shape requirements

TABLE 10-103 Cpdm definition

Aspect	Object-oriented program file	
Type	Source file	Header file
Content synonyms	Collection of program instructions including comments	Forward class declarations
Purpose	Compiled and linked to executable machine programs	Specifies how source files are obliged to use specified classes.
Architecture symbols		

10.23 Accounting in yet finer granularity

Programs are usually developed by in-house or subcontractor engineers, who most often are accounted per hour and not by any other unit. In mechanics and electronics departments, prices are not seldom calculated to several decimals, but in the programming department costs can be lumped together to a few very large sums. Such collapsed accounting makes it impossible to control and follow up development cost or compare efficiency of development.

Often this is a result of poorly documented *program partitioning*, which lacks units to account by, such as *solution alternatives, architecture ingredients*, or *white-box design items*.

It is highly recommended that *program* development costs are chopped down at least per *role items*, as shown so far. For huge *programs*, the *granularity* of *items* must be finer still, for example per *element, interface, requirement, use-case*, or *scenario* (see Table 10-104 below).

TABLE 10-104 White-box with interfaces or interfaces to environment estimated development work requisites hourly/fixed cost

Provider role item			Provider role item			Provider role item			Any kind of items	Any kind of reference
#	€/h	€	#	€/h	€	#	€/h	€		
									i100. Design item	A.a. Element
									i101. White-box item	A.B. White-box
									i102. Interface item	-A-1. Interface
									i103. Requirement item	A.R.1. Requirement
									i104. Restriction item	A.RR.1. Restriction requirement
									i105. Behavior item	A.RB.1. Behavior requirement
									i106. Scenario item	A.RB.S.1. Scenario
									i107. Use-case item	A.RB.S.U.1. Use-case

There is no inherent limit to lowering the accounting *granularity*. Instead of accounting for a total set of various *requirements*, they may be divided into separate *items*, for example, *restriction requirements, behavior requirements, use-cases*, and *scenarios*, which can be further divided between capturing and verifying (see Table 10-108, p. 582 and Table 10-109, p. 583).

10.24 EXAMPLE Phonebook: Finalize design items of product white-box with product requisites

10.24.1 EXAMPLE Phonebook: About this example

In the finalize design & requisites *phase* the *programs* and staffing grow rapidly. In the future, the phonebook is meant to be a very rich Windows application involving a lot of programmers, and from the start the *program* is already *partitioned* into a large number of *modules* to allow access of a large number of programmers in the future.

10.24.2 EXAMPLE Phonebook: Process schedule to use

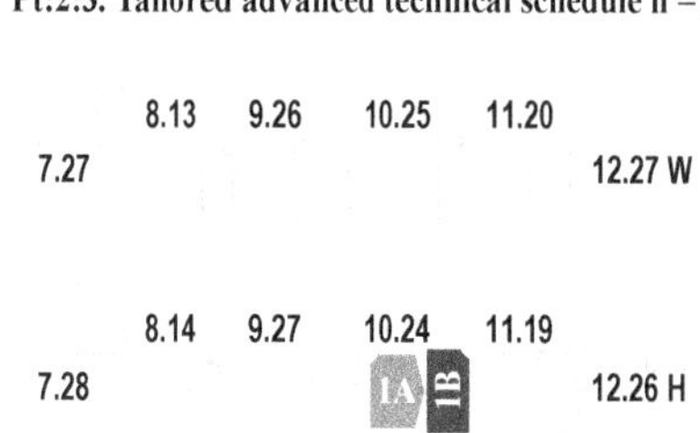

For the phonebook example, a special technical overall *schedule* is *tailored* (illustrated in Fig. 10-57, left):

- **Pt:2:3. Tailored advanced technical schedule n = 2,** page 169

In this *tailored schedule* the generic *schedule* to use is (filled-in symbols):

- **Pt.n.F. Finalize design of white-box with product requisites,** page 140.

FIGURE 10-57 Position of schedule and master to use (filled in) in overall schedule

EXAMPLE Phonebook: Finalize white-box detailed itemsIt is now time to prepare the design chart (see Fig. 10-58 below).

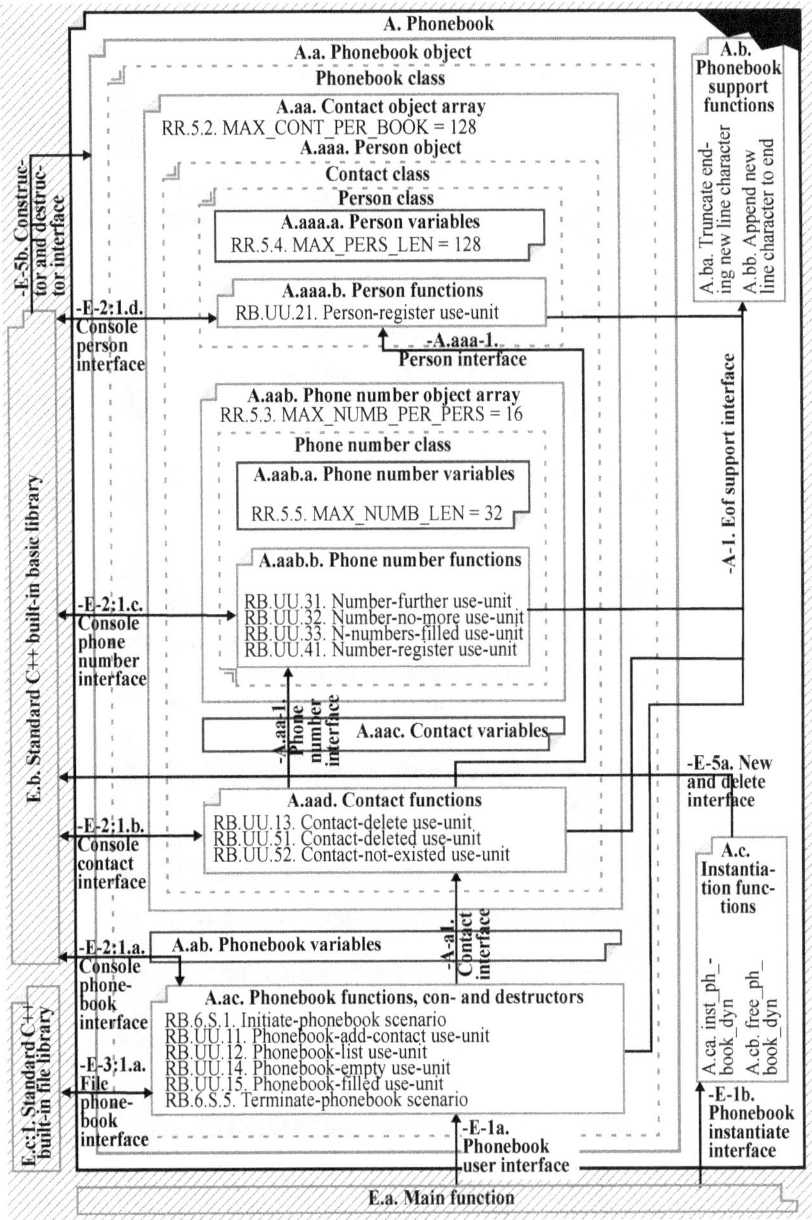

FIGURE 10-58 Phonebook design chart

EXAMPLE Phonebook: Finalize white-box detailed items

Based on the design chart above it is now ideal to *modularize* the phonebook into recognizable types of *program* files (see Fig. 10-59 below).

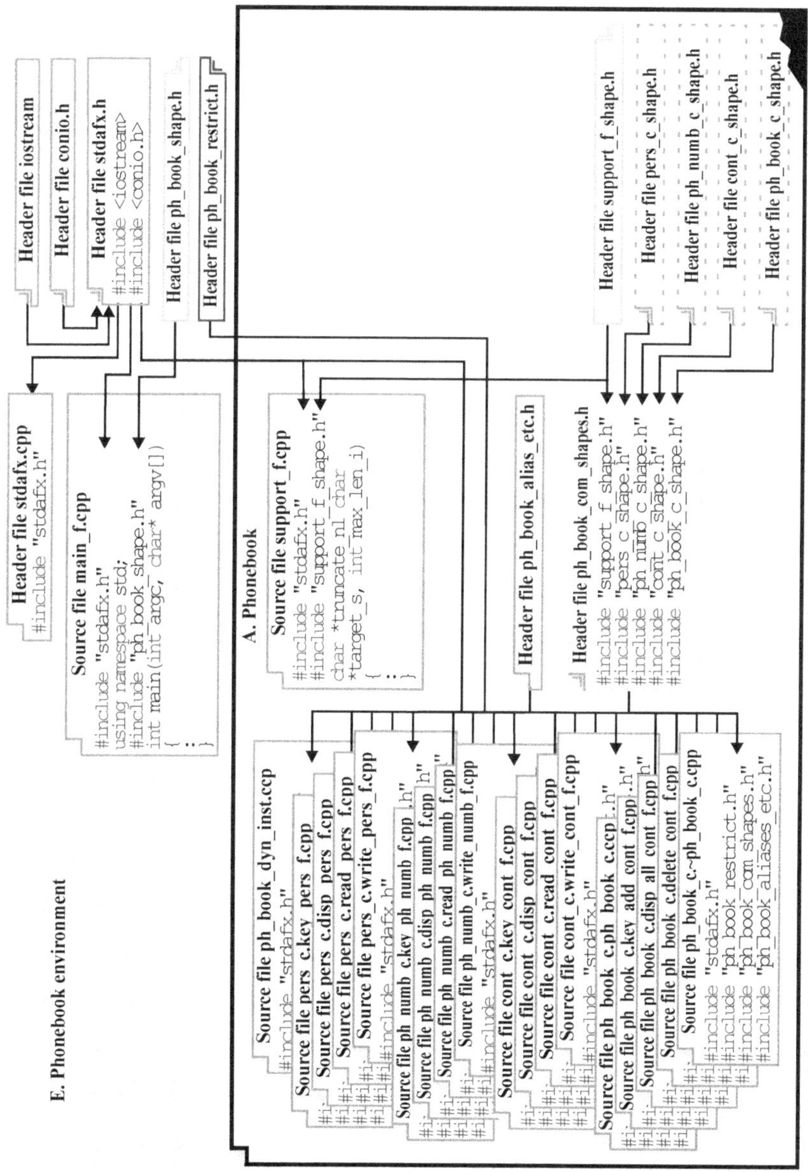

FIGURE 10-59 Phonebook file chart

Of course, charts and *program* files are not all developed in sequence during the finalize design & requisites *phase*. But great care must be taken to make all consistent in the end, needing substantial step-by-step documentation and reviews. Beware of *complexity* growth, and be careful when organizing files for large *programs*.

Observe that the debugger didn't need any *fixture* to surround the phonebook *white-box* during debugging. Very limited parameters are passed to the phonebook *functions* and phonebook instantiation *functions*, and they are easy to execute from within the debugger. Instead, the main *function source file* and *header files* are developed together with the *interface*s to the phonebook *environment* (see ch. 10.25.2, p. 586, and ch. 10.25.3, p. 587).

There is certainly no rational need to include all phonebook files in this book, but it is important to show that even if a lot of documentation is done, this so far small *program* is rather *complex* to understand and fully master.

10.24.3 EXAMPLE Phonebook header files: Finalize white-box detailed items

First there are some crucial restrictions limiting every corner of the *program* (see Figure 10-60 below).

```
                          Header file ph_book_restrict.h
#define MAX_NUMB_LEN   32
#define MAX_PERS_LEN   128
#define MAX_TITLE_LEN  50
#define MAX_CONTACTS   128
#define MAX_NUMB_PERS  16
```

FIGURE 10-60 Phonebook restriction specification C++ program

Header files specified for *functions* and *classes* are gathered into one common *header file* (see Fig. 10-61 below).

```
                          Header file ph_book_com_shapes.h
#include "support_f_shape.h"    // Own developed support function shape header file
#include "pers_c_shape.h"       // Person base class shape header file
#include "ph_numb_c_shape.h"    // Phone number class shape header file
#include "cont_c_shape.h"       // Contact class shape header file
#include "ph_book_c_shape.h"    // Phonebook class shape header file
```

FIGURE 10-61 Phonebook C++ program collecting function and class shapes

A *function shape header file* is prepared (see Fig. 10-61 below).

```
                          Header file support_f_shape.h
char *truncate_nl(char *, int);
char *append_nl(char *, int);
```

FIGURE 10-62 Phonebook support function specification C++ program

Aliases for some control characters are also defined (see Fig. 10-63 below).

```
                        Header file ph_book_alias_etc.h
#define END OF PH BOOK 2
#define END OF CONTACT 1
#define TRUNC PH BOOK  -1
#define NO ENDINGS      0
using namespace std;
```

FIGURE 10-63 Phonebook C++ program aliases

10.24.4 EXAMPLE Phonebook support functions: Finalize white-box detailed items

Now over to *source files*. The *source file* for the small support *functions* are pre-pared (see Fig. 10-64 below). They simply add or remove the new line character at the end of a string.

```
                        Source file support_f.cpp
#include "stdafx.h"
#include "support_f_shape.h"

    /* A.ba. Truncate ending new line character */

char *truncate_nl(char *target_s, int max_len_i)
{
    int char_numb_i;

    for (char_numb_i = 0; (target_s[char_numb_i] != '\n')
        && (target_s[char_numb_i] != 0) && (char_numb_i < max_len_i); char_numb_i++);
    target_s[char_numb_i] = 0;
    return(target_s);
}

    /* A.bb. Append new line character to end */

char *append_nl(char *target_s, int max_len_i)
{
    int char_numb_i;

    for (char_numb_i = 0; target_s[char_numb_i] != 0
        && (char_numb_i < max_len_i); char_numb_i++);
    target_s[char_numb_i] = '\n';
    target_s[char_numb_i + 1] = 0;
    return(target_s);
}
```

FIGURE 10-64 Phonebook support functions C++ program

10.24.5 EXAMPLE Phonebook instantiation functions: Finalize white-box detailed items

To hide instantiation mechanisms, a couple of support *functions* have been prepared to allocate and free up working memory space for the phonebook *variables*. The *header files* for these *functions* are specified as part of the *environment interfaces* (see ch. 10.25.3, p. 587).

```
Source file ph_book_dyn_inst.ccp

#include "stdafx.h"
#include "ph_book_restrict.h"
#include "ph_book_com_shapes.h"
#include "ph_book_aliases_etc.h"

ph_book_c *inst_ph_book_dyn()  /* A.ca. inst_ph_book_dyn */
{
    ph_book_c *ph_book_p;

        ph_book_p = new (nothrow) ph_book_c;
        if (ph_book_p == NULL)
            return (NULL);
        else
        {
        cout << " phonebook allocated size: " << sizeof(ph_book_c) << " Bytes\n";
        cout << " Version: fixed arrays, class inherited, multi-files, dynamic\n";
        return(ph_book_p);
        }
}

void free_ph_book_dyn(ph_book_c *ph_book_p)  /* A.cb. free_ph_book_dyn */
{
    delete ph_book_p;
}
```

FIGURE 10-65 Phonebook instantiation functions C++ program

10.24.6 EXAMPLE Phonebook object: Finalize white-box detailed items

Now let's go to more key *source files* and *header files* for the phonebook. Note that the *header files* capturing the *environment interfaces* are developed later as part of the *environment interfaces* (see ch. 10.25.2, p. 586).

First the phonebook *class shape header file* specifies the phonebook *objects, variables*, and *functions* (see Fig. 10-66 below). The phonebook is specified as an array of contacts, with an upper limit of MAX_CONTACTS. Further on, a *variable* is specified to hold the actual number of contacts. The phonebook *functions interfaces* are offered to be available for any *user* in the *environment*.

Note that a *class* can be interpreted as a *shape requirement* of *functions, variables* and nested *objects*, and such a *class* creates no *machine programs* to be executed. It is when these *shape requirements* are implemented by *programs* in *source files*, that the *result* will end up as executable *machine programs*.

```
                    Header file ph_book_c_shape.h
   class ph_book_c
   {
       /* A.aa. Contact object */
   private:
       cont_c cont_o[MAX_CONTACTS];


       /* A.ab. Phonebook variables */
   private:
       int    ph_book_cont_numb_i;


       /* A.ac. Phonebook functions, con- and destructors */
   public:
       ph_book_c (void);
       void key_add_cont_f (void);
       void disp_all_cont_f (void);
       void delete_cont_f (void);
       ~ph_book_c (void);
   };
```

FIGURE 10-66 Phonebook class specification C++ program

Now the *functions* will be prepared for the overall control of the phonebook. First a *function source file* is prepared, to be used when appending a contact to the phonebook if it is not already full (see Fig. 10-67 below). It is always important that all exceptions (like phonebook full) are reported to the *user*.

```
                 Source file ph_book_c.key_add_cont_f.cpp
#include "stdafx.h"
#include "ph_book_restrict.h"
#include "ph_book_com_shapes.h"
#include "ph_book_aliases_etc.h"

   void ph_book_c::key_add_cont_f (void)
   {
       if (ph_book_cont_numb_i >= MAX_CONTACTS)
           cout << "- Sorry, telephonebook full -";
       else
       {
           cout << "= Appending phonebook contact =\n";
           cont_o[ph_book_cont_numb_i].key_cont_f ();
           ph_book_cont_numb_i++;
           cout << " Used phonebook contacts " << ph_book_cont_numb_i
               << " out of " << MAX_CONTACTS << "\n";
       }
       return;
   }
```

FIGURE 10-67 Phonebook add contact function C++ program

When the phonebook begins to be used after the *program* is started, the phonebook *object* is instantiated from its *class*. If a constructor exists, this is executed and can establish the phonebook (see Fig. 10-68 below).

The phonebook tries to open and read the phonebook data file on hard disk. If there is no problem so far, this *function* repeatedly calls an *inward function* to read all contacts one after another.

```
                         Source file ph_book_c.ph_book_c.ccp
#include "stdafx.h"
#include "ph_book_restrict.h"
#include "ph_book_class_headers.h"
#include "ph_book_aliases_etc.h"

    ph_book_c::ph_book_c (void)
    {
        FILE *ph_book_f_p;
        char *line_f_p, title_s[MAX_TITLE_LEN + 1];
        int  cont_numb_i, result_i;

        ph_book_cont_numb_i = 0;
        ph_book_f_p = fopen("phonebook.txt", "r");
        if (ph_book_f_p == NULL)
            cout << "= No phonebook file found, a new one is created =\n";
        else
        {
            // read title line
            line_f_p = fgets(title_s, MAX_TITLE_LEN, ph_book_f_p);
            if (line_f_p == NULL)
                cout << "= phonebook file empty, start add contacts =\n";
            else
            {
                truncate_nl (title_s, MAX_TITLE_LEN);
                cout << "= Welcome to: " << title_s << " =\n";
                // read empty line after title line
                line_f_p = fgets(title_s, MAX_TITLE_LEN, ph_book_f_p);
                if (line_f_p == NULL)
                    cout << "= phonebook file empty, start add contacts =\n";
                else
                {
                    // read all contacts
                    result_i = cont_o[0].read_cont_f (ph_book_f_p);
                    for (cont_numb_i = 1; result_i == NO_ENDINGS; cont_numb_i++)
                    {
                        result_i = cont_o[cont_numb_i].read_cont_f (ph_book_f_p);
                    }
                    ph_book_cont_numb_i = cont_numb_i - 1; //end_of_file extra loop
                    if (result_i == TRUNC_PH_BOOK)
                    {
                        cout << "= File truncated, last contact " << cont_numb_i
                            << " not used =\n";
                        ph_book_cont_numb_i--;
                    }
                    cout << " Used phonebook contacts " << ph_book_cont_numb_i
                        << " out of " << MAX_CONTACTS << "\n";
                }
            }
            fclose(ph_book_f_p);
        }
    }
```

FIGURE 10-68 Phonebook constructor function C++ program

When dealing with hard disk files in above constructor, many things can go wrong, and it is as important as ever that the *user* of the *function* finds out if any exception has occurred.

The constructor *function* is recognized by having the same name as the *class*.

The next *function source file* is prepared to list all contacts for observation (see Fig. 10-69 below). Note that all these phonebook *functions* in turn may use *inward functions* in the phonebook *object* hierarchy. For example, the *function* disp_cont_f in *object* cont_o[cont_i] (see Fig. 10-74, p. 573) is called from the *program* below.

```
                    Source file ph_book_c.disp_all_cont_f.cpp
#include "stdafx.h"
#include "ph_book_restrict.h"
#include "ph_book_com_shapes.h"
#include "ph_book_aliases_etc.h"

   void ph_book_c::disp_all_cont_f (void)
   {
       int cont_i;

       if (ph_book_cont_numb_i == 0)
           cout << "- There are no contacts to display in phonebook -\n";
       else
       {
           cout << "= List of all contacts in phonebook =\n";
           for (cont_i = 0; cont_i < ph_book_cont_numb_i; cont_i++)
           {
               cout << " Contact number " << cont_i << "\n";
               cont_o[cont_i].disp_cont_f();
           }
       }
       return;

   }
```

FIGURE 10-69 Phonebook list all contacts function C++ program

There must also be a *function* for deleting unwanted contacts (see Fig. 10-70 below).

```
                      Source file ph_book_c.delete_cont_f.cpp
#include "stdafx.h"
#include "ph_book_restrict.h"
#include "ph_book_com_shapes.h"
#include "ph_book_aliases_etc.h"

   void ph_book_c::delete_cont_f (void)
   {

       char del_const_s[MAX_NUMB_LEN];
       int  del_cont_i, cont_i;


       if (ph_book_cont_numb_i == 0)
           cout <<"- There are no contacts to delete in phonebook -\n";
       else
       {
           disp_all_cont_f ();
           cout << "= What contact number to be deleted ? ";
           gets_s (del_const_s, MAX_NUMB_LEN);
           del_cont_i = atoi(del_const_s);
           if ((del_cont_i < 0) || (del_cont_i > ph_book_cont_numb_i))
               cout << "= There are no contacts with this number -\n";
           else
           {
               ph_book_cont_numb_i--;
               for (cont_i = del_cont_i; cont_i < ph_book_cont_numb_i; cont_i++)
               {
                   cont_o[cont_i] = cont_o[cont_i + 1];
               }
               cout << " Used phonebook contacts " << ph_book_cont_numb_i
                    << " out of " << MAX_CONTACTS << "\n";
           }
       }
   }
```

FIGURE 10-70 Phonebook delete contact function C++ program

When the phonebook *program* is about to terminate, a destructor *function* is called to take actions for saving the phonebook (see Fig. 10-71 below). Now the reverse of the constructor action will occur; the phonebook *variables* are fetched from the phonebook *object* in working memory, and saved back to the disk data file.

```
                    Source file ph_book_c.~ph_book_c.cpp
#include "stdafx.h"
#include "ph_book_restrict.h"
#include "ph_book_com_shapes.h"
#include "ph_book_aliases_etc.h"

    ph_book_c::~ph_book_c (void)
    {

        FILE *ph_book_f_p;
        int  result_i, cont_numb_i;


        ph_book_f_p = fopen("phonebook.txt", "w");
        if (ph_book_f_p == NULL)
            cout << "= phonebook file fails to be created =\n";
        else
        {
            result_i = fputs("Qpdm phonebook\n\n", ph_book_f_p);
            if (result_i < 0)
                cout << "= Nothing written to phonebook file =\n";
            else
            {
                for (cont_numb_i = 0; (cont_numb_i < ph_book_cont_numb_i)
                    && (result_i >= 0); cont_numb_i++)
                {
                    result_i = cont_o[cont_numb_i].write_cont_f (ph_book_f_p);
                    if (result_i >=0) result_i = fputs("\n", ph_book_f_p);
                }
                if (result_i < 0)
                    cout << "= phonebook file incompletely written =\n";
            }
        }
        fclose(ph_book_f_p);
        cout << "= Hit any key to leave console =\n";
        _getch ();

    }
```

FIGURE 10-71 Phonebook destructor function C++ program

10.24.7 EXAMPLE Phonebook contact object: Finalize white-box detailed items

The *class* cont_c, is now specified by a *class shape header file* (see Fig. 10-72 below) to capture *inward functions* needed by the *outward* phonebook *functions*. The *class* cont_c also inherits the *class* pers_c by the *program* line class cont_c : public pers_c.

It is specified that each contact shall hold a number of telephone numbers, with an upper limit of MAX_NUMB_PERS. A *variable* is also specified to indicate the actual numbers of telephone numbers in the contact.

```
                        Header file cont_c_shape.h

class cont_c : public pers_c    // Contact class inherits person class
{
       /* A.aab. Phone number object */
private:
    ph_numb_c ph_numb_o[MAX_NUMB_PERS];

       /* A.aac. Contact variables */
private:
    int        cont_ph_numb_numb_i;

       /* A.aad. Contact functions */
public:
       void key_cont_f (void);
       void disp_cont_f (void);
       int read_cont_f (FILE *);
       int write_cont_f (FILE *);
};
```

FIGURE 10-72 Contact class specification C++ program

A *function* is prepared to register the name of the contact and all phone numbers in a contact, provided that the limit is not reached (see Fig. 10-73 below). It uses the inherited *function* key_pers_f for registering the name and key_ph_numb_f to register each telephone number.

```
                    Source file cont_c.key_cont_f.cpp
#include "stdafx.h"
#include "ph_book_restrict.h"
#include "ph_book_com_shapes.h"
#include "ph_book_aliases_etc.h"

    void cont_c::key_cont_f (void)
    {
        int  ph_numb_numb_i;
        char menu_ch;

        key_pers_f();
        for(ph_numb_numb_i=0; ph_numb_numb_i<MAX_NUMB_PERS; ph_numb_numb_i++)
        {
            ph_numb_o[ph_numb_numb_i].key_ph_numb_f();
            cout <<" More phone numbers ? y = yes n = no \n";
            menu_ch = _getch ();
            switch(menu_ch)
            {
                case 'n': case 'N':
                cont_ph_numb_numb_i = ph_numb_numb_i + 1;
                return;
            }
        }
        cont_ph_numb_numb_i = MAX_NUMB_PERS;
        cout <<"- Sorry, contact numbers full -\n";
        return;
    }
```

FIGURE 10-73 Register contact function C++ program

This *source file* lists one contact with its telephone numbers (see Fig. 10-74 below). It was the *function* once used from the *function* in Figure 10-69, page 569.

```
                    Source file cont_c.disp_cont_f.cpp
#include "stdafx.h"
#include "ph_book_restrict.h"
#include "ph_book_com_shapes.h"
#include "ph_book_aliases_etc.h"

void cont_c::disp_cont_f (void)
   {
       int ph_numb_numb_i;

       cout << " Person : ";
       disp_pers_f();
       cout << "\n" << " Number(s) : ";
       for (ph_numb_numb_i=0; ph_numb_numb_i<cont_ph_numb_numb_i; ph_numb_numb_i++)
       {
           ph_numb_o[ph_numb_numb_i].disp_ph_numb_f();
           if (ph_numb_numb_i + 1 < cont_ph_numb_numb_i) cout << ", ";
       }
       cout << "\n";
       return;

   }
```

FIGURE 10-74 Display contact function C++ program

A *function* is prepared to read contacts from the disk file (see Fig. 10-75 below). This *function* in turn first calls a further *inward function* to read the person data of this contact, and then repeatedly calls another *inward function* to read all telephone numbers one after another of this contact.

```
                    Source file cont_c.read_cont_f.cpp
#include "stdafx.h"
#include "ph_book_restrict.h"
#include "ph_book_com_shapes.h"
#include "ph_book_aliases_etc.h"

   int cont_c::read_cont_f (FILE *ph_book_f_p)
   {
       int ph_numb_numb_i, result_i;

       cont_ph_numb_numb_i = 0;
       result_i = read_pers_f (ph_book_f_p);
       if ((result_i == END_OF_PH_BOOK) || (result_i == TRUNC_PH_BOOK))
           return (result_i); // File end possibly expected at this position
       result_i = ph_numb_o[0].read_ph_numb_f (ph_book_f_p);
       for (ph_numb_numb_i = 1; result_i == NO_ENDINGS; ph_numb_numb_i++)
       {
           result_i = ph_numb_o[ph_numb_numb_i].read_ph_numb_f (ph_book_f_p);
       }
       cont_ph_numb_numb_i = ph_numb_numb_i - 1;
       if (result_i == END_OF_CONTACT) return (NO_ENDINGS);
       return(result_i);

   }
```

FIGURE 10-75 Read contact function C++ program

The last *function* to prepare in the contact *object* is the one being called from the phonebook destructor, to take a contact and write it to the disk file (see Fig. 10-75 below). This is done by first calling the write_pers_f to write the person data to the disk and then repeatedly call the *function* write_numb_f to write each phone number to the disk.

```
                    Source file cont_c.write_cont_f.cpp
#include "stdafx.h"
#include "ph_book_restrict.h"
#include "ph_book_com_shapes.h"
#include "ph_book_aliases_etc.h"

    int cont_c::write_cont_f (FILE *ph_book_f_p)
    {
        int result_i, ph_numb_numb_i;


        result_i = write_pers_f (ph_book_f_p);
        for (ph_numb_numb_i = 0; (ph_numb_numb_i < cont_ph_numb_numb_i)
            && (result_i >= 0); ph_numb_numb_i++)
        {
            result_i = ph_numb_o[ph_numb_numb_i].write_numb_f (ph_book_f_p);
        }
        return(result_i);

    }
```

FIGURE 10-76 Write contract function C++ program

10.24.8 EXAMPLE Phonebook person object: Finalize white-box detailed items

Now over to the person *object*. First a *class shape header file* is prepared (see Fig. 10-77 below).

```
                    Header file pers_c_shape.h
    class pers_c
    {
        /* A.aaa.a. Person variables */
    private:
        char pers_data_s[5][MAX_PERS_LEN];

        /* A.aaa.b. Person functions */
    public:
        void key_pers_f (void);
        void disp_pers_f (void);
        int  read_pers_f (FILE *);
        int  write_pers_f (FILE *);

    };
```

FIGURE 10-77 Person class specification C++ program

First a *function* is prepared to register contact person data (see Fig. 10-78 below). These data are all stored in the string array pers_data_s.

```
                        Source file pers_c.key_pers_f.cpp
#include "stdafx.h"
#include "ph_book_restrict.h"
#include "ph_book_com_shapes.h"
#include "ph_book_aliases_etc.h"

    void pers_c::key_pers_f (void)
    {

        cout << "      Input persons name ? ";
        gets_s (pers_data_s[0], MAX_PERS_LEN);
        cout << "   Input persons address ? ";
        gets_s (pers_data_s[1], MAX_PERS_LEN);
        cout << "Input persons post_code ? ";
        gets_s (pers_data_s[2], MAX_PERS_LEN);
        cout << "      Input persons city ? ";
        gets_s (pers_data_s[3], MAX_PERS_LEN);
        cout << "   Input persons country ? ";
        gets_s (pers_data_s[4], MAX_PERS_LEN);

    }
```

FIGURE 10-78 Register person function C++ program

Then a *function* is prepared to list all contact person data (see Fig. 10-79 below).

```
                        Source file pers_c.disp_pers_f.cpp
#include "stdafx.h"
#include "ph_book_restrict.h"
#include "ph_book_com_shapes.h"
#include "ph_book_aliases_etc.h"

    void pers_c::disp_pers_f (void)
    {

        int data_numb_i;

        for (data_numb_i = 0; data_numb_i < 5 ;data_numb_i++)
        {
            cout << pers_data_s[data_numb_i];
            if (data_numb_i < 4) cout << ", ";
        }

    }
```

FIGURE 10-79 Display person function C++ program

This next *function* reads person data from the disk file into the working memory
(see Fig. 10-80 below). This *function* is called by the *function* read_cont_f as a a part
of the phonebook construction. The *function* returns the end-of-file character if
the file end was expected; otherwise, a failure indicator is returned.

Source file pers_c.read_pers_f.cpp

```cpp
#include "stdafx.h"
#include "ph_book_restrict.h"
#include "ph_book_com_shapes.h"
#include "ph_book_aliases_etc.h"

    int pers_c::read_pers_f (FILE *ph_book_f_p)
    {
        char *line_f_p;
        int data_numb_i;

        data_numb_i = 0;
        line_f_p = fgets(pers_data_s[0], MAX_PERS_LEN, ph_book_f_p);
        if (line_f_p == NULL) return(END_OF_PH_BOOK); // file end ok
        for (data_numb_i = 1; (data_numb_i < 5) && (line_f_p != NULL);
          data_numb_i++)
        {
            truncate_nl (pers_data_s[data_numb_i - 1], MAX_PERS_LEN);
            line_f_p = fgets(pers_data_s[data_numb_i], MAX_PERS_LEN, ph_book_f_p);
        }
        if (line_f_p == NULL) return(TRUNC_PH_BOOK);
        truncate_nl (pers_data_s[data_numb_i - 1], MAX_PERS_LEN);
        return(NO_ENDINGS);

    }
```

FIGURE 10-80 Read person function C++ program

As usual, a *function* must be used by the destructor hierarchy, writing back the person data from the working memory into the disk file (see Fig. 10-81 below).

Source file pers_c.write_pers_f.cpp

```cpp
#include "stdafx.h"
#include "ph_book_restrict.h"
#include "ph_book_com_shapes.h"
#include "ph_book_aliases_etc.h"

    int pers_c::write_pers_f (FILE *ph_book_f_p)
    {
        int result_i, data_numb_i;

        result_i = fputs(append_nl(pers_data_s[0], MAX_PERS_LEN), ph_book_f_p);
        for (data_numb_i = 1; (data_numb_i < 5) && (result_i >= 0); data_numb_i++)
        {
            result_i = fputs(append_nl(pers_data_s[data_numb_i], MAX_PERS_LEN),
                ph_book_f_p);
        }
        return(result_i);

    }
```

FIGURE 10-81 Write person function C++ program

10.24.9 EXAMPLE Phonebook phone number object: Finalize white-box detailed items

There is also an *object* taking care of all phone number *variables*, and the *class shape* and *function shape*s are shown in Figure 10-82 below.

```
                    Header file ph_numb_c_shape.h
    class ph_numb_c
    {
        /* A.aab.a. Phone number variables */
    private:
        char ph_numb_s[MAX_NUMB_LEN];

        /* A.aab.b. Phone number functions */
    public:
        void key_ph_numb_f(void);
        void disp_ph_numb_f (void);
        int read_ph_numb_f (FILE *);
        int write_numb_f(FILE *);
    };
```

FIGURE 10-82 Phone number class specification C++ program

The *function* in Figure 10-83 below shows the registration of each phone number. It is currently only a few lines in this first prototype, but a lot of *verification* and formatting of the telephone number is imaginable, such as dividing the phone number into country number part, and region and local parts, which are also specific for each country. It is also possible to register what kind of number it is, like personal or business phone, if it is a mobile number or land line, and so forth.

```
                Source file ph_numb_c.key_ph_numb_f.cpp
#include "stdafx.h"
#include "ph_book_restrict.h"
#include "ph_book_com_shapes.h"
#include "ph_book_aliases_etc.h"

    void ph_numb_c::key_ph_numb_f(void)
    {
        cout << " Input telephone number ? ";
        gets_s (ph_numb_s, MAX_NUMB_LEN);
    }
```

FIGURE 10-83 Register phone number function C++ program

The *function* to list a phone number is also currently short (see Fig. 10-84 below). A lot of extensions are imaginable here as well, such as the type of number, showing clearly the different parts of the phone number.

When choosing an object-orientation language, extensions are rather easy to do. Phone number handling is rather encapsulated and separated from the rest by a future-proof *architecture*, and it is easy to also extend the organization with a specific team that *programs* only the phone number part of the phonebook.

```
                    Source file ph_numb_c.disp_ph_numb_f.cpp
#include "stdafx.h"
#include "ph_book_restrict.h"
#include "ph_book_com_shapes.h"
#include "ph_book_aliases_etc.h"

    void ph_numb_c::disp_ph_numb_f (void)
    {

        cout << ph_numb_s;

    }
```

FIGURE 10-84 Display phone number function C++ program

The *function* in Figure 10-85 below reads a phone number from the disk file, and
returns a failure indicator if an unexpected file-end occurs. Otherwise, the *function*
returns a normal indicator of file ending, whether there are more contacts to read
or this was the last.

```
                    Source file ph_numb_c.read_ph_numb_f.cpp
#include "stdafx.h"
#include "ph_book_restrict.h"
#include "ph_book_com_shapes.h"
#include "ph_book_aliases_etc.h"

    int ph_numb_c::read_ph_numb_f (FILE *ph_book_f_p)
    {

        char *line_f_p;


        line_f_p = fgets(ph_numb_s, MAX_NUMB_LEN, ph_book_f_p);
        if (line_f_p == NULL) return (TRUNC_PH_BOOK);
        truncate_nl (ph_numb_s, MAX_NUMB_LEN);
        if (line_f_p[0] == 0) return (END_OF_CONTACT); // empty line end_of_contact
        return (NO_ENDINGS);

    }
```

FIGURE 10-85 Read phone number function C++ program

As usual, a *function* is also included in the destruction mechanism (see Fig. 10-86
below).

```
                    Source file ph_numb_c.write_numb_f.cpp
#include "stdafx.h"
#include "ph_book_restrict.h"
#include "ph_book_com_shapes.h"
#include "ph_book_aliases_etc.h"

    int ph_numb_c::write_numb_f(FILE *ph_book_f_p)
    {

        int result_i;


        result_i = fputs(append_nl(ph_numb_s, MAX_NUMB_LEN), ph_book_f_p);
        return(result_i);

    }
```

FIGURE 10-86 Write phone number function C++ program

10.24.10 EXAMPLE Phonebook start up and terminate message sequence diagram: Finalize white-box detailed items

The trickiest parts of the phonebook *program* may be its start-up and termination; at least, this is true for *managers* and other developers not used to reading *programs*. A good way to explain and share the *complex* sequence behind the simple *architecture* in Figure 9-64, page 428, is to use a *message sequence chart* (see Fig. 10-87 below).

- From the start, the phonebook *program* resides on the disk as a Windows executable file (exe file).
- The executable *program* is transferred to the working memory by the Windows operating system.
- Windows transfers the execution to the main *function*.
- Memory space for the phonebook *object* is first allocated from the Windows operating system by the instantiation support *functions* (containing the new *instruction*).
- After memory space is obtained, the instantiation *function* (the new *instructions*) transfers execution to the phonebook *object* constructor.
- The constructor reads phonebook data from the file and loads them to the *object* in working memory.
- Finally, the main *function* regains execution and shows the phonebook menu.
- When choosing quit in the menu, the phonebook termination is more or less as above, in reverse order.

A good question is what the best *granularity* is for a *message sequence chart*. For example, if every single message is illustrated, the *message sequence chart* gets very elongated, and the "trees get hidden behind the forest."

If an *element* in a *message sequence chart* asks another *element* a question, and the latter *element* is answering, maybe only the answer needs to be illustrated, because the question is implicit.

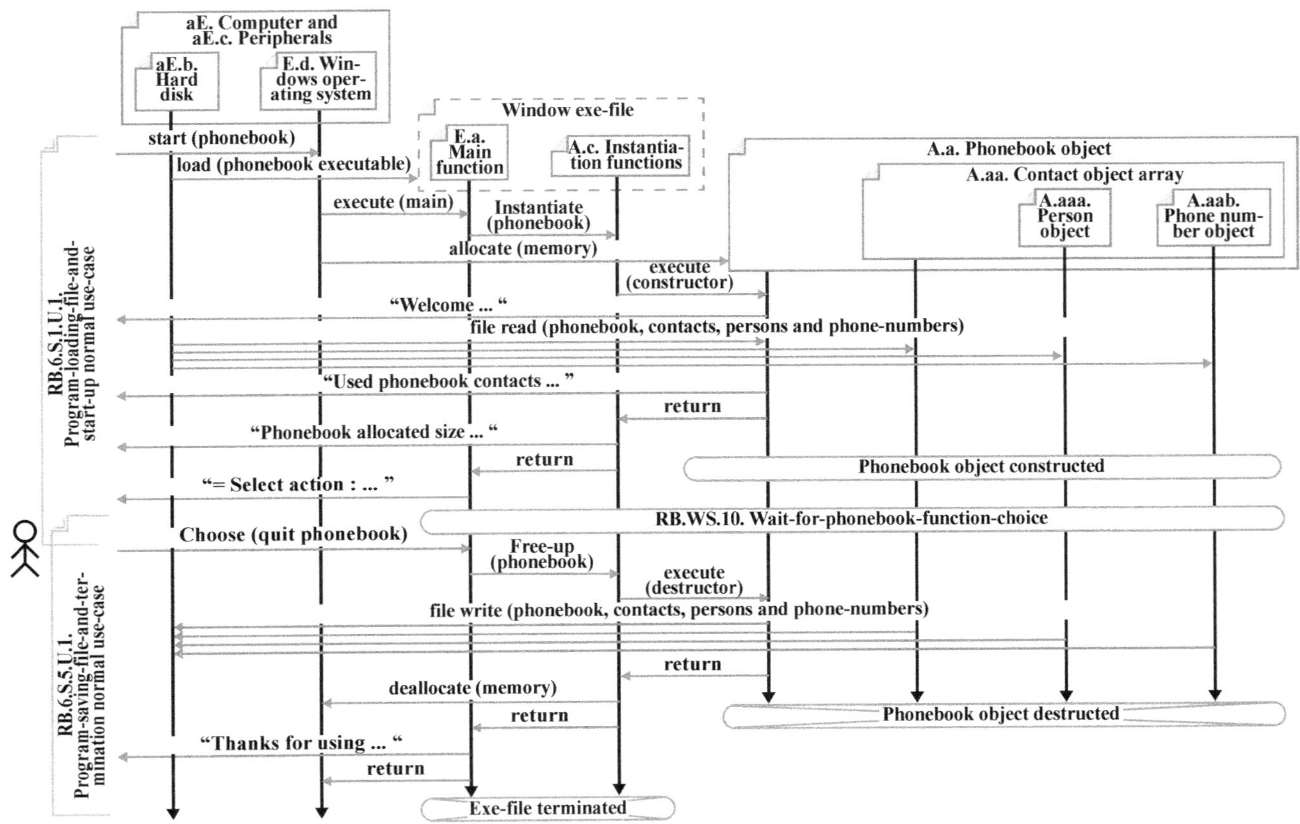

FIGURE 10-87 Phonebook constructor and destructor message sequence chart

10.24.11 EXAMPLE Phonebook: Finalize white-box detailed items list

Since the phonebook *program* is required to be heavily *modularized* from the start, the most elegant way to design is to let every single file be a *white-box design item* (see Table 10-105 below).

TABLE 10-105 Phonebook program white-box design items

Item	Type	LoC	White-box ingredient
i104.Header file ph_book_restrict.h	Restrictions header files	5	A. Phonebook
i105.Header file ph_book_alias_etc.h	Aliases etc. header file	5	A. Phonebook
i106.Header file ph_book_com_shapes.h	References to interfaces header file	5	A. Phonebook
i107.Source file ph_book_dyn_inst.ccp	Dynamic instantiation source file	20	A.c. Instantiation functions
i108.Header file support_f_shape.h	Support functions header file	2	-A-1. Eof support interface
i109.Source file support_f.cpp	Support functions source file	34	A.b. Phonebook support functions
i110.Header file ph_book_c_shape.h	Class shape header file	14	A.a. Phonebook object
i111.Source file ph_book_c.ph_book_c.ccp	Phonebook constructor source file	48	
i112.Source file ph_book_c.key_add_-cont_f.cpp	Phonebook function source file	16	A.ac. Phonebook functions, con- and destructors
i113.Source file ph_book_c.disp_all_-cont_f.cpp	Function source file	19	
i114.Source file ph_book_c.delete_cont_f.cpp	Function source file	29	
i115.Source file ph_book_c.~ph_book_c.cpp	Phonebook destructor source file	31	
i116.Header file cont_c_shape.h	Class shape header file	13	A.aa. Contact object
i117.Source file cont_c.key_cont_f.cpp	Function source file	24	
i118.Source file cont_c.disp_cont_f.cpp	Function source file	18	A.aad. Contact functions
i119.Source file cont_c.read_cont_f.cpp	Function source file	19	
i120.Source file cont_c.write_cont_f.cpp	Function source file	14	
i121.Header file pers_c_shape.h	Class shape header file	12	A.aaa. Person object
i122.Source file pers_c.key_pers_f.cpp	Function source file	15	
i123.Source file pers_c.disp_pers_f.cpp	Function source file	12	A.aaa.b. Person functions
i124.Source file pers_c.read_pers_f.cpp	Function source file	20	
i125.Source file pers_c.write_pers_f.cpp	Function source file	14	
i126.Header file ph_numb_c_shape.h	Class shape header file	12	A.aab. Phone number object
i127.Source file ph_numb_c.key_ph_-numb_f.cpp	Function source file	7	
i128.Source file ph_numb_c.disp_ph_-numb_f.cpp	Function source file	6	A.aab.b. Phone number functions
i129.Source file ph_numb_c.read_ph_-numb_f.cpp	Function source file	12	
i130.Source file ph_numb_c.write_-numb_f.cpp	Function source file	9	
Program lines sum		435	

The equipment may also need to be specified (see Table 10-106 below).

TABLE 10-106 Phonebook development system
white-box design items

Item	Product	Size	White-box ingredient
i131.Host computer	Well Windows PC	Working memory space 8 GB. Hard disc space 400 GB. CPU speed 3 GHz	Development equipment
i132.Development system	Micro hard	C++ compiler and linker with debugging possibility	

10.24.12 EXAMPLE Phonebook: Account for white-box development, realization, and manufacturing

The linker tool builds all *program elements* together. If the *architecture* and *white-box design items* are well prepared, this should be only a minor expense (see Table 10-107 below).

TABLE 10-107 Phonebook white-box,
estimated realization work requisites hourly/fixed/rental cost

h	€ /h	€	Item	Provider	Interface
20	45	900	i29. Configuration manager	PR.10. In-house	-A-1. Eof support interface -A-a1. Contact interface -A.aa-1. Phone number interface -A.aaa-1. Person interface
		900			

To be future-proof when scaling up the phonebook *project*, accounting is already made detailed at this point. Accounting for capturing and verifying is distributed across the *formal scenario* (see Table 10-108 below).

TABLE 10-108 Phonebook white-box,
capturing and verification estimated development work requisites hourly/fixed cost

Pt.n.Ra. Refine and capture requirements			Pt.n.V. Verify black-/white-box				
i20. Specifier			i31. Verifier				
#	€ /h	€	#	€ /h	€	Behavior item	Requirement
24	45	1 080	60	40	2 400	i1. Start scenario.	RB.6.S.1. Initiate-phonebook scenario
16	45	720	40	40	1 600	i2. Append scenario.	RB.6.S.2. Append-contact-to-phonebook scenario
8	45	360	40	40	1 600	i3. List scenario.	RB.6.S.3. List-phonebook-contacts scenario
8	45	360	40	40	1 600	i4. Delete scenario.	RB.6.S.4. Delete-phonebook-contact scenario
24	45	1 080	60	40	2 400	i5. Terminate scenario.	RB.6.S.5. Terminate-phonebook scenario
Sum		3 600			9 600		

White-box design items are divided into three competencies for each type of file (see Table 10-109 below).

TABLE 10-109 Phonebook white-box, estimated development work requisites hourly/fixed cost

Pt.n.Fa. Finalize white-box detailed items									
i25. Architect programmer			i26. Object-oriented programmer			i27. General C++ programmer			
#	€/h	€	#	€/h	€	#	€/h	€	File item
48	45	2 160				8	60	480	i106. Header file ph_book_com_shapes.h
						15	60	900	i104. Header file ph_book_restrict.h
						5	60	300	i105. Header file ph_book_alias_etc.h
12	45	540							i108. Header file support_f_shape.h
						30	60	1 800	i107. Source file ph_book_dyn_inst.ccp
						15	60	900	i109. Source file support_f.cpp
20	45	900	10	50	500				i110. Header file ph_book_c_shape.h
						32	60	1 920	i111. Source file ph_book_c.ph_book_c.ccp
						25	60	1 500	i112. Source file ph_book_c.key_add_cont_f.cpp
						10	60	600	i113. Source file ph_book_c.disp_all_cont_f.cpp
						32	60	1 920	i114. Source file ph_book_c.delete_cont_f.cpp
						30	60	1 800	i115. Source file ph_book_c.~ph_book_c.cpp
8	45	360	20	50	1 000				i116. Header file cont_c_shape.h
						36	60	2 160	i117. Source file cont_c.key_cont_f.cpp
						12	60	720	i118. Source file cont_c.disp_cont_f.cpp
						22	60	1 320	i119. Source file cont_c.read_cont_f.cpp
						20	60	1 200	i120. Source file cont_c.write_cont_f.cpp
8	45	360	16	50	800				i121. Header file pers_c_shape.h
						8	60	480	i122. Source file pers_c.key_pers_f.cpp
						8	60	480	i123. Source file pers_c.disp_pers_f.cpp
						15	60	900	i124. Source file pers_c.read_pers_f.cpp
						25	60	1 500	i125. Source file pers_c.write_pers_f.cpp
12	45	540	16	50	800				i126. Header file ph_numb_c_shape.h
						6	60	360	i127. Source file ph_numb_c.key_ph_numb_f.cpp
						4	60	240	i128. Source file ph_numb_c.disp_ph_numb_f.cpp
						14	60	840	i129. Source file ph_numb_c.read_ph_numb_f.cpp
						10	60	600	i130. Source file ph_numb_c.write_numb_f.cpp
Sum		4 860			3 100			22 920	

Some remaining development costs, not accounted for in Table 10-108 and Table 10-109 above, are accounted in Table 10-110 below.

TABLE 10-110 Phonebook white-box,
remaining estimated development work requisites hourly/fixed cost

h	€/h	€	Item	Provider	Workflows
40	80	3 200	i21. Product manager	PR.10. In-house	Pt.n.Ra. Refine and capture requirements.
16	70	1 120	i22. Development department management	PR.10. In-house	Pt.n.Re. Plan development roles and providers.
40	100	4 000	i23. Chief architect	PR.10. In-house	Pt.n.Pb. Solution alternatives Pt.n.Pd. Supplier opportunities
56	40	2 240	i24. Windows designer	PR.11. Zide Kft	Pt.n.Aa. Identify white-box architecture ingredients. Pt.n.Ac. Arrange ingredients and identify possible black-boxes. Pt.n.Ae. Lay out physical architecture.
36	50	1 800	i28. Development department management	PR.10. In-house	Pt.n.Fc. Account for white-box development, realization, and manufacturing.
20	70	1 400	i30. Certified verifier	PR.10. In-house	Pt.n.V. Verify black-/white-box.
		13 760			

Finally, the cost of development equipment needs to be accounted (see Table 10-111 below).

TABLE 10-111 Phonebook white-box,
estimated development equipment requisites purchase/rental cost

Pcs	€/pc	€	Item	Product	Supplier	Elements
2	1 500	3 000	i131. Host computer	Delly	Delly	Development equipment
2	Free downloadable	0	i132. Development system	Small&Softer	Internet	
		3 000				

See Table 10-111 below for summarizing the phonebook *white-box* accounting.

TABLE 10-112 Phonebook white-box,
summed/accumulated requisites estimates

Item	Reference	Cost €
Phonebook white-box, estimated realization work requisites hourly/fixed/rental cost	Table 10-107	900
Phonebook white-box, capturing and verification estimated development work requisites hourly/fixed cost	Table 10-108	3 600 9 600
Phonebook white-box, estimated development work requisites hourly/fixed cost	Table 10-109	4 860 3 100 22 920
Phonebook white-box, remaining estimated development work requisites hourly/fixed cost	Table 10-110	13 760
Phonebook white-box, estimated development equipment requisites purchase/rental cost	Table 10-111	3 000
Phonebook white-box, summed/accumulated requisites estimates		**61 740**

10.24.13 EXAMPLE Phonebook: Examine white-box success

The gate members had not expected the development cost to be this high. A *guru* programmer claimed that he could have realized this *white-box* for a third of this amount.

However, to continue to extend a phonebook *program* hacked together by a *guru* into a full-fledged rich Windows application would have been a nightmare. There would have been only the *monolithic program* to continue from, with no documentation of *requirements*, *architectures*, charts, or *verification* outcome.

The chair-leader of the gate had seen too many *products* developed by *gurus*, and was really impressed by being able to understand each corner of the development, by just reviewing all prepared *results*. He decided to continue developing the phonebook in this way.

10.24.14 EXAMPLE Phonebook: Proceed reading

To follow this example, these are the alternatives:

- Proceed to chapter 11.19, "EXAMPLE Phonebook: Procure items and linkable programs in order to realize white-box," page 655.
- Proceed to chapter 12.26, "EXAMPLE Phonebook: Verify black-/white-box," page 770.

10.25 EXAMPLE Phonebook environment: Finalize design of environment interfaces with product requisites

10.25.1 EXAMPLE Phonebook environment: Process schedule to use

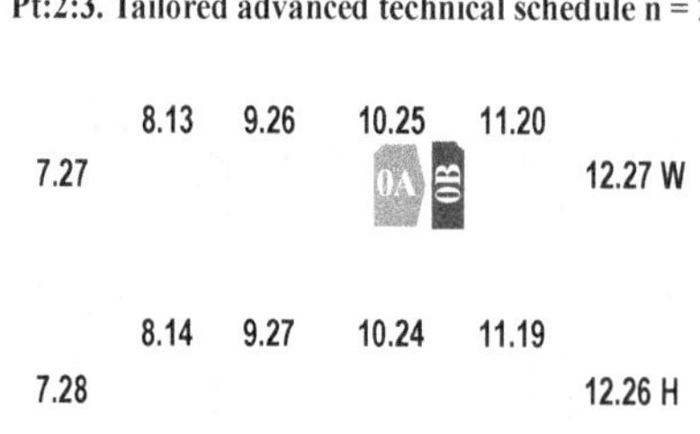

For the phonebook example, a special technical overall *schedule* is *tailored* (illustrated in Fig. 10-88, left):

- **Pt:2:3. Tailored advanced technical schedule n = 2,** page 169

In this *tailored schedule* the generic *schedule* to use is (filled-in symbols):

- **Pt.0.F. Finalize design of environment interfaces with product requisites,** page 118.

FIGURE 10-88 Position of schedule and master to use (filled in) in overall schedule

10.25.2 EXAMPLE Phonebook environment main function: Finalize white-box detailed items

To be able to execute the phonebook *program* as a stand-alone unit on a Windows computer, a simple *user interface* is developed (see Fig. 10-89 below).

```
                       Source file main_f.cpp
#include "stdafx.h"
#include <iostream>
#include <conio.h>
using namespace std;
#include "ph_book_shape.h"

int main(int argc, char* argv[]) /* E.a. Main function */
{
    ph_book_c   *ph_book_p;
    char    menu_ch = '\0';

    ph_book_p = inst_ph_book_dyn();
    if (ph_book_p == NULL)
    {
        cout << "No memory could be allocated for phonebook object\n";
        cout << "= Hit any key to leave console =\n";
        _getch ();
        return(false);
    }
    else
    {
        for(; (menu_ch != 'q') && (menu_ch != 'Q'); )
        {
            cout << "= Select action :\n";
            cout << " a = append phonebook contact\n";
            cout << " l = display all phonebook contacts\n";
            cout << " d = delete phonebook contact\n";
            cout << " q = quit phonebook program\n";
            menu_ch = _getch ();
            switch (menu_ch)
            {
                case 'a': case 'A':
                {
                    (*ph_book_p).key_add_cont_f ();
                    break;
                }
                case 'l': case 'L':
                {
                    (*ph_book_p).disp_all_cont_f ();
                    break;
                }
                case 'd': case 'D':
                {
                    (*ph_book_p).delete_cont_f ();
                    break;
                }
            }
        }
        free_ph_book_dyn(ph_book_p);
        cout << "= Thanks for using Cpdm phonebook =\n";
        cout << "= Hit any key to leave console =\n";
        _getch ();
        return (true);
    }
}
```

FIGURE 10-89 Phonebook main function C++ program

10.25.3 EXAMPLE Phonebook environment: Finalize environment interfaces detailed items

To be able to compile and link the prepared phonebook *white-box* to the newly developed *environment interfaces*, a specification *header file* is prepared (see Fig. 10-90 below).

```
                     Header file ph_book_shape.h

    /* -E-1a. Phonebook user interface */

    class ph_book_c
    {
    public:
        void key_add_cont_f (void);
        void disp_all_cont_f (void);
        void delete_cont_f (void);
    };


    /* -E-1b. Phonebook instantiate interface */

    ph_book_c   *inst_ph_book_dyn();
    void         free_ph_book_dyn(ph_book_c *ph_book_p);
```

FIGURE 10-90 Phonebook class and instantiate function specification C++ program header file

There are not too many *environment interfaces design item* (see Table 10-113 below).

TABLE 10-113 Phonebook interfaces to environment design items

Item	Type	Program lines	Interface ingredient
i133.Source file main_f.cpp	For verification of phonebook,	58	E.a. Main function
i134.Header file ph_book_shape.h	For programs using the phonebook programs	15	-E-1. Phonebook interface -E-5b. Constructor and destructor interface -E-5a. New and delete interface
i135.Header file iostream i136.Header file conio.h i137.Header file stdafx.cpp i138.Header file stdafx.h	Standard built-in development system library header files	?	-E-2:1. Console interface -E-3:1. File interface
Program lines sum		73	

10.25.4 EXAMPLE Phonebook environment: Account for environment interfaces development, realization, and manufacturing

The small number of *product requisites* can easily be explained since the *program* will not be produced and publicly delivered.

However, some *realization* work must yet be accounted (see Table 10-114 below).

TABLE 10-114 Phonebook environment,
estimated realization work requisites hourly/fixed/rental cost

h	€ /h	€	Item	Provider	Interface
40	55	2 200	i18. C++ development system analyst	PR.10. In-house	-E-1a. Phonebook user interface -E-1b. Phonebook instantiate interface -E-2:1.a. Console phonebook interface -E-2:1.b. Console contact interface -E-2:1.c. Console phone number interface -E-2:1.d. Console person interface -E-3:1.a. File phonebook interface -E-5a. New and delete interface -E-5b. Constructor and destructor interface
		2 200			

Then there are *environment interfaces design items*, see Table 10-115 below.

TABLE 10-115 Phonebook environment,
estimated development work requisites hourly/fixed cost

i16. C++ programmer PR.10. In-house **Pt.n.Fa. Finalize white-box detailed items**				
#	**€ /h**	**€**	**Design item**	**Element**
32	55	1 760	i133. Source file main_f.cpp	E.a. Main function
4	40	160	i134. Header file ph_book_shape.h	-E-1. Phonebook interface
Sum		**1 920**		

Cost of development work in other *phases* than finalize design & requisites in Table 10-115 above is shown in Table 10-116 below.

TABLE 10-116 Phonebook environment,
estimated development work requisites hourly/fixed cost

h	€ /h	€	Item	Provider	Schedule/task
48	55	2 640	i10. Product manager	PR.10. In-house	Pt.0.Ra. Explore restrictions
4	40	160	i11. Senior Windows graphic MMI designer	PR.11. Zide Kft	Pt.0.Ra. Explore restrictions.
12	80	960	i12. Development department management	PR.10. In-house	Pt.0.Rb. Plan development roles and providers.
24	60	1 440	i13. Windows programmer consultant	PR.11. Zide Kft	Pt.0.Pb. Interfaces from environment solution alternatives
12	80	960	i14. Development department management	PR.10. In-house	Pt.0.Pd. Environment demarcated by interface solutions with supplier opportunities
60	60	3 600	i15. Windows programmer consultant	PR.12. DeZine Kft	Pt.0.A. Constitute environment architecture.

TABLE 10-116 Phonebook environment,
estimated development work requisites hourly/fixed cost

h	€ /h	€	Item	Provider	Schedule/task
16	80	1 280	i17. Development department management	PR.10. In-house	Pt.0.Fc. Account for environment interfaces development, realization, and manufacturing.
0			i18. C++ development system analyst	PR.10. In-house	Pt.0.I. Realize interfaces to environment and insert/ await outermost white-box.
0			i19. C++ programmer	PR.10. In-house	Pt.0.V. Verify prototype in environment.
		11 040			

/Finally, the total costs for the phonebook are summarized in Table 10-116 below.

TABLE 10-117 Phonebook environment,
summed/accumulated requisites estimates

Item	Reference	Cost €
Phonebook white-box, summed/accumulated requisites estimates	Table 10-112, chapter 10.24	61 740
Phonebook environment, estimated development work requisites hourly/fixed cost	Table 10-115, this chapter	1 920
Phonebook environment, estimated development work requisites hourly/fixed cost	Table 10-116, this chapter	11 040
Phonebook environment, estimated realization work requisites hourly/fixed/rental cost	Table 10-114, this chapter	2 200
Phonebook environment, summed/accumulated requisites estimates		**76 900**

10.25.5 EXAMPLE Phonebook environment: Examine prototype success

The programming of both the *white-box* and the *environment interfaces* has been very structured and successful. It was decided to start up a new development iteration of the phonebook, to include more badly needed functionality.

10.25.6 EXAMPLE Phonebook: Proceed reading

To follow this example, these are the alternatives:

- Proceed to chapter 11.20, "EXAMPLE Phonebook environment: Realize interfaces to environment and insert/await outermost white-box," page 657.
- Proceed to chapter 12.27, "EXAMPLE Phonebook environment: Verify prototype in environment," page 780.
- Proceed to next chapter 10.26 below to leave the phonebook example and to dig deeper into the finalize design & requisites topic.

10.26 Scaling up programs

10.26.1 Language safety versus capability

Since the early days of programming, scaling up *programs* without losing control has been the major obstacle. The development of *program* languages has lately failed to be the necessary tool for this, and many companies ran into trouble with oversized *programs* (see Fig. 10-91 below).

Support for scaling up has failed to cure the everlasting program system growth.

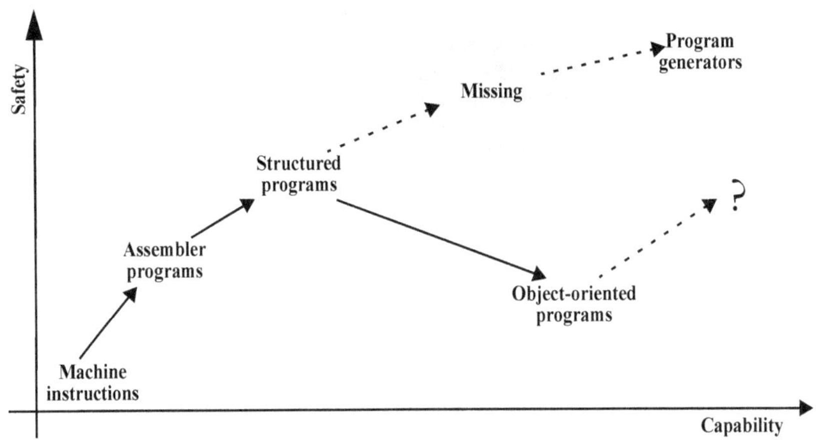

FIGURE 10-91 Development of program languages

This figure is not scientifically exact, but is intended to illustrate the evolution of *program* language safety and capability. Many people remember what a relief it was to get hold of a good assembler tool, instead of programming more or less with switches on the computer front panel. And let us not forget the relief of getting hold of a good compiler of a structured language, which made it a pleasure to throw out the beloved assembler tool. Great hope was created by the development of *object-oriented programs*, but not the same relief as before. Not everybody agrees, but object orientation has failed to support programmers as well as was expected.

Development of program languages has in the past been the most significant improvement in mastering program complexity.

10.26.2 Complexity caused by product capability growth

Integration of electronics has substantially contributed to mastery of *complexity* and support for growth of system size. Looking back into the electromechanics and electronics *product development*, the scale-up has progressed in powerful steps (see Fig. 10-92 below).

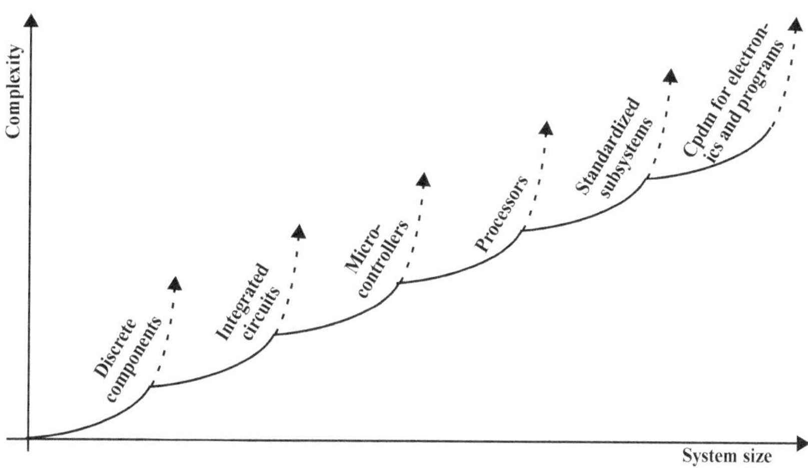

FIGURE 10-92 Mastering electronics complexity despite growth of system size

Interestingly, the steps to mitigate *complexity* are rather similar also for *programs*, when considering the earlier development of *program* languages (see Fig. 10-93 below), perhaps with the exception of lower usage of *program libraries* compared to electronic *components*.

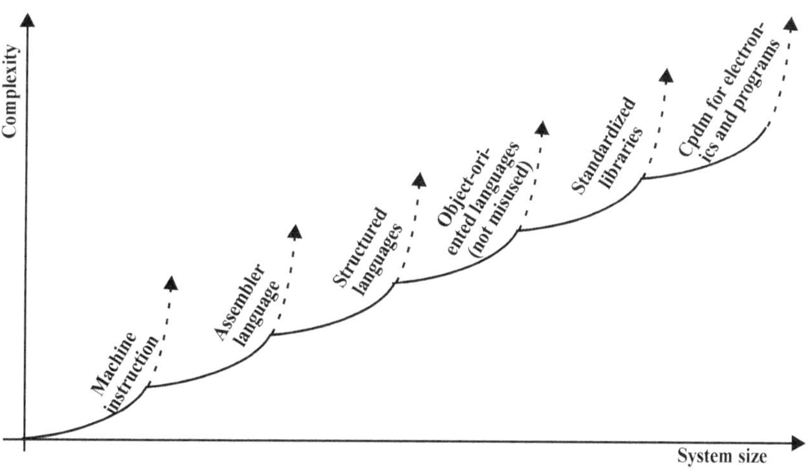

FIGURE 10-93 Mastering program complexity despite growth of system size

Most interesting is that *Cpdm*, with seamless *integration* of mechanics, electronics, and *programs*, and the use of box hierarchies, allows a new step in mastering *complexity*.

In addition, my next book "*Cpdm* technical overhead" will proceed from technical foundation to development overhead. This will explain *complexity* which occurs beyond the technical foundation, such as enlarging an already developed *product*, managing *line* and *projects*, managing *processes* and secure *quality* of a *product*.

Cpdm is the new support to master complexity. This book discusses development foundations and the next planned book will address development overhead

Realize & integrate white-box

If interfaces are properly designed by architects, integration will be a walk in the park.

Will the items fit a prototype ?

11.1 About realize & integrate white-box

Up to now, development *activities* have mainly produced technical documentation of various kinds, such as *requirements, architectures,* and *product requisites.* From now on, this documentation becomes the basis for realizing all prototype parts. Documentation will consist of reports of different kinds, describing the progress of *realization* work (see Fig. 6-6, Connection between methodology, own tailored process, and manufacturing process, p. 98).

Up to this phase, we have mainly developed descriptions, but from here the described tangibles will be realized.

11.1.1 Purpose of realization

Since *products* to be developed are *complex* and hard to understand only from its documentation, it is extremely rewarding to manually realize them in small prototype series in the laboratory, before the expensive and hard-to-change manufacturing lines are established. A prototype appearance is much easier to examine in the real world than in descriptions, and it can be manipulated for easy observation of its actual behavior.

Prototypes are constructed from *design items* developed in-house or obtained from *suppliers. Design items* are all the pieces needed for assembling the prototype (and later for manufacturing the *product*), such as pieces of mechanics, electronics, and *programs.*

If the prototype is unsatisfactory, it can be manipulated and improved by going back to the technical descriptions and optimizing them where development imagination obviously went wrong. In the **Pt.V. Verify black- & white-box** *phase*, all prototype boxes can be formally verified against their *requirements*, to ensure that the prototype complies with technical documentation in a consistent and comprehensible way.

In a *complex product* applying inside-out *incorporative integration* (see explanation of this further below), boxes are fully verified and detected failures are eliminated before being integrated into their *outward nesting level*, which in turn is fully verified, with failures detected and eliminated. This is an efficient way to gradually approve all *white-box* appearances and *black-box* behavior.

11.1.2 Integration workload depends on architecture effort

In this *phase*, the heaviest workload is to handle and assemble myriads of mechanics and electronics pieces (see Fig. 11-1 below). Linking together myriads of *program* pieces can also be tricky, especially when *porting programs*, but when everything has been set correctly, it is more or less an automatic work performed by a linker tool.

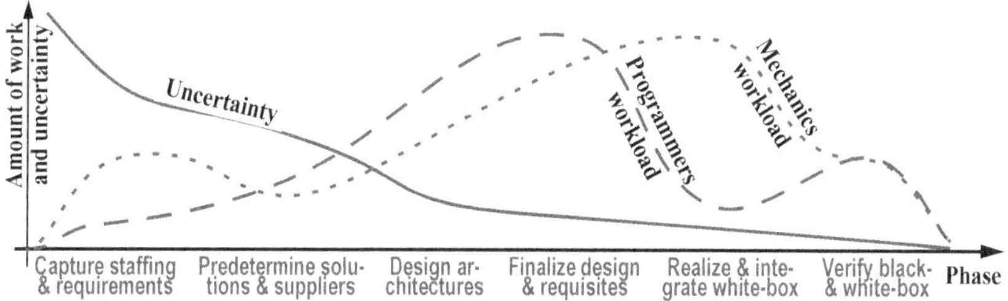

FIGURE 11-1 Amount of uncertainty and workload during the phases

11.1.3 Inside-out incorporative integration

Inside-out *incorporative integration* means that the deepest *innermost white-boxes* are procured and integrated first, and these are in turn integrated into *white-boxes* one *nesting level outwards*, and so on, until the *outermost white-box* is integrated into the *environment* (see Fig. 6-34, p. 162).

Inside-out integration of a product prototype is convenient, but not always possible.

Inside-out *incorporative integration* has drawbacks and benefits:

- It is time saving to execute **finalize design & requisites, realize & integrate white-box**, and **verify black- & white-box** in the same inside-out direction. These *phases* are mostly run in parallel, and the full department of engineers can be efficiently used.

- A problem with inside-out *incorporative integration* can be that the most detailed *white-boxes* are realized first. At this early time there is some risk that these details will not fit together when *integration* approaches outermost *black-/white-box nesting levels*. A well-done architectural work will mitigate this risk.

11.1.4 Outside-in invasive integration

Outside-in *invasive integration* means that the *interfaces* from the *product prototype* to the *environment* are first realized. The *realization* of these *interfaces* is done before the *outermost white-box* is developed. Then the time comes to invade and build the *outermost white-box* with *white-box design items*. When this is done, in turn the *inward nesting level white-boxes* can be invaded and built. This continues until every *innermost white-box* is invaded and built (see Fig. 6-38, p. 167).

When embedded black-boxes are large or heavy, it might be too cumbersome to first prepare them separately and then integrate them into outward white-boxes.

Outside-in *invasive integration* has drawbacks and benefits:

- Some *white-boxes* are big or heavy (for example, houses), which makes *realization* and stand-alone *verification* for later incorporation impossible, and they simply have to apply outside-in *invasive integration*.

- The realization will fit better together, because the surroundings to the *white-box* are always in place when opening and invading the *embedded black-boxes*.

- A risk with above bullet is that invaded and *embedded white-boxes* get too tightly attached to the surrounding *white-box*, making it hard to separately access them during later *verification*.

- *Full verification* cannot start until all *innermost white-boxes* are invaded. If a large nested hierarchy of *black-boxes* is to be invaded, the *verification* will start first after the *innermost white-boxes* are realized. Failures detected late are more expensive to correct, and a way to detect failures earlier would be beneficial. A *partial verification* can be obtained if *embedded black-/white-boxes* are temporarily developed as *fake black-/white-boxes*. For example, the outer insulation of a house can be verified (with simple fake rooms if needed), before the rooms are fully invaded.

- If *architectures* do not show and ensure that all *embedded black-boxes* fit together, the lack of trust implies that the *product* gets realized outside-in. Developers simply become afraid that the *product* will not fit together at the end if they apply inside-out *incorporative integration*, beginning *realization* with the most *inward white-boxes*.

- The worst drawback with outside-in *invasive integration* may be that the calendar *lead time* gets unacceptably long when more or less forced to run the *phases* **finalize design & requisites**, **realize & integrate white-box**, and **verify black- & white-box** in sequence (see Fig. 6-38, p. 167).

- A big risk with the long development calendar time is that the outside-in invasion is forced to start too early, before the architecture and final design of the innermost *nesting levels* are completed. Then there will be an obvious risk that the fit of the undeveloped innermost realization will be poor.

Since *invasive integration* starts from the outside, *embedded black-boxes* start empty before being invaded. As indicated above, it may be a good idea to temporarily fill these embedded empty *white-boxes* with fake *items*, since this makes it possible to verify the surrounding *white-box* containing these *fake black-/white-boxes*. A certain part of the *complexity* still resides in the surrounding *white-box*, and it can be good to partly verify and secure these parts. Such *verifications* are only intermediate or temporary, awaiting the *full verification* when all *fake black-/white-boxes* have been finally invaded and realized. These *verification results* are marked as "partial ver" in Figure 6-38, page 173.

In general, the outside-in *invasive integration* is the more difficult and circumstantial *process*, but sometimes there are no other ways, as with big construction *integration*, and sometimes it is obviously beneficial, as with *complex* big *programs*. If multiple *nesting levels* of *black-boxes* are used, each *nesting level* can be realized in any order, thus mixing outside-in *invasive integration* and inside-out *incorporative integration*. But beware of adding *artificial complexity*!

The necessary use of *configuration management* now enters more and more into the above discussion.

- During development of a large *product*, market risks can be considerably mitigated through a step-by-step *process* with continuous involvement of the *customers*. The large *product* is then divided into increments, which do not correspond to *architecture ingredients*, but represent small steps of behavior extensions repeatedly developed by quick execution of more or less the entire technical *process*. *Incremental development* is rather straightforward in itself, but when many increments are developed in parallel, strict *configuration management* must be applied.

- Large *products* containing different technologies can take very different development times, for example, computer mechanics and electronics versus operating system and many applications. Such parts must be developed separately to avoid large interdependent *complexity*, and development of these parts must constantly be synchronized by *configuration management*.

How to apply *configuration management* will be explained in my next book "*Cpdm* technical overhead".

11.1.5 EXAMPLE: Invasive integration with partial implicit verification

Since *invasive integration* and partial *implicit verification* are a bit tricky (but can be both unavoidable and advantageous), an example might be in place.

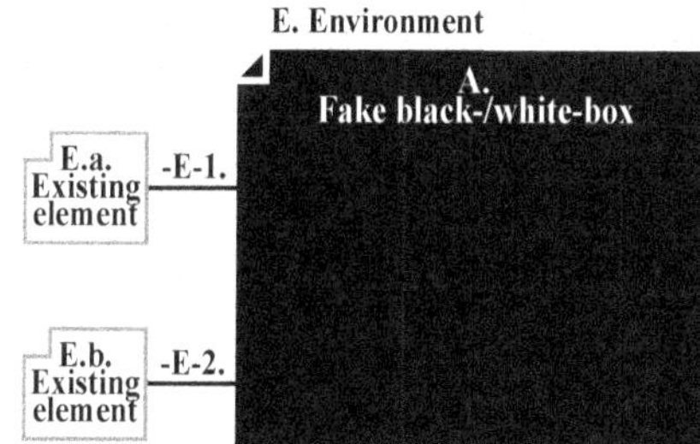

FIGURE 11-2 Product described by Pt.0.J.

Assume that the *product* in Figure 11-2 to the left is developed according to the *tailored process* in Figure 6-38, page 173. To make it possible to partially verify the newly designed and realized *interfaces* **-E-1.** and **-E-2.** to the *environment*, a *fake black-/white-box* **A.** may be integrated (for example, a manually controlled *white-box*). If desired, these *interfaces* can now be partially verified.

After *partial verification* of the *environment interfaces*, the *fake black-/white-box* **A.** is replaced with the intended *white-box* **A.** (see Fig. 11-3 to the right). To make it possible to partially verify the *white-box* **A.**, the identified two *black-boxes* **A.A.** and **A.B.** are integrated as *fake black-/white-boxes*, for example, with simple return *instructions* if they are *functions* in a *programs*.

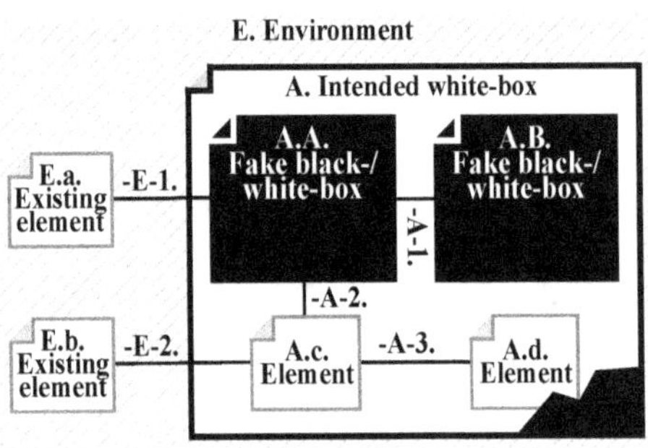

FIGURE 11-3 Designed product containing two fake black-boxes

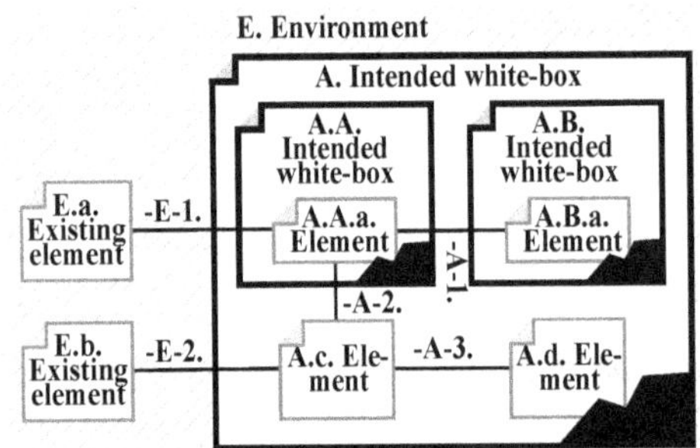

FIGURE 11-4 Completely designed product

After *partial verification* of the *white-box* **A.**, the two *fake black-/white-boxes* **A.A.** and **A.B.** are replaced with the two intended *white-boxes* (see Fig. 11-4, right). Now there are no more fakes in the *product*, and the entire *product* can now be fully implicitly verified inside-out on each *nesting level*.

As indicated before, *invasive integration* from outside can meet *incorporative integration* from inside. In this example, the final *innermost white-boxes* **A.A.** and **A.B.** may be fully developed separately, and after the *fake black-/white-boxes* are removed, they may be candidates for *incorporative integration* into the *white-box* **A.**

11.1.6 Cpdm definitions

In *Cpdm, received items deviation report* is defined according to Table 11-1 below.

TABLE 11-1 Cpdm definitions

Aspect	Received items deviation report
Definition	Deviation report on how received items (including linkable and possibly ported programs) match placed orders and other expectations
Synonyms	Receiving acknowledgement, unpacking control
Symbol	**TABLE 11-2** Items received deviation report | Deviation	Item	Item information	

11.1.7 About deviation report

The most important objective of a deviation report is to determine that the correct *design item* has been received, according to what has been documented and ordered in the *phase* **finalize design & requisites**. If wrong *design items* have been received, and this is not detected and eliminated, it becomes unpredictable what will arrive to manufacturing when this *design item* is ordered again from the same documentation.

11.1.8 Cpdm definitions

In *Cpdm, integration report* is defined according to Table 11-3 below.

TABLE 11-3 Cpdm definitions

Aspect	Integration report
Definition	Description of fulfillments and shortcomings when joining and assembling items, like wrongly designed interfaces, items not fitting together, and so forth
Synonyms	Prototype integration, prototype building
Symbol	**TABLE 11-4** Integration report | Fulfillments and shortcomings	Interfaces to realize	Responsible provider	

11.1.9 About integration report

This report can be a diary describing important events during prototype construction. It can also include a calendar with time plans, in order to follow up progress and monitor completion of the prototype in time for *verification*.

Interestingly, integrating and realizing a *white-box* is very much about how *design items* fit together, in other words, the fitness and *quality* of the *interfaces*. Consequently, the *integration report* should be centered around the fulfillments and shortcomings of *interfaces* (see examples below).

11.1.10 Cpdm definitions

In *Cpdm, start-up report* is defined according to Table 11-3 below.

TABLE 11-5 Cpdm definitions

Aspect	Start-up report
Definition	Report of obstacles and accomplishments when starting up a white-box or the environment interfaces
Synonyms	Prototype integration, prototype build
Symbol	**TABLE 11-6** Start-up report <table><tr><td>Accomplishments and obstacles</td><td>Troubling ingredient</td><td>Responsible provider</td></tr></table>

11.1.11 About start-up report

The purpose of a formal start-up of *white-boxes* or *product prototypes* is to make them useful for *verification*. There is no point in leaving anything for *verification* that requires a stand-by team of engineers to keep it running or restarting it all the time. This is simply a waste of time.

The question is, what are the criteria for a useful *white-box* or the *product prototype*? A simple rule of thumb is that at least the *normal use-cases* must execute without constantly failing and restarting. If there also are commonly occurring exceptions, these must also execute without obvious failures and restarts.

If a *white-box* or *product prototype* is found to not be ready for *verification*, the *schedule* in use must be traced *upstream* to find out which organization(s) have to fix the start-up problem(s). After location of the start-up problem is detected and eliminated, the *schedule* in use must be traced *downstream* again, to correct all consequences of the start-up problem (compare with ch. 12.3, "Troubleshooting," p. 665).

One beautiful benefit from *tailoring* a *process master* to *govern white-boxes*, is that such a *master* may now carefully describe how to pass the start-up criteria for all typical *white-boxes* being developed in the organization.

For outside-in *integration* (see Fig. 6-38, p. 167), the *full verification* takes place first when all *fake black-/white-boxes* are realized. However, the *white-box* with *fake black-/white-box* may be started for an intermediate *verification*.

11.2 EXAMPLE House environment: Realize interfaces to environment and insert/await outermost white-box

11.2.1 EXAMPLE House: About this example

For the house example, a special technical overall *schedule* has been *tailored* (see ch. Pt:2., "Cpdm generic advanced technical schedule," p. 104). This *process schedule* is here explicitly modified to a *view* better fitting outside-in *integration* with consolidated *verification* (see Fig. 11-5 and Fig. 11-6 below).

When it comes to integrating prototypes, it might seem tricky to apply this to a house development. There are some peculiarities to fit house development into traditional *integration*, such as:

Integrating design items to a prototype house. Translated to building vocabulary, this is more often referred to as simply building a house. A house is most often developed and built as a unique finished *product*, and not as a prototype to only be verified in order to secure it for later series manufacturing. Nevertheless, there is nothing wrong with verifying a single house prototype.

Integrated outside-in instead of inside-out. To hold people, major parts in a house are large, and thus realized by fixed and static *interfaces*, and there are no possibilities (even if one would prefer) to easily connect or disconnect *embedded white-boxes* before and after they are built (apart from *modularized* houses, constructed for this purpose). For example, most often static and fixed walls are the *boundary interfaces* between rooms, which can hardly be separately prepared and then integrated to the house, or be verified stand-alone.

Consolidated verification inside-out. Depending on the sheer size of a house, it cannot be integrated by incorporation of separately developed inner *black-boxes* into outer *white-boxes*. Consequently the house is built outside-in, and this has consequences for the *verification*. Invaded *black-/white-boxes* cannot be verified separately from their surroundings, but must be verified within the surroundings where it is inseparably built. When a failure is detected in an invaded *black-/white-box*, this failure can, in fact, be the result of a failure localized in the inseparable surroundings. (If a failure is detected in an *embedded black-/white-box* that is separable and yet not integrated in its surroundings when verified, a failure detected cannot be localized anywhere else than inside the *black-/white-box* itself). The consequence is that each failure detected can, of course, be located on the same *nesting level* where it is verified, but can also be located anywhere else *outwards*, thus making troubleshooting more *complex* and time-consuming.

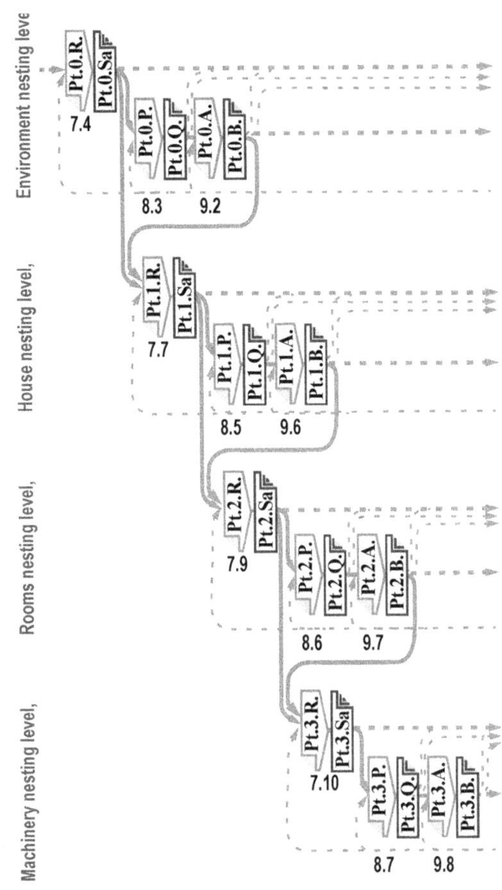

FIGURE 11-5 Outside-in development, house schedule n = 4

But since a house is so visible and tangible, it is very educational to also describe *realization* of a house prototype in this book. See it as a proof of *quality* that *Cpdm* easily manages development of a mostly static mechanical *product*.

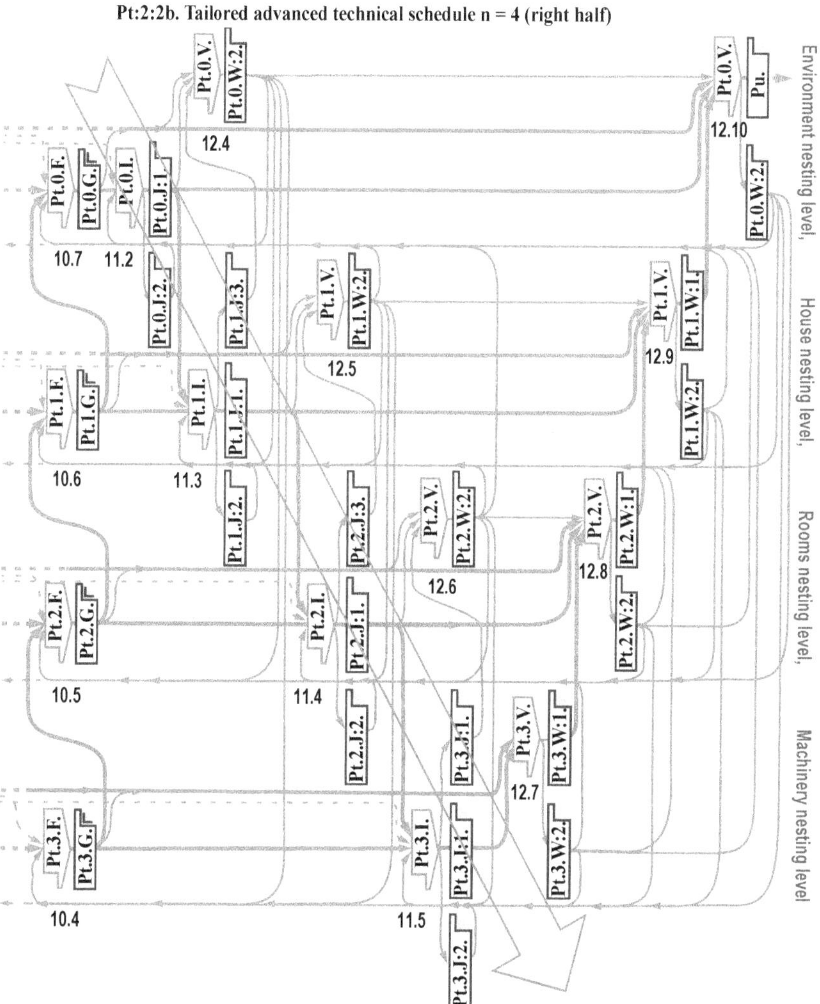

FIGURE 11-6 Outside-in invasive integration (the arrow), house schedule n = 4

11.2.2 EXAMPLE House environment: Process schedule to use

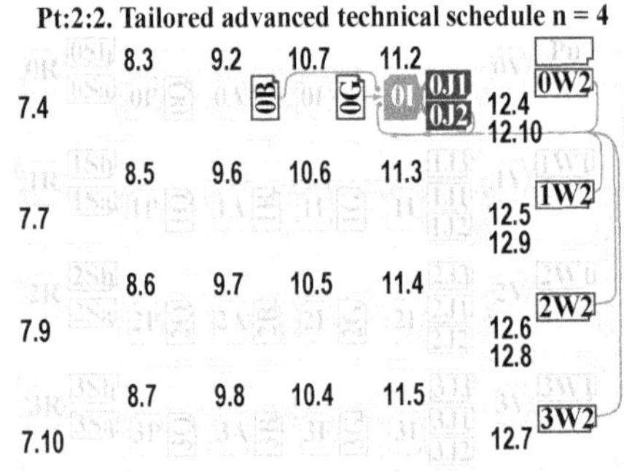

For the house example, a special technical overall *schedule* is *tailored* (illustrated in Fig. 11-7, left):

• **Pt:2:2. Tailored advanced technical schedule n = 4**, page 164

In this *tailored schedule* the generic *schedule* to use is (filled-in symbols):

• **Pt.0.I. Realize interfaces to environment and insert/await outermost white-box,** page 122.

FIGURE 11-7 Position of schedule and master to use (filled in) in overall schedule

Below *results* may be demanded (their *masters* are indicated *upstream* from the filled-in *schedule* in the figure above):

0B Environment existing ingredients and design of interfaces to the environment
Table 9-6, House environment architecture ingredients resumé, page 341
Figure 9-5, House environment logical architecture, page 341
Figure 9-6, House environment physical architecture, page 342

0G Environment interface design items and product requisites
Figure 10-14, House environment connections, page 479
Table 10-41, House interfaces to environment design items, page 479
Table 10-42, House environment estimated realization material requisites cost, page 480
Table 10-43, House environment estimated realization work requisites hourly/fixed/rental cost, page 480
Table 10-44, House environment estimated development work requisites hourly/fixed cost, page 481
Table 10-45, House environment summed/accumulated requisites estimates, page 481

H Picked in-house legacy reusable items

XaH Ordered suppliers' off-the-shelf reusable items

XbH Ordered suppliers' turnkey reusable items

For possible *failure localization* and *failure elimination*, below *results* may be demanded (their *masters* are indicated *upstream* from the filled-in *schedule* in the figure above or in greater detail in Fig. 11-5, p. 602 and Fig. 11-6, p. 603):

0J2 Table 11-9, House environment start-up report 1, page 606

0W2 List 12-1. House environment interfaces with fake house failure elimination report 1, page 674

0W2 Table 12-27, House environment verification report 1, page 701

Table 12-29, House environment reverification report 2, page 703

Table 12-28, House environment failure elimination report 1, page 702

[1W2] List 12-2. House with fake rooms failure elimination report 1, page 676

[1W2] Table 12-22, House verification report 1, page 697

Table 12-24, House verification report 2, page 698

Table 12-23, House failure elimination report 1, page 698

[2W2] List 12-3. House rooms with fake machinery failure elimination report 1, page 678

[2W2] Table 12-14, House kitchen verification report 1, page 688

Table 12-17, House kitchen verification report 2, page 691

Table 12-15, House kitchen failure elimination report 1, page 689

[3W2] Table 12-9, House kitchen machinery verification report 1, page 682

Table 12-11, House kitchen machinery verification report 2, page 683

Table 12-10, House kitchen machinery failure elimination report 1, page 683

11.2.3 Table 12-10, House kitchen machinery failure elimination report 1, page 683EXAMPLE House environment: Obtain items for interface to environment and ported linkable program

Only small adjustments were needed when *environment interfaces design items* were received (see Table 11-7 below).

TABLE 11-7 House environment items received deviation report 1

Deviation	Item	Sourcing	№	Unit	€/Unit	€
DR.1. Double thickness of gravel needed for heavy vehicles to enter property	i255. Gravel	Local	48	m³	12	576

11.2.4 EXAMPLE House environment: Complete and connect interfaces to environment

This is the first *task* to perform when "building a house," using outside-in *invasive integration* beginning with the *environment interfaces* (see Table 11-8 below).

TABLE 11-8 House environment integration report 1

Fulfillments and shortcomings	Interfaces to realize	Responsible provider
IR.1. The water hose was dug down to 60 cm. No bedrock found.	-E-2. Water	
IR.2. Canalization was also dug down together with water hose, and now waiting for connection.	-E-3. Sewage	
IR.3. Gas pipe was dug down from gas tank and waiting for connection.	-E-5. Gas	PR.13. House builder
IR.4. The final amount of gravel had been calculated only for small cars and not trucks. The double thickness of gravel was applied.	-E-6. Driveway	
IR.5. The land around the house was bulldozed without problems according to physical architecture.	E. House environment	
IR.6. The electrical cable was dug down to 60 cm. Some big boulders were bypassed. The drawing were accordingly updated.	-E-4. Electricity	PR.12. Elert AB

11.2.5 EXAMPLE House environment: Start interfaces to environment

To ensure that the *environment interfaces* are approved for use and ready for later *verification*, they should be started up (see Table 11-9 below).

TABLE 11-9 House environment start-up report 1

Accomplishments and obstacles	Troubling ingredient	Responsible provider
SR.1. Water was turned on at the street, and water became available at the building site.	-E-2. Water	
SR.2. Water from pipe drained in the sewage.	-E-3.Sewage	PR.13. House builder
SR.3. The truck with gravel used the road with no problems after the road surface was planed and rolled.	-E-6. Driveway	
SR.4. Electricity was turned on at the power pole and electricity became available at the building site.	-E-4. Electricity	PR.12. Elert AB

11.2.6 EXAMPLE House environment: Examine prototype reliability

The *architect* Robber HB and the house builder Mobber AB were called to the gate meeting held by the house proprietor. Some problems during road construction were reported, but all were solved. Responsible parties were reminded to update requisites with extra cost for gravel. The start-up was passed, and a *partial verification* of the *environment interfaces* (with a fake house) was decided upon.

11.2.7 EXAMPLE House: Proceed reading

To follow this example, these are the alternatives (see again Fig. 11-7, p. 604):

- Trace *upstream* in the overall *schedule* to understand how to fix a house environment that is found to not be ready for field *verification*.

- [0V] Proceed to chapter 12.4, "EXAMPLE House environment: Partially verify interfaces to environment with house fake," page 673.

- [0V] Proceed to chapter 12.10, "EXAMPLE House environment: Fully verify prototype in environment," page 699.

- [11] Proceed to chapter 11.3 "EXAMPLE House: Procure house items (and await invading room black-boxes) in order to build house" below.

11.3 EXAMPLE House: Procure house items (and await invading room black-boxes) in order to build house

11.3.1 EXAMPLE House: Process schedule to use

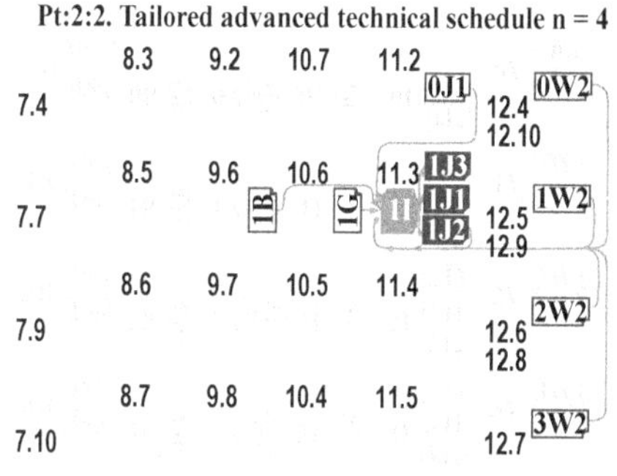

For the house example, a special technical overall *schedule* is *tailored* (illustrated in Fig. 11-8, left):

- **Pt:2:2. Tailored advanced technical schedule n = 4,** page 164

In this *tailored schedule* the generic *schedule* to use is (filled-in symbols):

- **Pt.n.I. Procure items, linkable programs, and integrate/await embedded black-boxes in order to realize white-box,** page 144.

FIGURE 11-8 Position of schedule and master to use (filled in) in overall schedule

Below *results* may be demanded (their *masters* are indicated *upstream* from the filled-in *schedule* in the figure above):

- [0J1] Usable prototype interfaces started in simulator or target

- [1B] House architectures with major ingredients and room black-boxes

    Table 9-13, House rooms black-boxes architecture ingredients resumé, page 360

    Table 9-14, House interfaces architecture ingredients resumé, page 361

    Table 9-15, House secondary space elements architecture ingredients resumé, page 362

    Figure 9-22, House logical architecture, page 363

Figure 9-23, House physical architecture, page 364

Figure 9-24, House façade physical architecture, page 364

1G Finalized house items design with product requisites

Figure 10-10, House boundary casing (upper part), page 467

Figure 10-11, House boundary casing (lower part), page 468

Figure 10-12, House connections, page 469

Table 10-32, House boundary white-box design items, page 470

Table 10-33, House connections white-box design items, page 471

Table 10-34, House boundary and connections white-box design items, page 471

Table 10-37, House estimated realization material requisites cost, page 472

Table 10-38, House estimated realization work requisites hourly/fixed/rental cost, page 474

H Picked in-house legacy reusable items

XaH Ordered suppliers' off-the-shelf reusable items

XbH Ordered suppliers' turnkey reusable items

For possible *failure localization* and *failure elimination*, below *results* may be demanded (their *masters* are indicated *upstream* from the filled-in *schedule* in the figure above):

1J2 Table 11-14, House start-up report 1, page 612

0W2 List 12-1. House environment interfaces with fake house failure elimination report 1, page 674

0W2 Table 12-27, House environment verification report 1, page 701

Table 12-29, House environment reverification report 2, page 703

Table 12-28, House environment failure elimination report 1, page 702

1W2 List 12-2. House with fake rooms failure elimination report 1, page 676

1W2 Table 12-22, House verification report 1, page 697

Table 12-24, House verification report 2, page 698

Table 12-23, House failure elimination report 1, page 698

2W2 List 12-3. House rooms with fake machinery failure elimination report 1, page 678

2W2 Table 12-14, House kitchen verification report 1, page 688

Table 12-17, House kitchen verification report 2, page 691

Table 12-15, House kitchen failure elimination report 1, page 689

3W2 Table 12-9, House kitchen machinery verification report 1, page 682

Table 12-11, House kitchen machinery verification report 2, page 683

Table 12-10, House kitchen machinery failure elimination report 1, page 683

11.3.2 EXAMPLE House: Obtain items including linkable program

The house estimated realization material requisites cost (see Table 10-37, p. 472) has been used to purchase all necessary *samples*. Only *samples* that caused problems are reported in Table 11-10 below.

TABLE 11-10 House items received deviation report 1

Deviation	Item	Sourcing	Producer	Amount	Unit	€/unit	Price €	Arch
DR.1. Weatherproof impregnated at same cost	i201. Roof truss	Local	NoPapp	22	Piece	410	9 020	]A]1.a.
DR.2. Weatherproof impregnated at € 1,9	Plank	Local	NoPapp	160	Meter	1.9	304	]A]1.d.
DR.3. Weatherproof impregnated at € 2,5	Plank	Local	NoPapp	27	Meter	2.5	67.5	]A]1.h.
DR.4. For barge board. Pressure impregnated spruce, missed in finalize design	Plank	Local	NoPapp	14	Meter	3	42	Missed
DR.5. Weatherproof impregnated at same cost	Plank	Local	NoPapp	27	Meter	11	297	]A]1.j.
DR.6. Joists over inner doors, missed in finalize design	Joist	AllBau		7	Meter	17.2	120.4	Missed
DR.7. Prefabricated superior mortar at € 86 per m³	Mortar	Local		5	m³	86	430	]A]1.o.
DR.8. Better found at € 3.3	Tiles	AllBau		1260	Piece	3.3	4 i58	]A]1.g.
DR.9. Oak handrail missed in finalize design	Stair		Pall&Papp	1	Piece	1 930	1 930	Missed
DR.10. Procured from No-Bau	Drain pipe	AllBau		8	Meter	6.7	54	-A-13.

11.3.3 EXAMPLE House: Integrate elements, including linkable program and embedded black-boxes

Before the house is realized, a fake house is arranged, in order to make some *partial verification* (see ch. 6.13.3, p. 166) of the *interfaces* to the *environment* (see Table 11-11 below).

TABLE 11-11 Fake house integration report 1

Fulfillments and shortcomings	Interfaces to realize	Responsible provider
IR.1. The fake house is realized by putting up strings between sticks knocked down in the ground, to show the house outer boundaries, IR.2. To fake pipes in the house, they are marked by dots painted on the ground.	-E- Interfaces to environment	PR.13. House builder
IR.3. To fake water consumption, a preliminary faucet was installed. This will also be used during house building, and then removed.	-E-2. Water	
IR.4. To fake electricity consumption, a preliminary electrical distribution box was connected without problem. This will also be used during building of the house, and then removed.	-E-4. Electricity	PR.12. Elert AB

After this fake is arranged, it is time to make the *partial verification* (see ch. 12.4.2, "EXAMPLE House environment: Process schedule to use," p. 673).

After the *partial verification* is done, the *task* is to build the real house, but not yet rooms and machinery (see the integration reports in Table 11-12 and Table 11-13 below).

TABLE 11-12 House static integration report 2

Fulfillments and shortcomings	Interfaces to realize		Responsible provider
IR.5. Preparation of ground Foundation was molded directly into trenches in the ground. Macadam was filled and the concrete slab was molded. Molding was delayed two months because of cold winter weather. The ground drainage was approved by a quality consultant.	]A[1. House boundary	]A]1.p. Water barrier]A]1.q. Foundation]A]1.r. Macadam]A]1.s. Drainage pipe]A]1.t. Concrete slab	PR.10. Proprietor
IR.6. Preparation of outer walls The hollow bricks were mortared without problem.The wall insulation was inspected by the energy consultant.		]A]1.i. Bracket]A]1.j. Joint beam]A]1.k. Opening joist]A]1.m. Inner brickwork]A]1.n. Wall insulation]A]1.o. Façade	PR.13. House builder
IR.7. Preparation of roof Roof trusses were of extraordinarily good quality. The insulation was no problem. The roof deck got 2 mm thinner than drawings, because proprietor accepted cheaper plank. Some 20 roof tiles were stored for possible future repairs.		]A]1.a. Roof truss]A]1.b. Truss insulation]A]1.c. Cross insulation]A]1.d. Cross space plank]A]1.e. Roof deck]A]1.f. Roof lath]A]1.h. Eave plank]A]1.g. Roof tile]A]1.l. Gypsum board	
IR.8. There were no problems in mortaring the inner wall hollow bricks.	-A-2. House inner walls		

TABLE 11-13 House connections integration report 1

Fulfillments and shortcomings	Interfaces to realize	Responsible provider
IR.9. Connections through outer wall Doors and windows were an inch smaller than specified, and a wider joint did occur, which was filled in with styrofoam. To lift the beams and brackets a crane was needed.	-E-11. Front door -E-8. Door section -E-10. T-window -E-9. K-window	
IR.10. Connections through inner walls The same beams as for the outer walls were used and lifted with the crane.	-A-3.E-door -A-4.Lock door -A-5.Open arch -A-6.Stairway -A-7.S-door -A-8.L-door	
IR.11. Preparation of house piping Sewer was forgotten to be connected before the concrete slab was molded. A groove was hammered out, and all house interfaces to the environment used this. The house builders insurance covered the extra cost. The floor heating was established, and photos were taken for the house manual before molding was done.	-A-9.Warm water -A-10.Cold water -A-11a.K-heating -A-11b.D-heating -A-11c.E-heating -A-11d.T-heating -A-11e.B-heating -A-12a.K-Wiring -A-12b.O-Wiring -A-13.Sewer -E-2.Water -E-6.Air outlet	PR.13. House builder
	-E-4.Electricity	PR.12. Elert AB

As shown in Table 8-13, House white-box supplier opportunities, page 303, the house builder is appointed to build the house. However, the excavation is done by the house proprietor, and not by the builder. (In addition to the table below, there are, of course, a lot of *project management* matters to report also, as in all other *activities*, but this is not a topic for this book.)

11.3.4 EXAMPLE House: Start up white-box

To ensure that the succeeding *verification* will be efficient and will not just take over the *integration* work, the assembled prototype must be started up before being invaded by rooms and later handed over for *verification*.

Since the house applies outside-in *invasive integration*, not all *interfaces* are connected to their absorbers, but can be started up for *partial verification*.

TABLE 11-14 House start-up report 1

Accomplishments and obstacles	Troubling ingredient	Responsible provider
SR.1. House ground-carrying capacity was inspected and approved by geology company. **SR.2.** Inner walls were masoned by hollow brick walls of approved type. Floors were made in reinforced concrete of approved type.	A. House	PR.13. House builder
SR.3. Feed piping for water, heating, sewer, and air outlet in place in ceiling, walls, and floor. Water available at the water scale, sewer connected to cesspit and air outlet piping in place. Windows and doors were in place, adjusted against cold draft, capable of being opened, and lockable.	A.aC. Kitchen A.aA. Dining room A.aB. Entrance A.aD. Toilet A.aE. Bed room	
SR.4. Wicket was in place.	A.b. Loft	
SR.5. Sewer cesspit was available and connected to all sewers.	A.c. Staircase space	
SR.6. Boiler placed on temporary stand, to be fastened on the wall after plastered. All water piping in place. Water pipe from environment was in place but not yet switched on.	A.cb. Boiler	
SR.7. Scale was in place on water supply pipe.	A.cc. Scale	
SR.8. Cesspit in place in floor. Piping from kitchen and toilet was in place, piping to environment in place.	A.cd. Cesspit	
SR.9. Main distribution box inserted in the wall of the staircase space and all electrical piping in place. Piping and wiring from environment and temporary electricity distribution box in place. Electricity was switched on and available in main box.	A.ca. Main distr. box	PR.12. Elert AB

11.3.5 EXAMPLE House: Examine integration robustness

A gate meeting was held with the house builder Mobber AB, the *architect* Robber HB, the quality inspector, and the house proprietor. Some weaknesses were discovered:

- Electricity had distribution boxes for each room; floor heating had separate thermostats for each room; in the staircase space, however, water piping to rooms had no separate valves to allow individual room *verification*. The house builder installed these for free.

- There was no extra piping to rooms for possible future needs. The electrician was phoned, and promised to apply 2 extra pipes to each room for the material cost only.

Since the house needs outside-in *invasive integration*, it is not yet really possible to fully verify it. The approval from the gate meeting means that *integration* of rooms into the house can be continued.

11.3.6 EXAMPLE House: Proceed reading

To follow this example, these are the alternatives (see again Fig. 11-8, p. 607):

- Trace *upstream* in the overall *schedule* to understand how to fix a house that is found to not be ready for *verification*.

[0V] Proceed to chapter 12.4, "EXAMPLE House environment: Partially verify interfaces to environment with house fake," page 673.

[1V] Proceed to chapter 12.5, "EXAMPLE House: Partially verify black-/white-box with embedded fakes," page 675.

[1V] Proceed to chapter 12.9, "EXAMPLE House: Fully verify house black-/white-box," page 693.

[2I] Proceed to chapter 11.4 "EXAMPLE House rooms: Procure room items (and await invading machinery black-boxes) in order to build rooms" below.

11.4 EXAMPLE House rooms: Procure room items (and await invading machinery black-boxes) in order to build rooms

11.4.1 EXAMPLE House kitchen: Process schedule to use

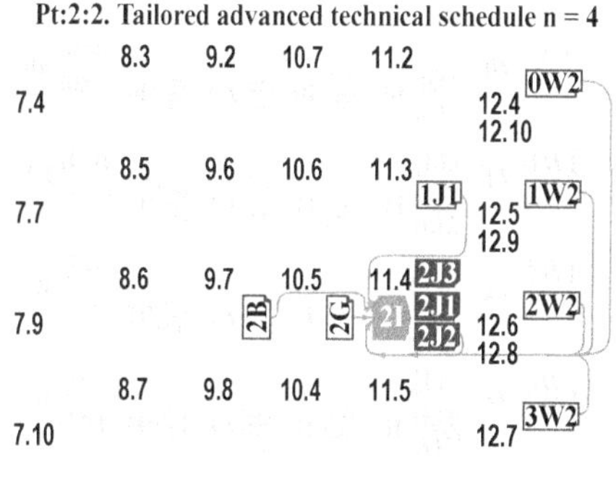

For the house example, a special technical overall *schedule* is *tailored* (illustrated in Fig. 11-9, left):

• **Pt:2:2. Tailored advanced technical schedule n = 4,** page 164

In this *tailored schedule* the generic *schedule* to use is (filled-in symbols):

• **Pt.n.I. Procure items, linkable programs, and integrate/await embedded black-boxes in order to realize white-box,** page 144.

FIGURE **11-9** Position of schedule and master to use (filled in) in overall schedule

Below *results* may be demanded (their *masters* are indicated *upstream* from the filled-in *schedule* in the figure above):

[1J] Completed usable house white-box

[2B] Room architectures with major ingredients and machinery black-boxes

Table 9-16, House kitchen elements architecture ingredients resumé, page 367

Table 9-17, House kitchen interfaces architecture ingredients resumé, page 368

Figure 9-27, House kitchen physical architecture, page 370

Figure 9-26, House kitchen logical architecture, page 369

[2G] Finalized room items design with product requisites

Figure 10-7, House kitchen interior drawing, page 455

Figure 10-8, House kitchen connections, page 456

Table 10-18, House kitchen electricity white-box design items, page 456

Table 10-19, House kitchen electricity white-box design items, page 457

Table 10-20, House kitchen interiors white-box design items (cabinets for machinery), page 457

Table 10-21, House kitchen interiors white-box design items (cabinets without machinery), page 458

Table 10-22, House kitchen interiors white-box design items (cabinet static interfaces), page 459

Table 10-23, House kitchen boundaries white-box design items, page 459

Table 10-26, House kitchen estimated realization material requisites cost, page 460

Table 10-27, House kitchen estimated realization work requisites hourly/fixed/rental cost, page 463

H Picked in-house legacy reusable items

XaH Ordered suppliers' off-the-shelf reusable items

XbH Ordered suppliers' turnkey reusable items

For possible *failure localization* and *failure elimination*, below *results* may be demanded (their *masters* are indicated *upstream* from the filled-in *schedule* in the figure above):

2J2 Table 11-18, House kitchen start-up report 1, page 617

0W2 List 12-1. House environment interfaces with fake house failure elimination report 1, page 674

0W2 Table 12-27, House environment verification report 1, page 701

Table 12-29, House environment reverification report 2, page 703

Table 12-28, House environment failure elimination report 1, page 702

1W2 List 12-2. House with fake rooms failure elimination report 1, page 676

1W2 Table 12-22, House verification report 1, page 697

Table 12-24, House verification report 2, page 698

Table 12-23, House failure elimination report 1, page 698

2W2 List 12-3. House rooms with fake machinery failure elimination report 1, page 678

2W2 Table 12-14, House kitchen verification report 1, page 688

Table 12-17, House kitchen verification report 2, page 691

Table 12-15, House kitchen failure elimination report 1, page 689

3W2 Table 12-9, House kitchen machinery verification report 1, page 682

Table 12-11, House kitchen machinery verification report 2, page 683

Table 12-10, House kitchen machinery failure elimination report 1, page 683

11.4.2 EXAMPLE House kitchen: Obtain items including linkable program

When starting the *invasive integration* of the rooms, such as the kitchen, the house must be about finished and inner walls well prepared. Walls, ceilings, and floors must, for example, be provided with holes needed for *interfaces* to feed the house. That way, pipes, hoses, and sockets can be inserted into the walls and ceilings, as well as heating pipes in the floors. Also human *interfaces* between rooms, such as doors and stairs, must be prepared as openings in the walls.

There is a huge number of *items* to assemble in a kitchen, as seen from Table 10-26, House kitchen estimated realization material requisites cost, page 460. The *product requisites* were the base for the large purchase, and delivery problems and deviations are reported in Table 11-15 below.

TABLE 11-15 :House kitchen items received deviation report 1

Deviation	Item	Sourcing	Producer	Type	Amt.	€/ unit	Cost €
DR.1. Delivery was delayed 2 months, mounted when arriving OK.	i124. Ventilation grid	K-Studio	MedMach	W34-G	1	35	35
DR.2. Delivery delayed 3 months, use pine door until arrival.	i142. Door	K-Studio	IKEY	C60-40	1	80	80
DR.3. Discontinued, use E40-60, change not visible, € 190.	i152. Housing	K-Studio	IKEY	D40-60	1	410	410
DR.4. Available in white only. Painted grey for + € 20.	i176. Revolving shelf	K-Studio	IKEY	UC40	2	70	140
DR.5. Sold out, delivery unknown. Purchased angle 180°, at price 52 apiece.	i183. Door hinges	K-Studio	IKEY		16	18	288
DR.6. Dark grey lighter than shown sample. Right color bought from Export AB	i191. Joint sealant	Import AB	Hazard	5C	5 kg	4.3	21.5

11.4.3 EXAMPLE House kitchen: Integrate elements, including linkable program and embedded black-boxes

Before the real rooms are realized, fake rooms are built in order to make some *partial verification* of the house (see *tailored schedule* in ch. 6.13.3, p. 166, and see Table 11-16 below).

TABLE 11-16 Fake house kitchen integration report 1

Fulfillments and shortcomings	Interfaces to realize	Responsible provider
IR.1. Kitchen drawings are used to fake piping and interiors with chalk crayon markings on the completed rough concrete floors and plastered walls.	-A- Interfaces inside the house	PR.17. K-Studio

After the *partial verification* is done (see ch. 12.5, p. 675), the real *invasive integration* of the rooms takes place Note that according to the architecture, the A.c. Staircase space and A.b. Loft belong to the house and are already in place.

As usual, the kitchen is the only room shown in this book example (see Table 11-17 below).

TABLE 11-17 House kitchen integration report 2

Fulfillments and shortcomings	Interfaces to realize		Responsible provider
IR.2. The floor heating hose was damaged by the proprietor and had to be exchanged at his cost. The floor was floated with putty without problem. The selected ceramics ran out and had to be imported again, and a delay of 10 weeks occurred which the house builder compensated for by finishing the ceramics for free.	[A.aC]4. Floor		PR.13. House builder
IR.3. The backsplash ceramics were mounted without checking on the hood. It had to be removed at the expense of the house builder.	[A.aC]5. Backsplash		
IR.4. Due to leaky electrical wall junctions, some of them let in plaster, and had to be fixed.	[A.aC]6. Wall and ceiling		
IR.5. Gas pipe was drawn directly from outside of house into the kitchen.	-E-5.Gas		
IR.6. Holes for bolts were predrilled from inside all cabinets. Ventilation holes were made later during machine integration. There were no problems arranging the floor socket and screwing it to the cabinets. The wall plugs and screws were too short, due to the hollowed brick, and were exchanged without cost. IR.7. Some doors used hinges, and some extra were bought.	-A.aC-2. Interior attachment	Bolt pairs Wood screw Wall plug Floor socket Outer shelf Door hinges Shelf pins	PR.17. K-Studio

TABLE 11-17 House kitchen integration report 2

Fulfillments and shortcomings	Interfaces to realize		Responsible provider
IR.8. Grooves were cut into the hollow bricks to hold all pipes. Brick nails were used to fix the pipes before the wall was plastered. Kitchen electrical distribution box was installed.	-A.aC-3. Machinery el. feed	-A.aC-3a.El. fridge	PR.12. Elert AB
		-A.aC-3b.El. micro	
		-A.aC-3c.El. warm	
		-A.aC-3d.El. bake	
		-A.aC-3e.El. hub	
		-A.aC-3f.El. hood	
		-A.aC-3g.El. adv. dish	
		-A.aC-3h.El. smp. dish	
		-A.aC-3i.El. lamp	
		-A.aC-3j.El. switch	

11.4.4 EXAMPLE House kitchen: Start up white-box

Note that only the start-up of the kitchen is shown in this book (see Table 11-18 below).

TABLE 11-18 House kitchen start-up report 1

Accomplishments and obstacles	Troubling ingredient	Responsible provider
SR.1. Flooring, plastering, and painting finished for usage. Gas pipe routed to heating area. Water available but not connected. Air outlet not connected. No kitchen machinery yet in place.	A.aC. Kitchen	PR.13. House builder
SR.2. Interiors in place and fixed to wall and floor.		PR.17. K-Studio
SR.3. Electrical wiring completed and wall sockets inserted.		PR.12. Elert AB

11.4.5 EXAMPLE House rooms: Examine integration robustness

K-Studio and Mobber AB were called to the gate meeting. There had been many conflicts between Mobber AB finishing the kitchen walls, floor, and ceiling and K-Studio installing the kitchen interiors. Unfortunately, the house proprietor had to invest much time and money to resolve these conflicts.

After sorting out all conflicts, the parties finally prepared a kitchen that could be verified.

11.4.6 EXAMPLE House: Proceed reading

To follow this example, these are the alternatives (see again Fig. 11-9, p. 613):

- Trace *upstream* in the overall *schedule* to understand how to fix rooms that are found to not be ready for *verification*.
- IV) Proceed to chapter 12.5, "EXAMPLE House: Partially verify black-/white-box with embedded fakes," page 675.

2V Proceed to chapter 12.6, "EXAMPLE House kitchen: Partially verify black-/white-box with embedded fakes," page 677.

2V Proceed to chapter 12.8, "EXAMPLE House kitchen: Fully verify room black-/white-box," page 684.

3I Proceed to chapter 11.5 "EXAMPLE House kitchen machinery: Procure machinery items in order to install machinery" below.

11.5 EXAMPLE House kitchen machinery: Procure machinery items in order to install machinery

11.5.1 EXAMPLE House kitchen machinery: Process schedule to use

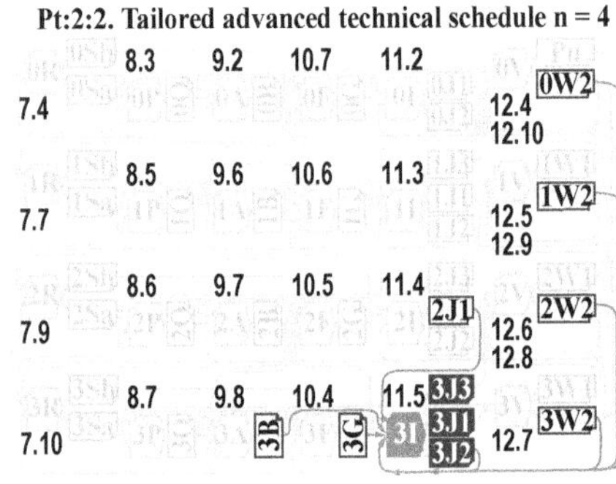

For the house example, a special technical overall *schedule* is *tailored* (illustrated in Fig. 11-10, left):

- **Pt:2:2. Tailored advanced technical schedule n = 4,** page 164

In this *tailored schedule* the generic *schedule* to use is (filled-in symbols):

- **Pt.n.I. Procure items, linkable programs, and integrate/await embedded black-boxes in order to realize white-box,** page 144.

FIGURE 11-10 Position of schedule and master to use (filled in) in overall schedule

Below *results* may be demanded (their *masters* are indicated *upstream* from the filled-in *schedule* in the figure above):

2J Completed usable room white-boxes

3B Machinery architectures with major ingredients but no black-boxes

Table 9-18, House kitchen machinery elements architecture ingredients resumé, page 373

Table 9-21, House kitchen machinery user interface architecture ingredients resumé, page 375

Table 9-19, House kitchen machinery connection interface architecture ingredients resumé, page 373

Table 9-20, House kitchen machinery boundaries interface architecture ingredients resumé, page 374

Figure 9-29, House kitchen machinery logical architecture, page 376

Figure 9-30, House kitchen machinery physical architecture, page 377

3G Finalized machinery items design with product requisites

Table 10-9, House kitchen machinery fake items, page 447

Table 10-10, House kitchen machinery white-box design items, page 447

Table 10-12, House kitchen machinery fakes, estimated realization work requisites hourly/fixed/rental cost, page 449

Table 10-13, House kitchen machinery estimated realization material requisites cost, page 449

Table 10-14, House kitchen machinery estimated realization work requisites hourly/fixed/rental cost, page 450

H Picked in-house legacy reusable items

XaH Ordered suppliers' off-the-shelf reusable items

XbH Ordered suppliers' turnkey reusable items

For possible *failure localization* and *failure elimination*, below *results* may be demanded (their *masters* are indicated *upstream* from the filled-in *schedule* in the figure above):

3J2 Table 11-22, House kitchen machinery start-up report 1, page 622

0W2 Table 12-27, House environment verification report 1, page 701

Table 12-29, House environment reverification report 2, page 703

Table 12-28, House environment failure elimination report 1, page 702

1W2 List 12-2. House with fake rooms failure elimination report 1, page 676

1W2 Table 12-22, House verification report 1, page 697

Table 12-24, House verification report 2, page 698

Table 12-23, House failure elimination report 1, page 698

2W2 List 12-3. House rooms with fake machinery failure elimination report 1, page 678

2W2 Table 12-14, House kitchen verification report 1, page 688

Table 12-17, House kitchen verification report 2, page 691

Table 12-15, House kitchen failure elimination report 1, page 689

3W2 Table 12-9, House kitchen machinery verification report 1, page 682

Table 12-11, House kitchen machinery verification report 2, page 683

Table 12-10, House kitchen machinery failure elimination report 1, page 683

11.5.2 EXAMPLE House kitchen machinery: Obtain items including linkable program

According to Table 8-17, House kitchen machinery white-box supplier opportunities, page 312, the K-Studio shall integrate the house kitchen machinery, specified in Table 10-13, House kitchen machinery estimated realization material requisites cost, page 449. All machine *items* are ordered by K-Studio and sent to the K-Studio storage waiting for *integration*. Problems with received items are reported in Table 11-19 below.

TABLE 11-19 :House kitchen machinery items received deviation report 1

Deviation	Item	Sourc-ing	Produc-er	Type	Piec e	€/ Piec e	Cost €
DR.1. WMic 8 sold out, re-placed with WMic 9	Micro oven	MedMach	Wirly	WMic 8	1	260	260
DR.2. Price changed to 85	Burner	MedMach	Buttan	Mall-2 Ex	1	120	120
DR.3. Delivery delayed 2 weeks	Hood	MedMach	Wirly	Vac 3411	1	90	90

11.5.3 EXAMPLE House kitchen machinery: Integrate elements, including linkable program and embedded black-boxes

Before the machinery is integrated, a house rooms machinery fake is prepared (see Table 11-20 below), in order to make some *partial verification* of the kitchen (see ch. 12.6, p. 677).

TABLE 11-20 House kitchen fake machinery integration report 1

Fulfillments and shortcomings	Interfaces to re-alize	Responsi-ble provider
IR.1. Machinery measurements and location were illustrated by marking their size and interfaces with a pen on the kitchen interiors.	-A.aC-2. Interior at-tachment (to ma-chinery)	PR.17. K-Studio

All cabinets (also for house rooms machinery) are already assembled in the kitchen, and now the machinery itself has to be installed (see Table 11-21 below).

TABLE 11-21 House kitchen real machinery integration report 2

Fulfillments and shortcomings	Interfaces to realize		Responsi-ble provider
IR.2. Connected to tin box.	-E-6. Air outlet	-E- Inter-faces to environ-ment	
IR.3. Pressure reducer connected to gas pipe. Armed hose for gas connects from valve to burner.	-E-5. Gas		
IR.4. Valve mounted on incoming water pipe. Connection hose mounted between valve and tap.	-A-9. Warm water		PR.13. House builder
IR.5. One-to-three branch mounted on incom-ing water pipe. One valve mounted on each output. Connection hoses mounted between valves and faucet and dishwashers.	-A-10. Cold water	-A- Inter-faces in-side the house	
IR.6. 3-inch water trap mounted to sewer pipe and connected to sink bowls. Jack on the wa-ter trap connected with a rubber hose to dish-washer.	-A-13. Sewer		

TABLE 11-21 House kitchen real machinery integration report 2

Fulfillments and shortcomings	Interfaces to realize		Responsible provider
IR.7. Back ventilation opening is sawn. Cover is predrilled for electrical wire. Feet are too short to be height-adjusted; an extra board was put underneath.	[A.aC.aA]1. Cold housing	[A.aC.a]. Machinery housing	PR.17. K-Studio
IR.8. Micro cover is mounted on top of oven cover. Covers are predrilled for electrical wires. Micro back ventilation and oven back ventilation opening were sawn. No problems in fastening the machines.	[A.aC.aB]1. Upper housing [A.aC.aB]2. Lower housing		
IR.9. Back ventilation opening sawn. Hole is predrilled for electrical wire.	[A.aC.aC]1. Oven housing		
IR.10. Stone countertop had customized hole fixed at delivery. Heat-resistant rubber gasket is glued between hub and stone countertop and between burner and stone countertop.	[A.aC.aD]1. Gasket seal		
IR.11. Tin box is mounted on the wall and connects to air outlet pipe. The hub is mounted to the tin box.	[A.aC.aE]1. Air tin box		
IR.12. Stone countertop had customized hole fixed at delivery. Water-resistant rubber gasket is glued between sink and stone countertop.	[A.aC.aF]1. Gasket seal		
IR.13. Covers are mounted side by side and get drilled for sewer hose, water hoses, and electrical wire.	[A.aC.aG]1. Left housing [A.aC.aG]2. Right housing		
IR.14. Fridge preset to +4 $^{\circ}$C and freezer to -18 $^{\circ}$C.	-A.aC-1a. Cold panel	-A.aC-1. Interior access	PR.17. K-Studio
IR.15. Adjusted to not leak out cold air.	-A.aC-1b. Cold door		
IR.16. Preset to English text.	-A.aC-1c. Micro panel		
IR.17. Changed to right handed.	-A.aC-1d. Micro door		
IR.18. Preset to English text.	-A.aC-1e. Warm panel		
IR.19. Preset to English text.	-A.aC-1g. Bake panel		
IR.20. Secured against child access.	-A.aC-1h. Bake door		
IR.21. Equipped with childproof controls.	-A.aC-1i. Hub panel		
IR.22. Equipped with childproof controls.	-A.aC-1j. Burner panel		
IR.23. Equipped with anti-slip cover.	-A.aC-1l. Handles		
IR.24. Preset to English text.	-A.aC-1m. Dish panel		
IR.25. Extra handle mounted.	-A.aC-1n. Dish door		
IR.26. Preset to English text.	-A.aC-1o. Dish panel		
IR.27. Extra handle mounted.	-A.aC-1p. Dish door		
IR.28. Owen configured for 3 phases with neutral connector. Perilex plug mounted on 5 x 2,5 mm² connection cable.	-A.aC-3d. El. bake	-A.aC-3. Machinery el. feed	PR.12. Elert AB
IR.29. Hub configured for 3 phases with neutral connector. Perilex plug mounted on 5 x 2,5 mm² connection cable.	-A.aC-3e. El. hub		

11.5.4 EXAMPLE House kitchen machinery: Start up white-box

Connecting and starting up kitchen machinery causes some difficulties (see Table 11-22 below).

TABLE 11-22 House kitchen machinery start-up report 1

Accomplishments and obstacles	Troubling ingredient		Responsible provider
SR.1. 220 V connected. Operating starts OK.	A.aC.aA.a. Combined fridge/freezer	A.aC.aA. Cold machinery	PR.17. K-Studio
SR.2. 220 V connected. A buzzing noise occurred when started. Resonance suppressed by a felt sheet.	A.aC.aB.a. Micro oven	A.aC.aB. Warm machinery	
SR.3. 220 V connected. Operation does not start. Repairman changes electronics board. Warranty covers cost.	A.aC.aB.b. Warming oven		
SR.4. Operation starts but oven not getting warm. Oven manual consulted to understand how to operate. Warming starts.	A.aC.aC.a. Wide oven	A.aC.aC. Bake machinery	
SR.5. The house builder stands on the hub and causes scratches. The house builder replaces hub at own cost.	AaC.aDa. Electrical hubs	A.aC.aD. Heat machinery	
SR.6. Gas connected. Stove lights OK.	A.aC.aD.b. Gas burner		
SR.7. 220 V connected. Operation starts and air blows outside.	A.aC.aE.a. Wide hood	A.aC.aE. Ventilation machinery	
SR.8. Water from faucet disappears in drain. No leakage detected.	A.aC.aF.a. Sink 2 bowls	A.aC.aF. Wet machinery	
SR.9. Water connected. Water flows at cold and warm positions.	A.aC.aF.b. Combined faucet		
SR.10. Water connected. 220 V connected. Operation starts and continues 10 minutes until fuse blows. The fuse is changed to higher ampere and operation OK. Kitchen machinery design items updated with appropriate current consumption.	A.aC.aG.a. Dishwasher coarse	A.aC.aG. Dish machinery	
SR.11. Water connected. 220 V connected. The fuse changed to higher ampere and operation OK.	A.aC.aG.b. Dishwasher gentle		

11.5.5 EXAMPLE House rooms machinery: Examine integration robustness

A gate meeting was held with K-Studio. There has been a lot of fixing to get all machinery to start. All these actions were included in the original price. The machinery is now started and ready for *verification*.

11.5.6 EXAMPLE House: Proceed reading

To follow this example, these are the alternatives (see again Fig. 11-10, p. 618):

- Trace *upstream* in the overall *schedule* to understand how to fix house rooms machinery that is found to not be ready for *verification*.

2V) Proceed to chapter 12.6, "EXAMPLE House kitchen: Partially verify black-/white-box with embedded fakes," page 677.

3V) Proceed to chapter 12.7, "EXAMPLE: House rooms machinery: Fully verify machinery black-/white-boxes," page 678.

- Proceed to chapter 11.6 below to leave the house example and to dig deeper into the realize & integrate white-box topic.

11.6 Completeness of prototypes

For many *complex products* it might be almost impossible to *finalize design items* all at once. *Complexity* emerging from all details might be impenetrable in the first go, and several prototypes must be developed to gradually get the understanding of how to *finalize design items* and how to optimize them.

When a prototype is complex, it might be necessary to prepare and verify several prototypes step by step.

A special *schedule* is *tailored* for this purpose (see List 6-35. Iterative development schedule for stepwise emerging prototypes, p. 163).

But be cautious. In these late *phases*, beware of starting *prototyping* (resorting to trial-and-error experimentation). Achieving a decided grade of *product quality* is not just the result of spending resources on *finalize design items*. If experimenting takes over, at the cost of *intellectual control*, only modest *product quality* can be expected. Heightening *quality* in a poor prototype by just preparing new prototypes is simply impossible.

Making many prototypes of gradually increased functionality leads into a methodology called *incremental development*. This is a powerful way to master *complexity*, if a *complex product* is developed for the first time or an unfamiliar technology is being introduced. My next book "*Cpdm* technical overhead" on *line* and *project* overhead will sort this out in detail.

Another reason to develop lesser grades of prototype completeness is having a prototype developed to only evaluate a concept and not be fully equipped for market validation. In the coming example "multiplication toy," one *variant* with hardwired multiplier will be a less complete prototype, without casing and full connection to its *target environment*. It will be evaluated only on correct behavior, and maybe the next prototype will have its multiplier exchanged to a customized ASIC circuit with casing and full connection to its *environment*.

11.7 EXAMPLE Multiplication toy with hard-wired multiplier: Procure items and linkable programs in order to realize white-box

11.7.1 EXAMPLE Multiplication toy: About this example

Note that *integration* in this and remaining examples is made according to the more common inside-out *incorporative integration* starting at innermost *nesting level* 1, in contradistinction to the outside-in *invasive integration* applied in the previous house example, starting at outermost *nesting level* 0. Also note that in the multiplication toy variant with hard-wired multiplier no *linkable program* is involved.

11.7.2 EXAMPLE Multiplication toy with hard-wired multiplier: Process schedule to use

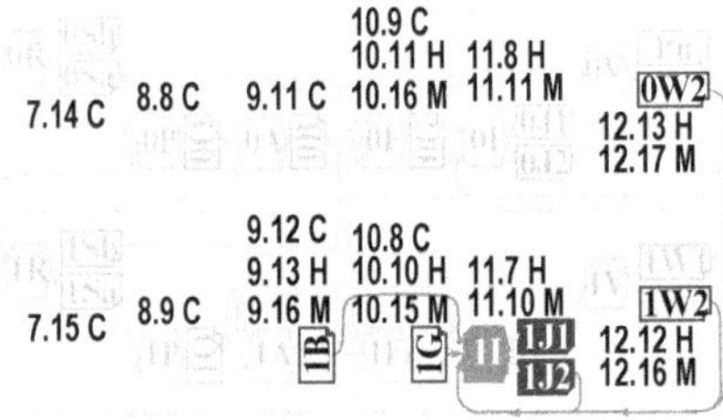

For the multiplication toy example, a special technical overall *schedule* is *tailored* (illustrated in Fig. 11-11, left):

• **Pt:2:3. Tailored advanced technical schedule n = 2,** page 169

In this *tailored schedule* the generic *schedule* to use is (filled-in symbols):

• **Pt.n.I. Procure items, linkable programs, and integrate/await embedded black-boxes in order to realize white-box,** page 144.

FIGURE 11-11 Position of schedule and master to use (filled in) in overall schedule

The very long list of below *results* needed in order to assemble such a small prototype *white-box* may be a surprise, but never underestimate *complexity*.

Below *results* may be demanded (their *masters* are indicated *upstream* from the filled-in *schedule* in the figure above):

1G Common finalized white-box items design with product requisites (only design items for prototype white-box)

Figure 10-16, Setting device diagram, page 484

Figure 10-17, Presenting device diagram, page 485

Figure 10-18, Power supply diagram, page 486

Table 10-46, Multiplication toy common electronics, white-box design items, page 487

Table 10-51, Multiplication toy white-box commonalities, estimated realization material requisites cost, page 490

IG Multiplication toy with hard-wired multiplier finalized white-box items design with product requisites (only design items for prototype white-box)

Figure 10-24, Hard-wired multiplier, arithmetic unit diagram, page 502

Figure 10-25, Hard-wired multiplier, full adder multiplier diagram, page 503

Figure 10-27, Hard-wired multiplier 3-module adder diagram, page 505

Figure 10-26, Hard-wired multiplier, Bin-BCD converter diagram, page 504

Figure 10-28, Multiplication toy with hard-wired multiplier prototype board drawing, page 505

Table 10-63, Multiplication toy with hard-wired multiplier electronics white-box design items, page 506

Table 10-64, Multiplication toy with hard-wired multiplier prototype board white-box design items, page 506

Table 10-65, Hard-wired multiplier in multiplication toy estimated realization material requisites cost, page 507

Table 10-66, Multiplication toy white-box with hard-wired multiplier estimated realization work requisites hourly/fixed/rental cost, page 507

IB Common architecture white-box with major ingredients but no embedded black-boxes

Table 9-25, Multiplication toy variants architecture ingredients resumé, page 388

Figure 9-38, Multiplication toy common logical architecture, page 389

Figure 9-39, Diversifying the multiplier element and circuit board interface, page 390

Figure 9-40, Multiplication toy common physical architecture, page 390

IB Hard-wired multiplier architecture white-box with major ingredients but no embedded black-boxes:

Table 9-26, Hard-wired multiplier architecture ingredients resumé, page 393

Figure 9-42, Hard-wired multiplier logical architecture, page 393

H Picked in-house legacy reusable items

XaH Ordered suppliers' off-the-shelf reusable items

XbH Ordered suppliers' turnkey reusable items

For possible *failure localization* and *failure elimination*, below *results* may be demanded (their *masters* are indicated *upstream* from the filled-in *schedule* in the figure above):

IJ2 Table 11-25, Multiplication toy with hard-wired multiplier start-up report 1, page 627

Table 11-33, Multiplication toy with microcontroller multiplier start-up report 2, page 636

IW2 Table 12-38, Multiplication toy with hard-wired multiplier failure elimination report 1, page 711

Table 12-47, Multiplication toy with microcontroller multiplier failure elimination report 1, page 723

OW2 Table 12-53, Multiplication toy environment (microcontroller multiplier variant) failure elimination report 1, page 728

From the above large list it is obvious that strict orderliness begins to pay off. One can imagine the disorder if *white-box design items* are sloppily handled, described, and documented. And note, this is a simple *product* when it comes to the number of *design items*.

Even if the procurement can be managed a bit sloppily, the trick will be to start the prototype after it has been put together. If the *architecture* isn't interrelated sufficiently, or the *white-box design items* are more or less guessed at, the prototype simply refuses to start. It is not seldom seen that it takes longer to start the prototype than to verify it, because both the design and the prototype contain so many failures to be eliminated before it starts. The saying "You made your bed, now lie in it" becomes very true here.

11.7.3 EXAMPLE Multiplication toy with hard-wired multiplier: Obtain items including linkable program

From unpacking of all ordered *white-box design items*, the below report in Table 11-23 is established.

TABLE 11-23 Multiplication toy with hard-wired multiplier items received deviation report 1

Deviation	Item	Sourcing	Producer	Pcs	€/pc	€/toy	Arch Id
DR.1. 1 000 price referred. Cost more per single piece	i100. Rotor switch	Gross Electro	Pointer	2	4	8	Ab.
DR.2. Replaced with discrete resistors	i102. Resistors 4SIL10k	Gross Electro	Resa	4	0.2	0.8	Ab.
DR.3. Replaced with discrete resistors	i105. Resistors 7SIL1k	Gross Electro	Texx	7	0.2	1.4	Ad.
DR.4. Exchanged to 3 x AA battery bin	i106. Battery bin	Gross Electro	Any	1	0.9	0.9	Ae.
DR.5. Exchanged to 3 x AA batteries	i107. AAA batteries	Gross Electro	Any	3	0.4	1.2	

The rise in cost is reported to the examine integration robustness gate referred to below.

11.7.4 EXAMPLE Multiplication toy with hard-wired multiplier: Integrate elements, including linkable program and embedded black-boxes

To manually wire all the logic capsules is quite an effort, and must be done extremely carefully to avoid mixing up the wires. A short report of the work can be seen in Table 11-24 below.

TABLE 11-24 Multiplication toy with hard-wired multiplier integration report 1

Fulfillments and shortcomings	Interfaces to realize	Responsible provider
IR.1. To hold all components, the printed circuit board was divided into one front board and three plug-in boards, tied together by 90-degree multi-pin connectors. IR.2. The circuit boards were equipped with logic component capsules without problem. Some brackets were added to secure mechanical stability. IR.3. Rotors were soldered without problem to front circuit board printed bus. IR.4. LEDs were mounted without problems on the front circuit board printed bus.	-A-1:1Prototype board -A-3. Operand bus -A-4. Operand bus -A-5. Product bus -Aa:1-1. Bin-BCD interface	PR.10. In-house project
IR.5. Screwed on rotor axes.	-aE-1.aa. Dial knob -aE-1.ab. Dial knob	
IR.6. Soldered to the front circuit board.	-A-6.a. On-off switch	
IR.7. The power is distributed on all circuit board printing.	-A-6. Power feed -A-7. Ground	

11.7.5 EXAMPLE Multiplication toy with hard-wired multiplier: Start up white-box

Finally, it is time to start the prototype (see Table 11-25 below).

TABLE 11-25 Multiplication toy with hard-wired multiplier start-up report 1

Accomplishments and obstacles	Troubling ingredient	Responsible provider
SR.1. At the initial start-up nothing happened on the presenting device when selectors were rotating, because power was not distributed to the arithmetic circuit boards. A wire was soldered.	-A-1:1Prototype board	PR.10. In-house project
SR.2. Toy is starting up, settings can be made on the rotation switches, and products are shown on the presentation device, but seem to be wrong.	Aa:1.a. Arithmetic unit	

11.7.6 EXAMPLE Multiplication toy with hard-wired multiplier: Examine integration robustness

A gate meeting concluded the realization *activities.*

The reported rise in cost doesn't matter, because this prototype will never result in any *products* on a competitive market.

The *start-up report* indicates problems with the arithmetic unit, but can also be a logic failure somewhere else in the path from the rotators to the displays. The prototype is decided to be ready for *verification*, to examine and eliminate such failures.

11.7.7 Multiplication toy: Proceed reading

To follow this example, these are the alternatives (see again Fig. 11-11, p. 624):

- Trace *upstream* through the overall *schedule* **Pt:2:3.** to understand how to fix a multiplication toy with hard-wired multiplier that is found to not be ready for *verification.*

- Proceed to chapter 12.12, "EXAMPLE Multiplication toy with hard-wired multiplier: Verify black-/white-box," page 706.

- To stay within the multiplication toy example, but not following the overall *schedule*, proceed to chapter 11.8 below.

11.8 EXAMPLE Multiplication toy environment (hard-wired multiplier variant): Realize interfaces to environment and insert/await outermost white-box

11.8.1 Multiplication toy environment (hard-wired multiplier variant): Process schedule to use

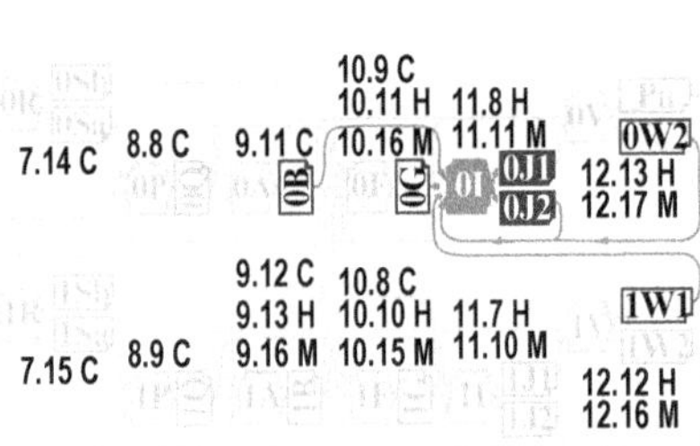

For the multiplication toy example, a special technical overall *schedule* is *tailored* (illustrated in Fig. 11-12, left):

- **Pt:2:3. Tailored advanced technical schedule n = 2,** page 169

In this *tailored schedule* the generic *schedule* to use is (filled-in symbols):

- **Pt.0.I. Realize interfaces to environment and insert/await outermost white-box,** page 122.

FIGURE 11-12 Position of schedule and master to use (filled in) in overall schedule

Below *results* may be demanded (their *masters* are indicated *upstream* from the filled-in *schedule* in the figure above):

1W Multiplication toy with hard-wired multiplier fully verified black-/white-box

0G Common environment interface design items and product requisites

Figure 10-21, Multiplication toy environment attachment, page 496

Table 10-56, Multiplication toy environment interfaces commonalities, interfaces to environment design items, page 496

Table 10-57, Multiplication toy environment interfaces commonalities, interfaces to environment design items, page 497

Table 10-59, Multiplication toy environment interfaces commonalities, estimated realization material requisites cost, page 497

Table 10-60, Multiplication toy environment interfaces commonalities, estimated realization material requisites cost, page 498

[0G] Multiplication toy with hard-wired multiplier environment interface design items and product requisites

Table 10-69, Multiplication toy environment (hard-wired multiplier variant) estimated realization work requisites hourly/fixed/rental cost, page 510

[0B] Common environment existing ingredients and design of interfaces to the environment

Table 9-23, Multiplication toy environment architecture ingredients resumé, page 383

Figure 9-35, Multiplication toy environment logical architecture, page 384

Figure 9-36, Multiplication toy environment physical architecture, page 384

[H] Picked in-house legacy reusable items

[XaH] Ordered suppliers' off-the-shelf reusable items

[XbH] Ordered suppliers' turnkey reusable items

For possible *failure localization* and *failure elimination*, below *results* may be demanded (their *masters* are indicated *upstream* from the filled-in *schedule* in the figure above):

[0J2] Table 11-28, Multiplication toy environment (hard-wired multiplier variant), start-up report 1, page 630

Table 11-36, Multiplication toy environment (microcontroller multiplier variant), start-up report 1, page 640

[0W2] Table 12-53, Multiplication toy environment (microcontroller multiplier variant) failure elimination report 1, page 728

11.8.2 EXAMPLE Multiplication toy environment (hard-wired multiplier variant): Obtain items for interface to environment and ported linkable program

Batteries were ordered and have been received (see Table 11-28 below).

TABLE 11-26 Multiplication toy environment (hard-wired multiplier variant), items received deviation report 1

Deviation	Item	Supplier	Element
DR.1. Batteries are received without deviations.	Capacity battery with extension cable	SU.18. In-house storage	Ae. Power supply

11.8.3 EXAMPLE Multiplication toy environment (hard-wired multiplier variant): Complete and connect interfaces to environment

The only *realization* work to be done is to attach the power battery and get it charged.

TABLE 11-27 Multiplication toy environment (hard-wired multiplier variant), integration report 1

Fulfillments and shortcomings	Interfaces to realize	Responsible provider
IR.1. The battery was connected without problems	-A-6. Power feed interface extended to environment	PR.12. Mechano

11.8.4 EXAMPLE Multiplication toy environment (hard-wired multiplier variant): Start interfaces to environment for evaluation

The prototype is started with a battery to feed it (see Table 11-28 below).

TABLE 11-28 Multiplication toy environment (hard-wired multiplier variant), start-up report 1

Accomplishments and obstacles	Troubling ingredient
SR.1. As expected, the prototype starts up correctly.	-A-6. Power feed interface extended to environment

11.8.5 EXAMPLE Multiplication toy environment (hard-wired multiplier variant): Examine prototype reliability

The multiplication toy with hard-wired multiplier is not intended for typical *users* in the *environment*, but is to be used for in-house *verification* of correct behavior of network of logic circuits. A short gate meeting was satisfied with the current incomplete *environment interfaces*.

11.8.6 EXAMPLE Multiplication toy: Proceed reading

To follow this example, these are the alternatives (see again Fig. 11-12, p. 628):

- Trace *upstream* through the overall *schedule* **Pt:2:3.** to understand how to fix the multiplication toy environment (hard-wired multiplier variant) that is found to not be ready for field *verification*.
- Proceed to chapter 12.13, "EXAMPLE Multiplication toy environment (hard-wired multiplier variant): Verify prototype in environment," page 712.
- To leave the multiplication toy example and to dig deeper into the chapter 11, "Realize & integrate white-box" topic, proceed to chapter 11.9 below.

11.9 Integrate assembler programs in a white-box

11.9.1 Integrate detailed process schedule

The *process* to procure executable *machine programs* from *assembler programs* is rather simple (see Fig. 11-13 below). This figure illustrates how the mechanics and electronics can also be integrated to constitute a complete *white-box*.

For a thorough explanation of a complete *integration*, see the *schedule* **Pt.n.I.** fully expanded in Figure 6-28, page 148.

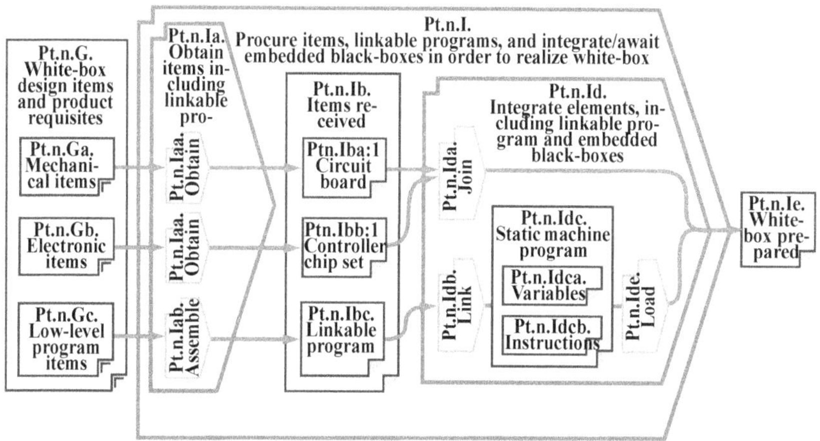

FIGURE 11-13 Process for assembling, linking, and loading microcontroller

11.10 EXAMPLE Multiplication toy with microcontroller multiplier: Procure items and linkable programs in order to realize white-box

11.10.1 EXAMPLE Multiplication toy: About this example

Note that the multiplication toy variant with microcontroller has both electronic *design items* and executable *machine programs* to be realized and integrated.

11.10.2 EXAMPLE Multiplication toy with microcontroller multiplier: Process schedule to use

For the multiplication toy example, a special technical overall *schedule* is *tailored* (illustrated in Fig. 11-14, left):

- **Pt:2:3. Tailored advanced technical schedule n = 2,** page 169

In this *tailored schedule* the generic *schedule* to use is (filled-in symbols):

- **Pt.n.I. Procure items, linkable programs, and integrate/await embedded black-boxes in order to realize white-box,** page 144 and expanded *schedule* page 147.

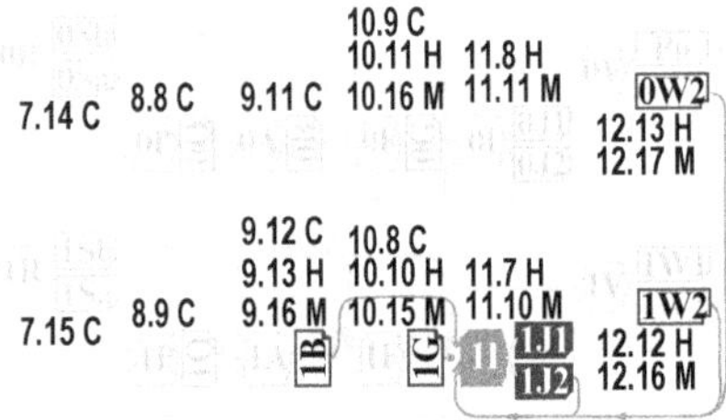

FIGURE 11-14 Position of schedule and master to use (filled in) in overall schedule

Below *results* may be demanded (their *masters* are indicated *upstream* from the filled-in *schedule* in the figure above):

1G Common finalized white-box items design with product requisites (only design items for prototype white-box)

Figure 10-16, Setting device diagram, page 484

Figure 10-17, Presenting device diagram, page 485

Figure 10-18, Power supply diagram, page 486

Table 10-46, Multiplication toy common electronics, white-box design items, page 487

Table 10-47, Multiplication toy with microcontroller multiplier casing, white-box design items, page 487

Table 10-51, Multiplication toy white-box commonalities, estimated realization material requisites cost, page 490

Table 10-52, Multiplication toy white-box commonalities, estimated realization material requisites cost, page 491

Table 10-53, Multiplication toy white-box commonalities, estimated realization material requisites cost, page 491

1G Multiplication toy with microcontroller multiplier finalized white-box items design with product requisites (only design items for prototype white-box)

Figure 10-31, Microcontroller multiplier diagram, page 520

Figure 10-32, Microcontroller multiplier assembler program with flow chart, page 521

Figure 10-33, Printed circuit board drawing, page 522

Figure 10-19, Multiplication toy casing, page 486

Table 10-74, Multiplication toy with microcontroller multiplier electronics white-box design items, page 522

Table 10-75, Multiplication toy with microcontroller multiplier printed board white-box design items, page 522

Table 10-78, Microcontroller multiplier for multiplication toy, estimated realization material requisites cost, page 523

Table 10-79, Multiplication toy with microcontroller multiplier, estimated realization work requisites hourly/fixed/rental cost, page 524

1B Common architecture white-box with major ingredients but no embedded black-boxes

Table 9-25, Multiplication toy variants architecture ingredients resumé, page 388

Figure 9-38, Multiplication toy common logical architecture, page 389

Figure 9-39, Diversifying the multiplier element and circuit board interface, page 390

Figure 9-40, Multiplication toy common physical architecture, page 390

1B Microcontroller multiplier architecture white-box with major ingredients but no embedded black-boxes

Table 9-28, Microcontroller multiplier refined architecture ingredients resumé, page 398

Figure 9-46, Microcontroller multiplier logical architecture, page 399

H Picked in-house legacy reusable items

XaH Ordered suppliers' off-the-shelf reusable items

XbH Ordered suppliers' turnkey reusable items

For possible *failure localization* and *failure elimination*, below *results* may be demanded (their *masters* are indicated *upstream* from the filled-in *schedule* in the figure above):

1J2 Table 11-25, Multiplication toy with hard-wired multiplier start-up report 1, page 627

Table 11-33, Multiplication toy with microcontroller multiplier start-up report 2, page 636

1W2 Table 12-38, Multiplication toy with hard-wired multiplier failure elimination report 1, page 711

Table 12-47, Multiplication toy with microcontroller multiplier failure elimination report 1, page 723

0W2 Table 12-53, Multiplication toy environment (microcontroller multiplier variant) failure elimination report 1, page 728

11.10.3 EXAMPLE Multiplication toy with microcontroller multiplier: Obtain items including linkable program

All items have been ordered and are now received and checked. Deviations from what is expected are reported (see Table 11-29 below). Assembled *programs* to be linked are also received, and found without deviations.

TABLE 11-29 Multiplication toy with microcontroller multiplier items received deviation report 1

| Volume | | | | | 1 000 | | 100 000 | | |
Deviation	Item	Sourc-ing	Produc-er	Piec-es	€/piece	Cost €/toy	€/piece	Cost €/toy	Id Arch
DR.1. Difficult to solder on printed board, must be improved.	Rotor switch	Gross Electro	Pointer	2	0.4	0.8	0.2	0.4	Ab.
DR.2. To be phased out by supplier	LED-segment, no dp.	Gross Electro	Texx	3	0.3	0.9	0.2	0.6	Ad.
DR.3. All ordinary bins too high. Order specialized immediately.	Battery bin	Gross Electro	Any	1	0.2	0.2	0.12	0.12	Ae.
DR.4. Up to 4-month delivery time. In the meantime, use manually produced fronts, backs, and flaps for prototypes.	Pressed front housing	In-house	In-house	1	0.2	0.2	0.1	0.1	]A[2.a.
	Front housing tool	In depth press	In depth press	1 per 50 000	5	5	0.1	0.1	]A[2.a.
	Back housing	In-house	In-house	1	0.1	0.1	0.06	0.06	]A[2.b.
	Back housing tool	In depth press	In depth press	1 per 50 000	5	5	0.1	0.1	]A[2.b.
DR.5. Produced by same tool as above housing	Cut battery flap	In-house	In-house	1	0.025	0.025	0.02	0.02	-aE-4.a.
DR.6. Maximum capacity 4 000 pieces / week	Display window	Gross Electro	Designer	1	0.02	0.02	0.015	0.015	-aE-1.ca.
DR.7. There might be a better microcontroller in this huge family.	PIC16F57	Pick-it	MiChip	1	0.7	0.7	0.4	0.4	Aa:2a.

TABLE 11-29 Multiplication toy with microcontroller multiplier items received deviation report 1

| Volume | | | | | 1 000 | | 100 000 | | |
Deviation	Item	Sourc-ing	Produc-er	Piec-es	€/ piece	Cost € / toy	€/ piece	Cost € / toy	Id Arch
DR.8. Up to 2-month delivery time. Use expensive fast-produced PCB for prototypes.	PCB	In-house	PCB inc.	1	0.4	0.4	0.2	0.2	-A-1:2
DR.9. Add probe area for automated tests.	PCB preparation	In-house	In-house	1	0.4	0.4	0.01	0.01	-A-1:2

11.10.4 EXAMPLE Multiplication toy with microcontroller multiplier: Integrate elements, including linkable program and embedded black-boxes

The prototype is now built (see report Table 11-30 below). Contrary to the multiplication toy variant with hard-wired multiplier, an ordinary printed board is now prepared, which is much easier to manually equip.

Since the casing is decided to belong to the *outermost black-/white-box*, the toy is provided with casing including all *user interfaces* on the casing front panel.

TABLE 11-30 Multiplication toy with microcontroller multiplier integration report 1

Fulfillments and shortcomings	Interfaces to realize	Responsible provider
IR.1. The circuit board was equipped and soldered without problems. IR.2. Rotors were soldered without problem to front circuit board printed bus. IR.3. LEDs were mounted without problems on the front circuit board printed bus.	-A-1:2Printed board -A-3. Operand bus -A-4. Operand bus -A-5. Product bus -Aa:1-1. Bin-BCD interface	PR.10. In-house project
IR.4. Riveted on rotator axis	-aE-1.aa.Dial knob	
IR.5. Soldered to the printed circuit board. Note, there are components on both sides of the PCB.	-A-6.a.On-off switch	
IR.6. Manually produced front mounted	]A[2.a.Toy front housing	
IR.7. Manually produced back mounted	]A[2.b.Toy back housing	
IR.8. Manually produced plexiglas mounted	-aE-1.ca.Display window	
IR.9. Manually produced flap mounted	-aE-4.a.Battery flap	
IR.10. Ordinary screw temporarily used	-aE-4.b.Flap screw	
IR.11. The power is distributed on all circuit board printing.	-A-6.Power feed -A-7.Ground	
IR.12. Loading the machine instructions to microcontroller. The transfer protocol is rather tricky and a specialist from the supplier was called in to get it right the first time.	-Aa:2-1.In-circuit programmable port	PR.15. MicroHard

11.10.5 EXAMPLE Multiplication toy with microcontroller multiplier: Start up white-box

The *assembler program* is written for the specific *target* microcontroller, so no *porting* of the *program* is necessary. After running the assembler and linker tool, the *machine program* can just be loaded to the in-circuit programmable port, and the toy is ready to be started (see Table 11-31 below).

TABLE 11-31 Multiplication toy with microcontroller multiplier
start-up report 1

Accomplishments and obstacles	Troubling ingredient	Responsible provider
SR.1. Toy is powered up but nothing seems to happen. The LEDs are not lighted up.	-A-1:2Printed board	PR.15. MicroHard

11.10.6 EXAMPLE Multiplication toy with microcontroller multiplier: Examine integration robustness

A gate meeting was gathered to decide what to do with the dead prototype. The prototype was sent back to the in-house project for an inspection of the equipped and soldered PCB.

11.10.7 EXAMPLE Multiplication toy with microcontroller multiplier: Integrate elements, including linkable program and embedded black-boxes

The PCB circuit plating was found to be slightly loose, and sometimes the toy stopped to work (see Table 11-32).

TABLE 11-32 Multiplication toy with microcontroller multiplier
integration report 2

Fulfillments and shortcomings	Interfaces to realize	Responsible provider
IR.1. The observed failure could be eliminated by scratching with a knife and inserting a jumper wire.	-A-1:2Printed board	PR.10. In-house project

11.10.8 EXAMPLE Multiplication toy with microcontroller multiplier: Again start up white-box

Now the outcome gets better (see Table 11-33 below).

TABLE 11-33 Multiplication toy with microcontroller multiplier
start-up report 2

Accomplishments and obstacles	Troubling ingredient	Responsible provider
SR.1. Toy is starting up, but wrong digits are shown on the presentation device. The presentation shifts incomprehensibly when different operands are selected.	Aa:2aa. Microcontroller program	PR.15. MicroHard

11.10.9 EXAMPLE Multiplication toy with microcontroller multiplier: Examine integration robustness

Now the gate meeting probably faced a *program* failure, and decided to send the prototype for *verification*.

11.10.10 EXAMPLE Multiplication toy: Proceed reading

To follow this example, these are the alternatives (see again Fig. 11-14, p. 632):

- Trace *upstream* through the overall *schedule* **Pt:2:3.** to understand how to fix a multiplication toy with microcontroller multiplier that is found to not be ready for *verification*.

- Proceed to chapter 12.16, "EXAMPLE Multiplication toy with microcontroller multiplier: Verify black-/white-box," page 717.

- To stay within the multiplication toy example, but not following the overall *schedule*, proceed to chapter 11.11 below.

11.11 EXAMPLE Multiplication toy environment (microcontroller multiplier variant): Realize interfaces to environment and insert/await outermost white-box

11.11.1 Multiplication toy environment (microcontroller multiplier variant): Process schedule to use

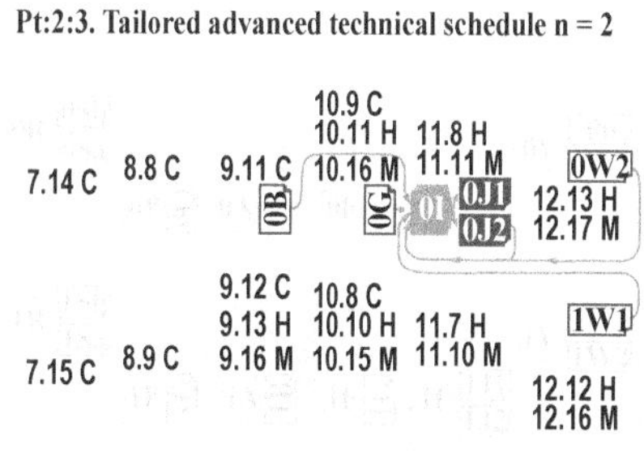

For the multiplication toy example, a special technical overall *schedule* is *tailored* (illustrated in Fig. 11-15, left):

- **Pt:2:3. Tailored advanced technical schedule n = 2,** page 169

In this *tailored schedule* the generic *schedule* to use is (filled-in symbols):

- **Pt.0.I. Realize interfaces to environment and insert/await outermost white-box,** page 122.

FIGURE 11-15 Position of schedule and master to use (filled in) in overall schedule

Below *results* may be demanded (their *masters* are indicated *upstream* from the filled-in *schedule* in the figure above):

1W   Multiplication toy with microcontroller multiplier fully verified black-/white-box

0G Common environment interface design items and product requisites (only
design items for prototype white-box)

Figure 10-21, Multiplication toy environment attachment, page 496

Table 10-56, Multiplication toy environment interfaces commonalities, interfaces to environment design items, page 496

Table 10-57, Multiplication toy environment interfaces commonalities, interfaces to environment design items, page 497

Table 10-59, Multiplication toy environment interfaces commonalities, estimated realization material requisites cost, page 497

Table 10-60, Multiplication toy environment interfaces commonalities, estimated realization material requisites cost, page 498

0G Multiplication toy with microcontroller multiplier environment interface
design items and product requisites (only design items for prototype white-
box)

Table 10-82, Multiplication toy environment (microcontroller multiplier variant) estimated realization work requisites hourly/fixed/rental cost, page
527

0B Common environment existing ingredients and design of interfaces to the
environment

Table 9-23, Multiplication toy environment architecture ingredients resumé,
page 383

Figure 9-35, Multiplication toy environment logical architecture, page 384

Figure 9-36, Multiplication toy environment physical architecture, page 384

H Picked in-house legacy reusable items

XaH Ordered suppliers' off-the-shelf reusable items

XbH Ordered suppliers' turnkey reusable items

For possible *failure localization* and *failure elimination*, below *results* may be demanded (their *masters* are indicated *upstream* from the filled-in *schedule* in the figure above):

0J2 Table 11-28, Multiplication toy environment (hard-wired multiplier variant),
start-up report 1, page 630

Table 11-36, Multiplication toy environment (microcontroller multiplier variant), start-up report 1, page 640

0W2 Table 12-53, Multiplication toy environment (microcontroller multiplier variant) failure elimination report 1, page 728

11.11.2 EXAMPLE Multiplication toy environment (microcontroller multiplier variant): Obtain items for interface to environment and ported linkable program

One of the received *environment interfaces design items* caused a deviation (see Table 11-34 below).

TABLE 11-34 Multiplication toy environment (microcontroller multiplier variant), items received deviation report 1

Deviation	Item	Sourcing	Producer	Count	Unit	€/Unit	€
DR.1. Sold out at Gross Hardware. Alternative supplier Standby HW cost € 0.32 / unit	Permanent rivet 2BS	Standby HW	Steel & Co	200	Pcs	0.25	50

The deviation report is forwarded to the coming gate meeting.

11.11.3 EXAMPLE Multiplication toy environment (microcontroller multiplier variant): Complete and connect interfaces to environment

The *environment interfaces* are now realized (see Table 11-35 below).

TABLE 11-35 Multiplication toy environment (microcontroller multiplier variant), integration report 1

Fulfillments and shortcomings	Interfaces to realize	Responsible provider
IR.1. Some cuddle multiplication toys are mounted to various cuddle toys for verification.	-E-1. Host attachments	PR.12. Mechano
IR.2. The host collar is easy to apply to any soft and unstable hosting toy and fits well to the multiplication toy.	-E-1a. Host collar frame -E-1c. Asymmetric holes (4x)	
IR.3. The gasket to prevent fluff to leak fits well between the cuddle toy boundary and the collar.	-E-1b. Fluff barrier	
IR.4. The rivets are easy to apply with a riveting hammer.	-E-1d. Permanent rivet (4x)	
IR.5. The batteries fit well in the bin.	-aE-4. Battery exchange	

11.11.4 EXAMPLE Multiplication toy environment (microcontroller multiplier variant): Start interfaces to environment

Finally, the *interfaces* to the multiplication toy environment are started up (see Table 11-36 below).

TABLE 11-36 Multiplication toy environment (microcontroller multiplier variant), start-up report 1

Accomplishments and obstacles	Troubling ingredient	Responsible provider
SR.1. The power switch is set to on, and the multiplication toy display illuminates nicely, and responds correctly when rotating switches are turned. The cuddly toy looks very nice.	None	None

11.11.5 EXAMPLE Multiplication toy environment (microcontroller multiplier variant): Examine prototype reliability

The multiplication toy was evaluated by the gate members. The environment interfaces seemed to work properly and the multiplication toy was decided to be sent for *verification*.

11.11.6 EXAMPLE Multiplication toy: Proceed reading

To follow this example, these are the alternatives (see again Fig. 11-14, p. 632):

- Trace *upstream* through the overall *schedule* **Pt:2:3.** to understand how to fix a multiplication toy environment (microcontroller multiplier variant) that is found to not be ready for field *verification*.

- Proceed to chapter 12.17, "EXAMPLE Multiplication toy environment (microcontroller multiplier variant): Verify prototype in environment," page 725.

- To leave the multiplication toy example and to dig deeper into the chapter 11, "Realize & integrate white-box" topic, proceed to chapter 11.12 below.

11.12 Integrate all kind of elements in a white-box

11.12.1 Integrate detailed process schedule

The *process* to procure executable *machine programs* from assemblers, compilers, and *program libraries* is rather simple to illustrate (see Fig. 11-16 below). For big *programs* with a lot of *program* files to keep track of, it might need automated *configuration management*.

All used technologies must integrate seamlessly with the complex prototype.

For a thorough explanation of a complete *integration*, see *schedule* **Pt.n.I.** fully expanded in Figure 6-28, page 148.

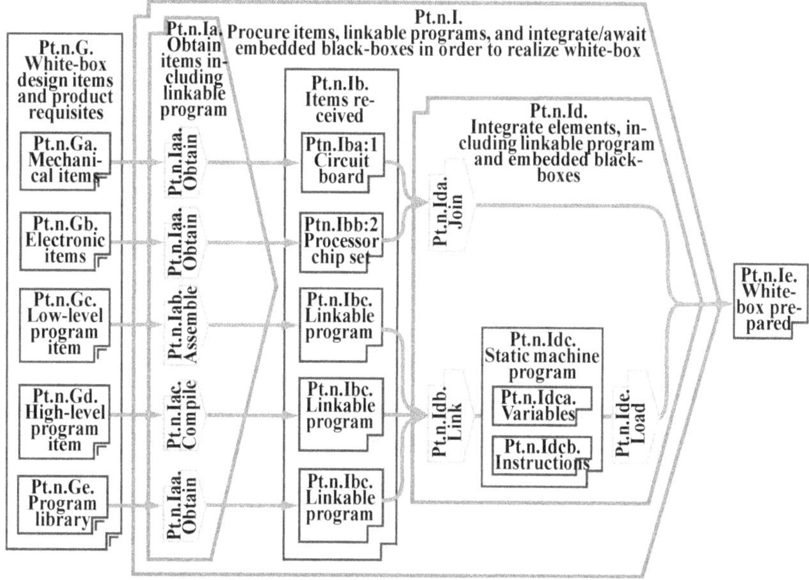

FIGURE 11-16 Process for assembling, linking, and loading microprocessor computer

11.13 Reuse and porting

11.13.1 Applying reuse

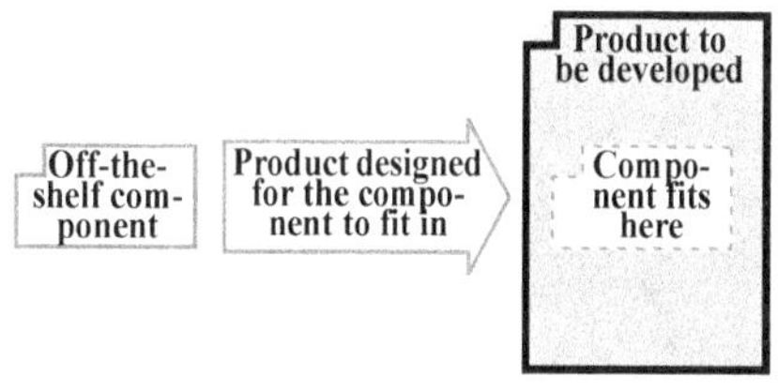

Many *elements* in traditional engineering are *reusable items*, such as windows, doors, and bricks for mechanics, and resistors and capacitors for electronics (see Fig. 11-17, left). Programming is still a young discipline, and has not yet formed any significant market for *reuse*, and it remains to be seen if this will happen or not.

FIGURE 11-17 Typical component reuse situation

11.13.2 Preparing for reuse

However, in the calc_logic example there is a decision to give *program reuse* a try, and the calc_logic *function* is decided to be built as a *component* for *reuse*. For the moment, calc_logic is decided to be offered only to one stand-alone calculator and to one Windows calculator, and in the future may be offered for public *reuse*.

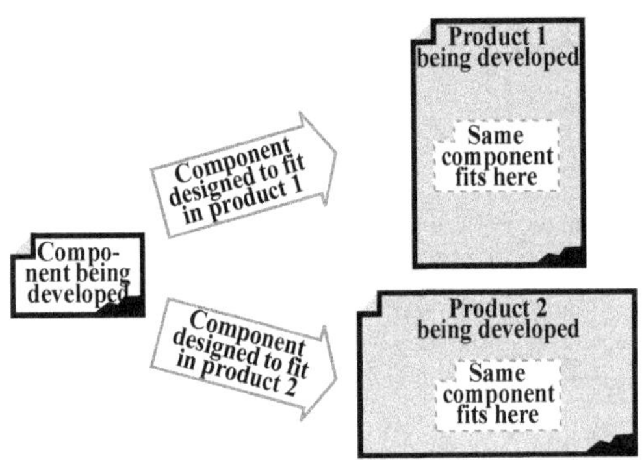

Since the stand-alone calculator and the Windows calculator are being developed in parallel with the calc_logic *reusable item*, there is a possibility for the *designers* to gain from each other.

Observe that the *environment interfaces* of the *component* for *reuse* are identical to the *target interface* to the *component* (see Fig. 11-18, left).

FIGURE 11-18 Develop the same component to fit in many different products

The conclusion is that the *reusable item* can be developed together with the *targets* for *reuse*, to achieve the best way to fit them both. This also means that the two *products'* environments to the *component* have the same *interfaces* and behavior. It may be decided in the future that the *reusable items* must be reusable also in other *target products*. If so, hopefully this increased generality can be achieved by the *component* with few if any updates.

If the *reusable item* is a *program*, one of the constraints might be how it fits to a *target product*, having different controller or processor chip sets or using different *program development systems*. This is about portability (see next chapter).

11.13.3 Host, simulator, and target

Often, the *environment interfaces* of a *product* are developed by the same *development system* as the *product* itself (*host* and *target* are the same), and the development of the *product* just continues seamlessly from innermost *nesting levels* to environment *nesting level*.

When preparing a compiled and linked *program* to a stand-alone *target*, having a lot of interaction with surrounded electronics, it is often far too big a leap to get the prototype reliably started up on the *target*.

Development and *realization* of a prototype might be far too *complex* to perform from scratch in the *target environment*. Most often and when possible, the *realization* of a prototype is performed in a *hosted environment*, where the context is much more capable and convenient. This is specially true for *target realization* of tiny electronics with real-time *programs*.

When the *realization* of a *program* is done in a *host environment*, provided with powerful *development systems* with compilers and *program libraries*, it is at the end *ported* to the final *target environment*. If the *program* interaction is *complex* and/ or the *target* scanty, this *process* can be divided into several steps to gradually approach the *target* (see Fig. 11-20 below).

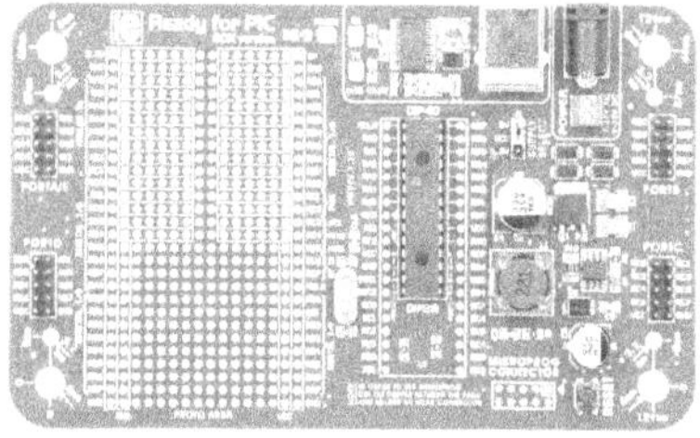

Programs can first be developed on a debugger in the *host* for finding elementary failures in the *program* logic. Then it can be further *ported* to a simulator, emulator, or *development board*, sometimes provided from the microcontroller vendor (see Fig. 11-19, left).

FIGURE 11-19 Development board "Ready for PIC" ™ supplied by MikroElektronika

The *development board* is not only practical for *target integration* of *programs* and electronics. It often has an area for custom-defined electronics which can be used for *prototyping* to gradually understand the microcontroller, in particular if used together with an in-circuit debugging tool.

When the *program* has been started and verified in the simulator, it can finally be *ported* to the *target* prototype. Now the *program* is almost bug-free, and hopefully only few and easy problems remain in the final cramped *target* prototype.

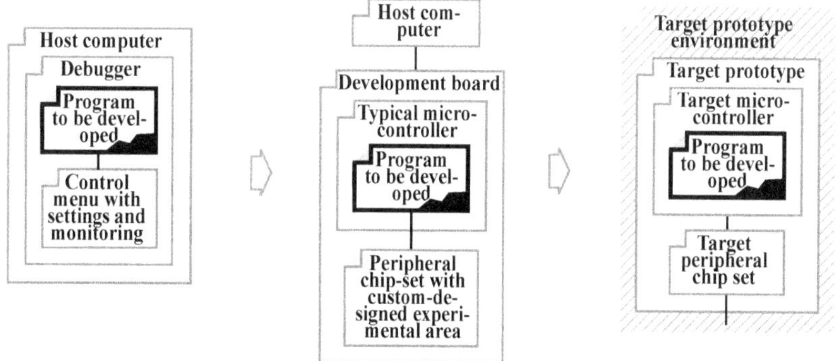

FIGURE 11-20 Porting programs stepwise to reach target environment

Developing and moving a *white-box* outside the *host* in three steps can be supported by a *tailored schedule* for this purpose, as shown in **Pt:2:3b. Tailored advanced technical schedule n = 2, porting to 3 targets (right half),** page 170.

11.14 EXAMPLE: Hosted calc_logic for reuse: Procure items and linkable programs in order to realize white-box

11.14.1 EXAMPLE Calc_logic: About this example

When *porting programs* to different *targets*, it is not only the *environment interfaces* that change. The processor in a Windows computer, the microcontroller in a *development board*, and the microcontroller in the *target* pocket calculator differ a lot, which means that all *programs* must be recompiled and linked again before they can execute on the different machines.

11.14.2 EXAMPLE Hosted calc_logic for reuse: Process schedule to use

For the calc_logic example, a special technical overall *schedule* is *tailored* (illustrated in Fig. 11-21, below):

- Pt:2:3b. *Tailored advanced technical schedule n = 2, porting to 3 targets (right half)*, page 170

In this *tailored schedule* the generic *schedule* to use is (filled-in symbols below):

- Pt.n.I. *Procure items, linkable programs, and integrate/ await embedded black-boxes in order to realize white-box*, page 144.

- Also see the *schedule* **Pt.n.I.** fully expanded in Figure 6-28, page 148.

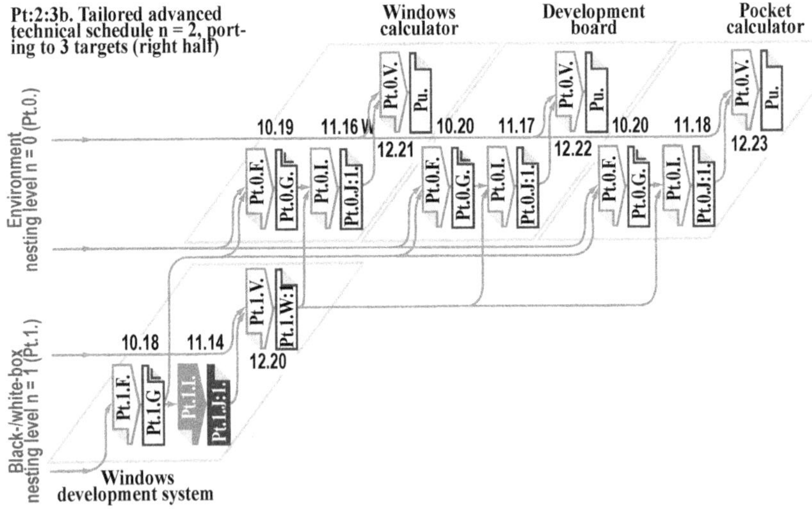

FIGURE 11-21 Position of schedule and master to use (filled in) in overall schedule

11.14.3 EXAMPLE: Hosted calc_logic for reuse: Obtain items including linkable program

The calc_logic file package was compiled by the in-house project "dg3" (see Table 11-37 below).

TABLE 11-37 Calc_logic for reuse items received deviation report 1

Deviation	Item	Sourcing
DR.1. Received compiled file package without problem.	i100. Hosted portable calc_logic programs on CD disc	SU.11. In-house project "dg3"
	i101. Adapted header files for Windows target	SU.11. In-house project "dg3"

11.14.4 EXAMPLE: Hosted calc_logic for reuse: Integrate elements, including linkable program and embedded black-boxes

The *linkable program* files shall now be integrated with *development system libraries* for standard *functions* and debug possibilities (see Table 11-38 below).

TABLE 11-38 Calc_logic for reuse integration report 1

Fulfillments and shortcomings	Interfaces to realize	Responsible provider
IR.1. There are few interfaces within the calc_logic program. The linker integrated the compiled calc_logic with development system libraries without problems.	E.b. Built-in standard library interface	PR.12. In-house project "dg3"

11.14.5 EXAMPLE Hosted calc_logic for reuse: Start up white-box

The calc_logic *machine program* is now residing in the *host* debugger, which can be started (see Table 11-39 below).

TABLE 11-39 Calc_logic for reuse start-up report 1

Accomplishments and obstacles	Troubling ingredient	Responsible provider
SR.1. The main development of the calc_logic program was made in Windows standard C development tool, and the program starts without problems in this debugger.	A. Calc_logic for reuse	PR.12. In-house project "dg3"

11.14.6 EXAMPLE Hosted calc_logic for reuse: Examine integration robustness

So far there is no trouble with calc_logic. It starts in the *host* debugger and was decided by the gate members to be sent for *verification*.

11.14.7 EXAMPLE Calc_logic for reuse: Proceed reading

To follow this example, these are the alternatives:

- Go back through the calc_logic example chapters to understand how to fix a *hosted* calc_logic for reuse that is not ready for *verification*.
- Proceed to chapter 12.20, "EXAMPLE Hosted calc_logic for reuse: Verify black-/white-box," page 733.
- Proceed to chapter 11.15 "EXAMPLE: Hosted calc_logic for reuse: Porting to Windows application, development board, and pocket calculator" below.

11.15 EXAMPLE: Hosted calc_logic for reuse: Porting to Windows application, development board, and pocket calculator

The *reusable item* calc_logic development is *hosted* in a PC with Windows operating system and C *development system*. The *reusable item* calc_logic *program* should be *ported* and used by two different *products*, one Windows application and one stand-alone pocket calculator. *Porting* to a Windows application is easy, since the *reusable item* calc_logic *program* is already *hosted* on a Windows computer. *Porting* to the *target* pocket calculator is more difficult, and it is planned to first *port* the *program* to a *development board* before finally being *ported* to the *target* pocket calculator (see Fig. 11-22 below).

To be reliable and safe for *reuse*, the behavior must be exactly the same regardless of where calc_logic is *ported*. However, there might be differences in the *products* that *reuse* calc_logic, such as different processors or different *development systems* with different compilers and *development system libraries*.

To adapt to these differences, a dedicated header file can be prepared and changed for the different *portings*. Since the *reusable item* is a *source system*, it is also imaginable to insert compiler switches in the *reusable item program instructions*, to handle differences in the *products* reusing the *component*. But beware, when changing the *program reusable item*, it might very quickly lose its generality and reusability may get very limited.

11.16 EXAMPLE Calc_logic environment (the Windows calculator application): Realize interfaces to environment and insert/await outermost white-box

11.16.1 Calc_logic environment (the Windows calculator application): Process schedule to use

For the calc_logic example, a special technical overall *schedule* is *tailored* (illustrated in Fig. 11-22, below):

- **Pt:2:3b. Tailored advanced technical schedule n = 2, porting to 3 targets (right half),** page 170

In this *tailored schedule* the generic *schedule* to use is (filled-in symbols below):

- **Pt.0.I. Realize interfaces to environment and insert/await outermost white-box,** page 122.

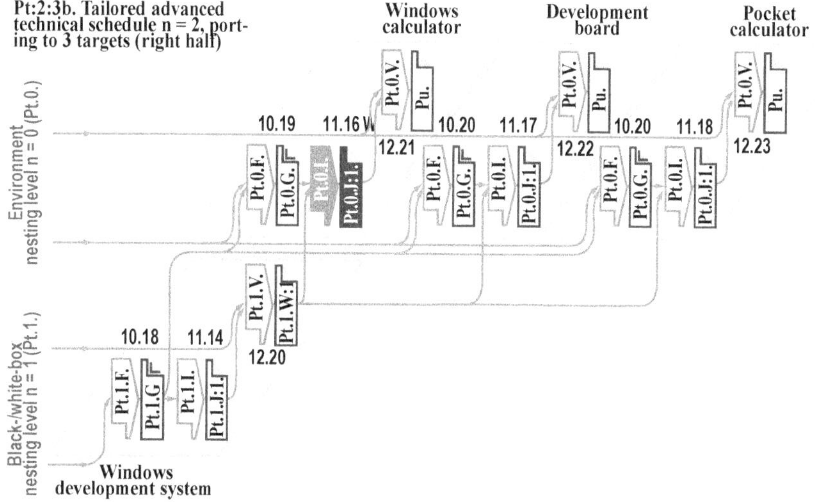

FIGURE 11-22 Position of schedule and master to use (filled in) in overall schedule

Note that the calc_logic is only a small *program* used by the large Windows calculator application *program*, which resides in the *target* Windows computer *development system* (see Fig. 9-58, calc_logic environment logical architecture, p. 417).

11.16.2 EXAMPLE Calc_logic environment (the Windows calculator application): Obtain items for interface to environment and ported linkable program

The calc_logic *program* is developed by the in-house project "dg3" *hosted* on a Windows *development system*, which is the same *development system* as is used by in-house project "dw2" to develop the calculator Windows application. Consequently, it is a very small *porting* effort to receive calc_logic and include it in the in-house project "dw2" calculator *development system* (see Table 11-40 below)

TABLE 11-40 Calc_logic environment (the Windows calculator application) items received deviation report 1

Deviation	Item	Sourcing
DR.1. The Windows calculator application is prepared for many more buttons and far more arithmetic than calc_logic is currently specified to perform. Delivered version will have no problem receiving any unspecified combinations of stimuli, and will simply return a null pointer if this happens. **DR.2. No deviations are noted when recompiling the calc_logic file package in the Windows calculator development system.**	i100. Hosted portable calc_logic programs on CD disc	SU.11. In-house project "dg3"
	i101. Adapted header files for Windows target	SU.11. In-house project "dg3"

11.16.3 EXAMPLE Calc_logic environment (the Windows calculator application): Complete and connect interfaces to environment

When calc_logic is in place at the calculator application *development system*, there is no big problem in linking it to the application (see Table 11-41 below).

TABLE 11-41 Calc_logic environment (the Windows calculator application) integration report 1

Fulfillments and shortcomings	Interfaces to realize	Responsible provider
IR.1. Within Windows Visual C/C++, there was no problem in integrating and linking the calc_logic program into the Windows pocket calculator application.	-E-1. Calc_logic interface to Windows calculator program	PR.18. In-house project "dw2"

11.16.4 EXAMPLE Calc_logic environment (the Windows calculator application): Start interfaces to environment

The calc_logic *program* has already been executed a lot when verified as a *black-/white-box*, and has no problem functioning in the *target* calculator application (see Table 11-42 below).

TABLE 11-42 Calc_logic environment (the Windows calculator application) start-up report 1

Accomplishments and obstacles	Troubling ingredient	Responsible provider
SR.1. For the in-house project "dw2" it was a bit challenging to start up the Windows calculator with its graphic man-machine-interface. **When that was done, it was no problem to start the Windows calculator, and make correct calls to calc_logic.**	-E-1. Calc_logic interface to Windows calculator program	PR.18. In-house project "dw2"

11.16.5 EXAMPLE Calc_logic environment (the Windows calculator application): Examine prototype reliability

After the Windows calculator application was adapted to calc_logic, it integrated perfectly, and there were no problems making calls to calc_logic from its surroundings.

11.16.6 EXAMPLE Calc_logic for reuse: Proceed reading

To follow this example, these are the alternatives:

- Go back through the calc_logic example chapters to understand how to fix a calc_logic environment (the windows calculator application) that is not ready for field *verification.*
- Proceed to chapter 12.21, "EXAMPLE Calc_logic environment (the Windows calculator application): Verify prototype in target environment," page 758.
- Proceed to chapter 11.17 "EXAMPLE Calc_logic environment (the development board fixture firmware): Realize interfaces to environment and insert/ await outermost white-box" below.

11.17 EXAMPLE Calc_logic environment (the development board fixture firmware): Realize interfaces to environment and insert/await outermost white-box

11.17.1 EXAMPLE Calc_logic: About this example

The microcontroller *development board* has been used as a facilitating middle station between development of calc_logic on a Windows PC and verifying it in the *target* pocket calculator. The in-house project "de1" included some effort into preparing the experimental area of the *development board* with 16 buttons and 8 LED displays.

The calc_logic *program* is surrounded by a small *assembler program fixture*, which is reading the buttons, calling calc_logic, and showing the result on the display, to verify that the calc_logic *program* is working.

11.17.2 EXAMPLE Calc_logic environment (the development board fixture firmware): Process schedule to use

For the calc_logic example, a special technical overall *schedule* is *tailored* (illustrated in Fig. 11-23, below):

- **Pt:2:3b. Tailored advanced technical schedule n = 2, porting to 3 targets (right half),** page 170

In this *tailored schedule* the generic *schedule* to use is (filled-in symbols below):

- **Pt.0.I. Realize interfaces to environment and insert/await outermost white-box,** page 122.

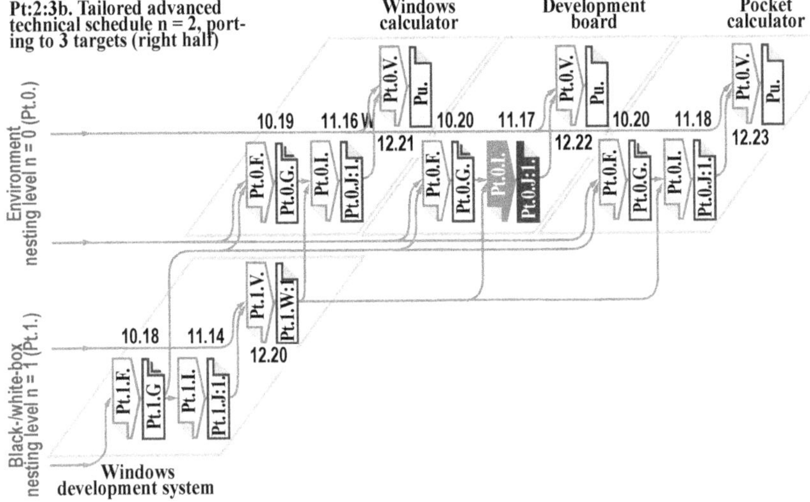

 Position of schedule and master to use (filled in) in overall schedule

Note that the calc_logic *environment* is the pocket calculator *program* surrounding it (see Fig. 9-54, calc_logic environment logical architecture, p. 410).

11.17.3 EXAMPLE Calc_logic environment (the development board fixture firmware): Obtain items for interface to environment and ported linkable program

To obtain *linkable program*, the calc_logic C-*program* is delivered to the in-house project "de1" to be recompiled with the *development board* microcontroller C-compiler (see Table 11-43 below).

TABLE 11-43 Calc_logic environment (the pocket calculator firmware), items received deviation report 1

Deviation	Item	Sourcing
DR.2. Some naming problems with the built-in standard library were identified and rather easily solved by a change in calc_logic, without losing its generality.	i100. Hosted portable calc_logic programs on CD disc	SU.11. In-house project "dg3"

11.17.4 EXAMPLE Calc_logic environment (the development board fixture firmware): Complete and connect interfaces to environment

Now the microcontroller linker should integrate the recompiled calc_logic *program* with the *fixture program* (see Table 11-44 below).

TABLE 11-44 Calc_logic environment (the pocket calculator firmware), integration report 1

Fulfillments and shortcomings	Interfaces to realize	Responsible provider
IR.1. Linker for microcontroller had problems linking the calc_logic linkable program with the program fixture linkable program. The supplier of the microcontroller assembler tool supported a correct solution.	-E-1. Calc_logic interface to pocket calculator program	PR.17. In-house project "de1" PR.16. Chk&Go

11.17.5 EXAMPLE Calc_logic environment (the development board fixture firmware): Start interfaces to environment

The calculator *program* for the *development board* now has been linked, and the *machine instructions* can be transferred to the *development board*, which can be started (see Table 11-45 below).

TABLE 11-45 Calc_logic environment (the pocket calculator firmware), start-up report 1

Accomplishments and obstacles	Troubling ingredient	Responsible provider
SR.1. The linked program was loaded to the development board. When the calculator program was started, it seemed to get stuck just after any button was pressed for the first time. Only after reset it started again, but got stuck in the same way.	-E-1. Calc_logic interface to pocket calculator program	PR.17. In-house project "de1" PR.16. Chk&Go

11.17.6 EXAMPLE Calc_logic environment (the development board fixture firmware): Examine prototype reliability

The readiness gate members were not satisfied with the start-up, and clearly, calc_logic was not executing well enough to be verified. The prototype was rejected and sent back for redesign.

11.17.7 EXAMPLE Calc_logic environment (the development board fixture firmware): Complete and connect interfaces to environment

The assembler worked without calling calc_logic, so the problem could not be within calc_logic, but probably in the *interface* to calc_logic. The integrating engineers had a first look at the problem before sending it further back for redesign (see Table 11-46 below).

TABLE 11-46 Calc_logic environment (the pocket calculator firmware)
on development board, integration report 2

Fulfillments and shortcomings	Interfaces to realize	Responsible provider
IR.1. C-manuals and assembler manuals were checked and it became apparent that the surrounding assembled program did not follow the C convention when calling C function. The stack was set up wrong before calling, and the return address was missed when executing the return at the end of calc_logic. When the problem was finally understood, the calculator program failure was easily eliminated, and recompiled, and relinked on the host.	-E-1. Calc_-logic interface to pocket calculator program	PR.17. In-house project "de1" PR.16. Chk&Go

11.17.8 EXAMPLE Calc_logic environment (the development board fixture firmware): Start interfaces to environment

The calculator *machine program* can once again be transferred to the *development board* and started (see Table 11-47 below).

TABLE 11-47 Calc_logic environment (the pocket calculator firmware) on
development board, start-up report 2

Accomplishments and obstacles	Troubling ingredient	Responsible provider
SR.1. The calculator program could now be started without problems with incorrect calls and returns to the calc_logic program. The pocket calculator showed that the key pressed digits on the display, but occasionally the calculator got a wrong calculation.	A. Calc_logic for reuse in pocket calculator on development board	PR.17. In-house project "de1" PR.16. Chk&Go

11.17.9 EXAMPLE Calc_logic environment (the development board fixture firmware): Examine prototype reliability

The decision was made that calc_logic design needed no further changes to be a sufficiently *reusable item* to be verified. It was also decided that a portability chapter should be included in the calc_logic manual.

11.17.10 Calc_logic for reuse: Proceed reading

To follow this example, these are the alternatives:

- Go back through the calc_logic example chapters to understand how to fix a calc_logic environment (the pocket calculator firmware) on a *development board* that is not ready for field *verification*.

- Proceed to chapter 12.22, "EXAMPLE Calc_logic environment (the pocket calculator firmware) on development board: Verify prototype in environment," page 761.

- Proceed to chapter 11.18 "EXAMPLE Calc_logic environment (the pocket calculator firmware): Realize interfaces to environment and insert/await outermost white-box" below.

11.18 EXAMPLE Calc_logic environment (the pocket calculator firmware): Realize interfaces to environment and insert/await outermost white-box

11.18.1 Calc_logic environment (the pocket calculator firmware): Process schedule to use

For the calc_logic example, a special technical overall *schedule* is *tailored* (illustrated in Fig. 11-24, below):

- **Pt:2:3b. Tailored advanced technical schedule n = 2, porting to 3 targets (right half),** page 170

In this *tailored schedule* the generic *schedule* to use is (filled-in symbols below):

- **Pt.0.I. Realize interfaces to environment and insert/await outermost white-box,** page 122.

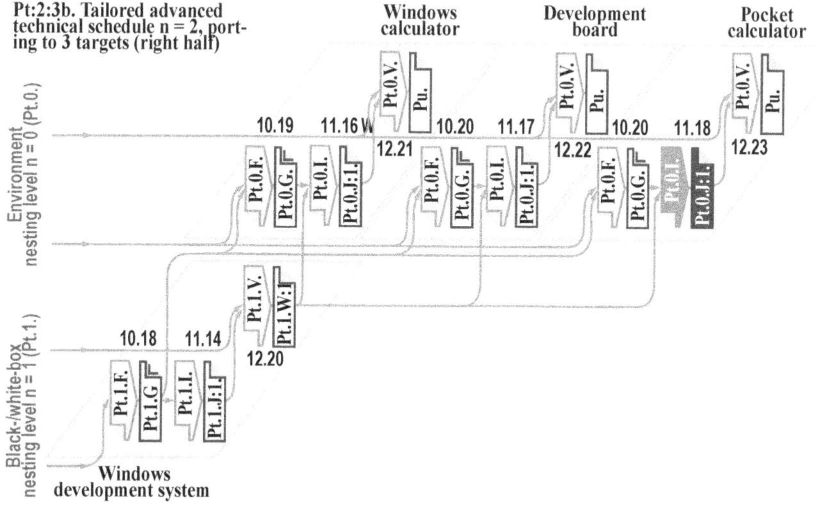

FIGURE 11-24 Position of schedule and master to use (filled in) in overall schedule

11.18.2 EXAMPLE Calc_logic environment (the pocket calculator firmware): Obtain items for interface to environment and ported linkable program

The *development board* had a more general microprocessor than the cheaper microprocessor used in the *target* pocket calculator. The assembler pocket calculator *program* must be assembled with some different control parameters to supply the *target* pocket calculator microprocessor (see Table 11-48 below).

TABLE 11-48 Calc_logic environment (the pocket calculator firmware)
items received deviation report 1

Deviation	Item	Sourcing
DR.1. Without deviations, calc_logic was recompiled with the microcontroller compiler for the particular target pocket calculator microcontroller.	i100. Hosted portable calc_logic programs on CD disc	SU.11. In-house project "dg3"

11.18.3 EXAMPLE Calc_logic environment (the pocket calculator firmware): Complete and connect interfaces to environment

The newly compiled calc_logic *program* must now be linked with the pocket calculator *program* and *program libraries*. The *target* microprocessor is cheaper and simpler, but should not cause any major problem (see Table 11-49 below).

TABLE 11-49 Calc_logic environment (the pocket calculator firmware)
integration report 1

Fulfillments and shortcomings	Interfaces to realize	Responsible provider
IR.1. Since all serious problems were identified and eliminated on the development board, there were no problems with the calc_logic integration and linking with the pocket calculator program.	-E-1. Calc_logic interface to pocket calculator program	PR.17. In-house project "de1" PR.16. Chk&Go

11.18.4 EXAMPLE Calc_logic environment (the pocket calculator firmware): Start interfaces to environment

The new pocket calculator *machine program* from the linker is loaded to the *target* stand-alone pocket calculator, which is tried for a start-up (see Table 11-50 below).

TABLE 11-50 Calc_logic environment (the pocket calculator firmware)
start-up report 1

Accomplishments and obstacles	Troubling ingredient	
SR.1. The in-house project "de1" had problem lighting up the pocket calculator LED digits, and problems with the assembler program addressing correct ports were solved. SR.2. When buttons were pressed, nothing happened. Diodes were shown to be mounted and soldered in wrong direction. SR.3. When all above failures were eliminated, the stand-alone calculator program started, and the calc_logic program could be called without causing problems.	E. Calc_logic environment (the pocket calculator firmware)	PR.17. In-house project "de1"

11.18.5 EXAMPLE Calc_logic environment (the pocket calculator firmware): Examine prototype reliability

To understand the reusability of calc_logic, the in-house project "dg3" engineers were also present at the *porting* and *integration* of calc_logic *program* with the *target* pocket calculator. The decision was made that calc_logic design needed no further changes to be a sufficiently *reusable item*. It was also noted how important it is to extend the portability chapter in the calc_logic manual.

11.18.6 EXAMPLE Calc_logic for reuse: Proceed reading

To follow this example, these are the alternatives:

- Go back through the calc_logic example chapters to understand how to fix a calc_logic environment (the pocket calculator firmware) that is not ready for field *verification*.

- Proceed to chapter 12.23, "EXAMPLE Calc_logic environment (the pocket calculator firmware): Verify prototype in environment," page 767.

- Proceed to chapter 11.9 below, to leave the calc_logic example and to enter the phonebook example.

11.19 EXAMPLE Phonebook: Procure items and linkable programs in order to realize white-box

11.19.1 EXAMPLE Phonebook: Process schedule to use

For the phonebook example, a special technical overall *schedule* is *tailored* (illustrated in Fig. 11-25, left):

- **Pt:2:3. Tailored advanced technical schedule n = 2,** page 169

In this *tailored schedule* the generic *schedule* to use is (filled-in symbols):

- **Pt.n.I. Procure items, linkable programs, and integrate/await embedded black-boxes in order to realize white-box,** page 144.

FIGURE 11-25 Position of schedule and master to use (filled in) in overall schedule

11.19.2 EXAMPLE Phonebook: Obtain items including linkable program

The *program* will integrate and run on the same *target environment* where it was developed. Because of this, there is only a need to recompile the files (see Table 11-51 below) with debugger request turned on.

TABLE 11-51 Phonebook items received deviation report 1

Deviation	Item	Sourcing
DR.1. Modularization of program files was ordered, but this exceeds expectations. Well done.	Source file ph_book_dyn_inst.ccp	SU.15. In-house development
	Source file support_f.cpp	
	Source file ph_book_c.ph_book_c.ccp	
	Source file ph_book_c.key_add_-cont_f.cpp	
	Source file ph_book_c.disp_all_-cont_f.cpp	
	Source file ph_book_c.delete_cont_f.cpp	
	Source file ph_book_c.~ph_book_c.cpp	
	Source file cont_c.key_cont_f.cpp	
	Source file cont_c.disp_cont_f.cpp	
	Source file cont_c.read_cont_f.cpp	
	Source file cont_c.write_cont_f.cpp	
	Source file pers_c.key_pers_f.cpp	
	Source file pers_c.disp_pers_f.cpp	
	Source file pers_c.read_pers_f.cpp	
	Source file pers_c.write_pers_f.cpp	
	Source file ph_numb_c.key_ph_-numb_f.cpp	
	Source file ph_numb_c.disp_ph_-numb_f.cpp	
	Source file ph_numb_c.read_ph_-numb_f.cpp	
	Source file ph_numb_c.write_-numb_f.cpp	

11.19.3 EXAMPLE Phonebook: White-box integration report

Building the phonebook from all its *linkable program* items is made quite simple by the linker tool (see Table 11-52 below). As the *fixture* to the phonebook serve as *stimuli* injector and *response* visualizer during *verification*, the *environment* source file main_f.cpp, page 587, is prepared in advance.

TABLE 11-52 Phonebook integration report 1

Fulfillments and shortcomings	Interfaces to realize		Responsible provider
IR.1. The phonebook linked without failures.	-A-1. Eof support interface	Header file support_f_shape.h	
	-A-a1. Contact interface	Header file cont_c_shape.h	PR.10. In-house
	-A.aaa-1. Person interface	Header file pers_c_shape.h	
	-A.aa-1. Phone number interface	Header file ph_-numb_c_shape.h	

11.19.4 EXAMPLE Phonebook: White-box start-up report

The phonebook with its *fixture* starts nicely in the debugger tool (see Table 11-53 below).

TABLE 11-53 Phonebook start-up report 1

Accomplishments and obstacles	Troubling ingredient
SR.1. Within the fixture, the program starts without problems in the development system debugger.	A. Phonebook

11.19.5 EXAMPLE Phonebook: Examine integration robustness

The gate decision was simply to send the phonebook for *verification*.

11.19.6 EXAMPLE Phonebook: Proceed reading

To follow this example, these are the alternatives:

- Go back through the phonebook example chapters to understand how to fix a phonebook that is not ready for *verification*.
- Proceed to chapter 12.26, "EXAMPLE Phonebook: Verify black-/white-box," page 770.
- Proceed to chapter 11.20 "EXAMPLE Phonebook environment: Realize interfaces to environment and insert/await outermost white-box" below.

11.20 EXAMPLE Phonebook environment: Realize interfaces to environment and insert/await outermost white-box

11.20.1 EXAMPLE Phonebook: About this example

This first *variant* of the phonebook *program* is only for evaluation, and the *target environment* is not yet planned and developed.

The items located in the *environment* are already prepared and used as the *fixture* to the phonebook, when verified in the debugger. The only thing needed to complete the phonebook is to make it a stand-alone *program*, executable on a Windows PC without the debugger tool.

11.20.2 EXAMPLE Phonebook environment: Process schedule to use

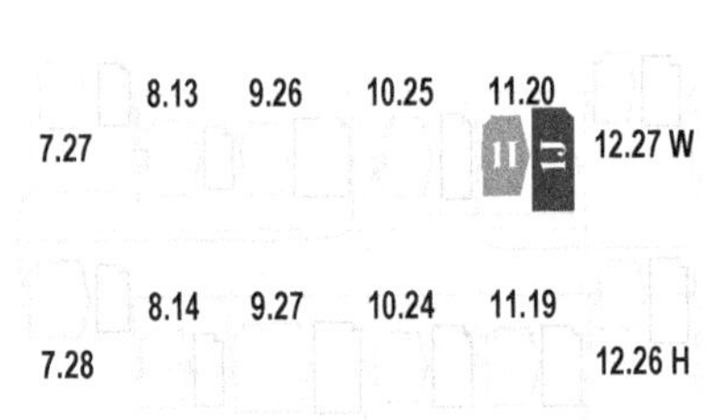

For the phonebook example, a special technical overall *schedule* is *tailored* (illustrated in Fig. 11-26, left):

- **Pt:2:3. Tailored advanced technical schedule n = 2,** page 169

In this *tailored schedule* the generic *schedule* to use is (filled-in symbols):

- **Pt.0.I. Realize interfaces to environment and insert/await outermost white-box,** page 122.

FIGURE 11-26 Position of schedule and master to use (filled in) in overall schedule

11.20.3 EXAMPLE Phonebook environment: Received items and ported linkable programs deviation report

The limited *environment* items were already prepared when the phonebook *black-box* was integrated, and there is nothing more to receive.

11.20.4 EXAMPLE Phonebook environment: Complete and connect interfaces to environment

The debug settings are replaced by stand-alone setting for the compiler and linker. All that remains to be done is recompile and relink the phonebook issues.

11.20.5 EXAMPLE Phonebook environment: Start interfaces to environment

As expected, the phonebook starts up as a stand-alone application.

11.20.6 EXAMPLE Phonebook environment: Pt.0.Ii. Examine prototype reliability

The gate concluded that the stand-alone phonebook application is ready for *verification.*

11.20.7 EXAMPLE Phonebook: Proceed reading

To follow this example, these are the alternatives:

- Go back through the phonebook example chapters to understand how to fix a phonebook environment that is not ready for field *verification.*
- Proceed to chapter 12.27, "EXAMPLE Phonebook environment: Verify prototype in environment," page 780.

## CHAPTER 12

Verify black- & white-box

"If the verification effort seems to get significant, abandon hope for affordable quality outcome."

Can even be automated.

12.1 About verification

12.1.1 Purpose

Verification is a bit special in the sense that the *result* from *verification* is not something developed further on, but something that is already developed and must be changed, improved, or removed.

With some pain, *requirements* have been specified, and with yet more pain *white-boxes* have been populated, in a belief that all this will perfectly match and even surpass *stakeholder* expectations. However, it remains to be proved that this is the case. Development is made by humans, and they make mistakes. Also, the original *requirements* have been developed by humans, and may be faulty.

> Verification confirms that the prepared black-/white-box is able to satisfy the requirements (or it reveals that it was wrong).

Requirements for the *black-/white-box* must again be retrieved, in order to allow checking that the design has specified restrictions and behavior. Identified deviations from the *black-/white-box* against correct *requirement* expectations are simply called failures and must be eliminated.

To verify a *black-/white-box* directly against its *requirements* is often too big a leap to make in one step. To bridge between specified *requirements* and the realized *black-/white-box*, *test-cases* need to be developed that refine and concretize the specified *requirements* to suit the work of verifying the realized *black-/white-box*.

12.1.2 Verification workload depends on earlier carefulness

During *verification*, the uncertainty of deviations between a *black-/white-box* and its *requirements* can get as close to zero as desired, just through an increase in the coverage and thereby the number of *test-cases*. The *verification* amount of work is again rather similar between programmers and mechanics (see Fig. 12-1 below).

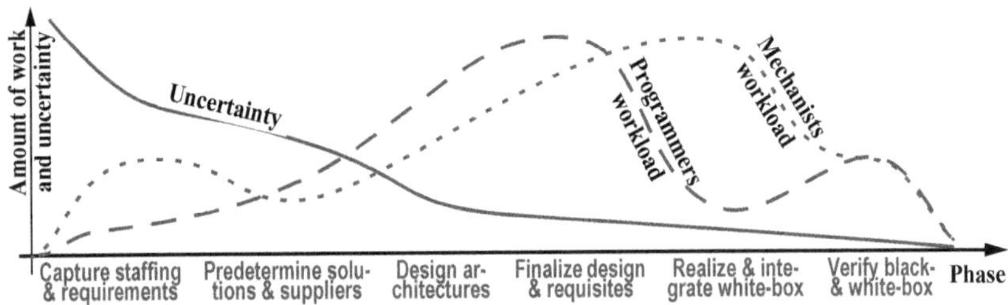

FIGURE 12-1 Amount of uncertainty and workload during the phases

12.1.3 Cpdm definition

In *Cpdm*, *verification* is defined according to Table 12-1 below

TABLE 12-1 Cpdm definition

Aspect	Verification
Definition	To confirm that a realized black-/white-box or a prototype in the environment fulfills requirements on design and behavior
Synonyms	Test, examination, try-out, check

12.1.4 Verification is based on all available requirements

All kinds of *requirements* to the *black-/white-box* intended to be verified must now be retrieved and used as the basis for producing *test-cases* for *verification* (see Fig. 12-2). Provided that *requirements* match all *stakeholder* desires, this will ensure that the verified *black-/white-box* complies with these desires from *customers* and other *stakeholders*.

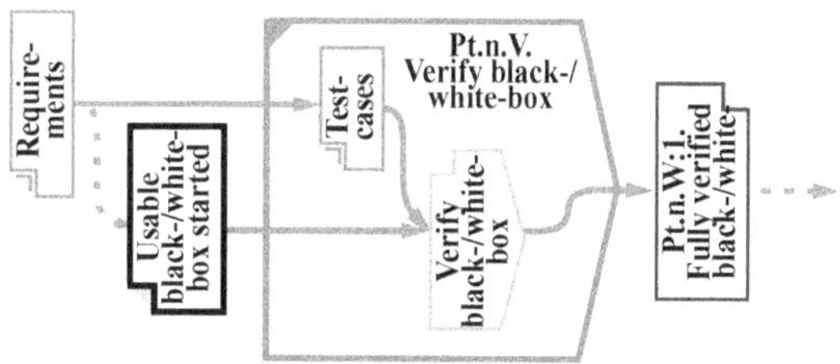

FIGURE 12-2 Principle of black-/white-box verification

When the launch of a *product* approaches, the development organizations often get anxious for *quality* and quite willing to produce serious and really stressful tests. It is not seldom seen that *test-cases* gradually become of higher *quality* and of better coverage than the *requirements* specified long before. This is problematic, because something else is then verified than what originally was specified. A lot of problematic consequences result from this, for example:

A poor match between requirements and test-cases is always a serious problem.

- *Test-cases* which cannot possibly refer to any *requirement* verify something not necessarily required by *stakeholders* and *customers*.

- Having *requirements* without *test-cases* implies that *stakeholder* desires are not verified and may be missed in the *product*.

- Overlaps between *test-cases* and *requirements* are not problematic in themselves, but what is overlapped with what may not be known. When the *verification* is done, nobody can tell exactly what *requirements* the prototype fulfills.

- If *requirements* and *test-cases* are both missing, it is not easy to tell how the *black-/white-box* will behave at all, or how well it complies with restrictions. It also becomes impossible to write a good manual. Not seldom manual writers run around in the organization disturbing developers, to find out how the *black-/white-box* works in order to understand and document its capability.

The best way to solve above mismatches is to eliminate them, which means going back through the *phases* and rewrite, add, and remove *requirements,* and then possibly complement them with new *test-cases* in order to achieve a full match between *requirements* and *test-cases.* All of this must then be reviewed by *stakeholders* before *verification* starts. This update may disclose shortages in the *black-/white-box,* but it is better to know them this late than to proceed with a wrong or insufficient *black-/white-box,* and ending up with an unattractive *product.*

12.1.5 Verify outside-in or inside-out

The *innermost black-/white-boxes* of the developed *product prototype* are always the simplest to verify, including tracing detected failures to their locations, because by definition, *innermost black-/white-boxes* don't contain any more *embedded black-/white-boxes* containing failures. The *outermost black-/white-box* is the most *complex* because all *embedded black-/white-boxes* are contained in it. Thus *verification* most often is made inside-out (see Fig. 6-34, p. 162 and ch. 11.1.3, p. 595).

Beginning verification of innermost black-boxes is most natural, since these are fractions, and least complex.

12.1.6 Relation between failure detection and failure location

As difficult as it may be to find the illness from a symptom of a human, it is the same type of difficulty to find the failure location in a *product prototype* after a failure is detected. It may be an intricate and tedious investigation to find the source location of a failure based on how the *product prototype* behaves.

The most probable is to find the failure on about the *nesting level* where it was detected, because most failures are located close to where they show up, like pain in the foot probably comes from an illness in the foot. However, this is not always the case, like pain in the foot may be caused by a narrowing of the spinal cord in a neck vertebra. Similarly, a wrong behavior detected on one *interface* of a *product prototype* can be caused by a very remote failing *element.* In the same way, the symptom and illness can be of very different depths in the *product prototype.* A nasty failure may be located deep in a *product prototype* when performing field *verification* (see example ch. 12.3.2, p. 666), or conversely (after *invasive integration* has been applied), a failure may be located on *environment interfaces* when verifying a deeply *embedded black-/white-box.*

It certainly is a considerable effort to develop *test-cases* and execute them, but often the biggest effort during *verification* is to find the location of detected failures.

12.1.7 Explicitly verify black-/white-boxes

The most straightforward way of *verification* is to fully verify a *black-/white-box* separately in any kind of *test-case* executor and then correct all detected failures before integrating the *black-boxes* into the next *outward nesting level white-box* or into the *environment*. In *Cpdm*, this kind of *verification* is called *explicit verification*.

Advantage of *explicit verification*:

* When practicing *explicit verification* throughout a whole *product prototype*, the *verification* begins with the *innermost black-/white-boxes* not containing *embedded black-/white-boxes* with possible failures. When going *outwards* with *verification*, inner *black-/white-boxes* always have been verified and bring almost no additional failures. Thus, inside-out *integration* of each *nesting level* does not result in any troublesome accumulation of failures, and can continue smoothly *outwards* all the way to the *environment nesting level*, with the few exceptions of deeply located failures detected during *verification* of *outward nesting levels*.

Drawback of *explicit verification*:

* When applying *explicit verification* on all *nesting levels*, for a very long time *integration* is done on inner *embedded black-/white-boxes*, and the full *product prototype* is the last to be integrated. Thus, the *product prototype* cannot be validated until the development is almost completed, which scares many development companies and their *customers*. *Implicit verification* in the next chapter may be a solution to this.

12.1.8 Implicitly verify black-/white-boxes

An alternative way of *verification* is to fully verify a *black-/white-box* after it has been integrated into its *outward nesting level black-/white-box*, which now is acting as part of the *test-case* executor. In *Cpdm*, this kind of *verification* is called *implicit verification*.

There are advantages in *implicit verification*.

* One obvious reason is that some *product*s are big or heavy, implying there is no way to build detachable *black-/white-boxes*, which can be explicitly verified, and the only way is to perform *implicit verification*.

* Another reason may be when there is a clear *requirement* to build the *product prototype* outside-in, which typically can occur when the *customer* wants to get closely involved during development. It is much simpler for *customers* to gradually validate a *product prototype* growing outside-in, because then they starts evaluating the usage of the *product*, and can gradually see how the *product prototype* grows more and more capable of satisfying them. This is called iterative development, and is sometimes advantageously practiced in development of *programs* for dedicated *customers*.

However, there are also serious drawbacks in *implicit verification*.

- Since the *black-/white-box* to be verified is not detached from its surroundings, its *interfaces* may be hidden and not available to inject *stimuli* and detect *responses*. This can be solved by arranging observation posts (see ch. 9.4.4, p. 350) on *interfaces* crossing the borders to the *black-/white-box*. In *programs*, this is a smaller problem, because modern *development systems* have excellent debugger tools, where all *interfaces* are available and can be exposed.

- Another serious problem is that a detected failure can be hard to locate, since it can occur not only in the *black-/white-box* being verified, but anywhere *outwards* in the attached surroundings. If there are many failures not yet detected, these can interact with the failure being searched for, which creates a very *complex* failure localization.

- If through necessity or desire a *product prototype* is integrated outside-in (by invasion or incorporation), a *full verification* cannot start until the *product prototype* is completely integrated all the way to innermost *nesting level*, and this with all failures accumulated from each *nesting level*. Even when *implicit verification* is done stepwise inside-out, starting with the *innermost black-/white-boxes*, a very turbulent *verification* will probably take place, since locating and eliminating each failure constantly will block proceeding the *verification*. A way to solve this is to fake the *embedded black-/white-boxes* and during *integration* perform partial *implicit verification* on each *nesting level* (see next chapter).

12.1.9 Partially verify black-/white-boxes with embedded fakes

To get started earlier with *implicit verification* during *invasive integration*, *inwards embedded black-/white-boxes* on each *nesting level* may temporarily be replaced with fakes to allow *partial verification* (see the house example *schedule* in Fig. 6-38, p. 167). After the *implicit verification* is performed, the fake is removed and the *embedded black-/white-boxes* is invaded (or incorporated).

The implicit *partial verification* can be more or less complete, depending on how well the fake reflects the coming *black-/white-box*. In some cases the implicit *partial verification* is only a few spot-checks, and in the best case it is almost a *full verification*. However, in all cases it is a good practice to perform early checks, to avoid later expensive rework.

After the *product prototype* is completely integrated, the full inside-out *implicit verification* takes place according to chapter 12.1.8 above.

12.2 EXAMPLE Lake: Size of meshes in fishing nets

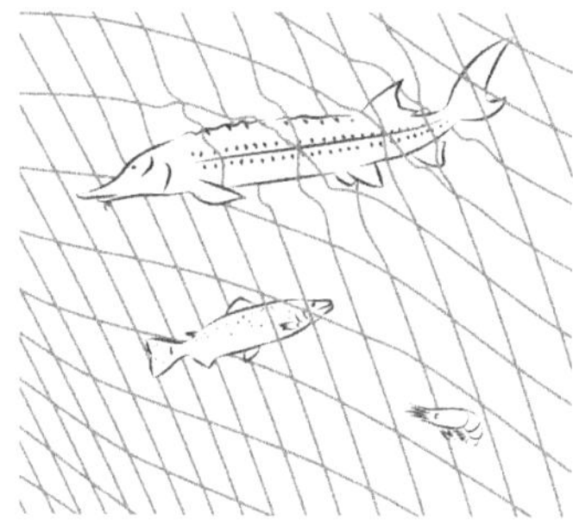

How can fishing be a metaphor for *verification?* Actually, a lot of catchy similarities can be found, as shown in the following sequence of lake examples. Just imagine that catching fish is catching failures. (Maybe fish are more valuable than failures, but never mind.)

This example illustrates that a large number of well-distributed and detailed *test-cases* have the effect of making each and every mesh smaller, resulting in more fish getting stuck.

FIGURE 12-3 A fishing net catches the big fish, but lets the smaller through.

12.3 Troubleshooting

12.3.1 Tracing upstream and inwards for failure localization

When a *response* does not match the expectation specified in a *test-case*, a failure is detected, which must be located in the *realization* being verified, and this localization starts in the *process* where the initial detection of the failure has been made.

If *invasive integration* has been applied (which means that all *interfaces* are inseparable at *verification*), note that failures can be located both *inwards* and *outwards* from the location where they were first detected.

When a *white-box verification* is performed, *white-boxes* are already open and the failure is more or less located when identified.

When a *black-box verification* is performed, it can take considerable time and effort to trace from the *interface* which responded incorrectly, *upstream* through the used *schedule*, and *inwards* through the structure of the *product prototype*, to find the failure location. This detective work can most often not be done only by verifiers, who have limited experience with the prototype structure. It is preferable also for attitude reasons that developers of the prototype themselves perform *failure localizations* to locate problems they have caused.

When tracing *inwards* in the *realization* looking for the failure, it easily gets lost. In the realization itself, there are seldom any explanations of the found design, and the detective work must move over to design documentation, which is the origin for the *realization.* When it comes to tracing in the documentation, it is also beneficial to trace the used *process schedule* to understand how and when the failure was made.

Manufacturing never works like copying a prototype. Instead, the design documentation is the origin of the *product* manufacturing. If only failures in the prototype are eliminated, and no corrections are made in the documentation, the failure will most probably reappear when *product* manufacturing starts.

12.3.2 EXAMPLE House: Find a failure localization

Recall the house example applying the overall *schedule* **Pt:2:2.**, page 164, shown in a smaller and simpler *view* in Figure 12-4 below.

EXAMPLE: Pt:2:2. Tailored advanced technical schedule n = 4, page 163

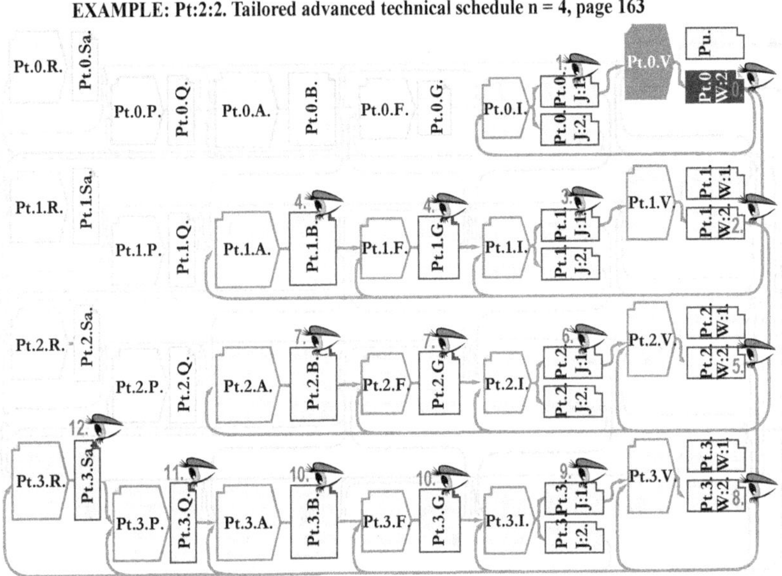

FIGURE 12-4 EXAMPLE: Failure localization in process schedule

To exemplify how tricky troubleshooting may be, the example is maybe a bit extreme, because many *inward verifications* have in one way or another missed detecting a residual current failure on the cable connection to the house. The reason may be as simple as a lack of competency in this field.

The eyes in the picture symbolize locations where the failure is first detected, and then where the failure is searched for and located. The numbers refer to the list of actions in the list below. Also note that all arrows starting from the *"masters governing* partially verified *results* **W:2.**" and then bending to proceed left, indicated the *upstream* direction to trace for localizing failures.

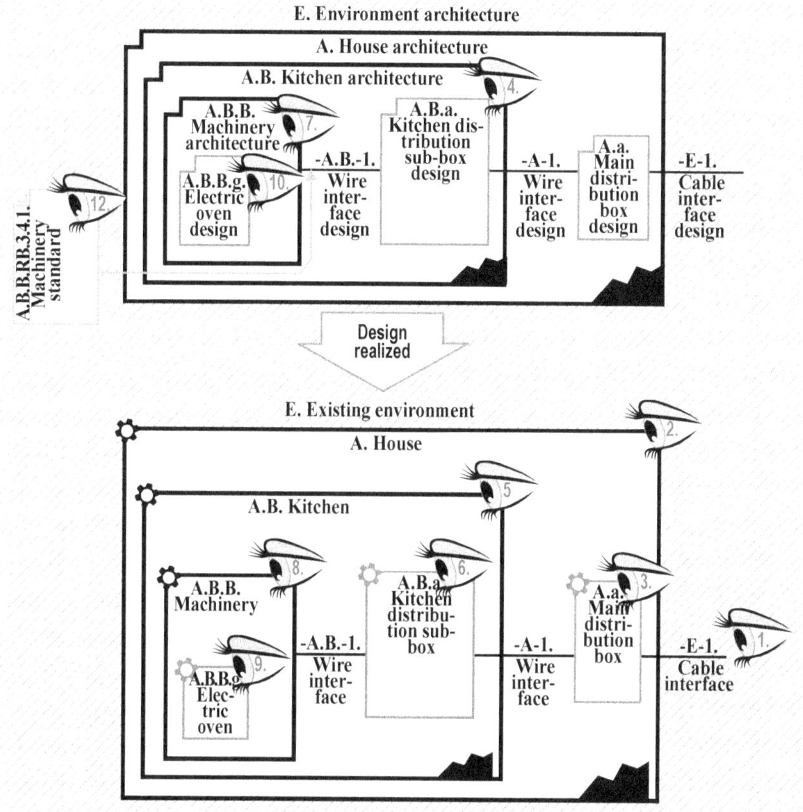

FIGURE 12-5 EXAMPLE: Failure localization in documentation and prototype

The below example is a detailed explanation of finding the failure VR.1.3. in Table 12-27, page 701. The explanation list below refers constantly to Figure 12-4 and Figure 12-5 above.

0. Assume a residual current leakage failure is detected in the **-E-1. Cable interface** (in Fig. 12-5 above) *governed* by *master* **Pt.0.W:2.** during execution of *schedule* **Pt.0.V.** (both in Fig. 12-4 above) by assistance from the **SU.11. Electricity supplier** (in Table 8-7, p. 293).

1. First check is done by again executing *upstream schedule* **Pt.0.I.**, to see if the failure emanates somewhere along the **-E-1. Cable interface** (now *governed* by *master* **Pt.0.J:1.**) in the *environment*. The main power switch in **A.a. Main distribution box** is turned on and off, and the conclusion is drawn that the **-E-1. Cable interface** does not cause the residual current problems.

2. Next assumption is that the failure must be localized *inwards* in the **A. House** (*governed* by *master* **Pt.1.W:2.**) on *nesting level* = 1. Observe that the house should already have been properly verified when *upstream schedule* **Pt.1.V.** was

executed, and possible failures should have been eliminated by now. But say that the current *test-cases* didn't cover this *verification* because it is seldom done in a private house.

3. The **A.a. Main distribution box** (*governed* by *master* **Pt.1.J:1.**) is revisited by *upstream schedule* **Pt.1.I..** No failures can be localized in this box, and when all fuses are released, the residual current failure disappears. Consequently, the failure must be on one or more connections to the main distribution box.

4. The *upstream schedules* **Pt.1.A.** and **Pt.1.F.** have been executed for several rooms which are described in architectures on *nesting level* = 2. In the architecture and final design the **-A.B.-1. Wire interface design** can be identified.

5. By executing *schedule* **Pt.2.V.**, the verifiers localize all distribution sub-boxes, and assume the failure is located in the **A.B. Kitchen** (*governed* by *master* **Pt.2.W:2.**). As for the house, the kitchen should also already have been verified and failure eliminated when the *upstream schedule* **Pt.2.V.** was executed. In this example, assume that there was no *test-case* ordering such strange thing as a residual current measurement on electricity feed to the newly purchased kitchen.

6. Now check the **A.B.a. Kitchen distribution sub-box** (installed by the *schedule* **Pt.2.I.** and *governed* by *master* **Pt.2.J:1.**) to find out if the failure may be localized here. However, this box has no residual current problems, and the failure must be searched for on further *inward* connections. In the kitchen architecture, the **-A.B.-1. Wire interface** can be seen, describing the different wires to each machine.

7. The **A.B. Kitchen architecture** (drawn by executing *schedule* **Pt.2.B.** and *governed* by *master* **Pt.2.B.**) contains the *embedded black-box* **A.B.B. Machinery architecture** on *nesting level* 3 connected with the **-A.B.-1. Wire interface.** The **A.B.B. Machinery** is identified, and by disconnecting one machine at a time it is possible to localize where the residual current problem is.

8. Assume the failure is located in the **A.B.B.g. Electric oven** (*governed* by *master* **Pt.3.W:2.**). Like all kitchen machinery, the electric oven is assumed to be delivered without residual current problems. It is maybe a bit outrageous that this failure has not already been identified when executing the *verification schedules* **Pt.1.V.**, **Pt.2.V..** and **Pt.3.V.**, but deeply buried failures do sometimes show up first when the entire *product is* tested in its *target environment.*

9. Attached on the back of the **A.B.B.g. Electric oven** (*governed* by *master* **Pt.3.J:1.**) is a power connection manual, revealing that a wrong type of oven was used when executing the *realization schedule* **Pt.3.I.**, which creates residual current when connected.

10. It is a question why that type of electric oven is chosen when executing *schedule* **Pt.3.G.** drawing the **A.B.B. Machinery architecture** (*governed* by *master* **Pt.3.B.**).

11. The failure has already been introduced when *schedule* **Pt.3.P.** was executed, proposing this kind of *sample* in their *solution alternatives* (*governed* by *master* **Pt.3.Q.**). Assume that this proposed oven was manufactured for a market without demand for residual current correctness.

12. When continuing to trace *upstream* for the failure, the **A.B.B.RB.3.4.1. Machinery standard** (*governed* by *master* **Pt.3.Sa.**) was found outdated because a new law mandates residual current correctness in all houses. This is obviously the primary failure. Assume that when the *schedule* **Pt.3.R.** was executed, the *speci-*

fiers didn't know the new demand for residual current detection, but the electricity *supplier* nowadays regularly measures residual currency correctness.

12.3.3 Tracing downstream and outwards for failure elimination and reverification

Now the primary failure is localized, and the *failure elimination activities* begins. The most natural way to eliminate failures is to retrieve the used development *process schedule*, and to start eliminating the primary failure and then trace *downstream* to put right all the secondary consequences from the primary failure, until all secondary failures are also eliminated. Notice that sometimes the failure is too difficult or too expensive to correct, and a walk-around is chosen, which is to let the failure stay and make another change that eliminates it. Then secondary consequences from this walk-around change should instead be taken care of. Among the elimination *activities*, assistance from the developers who originally caused the failures and secondary failures is most often desirable and needed.

After a possibly time-consuming *failure localization*, it might be easy to forget to eliminate the primary failure and all its *downstream* consequences, and only fix the prototype for a quick *reverification*. But as said earlier, if only the prototype is failure-free, and the design documentation is left uncorrected, the failure will probably reappear in later manufacturing, which will use the uncorrected documentation.

Elimination of failures is far from risk-free. If not very carefully performed, the risk is great that other failures are introduced during *failure elimination* of already localized failures. Because of this risk, *reverification* must always be made after failure eliminations. If *failure eliminations* are substantial, a complete *regression verification* must be done.

12.3.4 EXAMPLE House: Trace downstream for failure elimination and reverifying.

Now continue the previous example to show *failure elimination* and *reverification* (see Fig. 12-6 below). The example below is a detailed explanation of eliminating failure VR.1.3. in Table 12-27, page 701.

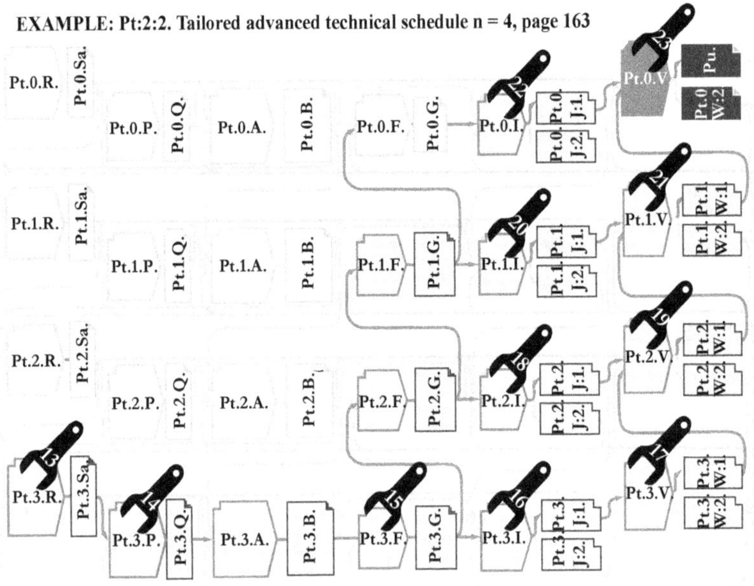

FIGURE 12-6 Eliminating failures downstream

In the previous example chapter, the primary failure was localized in the *requirement* **A.B.B.RB.3.4.1. Machinery standard**.

13. The primary failure of the outdated *requirement* (*governed* by *master* **Pt.3.Sa.**) must be eliminated by *schedule* **Pt.3.R.** *activities*. The *requirements* are supplemented with the updated regulation of residual current detection.

14. The outdated *requirement* **A.B.B.RB.3.4.1.** caused a failure in the proposed oven *solution alternatives* (*governed* by *master* **Pt.3.Q.**). This failure in the documentation must be eliminated by executing *schedule* **Pt.3.P.** *activities*. The *supplier* offers another oven *sample* approved for this market.

15. Because of the failure in the oven *solution alternatives*, the **A.B.B.g. Electric oven design** (*governed* by *master* **Pt.3.G.**) also became wrong, and must also be eliminated by *schedule* **Pt.3.F.** *activities*. Change this *white-box design item* to an approved oven with the right connection type.

16. The correction of the documented electric oven design caused a need for reordering of the *white-box design item* for the *element* **A.B.B.g. Electric oven design** by a *schedule* **Pt.3.I.** *activity*. Fortunately, the correct oven type could be found and changed out off the shelf, without troublesome delays.

17. According to the failure elimination *requirement* a new machinery *test-case* was added by executing *verification schedule* **Pt.3.V.**. The new *supplier* could help verify that the new oven had no residual current problems (now *governed* by *master* **Pt.3.W:1.** indicating failure-free machinery).

18. The verified *white-box* **A.B.B. Machinery** is fitted to the **A.B. Kitchen** by *schedule* **Pt.2.I.** *activities* without start-up problems (*governed* by *master* **Pt.2.J:1.**).

19. Since some other failures were also eliminated, the kitchen was regression-re-verified by executing *schedule* **Pt.2.V.** using all old and some newly prepared *test-cases*, without any *failure detection* (*governed* by *master* **Pt.2.W:1.**, indicating a failure-free kitchen).

20. All kitchen changes could be easily integrated and started in the **A. House** by *schedule* **Pt.1.I.** *activities.*

21. The house is now reverified with selected *test-cases* by *schedule* **Pt.1.V.** *activities.*

22. The reverified house prototype was reconnected to the *environment* by executing *schedule* **Pt.0.I..**

23. At the field *verification* by executing *schedule* **Pt.0.V.** (assisted by the electricity *supplier*) assume that the residual current failure was the last remaining failure to be eliminated (*governed* by *master* **Pu. Field-verified prototype**).

It has been said that *failure detection* deep in a prototype gets very expensive and time consuming for *failure eliminations* at the end of development. The above example cannot emphasize the truth of this statement any more clearly.

The *downstream* tracing noted above doesn't differ much from ordinary development. The only thing new is that the normally used *schedules* are traced for each failure, and not only progress as in normal development. Tracing is discussed a bit more in chapter 12.15.4, page 715 below.

12.3.5 Cpdm definition

In *Cpdm*, *test-case* is defined according to Table 12-3 below.

TABLE 12-2 Cpdm definition

Aspect	Test-case
Definition	Concretization of requirements in order to be a more efficient base for performing all kinds of verification
Synonyms	Test suite, test buffer, conditions for test
Symbol	**TABLE 12-3 Test-cases** <table><tr><td>Check design and/or check behavior</td><td>Expected outcome with pass/fail criteria</td><td>Require-ment</td></tr></table>

As said before, it is crucial that *test-cases* match *requirements*, and consequently each and every *test-case* references the *requirements* as the source for the *test-case*. For *complex requirements*, there may be many *test-cases* referring to the same *requirement*.

Requirements that have not resulted in any *test-case* at all, despite an effort to produce them, are problematic. A *requirement* that is not verifiable is a sign of poor technical understanding or insufficient *formalism* applied when such *requirement* was specified. A good habit when capturing *requirements*, apart from involving all *stakeholders*, is to also involve engineers within the technical field, as well as verifiers to check testability.

Additionally, each *test-case* has a description of the procedure for *verification*. This can be about how to prepare the *black-/white-box*, and how and where to inject *stimuli*, or what design to judge and how to judge it. Finally, a *test-case* contains information on what is expected from the checking procedure, and which expectations are considered to be a failure.

Who knows the expected outcome and pass/fail judgment? Sometimes this can be calculated automatically, but it is difficult to build and rely on an alternative judgment system that is better than the one that has been developed. An easier way is to go back to the *requirement*, find out what it specifies, and thereby find how to judge, though this may take a lot of time. This is a classic problem, and the source for the expected outcome and pass/fail judgments is sometimes called the *test-case* oracle (see pass/fail judgment in Fig. 12-27, p. 743).

In *Cpdm, verification report* is defined according to Table 12-4 below.

TABLE 12-4 Cpdm definition

Aspect	Verification report
Definition	Report of pass/fail conclusions from each executed test-case
Synonyms	Verification outcome, test report, test outcome, pass/fail report
Symbol	**TABLE 12-5 Verification report** | **Actual outcome and pass/fail conclusion**	Expected outcome and pass/fail criteria	Require-ment
---	---	---	

In *Cpdm, failure elimination report* is defined according to Table 12-6 below.

TABLE 12-6 Cpdm definition

Aspect	Failure elimination report
Definition	Report of each detected failure, explaining where its source is located, and proposing how to eliminate it in the optimal way
Syno-nyms	Debugging plan, failure firefighting
Symbol	**TABLE 12-7 Failure elimination report** | Failure location	Failure elim-ination plan	Actual outcome and pass/fail criteria	Expected outcome and pass/fail criteria	Require-ment
---	---	---	---	---	

12.4 EXAMPLE House environment: Partially verify interfaces to environment with house fake

12.4.1 EXAMPLE House: About this example

It takes considerable time to fully realize (build) a house by outside-in *invasive integration*, with *full verification* impossible until the prototype is finally completed. However, *partial verification* can be done after each *nesting level* of *inward* invasion, which gives a quick check to ensure that the *invasive integration* is progressing as planned and with expected *quality* of *results*. First, the newly realized *environment interfaces* will be partially verified.

All *partial verifications* are decided to be quite informal, and contain only a bulleted documentation of the outcome. However, there is nothing preventing a formal and strict *partial verification* based on formal and strict *test-cases*.

12.4.2 EXAMPLE House environment: Process schedule to use

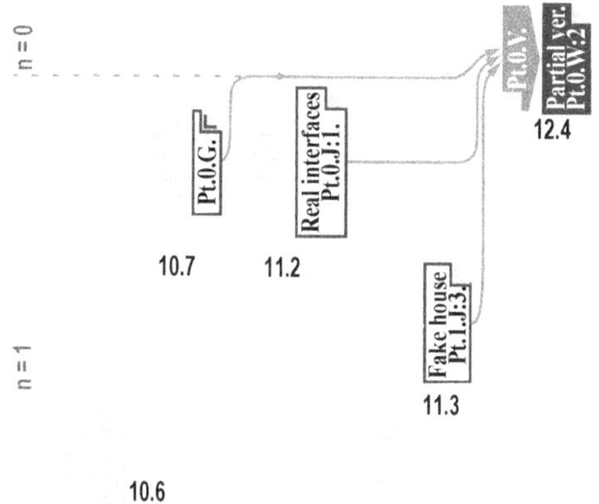

For the house *invasive integration* example, a special technical overall *schedule* is *tailored* (parts of right half illustrated in Fig. 12-7, left):

- **Pt:2:2b. Tailored advanced technical schedule n = 4 (right half),** page 167

In this *tailored schedule* the generic *schedule* to informally use is (filled-in symbols):

- **Pt.0.V. Verify prototype in environment,** page 125.

FIGURE 12-7 Position of schedule and master to use (filled in) in overall schedule

Below *results* may be demanded (their *masters* are indicated *upstream* from the filled-in *schedule* in the figure above and in Fig. 11-6, p. 603):

1J3 Fake house
Figure 11-11, Fake house integration report 1, page 609

0J1 Usable prototype interfaces started in simulator or target
Table 11-9, House environment start-up report 1, page 606

0G Environment interface design items and product requisites
Figure 10-14, House environment connections, page 479

0B Environment existing ingredients and design of interfaces to the environment
Figure 9-5, House environment logical architecture, page 341

0Q Solution alternatives of interfaces to environment with supplier opportunities
Figure 8-5, House environment solution alternatives of interfaces to environment, page 292

0Sa Environment restriction requirements
Table 7-5, House environment restriction requirement, page 182

The real *interfaces* to the *environment* are built according to *schedule* **Pt.0.I.** and a fake house is prepared according to *schedule* **Pt.1.I.** (see Fig. 6-38, p. 167).

This means that now the real *environment interfaces* are realized, and a fake house is temporarily prepared, to allow the *environment interfaces* to be partly verified by **Pt.0.V.** For a description of the fake, see Table 11-11, page 609. No formal *test-cases* were decided to be created. A *failure elimination report* is documented (see List 12-1 below).

LIST 12-1 House environment interfaces with fake house failure elimination report 1

1. The driveway was checked to ensure that it connected well to the garage, and it had enough space to reverse an ordinary car.
2. The entrance walkway was also checked for proper accessibility for wheel-chairs.
3. The water scale at the street was checked to indicate no consumption when the preliminary faucet was shut. From the preliminary faucet five 20-liter buckets were filled and flushed into the sewage to check that the water scale did measure this correctly and there were no obstructions along the sewage piping. However, it showed that the scale had measured 133 liters. A *failure elimination activity* was started, and finally the failure was found to be that the faucet was leaking after being too tightly closed. Some lubrication eliminated the failure.
4. The electricity meter at the street was checked to indicate no electricity consumption when the temporary distribution box was switched off. A 2 kW electrical heater was connected for one hour, and it was checked that the meter was adding 2 kWh.
5. The incoming piping was checked to connect well to the piping marks within the strings marking the house.

No failures were detected.

12.4.3 EXAMPLE House: Process schedule to use

To follow this example, these are the alternatives (see Pt:2:2b., p. 167):

- Trace *upstream* in the overall *schedule* to understand how and where detected failures are eliminated. Since the house is built outside-in, a failure can be located both *inwards* and *outwards* with respect to where it is detected.

0V Proceed to chapter 12.10, "EXAMPLE House environment: Fully verify prototype in environment," page 699.

- To stay within the house example, but not following the overall *schedule*, proceed to chapter 12.5 below.

12.5 EXAMPLE House: Partially verify black-/white-box with embedded fakes

12.5.1 About the house example

Now *partial verification* should be done for the realized house, to quickly spot-check that the *invasive integration* is maintaining an expected *quality*.

12.5.2 EXAMPLE House: Process schedule to use

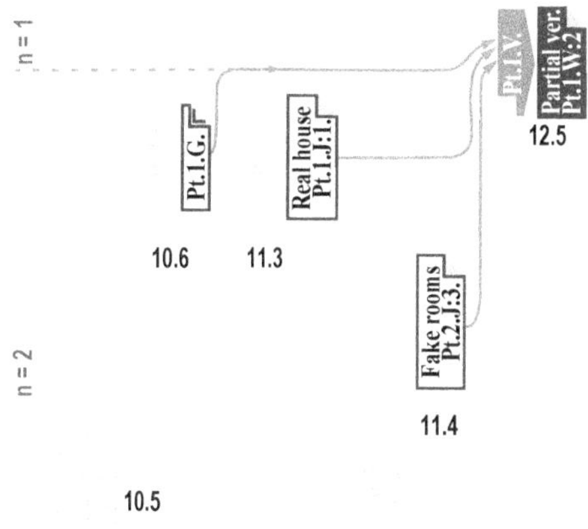

For the house *invasive integration* example, a special technical overall *schedule* is *tailored* (parts of right half illustrated in Fig. 12-8, left):

- **Pt:2:2b. Tailored advanced technical schedule n = 4 (right half)**, page 167

In this *tailored schedule* the generic *schedule* to informally use is (filled-in symbols):

- **Pt.n.V. Verify black-/white-box**, page 152.

FIGURE 12-8 Position of schedule and master to use (filled in) in overall schedule

Below *results* may be demanded (their *masters* are indicated *upstream* from the filled-in *schedule* in the figure above and in Fig. 11-6, p. 603):

2J3 Fake rooms

Figure 11-16, Fake house kitchen integration report 1, page 616

1J1 Completed usable house white-box

Table 11-14, House start-up report 1, page 612

1G Finalized house items design with product requisites

Figure 10-10, House boundary casing (upper part), page 467

Figure 10-11, House boundary casing (lower part), page 468

    Figure 10-12, House connections, page 469

1B House architectures with major ingredients and room black-boxes

    Figure 9-22, House logical architecture, page 363

    Figure 9-23, House physical architecture, page 364

1Q House solution alternatives with supplier opportunities

    Figure 8-12, House white-box solution alternatives, page 301

1Sa House prioritized requirements

    Table 7-9, House black-box compound requirements (small selection),
       page 189

The house is built according to *schedule* **Pt.1.I.** and fake rooms are prepared according to *schedule* **Pt.2.I.** (also see overall Fig. 6-38, p. 167).

For a description of the fake rooms, see Table 11-16, page 616. No formal *test-cases* were deemed necessary and the List 12-2 below was identified during the informal *partial verification*:

    LIST 12-2 House with fake rooms failure elimination report 1

1. The locations of the house windows and doors were measured and also checked for size. No big deviations were found.
2. The kitchen interiors were generally fitting well. On some occasions, where small gaps between interiors and walls were unavoidable, some cover strips were planned.
3. It became obvious from interior markings that the counter (A.aC.j. Preparation counter) would reach higher than the bottom of the kitchen window (-E-9. K-window). The current window had to be exchanged for a shorter one, and placed higher in the niche.
4. Piping from the kitchen floor and walls was checked, and all kitchen connections ended correctly inside the marked interiors.

Above *failure eliminations* were distributed to developers responsible within the faulty areas. No reverifying was needed.

12.5.3 EXAMPLE House: Process schedule to use

To follow this example, these are the alternatives (see Pt:2:2b., p. 167):

- Trace *upstream* in the overall *schedule* to understand how and where detected failures are eliminated. Since the house is built outside-in, a failure can be located both *inwards* and *outwards* with respect to where it is detected.

1V Proceed to chapter 12.9, "EXAMPLE House: Fully verify house black-/white-box," page 693.

- To stay within the house example, but not following the overall *schedule*, proceed to chapter 12.6 below.

12.6 EXAMPLE House kitchen: Partially verify black-/white-box with embedded fakes

12.6.1 About the house example

Now *partial verification* is done for the kitchen, to quickly spot-check that the *invasive integration* is maintaining the expected *quality*.

12.6.2 EXAMPLE House kitchen: Process schedule to use

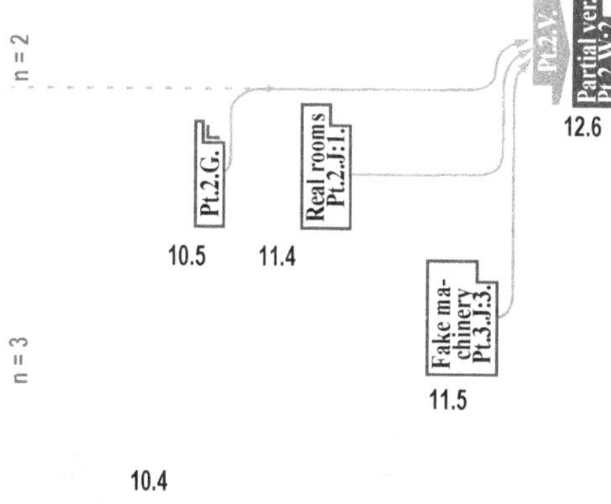

For the house *invasive integration* example, a special technical overall *schedule* is *tailored* (parts of right half illustrated in Fig. 12-8, left):

- **Pt:2:2b. Tailored advanced technical schedule n = 4 (right half),** page 167

In this *tailored schedule* the generic *schedule* to informally use is (filled-in symbols):

- **Pt.n.V. Verify black-/white-box,** page 152.

FIGURE 12-9 Position of schedule and master to use (filled in) in overall schedule

Below *results* may be demanded (their *masters* are indicated *upstream* from the filled-in *schedule* in the figure above and in Fig. 11-6, p. 603):

3J3 Fake machinery
Figure 11-20, House kitchen fake machinery integration report 1, page 620

2J1 Completed usable room white-box
Table 11-18, House kitchen start-up report 1, page 617

2G Finalized room items design with product requisites
Figure 10-7, House kitchen interior drawing, page 455
Figure 10-8, House kitchen connections, page 456

2B Room architectures with major ingredients and machinery black-boxes
Figure 9-26, House kitchen logical architecture, page 369
Figure 9-27, House kitchen physical architecture, page 370

2Q Room solution alternatives with supplier opportunities
Figure 8-14, House kitchen white-box solution alternatives, page 306

2Sa Rooms prioritized requirements
   Table 7-11, House kitchen black-box compound requirements (small selection), page 195

The rooms are built according to *schedule* **Pt.2.I.** and fake machinery is prepared according to *schedule* **Pt.3.I.** (also see overall Fig. 6-38, p. 167).

For a description of the fake machinery realization, see Table 11-20, page 620. The rooms with their interiors can now be verified (see List 12-3 below).

LIST 12-3 House rooms with fake machinery failure elimination report 1

1. The machinery fits into planned interiors.
2. The gas pipe drawn directly from outside the house protruded too far into the marked dish machinery. A sharper bend had to be mounted on the gas pipe before installing the dish machinery.
3. Correct gaps had to be made to some interiors to let electrical cables pass from the sockets to their machinery.

Appropriate documentation was updated and distributed to those involved.

12.6.3 EXAMPLE House: Process schedule to use

To follow this example, these are the alternatives (see Pt:2:2b., p. 167):

*   Trace *upstream* in the overall *schedule* to understand how and where detected failures are eliminated. Since the house is built outside-in, a failure can be located both *inwards* and *outwards* with respect to where it is detected.

2V Proceed to chapter 12.8, "EXAMPLE House kitchen: Fully verify room black-/white-box," page 684.

*   To stay within the house example, but not following the overall *schedule*, proceed to chapter 12.7 below.

Proceed to chapter 12.3 below to leave the house example and to dig deeper into the verify black- & white-box topic.

12.7 EXAMPLE: House rooms machinery: Fully verify machinery black-/white-boxes

12.7.1 EXAMPLE House: About this example

It may be hard to verify rooms as stand-alone parts of a house, because houses are developed according to *invasive integration*, which unfortunately results in *interfaces* that most often get hidden in walls, floors, and ceilings. If it is a strong *requirement* to be able to verify rooms as stand-alone parts (or a flat in an apartment building), these *interfaces* must be designed for *verification* to become easily accessible, such as providing *observation posts* in each room (apartment) for gas, electricity, water, sewage, and so forth.

Also note that when verifying on an innermost machinery *nesting level* after *invasive integration*, the failure can be located on any *nesting level*, because the house is now totally completed. For example, if the gas burner doesn't light, it can be due to failures in the burner itself, to clogged pipes or damaged valves in the kitchen or elsewhere in the house, or even to a gas shortage from the *supplier* in the *environment*. If the burner had been verified separately, gas would have been provided in the *verification fixture*, and the failure could not have been anywhere else than in the burner.

Since a house is visible and tangible and easy to understand, it may be important to also apply the *verification phase* of the *process schedule*. Dear reader, have some mercy if the example should occasionally get somewhat strained.

12.7.2 EXAMPLE House kitchen machinery: Process schedule to use

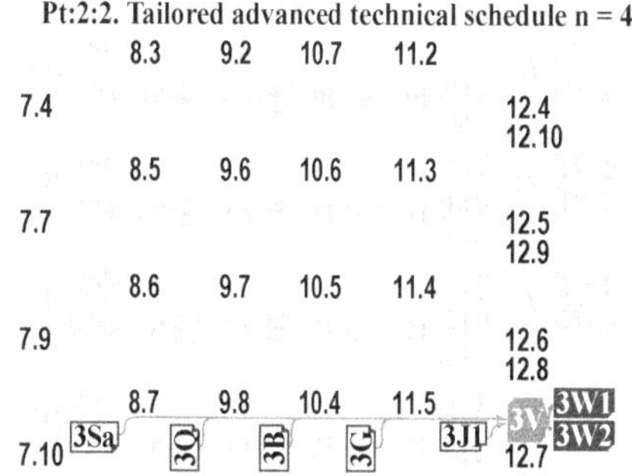

For the house example, a special technical overall *schedule* is *tailored* (illustrated in Fig. 12-10, left):

- **Pt:2:2. Tailored advanced technical schedule n = 4,** page 164

In this *tailored schedule* the generic *schedule* to use is (filled-in symbols):

- **Pt.n.V. Verify black-/white-box,** page 152.

FIGURE 12-10 Position of schedule and master to use (filled in) in overall schedule

Below *results* may be demanded (their *masters* are indicated *upstream* from the filled-in *schedule* in the figure above):

3Jt Completed usable machinery white-boxes
Table 11-22, House kitchen machinery start-up report 1, page 622

3G Finalized machinery items design with product requisites
Table 10-10, House kitchen machinery white-box design items, page 447

3B Machinery architectures with major ingredients but no black-boxes
Table 9-18, House kitchen machinery elements architecture ingredients resumé, page 373

Table 9-21, House kitchen machinery user interface architecture ingredients resumé, page 375

Table 9-19, House kitchen machinery connection interface architecture ingredients resumé, page 373

Table 9-20, House kitchen machinery boundaries interface architecture ingredients resumé, page 374

Figure 9-29, House kitchen machinery logical architecture, page 376

Figure 9-30, House kitchen machinery physical architecture, page 377

[3Q] Machinery solution alternatives with supplier opportunities

Table 8-12, House white-box solution alternatives, page 301

Table 8-13, House white-box supplier opportunities, page 303

[3Sa] Machinery prioritized requirements

Table 7-13, House kitchen machinery black-box compound requirements, page 200

12.7.3 EXAMPLE House kitchen machinery: Develop test-cases, also for fixtures and debugger

Requirements specified in Table 7-13, page 200 are the most important *results* to *demand* when developing *test-cases* (see Table 12-8 below).

TABLE 12-8 House kitchen machinery test-cases

Check design and/or check behavior	Expected outcome with pass/fail criteria	Requirement
A.aC.aA. Cold machinery		
CH.1. Measure volume by decimeter ruler. Apply a min/max thermometer. Measure one day after setting to 4 °C and 18 °C (room-temperature above 18 °C).	**XP.1. Pass if volume bigger than required and temperature maintained for 24 hours within 2 degree from set value.**	A.aC.aA.R.6.4.6.1. Fridge shall hold 4 bags (of 30 liters each) of chilled groceries.
CH.1. Measure volume by decimeter ruler. Apply a max/min thermometer. Measure one day after setting to -20 °C.	**XP.1. Pass if volume bigger than requested and temperature within 3 degree from set value.**	A.aC.aA.R.6.4.6.2. Freezer shall hold 1 bag of 30 liters frozen food.
CH.1. Leave the freezer with open door 24 hours to create a lot of frost, and then shut the freezer.	**XP.1. Pass if defrosted after 24 hours after door being shut.**	A.aC.aA.R.6.4.6.3. Fridge and freezer shall be maintenance-free.
A.aC.aB. Warm machinery		
CH.1. Start oven at room temperature and then set to +100 °C. Measure temperature with a thermometer.	**XP.1. Pass if set temperature achieved within 15 minutes and maintained for 2 hours within 10 degrees of set value.**	A.aC.aB.R.6.3.1.1. Equipment shall keep cooked food warm for 2 hours.
CH.1. Place 100-gram ice cubes (~ 2 cm sides) in a plastic box and turn on thawing.	**XP.1. Pass if all parts of the ice melt in less than 2 minutes.**	A.aC.aB.R.6.6.1.1 Thawing shall work on simple food in plastic containers.
CH.1. After above check is made, also check that nothing in the oven has been warmed up.	**XP.1. Pass if everything in oven is still below body temperature.**	A.aC.aB.R.6.6.1.2. When thawing, no parts of the food may get cooked.
CH.1. Check that there is a control for thawing speed.	**XP.1. Pass if there is such control.**	A.aC.aB.R.6.6.1.3. Thawing speed must be adjustable.
CH.1. Check the time of warming 1 liter of water from 20 °C to 70 °C in a crockery bowl.	**XP.1. Pass if it takes less than 4 minutes.**	A.aC.aB.R.6.6.2.1 Ready-to-serve food shall be warmed on heat-proof plates.
A.aC.aC. Bake machinery		

TABLE 12-8 House kitchen machinery test-cases

Check design and/or check behavior	Expected outcome with pass/fail criteria	Requirement
CH.1. Let a competent chef prepare a soufflé and bake it in the oven.	XP.1. Pass if the soufflé rises as expected and looks good.	A.aC.aC.R.6.6.6.1 Advanced baking shall be possible.
CH.1. Prepare grilled sausages and crusted Christmas ham.	XP.1. Pass if tasting and looking good.	A.aC.aC.R.6.6.6.2. Warming, grilling, and crusting shall be possible.
A.aC.aD. Heat machinery		
CH.1. Check the time from 20 °C to boiling of 1 liter of water in a metal pan.	XP.1. Pass if it takes less than 4 minutes.	A.aC.aD.R.6.6.2.1. Fast warming when cooking shall work on food in metal pots.
CH.1. Check to boil water in porcelain pan, metal pan, and wok pan.	XP.1. Pass if possible to boil in all pans (without damaging the pan) on at least one heat machinery.	A.aC.aD.R.6.6.3.1 Boiling food shall work with all kinds of pans.
CH.1. Check that all heating machineries have adjustable heat supply in more than 5 steps.	XP.1. Pass if heat supply can be easily adjusted continuously or at least in 5 steps.	A.aC.aD.R.6.6.3.2. Heat supply shall be adjustable and easy to control.
CH.1. Use a 0,5 liter pan with 2 dl water and adjust to barely boil (steaming but not bubbling).	XP.1. Pass if possible to simmer one hour without losing more than 1 dl.	A.aC.aD.R.6.6.3.3. Power shall be down-adjustable for small pans to simmer.
CH.1. Fry thin bacon in a 12-inch flat-bottom pan.	XP.1. Pass if the bacon fries in a controlled way on at least one heat machine.	A.aC.aD.R.6.6.4.1 Frying shall work on wide- and flat-bottom pans
CH.1. Use a pan with 2 dl water and adjust to barely boil (clearly bubbling but not rolling boil).	XP.1. Pass if possible to reduce to below 1 dl within one hour.	A.aC.aD.R.6.6.4.2. Power shall be down-adjustable, e.g., for thickening a sauce.
CH.1. Use a 30 cm wok pan on burner and fill with 1 liter of 20 °C water.	XP.1. Pass if water boils within half a minute.	A.aC.aD.R.6.6.5.1 Wokking shall be possible in cupped-bottom pans.
CH.1. Use a 30 cm wok pan on burner and fill with 3 dl cooking oil.	XP.1. Pass if oil boils within 15 seconds.	A.aC.aD.R.6.6.5.2. Heat shall bring 3 dl oil to almost instant boil.
A.aC.aE. Ventilation machinery		
CH.1. Generate smoke under the kitchen arch. Start the hood.	XP.1. The smoke shall move in direction of the hood.	A.aC.aE.R.6.7.1.1 Air evacuation from kitchen shall be stronger than air inflow.
A.aC.aF. Wet machinery		
CH.1. Test to fill a 12-liter bucket using only 1 hand. Test to fill ordinary 75 cl wine bottle.	XP.1. Pass if possible under 1 minute, without tools and without any water spilling outside.	A.aC.aF.R.6.4.1.1. Cold- and warm-water faucet shall be easily accessible.
CH.1. Run exclusively warm water on highest speed for 1 minute and then measure temperature.	XP.1. Pass if between 50 °C and 60 °C.	A.aC.aF.R.6.4.1.2. Water temperature should be efficient but not harmful.
CH.1. Check that there is running warm and cold water with drainage.	XP.1. Pass if any temperature between warm and cold can be settled and obtained.	A.aC.aF.R.6.5.3.1 Dishwashing by hand shall be possible.
A.aC.aG. Dish machinery		
CH.1. Wash the flat-bottom pan from above bacon frying in the coarse dishwashing machine. Wash used wine glasses in the fine machine.	XP.1. Pass if all food and grease stains are washed away.	A.aC.aG.R.6.5.1.1 One dishwasher for dirty utensils, one for cleaning tableware.
CH.1. Use a rack for wine glasses with dimensions 20 cm high, 7 cm wide and 3 dl in volume.	XP.1. Pass if possible to wash 25 of these glasses in dishwasher simultaneously without risk of breakage.	A.aC.aG.R.6.5.1.2. Glasses shall be loadable in a rack.

The golden rule is to always keep *requirements* consistent with *test-cases.*

EXAMPLE House kitchen machinery: Verify black-/white-box

The documentation of the *verification* outcome can be more or less abundant. In this example, only "Pass" is specified when the outcome is approved. However, if a failure is detected, it should be much better described to be understood by those engineers who will locate and eliminate the failure (see Table 12-9 below).

TABLE 12-9 House kitchen machinery verification report 1

Actual outcome and pass/fail conclusion	Expected outcome and pass/fail criteria	Requirement
VR.1. Pass	XP.1.	A.aC.aA.R.6.4.6.1.
VR.1. Pass	XP.1.	A.aC.aA.R.6.4.6.2.
VR.1.1. Fail. After door opened, freezer got too warm and no frost was created to test defrosting.	XP.1. Pass if defrosted after 24 hours after door being shut..	A.aC.aA.R.6.4.6.3. Fridge and freezer shall be maintenance-free.
VR.1. Pass	XP.1.	A.aC.aB.R.6.3.1.1.
VR.1. Pass	XP.1.	A.aC.aC.R.6.6.6.1
VR.1. Pass	XP.1.	A.aC.aC.R.6.6.6.2.
VR.1. Pass	XP.1.	A.aC.aD.R.6.6.3.1
VR.1. Pass	XP.1.	A.aC.aD.R.6.6.3.2.
VR.1. Pass	XP.1.	A.aC.aD.R.6.6.3.3.
VR.1. Pass	XP.1.	A.aC.aD.R.6.6.4.1
VR.1. Pass	XP.1.	A.aC.aD.R.6.6.4.2.
VR.1. Pass	XP.1.	A.aC.aD.R.6.6.5.1
VR.1.2. Fail. To boil 3 dl oil took 25 seconds.	XP.1. Pass if oil boils within 15 seconds.	A.aC.aD.R.6.6.5.2. Heat shall bring 3 dl oil to almost instant boil.
VR.1. Pass	XP.1.	A.aC.aE.R.6.7.1.1
VR.1. Pass	XP.1.	A.aC.aF.R.6.4.1.1.
VR.1.3. Fail.Temperature at 69 °C.	XP.1. Pass if between 50 °C and 60 °C..	A.aC.aF.R.6.4.1.2. Water temperature should be efficient but not harmful.
VR.1. Pass	XP.1.	A.aC.aG.R.6.5.1.1
VR.1. Pass	XP.1.	A.aC.aG.R.6.5.1.2.

12.7.4 EXAMPLE House kitchen machinery: Locate failure source and evaluate optimal elimination

The four failures detected must have their sources located in the prototype, and it must be investigated where in the *process schedule* the failure arose, to know where to send the prototype for failure elimination. To find failure locations in prototypes and *process schedules* may require a substantial effort, and developers from all around the company may be asked for assistance. The location of the failures must be documented (see column "failure location" in Table 12-10 below).

This table also serves as documentation for the gate meeting, which should review and approve the action for *failure elimination.* Those decisions are documented in column "failure elimination plan" in Table 12-10 below.

TABLE 12-10 House kitchen machinery failure elimination report 1

Failure loca-tion	Failure elimination plan	Actual outcome and pass/fail criteria	Expected outcome and pass/fail criteria	Requirement
FER.1.1. Incorrect test-case in activity verify kitchen machinery 3V	FER.1.1.a.Test-case check incorrect. Update test-case to leave door 2 cm open. FER.1.1.b. Reverify.	VR.1.1. Fail. After door opened, freezer got too warm and no frost was created to test defrosting.	XP.1. Pass if defrosted after 24 hours after door being shut..	A.aC.aA.R.6.4.6.3. Fridge and freezer shall be maintenance-free.
FER.1.2. Wrong item selected in finalize kitchen machinery 3F	FER.1.2.a.Update requisite with more powerful burner. FER.1.2.b. Procure burner at K-Studio cost. FER.1.2.c. Reverify.	VR.1.2. Fail. To boil 3 dl oil took 25 seconds.	XP.1. Pass if oil boils within 15 seconds.	A.aC.aD.R.6.6.5.2. Heat shall bring 3 dl oil to almost instant boil.
FER.1.3. Missing requirement at capture kitchen machinery 3R	FER.1.3.a.This is a requirement added after contracting the builder. FER.1.3.b. Update finalize kitchen machinery with a warm/cold water mixer. FER.1.3.c. Install mixer at proprietor cost. FER.1.3.d. Set mixer to 55 °C. FER.1.3.e. Reverify.	VR.1.3. Fail.Temperature at 69 °C.	XP.1. Pass if between 50 °C and 60 °C..	A.aC.aF.R.6.4.1.2. Water temperature should be efficient but not harmful.

12.7.5 EXAMPLE House kitchen machinery: Verify black-/white-box

The list is sent out to those organizations that caused the failures, to be eliminated according to the elimination decisions. After all failures have been reported eliminated, the house kitchen machinery needs to be reverified, in order ensure that the failures are totally eliminated. The *task* **Pt.n.Vc. Verify black-/white-box** used at the first *verification* number 1 may be entered once again.

The four *test-cases* that failed are verified again in the prototype and a new re*verification report* is established (see Table 12-11 below).

TABLE 12-11 House kitchen machinery verification report 2

Actual outcome and pass/fail conclusion	Expected outcome and pass/fail criteria	Requirement
VR.2. Pass	XP.1.	A.aC.aA.R.6.4.6.3.
VR.2. Pass	XP.1.	A.aC.aD.R.6.6.5.2.
VR.2. Pass	XP.1.	A.aC.aF.R.6.4.1.2.

As can be seen, all failures are now eliminated.

12.7.6 EXAMPLE House kitchen machinery: Prepare product manual

In fact, it was rather easy to prepare the *user* manual, because all the machinery was delivered with manuals attached (see Table 12-12 below).

TABLE 12-12 House kitchen machinery manual table of contents

Manual name	Producer and type	Element / interface
MA.1. W34-76 user manual (eng)	Wirly W34-76	A.aC.aA.a. Combined fridge/freezer
MA.2. WMic 7/8/9 user manual (eng)	Wirly WMic 8	A.aC.aB.a. Micro oven
MA.3. W-series user manual (eng)	Wirly W+ 4800	A.aC.aB.b. Warming oven
MA.4. WZ-1000/2000 operating (eng)	Wirly WZ-2000	A.aC.aC.a. Wide oven
MA.5. E4 W2 my manual (eng)	Wirly E4 W2	AaC.aDa. Electrical hubs
MA.6. Use your Ex burner (eng)	Buttan Mall-2 Ex	A.aC.aD.b. Gas burner
MA.7. Vac 3411 user manual (eng)	Wirly Vac 3411	A.aC.aE.a. Wide hood
MA.8. Clean your plating (eng)	Plating K2 Fe	A.aC.aF.a. Sink 2 bowls
MA.9. No user manual issued	Plating Z34-Px	A.aC.aF.b. Combined faucet
MA.10. Operate Bull 4 (eng)	Teller Inc Bull-4-30	A.aC.aG.a. Dishwasher coarse
MA.11. Operate Cat 6 (eng)	Teller Inc Cat-6-40	A.aC.aG.b. Dishwasher gentle

12.7.7 EXAMPLE House: Proceed reading

To follow this example, these are the alternatives (see again Fig. 12-10, p. 679):

- Trace *upstream* in the overall *schedule* to understand how and where detected failures are eliminated. Since the house is built outside-in, a failure can be located both *inwards* and *outwards* in respect to where it is detected.

- 2V Proceed to chapter 12.8, "EXAMPLE House kitchen: Fully verify room black-/white-box" below.

12.8 EXAMPLE House kitchen: Fully verify room black-/white-box

12.8.1 About the house example

The *verification* has been finished on the *innermost black-/white-boxes*, which now are found to be failure-free. This means that:

- No *test-case* results in unexpected outcome. However, it is not entirely certain that prepared *test-cases* reflect the *requirements* correctly or fully. Meticulous reviews by developers are needed to secure a correct relationship between *requirements* and *test-cases*.

- *Requirements* can be incorrect, ambiguous, or inadequate. Meticulous reviews by *stakeholders* are needed to ensure that *requirements* correctly reflect their desires.

It is now time to continue with *black-/white-boxes* on the next *nesting level outwards.*

12.8.2 EXAMPLE House kitchen: Process schedule to use

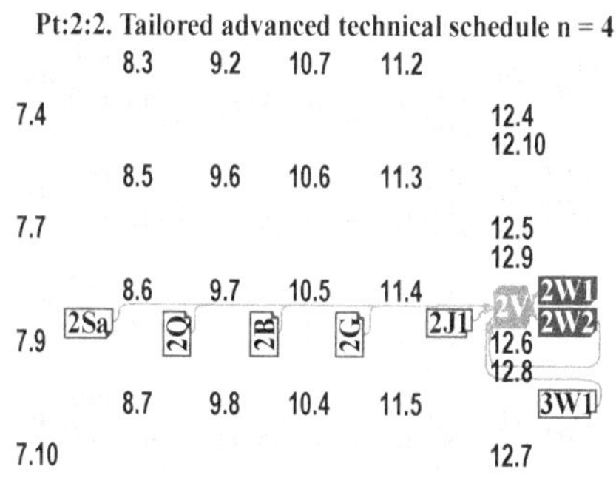

For the house example, a special technical overall *schedule* is *tailored* (illustrated in Fig. 12-11, left):

- **Pt:2:2. Tailored advanced technical schedule n = 4,** page 164

In this *tailored schedule* the generic *schedule* to use is (filled-in symbols):

- **Pt.n.V. Verify black-/white-box,** page 152.

FIGURE 12-11 Position of schedule and master to use (filled in) in overall schedule

Below *results* may be demanded (their *masters* are indicated *upstream* from the filled-in *schedule* in the figure above or greater detail in Fig. 11-5, p. 602 and Fig. 11-6, p. 603):

3W1 Fully verified machinery black-/white-boxes

    Table 12-9, House kitchen machinery verification report 1, page 682

    Table 12-11, House kitchen machinery verification report 2, page 683

2W2 Partially verified room black-/white-boxes

    List 12-3. House rooms with fake machinery failure elimination report 1, page 678

2J1 Completed usable room white-boxes

    Table 11-18, House kitchen start-up report 1, page 617

2G Finalized room items design with product requisites

    Figure 10-7, House kitchen interior drawing, page 455

    Figure 10-8, House kitchen connections, page 456

    Table 10-18, House kitchen electricity white-box design items, page 456

    Table 10-19, House kitchen electricity white-box design items, page 457

    Table 10-20, House kitchen interiors white-box design items (cabinets for machinery), page 457

    Table 10-21, House kitchen interiors white-box design items (cabinets without machinery), page 458

    Table 10-22, House kitchen interiors white-box design items (cabinet static interfaces), page 459

    Table 10-23, House kitchen boundaries white-box design items, page 459

2B Room architectures with major ingredients and machinery black-boxes

Table 9-16, House kitchen elements architecture ingredients resumé, page 367

Table 9-17, House kitchen interfaces architecture ingredients resumé, page 368

Figure 9-27, House kitchen physical architecture, page 370

Figure 9-26, House kitchen logical architecture, page 369

[2Q] Room solution alternatives with supplier opportunities

Table 8-14, House kitchen white-box solution alternatives, page 306

Table 8-15, House kitchen white-box supplier opportunities, page 308

[2Sa] Rooms prioritized requirements

Table 7-11, House kitchen black-box compound requirements (small selection), page 195

12.8.3 EXAMPLE House kitchen: Develop test-cases, also for fixtures and debugger

As usual, the *test-cases* are documented (see Table 12-13 below).

TABLE 12-13 House kitchen test-cases

Check design and/or check behavior	Expected outcome with pass/fail criteria	Requirement
CH.1. Place the chef at the food preparation area.	XP.1. Pass if the chef can see the lake through the window and still reach the preparation area with one hand.	A.aC.R.1.1.1. The lake shall be visible from kitchen food preparation area.
CH.1. Check if a person can move around freely in a wheelchair.	XP.1. Pass in either case.	A.aC.R.2.2.1. Kitchen may be inaccessible by disabled guests.
CH.1. Building engineer checks kitchen against norms, regulations, and standards.	XP.1. Pass if inspector finds that norms, regulations, and standards are met.	A.aC.R.3.1.1. Kitchen shall follow established norms, regulations, and standards.
CH.1. Check surfaces on all interiors, including machinery.	XP.1. Pass if no scratches, stains or any other damages can be detected.	A.aC.R.3.1.2. Kitchen interiors may not have any damage, scratches, or stains.
CH.1. Proprietor checks kitchen interiors and machinery appearance.	XP.1. Pass if proprietor has chosen all interior appearances.	A.aC.R.3.1.3. Kitchen interior appearance must be chosen by proprietor.
CH.1. Check the requisites cost summary of the kitchen room boundaries, interiors, and machines.	XP.1. Pass if the sum is below € 20 000.	A.aC.R.4.1.1. Kitchen including interiors shall cost maximally 20 000 € .
CH.1. Check how many suppliers have delivered interiors.	XP.1. Pass if there are three or fewer major interior suppliers.	A.aC.R.4.1.2. Purchase only ordinary interiors from few suppliers to keep prices down.
CH.1. Identify the equipment to keep one course warm.	XP.1. Pass if cost less than € 100. If it can be used for other cooking, pass if less than € 200.	A.aC.R.6.3.1. Cheap equipment keeps one course warm during eating.
CH.1. Identify the preparation area and the water supply.	XP.1. Pass if preparation-free area bigger than 1 m² and if chef can reach minimum 50 x 50 cm preparation area with one hand, while simultaneously operating warm- and cold-water faucets with the other hand.	A.aC.R.6.4.1. Include large food preparation area with hot and cold water.

TABLE 12-13 House kitchen test-cases

Check design and/or check behavior	Expected outcome with pass/fail criteria	Requirement
CH.1. Store 12 pans holding 36 liters in total near cooking area.	XP.1. Pass if possible to store 12 pans of 36 liters at the same time as all other storage is verified.	A.aC.R.6.4.2. Storage shall hold 12 pans of 3 liters near cooking area.
CH.1. Store 6 plates for baking oven and store 6 bowls at totally 12 liters near baking area.	XP.1. Pass if possible to store 6 plates and 6 bowls of 12 liters at the same time as all other storage is verified.	A.aC.R.6.4.3. Storage shall hold 6 plates and 6 bowls near baking area.
CH.1. Identify the knife hanger.	XP.1. Pass if chef can reach knives with one hand, and a preparation area of minimally 50 x 50 cm with the other hand, and if a 6-year-old child standing on the floor cannot reach any of them.	A.aC.R.6.4.4. Knife storage shall be child-safe and near preparation area.
CH.1. Store needed types of utensils for cooking 5-course menu.	XP.1. Pass if utensils of needed type can be stored, and can be reached by chef with one hand, and simultaneously the chef's other hand can reach preparation area of minimum 50 x 50 cm.	A.aC.R.6.4.5. Storage shall hold all kinds of cutlery and kitchen utensils which shall be easily accessible from preparation area.
CH.1. Fill 3 grocery bags of 30 liters with room-temperature food, 4 bags with chilled food, and 1 bag with frozen food.	XP.1. Pass if possible to store the 8 grocery bags of 30 litres at the same time as all other storage is verified.	A.aC.R.6.4.6. Storage shall hold 8 bags (of 30 liters each) of groceries, 4 of which are chilled, and 1 frozen.
CH.1. Identify dishing equipment. At the same time, load these with dirty and barely stained tableware, with sensitive barely stained tableware and with dirty utensils.	XP.1. Pass if everything can be washed without being moved around during the washing, and is perfectly cleaned and not damaged.	A.aC.R.6.5.1. Dishwashing shall clean 5-course tableware (5 x 8 plates, cutlery, glasses, and cooking utensils) in same run.
		A.aC.R.6.5.2. Gentle dishwashing shall support fragile tableware.
CH.1. Identify the safe storage.	XP.1. Pass if 6-year-old child cannot reach any chemicals.	A.aC.R.6.5.3. Child-safe storage shall hold chemicals for dishwashing etc.
CH.1. Identify verified thawing equipment.	XP.1. Pass if equipment verified and chef can thaw food with one hand and simultaneously reach a preparation area of minimally 50 x 50 cm with the other hand.	A.aC.R.6.6.1. Food thawing shall be near preparation area.
CH.1. Identify verified fast-warming area.	XP.1. Pass if equipment verified and chef can heat food with one hand and simultaneously reach minimum 50 x 50 cm preparation area with the other hand.	A.aC.R.6.6.2. Warming of all kinds of food shall be near preparation area.
CH.1. Identify verified boiling equipment.	XP.1. Pass if equipment verified and chef can boil food with one hand and simultaneously reach minimum 50 x 50 cm preparation area with the other hand.	A.aC.R.6.6.3. Food boiling shall be near preparation area.
CH.1. Identify verified frying equipment.	XP.1. Pass if equipment verified and chef can fry food with one hand and simultaneously reach minimum 50 x 50 cm preparation area with the other hand.	A.aC.R.6.6.4. Food frying shall be near preparation area.

TABLE 12-13 House kitchen test-cases

Check design and/or check behavior	Expected outcome with pass/fail criteria	Requirement
CH.1. Identify verified wokking equipment.	XP.1. Pass if equipment verified and chef can wok food with one hand and simultaneously reach minimum 50 x 50 cm preparation area with the other hand.	A.aC.R.6.6.5. Food wokking shall be near preparation area.
CH.1. Identify verified baking equipment.	XP.1. Pass if equipment verified and chef can bake food with one hand and simultaneously reach minimum 50 x 50 cm preparation area with the other hand.	A.aC.R.6.6.6. Food baking shall be near preparation area.
CH.1. Start the hood at maximum suction. Release a "stink bomb" (ammonium sulfide) at the stove top	XP.1. Pass if no significant odors occur in the dining room.	A.aC.R.6.7.1. Odor from cooking may not spread to other rooms.
CH.1. Release an ammonium sulfide "stink bomb" in the garbage.	XP.1. Pass if no significant "stink bomb" odor occurs in the kitchen.	A.aC.R.6.8.1. Garbage odor may not be recognizable in kitchen.
CH.1. Fill 2 grocery bags of 30 liters with snack meals.	XP.1. Pass if possible at the same time that all other storage is verified.	A.aC.R.7.3.1. Storage shall hold 24 snack meals (2 grocery bags).
CH.1. Check if kitchen holds child-safe storage.	XP.1. Pass if 6-year-old child cannot reach storage.	A.aC.R.9.1.1. Kitchen shall hold family medicines in child-safe way.

12.8.4 EXAMPLE House kitchen: Verify black-/white-box

The *verification* is performed and documented (see Table 12-14 below).

TABLE 12-14 House kitchen verification report 1

Actual outcome and pass/fail conclusion	Expected outcome and pass/fail criteria	Requirement
VR.1. Pass	XP.1.	A.aC.R.1.1.1.
VR.1. Pass	XP.1.	A.aC.R.2.2.1.
VR.1. Pass	XP.1.	A.aC.R.3.1.1.
VR.1. Pass	XP.1.	A.aC.R.3.1.2.
VR.1. Pass	XP.1.	A.aC.R.3.1.3.
VR.1. Pass	XP.1.	A.aC.R.4.1.1.
VR.1.1. Fail. far too many different suppliers of interiors found	XP.1. Pass if there are three or fewer major interior suppliers.	A.aC.R.4.1.2.Purchase only ordinary interiors from few suppliers to keep prices down.
VR.1. Pass	XP.1.	A.aC.R.6.3.1.
VR.1. Pass	XP.1.	A.aC.R.6.4.1.
VR.1.2. Fail. Only 10 pans possible	e	A.aC.R.6.4.2. Storage shall hold 12 pans of 3 liters near cooking area.
VR.1.3. Fail. Only 4 bowls possible	XP.1. Pass if possible to store 6 plates and 6 bowls of 12 liters at the same time as all other storage is verified.	A.aC.R.6.4.3. Storage shall hold 6 plates and 6 bowls near baking area.
VR.1. Pass	XP.1.	A.aC.R.6.4.4.
VR.1. Pass	XP.1.	A.aC.R.6.4.5.
VR.1. Pass	XP.1.	A.aC.R.6.4.6.

TABLE 12-14 House kitchen verification report 1

Actual outcome and pass/fail conclusion	Expected outcome and pass/fail criteria	Requirement
VR.1. Pass	XP.1.	A.aC.R.6.5.1.
VR.1. Pass	XP.1.	A.aC.R.6.5.2.
VR.1. Pass	XP.1.	A.aC.R.6.5.3.
VR.1. Pass	XP.1.	A.aC.R.6.6.1.
VR.1. Pass	XP.1.	A.aC.R.6.6.2.
VR.1. Pass	XP.1.	A.aC.R.6.6.3.
VR.1. Pass	XP.1.	A.aC.R.6.6.4.
VR.1. Pass	XP.1.	A.aC.R.6.6.5.
VR.1. Pass	XP.1.	A.aC.R.6.6.6.
VR.1. Pass	XP.1.	A.aC.R.6.7.1.
VR.1.4. Fail. Odor was immediately spread through kitchen	XP.1. Pass if no significant "stink bomb" odor occurs in the kitchen.	A.aC.R.6.8.1. Garbage odor may not be recognizable in kitchen.
VR.1. Pass	XP.1.	A.aC.R.7.3.1.
VR.1. Pass	XP.1.	A.aC.R.9.1.1.

12.8.5 EXAMPLE House kitchen: Locate failure source and evaluate optimal elimination

All detected failures in the kitchen *verification report* are investigated by developers, and all sources are located (see Table 12-15 below). A gate meeting is arranged to review the kitchen *verification report*, and decide how to eliminate the located failures.

TABLE 12-15 House kitchen failure elimination report 1

Failure location	Failure elimination plan	Actual outcome and pass/fail criteria	Expected outcome and pass/fail criteria	Requirement
FER.1.1. Too many suppliers at procure kitchen machinery [31]	FER.1.1.a. Try to reduce to three suppliers. FER.1.1.b. K-Studio contract questioned. Claim 10% price reduction on interiors.	VR.1.1. Fail. far too many different suppliers of interiors found	XP.1. Pass if there are three or fewer major interior suppliers.	A.aC.R.4.1.2. Purchase only ordinary interiors from few suppliers to keep prices down.

TABLE 12-15 House kitchen failure elimination report 1

Failure location	Failure elimination plan	Actual outcome and pass/fail criteria	Expected outcome and pass/fail criteria	Requirement
FER.1.2. Too small storage places at kitchen verification 2V	FER.1.2.a.Update test-case to allow mixing large utensils with baking bowls, and also use and mix storage with small utensils storage place. FER.1.2.b. Reverify.	VR.1.2. Fail. Only 10 pans possible.	XP.1. Pass if possible to store 12 pans of 36 liters at the same time as all other storage is verified.	A.aC.R.6.4.2. Storage shall hold 12 pans of 3 liters near cooking area.
		VR.1.3. Fail. Only 4 bowls possible	XP.1. Pass if possible to store 6 plates and 6 bowls of 12 liters at the same time as all other storage is verified.	A.aC.R.6.4.3. Storage shall hold 6 plates and 6 bowls near baking area.
FER.1.3. Insufficient construction of garbage storage at kitchen machinery architecture 3A ·	FER.1.3.a.Update kitchen machinery architecture with a small (low noise) 220 V wall fan. FER.1.3.b. Update dependent architecture and finalize design of kitchen machinery, kitchen, house, and environment. FER.1.3.c. Ask Elert AB to mount another AC/DC plug behind the garbage storage according to updated drawings. FER.1.3.d. Ask Mobber AB to drill a hole through the outer wall according to updated drawings. FER.1.3.e. Ask K-Studio to mount the fan. FER.1.3.f. Reverify.	VR.1.4. Fail. Odor was immediately spread through kitchen	XP.1. Pass if no significant "stink bomb" odor occurs in the kitchen..	A.aC.R.6.8.1. Garbage odor may not be recognizable in kitchen.

12.8.6 EXAMPLE House kitchen: Verify black-/white-box

Sometimes a *test-case* is shown to demand more than the *requirements* actually capture, or a *requirement* must be compromised a bit, especially if it is not of high priority.

If a *test-case* is to be compromised, the *requirement* should also be changed to reflect this. If the change to a *test-case* is only minor (see Table 12-16 below), at least the **Pt.n.Vj. Evaluate optimal elimination** gate meeting should be notified for approval of the change.

TABLE 12-16 Updated house kitchen test-cases

Check design and/or check behavior	Expected outcome with pass/fail criteria	Requirement
CH.1. Store 12 pans holding totally 36 liters, store 6 plates for baking oven, and store 6 bowls at totally 12 liters,in all storage places near baking and cooking areas.	XP.1. Pass if specified number of pans, plates, and bowls can be practically stored near baking and cooking areas.	A.aC.R.6.4.2. Storage shall hold 12 pans of 3 liters near cooking area.
		A.aC.R.6.4.3. Storage shall hold 6 plates and 6 bowls near baking area.
		A.aC.R.6.4.5. Storage shall hold all kinds of cutlery and kitchen utensils which shall be easily accessible from preparation area.

After the update of the *test-case*, a *reverification* was done (see Table 12-17 below). Still the A.aC.R.4.1.2. *requirement* was not satisfied.

TABLE 12-17 House kitchen verification report 2

Actual outcome and pass/fail conclusion	Expected outcome and pass/fail criteria	Requirement
VR.2.1. Fail. Four different suppliers of machinery still found	XP.1. Pass if there are three or fewer major interior suppliers.	A.aC.R.4.1.2.Purchase only ordinary interiors from few suppliers to keep prices down.
VR.2. Pass	XP.1. Pass if specified number of pans, plates, and bowls can be practically stored near baking and cooking areas.	A.aC.R.6.4.2. Storage shall hold 12 pans of 3 liters near cooking area.
		A.aC.R.6.4.3. Storage shall hold 6 plates and 6 bowls near baking area.
		A.aC.R.6.4.5. Storage shall hold all kinds of cutlery and kitchen utensils which shall be easily accessible from preparation area.
VR.2. Pass	XP.1. Pass if no significant "stink bomb" odor occurs in the kitchen.	A.aC.R.6.8.1. Garbage odor may not be recognizable in kitchen.

12.8.7 EXAMPLE House kitchen: Locate failure source and evaluate optimal elimination

There is no need to investigate the repeated failure for a second time (see Table 12-18 below).

TABLE 12-18 House kitchen machinery failure elimination report 2

Failure location	Failure elimination plan	Actual outcome and pass/fail criteria	Expected outcome and pass/fail criteria	Requirement
FER.2.1. Still too many suppliers after a few items are exchanged	FER.2.1.a.K-Studio contract is proposed to be canceled. FER.2.1.b. Claim 5% price reduction on interiors. FER.2.1.c. Accepted by K-Studio, no reverification needed	VR.2.1. Fail. Four different suppliers of machinery still found	XP.1. Pass if there are three or fewer major interior suppliers.	A.aC.R.4.1.2.Purchase only ordinary interiors from few suppliers to keep prices down.

The gate meeting concluded that the first time it was rather easy to just exchange some cheaper *samples* to reduce the number of *suppliers*. But this time it is much more expensive to reduce by one more. The failure was decided to be resolved without exchanging more *samples*, and the *supplier* accepted to reduce 5% of the price, so there is no need for a second *reverification*.

12.8.8 EXAMPLE House kitchen: Prepare product manual

To write the kitchen manual is not as convenient as it was to do the house rooms machinery manual, since the kitchen is mostly assembled and developed to a unique room, and not bought as a whole from *suppliers*.

Only the table of contents is included in this book, because many of the manuals might be thicker than this entire book.

TABLE 12-19 House kitchen manual table of contents

Manual name	Producer and type	Element / interface
MA.1. Kitchen architectures	Robber HB	Drawings
MA.2. Kitchen interior architectures and ergonomic study	K-Studio	Drawings
MA.3. Kitchen finalized architectures	Mobber AB	Drawings
MA.4. Kitchen electricity drawings	Elert AB	Drawings
MA.5. Maintaining our cabinets (Eng)	IKEY	Cabinets
MA.6. Extending our hinges' life-span (Eng)	IKEY	Cabinet hinges
MA.7. Kitchen fuses	Elert AB	Distribution box
MA.8. Maintain kitchen (plaster type, paint types, color codes, etc.)	Mobber AB	Wall and ceiling

TABLE 12-20 Staircase space, loft, and other rooms user manual (table of contents)

Manual name	Producer and type	Element / interface
MA.1. Room architectures	Robber HB	Drawings
MA.2. Room finalized architectures	Mobber AB	Drawings
MA.3. Room electricity drawings	Elert AB	Drawings
MA.4. Reading the water scale	Wat Son	Scale
MA.5. Operating the boiler	Wat Son	Boiler
MA.6. House fuses	Elert AB	Main distr. box

12.8.9 EXAMPLE House: Proceed reading

To follow this example, these are the alternatives (see again Fig. 12-11, p. 685):

- Trace *upstream* in the overall *schedule* to understand how and where detected failures are eliminated. Since the house is built outside-in, a failure can be located both *inwards* and *outwards* with respect to where it is detected.

IV Proceed to chapter 12.9, "EXAMPLE House: Fully verify house black-/ white-box" below.

12.9 EXAMPLE House: Fully verify house black-/ white-box

12.9.1 EXAMPLE House: Process schedule to use

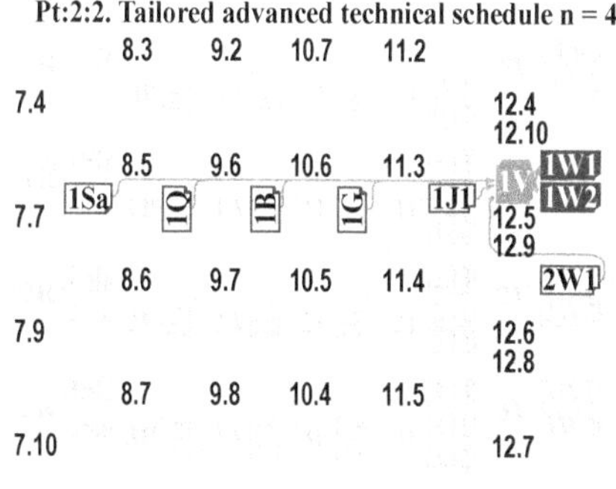

For the house example, a special technical overall *schedule* is *tailored* (illustrated in Fig. 12-12, left):

- **Pt:2:2. Tailored advanced technical schedule n = 4,** page 164

In this *tailored schedule* the generic *schedule* to use is (filled-in symbols):

- **Pt.n.V. Verify black-/white-box,** page 152.

FIGURE 12-12 Position of schedule and master to use (filled in) in overall schedule

Below *results* may be demanded (their *masters* are indicated *upstream* from the filled-in *schedule* in the figure above or in greater detail in Fig. 11-5, p. 602 and Fig. 11-6, p. 603):

2W1 Fully verified room black-/white-boxes

Table 12-14, House kitchen verification report 1, page 688
Table 12-17, House kitchen verification report 2, page 691

1W2 Partially verified house black-/white-boxes

List 12-2. House with fake rooms failure elimination report 1, page 676

1J1 Completed usable house white-box

Table 11-14, House start-up report 1, page 612

1G Finalized room items design with product requisites

Figure 10-10, House boundary casing (upper part), page 467
Figure 10-11, House boundary casing (lower part), page 468
Figure 10-12, House connections, page 469
Table 10-32, House boundary white-box design items, page 470
Table 10-33, House connections white-box design items, page 471
Table 10-34, House boundary and connections white-box design items, page 471

1B House architectures with major ingredients and room black-boxes

Table 9-13, House rooms black-boxes architecture ingredients resumé, page 360
Table 9-14, House interfaces architecture ingredients resumé, page 361
Table 9-15, House secondary space elements architecture ingredients resumé, page 362
Figure 9-22, House logical architecture, page 363
Figure 9-23, House physical architecture, page 364
Figure 9-24, House façade physical architecture, page 364

1Q House solution alternatives with supplier opportunities

Table 8-14, House kitchen white-box solution alternatives, page 306
Table 8-15, House kitchen white-box supplier opportunities, page 308

1Sa House prioritized requirements

Table 7-9, House black-box compound requirements (small selection), page 189

12.9.2 EXAMPLE House: Develop test-cases, also for fixtures and debugger

A table is used for documenting the *test-cases* (see Table 12-21 below). Normally most house *verifications* are very technically oriented, but in this book I also like to complement with a more user-oriented *view* on houses.

TABLE 12-21 House test-cases

Check design and/or check behavior	Expected outcome with pass/fail criteria	Requirement
CH.1. Place the chef at the cooking and dining areas.	XP.1. Pass if the chef can enjoy the lake from at least one of these places.	A.R.1.1. During cooking and dining, the lake shall be as visible as possible.
CH.1. Ask the proprietor if there are any façade elements not chosen by him.	XP.1. Pass if all major façade elements are chosen by the proprietor.	A.R.1.2. Façade and roofing material appearance shall be chosen by proprietor.
CH.1. Show the area to a landscape architect for a comment on house style.	XP.1. Pass if the architect doesn't identify this particular house to significantly differ from norm	A.R.2.1. The house style shall fit in with neighbors' houses.
CH.1. Enter house with wheelchair, move around in the house (except in the kitchen), and use the toilet.	XP.1. Pass if possible to enter, move around everywhere (except in the kitchen), and use toilet.	A.R.2.2. The house shall be accessible for disabled persons.
CH.1. Check if a permit application has been made before house construction began.	XP.1. When this verification takes place, the permit must have been received long before.	A.R.2.3. A building permit is mandatory before building starts.
CH.1. Have house engineer inspect the house.	XP.1. Pass, if house inspector finds the house in compliance with proven methods, regulations, and standards.	A.R.3.1. The house shall follow proven methods, regulations, and standards.
CH.1. Connect a gas meter. Measure the gas consumption for one day and night at calm weather and outside temperature 7° C (climate average temperature).	XP.1. Pass if gas consumption for one day and night is less than 1,35 Nm³ (corresponds to 90 kWh/m²a for the 60 m² house).	A.R.3.2. The house shall have an economical energy consumption.
CH.1. Have a bank appraiser inspect the house.	XP.1. Pass if the appraiser finds the house attractive for potential buyers.	A.R.4.1. The house shall be attractive to allow maximum bank mortgage.
CH.1. Check the utility room of the house.	XP.1. Pass if all connections are well arranged in the utility room in a secondary space.	A.R.5.1. Provide the house with a utility room for all public utilities.
CH.1. Release an ammonium sulfide "stink bomb" in the middle of the kitchen.	XP.1. Pass if possible to evacuate kitchen air, so no significant smell can be detected in dining room.	A.R.6.7. Smell from cooking shall be eliminated.
CH.1. Check if the dining room is almost noiseless, with soft music and dimmed light.	XP.1. Pass if chef finds the dining room sufficiently relaxing.	A.R.6.1. The atmosphere during dining shall be relaxed.
CH.1. Check if the dining room easily holds 8 sitting guests.	XP.1. Pass if easily possible.	A.R.6.2. The dining space shall support up to 8 guests.
CH.1. Check if there is good serving space away from the dinner table to accommodate storage for one course.	XP.1. Pass if one course is easily served and stored during eating.	A.R.6.3. Serving support shall exist for one course at a time.

TABLE 12-21 House test-cases

Check design and/or check behavior	Expected outcome with pass/ fail criteria	Requirement
CH.1. The kitchen room implements these house requirements almost entirely, and it is sufficient for the house if the kitchen verification passes.	XP.1. Pass if kitchen test-cases have been passed.	A.R.6.4. Storing, cooking, and dining support shall allow a 5-course French menu for 8 persons.
		A.R.6.5. Dishwashing support needed, shall hold complete dinner.
		A.R.6.6. Cooking shall support warming, boiling, frying, wokking, and baking.
		A.R.6.8. Waste and garbage from cooking shall be odorlessly disposed of.
CH.1. Check if dining room chairs can be aligned at the walls, and check dancing or mingling space after the table is removed.	XP.1. Pass if 12 persons can sit and 12 persons can dance or mingle.	A.R.7.1. Good space shall exist for 24 party guests, half standing, half sitting.
CH.1. Check if dining room allows small table for snacks while previous test-case is applied.	XP.1. Pass if possible.	A.R.7.2. Serving snack meals shall support 12 party guests.
CH.1. Storage is in kitchen (see kitchen refined test-case).	XP.1. Pass if refined kitchen test-case passed.	A.R.7.3. Storing capacity shall exist for 24 party guests' snack meals.
CH.1. Check that two persons can use the toilet and bathroom at the same time.	XP.1. Pass if at least 1 person can lock and use the water closet, while another person refreshes make-up.	A.R.7.4. Toileting capacity shall exist for 2 guests simultaneously.
CH.1. Check the capacity for temporary storage of outer clothes.	XP.1. Pass if easily possible to store outer clothes for 24 guests.	A.R.7.5. Hanger capacity shall exist for 24 coats and such.

The outdated *requirement* A.R.2.3. was first discussed. It was specified by the *stakeholders* with the best intentions, but should have been a directive to the builder. This *test-case* was excluded.

12.9.3 EXAMPLE House: Verify black-/white-box

Many readers will claim that many of the *user requirements* are too soft and vague to be useful in a formal *verification*. Maybe the failure list is too weak for guarantee claims or a *process* in court, but they are a strong tool for ordinary people struggling for a home designed for convenience. Also note that there are many more *requirements* for a house *nesting level* than fit into this book.

Now the *verification* takes place (see Table 12-22 below).

TABLE 12-22 House verification report 1

Actual outcome and pass/fail conclusion	Expected outcome and pass/fail criteria	Requirement
VR.1. Pass	XP.1.	A.R.1.1.
VR.1. Pass	XP.1.	A.R.1.2.
VR.1. Pass	XP.1.	A.R.2.1.
VR.1. Pass	XP.1.	A.R.2.2.
VR.1. Pass	XP.1.	A.R.2.3.
VR.1. Pass	XP.1.	A.R.3.1.
VR.1.1. Fail. The gas consumption was 1,47 Nm³.	XP.1. Pass if gas consumption for one day and night is less than 1,35 Nm³ (corresponds to 90 kWh/m²a for the 60 m² house).	A.R.3.2. The house shall have an economical energy consumption.
VR.1. Pass	XP.1.	A.R.4.1.
VR.1. Pass	XP.1.	A.R.5.1.
VR.1. Pass	XP.1.	A.R.6.7.
VR.1.2. Fail. Telephone and doorbell far too noisy	XP.1. Pass if chef finds the dining room sufficiently relaxing.	A.R.6.1. The atmosphere during dining shall be relaxed.
VR.1. Pass	XP.1.	A.R.6.2.
VR.1. Pass	XP.1.	A.R.6.3.
VR.1. Pass	XP.1.	A.R.6.4., A.R.6.5., A.R.6.6., A.R.6.7.
VR.1. Pass	XP.1.	A.R.7.1.
VR.1. Pass	XP.1.	A.R.7.2.
VR.1. Pass	XP.1.	A.R.7.3.
VR.1. Pass	XP.1.	A.R.7.4.
VR.1.3. Fail. There is no equipment for this.	XP.1. Pass if easily possible to store outer clothes for 24 guests.	A.R.7.5. Hanger capacity shall exist for 24 coats and such.

12.9.4 EXAMPLE House: Locate failure source and evaluate optimal elimination

Detected failures are both in technical areas and in user-friendly areas, which is a good example of diversifying *stakeholder views* on what is important (see Table 12-23 below).

TABLE 12-23 House failure elimination report 1

Failure location	Failure elimination plan	Actual outcome and pass/fail criteria	Expected outcome and pass/fail criteria	Requirement
FER.1.1. Sloppy procured house outer interface elements, and sloppy realization of house outer shell at house procurement and realization ⬡11	FER.1.1.a.Scan the house with thermal camera. FER.1.1.b. Isolate cold drafts, for example at outer doors and windows. FER.1.1.c. Improve sloppy insulation around the house. Reverify.	VR.1.1. Fail. The gas consumption was 1,47 Nm³.	XP.1. Pass if gas consumption for one day and night is less than 1,35 Nm³ (corresponds to 90 kWh/m²a for the 60 m² house).	A.R.3.2. The house shall have an economical energy consumption.
FER.1.2. Wrong bells selected at entry hall procurement ⬡21	FER.1.2.a.Update dining room requisite with noiseless (light) bells. FER.1.2.b. Change bells in entry. FER.1.2.c. Reverify	VR.1.2. Fail. Telephone and doorbell far too noisy	XP.1. Pass if chef finds the dining room sufficiently relaxing.	A.R.6.1. The atmosphere during dining shall be relaxed.
FER.1.3. No hanger place designed at finalizing entry hall ⬡2F	FER.1.3.a.Acquire a fold-up stand with clothes hangers. FER.1.3.b. Reverify.	VR.1.3. Fail. There is no equipment for this.	XP.1. Pass if easily possible to store outer clothes for 24 guests.	A.R.7.5. Hanger capacity shall exist for 24 coats and such.

12.9.5 EXAMPLE House: Verify black-/white-box

This *reverification* is to repeat the *activity verify prototype in target environment* already presented in the house *verification* above (see Table 12-24 below).

TABLE 12-24 House verification report 2

Actual outcome and pass/fail conclusion	Expected outcome and pass/fail criteria	Requirement
VR.2. Pass	XP.1.	R.3.2.
VR.2. Pass	XP.1.	R.6.1.
VR.2. Pass	XP.1.	R.7.5.

Failures to reverify were fairly harmless and could be quickly eliminated.

12.9.6 EXAMPLE House: Prepare product manual

The house manuals are rather similar to the room manuals, but one *nesting level outwards.*

TABLE 12-25 House manual table of contents

Manual name	Producer and type	Element / interface
MA.1. House architectures	Robber HB	Drawings
MA.2. House finalized architectures	Mobber AB	Drawings
MA.3. House electricity drawings	Elert AB	Drawings
MA.4. Weatherproof and impregnation certificate	NoPapp	Plank
MA.5. Galvanization certificate	IronWare	Bracket
MA.6. High-pressure water cleaning guidance	Bricket	Façade Roof tile
MA.7. How to open for cleaning	Pall&Papp	Windows

12.9.7 EXAMPLE House: Proceed reading

To follow this example, these are the alternatives (see again Fig. 12-12, p. 693):

- Trace *upstream* in the overall *schedule* to understand how and where detected failures are eliminated. Since the house is built outside-in, a failure can be located both *inwards* and *outwards* with respect to where it is detected.

- **0V)** Proceed to chapter 12.10, "EXAMPLE House environment: Fully verify prototype in environment" below.

12.10 EXAMPLE House environment: Fully verify prototype in environment

12.10.1 About the house example

Now it is time to verify the *environment interfaces* of the house. A house may be built as a scale model, and partly verified in a simulator. This is not the case in this example, so *verification* will take place in the real *target environment.*

12.10.2 EXAMPLE House environment: Process schedule to use

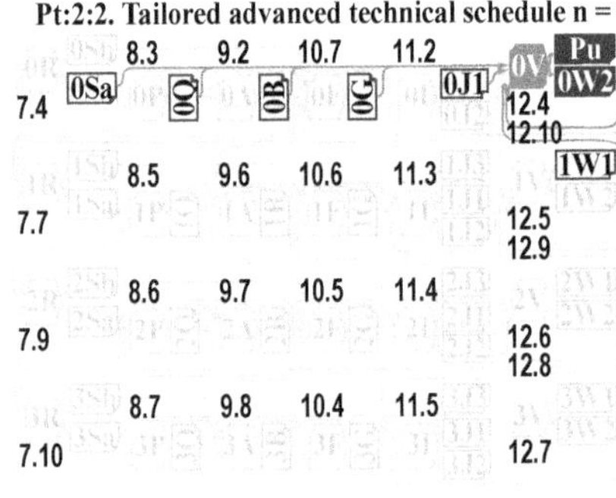

For the house example, a special technical overall *schedule* is *tailored* (illustrated in Fig. 12-13, left):

- **Pt:2:2. Tailored advanced technical schedule n = 4,** page 164

In this *tailored schedule* the generic *schedule* to use is (filled-in symbols):

- **Pt.0.V. Verify prototype in environment,** page 125.

FIGURE 12-13 Position of schedule and master to use (filled in) in overall schedule

Below *results* may be demanded (their *masters* are indicated *upstream* from the filled-in *schedule* in the figure above or in greater detail in Fig. 11-5, p. 602 and Fig. 11-6, p. 603):

1W1 Fully verified house black-/white-box

    Table 12-22, House verification report 1, page 697

    Table 12-24, House verification report 2, page 698

0W2 Partially field-verified prototypes

    List 12-1. House environment interfaces with fake house failure elimination report 1, page 674

0J1 Usable prototype interfaces started in simulator or target

    Table 11-9, House environment start-up report 1, page 606

0G Environment interface design items and product requisites

    Figure 10-14, House environment connections, page 479

    Table 10-41, House interfaces to environment design items, page 479

0B Environment existing ingredients and design of interfaces to the environment

    Table 9-6, House environment architecture ingredients resumé, page 341

    Figure 9-5, House environment logical architecture, page 341

    Figure 9-6, House environment physical architecture, page 342

0Q Solution alternatives of interfaces to environment with supplier opportunities

    Table 8-5, House environment solution alternatives of interfaces to environment, page 292

    Table 8-6, House environment solutions excluded, page 293

    Table 8-7, House environment solutions included with supplier opportunities, page 293

0Sa Environment restriction requirements

Table 7-5, House environment restriction requirement, page 182

12.10.3 EXAMPLE House environment: Develop field test-cases for target or simulator

A table is again used for documenting the *test-cases* (see Table 12-26 below).

TABLE 12-26 House environment test-cases

Check design and/or check behavior	Expected outcome with pass/fail criteria	Requirement
CH.1. Have environment authority inspect and approve the house.	XP.1. Pass if environment authority inspector approves the house.	E.R.1. Swedish environmental code, 7th chapter
CH.1. Have Robber HB ask building authority to inspect and approve compliance with building permit.	XP.1. Pass if authority inspector approves compliance with building permit and approves use of the house.	E.R.2. Malmö urban planning and local construction ordinance
CH.1. Check if building permit is obtained.	XP.1. Pass if building permit obtained.	E.R.3. Swedish national building regulations
CH.1. Have bank evaluate loan approval.	XP.1. Pass if bank accepts a request for maximum allowed loan.	E.R.4. Swedish Financial Supervisory Authority
CH.1. Check all public utilities to the house.	XP.1. Pass if all available public utilities are used and approved by the suppliers.	E.R.5. All public utilities available shall be used and comply with building standard 2011:45

12.10.4 EXAMPLE House environment: Verify prototype in target environment

The *verification* is performed and shows a significant failure. The hydrant requested by the authorities has been forgotten by the builder since it was not included in the *architecture* and *environment interfaces design items*.

TABLE 12-27 House environment verification report 1

Actual outcome and pass/fail conclusion	Expected outcome and pass/fail criteria	Requirement
VR.1. Pass	XP.1.	E.R.1.
VR.1.1. Fail. Outdoor hydrant missing	XP.1.	E.R.2. Malmö urban planning and local construction ordinance
VR.1.2. Fail. Outdoor staircase not safe without handrail	Pass if authority inspector approves compliance with building permit and approves use of the house.	
VR.1. Pass	XP.1.	E.R.3.
VR.1. Pass	XP.1.	E.R.4.
VR.1.3. Fail. Electricity supplier found residual current on cable connection to the house.	XP.1. Pass if all available public utilities are used and approved by the suppliers.	E.R.5. All public utilities available shall be used and comply with building standard 2011:45

12.10.5 EXAMPLE House environment: Locate failure source and evaluate prototype completeness

The missing hydrant is an expensive failure with a lot of excavation in the finished garden. An extra inspection of the *architecture* and *environment interfaces design items* would have cost only a fraction of the expense of eliminating this failure.

TABLE 12-28 House environment failure elimination report 1

Failure loca-tion	Failure elimination plan	Actual outcome and pass/ fail crite-ria	Expected outcome and pass/ fail criteria	Require-ment
FER.1.1. Absent in architecture constitution 0A	FER.1.1.a.Update architecture constitution and all dependent documentation. FER.1.1.b. Ask Mobber AB to build hydrant and submit water pressure certificate to authorities. FER.1.1.c. Reinspect compliance with building permit. FER.1.1.d. Have lawyer claim damage from Robber HB for missed hydrant in architecture.	VR.1.1. Fail. Outdoor hydrant missing	XP.1. Pass if authority inspector approves compliance with building permit and approves use of the house.	E.R.2. Malmö urban planning and local construction ordinance
FER.1.2. Absent in environment finalized design 0F	FER.1.2.a.Ask Pall & Papp to put up a handrail. FER.1.2.b. Have lawyer claim damage from Robber HB for missed handrail in architecture.	VR.1.2. Fail. Outdoor staircase not safe without handrail		
FER.1.3. Specifiers (and anybody else) were not aware of a new standard demanding absence of residual current in private houses 3Sa (For a detailed explanation see ch. 12.3.2, p. 666)	FER.1.3.a. Change the failures in machinery specification, solution alternatives, architecture and final design. FER.1.3.b. Change the electrical oven to a correct one without residual current problems. FER.1.3.c. Prepare a new test-case to verify that the new oven creates no residual current problems FER.1.3.d. If more failures detected in the kitchen, perform a regression test. FER.1.3.e. Verify house with selected test-cases. FER.1.3.f. Let electricity supplier approve electrical connection to house. (For a detailed explanation, see ch. 12.3.4, p. 669)	VR.1.3. Fail. Electricity supplier found residual current on cable connection to the house.	XP.1. Pass if all available public utilities are used and approved by the suppliers.	E.R.5. All public utilities available shall be used and comply with building standard 2011:45

12.10.6 EXAMPLE House environment: Verify prototype in target environment

The missing handrail was a surprise for the gate meeting. How come this failure was not detected when verifying the house? The simple answer is that such an inspection cannot be ordered before the house is finished, including its *interfaces* to the *environment*.

The two failures were eliminated by the *suppliers*, and to avoid tedious and time-consuming legal measures, the house proprietor compromised and paid half the cost.

The residual current failure was also a surprise, but the builder took on the responsibility to be aware of latest regulations. He fixed it all without extra cost.

TABLE 12-29 House environment reverification report 2

Actual outcome and pass/fail conclusion	Expected outcome and pass/fail criteria	Requirement
VR.2. Pass	XP.1.	E.R.2.
VR.2. Pass		
VR.2. Pass	XP.1.	E.R.5.

12.10.7 EXAMPLE House environment: Finalize product manual

Both technical and legal matters are pushing preparation of the manual (see Table 12-30 below).

TABLE 12-30 House environment product manual

Manual name	Producer and type	Element / interface
MA.1. House connections	Mobber AB	Drawing
MA.2. Environment architecture	Robber HB	Drawing
MA.3. House connection wiring	Elert AB	Drawing
MA.4. Pressure certificate	Ion Gas	Gas tank
MA.5. Building permit	Authorities	Building permit
MA.6. Authority inspections 1 and 2	Authorities	Inspection of house use

12.10.8 EXAMPLE House: Proceed reading

To follow this example, these are the alternatives (see again Fig. 12-13, p. 700):

- Trace *upstream* in the overall *schedule* to understand how and where detected failures are eliminated. Since the house is built outside-in, a failure can be located both *inwards* and *outwards* with respect to where it is detected.
- The house example has now terminated and market validation is supposed to take place (see ch. 6.6, "P. Cpdm generic development schedule," p. 99). Pull the cork off a champagne bottle, and cheers to both the house proprietor and house developers as well as to yourself for reading so far in this book.

12.11 Verification depends on type of requirements

In the previous house example, no distinction was made between *restriction requirements* and *behavior requirements*. However, when entering *product*s that are not as visible as a house, the *formalism* can be heightened to keep up mastering *complexity*.

Separating *requirements* into either *restriction requirements* or *behavior requirements* improves engineer awareness, and *requirements* and *verification* become stricter. These differences have greatest impact on *requirements*, which become completely different, but *verification* also differs a lot between restrictions and behavior (see Fig. 12-14 below).

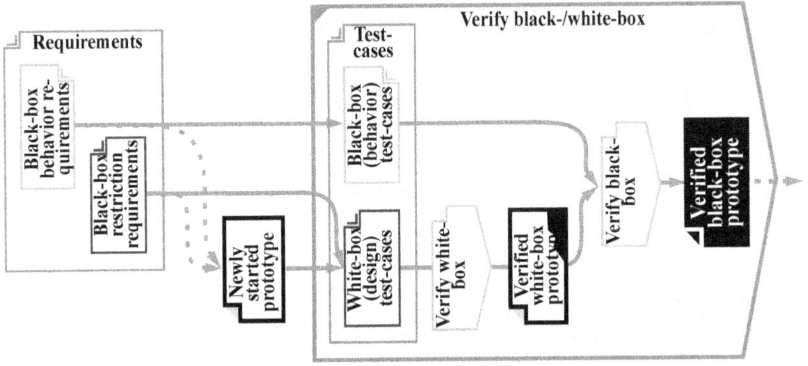

FIGURE 12-14 Principle of black-/white-box verification

12.11.1 EXAMPLE Lake: Fish behavior and appearance

The example of lake fishing continues to illustrate the difference between *verification* of a *black-box* and *verification* of a *white-box*.

A *failure detection* in the *black-box* (a fish in the lake) is done by its behavior from outside the *black-box* (see Fig. 12-15, right).

FIGURE 12-15 Fishing in the black-box

A *failure detection* in a *white-box* (another fish in the lake) is done by opening the *black-box* and looking into the exposed *white-box* (see Fig. 12-15, left).

FIGURE 12-16 Fishing in the white-box

12.11.2 Cpdm definitions

In *Cpdm*, *white-box test-case* and *black-box test-case* are defined according to Table 12-31 below.

TABLE 12-31 Cpdm definitions

Aspect	White-box test-case	Black-box test-case
Definition	Concretization of restriction requirements, to be an efficient base for verification of white-box design	Concretization of behavior requirements, to be a practical base for verification of black-box behavior
Synonyms	Test-case, test, design test, design check, inspection, review, design review	Test-case, test, behavioral test
Symbol	**TABLE 12-32** White-box test-cases 	White-box design check	Expected appearance with pass/fail criteria	Restriction requirement
				TABLE 12-33 Black-box test-cases | Black-box stimulus injection	Expected response with pass/fail criteria	Behavior requirement

There is no strange magic in the above definitions. If a *restriction requirement* is specified for a *black-box*, one has to open it and check if the *restriction requirement* is satisfied. If a behavior is specified on a *black-box interface*, one has to stimulate the *interface* and check the *response*.

However, performing a *restriction requirements verification* on the *white-box* will be closer to the design than verifying behavior on a *black-box interface*. Thus, the *restriction requirements verification* often involves *designers* to explain the checked design.

12.11.3 Responsibility for test

The question often raised is whether to have a separate organization for performing *verifications.* A separate organization for *verification* brings some extra overhead and gets a bit more expensive, that is true. On the other hand, there are big benefits, especially if you count independence and formal control. Letting *designers* do their own *verification* brings the risk that *test-cases* will be biased and personal, failures will be hidden, and *verification reports* will be incomplete.

From the Cpdm advanced technical schedule, organizations can be partitioned according to phases, or according to nesting levels, or to both, or to neither of them.

Leaving the responsibility to a separate *verification* organization is to separate *performers* and inspectors. This is often used by authorities to fight corruption, and it is a strong tool also in development.

In a separate *verification* organization, the verifiers define what *test-cases* to create and they conduct the *verification.* Whenever needed, for example, at *white-box verification* and localization of behavior failures, the verifiers call in *designer* experts for support and to explain their technical constructions.

12.12 EXAMPLE Multiplication toy with hard-wired multiplier: Verify black-/white-box

12.12.1 Multiplication toy: About this example

The *verification* of the multiplication toy with hard-wired multiplier is rather straightforward. *Complexity* is buried in the logic gate network, to get it theoretically right, and to connect all prototype gate capsules in the right way.

12.12.2 Multiplication toy with hard-wired multiplier: Process schedule to use

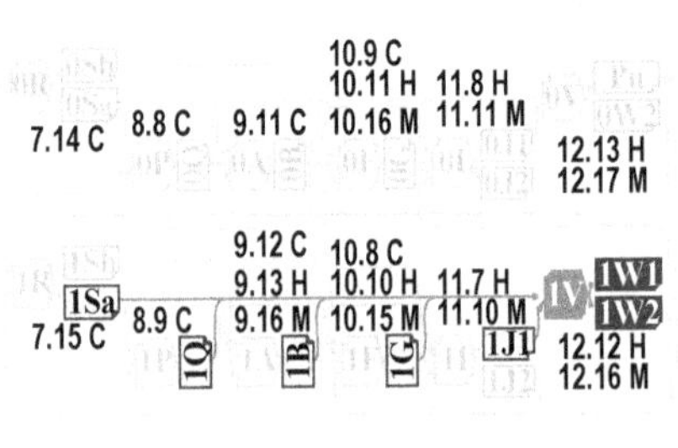

For the multiplication toy example, a special technical overall *schedule* is *tailored* (illustrated in Fig. 12-17, left):

- **Pt:2:3. Tailored advanced technical schedule n = 2,** page 169

In this *tailored schedule* the generic *schedule* to use is (filled-in symbols):

- **Pt.n.V. Verify black-/white-box,** page 152.

FIGURE 12-17 Position of schedule and master to use (filled in) in overall schedule

Below *results* may be demanded (their *masters* are indicated *upstream* from the filled-in *schedule* in the figure above):

[1J1] Completed usable white-box in target, fixture, or debugger

[1G] Common finalized white-box items design with product requisites
Figure 10-16, Setting device diagram, page 484
Figure 10-17, Presenting device diagram, page 485
Figure 10-18, Power supply diagram, page 486
Table 10-46, Multiplication toy common electronics, white-box design items, page 487

[1G] Multiplication toy with hard-wired multiplier finalized white-box items design with product requisites
Figure 10-24, Hard-wired multiplier, arithmetic unit diagram, page 502
Figure 10-25, Hard-wired multiplier, full adder multiplier diagram, page 503
Figure 10-27, Hard-wired multiplier 3-module adder diagram, page 505
Figure 10-26, Hard-wired multiplier, Bin-BCD converter diagram, page 504
Figure 10-28, Multiplication toy with hard-wired multiplier prototype board drawing, page 505
Table 10-63, Multiplication toy with hard-wired multiplier electronics white-box design items, page 506
Table 10-64, Multiplication toy with hard-wired multiplier prototype board white-box design items, page 506

[1B] Common architecture white-box with major ingredients but no embedded black-boxes
Table 9-25, Multiplication toy variants architecture ingredients resumé, page 388
Figure 9-38, Multiplication toy common logical architecture, page 389
Figure 9-39, Diversifying the multiplier element and circuit board interface, page 390
Figure 9-40, Multiplication toy common physical architecture, page 390

[1B] Multiplication toy with hard-wired multiplier architecture white-box with major ingredients but no embedded black-boxes
Table 9-26, Hard-wired multiplier architecture ingredients resumé, page 393
Figure 9-42, Hard-wired multiplier logical architecture, page 393

[1Q] Solution alternatives and supplier opportunities
Table 8-21, Multiplication toy white-box solution alternatives, page 317
Table 8-22, Multiplication toy white-box supplier opportunities, page 318

[1Sa] Prioritized requirements
Table 7-25, Multiplication toy restriction requirement, page 216

12.12.3 EXAMPLE Multiplication toy with hard-wired multiplier: Develop test-cases, also for fixtures and debugger

The prototype with hard-wired multiplier is developed and verified to secure the function of the electronic logic gate network, before transferring this gate network into an ASIC. An ASIC development is extremely expensive in investment for chip layout, but cheap apiece to produce in large series. Even minor failures might cause a new round of expensive chip masks investments.

Most of the *white-box test-cases* for the multiplication toy with hard-wired multiplier are design-specific, but some of the *white-box test-cases* developed for the multiplication toy with microcontroller multiplier can be *reused* (see Table 12-34 below).

TABLE 12-34
Multiplication toy with hard-wired multiplier white-box test-cases

White-box design check	Expected appearance with pass/fail criteria	Restriction requirement
Not applicable		A.RR.1.1.
Not applicable		A.RR.1.2.
CH.1. Check in the architecture and design graphs how the interface to the multiplier is designed. Check in the prototype that exactly these interfaces are kept.	XP.1. Pass, if the prototype has realized all interfaces to the multiplier in accordance with the architecture and design graphs.	A.RR.5.1.The multiplication calculation shall be allocated to a separate element that can be redesigned with minimal rework.
CH.1. The multiplication toy with hard-wired multiplier is not intended to be a product with final casing until the ASIC is developed. Transfer this requirement to the possible coming ASIC product development, and make this verification on both products.	XP.1. Pass so far.	A.RR.5.2.The casing of the hosting toy shall be exactly the same for all products independent of technology used.
Not applicable		A.RR.6.1.
CH.1. When the multiplication toy with ASIC is developed, check the manufacturing requisites for total accumulated cost. Have manufacturing experts review these cost estimates.	XP.1. Pass for now.	A.RR.7.1.If the toy is produced, it shall be made in such volume that the price per unit is less than € 5.

The most important *verification* for the multiplication toy with hard-wired multiplier is to ensure that the gate design has the correct behavior, which is verified with the *black-box test-cases* in Table 12-35 below.

TABLE 12-35
Multiplication toy with hard-wired multiplier black-box test-cases

Black-box stimulus injection	Expected response with pass/fail criteria	Behavior requirement
CH.1. Check all possible settings on one operand multiplied by all possible settings on the other operand.	XP.1. Pass if all product values are calculated correctly.	A.RB.8.1.User amusement shall be to multiply two operand values into one product value.
CH.1. Check that both multiplicand switches can be set to different values..	XP.1. Pass if both operands can be set from 0 to at least 10.	A.RB.8.2.The toy shall have one input device for each multiplicand value. Multiplicand values shall be at least between 0 and 10. The multiplicands must be easy for a child to settle.
CH.2. Set one operand switch at value 1 and manually turn the other operand switch through all positions. Reverse operands above..	XP.2. Pass if one operand is set to 1 and presentation shows the value on the other setting, and if the same occurs when swapping operands.	
CH.1. Check the product when both operands are set to 10.	XP.1. Pass if value 100 shown.	A.RB.8.3.The toy shall have at least a 3-digit output device for the product value. The product must be easy to read for a child.
CH.1. Operate the operand switches manually as fast as possible..	XP.1. Pass if every product value is shown without a perception of lagging.	A.RB.8.4.The toy calculates and shows the product value practically instantly after either multiplicand is settled.

12.12.4 EXAMPLE Multiplication toy with hard-wired multiplier: Verify white-box

With separated *test-cases* for *restriction requirements* and *behavior requirements*, the *verification* can now be separated. A *restriction requirement verification* assumes that the transparent prototype *white-box* design is checked (see Table 12-36 below).

TABLE 12-36
Multiplication toy with hard-wired multiplier
verification report 1

Actual outcome and pass/ fail conclusion	Expected outcome and pass/fail criteria	Requirement
VR.1.1. Fail. **Operand and product buses not kept explicitly separated in prototype**	XP.1. Pass, if the prototype has realized all interfaces to the multiplier in accordance with the architecture and design graphs.	A.RR.5.1.The multiplication calculation shall be allocated to a separate element that can be redesigned with minimal rework.

12.12.5 EXAMPLE Multiplication toy with hard-wired multiplier: Reverify black-box

A behavior *verification* assumes that the *black-box* remains closed, and the *verification* is made on its *interfaces*. As the casing is defined as belonging to the *outermost black-/white-box*, the two operand inputs are the rotating switches set by hand and one display read by eyes (see Table 12-37 below).

TABLE 12-37
Multiplication toy with hard-wired multiplier
verification report 2

Actual outcome and pass/ fail conclusion	Expected outcome and pass/fail criteria	Requirement
VR.2.1. Fail. Most products were wrongly calculated.	XP.1. Pass if all product values are calculated correctly.	A.RB.8.1.User amusement shall be to multiply two operand values into one product value.
VR.2. Pass		A.RB.8.2.
VR.2. Pass		A.RB.8.2.
VR.2. Pass		A.RB.8.3.
VR.2. Pass		A.RB.8.4.

After both *verification*s above, the sources of detected failures are located by developers and documented

12.12.6 EXAMPLE Multiplication toy with hard-wired multiplier: Locate failure source

The VR.1.1. failure was rather easy to find, and it was obvious that the prototype engineer had not kept the *interfaces* separate and available in the prototype.

The VR.2.1. failure was much more difficult to find. Developers had to follow the detected failure from the presenting device and upstream along the calculation flow. At the output of the hard-wired multiplier, the calculation was measured correctly, so the failure must be in the bin-bcd converter. After great effort, it was found that certain inputs to the gate logic were not connected anywhere at all.

Both failures were documented in the failure location column of Table 12-38 in the next chapter.

12.12.7 EXAMPLE Multiplication toy with hard-wired multiplier: Evaluate optimal elimination

A gate meeting was arranged and ways to eliminate detected failures were discussed and decided (see Table 12-38 below).

TABLE 12-38
Multiplication toy with hard-wired multiplier
failure elimination report 1

Failure location	Failure elimination plan	Actual outcome and pass/fail criteria	Expected outcome and pass/fail criteria	Restriction requirement
FER.1.1. Prototype board designers did not understand the architecture drawing in Figure 9-38, page 389.	FER.1.1.a.Train board designers in architecture. FER.1.1.b. Order redesign of new prototype boards.	VR.1.1. Fail. Operand and product buses not kept explicitly separated in prototype	XP.1. Pass, if the prototype has realized all interfaces to the multiplier in accordance with the architecture and design graphs.	A.RR.5.1.The multiplication calculation shall be allocated to a separate element that can be redesigned with minimal rework.
FER.1.2. Open inputs a_3 in detail design drawing Figure 10-26, page 504.	FER.1.2.a.Solder small jumpers on prototype board. FER.1.2.b. Update drawing. FER.1.2.c. Correct layout on new boards.	VR.2.1. Fail. Most products were wrongly calculated.	XP.1. Pass if all product values are calculated correctly.	A.RB.8.1.User amusement shall be to multiply two operand values into one product value.

12.12.8 EXAMPLE Multiplication toy with hard-wired multiplier: Verify black-/white-box

After the failures were eliminated, the *reverification* was successful (see Table 12-39 below).

TABLE 12-39
Multiplication toy with hard-wired multiplier
verification report 3

Actual outcome and pass/fail conclusion	Expected outcome and pass/fail criteria	Requirement
VR.3. Pass		A.RR.5.1.
VR.3. Pass		A.RB.8.1.

12.12.9 EXAMPLE Multiplication toy with hard-wired multiplier+: Manual table of contents

Since the hard-wired prototype is to ensure that the technical concept is useful to *port* to an ASIC, there is no need to make a manual for an *end user* of the toy.

TABLE 12-40
Multiplication toy with hard-wired multiplier manual table of contents

Manual name	Producer and type	Element / interface
MA.1. Prerequisites for ASIC porting	Electronic designers	Electronic drawings

12.12.10 Multiplication toy: Proceed reading

To follow this example, these are the alternatives (see again Fig. 12-17, p. 706):

- Trace *upstream* through the overall *schedule nesting level* = 1 to understand how to eliminate a detected failure.
- Proceed to chapter 11.8, "EXAMPLE Multiplication toy environment (hard-wired multiplier variant): Realize interfaces to environment and insert/await outermost white-box," page 628.
- To stay within the multiplication toy example, but not following the overall *schedule*, proceed to chapter 12.13 below.

12.13 EXAMPLE Multiplication toy environment (hard-wired multiplier variant): Verify prototype in environment

12.13.1 Multiplication toy: About this example

The hard-wired multiplier prototype is developed to ensure correct functionality of the hard-wired multiplier, before the gate design is allowed to be *ported* to an ASIC circuit. Therefore, there is no need to connect the multiplication toy to any hosting toys for *verification* by *users* in the *target environment*.

12.13.2 Multiplication toy: Proceed reading

To follow this example, these are the alternatives:

- The multiplication toy (hard-wired multiplier variant) example ends here, and a new *project* will be started with how to *port* the hard-wired multiplier into an ASIC, to achieve a much lower cost apiece. However, this will not be presented in this book, so this example ends here. Pull another bottle of champagne:-).

12.14 Prepare verification during manufacturing

In the above chapters, *verification* is described for verifying prepared prototypes, and not for verifying *products* after being assembled by manufacturing. Regular manufacturing *verification* certainly implies that the prototype has been successfully verified during development, but in addition that the *product* is designed to be easy to verify at manufacturing.

The prototype must be designed to be easily verified during development, but more important, the product must be easily verified during manufacturing.

However, to make automated manufacturing tests possible, almost always the development *phases* must include significant supportive design. Since development from automated design must take place, it is also natural to verify manufacturing *verification* during this prototype *verification* (for details see Table 12-41, p. 719).

12.15 Making verification efficient

12.15.1 Verification formalism

Many development departments are heavily stressed, because of market need for launch of new *products* before competitors offer the same or better. The stress causes a lot of failures, and this most often results in lack of time for *verification formalism* or even for sufficient *verification* at all. Too often, *verification* with endless updates and releases is left to be the problem of the *customers*.

Good order and formalism during verification have the perplexing result that more failure problems are detected

Not being formal enough unfortunately brings on a negative blame game, because most of the failures result from too little *formalism* applied before *verification*. If true *stakeholders* have good order in their *requirements*, and if architects bring a good order of *product structure*, that will dramatically bring down severe failures in the finalization *phase* and make *integration* smooth and swift. The saying, "You made your bed, now lie in it," is awfully true in this context.

However it is unavoidable that some bugs will creep into the prototypes, but even here good order in the *requirements* and *architecture* is rewarding. First, it is much easier to locate a failure in a good, structured documentation and prototype, and second, when you eliminate the detected failure you have much better control, and it is much easier to avoid secondary failures if you understand the context around the failure location.

The simple conclusion is that *formalism* takes some extra effort and time in the beginning of the development, but this is earned back in the end in time and better *quality*.

Without order and structure one has a messy *product* that is hard for outsiders (and even for insiders) to understand, maintain, or extend. With well-structured documentation, engineers can be replaced without loss of productivity from day one, and they will keep a high efficiency and desired quality. Without structure and documentation, even with reconstruction when needed, the risk is that extensions bring more and more *complexity* (with added *artificial complexity*), which gets harder and harder to master. In the worst case, a prototype, possibly extended many times, might finally collapse and never again function properly.

12.15.2 EXAMPLE Lake: Make design for efficient fishing

In the fishing example, it is very obvious that efficient equipment results in catching a lot more fish.

This can be compared with automated *verification*, through development of a clever design both in the program and in the surrounding *fixtures*.

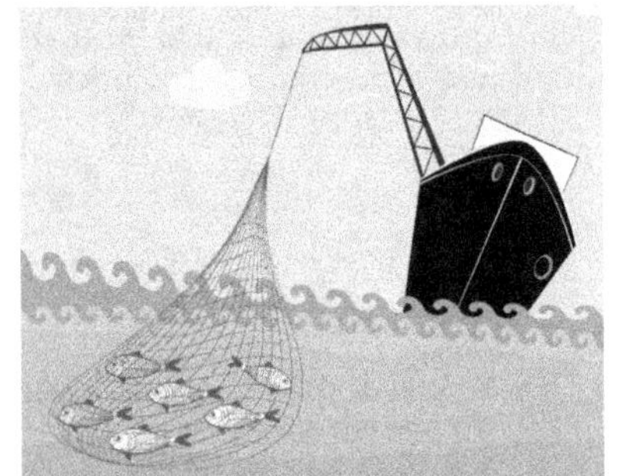

FIGURE 12-18 Efficient fishing tools

12.15.3 Automated verification

Even if the *product* is developed with care and sufficient *formalism*, modern *products* get more and more *complex*. Modern *products* are often very user-permissive, which demands *complex* man-machine *interfaces*, with a lot of *complex* exception handling, which often is underestimated.

Automated verification must already be decided when capturing requirements.

Consequently, if a prototype has a lot of *interface* interaction, it takes a lot of time to verify it.

Automated tests can be a very good investment and *quality* improvement in these cases. Maybe the biggest benefit is that the automated *verification*s used during development can also be easily moved to manufacturing *verification* when development is finished. But don't take out the victory in advance. Automated *verification* must be planned as early as in the **Pt.R. Capture staffing & requirements** *phase*. Additionally, it is very important during make/buy planning to ensure that *white-box design items* and *black-/white-boxes* can be machine connectable for automated *verification*. The architects must identify supportive test points to *black-boxes*, and engineers must design them into the construction of the *black-/white-boxes*. *Samples* must be ordered according to test points and prototype and even the final *product* must be provided with sufficient additional test points.

The paradox here is that only well-designed prototypes will allow automatic *verification*. Automated *verification* is often asked for as a cure to speed up *verification* after sloppy development has caused far too many failures. But in such cases, the deceased is too far gone for this kind of medicine.

When automated tests are applied, regression tests are almost free. It takes only some monitoring of the execution and outcome to check that old failures never reappear, which preserves the prototype on a continuously rising *quality* grade during *verification*.

12.15.4 The process schedule for repetitive manual or automated verification

A located *program* failure can be eliminated in the *activity finalize white-box detailed items*, and after that there are no really time-consuming *activities* until the *activity verify black-box* (see Fig. 12-19 below). It might take only a couple of hours to localize a failure, eliminate the failure, and start up the *white-box* to verify a successful *failure elimination*. The *verification*-elimination-*reverification* forms a closed loop of *activities*. The time it takes from when the elimination is made until the elimination is reverified is sometimes called (*verification*) *turnaround time* (see Fig. 12-19 below).

Many *managers* and nontechnicians interpret a *turnaround time* of a couple of hours to be the time to develop that piece of a *program*. Many *guru* developers support such misunderstanding, in order to demonstrate how smart they are to develop *programs*.

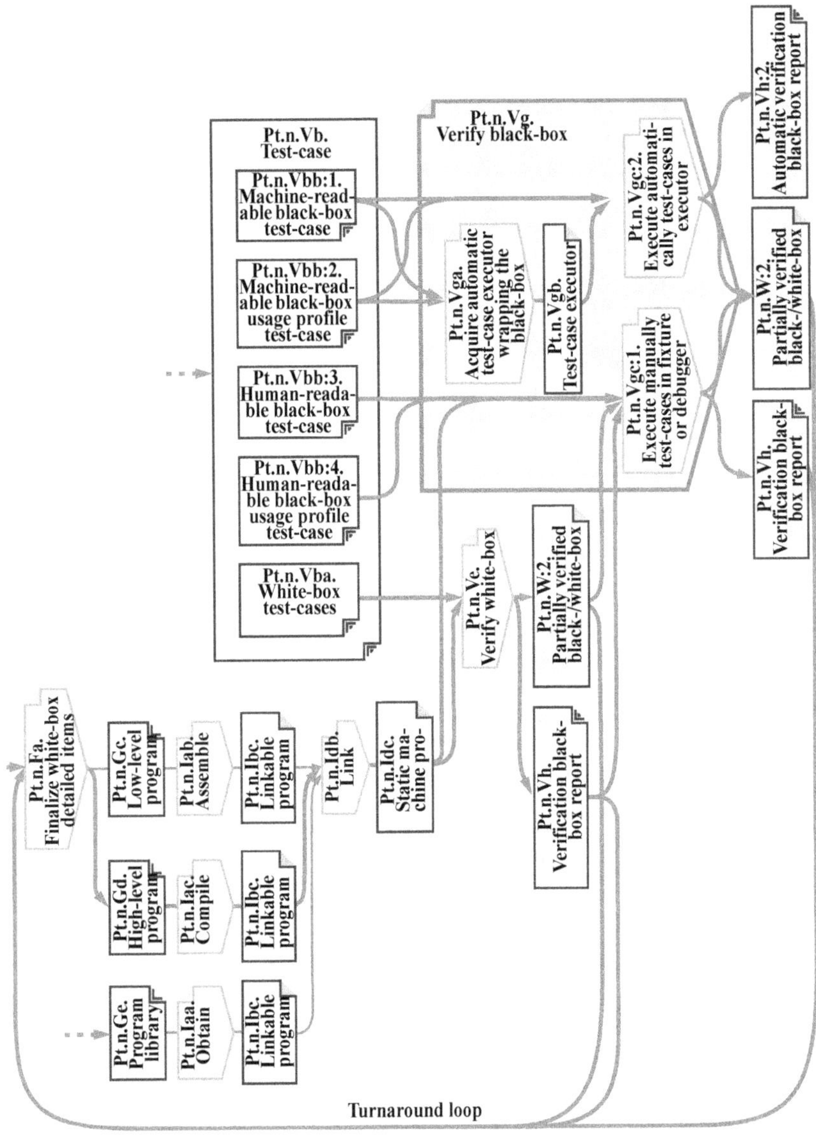

FIGURE 12-19 Repetitive verification process schedule

If the failure is found to be introduced prior to the *activity finalize white-box detailed items*, the loop must be enlarged until the *activity* for the failure localization is included and the failure being eliminated, but the regression principle remains. For example, if a failure depends on wrong *requirements* or wrong design, the elimination loop must include also these *activities*.

12.15.5 Regression test

Even if one has good, documented, and structured prototypes, and even if a failure is eliminated with a careful hand, it might be hard to avoid some secondary failures. If *configuration management* is poor, a properly eliminated failure might reappear, because of an unintended exchange of the corrected prototype for the original prototype that contained the failure. In *programs* it is rather common that one *failure elimination* is not the origin of the next *failure elimination*, and the first *failure elimination* is replaced by the faulty one again.

Test-cases should be planned to be easily repeatable on the same prototype. Frequently these *test-cases* should be executed to verify that no cured failures or secondary failures have reemerged. To execute the same *test-cases* for a second time is called a *regression verification*. New *verifications* that are performed are always added to the regression tests.

12.15.6 EXAMPLE Lake: Let the nets stay and continue fishing

In the fishing example, imagine that the fishing nets laid out in a lake will stay in the water until the fishing season is over. For little effort and low cost, these will always catch old fish not yet caught, as well as possibly new fish that have popped up in the lake.

To catch all fish, let the nets stay, to be emptied over and over again.

12.16 EXAMPLE Multiplication toy with microcontroller multiplier: Verify black-/white-box

12.16.1 Multiplication toy with microcontroller multiplier: About this example

This example replaces the hard-wired multiplier with a microcontroller, to introduce *program* development. The changes of *black-box verification* are minor, but of course, the *white-box verification* now *targets* a very different technique.

12.16.2 Multiplication toy with microcontroller multiplier: Process schedule to use

Pt:2:3. Tailored advanced technical schedule n = 2

10.9 C
10.11 H 11.8 H
7.14 C 8.8 C 9.11 C 10.16 M 11.11 M
12.13 H
12.17 M

9.12 C 10.8 C
9.13 H 10.10 H 11.7 H
1Sa
7.15 C 8.9 C 9.16 M 10.15 M 11.10 M
1Q 1B 1G 1J1 12.12 H
12.16 M
1W1
1W2

For the multiplication toy example, a special technical overall *schedule* is *tailored* (illustrated in Fig. 12-20, left):

• **Pt:2:3. Tailored advanced technical schedule n = 2,** page 169

In this *tailored schedule* the generic *schedule* to use is (filled-in symbols):

• **Pt.n.V. Verify black-/white-box,** page 152.

FIGURE 12-20 Position of schedule and master to use (filled in) in overall schedule

Below *results* may be demanded (their *masters* are indicated *upstream* from the filled-in *schedule* in the figure above):

[1J1] Completed usable white-box in target, fixture, or debugger

[1G] Common finalized white-box items design with product requisites

> Figure 10-16, Setting device diagram, page 484

> Figure 10-17, Presenting device diagram, page 485

> Figure 10-18, Power supply diagram, page 486

> Table 10-46, Multiplication toy common electronics, white-box design items, page 487

[1G] Multiplication toy with microcontroller multiplier finalized white-box items design with product requisites

> Figure 10-31, Microcontroller multiplier diagram, page 520.

> Figure 10-32, Microcontroller multiplier assembler program with flow chart, page 521

> Figure 10-33, Printed circuit board drawing, page 522

> Figure 10-19, Multiplication toy casing, page 486

> Table 10-74, Multiplication toy with microcontroller multiplier electronics white-box design items, page 522

> Table 10-75, Multiplication toy with microcontroller multiplier printed board white-box design items, page 522

[1B] Common architecture white-box with major ingredients but no embedded black-boxes

> Table 9-25, Multiplication toy variants architecture ingredients resumé, page 388

> Figure 9-38, Multiplication toy common logical architecture, page 389

> Figure 9-39, Diversifying the multiplier element and circuit board interface, page 390

> Figure 9-40, Multiplication toy common physical architecture, page 390

☐1B Multiplication toy with microcontroller multiplier architecture white-box with major ingredients but no embedded black-boxes

Table 9-28, Microcontroller multiplier refined architecture ingredients resumé, page 398

Figure 9-46, Microcontroller multiplier logical architecture, page 399

☐1Q Solution alternatives and supplier opportunities

Table 8-21, Multiplication toy white-box solution alternatives, page 317

Table 8-22, Multiplication toy white-box supplier opportunities, page 318

☐1Sa Prioritized requirements

Table 7-25, Multiplication toy restriction requirement, page 216

12.16.3 EXAMPLE Multiplication toy with microcontroller multiplier: Develop test-cases, also for fixtures and debugger, and even for manufacturing

There is a strong demand to keep manufacturing prices of the multiplication tool low apiece. One way to save money is to verify automatically in manufacturing. Most often, not everything can be automatically verified, such as pushing buttons and reading a display. But the signal flow in between, starting with the connector to the rotation switches (-A-7a., p. 484 and -A-7b., p. 485) and ending at the LED digits (-A-7c., -A-7d. and -A-7e., p. 485), can be automatically verified (see Table 12-41 below). Automatic *verification* must most often be prepared by design for it, and now it can already be verified that these *verifications* will function in manufacturing

TABLE 12-41 Multiplication toy with microcontroller multiplier, manufacturing black-box test-cases

Black-box stimulus injection	Expected response with pass/fail criteria	Behavior requirement
CH.1. Check to connect a preliminary manufacturing fixture of injection test probes to all printed circuit board injection points (marked in finalized design in Fig. 10-16, p. 484) and detecting points (marked in Fig. 10-17, p. 485).	XP.1. Pass if possible to inject circuit board probes with all possible operands stimuli, and for each injection that it is possible to measure digital responses at the detecting probes.	A.RB.8.1.User amusement shall be to multiply two operand values into one product value.
CH.1. Check to connect between test point 1 and test point 2 at printed circuit board lamp test (marked in finalized design in Fig. 10-17, p. 485.)	XP.1. Pass if possible to manually connect and manually check which segments are illuminated.	A.RB.8.3.The toy shall have at least a 3-digit output device for the product value. The product must be easy to read for a child.

Now we start preparing *white-box test-cases* for the realization (see Table 12-42 below). Since the casing has been decided to belong to the *outermost black-/white-box* and not to the *environment*, it will be verified at this *nesting level*. A youth recreation center near the development site has been contracted to appoint children to assist with *verification*.

TABLE 12-42 Multiplication toy with microcontroller multiplier white-box test-cases

White-box design check	Expected appearance with pass/fail criteria	Restriction requirement
CH.1. Try to break off anything from the front panel with bare hands.	XP.1. Pass, if verifiers cannot break off anything from the front panel.	A.RR.1.1.Everything accessible from the outside of the multiplication toy shall be impossible for a child to remove.
CH.2. Ask four 10-year-old children at the selected youth recreation center, to try to break off anything from the front panel, even with the help of simple tools they are used to.	XP.2. Pass, if children cannot remove anything from the front panel.	
CH.1. Identify all safety instructions for all used coating agents and surface treatments. Check if any used agents or treatments are hazardous for humans after the treatment is completed.	XP.1. Pass, if no coating and treatments used have any hazard warnings.	RR.1.2.Everything accessible from the outside of the multiplication toy shall be nontoxic for a licking child.
CH.1. Check in the architecture and design graphs how the interface to the multiplier is designed. Check in the prototype that exactly these interfaces are kept.	XP.1. Pass, if the prototype has realized all interfaces to the multiplier in accordance with the architecture and design graphs.	A.RR.5.1.The multiplication calculation shall be allocated to a separate element that can be redesigned with minimal rework.
CH.1. The multiplication toy with hard-wired multiplier is not yet a product with an ASIC. When the product with an ASIC is ready for verification, make this verification on both products.	XP.1. Pass, so far.	A.RR.5.2.The casing of the hosting toy shall be exactly the same for all products independent of technology used.
CH.1. Try to open the prototype power supply with the hands.	XP.1. Pass, if verifiers find it impossible to open the power supply.	A.RR.6.1.The toy shall have an internal power supply, not possible for a child to operate.
CH.2. Ask four 10-year-old children at the selected youth recreation center to try to open the prototype power supply, even with the help of simple tools they are used to.	XP.2. Pass, if verifiers and children shall find it impossible to open the power supply	
CH.1. Check the manufacturing requisites for total accumulated cost. Have manufacturing experts review these cost estimates.	XP.1. Pass, if estimated sum does not exceed € 5. and manufacturing experts find the estimate reliable. If cost exceeds €5, make a cost reduction with new manufacturing requisites and reverify.	A.RR.7.1.If the toy is produced, it shall be made in such volume that the price per unit is less than € 5.

Now we start preparing *black-box test-cases* for the *verification* of the multiplication toy *black-box* (see Table 12-43 below).

TABLE 12-43 Multiplication toy with microcontroller multiplier black-box test-cases

Black-box stimulus injectio- nion	Expected response with pass/fail criteria	Behavior requirement
CH.1. Check all possible settings on one operand multiplied by all possible settings on the other operand.	XP.1. Pass if all product values are calculated correctly.	A.RB.8.1.User amusement shall be to multiply two operand values into one product value.
CH.1. Check that both multiplicand switches can be set to different values.	XP.1. Pass if both operands can be set from 0 to at least 10.	A.RB.8.2.The toy shall have one input device for each multiplicand value. Multiplicand values shall be at least between 0 and 10. The multiplicands must be easy for a child to settle.
CH.2. Set one operand switch at value 1 and manually turn the other operand switch through all positions. Reverse operands above.	XP.2. Pass if one operand is set to 1 and presentation shows the value on the other setting, and if the same occurs when swapping operands.	
CH.3. Ask four 10-year-old children at a youth recreation center to input values (without more explanations).	XP.3. Pass if most children find it easy to input values.	
CH.1. Check the product when both operands are set to 10.	XP.1. Pass if value 100 shown.	A.RB.8.3.The toy shall have at least a 3-digit output device for the product value. The product must be easy to read for a child.
CH.2. Ask four 10-year-old children at a youth recreation center to read the display.	XP.2. Pass if most children find it easy to read the display.	
CH.1. Operate the operand switches manually as fast as possible.	XP.1. Pass if every product value is shown without a perception of lagging.	A.RB.8.4.The toy calculates and shows the product value practically instantly after either multiplicand is settled.

12.16.4 EXAMPLE Multiplication toy with microcontroller multiplier: Verify white-box in simulated manufacturing

The multiplication toy *product prototype* is not yet mounted in any *environment* hosting toy, but is now mounted in a preliminary manufacturing test *fixture* (similar to what will be used during manufacturing), which automatically connects to circuit board test points. For the *verification report*, see Table 12-44 below.

TABLE 12-44 Multiplication toy with microcontroller multiplier manufacturing verification report 1

Actual outcome and pass/ fail conclusion	Expected outcome and pass/fail criteria	Requirement
VR.1. Pass	XP.1.	A.RB.8.1.
VR.1. Pass	XP.1.	A.RB.8.3.

12.16.5 EXAMPLE Multiplication toy with microcontroller multiplier: Verify white-box

The *verification* of the *white-box* is performed next (see Table 12-45 below).

TABLE 12-45 Multiplication toy with microcontroller multiplier verification report 2

Actual outcome and pass/ fail conclusion	Expected outcome and pass/fail criteria	Requirement
VR.2. Pass	XP.1.	A.RR.1.1.
VR.2. Pass	XP.2.	A.RR.1.1.
VR.2. Pass	XP.1.	A.RR.1.2.
VR.2. Pass	XP.1.	A.RR.5.1.
VR.2. Pass	XP.1.	A.RR.5.2.
VR.2. Pass	XP.1.	A.RR.6.1.
VR.2. Pass	XP.2.	A.RR.6.1.
VR.2. Pass	XP.1.	A.RR.7.1.

12.16.6 EXAMPLE Multiplication toy with microcontroller multiplier: Verify black-box

Finally, the *verification* of the *black-box* is performed (see Table 12-46 below).

TABLE 12-46 Multiplication toy with microcontroller multiplier verification report 3

Actual outcome and pass/ fail conclusion	Expected outcome and pass/fail criteria	Requirement
VR.3.1. Fail. Products calculated as (operand + 1)	XP.1. Pass if all product values are calculated correctly.	A.RB.8.1.User amusement shall be to multiply two operand values into one product value.
VR.3. Pass	XP.1.	A.RB.8.2.The toy shall have one input device for each multiplicand value. Multiplicand values shall be at least between 0 and 10. The multiplicands must be easy for a child to settle.
VR.3.2. Fail. Calculations are done, but product seems to be calculated incorrectly as above. VR.3.3. Fail. Calculations also incorrect when reversing operands	XP.2. Pass if one operand is set to 1 and presentation shows the value on the other setting, and if the same occurs when swapping operands..	
VR.3. Pass	XP.3.	
VR.3. Pass	XP.1.	A.RB.8.3.
VR.3. Pass	XP.2.	
VR.3. Pass	XP.1.	A.RB.8.4.

12.16.7 EXAMPLE Multiplication toy with microcontroller multiplier: Locate failure source

Some failures were detected during the *black-box verification*, but all of them seemed to originate from the same location. When programming, a *variable* may easily be incremented or decremented too early or too late, which results in many secondary failures, as detected here.

How to localize the annoying failures was discussed a lot by the gate members. It was easy to measure the static settings on the microcontroller pinning for a *failure localization* of the *program*. The conclusion was that the electronics were correct.

Therefore, the *program* must contain the failure. There is the option of buying *program* debugger tools, but is it worth the cost? Masking away different parts of the *program*, to locate the failure more closely, can also be considered.

Since the *program* is fairly simple, it was decided to first try to carefully review the *program* to find the location of the failure (see column failure location; see Table 12-47 next chapter).

12.16.8 EXAMPLE Multiplication toy with microcontroller multiplier: Evaluate optimal elimination

When the failure is located, it becomes very simple to correct the *program* and eliminate the failure (see Table 12-47 below).

TABLE 12-47 Multiplication toy with microcontroller multiplier failure elimination report 1

Failure location	Failure elimination plan	Actual outcome and pass/fail criteria	Expected outcome and pass/fail criteria	Behavior requirement
FER.1.1. Correct values on input operand buses implying wrong values on product bus, indicate that the program in Figure 10-32, page 521 is calculating the product incorrectly.	FER.1.1.a.The program is simple enough to be debugged by thorough review. FER.1.1.b. Correct the program. FER.1.1.c. Reverify.	VR.3.1. Fail. Products calculated as (operand + 1). VR.3.2. Fail. Calculations are done, but product seems to be calculated incorrectly as above. VR.3.3. Fail. Calculations also incorrect when reversing operands	XP.1. Pass if all product values are calculated correctly. XP.2. Pass if one operand is set to 1 and presentation shows the value on the other setting, and if the same occurs when swapping operands.	A.RB.8.1.User amusement shall be to multiply two operand values into one product value. A.RB.8.2.The toy shall have one input device for each multiplicand value. Multiplicand values shall be at least between 0 and 10. The multiplicands must be easy for a child to settle.

12.16.9 EXAMPLE Multiplication toy with microcontroller multiplier: Verify black-/white-box regressively

After some review effort and assembler manual reading, the failure was localized and eliminated for reverifying (see Table 12-48 below). Since the *test-cases* are rather few, a regression test is decided on, which means that all *test-cases* should be used again by this *verification*, regardless of which were used in previous *verification*.

TABLE 12-48 Multiplication toy with microcontroller multiplier
re-verification report 4

Actual outcome and pass/ fail conclusion	Expected outcome and pass/fail criteria	Requirement
VR.4. Pass	XP.1.	A.RR.1.1.
VR.4. Pass	XP.2.	A.RR.1.1.
VR.4. Pass	XP.1.	A.RR.1.2.
VR.4. Pass	XP.1.	A.RR.5.1.
VR.4. Pass	XP.1.	A.RR.5.2.
VR.4. Pass	XP.1.	A.RR.6.1.
VR.4. Pass	XP.2.	A.RR.6.1.
VR.4. Pass	XP.1.	A.RR.7.1.
VR.4. Pass	XP.1.	A.RB.8.1.
VR.4. Pass	XP.1.	A.RB.8.2.
VR.4. Pass	XP.2.	
VR.4. Pass	XP.3.	
VR.4. Pass	XP.1.	A.RB.8.3.
VR.4. Pass	XP.2.	
VR.4. Pass	XP.1.	A.RB.8.4.

Now the failure is eliminated, and no other introduced.

12.16.10 EXAMPLE Multiplication toy with microcontroller multiplier: Prepare product manual

The multiplication toy with microcontroller multiplier will be widely marketed, and a good use manual is essential (see Table 12-49 below). For internal use, a *reuse* and portability manual is prepared.

TABLE 12-49 Multiplication toy with microcontroller multiplier
manual table of contents

Manual name	Requirement
MA.1. Internal reuse and portability guide	A.RR.5.1.The multiplication calculation shall be allocated to a separate element that can be redesigned with minimal rework.
MA.2. User playing with the toy	A.RB.8.1.User amusement shall be to multiply two operand values into one product value.
MA.3. User playing with the toy	A.RB.8.2.The toy shall have one input device for each multiplicand value. Multiplicand values shall be at least between 0 and 10. The multiplicands must be easy for a child to settle.
MA.4. User playing with the toy	A.RB.8.3.The toy shall have at least a 3-digit output device for the product value. The product must be easy to read for a child.
MA.5. User guardian MA.6. changing batteries	A.RR.6.1.The toy shall have an internal power supply, not possible for a child to operate.

12.16.11 EXAMPLE Multiplication toy: Proceed reading

To follow this example, these are the alternatives (see again Fig. 12-20, p. 718):

- Trace *upstream* through the overall *schedule nesting level* = 1 to understand how to eliminate a detected failure.

- Proceed to chapter 11.11, "EXAMPLE Multiplication toy environment (microcontroller multiplier variant): Realize interfaces to environment and insert/await outermost white-box," page 637.

- To stay within the multiplication toy example, but not following the overall *schedule*, proceed to chapter 12.17 below.

12.17 EXAMPLE Multiplication toy environment (microcontroller multiplier variant): Verify prototype in environment

12.17.1 EXAMPLE Multiplication toy environment (microcontroller multiplier variant): Process schedule to use

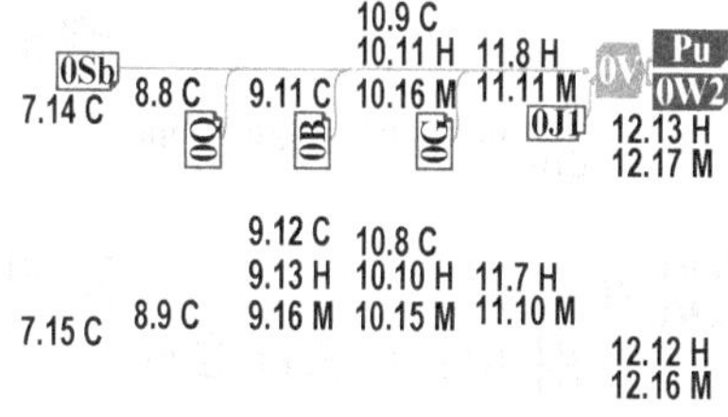

For the multiplication toy example, a special technical overall *schedule* is *tailored* (illustrated in Fig. 12-21, left):

- **Pt:2:3. Tailored advanced technical schedule n = 2,** page 169

In this *tailored schedule* the generic *schedule* to use is (filled-in symbols):

- **Pt.0.V. Verify prototype in environment,** page 125.

FIGURE 12-21 Position of schedule and master to use (filled in) in overall schedule

Below *results* may be demanded (their *masters* are indicated *upstream* from the filled-in *schedule* in the figure above):

0J1 Usable prototype interfaces started in simulator or target

0G Environment interface design items and product requisites

  Table 10-21, Multiplication toy environment attachment, page 496

  Table 10-56, Multiplication toy environment interfaces commonalities, interfaces to environment design items, page 496

0B Environment existing ingredients and design of interfaces to the environment

  Table 9-23, Multiplication toy environment architecture ingredients resumé, page 383

  Figure 9-35, Multiplication toy environment logical architecture, page 384

Figure 9-36, Multiplication toy environment physical architecture, page 384

0Q Solution alternatives of interfaces to environment with supplier opportunities

Table 8-18, Multiplication toy environment solution alternatives of interfaces to environment, page 314

Table 8-19, Multiplication toy environment solutions excluded, page 315

Table 8-20, Multiplication toy environment solutions included with supplier opportunities, page 315

0Sb Environment restriction requirements

Table 7-23, Multiplication toy environment restriction requirement, page 213

12.17.2 EXAMPLE Multiplication toy environment (microcontroller multiplier variant): Develop field test-cases for target or simulator

Field *environment verification test-cases* are now prepared (see Table 12-50 below).

TABLE 12-50 Multiplication toy environment (microcontroller multiplier variant) white-box test-cases

White-box design check	Expected appearance with pass/fail criteria	Restriction requirement
CH.1. Submit the multiplication toy prototype to specialized companies, accredited to issue certification for compliance with these regulations.	XP.1. Pass, if certified.	E.RR.1.The multiplication toy shall be childproof according to European Community regulations.
CH.1. Mount the multiplication toy prototype with gasket and collar on an airtight plastic bag, seal the bag, and blow it up until it barely loses all its wrinkles.	XP.1. Pass, if the bag doesn't significantly lose its air tension within a day.	E.RR.2.The multiplication toy attachment to the hosting toy may not leak padding from the inside of the hosting toy.
CH.1. Check interface design descriptions.	XP.1. Pass, if interfaces to the environment are documented according to process.	E.RR.3.The multiplication toy shall be easily reusable in a safe way, by all kinds of different hosting toys.
CH.2. Check the change control history of the interface.	XP.2. Pass, if change control history has been properly managed.	
CH.3. With help only from the manual, have 10 potential customers mount the multiplication toy prototype in some of their cuddling toys.	XP.3. Pass, if all customers have mounted the multiplication toy correctly.	
CH.4. Try to mount the multiplication toy into the hosting toy in a wrong way.	XP.4. Pass, if it is difficult to mount upside down, inside out, or in any other wrong way.	
CH.5. Try to remove a mounted multiplication toy from its host toy.	XP.5. Pass, if impossible to demount it by hand or by simple child tools.	
Excluded from development responsibility		E.RR.4.No legal responsibility is taken for incorrect mounting of the multiplication toy into the hosting toy by companies buying the multiplication toy.

12.17.3 EXAMPLE Multiplication toy environment (microcontroller multiplier variant): Verify prototype in field simulator environment

The European Union standards have been very strict on good design, to avoid a bad reputation for dangerous children's toys. Because of that, the toy was sent to the institute accredited to certify by European standards. The certification didn't require the multiplication toy to be mounted into any *target* host toy.

The certification was received with a wide margin (see Table 12-51 below).

TABLE 12-51 Multiplication toy environment (microcontroller multiplier variant) verification report 1

Actual outcome and pass/fail conclusion	Expected outcome and pass/fail criteria	Requirement
VR.1. Certified	XP.1.	E.RR.1.

12.17.4 EXAMPLE Multiplication toy environment (microcontroller multiplier variant), (certified): Verify prototype in target environment

The multiplication toy was sent out to prospective hosting toy *customers*, for evaluation of mounting the multiplication toy to their hosting toys.

The pass/fail list was filled in after the *customers* had made their trial mounting.

TABLE 12-52 Multiplication toy environment (microcontroller multiplier variant) verification report 2

Actual outcome and pass/fail conclusion	Expected outcome and pass/fail criteria	Requirement
VR.2. Pass	XP.1.	
VR.2. Pass	XP.1.	
VR.2. Pass	XP.2.	
VR.2.4. Fail. Some hosting toy customers didn't understand how to mount the multiplication toy in their cuddling toys.	XP.3. Pass, if all customers have mounted the multiplication toy correctly.	E.RR.3. The multiplication toy shall be easily reusable in a safe way, by all kinds of different hosting toys.
VR.2. Pass	XP.4.	
VR.2. Pass	XP.5.	

Surprisingly, some of the hosting toy *customers* didn't succeed in mounting the multiplication toy.

12.17.5 EXAMPLE Multiplication toy environment (microcontroller multiplier variant): Locate failure source

The failed mounting of the multiplication toy was carefully examined and the hosting toy *customers* were carefully interviewed.

It showed that the mounting skill of the hosting toys *customers* had been overrated. These *customers* seldom worked with hard plates when producing cuddle toys, and they got very confused on how to do it. Also, many of the small *customers* were not used to reading long textual descriptions explaining the work flow. These failure locations are documented in Table 12-53 in the next chapter.

12.17.6 EXAMPLE Multiplication toy environment (microcontroller multiplier variant): Evaluate prototype completeness

To eliminate the failure, the manual needs to be improved (see Table 12-53 below).

TABLE 12-53 Multiplication toy environment (microcontroller multiplier variant) failure elimination report 1

Failure location	Failure elimination plan	Actual outcome and pass/fail criteria	Expected outcome and pass/fail criteria	Restriction requirement
FER.1.2. For small hosting toy producers, it is too complicated to interface their hosting toy with the multiplication toy.	FER.1.2.a. How to use the collar was insufficiently described in the manual. FER.1.2.b. Insert some pictures and illustrations in the manual. FER.1.2.c. Reverify.	VR.2.4. Fail. Some hosting toy customers didn't understand how to mount the multiplication toy in their cuddling toys..	XP.3. Pass, if all customers have mounted the multiplication toy correctly.	E.RR.4.No legal responsibility is taken for incorrect mounting of the multiplication toy into the hosting toy by companies buying the multiplication toy.

12.17.7 EXAMPLE Multiplication toy environment (microcontroller multiplier variant): Reverify prototype in target environment

Multiplication toys were again sent out to prospective *customers* for their evaluation, and also attached was the newly improved manual. This time the *verification* went well (see Table 12-54 below).

TABLE 12-54 Multiplication toy environment (microcontroller multiplier variant) reverification report 3

Actual outcome and pass/fail conclusion	Expected outcome and pass/fail criteria	Requirement
VR.3. Pass	XP.3.	E.RR.4.

12.17.8 EXAMPLE Multiplication toy environment (microcontroller multiplier variant): Finalize product manual

The previous *verification* has shown that the current manual must be improved a lot, to better support *customers* mounting the multiplication toy (see Table 12-55 below).

TABLE 12-55
Multiplication toy environment (microcontroller multiplier variant) product manual

Manual name	Requirement
MA.1. Include the European Union certificate in the manual.	E.RR.1.The multiplication toy shall be childproof according to European Community regulations.
MA.2. Explain hosting toy customer mounting guidelines with work flow by photos.	E.RR.4.No legal responsibility is taken for incorrect mounting of the multiplication toy into the hosting toy by companies buying the multiplication toy.

12.17.9 EXAMPLE Multiplication toy: Proceed reading

To follow this example, these are the alternatives (see again Fig. 12-21, p. 725):

- Trace *upstream* through the overall *schedule nesting level* = 0 to understand how to eliminate a localized failure.

- The multiplication toy (microcontroller multiplier variant) example ends here, and market validation takes place (see ch. 6.6, "P. Cpdm generic development schedule," p. 99). If still enthusiastic about *Cpdm*, pull yet another bottle of champagne :-).

12.18 EXAMPLE Lake: Which fish to catch first ?

As with*requirements*, there is no time or personnel to make any number of *test-cases*, implying that a prioritization is needed for which kind of failures to catch first and which to proceed with if you have more capacity. Compare with fishing—which are the most important fish to catch first (see Fig. 12-22 below)?

Most often the big fish are targeted first, because they are most valuable. The problem is that they are also scaring away the smaller fish, which disappear and can't be found. Compare with fatal failures that halt the *verification* or cause it to deviate in unexpected ways; the subsequent failures after the break are impossible to find until the fatal failure is eliminated.

FIGURE 12-22 Which are the most attractive fish to catch?

12.19 Usage profile modeling

12.19.1 The problem

Sooner or later *verification* organizations ask themselves, what set of *test-cases* should we develop? How many *test-cases* should we have? What parts are most important to verify? Is it the most commonly used, or the most difficult to develop, or something else? And behind all of these questions—who in the organization decides this?

The numbers of test-cases that can be developed is limited—which have the highest priority?

There is an elegant and knowledgeable solution to all these questions, called *usage profile* modeling. It is somewhat theoretical and academic, but if applied in a realistic way, the benefits are great.

12.19.2 EXAMPLE: Imagine how to choose test-cases

It is hard for verifiers and engineers to predict how normal *users* will handle the *product*, or in other words, what *stimuli* are entered and what *responses* are expected. Developers of the *product* have (hopefully) good inside understanding of the *product*, but are often very poor at putting themselves in the place of a target *end user*. Even if *use-case requirements* are used to create and develop *test-cases*, they often become biased by too much or too specialized internal knowledge of the *product*.

Imagine that somebody instead of deciding how to use a *product*, throws a die to decide how to use that *product*. Recall the calculator example, and imagine that the different die dots correspond to different *stimuli*. Imagine "one dot" to be a number to be pressed, "two dots" to be addition, "three dots" to be subtraction, "four dots'" to be multiplication, "five dots" to be division and "six dots" to be pressing the equal sign.

> To avoid biased humans, imagine for a short moment that a die determines which stimuli arrive to a calculator.

The above example would certainly generate too few digits and too many operators, since each of them has a 1/6 probability. But the problem is now reduced to figuring out what the probability is for an average *user* to push a certain button on the calculator. For this a biased *stakeholder* is not needed.

12.19.3 Automate generation of a usage profile

If *requirements* are made as *formal use-cases*, it is not very difficult to see how a normal *user* will operate a calculator. Every transition from a *wait-state* in a *formal scenario*, can be assigned probabilities for what the *user* might do. Since different *users* have different ways to use a calculator, some different kinds of *usage profiles* can be developed.

> Throwing a die is too random to describe stimuli from a normal user; instead, assign probabilities for the stimuli from that user.

After each *usage profile* is populated with probabilities, the *test-case stimuli* sequence can be determined according to their probabilities by a die or any kind of random generator. A simple *program* or a complete *test-case* tool can even be built to generate the *test-case stimuli* sequence based on the usage of the *product*.

After a *test-case stimuli* sequence is generated, it must be supplemented with expected *responses* from the prototype and a criterion for how to judge the *response*, whether it is sufficient for a pass/fail judgment. In chapter 12.20, page 733 this will be further explained.

Assume that statistical *test-cases* have been generated to correspond to 1 000 hours of usage of a *product*, and four failures from these *test-cases* are found at *verification*. If the usage being modeled really matches realistic usage, the conclusion can be drawn that the mean time between failures (MTBF) would be 250 hours. What other *verification* method can provide such a reliable prediction?

12.19.4 Cpdm definition

Since a *usage profile* is the core of this *verification* method, a formal *Cpdm* definition would be great.

TABLE 12-56 Cpdm definition

Aspect	Usage profile
Definition	Every stimulus causing departure from every wait-state is assigned estimated probabilities, to be used by black-box test-cases.
Synonyms	Probability test-case, statistical usage
Symbol	*(see Table 12-57 below)*

TABLE 12-57 Probabilities for different usage

Wait-state-machine with input	Usage profiles, estimates in %			Next state
	Profile 1	Profile 2	Profile 3	
A.SM initial	100%	100%	100%	
WS 1				
A.SM Stimulus 1	60%	30%	10%	WS 1
A.SM Stimulus 2	40%	30%	10%	WS 2
A.SM Stimulus 4		40%	80%	WS 2
Sum	100%	100%	100%	
WS 2				
A.SM Stimulus 4	50	80	100%	WS 1
A.SM Stimulus 4 / A.SM Stimulus 3	20 / 30	10 / 10		WS 2
Sum	100%	100%	100%	

12.19.5 Process schedule for usage profile modeling

The normal *process schedule* **Pt.n.Va. Develop test-cases, also for fixtures and debugger,** page 152, can be specialized and detailed for *usage profile* modeling (see Fig. 12-23 below).

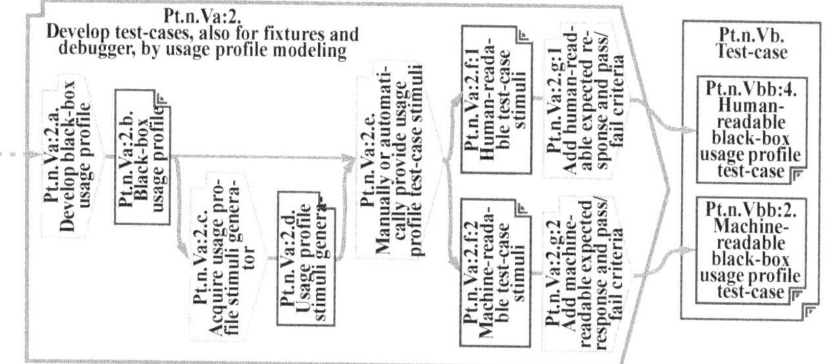

FIGURE 12-23 Develop test-cases specialized for usage profile modeling

12.20 EXAMPLE Hosted calc_logic for reuse: Verify black-/white-box

12.20.1 Calc_logic: About this example

In this example, the calc_logic *program* will be *ported* from development at *host* to various *targets*, and because of that the *tailored schedule* for *porting* will be used.

In addition, the *verification* should prepare *black-box test-cases* useful for *usage profile* modeling. It means to develop *test-cases* from a *wait-state-machine*, which have been defined when capturing *behavior requirements* for calc_logic.

12.20.2 Hosted calc_logic for reuse: Process schedule to use

For the calc_logic example, a special technical overall *schedule* is *tailored* (illustrated in Fig. 12-24, below):

- **Pt:2:3b. Tailored advanced technical schedule n = 2, porting to 3 targets (right half),** page 170

In this *tailored schedule* the generic *schedule* to use is (filled-in symbols):

- **Pt.n.V. Verify black-/white-box,** page 152.

To support *usage profile* modeling, the first *task* **Pt.n.Va.,** page 152 in this *schedule* is expanded to *schedule version* **Pt.n.Va:2. Develop test-cases, also for fixtures and debugger, by usage profile modeling,** page 158.

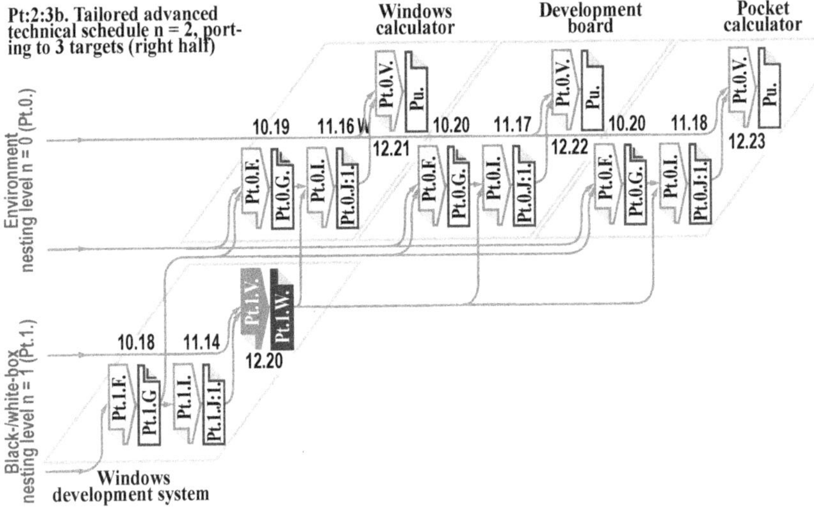

FIGURE 12-24 Position of schedule and master to use (filled in) in overall schedule

12.20.3 EXAMPLE Hosted calc_logic for reuse: Develop test-cases, also for fixtures and debugger

Before starting with *usage profile* modeling, ordinary *white-box test-cases* are prepared to be used for *verification* on the *host* Windows *development system* (see Table 12-58 below).

TABLE 12-58 Calc_logic for reuse white-box test-cases

White-box design check	Expected appearance with pass/fail criteria	Restriction requirement
CH.1. Check that calc_logic is able to receive 64 different stimuli types (buttons).	XP.1. Pass if the calc_logic program is prepared to receive 64 different stimuli types (buttons).	A.RR.1.1. The calc_logic element shall be able to process 16-button stimuli at a minimum and be capable of processing many more future button stimuli.
CH.1. Check that calc_logic is able to return 128 digits.	XP.1. Pass if calc_logic is prepared to return 128 digits.	A.RR.2.1. The calc_logic element shall be able to process 8-digit responses, and be capable of processing a much larger future number of digits.
CH.1. Check if calc_logic program uses math-coprocessor.	XP.1. Pass if math-coprocessor is not used and not needed.	A.RR.3.1. To reduce complexity from the unpredictable existence of coprocessor, math coprocessors shall not be used at all.

TABLE 12-58 Calc_logic for reuse white-box test-cases

White-box design check	Expected appearance with pass/fail criteria	Restriction requirement
CH.1. Check that all used development systems are capable of supporting all specified targets, and that calc_logic is capable of porting to specified targets without redesign.	XP.1. Pass if all development systems support all targets, and if calc_logic can be ported to all targets without redesign.	A.RR.5.1. Calc_logic for reuse shall be portable to the Windows calculator application, to development boards, and to the pocket stand-alone calculator.
CH.1. Check the black-box requirements for calc_logic.	XP.1. Pass if requirements are using wait-states.	A.RB.6.S1. Calc_logic for reuse, behavior requirement 6, scenario 1

12.20.4 EXAMPLE Hosted calc_logic for reuse: Verify white-box

It is decided to perform the *white-box verification* before dealing with the *usage profile* modeling (see Table 12-59 below).

TABLE 12-59 Calc_logic for reuse verification report 1

Actual outcome and pass/fail conclusion	Expected outcome and pass/fail criteria	Requirement
VR.1. Pass	XP.1.	A.RR.1.1.
VR.1. Pass	XP.1.	A.RR.2.1.
VR.1. Pass	XP.1.	A.RR.3.1.
VR.1. Pass	XP.1.	A.RR.4.1.
VR.1. Pass	XP.1.	A.RR.5.1.
VR.1. Pass	XP.1.	A.RB.6.S1.

12.20.5 EXAMPLE Hosted calc_logic for reuse: Develop test-cases, also for fixtures and debugger, by usage profile modeling

Now it is about time to start with the *usage profile* modeling. First the *black-box test-casees* are prepared (see Table 12-60 below).

TABLE 12-60 Calc_logic for reuse usage profile black-box test-cases

Black-box stimulus injection	Expected response with pass/fail criteria	Behavior requirement
CH.1. Minimum of 200 stimuli automatically is generated for the basic usage profile. Verify and eliminate all failures before verifying with below advanced usage profile.	XP.1. Expect a relative error criterion of nothing else than 0 for integer responses, and less than 1e-16 relative error criteria for fraction responses.	A.RB.6.S1. Calc_logic for reuse, behavior requirement 6, scenario 1
CH.2. Minimum of 200 stimuli is automatically generated for advanced usage profile. Verify and eliminate all failures before verifying with below stressed usage profile.		A.RR.4.1. A maximum calculation precision of $\approx$ 15 digits during host development of the calc_logic program is sufficient.
CH.3. Minimum of 200 stimuli automatically generated for stressed usage profile.		

The basic usage profile is intended to imitate a novice calculator *user*, making only very safe and simple calculations that have been specified. The advanced usage profile uses all kinds of specified calculating sequences, including repetitive calculations without using equal signs in between. The stressed usage profile also specifies unexpected *stimuli* sequences, many of them probably not specified at all.

12.20.6 EXAMPLE Hosted calc_logic for reuse: Develop black-box usage profile

The *process schedule* to use is described in above chapter 12.19, "Usage profile modeling," page 730 and in chapter 6.12.9, "Detailed Pt.n.Va:2. Develop test-cases, also for fixtures and debugger, by usage profile modeling," page 158.

Usage probabilities now must be built from the *formal scenario* in Figure 7-28, page 247. It is also neat to translate this *formal scenario* to a *wait-state-machine* (see left of Fig. 12-25 below). It has already been decided to create three different *usage profiles* which are noted in separate columns (see Table 12-61 below).

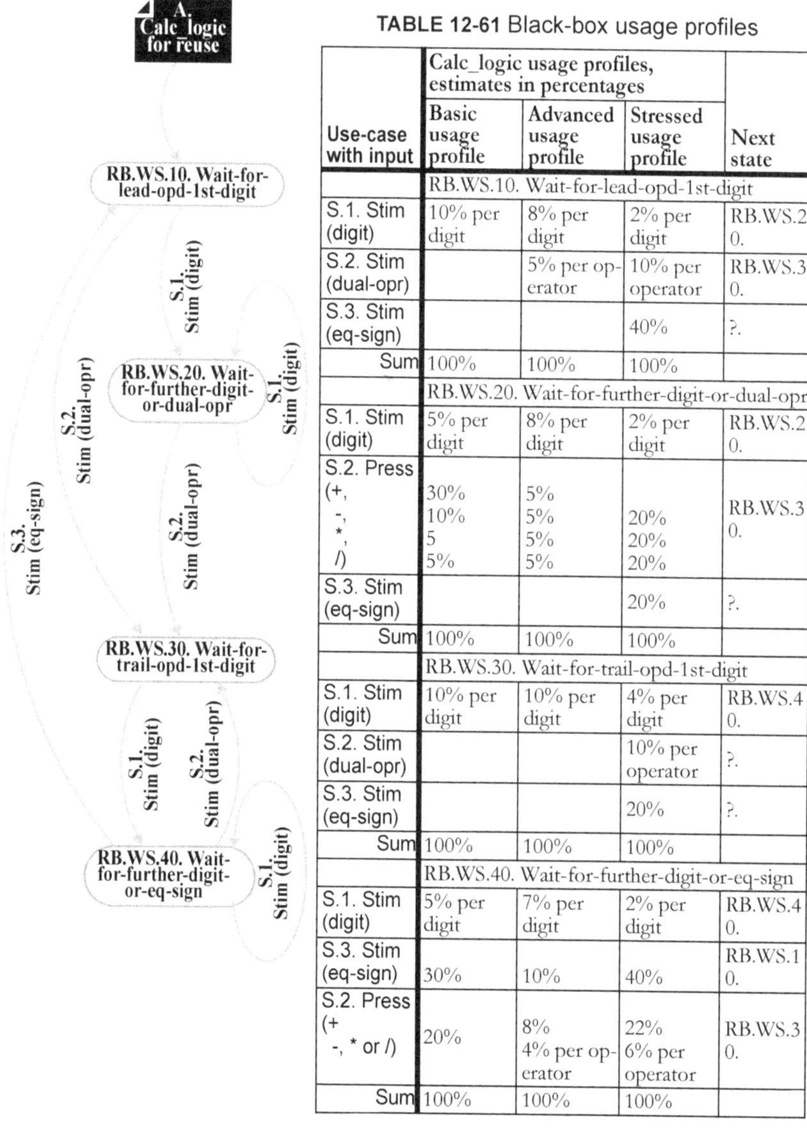

TABLE 12-61 Black-box usage profiles

Use-case with input	Calc_logic usage profiles, estimates in percentages			Next state
	Basic usage profile	Advanced usage profile	Stressed usage profile	
	RB.WS.10. Wait-for-lead-opd-1st-digit			
S.1. Stim (digit)	10% per digit	8% per digit	2% per digit	RB.WS.20.
S.2. Stim (dual-opr)		5% per operator	10% per operator	RB.WS.30.
S.3. Stim (eq-sign)			40%	?.
Sum	100%	100%	100%	
	RB.WS.20. Wait-for-further-digit-or-dual-opr			
S.1. Stim (digit)	5% per digit	8% per digit	2% per digit	RB.WS.20.
S.2. Press (+, -, *, /)	30% 10% 5 5%	5% 5% 5% 5%	20% 20% 20%	RB.WS.30.
S.3. Stim (eq-sign)			20%	?.
Sum	100%	100%	100%	
	RB.WS.30. Wait-for-trail-opd-1st-digit			
S.1. Stim (digit)	10% per digit	10% per digit	4% per digit	RB.WS.40.
S.2. Stim (dual-opr)			10% per operator	?.
S.3. Stim (eq-sign)			20%	?.
Sum	100%	100%	100%	
	RB.WS.40. Wait-for-further-digit-or-eq-sign			
S.1. Stim (digit)	5% per digit	7% per digit	2% per digit	RB.WS.40.
S.3. Stim (eq-sign)	30%	10%	40%	RB.WS.10.
S.2. Press (+ -, * or /)	20%	8% 4% per operator	22% 6% per operator	RB.WS.30.
Sum	100%	100%	100%	

FIGURE 12-25 Usage profile wait-state-machine with transition probabilities

Since the calc_logic specification in Figure 7-28, page 247 is not complete (specifications can be but seldom are), it must be decided how to handle *wait-state* transitions, when *stimuli* arrive that are not expected and specified by the *wait-state-*

machine (see question marks in *wait-state* column in Table 12-61 above). For example, it is important to know how to find out which *wait-state* to enter after two equal signs appear immediately after each other.

There are mainly two solutions:

- Go back to *requirement* specification and make the *requirements* complete for all *exception use-cases* before starting to generate the *stimuli*.
- Decide what will happen to an unexpected and unspecified *stimulus*. In this example for the *stimuli* generator below, it is decided on the spot that whatever unspecified *stimulus* is received shall not cause leaving the current *wait-state*.

12.20.7 EXAMPLE Hosted calc_logic for reuse: Acquire usage profile stimuli generator

The *process schedule* to use is described in above chapter 12.19, "Usage profile modeling," page 730and in chapter 6.12.9, "Detailed Pt.n.Va:2. Develop test-cases, also for fixtures and debugger, by usage profile modeling," page 158.

Since the calc_logic *program* is developed by a C/C++ windows *development system*, it is convenient to also develop the *stimuli* generator by this system.

Note that the below *stimuli* generator *program* is not presented as a full example of *Cpdm* development, but it is nice to present the *program* Figure 12-26 below, to show that it is not overly *complex* to build such a generator.

The *program* below and all other *programs* presented in this book, are also available at the website www.cpdm.com.

```
/* A. Usage profile stimuli generator */

        /* A.a. #include, #define etc. */
#include "stdafx.h"
#include <stdlib.h>
#include <stdio.h>
#include <time.h>
#include <tchar.h>
#include <iostream>
#include <conio.h>
using namespace std;

#define PRESS_BUTT_COL 0
#define PERCENT_BASIC_COL 1
#define PERCENT_ADVANCED_COL 2
#define PERCENT_STRESSED_COL 2
#define NEXT_STATE_COL 4

#define WAIT_FOR_LEADH_OPD_1ST_DIGIT 0

#define WAIT_FOR_FURTHER_DIGIT_OR_DUAL_OPR 1
#define WAIT_FOR_TRIAL_OPD_1ST_DIGIT 2
#define WAIT_FOR_FURTHER_DIGIT_OR_EQ_SIGN 3

        /*A.b.Probabilities table */
// organized as value [state][row][col] */
static char value[4][15][5] =
{
```

```
   {
// wait_state 0 (S.10. WAIT FOR LEAD OPD 1ST DIGIT)
       '0', 10,  8,  2,   WAIT FOR FURTHER DIGIT OR DUAL OPR,
       '1', 10,  8,  2,   WAIT FOR FURTHER DIGIT OR DUAL OPR,
       '2', 10,  8,  2,   WAIT FOR FURTHER DIGIT OR DUAL OPR,
       '3', 10,  8,  2,   WAIT FOR FURTHER DIGIT OR DUAL OPR,
       '4', 10,  8,  2,   WAIT FOR FURTHER DIGIT OR DUAL OPR,
       '5', 10,  8,  2,   WAIT FOR FURTHER DIGIT OR DUAL OPR,
       '6', 10,  8,  2,   WAIT FOR FURTHER DIGIT OR DUAL OPR,
       '7', 10,  8,  2,   WAIT FOR FURTHER DIGIT OR DUAL OPR,
       '8', 10,  8,  2,   WAIT FOR FURTHER DIGIT OR DUAL OPR,
       '9', 10,  8,  2,   WAIT FOR FURTHER DIGIT OR DUAL OPR,
       '+',  0,  5, 10,   WAIT FOR LEAD OPD 1ST DIGIT,
       '-',  0,  5, 10,   WAIT FOR LEAD OPD 1ST DIGIT,
       '*',  0,  5, 10,   WAIT FOR LEAD OPD 1ST DIGIT,
       '/',  0,  5, 10,   WAIT FOR LEAD OPD 1ST DIGIT,
       '=',  0,  0, 40,   WAIT FOR LEAD OPD 1ST DIGIT,
   },
// Wait_state 1 (S.20. WAIT_FOR_FURTER_DIGIT_OR_DUAL_OPR)
   {
       '0',  5,  8,  2,   WAIT FOR FURTHER DIGIT OR DUAL OPR,
       '1',  5,  8,  2,   WAIT FOR FURTHER DIGIT OR DUAL OPR,
       '2',  5,  8,  2,   WAIT FOR FURTHER DIGIT OR DUAL OPR,
       '3',  5,  8,  2,   WAIT FOR FURTHER DIGIT OR DUAL OPR,
       '4',  5,  8,  2,   WAIT FOR FURTHER DIGIT OR DUAL OPR,
       '5',  5,  8,  2,   WAIT FOR FURTHER DIGIT OR DUAL OPR,
       '6',  5,  8,  2,   WAIT FOR FURTHER DIGIT OR DUAL OPR,
       '7',  5,  8,  2,   WAIT FOR FURTHER DIGIT OR DUAL OPR,
       '8',  5,  8,  2,   WAIT FOR FURTHER DIGIT OR DUAL OPR,
       '9',  5,  8,  2,   WAIT FOR FURTHER DIGIT OR DUAL OPR,
       '+', 30,  5,  0,   WAIT FOR TRIAL OPD 1ST DIGIT,
       '-', 10,  5, 20,   WAIT FOR TRIAL OPD 1ST DIGIT,
       '*',  5,  5, 20,   WAIT FOR TRIAL OPD 1ST DIGIT,
       '/',  5,  5, 20,   WAIT FOR TRIAL OPD 1ST DIGIT,
       '=',  0,  0, 20,   WAIT FOR FURTHER DIGIT OR DUAL OPR,
   },
// wait_state 2 (S.30. WAIT_FOR_TRIAL_OPD_1ST_DIGIT)
   {
       '0', 10, 10,  4,   WAIT FOR FURTHER DIGIT OR EQ SIGN,
       '1', 10, 10,  4,   WAIT FOR FURTHER DIGIT OR EQ SIGN,
       '2', 10, 10,  4,   WAIT FOR FURTHER DIGIT OR EQ SIGN,
       '3', 10, 10,  4,   WAIT FOR FURTHER DIGIT OR EQ SIGN,
       '4', 10, 10,  4,   WAIT FOR FURTHER DIGIT OR EQ SIGN,
       '5', 10, 10,  4,   WAIT FOR FURTHER DIGIT OR EQ SIGN,
       '6', 10, 10,  4,   WAIT FOR FURTHER DIGIT OR EQ SIGN,
       '7', 10, 10,  4,   WAIT FOR FURTHER DIGIT OR EQ SIGN,
       '8', 10, 10,  4,   WAIT FOR FURTHER DIGIT OR EQ SIGN,
       '9', 10, 10,  4,   WAIT FOR FURTHER DIGIT OR EQ SIGN,
       '+',  0,  0, 10,   WAIT FOR TRIAL OPD 1ST DIGIT,
       '-',  0,  0, 10,   WAIT FOR TRIAL OPD 1ST DIGIT,
       '*',  0,  0, 10,   WAIT FOR TRIAL OPD 1ST DIGIT,
       '/',  0,  0, 10,   WAIT FOR TRIAL OPD 1ST DIGIT,
       '=',  0,  0, 20,   WAIT FOR TRIAL OPD 1ST DIGIT,
   },
// wait_state 3 (S.40. WAIT_FOR_FURTHER_DIGIT_OR_EQ_SIGN)
   {
       '0',  5,  7,  2,   WAIT FOR FURTHER DIGIT OR EQ SIGN,
       '1',  5,  7,  2,   WAIT FOR FURTHER DIGIT OR EQ SIGN,
       '2',  5,  7,  2,   WAIT FOR FURTHER DIGIT OR EQ SIGN,
       '3',  5,  7,  2,   WAIT FOR FURTHER DIGIT OR EQ SIGN,
       '4',  5,  7,  2,   WAIT FOR FURTHER DIGIT OR EQ SIGN,
       '5',  5,  7,  2,   WAIT FOR FURTHER DIGIT OR EQ SIGN,
       '6',  5,  7,  2,   WAIT FOR FURTHER DIGIT OR EQ SIGN,
       '7',  5,  7,  2,   WAIT FOR FURTHER DIGIT OR EQ SIGN,
       '8',  5,  7,  2,   WAIT FOR FURTHER DIGIT OR EQ SIGN,
       '9',  5,  7,  2,   WAIT FOR FURTHER DIGIT OR EQ SIGN,
       '=', 30, 10, 40,   WAIT FOR LEAD OPD 1ST DIGIT,
       '+', 20,  8, 22,   WAIT FOR TRIAL OPD 1ST DIGIT,
       '-',  0,  4,  6,   WAIT FOR TRIAL OPD 1ST DIGIT,
       '*',  0,  4,  6,   WAIT FOR TRIAL OPD 1ST DIGIT,
       '/',  0,  4,  6,   WAIT FOR TRIAL OPD 1ST DIGIT,
   },
};

/* Ac. main.c */
```

```cpp
int _tmain(int argc, _TCHAR* argv[])
{
                        /* A.ca. Local variables */
    int  accu_percent, accu_novis_percent;
    int  accu_advanced_percent, accu_STRESSED_percent;
    int  rand_gen, gen_press, row, state, press_desired;
    int  usage_desired, usage_column;
    char pressed_butt, m;
    FILE *gen_values_comma_sep_p, *gen_values_line_sep_p;

                        /* A.cb. Check consistency */
// Check that probability percents sum up to 100
    for (state = 0; state<=3; state++)
    {
        accu_novis_percent = 0;
        accu_advanced_percent = 0;
        accu_stressed_percent = 0;
        for (row = 0; row <= 14; row++)
        {
            accu_novis_percent += value[state][row][PERCENT_BASIC_COL];
            accu_advanced_percent += value[state][row][PERCENT_ADVANCED_COL];
            accu_stressed_percent += value[state][row][PERCENT_STRESSED_COL];
        }
        if ((accu_novis_percent != 100) ||
            (accu_advanced_percent != 100) ||
            (accu_stressed_percent != 100))
        {
            printf ("Sum not equal to 100 percent in column of state %d\n", state);
            cout << "Hit any key to quit\n";
            m = _getch();
            return(-1);
        }
    }

                        /* A.cc. Initiate program */
// Input some control parameters
    cout << "Test-case generator for Basic, Advanced or Stressed\n";
    cout << "Which usage should be generated? Basic=1,Advanced=2,Stressed=3\n";
    cin >> usage_desired;
    switch(usage_desired)
    {
        case 1 :
            gen_values_comma_sep_p = fopen("Test-case-stimuli-Basic-c.txt", "w");
            gen_values_line_sep_p = fopen("Test-case-stimuli-Basic-l.txt", "w");
            break;
        case 2 :
            gen_values_comma_sep_p = fopen("Test-case-stimuli-Advanced-c.txt", "w");
            gen_values_line_sep_p = fopen("Test-case-stimuli-Advanced-l.txt", "w");
            break;
        case 3 :
            gen_values_comma_sep_p = fopen("Test-case-stimuli-Stressed-c.txt", "w");
            gen_values_line_sep_p = fopen("Test-case-stimuli-Stressed-l.txt", "w");
            break;
        default :
            printf ("Answer with either of 1, 2 or 3\n");
            cout << "Hit any key to quit\n";
            m = _getch();
            return(-1);
    }
    usage_column = usage_desired;
    cout << "How many button pressings should be generated ?\n";
    cin >> press_desired;
    // Set random start by using time
    srand( (unsigned)time( NULL ) );

                        /* A.cd. Generate stimuli */
    // For each wait-state scan value table row by row,
    // and match when rand_gen >= accumulated percentage + 1 accumulation
    cout << "Generated button pressings\n";
    state = WAIT_FOR_LEAD_OPD_1ST_DIGIT;

                        /* A.cd.a Generate one char */
```

```
for (gen_press = 0; gen_press < press_desired; gen_press++)
{
    rand_gen = (rand() % 100 );
    accu_percent = 0;


    for(row = 0;rand_gen>=accu_percent+value[state][row][usage_column];row++)
    {
        accu_percent = accu_percent + value[state][row][usage_column];
    }

    pressed_butt = value[state][row][PRESS_BUTT_COL];
    printf("%c\n", pressed_butt);
    fprintf(gen_values_comma_sep_p, "%c, ", pressed_butt);
    fprintf(gen_values_line_sep_p, "%c\n", pressed_butt);
    state = value[state][row][NEXT_STATE_COL];
}


    fclose(gen_values_comma_sep_p);
    fclose(gen_values_line_sep_p);
    cout << "Hit any key to quit\n";
    m = _getch();
    return 0;
```

FIGURE 12-26 Program generating stimuli according to black-box test-cases

12.20.8 EXAMPLE Hosted calc_logic for reuse: Manually or automatically provide usage profile test-case stimuli

As said, the *process schedule* to use is described in above chapter 12.19, "Usage profile modeling," page 730, and in chapter 6.12.9, "Detailed Pt.n.Va:2. Develop test-cases, also for fixtures and debugger, by usage profile modeling," page 158.

Below are shown 400 *stimuli* generated by the *stimuli* generator *program* above presented for the three decided *usage profiles*. Note that the lists below are presented as comma-separated *stimuli*, since this is more compact for a book. Later when *test-cases* are presented, each *stimulus* occupies at minimum a full row.

```
8, 7, +, 4, =, 5, +, 5, 9, 7, +, 2, 8, 7, =, 6, +, 5, =, 3, +, 8, +, 8, =, 7, /, 2, 0,
8, =, 9, 3, -, 2, +, 1, 0, 7, 3, 2, 6, 4, 8, =, 8, 2, +, 8, 9, 4, =, 6, -, 5, +, 7, =,
0, +, 3, =, 9, 2, 1, 0, 0, 8, +, 4, =, 6, 6, *, 7, =, 5, +, 9, 7, 8, 3, 8, 6, 7, 9, 9,
9, 5, 4, 2, =, 2, /, 6, =, 4, 2, 4, 6, 3, 0, +, 7, 5, 2, 4, =, 7, 2, 7, 8, +, 0, +, 1,
+, 4, =, 9, 9, 4, 4, -, 4, 5, +, 1, 5, 7, 2, 9, =, 2, 7, 1, 5, 4, *, 7, =, 1, 1, +, 2,
=, 3, *, 1, 7, 0, =, 3, 0, +, 8, 4, 1, 6, 3, =, 2, -, 0, 8, =, 2, +, 1, 4, =, 4, 9, 7,
1, 6, +, 2, 8, 6, =, 7, 8, +, 5, =, 4, 8, +, 7, +, 7, 9, 4, =, 3, 4, 9, 7, +, 0, +, 6,
=, 8, +, 3, =, 1, /, 6, +, 9, 8, +, 3, 5, =, 2, /, 6, 8, =, 2, +, 6, =, 9, -, 1, 9, =,
3, +, 8, 5, 0, =, 4, 5, 1, /, 5, 7, 6, 9, 9, +, 2, 0, =, 8, +, 2, 8, =, 0, /, 2, =, 6,
+, 4, +, 0, =, 0, 1, 6, +, 3, +, 6, 6, 7, +, 8, 7, 0, 1, =, 8, -, 1, 0, =, 4, 0, -, 3,
+, 5, 2, 9, 8, +, 5, =, 8, +, 9, =, 6, 9, 4, 5, 6, 6, 1, +, 5, 5, =, 3, +, 6, 2, 5, +,
8, +, 4, 2, 4, =, 7, +, 6, +, 0, =, 1, 3, +, 7, 7, 8, 4, =, 5, +, 7, 5, =, 2, 0, 3, +,
3, =, 5, 4, *, 6, 6, =, 1, 2, +, 9, 6, 1, 3, 5, 8, 7, =, 0, 3, +, 1, 2, 5, 7, 2, 0, =,
3, /, 5, +, 1, 6, =, 2, 2, +, 8, +, 8, +, 6, 0, 5, 7, 8, +, 3, 4, =
```

LIST 12-4 Basic usage profile machine-readable test-case stimuli (comma-separated)

```
5, 7, +, 6, /, 9, 6, 6, 1, 6, +, 8, 0, -, 3, 3, 6, 7, 3, 8, 0, 9, 9, =, /, 0, 8, 9, 9,
9, +, 9, /, 5, 3, 0, 6, 7, 8, 6, +, 5, 8, 0, 3, 6, 2, +, 4, 4, 7, +, 4, +, 8, 0, 5, 4,
2, =, 8, 3, -, 6, 5, 5, =, -, 9, 8, 4, *, 7, 4, *, 3, =, +, 5, 1, 5, 4, 8, 8, 3, 5, 6,
9, 2, 2, 7, 3, 1, 3, *, 7, =, 1, +, 2, 4, 4, 1, 0, 7, 4, 9, -, 5, 6, 1, 9, 2, -, 0, 8,
```

```
-, 8, 4, -, 9, 8, 4, 3, +, 1, 1, 3, -, 4, +, 4, 1, 2, 2, +, 4, 7, 5, 0, 0, 7, 2, *, 3,
4, 2, +, 7, 2, 3, 9, 1, 1, +, 1, 2, 6, 8, 0, 8, +, 9, 0, 0, 2, 6, 8, 2, 4, 0, -, 2, =,
+, 1, +, 3, 0, 6, 4, 2, 0, 9, 3, 3, 8, 1, 7, 8, 6, 5, *, 1, 9, 8, 3, *, 1, 6, 2, 8, 1,
=, 3, 6, +, 5, 7, 6, 6, *, 2, 3, 4, +, 0, 4, 6, 0, 6, 5, 3, /, 2, 4, 3, 4, 3, +, 2, 0,
+, 6, =, 0, 4, 2, 8, 8, 5, 2, 2, -, 4, -, 8, 7, 9, =, 9, 4, 4, 2, +, 6, 6, 1, 4, 6, 1,
9, =, 4, *, 1, =, 2, 3, +, 9, 5, +, 6, =, 4, 0, 1, 9, 0, 6, 1, -, 5, 6, 3, 7, *, 1, 7,
4, =, 3, 4, 4, 3, 7, +, 9, 7, 8, 5, 5, +, 3, 8, 2, 3, 7, 7, =, 3, 6, 7, -, 9, 2, 7, 8,
=, 4, *, 6, 8, 4, 4, 5, 6, 9, 9, *, 1, 6, 3, 8, =, 7, 4, 6, 5, 9, 9, 1, 0, 2, 9, 5, 0,
9, 6, 0, 2, 0, 3, 8, 7, *, 4, 5, 0, 7, *, 3, 0, 3, =, 6, 2, 8, -, 1, 3, 3, +, 1, 5, 5,
7, 0, 9, 5, 3, 3, 4, 9, 0, *, 6, 8, 7, =
```

LIST 12-5 Advanced usage profile machine-readable test-case stimuli (comma-separated)

```
2, -, =, -, *, -, *, 7, =, 8, =, *, *, +, 7, =, 0, =, /, =, =, =, =, =, 8, -, +, 1, +,
8, =, 4, 2, -, *, /, /, +, 7, =, *, =, =, +, *, =, =, 4, *, =, 9, =, =, =, *, 0, 0, /,
=, =, 2, -, =, =, =, =, =, =, /, /, -, =, =, /, +, =, =, =, *, =, +, =, =, /, =, 5, *,
*, =, =, =, -, =, 8, =, 2, *, 5, =, /, *, =, 4, *, =, 0, /, =, =, +, -, +, =, 5, /, 0,
+, 1, =, /, 8, *, -, +, 7, 2, 7, 6, =, =, =, 2, -, 6, =, =, +, *, +, =, =, =, *, =, =,
*, -, 5, -, 6, =, -, -, 4, *, 0, +, -, *, +, =, 0, /, =, -, /, *, 1, -, *, 1, =, =, =,
/, +, =, 5, -, 1, +, 5, +, 8, =, =, =, 1, 5, -, =, =, =, /, =, 2, 4, /, =, =, 2, -, +,
*, 3, =, =, =, =, /, =, =, /, /, *, -, =, *, /, =, 6, -, 3, =, *, =, 7, *, 1, =, +, 5,
=, 1, -, *, *, 1, /, -, +, +, =, =, /, *, /, 4, =, =, /, =, 0, =, =, =, =, *, -, 0, 4,
8, =, 5, *, /, 6, +, *, 4, 3, =, =, =, =, =, +, /, =, /, =, +, =, =, =, 0, -, *, 7, 8,
7, 7, /, *, -, 3, *, 4, +, 8, /, =, 3, /, =, /, 8, +, =, /, 6, +, -, =, *, -, -, -, 2,
=, -, -, =, =, =, =, =, =, 3, 0, *, +, -, *, 8, =, -, =, =, =, =, /, +, *, =, 2, *, *,
/, 9, /, =, =, =, 7, =, *, +, =, -, 3, -, *, =, =, +, /, +, 5, =, *, 6, =, =, 0, 6, -,
=, -, 5, =, =, -, =, +, -, 6, /, 9, =, =, +, =, 6, *, *, 6, =, -, =
```

LIST 12-6 Stressed usage profile machine-readable test-case stimuli (comma-separated)

12.20.9 EXAMPLE Hosted calc_logic for reuse (basic usage profile stimuli): Add machine-readable expected response and pass/fail criteria

As stated, the *process schedule* to use is described in above chapter 12.19, "Usage profile modeling," page 730 and in chapter 6.12.9, "Detailed Pt.n.Va:2. Develop test-cases, also for fixtures and debugger, by usage profile modeling," page 158.

The List 12-7 below, is an extraction from the beginning of the basic *usage profile stimuli* presented above. Each row contains the originally generated *stimuli*, which have optionally been extended with expected *responses* and with maximum relative error criteria.

The farthest left character of the below *test-case* contains the *stimuli*. If there is nothing more on the row, no *response* is expected. If there is more information after the *stimuli*, the first is the expected *response*, and the second is the maximal allowed relative error between the expected and actual *response* that is accepted.

```
8
7
+
4
=; 91 0
5
+
5
9
7
+; 602 0
2
```

```
8
7
=; 889 0
6
+
5
=; 11 0
3
+
8
+; 11 0
8
=; 19 0
7
/
2
0
8
=; 0.033653846153846153846153846153846 1e-16
9
3
-
2
+; 91 0
1
0
7
3
2
6
4
8
=; 10732739 0
```

LIST 12-7 Beginning of basic usage profile completed test-case

12.20.10 EXAMPLE Hosted calc_logic for reuse: Acquire automatic test-case executor wrapping the black-box calc_logic

When executing *verification* automatically, *activity* Pt.n.Vg. can be expanded to a full *schedule* (see Fig. 6-30, p. 156). Accordingly, a *test-case* executor was easily developed in-house (see its *logical architecture* in Fig. 12-27 below).

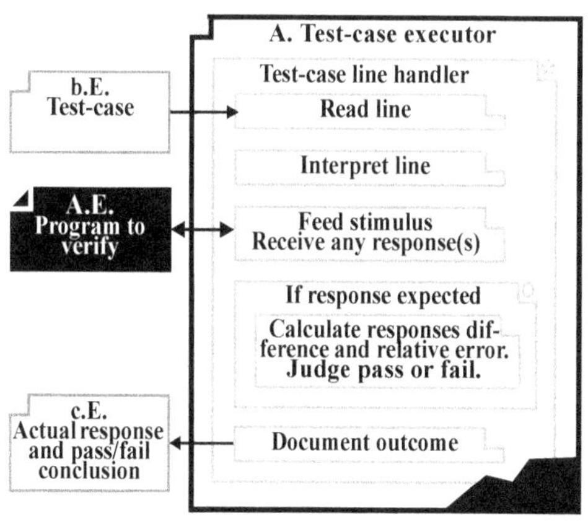

A nasty problem appeared when calculating the actual *responses* and expected *response* difference and relative error. The maximum accuracy of the *test-case* executor was the same as the maximum allowed relative error, which means that the tool was not better than the program for *verification* and useless to calculate and judge pass or fail. The *test-case* executor was provided with math *program libraries* of quite sufficient accuracy, and the problem was solved.

FIGURE 12-27 Simplified architecture of a test-case executor

12.20.11 EXAMPLE Hosted calc_logic for reuse: Execute automatically test-cases in executor for basic usage profile

Since three *usage profiles* are defined which are intentionally harder and harder for the calc_logic *program* to execute correctly, it might be wise to verify one at a time. It is better to start with the simplest *usage profile*, and eliminate the easy failures first, before coping with the difficult ones. For the *verification report*, see Table 12-62 below.

TABLE 12-62 Calc_logic for reuse
basic usage profile verification report 2

Actual outcome and pass/fail conclusion	Expected outcome and pass/fail criteria	Use-case require-ment	Scenario re-quirement
VR.2.1. Fail. **Stimulus +;** **Response 0.59999999999999998;** **Expected 0.6;** **Rel. error 0.0000000000000003;** **Max allowed 0;** • Line 383 in List 12-8, page 744	CH.1.XP.1. Expect a relative error criterion of nothing else than 0 for integer responses, and less than 1e-16 relative error criteria for fraction responses.	A.RB.UU.43. Do-conclu-sion-with-continuation	A.RB.6.S1. Calc_logic for reuse, behavior requirement 6, scenario 1 A.RR.4.1. A maximum calcu-lation precision of ≈ 15 digits during host de-velopment of the calc_logic program is suffi-cient.
VR.2.2. Fail. Secondary to above. **Stimulus =;** **Response 16.600000000000001; Expected 16.6;** **Rel. error 0.0000000000000006;** **Max allowed 0;** • Line 386 in List 12-8, page 744		A.RB.UU.42. Do-conclu-sion	

Observe that extracted fail lists must be retrieved by starting from an identifiable *wait-state* as shown below. This is because it gets very difficult to judge the failure reason, and hard to repeat the failure, if it is uncertain where to start reverifying. Of course, it is always possible to reverify from the beginning of the *test-case*, but it might be tedious and time consuming if the failure is way down in the *test-case*. (Another bizarre consequence of this phenomenon, in which the novice *users* lose in which *wait-state* they are during calculation, is that not seldom such *users* switch off their calculators to reenter the secure *idle wait-state*, causing the power switch to get worn out : -).

```
000 ! ********* Verification
001 ! Test-case Basic usage of calc_logic
002 ! Stimulus; Response; Expected-response; Relative error; Max-diff-criteria;
    :
    :
380 Stimulus 3; Response 3
381 Stimulus /
382 Stimulus 5; Response 5
383 Stimulus +; Response 0.59999999999999998; Expected 0.6;
    Rel. error  0.0000000000000003; Max allowed 0; FAIL
384 Stimulus 1; Response 1
385 Stimulus 6; Response 16
386 Stimulus =; Response 16.600000000000001; Expected 16.6;
    Rel. error  0.0000000000000006; Max allowed 0; FAIL
387 ! ********* Total number of failures 2
    :
    :
```

LIST 12-8 Basic usage profile automatic verification black-box report V1

12.20.12 EXAMPLE Hosted calc_logic for reuse: Locate failure source

When a behavior fault is identified in *verification*, as already noted, it might be hard to find its exact location in the *program*. Modern *development systems* provide very good support to run *programs* in a debugger interactively with the developer watching how the *program* is executed.

The key tools in a modern *development system* debugger are "break points" and "watch points." When these are set, the *program* executes until the break point is reached, and required watch points show actual values of any *variables*.

The log below is a watch point documentation list from execution of calc_logic in Figure 10-42, page 541. In the list below, the watch points are named WPn:, and in the referred figure, corresponding watch points are marked out by □ squares in the *program*.

```
Press 3, WP1:entrystate 1, WP2:digit 3,main operand 3,exitstate 2, Shown 3
Press /, WP1:entrystate 2, WP6:operator /,prev operator /,
  main operand 3 to side operand 3,exitstate 3, No change shown
Press 5, WP1:entrystate 3, WP8:digit 5,side operand 5,exitstate 4, Shown 5
Press +, WP1:entrystate 4, WP12:conclusion +,calculated +,
  main operand result 0.59999999999999998,prev operator +,exitstate 3, Shown
  0.59999999999999998
Press 1, WP1:entrystate 3, WP8:digit 1,side operand 1,exitstate 4, Shown 1
Press 6, WP1:entrystate 4, WP10:digit 6,side operand 16,exitstate 4, Shown 16
Press =, WP1:entrystate 4, WP11:conclusion =,calculated +,
  main operand result 16.600000000000001,exitstate 1, Shown 16.600000000000001
```

LIST 12-9 Watch point listing from repeating basic usage profile failures

12.20.13 EXAMPLE Hosted calc_logic for reuse: Evaluate optimal elimination

The decision gate members were at first a bit surprised by this failure. How can a division of 3 by 5 go this wrong?

The explanation is that floating-point numbers are implemented by the compiler and its *program libraries* in a scientific notation, with the sign stored in one bit, value fraction stored as binary in the mantissa part, and the value size (the decimal position) stored as binary in the exponent part. A binary value is built by a base of 2, which often cannot exactly represent values as built by a base of 10.

However, floating point is a simple and efficient notation with good understanding from advanced and scientific *users*. The problem is the basic *users*, who most often deal with integers without *complex* fractions, and for these *users* their simple integers must be treated without incomprehensible incorrectnesses.

After the floating-point briefing among the gate members, the *failure elimination* decisions were made (see Table 12-63 below).

TABLE 12-63 Calc_logic for reuse basic usage profile failure elimination report 1

Failure location	Failure elimination plan	Actual outcome and pass/fail criteria	Expected outcome and pass/fail criteria	Use-case requirement
FER.1.1. The failure emerges from the compiler's and libraries' way of handling fixed point variables. The compiler uses normal fixed-point rules, which may imprecisely represent integers.	FER.1.1.a.Accept a deteriorated relative error of the pass/fail criteria (see Table 12-64 below), and round off last decimal in the program to get a truer integer value (see Fig. 12-28 below).	VR.2.1. Fail. Stimulus +; Response 0.59999999999999998; Expected 0.6; Rel. error 0.00000000000000003; Max allowed 0;	CH.1.XP.1. Expect a relative error criterion of nothing else than 0 for integer responses, and less than 1e-16 relative error criteria for fraction responses.	A.RB.UU.43. Do-conclusion-with-continuation
		VR.2.2. Fail. Secondary to above. Stimulus =; Response 16.600000000000001; Expected 16.6; Rel. error 0.00000000000000006; Max allowed 0;		A.RB.UU.42. Do-conclusion

The calc_logic for reuse *test-case* pass/fail criteria was redefined (see Table 12-64 below).

TABLE 12-64 Calc_logic for reuse usage profile black-box test-cases

Black-box stimulus injection	Expected response with pass/fail criteria	Behavior requirement
CH.1. Minimum of 200 stimuli automatically is generated for the basic usage profile. Verify and eliminate all failures before verifying with below advanced usage profile.	XP.2. Expect a relative error criterion of 0 for integer responses, and less than 1e-15 relative error criteria for fraction responses.	A.RB.6.S1. Calc_logic for reuse, behavior requirement 6, scenario 1
CH.2. Minimum of 200 stimuli is automatically generated for advanced usage profile. Verify and eliminate all failures before verifying with below stressed usage profile.		A..RR.4.1. A maximum calculation precision of $\approx$ 15 digits during host development of the calc_logic program is sufficient.
CH.3. Minimum of 200 stimuli automatically generated for stressed usage profile.		

The machine-readable *test-cases* were retrieved and exchanged for the new pass/fail value.

Precision of calculated floating-point values may be set by a format string within a C language "scanf" *program instruction*. To be easy to change when using other compilers and to easily change between various precisions, it is placed into the *pro-*

gram header "calc_logic_interface.h" (see below *program* listing) .

```
#define OP MAX LEN 100
char     format[] = "%.16g";   // changed from "%.17g" to cure basic profile fail 1
char     *calc_logic (char);
```

FIGURE 12-28 Calc_logic program changes to cure failure 1

12.20.14 EXAMPLE Hosted calc_logic for reuse: Execute automatically test-cases in executor for basic usage profile

Now the *verification* is repeated, and the failure has been eliminated (see Table 12-65 below).

TABLE 12-65 Calc_logic for reuse
basic usage profile verification report 3

Actual outcome and pass/fail conclusion	Expected outcome and pass/fail criteria	Use-case requirement	Scenario requirement
VR.3. Pass	CH.1.XP.2.	A.RB.UU.43.	
VR.3. Pass	Expect a relative error criterion of 0 for integer responses, and less than 1e-15 relative error criteria for fraction responses.	A.RB.UU.42.	A.RB.6.S1.

12.20.15 EXAMPLE Hosted calc_logic for reuse (advanced usage profile stimuli): Add machine-readable expected response and pass/fail criteria

There were no real big problems in entering expected *responses* and pass/fail criteria into the generated advanced *usage profile stimuli*. The only obstacle was performing some high-precision integer number calculations, but paper and pen are still going strong as unlimited precision tools for the four operations.

12.20.16 EXAMPLE Hosted calc_logic for reuse: Execute automatically test-cases in executor for advanced usage profile

Now it is time to verify with the advanced *usage profile*. For the *verification report*, see Table 12-66 below.

TABLE 12-66 Calc_logic for reuse
advanced usage profile verification report 4

Actual outcome and pass/fail conclusion	Expected outcome and pass/fail criteria	Use-case requirement	Scenario requirement
VR.4.1. Fail. **Stimulus =;** **Response 3.608418498217317e+016;** **Expected 36084184982173167;** **Rel. error 0.00000000000000008;** **Max allowed 0;** • Line 101 in List 12-10, page 749		A.RB.UU.42. Do-conclusion	
VR.4.2. Fail. **Stimulus *;** **Response 6.076542331639483e+017;** **Expected 607654233163948245;** **Rel. error 0.00000000000000009;** **Max allowed 0;** • Line 202 in List 12-11, page 749	CH.2.XP.2. Expect a relative error criterion of 0 for integer responses, and less than 1e-15 relative error criteria for fraction responses.	A.RB.UU.43. Do-conclusion-with-continuation	A.RB.6.S1. Calc_logic for reuse, behavior requirement 6, scenario 1
VR.4.3. Fail. **Stimulus =;** **Response 9.893218570142241e+021;** **Expected 9893218570142241376845;** **Rel. error 0.00000000000000003;** • Line 208 in List 12-11, page 749		A.RB.UU.42. Do-conclusion	
VR.4.4. Fail. **Stimulus *;** **Response 3.364922156999978e+023;** **Expected 336492215699997763884209;** **Rel. error 0.0000000000000001;** V.0.1. Line 366 in List 12-12, page 750		A.RB.UU.43. Do-conclusion-with-continuation	
VR.4.5. Fail. **Stimulus =;** **Response 1.019571413570993e+026;** **Expected 101957141357099322456915327;** **Rel. error 0.0000000000000002;** **Max allowed 0;** • Line 370 in List 12-12, page 750		A.RB.UU.42. Do-conclusion	

The listings below are extracts of failures from the automatic executor of the advanced *usage profile test-case*.

```
   :
   :
081 Stimulus =; Response -345432; Expected -345432; Rel. error  0.0; Max allowed 0; PASS
082 Stimulus +
083 Stimulus 5; Response 5
084 Stimulus 1; Response 51
085 Stimulus 5; Response 515
086 Stimulus 4; Response 5154
087 Stimulus 8; Response 51548
088 Stimulus 8; Response 515488
089 Stimulus 3; Response 5154883
```

```
090 Stimulus 5; Response 51548835
091 Stimulus 6; Response 515488356
092 Stimulus 9; Response 5154883569
093 Stimulus 2; Response 51548835692
094 Stimulus 2; Response 515488356922
095 Stimulus 7; Response 5154883569227
096 Stimulus 3; Response 51548835692273
097 Stimulus 1; Response 515488356922731
098 Stimulus 3; Response 5154883569227313
099 Stimulus *; Response 5154883568881881; Expected 5154883568881881;
    Rel. error  0.0; Max allowed 0; PASS
100 Stimulus 7; Response 7
101 Stimulus =; Response 3.608418498217317e+016; Expected 36084184982173167;
    Rel. error  0.00000000000000008; Max allowed 0; FAIL
    :
    :
```

LIST 12-10 Advanced usage profile automatic verification black-box report V3.1

```
    :
    :
178 Stimulus =; Response 10852951649; Expected 10852951649;
    Rel. error  0.0; Max allowed 0; PASS
179 Stimulus +
180 Stimulus 1; Response 1
181 Stimulus +; Response 10852951650; Expected 10852951650;
    Rel. error  0.0; Max allowed 0; PASS
182 Stimulus 3; Response 3
183 Stimulus 0; Response 30
184 Stimulus 6; Response 306
185 Stimulus 4; Response 3064
186 Stimulus 2; Response 30642
187 Stimulus 0; Response 306420
188 Stimulus 9; Response 3064209
189 Stimulus 3; Response 30642093
190 Stimulus 3; Response 306420933
191 Stimulus 8; Response 3064209338
192 Stimulus 1; Response 30642093381
193 Stimulus 7; Response 306420933817
194 Stimulus 8; Response 3064209338178
195 Stimulus 6; Response 30642093381786
196 Stimulus 5; Response 306420933817865
197 Stimulus *; Response 306431786769515; Expected 306431786769515;
    Rel. error  0.0; Max allowed 0; PASS
198 Stimulus 1; Response 1
199 Stimulus 9; Response 19
200 Stimulus 8; Response 198
201 Stimulus 3; Response 1983
202 Stimulus *; Response 6.076542331639483e+017; Expected 607654233163948245;
    Rel. error  0.00000000000000009; Max allowed 0; FAIL
203 Stimulus 1; Response 1
204 Stimulus 6; Response 16
205 Stimulus 2; Response 162
206 Stimulus 8; Response 1628
207 Stimulus 1; Response 16281
208 Stimulus =; Response 9.893218570142241e+021; Expected 9893218570142241376845;
    Rel. error  0.00000000000000003; Max allowed 0; FAIL
    :
    :
```

LIST 12-11 Advanced usage profile automatic verification black-box report V3.2 and V3.3

```
    :
    :
341 Stimulus 7; Response 7
342 Stimulus 4; Response 74
343 Stimulus 6; Response 746
344 Stimulus 5; Response 7465
345 Stimulus 9; Response 74659
```

```
346 Stimulus 9; Response 746599
347 Stimulus 1; Response 7465991
348 Stimulus 0; Response 74659910
349 Stimulus 2; Response 746599102
350 Stimulus 9; Response 7465991029
351 Stimulus 5; Response 74659910295
352 Stimulus 0; Response 746599102950
353 Stimulus 9; Response 7465991029509
354 Stimulus 6; Response 74659910295096
355 Stimulus 0; Response 746599102950960
356 Stimulus 2; Response 7465991029509602
357 Stimulus 0; Response 74659910295096020
358 Stimulus 3; Response 746599102950960203
359 Stimulus 8; Response 7465991029509602038
360 Stimulus 7; Response 74659910295096020387
361 Stimulus *
362 Stimulus 4; Response 4
363 Stimulus 5; Response 45
364 Stimulus 0; Response 450
365 Stimulus 7; Response 4507
366 Stimulus *; Response 3.364922156999978e+023; Expected 33649221569999977763884209;
  Rel. error  0.0000000000000001; Max allowed 0; FAIL
367 Stimulus 3; Response 3
368 Stimulus 0; Response 30
369 Stimulus 3; Response 303
370 Stimulus =; Response 1.019571413570993e+026; Expected 101957141357099322456915327;
  Rel. error  0.0000000000000002; Max allowed 0; FAIL
  ⋮
  ⋮
```

LIST 12-12 Advanced usage profile automatic verification black-box report V3.4 and V3.5

12.20.17 EXAMPLE Hosted calc_logic for reuse: Locate failure source

Here another surprising failure shows up. When multiplying two long integers, the product may get too long to be kept by the floating point *variable*. The failure is not in any *program instruction*, but in unreasonable expectations of the calc_logic capability (see Table 12-67 below).

12.20.18 EXAMPLE Hosted calc_logic for reuse: Evaluate optimal elimination

A decision is made to lower the expectation of calc_logic (see Table 12-67 below).

TABLE 12-67 Calc_logic for reuse advanced usage profile failure elimination report 2

Failure location	Failure elimination plan	Actual outcome and pass/fail criteria	Expected outcome and pass/fail criteria	Use-case requirement
FER.2.1. Used double floating-point precision is limited to ≈ 17 digits. Integers with more digits consequently get imprecise. Note, these integers are not truncated to ≈ 17 digits, only thus limited in precision.	FER.2.1.a. **Accept that integers larger than ≈ 17 digits get imprecise.** FER.2.1.b. **Explain this in manual.** FER.2.1.c. In reverification, accept relative error of 1e-15 on integers larger than 16 digits.	VR.4.1. Fail. Stimulus =; Response 3.608418498217317e+016; Expected 36084184982173167; Rel. error 0.00000000000000008; Max allowed 0; VR.4.2. Fail. Stimulus *; Response 6.076542331639483e+017; Expected 607654233163948245; Rel. error 0.00000000000000009; Max allowed 0; VR.4.3. Fail. Stimulus =; Response 9.893218570142241e+021; Expected 9893218570142241376845; Rel. error 0.00000000000000003; VR.4.4. Fail. Stimulus *; Response 3.364922156999978e+023; Expected 336492215699997763884209; Rel. error 0.0000000000000001; VR.4.5. Fail. Stimulus =; Response 1.019571413570993e+026; Expected 101957141357099322456915327; Rel. error 0.0000000000000002; Max allowed 0;	XP.2. Expect a relative error criterion of 0 for integer responses, and less than 1e-15 relative error criteria for fraction responses.	A.RB.6.S1. Calc_logic for reuse, behavior requirement 6, scenario 1

The calc_logic for reuse *test-case* pass/fail criteria was again redefined (see Table 12-64 below).

TABLE 12-68 Calc_logic for reuse usage profile black-box test-cases

Black-box stimulus injection	Expected response with pass/fail criteria	Behavior requirement
CH.2. Minimum of 200 stimuli is automatically generated for advanced usage profile. Verify and eliminate all failures before verifying with below stressed usage profile.	XP.3. Expect a relative error criterion of 0 for integers not longer than 16 digits responses, and less than 1e-15 relative error criteria for all other responses.	A.RB.6.S1. Calc_logic for reuse, behavior requirement 6, scenario 1
CH.3. Minimum of 200 stimuli automatically generated for stressed usage profile.		

12.20.19 EXAMPLE Hosted calc_logic for reuse: Execute automatically test-cases in executor for advanced usage profile with basic usage profile regressively

After mitigating the fail criteria in the *test-case*, no failures were detected (see Table 12-69 below.)

TABLE 12-69 Calc_logic for reuse
advanced usage profile verification report 5

Actual outcome and pass/fail conclusion	Expected outcome and pass/fail criteria	Requirement
VR.5. Pass.	CH.2.XP.3. Expect a relative error criterion of 0 for integers not longer than 16 digits responses, and less than 1e-15 relative error criteria for all other responses.	A.RB.6.S1. Calc_logic for reuse, behavior requirement 6, scenario 1

When eliminating failures in *programs*, secondary faults are often introduced. To avoid that these failures remain in the *program*, all *test-cases* are executed again, called regression testing (see Table 12-70 below).

TABLE 12-70 Calc_logic for reuse
basic usage profile regression verification report 6

Actual outcome and pass/fail conclusion	Expected outcome and pass/fail criteria	Requirement
VR.6. Pass	CH.1.XP.2. Expect a relative error criterion of 0 for integer responses, and less than 1e-15 relative error criteria for fraction responses.	A.RB.6.S1. Calc_logic for reuse, behavior requirement 6, scenario 1

12.20.20 EXAMPLE Hosted calc_logic for reuse: Add machine-readable expected response and pass/fail criteria

This time it turns out to be really difficult to add expected *responses* to the generated *stimuli*, since some of the *stimulus* sequences are not specified by the *formal scenario* in Figure 7-28, Calc_logic for reuse, behavior requirement 6, scenario 1, page 247.

The problem is similar to the problem when building the probability table in Figure 12-25, page 737. As said before, there are mainly two ways out of this:

- Go back to *requirement* specification and make the *requirements* complete for all *exception use-cases* before generating *stimuli*.

- Decide what will happen to an unexpected *stimulus*. In this example, for the *stimuli* generator below, it is decided on the spot that not every unexpected *stimulus* causes leaving of the current *wait-state*.

The expected *responses* were chosen as well as possible, but with anxiety about whether the absence of *requirements* resulted in inconsistent expected *responses*.

12.20.21 EXAMPLE Hosted calc_logic for reuse: Execute automatically test-cases in executor for stressed usage profile

The stressed *usage profile* is now used to verify calc_logic. For the *verification report* see Table 12-71 below.

TABLE 12-71 Calc_logic for reuse
stressed usage profile verification report 7

Actual outcome and pass/fail conclusion	Expected outcome and pass/fail criteria	Use-case requirement	Scenario requirement
VR.7.1. Fail. Stimulus +; Response -1.#IND; See line 118 in List 12-13 below.	CH.3.XP.3. Expect a relative error criterion of 0 for integers not longer than 16 digits responses, and less than 1e-15 relative error criteria for all other responses.	A.RB.UU.42. Do-conclusion A.RB.UU.43. Do-conclusion-with-continuation	A.RB.6.S1. Calc_logic for reuse, behavior requirement 6, scenario 1
VR.7.2. Fail. A total of 24 failed responses reported from executing stressed usage profile test-case	Expected response from many stimuli sequences hard to determine without complete specification.	Incomplete	

The listing below is an extraction of the failures from the automatic executor of the stressed *usage profile test-case.*

```
115 Stimulus 5; Response 5
116 Stimulus /; Response 0
117 Stimulus 0; Response 0
118 Stimulus +; Response -1.#IND; FAIL
119 Stimulus 1; Response 1
120 Stimulus =; Response 0
```

LIST 12-13 Stressed usage profile automatic verification black-box report V3

12.20.22 EXAMPLE Hosted calc_logic for reuse: Locate failure source

The failure was initially a problem to understand. However, a detailed reading of the automatic *verification* shows it to be the well-known divide-by-zero failure. The problem was not that there was no silly calculation done when dividing by zero, but that there was no clear message when it happened (see Table 12-72, p. 754).

The following watch point listing was made to figure out how the strange divide-by-zero *response* was formed.

```
Press 5, WP1:entrystate 1, WP2:digit 5,main operand 5,exitstate 2, Shown 5
Press /, WP1:entrystate 2, WP6:operator /,prev operator /,
  main operand 5 to side operand 5,exitstate 3, No change shown
Press 0, WP1:entrystate 3, WP8:digit 0,side operand 0,exitstate 4, Shown 0
Press +, WP1:entrystate 4, WP12:conclusion +,calculated +,
  main operand result 1.#INF,prev operator +,exitstate 3, Shown 1.#INF
Press 1, WP1:entrystate 3, WP8:digit 1,side operand 1,exitstate 4, Shown 1
Press =, WP1:entrystate 4, WP11:conclusion =,calculated +,
  main operand result 2,exitstate 1, Shown 2
```

LIST 12-14 Watch point listing from repeating stressed usage profile failure 3

12.20.23 EXAMPLE Hosted calc_logic for reuse: Evaluate optimal elimination

The decision gate members were satisfied with the *verification* performed and passed off basic *usage profiles* and advanced *usage profile*. However, without passing stressed *usage profile verification* with a consistent expected result, a calculator is not commercially viable, so a decision was made on how to fulfill the calculator.

The calc_logic prototype was now judged very promising, and it was decided that this time the *requirements* must be extended to also cover the divide-by-zero exception. Seen in retrospect, it is a bit embarrassing that this well-known exception was not identified and taken care of much earlier, even at specification time.

TABLE 12-72 Calc_logic for reuse advanced usage profile failure elimination report 3

Failure location	Failure elimination plan	Actual outcome and pass/fail criteria	Expected outcome and pass/fail criteria	Use-case requirement
FER.3.1. Not specified by exception use-case and thus not implemented in program	FER.3.1.a. Capture exception use-case for divide by zero. FER.3.1.b. Add pass/fail criteria for this failure. FER.3.1.c. Implement division by zero return. FER.3.1.d. Reverify.	VR.7.1. Fail. Stimulus +; Response -1.#IND; See line 118 in List 12-13 below.	CH.3.XP.3. Expect a relative error criterion of 0 for integers not longer than 16 digits responses, and less than 1e-15 relative error criteria for all other responses.	A.RB.6.S1. Calc_logic for reuse, behavior requirement 6, scenario 1
FER.3.2. Inconsistent test-case due to incomplete requirements	FER.3.2.a. Cpdm book reader completes calc_logic normal and exception use-cases.	VR.7.2. Fail. A total of 24 failed responses reported from executing stressed usage profile test-case.		

The divide-by-zero *exception use-case* is defined in Figure 12-29 below.

Addition to "Calc_logic for reuse, behavior requirement 6, scenario 1", page 245

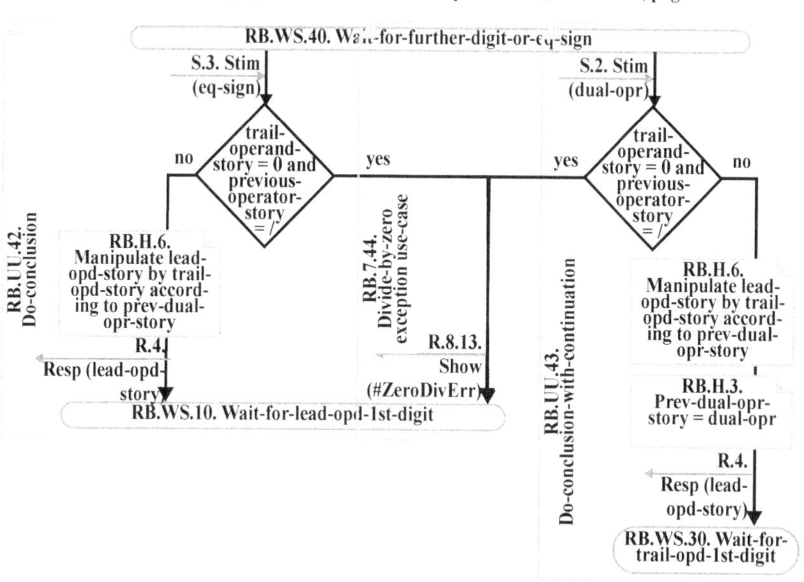

FIGURE 12-29 Divide-by-zero exception use-case

A pass/fail criterion was added to the *use-case* for the stressed usage profile (see Table 12-73 below).

TABLE 12-73 Calc_logic for reuse usage profile black-box test-cases

Black-box stimulus injection	Expected response with pass/fail criteria	Behavior requirement
CH.3. Minimum of 200 stimuli automatically generated for stressed usage profile.	XP.4. Pass if response is "#ZeroDivErr".	Divide-by-zero exception use-case

As decided, the non-numeric *response* from calc_logic *program* is implemented (see Fig. 12-14 below). To be simple, this kind of *response* has a leading '#' to be easily distinguished from a numerical value.

```c
/* A.cec.d. Divide */

case '/':
// Inserted to cure stress usage profile failure
if(strtod(trial_operand,'\0') == 0)
{
    wait_state = WAIT_FOR_LEAD_OPD_1ST_DIGIT;
    return("#ZeroDivErr");
}
sprintf(lead_operand, format, strtod(lead_operand,'\0')
    / strtod(trial_operand,'\0'));
break;
```

LIST 12-15 Program addition to cure divide-by-zero failure

12.20.24 EXAMPLE Hosted calc_logic for reuse: Execute automatically test-cases in executor for stressed usage profile, and execute advanced usage profile and basic usage profile regressively

First the *verification* is repeated for the stressed *usage profile* (see the *verification report* in Table 12-74 below). The zero-divide exception now works fine.

TABLE 12-74 Calc_logic for reuse
stressed usage profile verification report 8

Actual outcome and pass/fail conclusion	Expected outcome and pass/ fail criteria	Requirement
VR.8. Pass	CH.3.XP.4. Pass if response is "#ZeroDivErr".	RB.7.44. Divide-by-zero exception use-case, page 755

To be sure that no secondary failures have been introduced, the advanced *usage profile* is regressively verified (for the *verification report* see Table 12-75 below).

TABLE 12-75 Calc_logic for reuse
advanced usage profile regression verification report 9

Actual outcome and pass/fail conclusion	Expected outcome and pass/ fail criteria	Requirement
VR.9. Pass	CH.2.XP.3. Expect a relative error criterion of 0 for integers not longer than 16 digits responses, and less than 1e-15 relative error criteria for all other responses.	A.RB.6.S1. Calc_logic for reuse, behavior requirement 6, scenario 1

To be sure that no other secondary failures have been introduced, the basic *usage profile* is regressively verified (for the *verification report* see Table 12-75 below).

TABLE 12-76 Calc_logic for reuse
basic usage profile regression verification report 10

Actual outcome and pass/fail conclusion	Expected outcome and pass/ fail criteria	Requirement
VR.10. Pass	CH.1.XP.2. Expect a relative error criterion of 0 for integer responses, and less than 1e-15 relative error criteria for fraction responses.	A.RB.6.S1. Calc_logic for reuse, behavior requirement 6, scenario 1

12.20.25 EXAMPLE Hosted calc_logic for reuse: Manual table of contents

A manual explaining limitations of calc_logic is prepared (see Table 12-77 below).

TABLE 12-77 Calc_logic for reuse limitations manual table of contents

Manual name	Limitation	Requirement
MA.1. About floating-points	Maximum relative error criterion has been lowered to 1e-15 (originally 1e-16).	A.RR.4.1. A maximum calculation precision of ≈ 15 digits during host development of the calc_logic program is sufficient.
MA.2. About managing integers	Integers longer than 16 digits have been allowed a relative error of maximum 1e-15 (originally 0).	
MA.3. Exceptions handling	Non-numerical return from calc_logic.	Divide-by-zero exception use-case

A manual explaining proper usage of calc_logic is prepared (see Table 12-78 below).

TABLE 12-78 Calc_logic for reuse behavior manual table of contents

Manual name	Use-case	Requirement
MA.1. Initiation of calc_logic	RB.UU.01. Start up	A.RB.6.S1. Calc_logic for reuse, behavior requirement 6, scenario 1
MA.2. Input of first operand	RB.6.S1.U1. Lead-operand-input normal use-case	
MA.3. Input of dual operator	RB.UU.22. Input-dual-opr	
MA.4. Input of second operand	RB.6.S1.U2. Trail-operand-input normal use-case	
MA.5. Calculate result from input operands and operator	RB.UU.42. Do-conclusion or RB.UU.43. Do-conclusion-with-continuation	
MA.6. Completing calculation	RB.6.S1.U3. Conclusion normal use-case	
MA.7. Repetitive calculation	RB.6.S1.U4. Repeating normal use-case	
MA.8. Completing yet repetitive calculation	RB.6.S1.U5. Repeating-after-conclusion normal use-case	

12.20.26 EXAMPLE Calc_logic for reuse: Proceed reading

To follow this example, these are the alternatives:

- Trace back through the calc_logic example chapters to understand how to eliminate a detected failure.
- Proceed to chapter 11.17, "EXAMPLE Calc_logic environment (the development board fixture firmware): Realize interfaces to environment and insert/await outermost white-box," page 649.
- Proceed to chapter 11.18, "EXAMPLE Calc_logic environment (the pocket calculator firmware): Realize interfaces to environment and insert/await outermost white-box," page 653.
- Proceed to chapter 11.18, "EXAMPLE Calc_logic environment (the pocket calculator firmware): Realize interfaces to environment and insert/await outermost white-box," page 653.

- Proceed to chapter 12.21, "EXAMPLE Calc_logic environment (the Windows calculator application): Verify prototype in target environment" below.

12.21 EXAMPLE Calc_logic environment (the Windows calculator application): Verify prototype in target environment

12.21.1 EXAMPLE Calc_logic environment (the Windows calculator application): Process schedule to use

For the calc_logic example, a special technical overall *schedule* is *tailored* (illustrated in Fig. 12-30, below):

- **Pt:2:3b. Tailored advanced technical schedule n = 2, porting to 3 targets (right half),** page 170

In this *tailored schedule* the generic *schedule* to use is (filled-in symbols below):

- **Pt.0.V. Verify prototype in environment,** page 125.

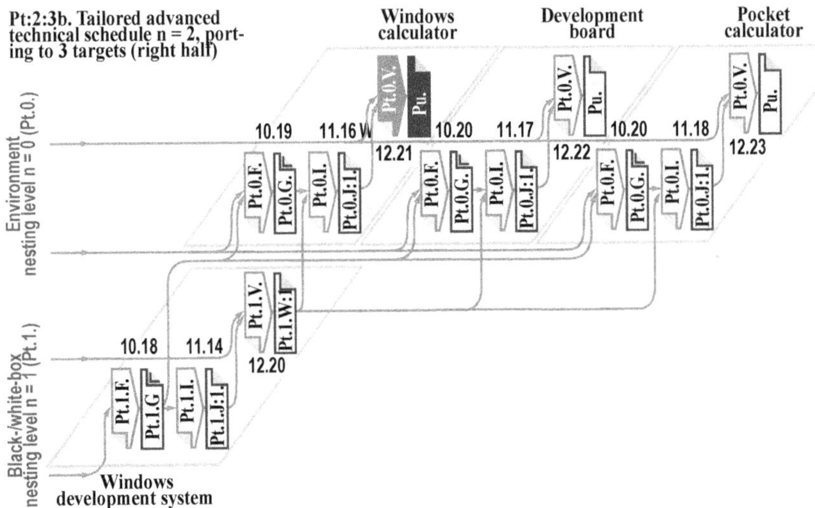

FIGURE 12-30 Position of schedule and master to use (filled in) in overall schedule

The calc_logic *white-box* has been *hosted* for development and *verification* in a Windows *development system*, which is very similar to the *development system* used for the Windows calculator application development. The calc_logic *program* only needs to be moved to the Windows calculator *development system* for *reuse* and *integration* into the *target* Windows calculator.

12.21.2 EXAMPLE Calc_logic environment (the Windows calculator application): Develop field test-cases for target or simulator

After the statistical usage testing of the calc_logic *white-box*, there shouldn't be too much that can go wrong with this *program* after *integration* into the Windows calculator application.

Most of the *verification* of the Windows calculator should be performed by the in-house project "dw2". However, as this is the first time calc_logic is *reused* by any *program product*, it is beneficial for all that the in-house project "dg3" verifiers are present doing some basic *verification*s to ensure that calc_logic is a highly *reusable item* and works well in its *target product.*

Test-cases are prepared for the Windows calculator *environment restriction requirements* (see Table 12-79 below).

TABLE 12-79 Calc_logic environment (the Windows calculator application) white-box test-cases

White-box design check	Expected appearance with pass/fail criteria	Restriction requirement
CH.1. Press digits in ascending order. Between digits press buttons not specified as stimulus by calc_logic.	XP.1. Pass, if the display shows the ascending digits, with no glitches or other malfunction between the digits. (up to ≈ 15 digits).	E.RR.7. Since a window in Microsoft Windows can hold almost any number of buttons, a large number of buttons shall be possible stimuli for calc_logic.
CH.1. Type 1, /, 3, = to divide 1 by 3.	XP.1. Pass, if the number 0.33333 ... is displayed with at least double precision (≈ 15 digits).	E.RR.8. Since a window field in Microsoft Windows can show almost any number of alphanumeric characters, the floating-point precision in calc_logic shall be as high as possible.
CH.2. Type 1, /, 0, = to divide one by zero.	XP.2. Pass, if the illegal operation is presented in an understandable text.	E.RR.10. There are a lot of compilers for Windows, with a lot of different kinds of mathematical calculation support. Standard Windows C development systems with library supports have float (≈ 7 digits), double-float (≈ 15 digits) and long double-float (≈ 31 digits) precision. However, long double-float variables most often are reduced to the precision of double variables. As Windows 64-bits version gets more popular, in the near future many compilers will probably offer no undiminished support for long double-floating-point (≈ 31 digits) precision. RB.7.44. Divide-by-zero exception use-case, page 755.
CH.1. Check with debugger that parameter passing mechanism is correct between calc_logic and its environment.	XP.1. Pass, if parameter passing occurs in a C standard way.	E.RR.12. Calc_logic for reuse developed on a host system must be portable to the target Windows disk and input/output system.

Note that some of the *environment restriction requirements* are not applicable anymore.

12.21.3 EXAMPLE Calc_logic environment (the Windows calculator application): Verify prototype in target environment

The *verification* is performed with the *verification report* in Table 12-80 below.

TABLE 12-80 Calc_logic environment (the Windows calculator application) verification report 1

Actual outcome and pass/fail conclusion	Expected outcome and pass/fail criteria	Requirement
VR.1. Pass	XP.1.	E.RR.7.
VR.1. Pass	XP.1.	E.RR.8.
VR.1. Pass	XP.2.	E.RR.10.
VR.1. Pass	XP.1.	E.RR.12.

12.21.4 EXAMPLE Calc_logic environment (the Windows calculator application): Finalize product manual

Both the calc_logic behavior and portability need an extensive description (see Table 12-81 below).

TABLE 12-81 Porting calc_logic to Windows product manual

Manual name	Behavior / restriction	Requirement / design
MA.1. The calc_logic behavior	Stimuli to calc_logic	RB.6.S1. Calc_logic for reuse, behavior requirement 6, scenario 1
	Responses from calc_logic	
MA.2. Porting the calc_logic program to Windows development systems	Use of float, strings etc.	E.b. Development system built-in standard library

12.21.5 Calc_logic for reuse: Proceed reading

To follow this example, these are the alternatives:

- Trace back through the calc_logic example chapters to understand how to eliminate a detected failure.
- Here the story about the windows calc_logic *program* is finished, and market validation takes place (see ch. 6.6, "P. Cpdm generic development schedule," p. 99). Please visit the *Cpdm* website for more information, and open a half bottle of champagne.

12.22 EXAMPLE Calc_logic environment (the pocket calculator firmware) on development board: Verify prototype in environment

12.22.1 EXAMPLE Calc_logic environment (the pocket calculator firmware) on development board: Process schedule to use

For the calc_logic example, a special technical overall *schedule* is *tailored* (illustrated in Fig. 12-31, below):

- **Pt:2:3b. Tailored advanced technical schedule n = 2, porting to 3 targets (right half),** page 170

In this *tailored schedule* the generic *schedule* to use is (filled-in symbols below):

- **Pt.0.V. Verify prototype in environment,** page 125.

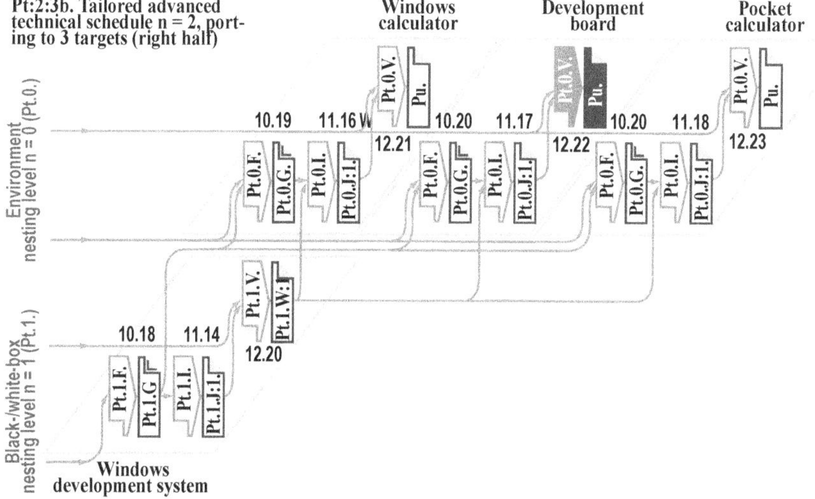

FIGURE 12-31 Position of schedule and master to use (filled in) in overall schedule

12.22.2 EXAMPLE Calc_logic environment (the pocket calculator firmware) on development board: Develop field test-cases for target or simulator

The pocket calculator is a *product* quite different from Windows, where calc_logic is first *hosted*. The calc_logic *program black-box* is definitely well verified with statistical *usage profile verification* with Windows *development system* tools, but there might be more real-time complications when verifying calc_logic on the *development board*. Most problems are expected in the *interface* between the pocket calculator assembled *program* and the calc_logic compiled *program*. For the test cases, see Table 12-82 below.

TABLE 12-82 for Calc_logic environment (the pocket calculator firmware) on development board white-box test-cases

White-box design check	Expected appearance with pass/fail criteria	Restriction requirement
CH.1. Check with in-circuit debugger that parameter passing mechanism is correct between calc_logic and its environment.	**XP.1. Pass, if parameter passing occurs in a C standard way (to make calc_logic automatically reusable in C targets).**	E.RR.6. Calc_logic for reuse developed on a host system must be portable to development boards and the target pocket calculator.
CH.1. Press digits in ascending order.	**XP.1. Pass, if the display shows the ascending digits (up to ≈ 7 digits).**	E.RR.1. To fit market cost of an affordable everyday calculator, the number of buttons shall be no greater than 16.
CH.1. Type 1, /, 3, = to divide 1 by 3.	**XP.1. Pass, if the number 0.33333 ... is displayed with at least double precision (≈ 7 digits).**	E.RR.2. To fit market cost of an affordable calculator, the stand-alone calculator display shall hold a maximum of 8 digits.
CH.2. Type 1, /, 0, = to divide 1 by 0.	**XP.2. Pass, if the illegal operation is presented in an understandable way.**	E.RR.4. Standard compilers to microcontrollers seldom have support for extensive mathematical arithmetic, but there are some microcontroller development systems with floating-point library support for float variables (≈ 7 digits). RB.7.44. Divide-by-zero exception use-case, page 755.

12.22.3 EXAMPLE Calc_logic environment (the pocket calculator firmware) on development board: Verify prototype in field simulator environment

The calc_logic has been integrated into the pocket calculator *program* and is started on the *development board*. The *verification* can begin (see Table 12-83 below).

TABLE 12-83 Calc_logic environment (the pocket calculator firmware) on development board verification report 1

Actual outcome and pass/fail conclusion	Expected outcome and pass/fail criteria	Requirement
VR.1.1. **Fail. The prototype doesn't get stuck, but pressing a button may bring up random digits on the display.**	XP.1. Pass, if parameter passing occurs in a C standard way (to make calc_logic automatically reusable in C targets).	E.RR.6. Calc_logic for reuse developed on a host system must be portable to development boards and the target pocket calculator.

12.22.4 EXAMPLE Calc_logic environment (the pocket calculator firmware) on development board: Locate failure source

The *development board* offers an easy way to connect an in-circuit debugger tool to locate the failure (see Table 12-84 below).

12.22.5 EXAMPLE Calc_logic environment (the pocket calculator firmware) on development board: Evaluate optimal elimination

The board members were very satisfied that this fatal failure was prioritized first by the verifiers. It is obvious that none of the other *test-cases* would have worked and a lot of time would have been wasted.

The failure in the parameter passing must simply be eliminated (see Table 12-84 below).

TABLE 12-84 Calc_logic environment (the pocket calculator firmware) on development board failure elimination report 1

Failure location	Failure elimination plan	Actual outcome and pass/fail criteria	Expected outcome and pass/fail criteria	Restriction requirement
FER.1.1. Typically, the parameters are still not correctly passed on the interface to calc_logic.	FER.1.1.a.Work with the debugger to fully understand the failure. FER.1.1.b. Study the C convention on parameter passing. FER.1.1.c. Eliminate the failure and be careful to comply with the C-convention. FER.1.1.d. Reverify.	VR.1.1. Fail. The prototype doesn't get stuck, but pressing a button may bring up random digits on the display.	XP.1. Pass, if parameter passing occurs in a C standard way (to make calc_logic automatically reusable in C targets).	E.RR.6. Calc_logic for reuse developed on a host system must be portable to development boards and the target pocket calculator.

12.22.6 EXAMPLE Calc_logic environment (the pocket calculator firmware) on development board: Verify prototype in field simulator environment

Now reverify the E.RR.6.CH1. *test-case*, and verify remaining *test-cases* not yet used (see the *verification report* in Table 12-85 below).

TABLE 12-85 Calc_logic environment (the pocket calculator firmware)
on development board verification report 2

Actual outcome and pass/fail conclusion	Expected outcome and pass/fail criteria	Requirement
VR.2. Pass	XP.1.	E.RR.6.
VR.2. Pass	XP.1.	E.RR.1.
VR.2. Pass	XP.1.	
VR.2.1. Fail After each zero division, the display showed the same weird segment pattern.	XP.2.	E.RR.2. E.RR.4. RB.7.44.

12.22.7 EXAMPLE Calc_logic environment (the pocket calculator firmware) on development board: Locate failure source

The *response* from calc_logic after divide by zero is the non-numeric #, which cannot be handled by the *development board* (see Table 12-86 next chapter).

12.22.8 EXAMPLE Calc_logic environment (the pocket calculator firmware) on development board: Evaluate optimal elimination

It was embarrassing that the *response* from calc_logic contained a long text string to indicate a divide-by-zero exception. It was easy to locate the failure but the verifying decision gate members had a difficult time agreeing on the solution (see Table 12-86 below).

TABLE 12-86 Calc_logic environment (the pocket calculator firmware) on development board failure elimination report 2

Failure location	Failure elimination plan	Actual outcome and pass/fail criteria	Expected outcome and pass/fail criteria	Restriction requirement
FER.2.1. calc_logic responds with text, which the pocket calculator can't handle and show in an understandable way.	FER.2.1.a.Change the calc_logic divide-by-zero exception response from long text string to simple codes. FER.2.1.b. Use '#' as code for exception has occurred, and 'z' as divide-by-zero code. (The full divide-by-zero response code then becomes '#z'.) FER.2.1.c. Tell the in-house project "de1" to decode response '#' as exception after calling calc_logic, and to handle next character as the exception type. FER.2.1.d. Tell in-house project "dw2" to do the same in the Windows calculator. FER.2.1.e. Ask the calculator stakeholders what shall be shown on exceptions. FER.2.1.f. Update requirements and all succeeding documentation. FER.2.1.g. Reverify both Windows and pocket calculator.	VR.2.1. Fail After each zero division, the display showed the same weird segment pattern.	XP.2. Pass, if the illegal operation is presented in an understandable way.	E.RR.2. To fit market cost of an affordable calculator, the stand-alone calculator display shall hold a maximum of 8 digits. RB.7.44.

Typically, this careless failure, detected in the very last moment of calculator development, causes a lot of rework to be done. Fortunately, the Windows calculator works as it is, and needs not be withdrawn and calc_logic has not been delivered to any *customers*, and does not need to be withdrawn either.

12.22.9 EXAMPLE Calc_logic environment (the pocket calculator firmware) on development board: Verify prototype in field simulator environment

Now reverify the divide-by-zero *test-case* (for the *verification report* see Table 12-87 below).

TABLE 12-87 Calc_logic environment (the pocket calculator firmware) on development board verification report 3

Actual outcome and pass/fail conclusion	Expected outcome and pass/fail criteria	Requirement
VR.3. Pass	XP.2.	E.RR.2. RB.7.44.

12.22.10 EXAMPLE Calc_logic environment (the pocket calculator firmware) on development board: Finalize product manual

How to do the *porting* to the *development board* needs to be explained to potential *customers* in the manual for *reuse* (see Table 12-88 below).

TABLE 12-88 Porting calc_logic to PIC development board product manual

Manual name	Restriction	Design
MA.1. Porting calc_logic to PIC-ready development board	Use of float, strings etc.	E.b. Development system built-in standard library

12.22.11 EXAMPLE Calc_logic for reuse: Proceed reading

To follow this example, these are the alternatives:

- Trace back through the calc_logic example chapters to understand how to eliminate a detected failure.
- Proceed to chapter 12.23, "EXAMPLE Calc_logic environment (the pocket calculator firmware): Verify prototype in environment" below.

12.23 EXAMPLE Calc_logic environment (the pocket calculator firmware): Verify prototype in environment

12.23.1 Calc_logic for reuse: Process schedule to use

For the calc_logic example, a special technical overall *schedule* is *tailored* (illustrated in Fig. 12-32, below):

- **Pt:2:3b. Tailored advanced technical schedule n = 2, porting to 3 targets (right half),** page 170

In this *tailored schedule* the generic *schedule* to use is (filled-in symbols below):

- **Pt.0.V. Verify prototype in environment,** page 125.

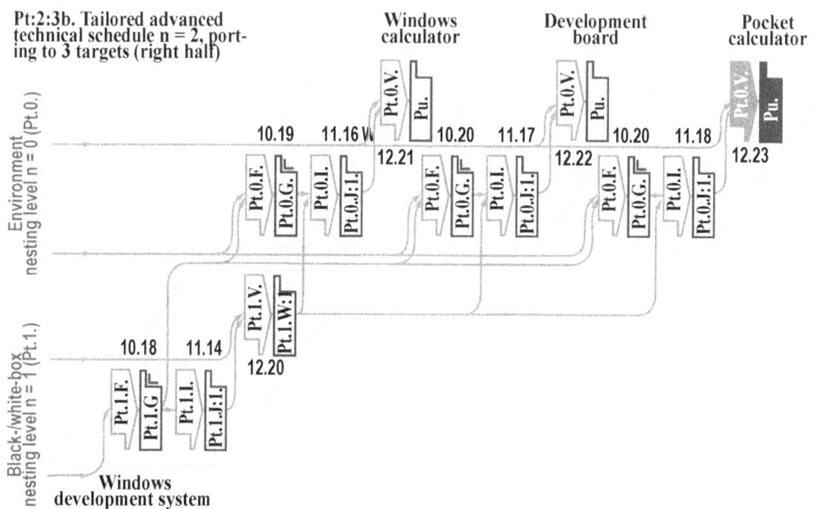

FIGURE 12-32 Position of schedule and master to use (filled in) in overall schedule

12.23.2 EXAMPLE Calc_logic environment (the pocket calculator firmware): Develop field test-cases for target or simulator

When *porting* from the *development board* to the real *target* pocket calculator prototype, the major issue is that the *target* microcontroller is a cheap and slimmed-down *version* compared to the more general microcontroller on the *development board.* The calc_logic file packet and the *assembler program* have been recompiled, relinked, and reloaded on *target* for this reason. As usual, some in-house project "dg3" engineers assist with *test-cases* to check that calc_logic fits its *environment*, the pocket calculator surroundings. For the prepared *test-cases*, see Table 12-87 below.

Also, the surrounding electronics to the microcontroller might contain failures, but these are verified and eliminated by the in-house project "de1".

TABLE 12-89 Calc_logic environment (the pocket calculator firmware) white-box test-cases

White-box design check	Expected appearance with pass/fail criteria	Restriction requirement
CH.1. Press digits in ascending order.	XP.1. Pass, if the display shows the ascending digits (up to ≈ 7 digits).	E.RR.1. To fit market cost of an affordable everyday calculator, the number of buttons shall be no greater than 16.
CH.1. Type 1, /, 3, = to divide 1 by 3.	XP.1. Pass, if the number 0.33333 ... is displayed with at least double precision (≈ 7 digits).	E.RR.2. To fit market cost of an affordable calculator, the stand-alone calculator display shall hold a maximum of 8 digits. E.RR.4.
CH.1. Type 1, /, 0, = to divide 1 by 0.	XP.2. Pass, if the illegal operation is presented in an understandable way.	Standard compilers to microcontrollers seldom have support for extensive mathematical arithmetic, but there are some microcontroller development systems with floating-point library support for float variables (≈ 7 digits). RB.7.44. Divide-by-zero exception use-case, page 755

12.23.3 EXAMPLE Calc_logic environment (the pocket calculator firmware): Verify prototype in target environment

Verification of the calc_logic *interfaces* to the *environment*, which is the *target* pocket calculator, is performed (see Table 12-90 below).

TABLE 12-90 Calc_logic environment (the pocket calculator firmware) verification report 1

Actual outcome and pass/fail conclusion	Expected outcome and pass/ fail criteria	Requirement
VR.1. Pass	XP.1.	E.RR.1.
VR.1. Pass	XP.1.	E.RR.2.
VR.1. Pass	XP.2.	E.RR.4.

12.23.4 EXAMPLE Calc_logic environment (the pocket calculator firmware): Finalize product manual

Documentation of *porting* to this *target* is described in a technical manual (see Table 12-91 below).

TABLE 12-91 Porting calc_logic to PIC 16F57 product manual

Manual name	Restriction	Design
MA.1. Porting calc_logic to PIC16F57 microprocessor target	Use of float, strings etc.	E.b. Development system built-in standard library

12.23.5 Calc_logic for reuse: Proceed reading

To follow this example, these are the alternatives:

- Trace back through the calc_logic example chapters to understand how to eliminate a detected failure.

- Here the story about the pocket calculator use of the calc_logic *program* is finished. Please visit the *Cpdm* website for more information. And open another half bottle of champagne.

12.24 EXAMPLE Lake: Fishing in same lake with changed environment

If the *environment* for the lake has changed, catching fish may also change. Some fish may have disappeared, some remain but bite better, and there might be some new fish as well.

This is analogous to a *product* that seems to work perfectly in one *environment*, and may have some failures in another *environment*.

FIGURE 12-33 The same lake but in another environment

12.25 About human-decided verification coverage

In the previous calc_logic example, a *usage profile* model was used to generate *test-case stimuli* without being biased by any human (other than proposing probabilities for the usage). In this example, the knowledge about specified *formal scenarios* is used actively to develop *test-cases*.

The benefit of human-selected *test-cases* is that *complex* behavior can be quickly identified, and *verification* can be focused on such behavior. With *usage profile* modeling, simple and *complex* behavior will be covered according to probability, which not often corresponds to *complexity*. On the other hand, humans tend to forget *verification* of simple but user-important behavior, which then gets poorly verified, and difficult but seldom used behavior will be much harder to verify.

Some failures might be very hard to locate without knowing how the *product* is structured. Such failures can, of course, also be located by statistical *test-case* generation, but very long *stimuli* sequences might be needed before the crucial location is reached. Or in other words, in real use of a *program*, special crucial locations might be reached much faster than the probability usage transition model would reflect.

To identify *complex* interaction behavior is, in fact, easier than first imagined. In interaction between *environment users* (human or machine) and the *product, stimuli* often have a range of values, which is expected and treated as normal by the receiving *product*. Any allowed range always has at least one lower and one upper end, and outside these, the *product* protects itself by rejecting the *stimuli* and responds with an exception message. The highest *complexity* appears at exactly these location ends, where the range shifts between normal and exception. When numeric, one value is the last of the normal case, and the adjacent value is the first of the exception case. To catch potential failures of this kind, at least verify these parameter end values of each *stimulus*. This is shown in the phonebook example below.

12.26 EXAMPLE Phonebook: Verify black-/white-box

12.26.1 About this example

Among others, this example shows how to verify *complexity* caused by unexpected *stimuli*, or unexpected *stimuli* parameters, received by a *product*.

For example, when a file is read from disk and unexpectedly terminates, this will be detected in many locations of the *product* design. All these locations must be able to protect themselves from unexpected file terminations. To verify such black-box behavior, the exception locations of the *product* are not visible, but it is possible to detect where in the *formal scenario* this might occur. Then a set of *test-cases* can be developed to verify that each *formal scenario* path handles the unexpected file termination properly.

12.26.2 EXAMPLE Phonebook: Process schedule to use

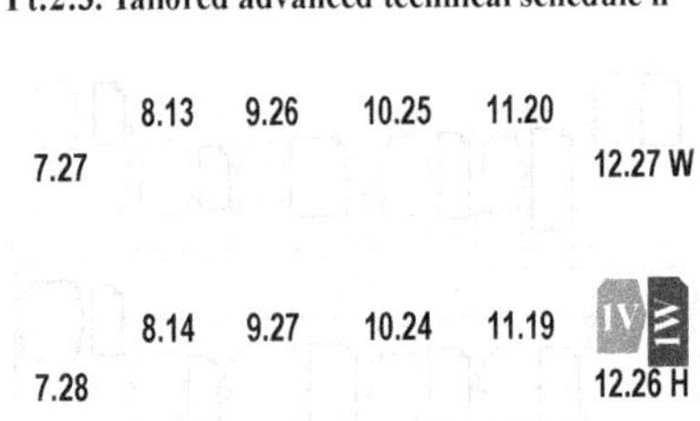

For the phonebook example, a special technical overall *schedule* is *tailored* (illustrated in Fig. 12-34, left):

- **Pt:2:3. Tailored advanced technical schedule n = 2,** page 169

In this *tailored schedule* the generic *schedule* to use is (filled-in symbols):

- **Pt.n.V. Verify black-/white-box,** page 152.

FIGURE 12-34 Position of schedule and master to use (filled in) in overall schedule

12.26.3 EXAMPLE Phonebook: Develop test-cases, also for fixtures and debugger

The major phonebook design *complexity* is its future-proof architecture. Thus, the *white-box test-cases* will verify crucial structures (see Table 12-92 below).

TABLE 12-92 Phonebook white-box test-cases

White-box design check	Expected appearance with pass/fail criteria	Restriction requirement
CH.1. Check which compiler is used.	XP.1. Pass, if used compiler is widely available, considered trustworthy, and inexpensive.	A.RR.1.1. The phonebook program shall be developed for a standard Windows computer.
CH.1. Check how user interaction is solved.	XP.1. Pass.	A.RR.2.1. First phonebook program shall use simple text-oriented console interaction.
CH.1. Start the phonebook, add a contact and terminate the program.	XP.1. Pass, if the added contact is saved in a phonebook file, and reads this file on a restart.	A.RR.3.1. The program shall load phonebook data at start-up and save them on program termination.
CH.1. Have a senior programmer check the phonebook program.	XP.1. Pass, if the selected language is object-oriented and if the program is developed according to designed architecture and if the program is highly modularized in a competent way.	A.RR.4.1. Choose an object-oriented program development system, design a strict modularization, and keep user interaction together with used objects.
CH.1. Check the phonebook program.	XP.1. Pass, if the program uses fixed arrays according to the requirement for variable storages.	A.RR.5.1. The first phonebook program version shall use only compiler built-in fixed-size variable storage. The phonebook file shall hold an array of contacts with contact data, one of which is an array of phone numbers with phone number data.
CH.1. Register contacts to the phonebook.	XP.1. Pass, if the phonebook can hold at least 128 contacts.	A.RR.5.2. MAX_CONT_PER_BOOK = 128
• Compared to maximum length, key in person name with: CH.1. one character less than maximum CH.2. equal number of characters CH.3. one character more	XP.1. Pass, if oversized person data and phone numbers get truncated at their max length, and if the program continues to execute normally, and if these data keep after termination and restart.	A.RR.5.4. MAX_PERS_LEN = 128
• Compared to maximum length, key in phone numbers with: CH.4. one character less CH.5. equal number of characters CH.6. with one character more		A.RR.5.5. MAX_NUMB_LEN = 32

When it comes to black-box behavior, much of the *complexity* resides in exception handling of received *stimuli*. Each *scenario requirement* has its own set of *test-*

cases (see Table 12-93 to Table 12-97 below).

TABLE 12-93 Phonebook black-box test-cases
for initiate-phonebook scenario

Black-box stimulus injection	Expected response with pass/fail criteria	Behavior requirement
CH.1. Use a correct half-full phonebook file. Start the phonebook program.	XP.1. Pass, if the console window opens and if the welcome message is displayed	A.RB.6.S.1.U.1. Program-loading-file-and-start-up normal use-case
CH.1. Delete the phonebook file. Start the phonebook program.	XP.1. Pass, if a warning message is displayed, and thereafter if a welcome message is displayed.	A.RB.6.S.1.X.2. No-file-found Exception use-case
• During start-up of the program, use various phonebook files, which are: CH.1. An existing but completely empty file CH.2. With title but without new line character CH.3. With title with new line character CH.4. With title with two new line characters	XP.1. Pass, if a warning message is displayed (for CH.1 to CH.3.), and if then a welcome message is displayed, and if the phonebook file is correct after termination.	A.RB.6.S.1.X.3. Found-file-empty exception use-case
• Use phonebook files which are: CH.1. Correct but empty of contacts CH.2. Half full of intact contacts • Then end this file incorrectly in various ways (see Fig. 9-67, p. 433) like: CH.3.a. - CH.3.j. File ending after each person data string, with and without new line character CH.3.k. - CH.3.q. File ending after no phone number (equal to last case in previous paragraph), one phone number, one ending after less than one maximal number and one ending after maximal numbers, with and without new line character CH.3.r. File ending after the full contact with only one new line character (equal to last case in previous paragraph)	XP.1. Pass, if a warning message is displayed and thereafter if a welcome message is displayed, and if all intact contacts are read and available in the phonebook.	A.RB.6.S.1.X.4. Found-file-unexpected-end exception use-case

TABLE 12-94 Phonebook black-box test-cases for append-contact-to-phonebook scenario

Black-box stimulus injection	Expected response with pass/fail criteria	Behavior requirement
• Use a phonebook file which has various numbers of existing contacts, described below. Start the phonebook program, and append one contact. CH.1. Phonebook file correct, but without contacts CH.2. Phonebook file approximately half full CH.3. Phonebook file still having place for one more contact • For each of the above ways of registering contacts, also register phone numbers in the five various ways. CH.4.a. One phone number CH.4.b. Half full of phone numbers CH.4.c. Next to full of phone numbers CH.4.d. Full of phone numbers	XP.1. Pass, if the appended contact and phone numbers are present in the phonebook at a listing of contacts.	A.RB.6.S.2.U.1. Append-contact-to-phonebook normal use-case
CH.1. Use a phonebook file which has no more place for contacts. Start the phonebook program, and append one contact.	XP.1. Pass, if the program displays an error message, and if a listing doesn't contain anything from the appended contact, and if the program still executes normally.	A.RB.6.S.2.X.2. Append-contact-to-full-phonebook exception use-case
CH.1. Use a phonebook file which is half full of contacts. Start the phonebook program. CH.2. Append one contact and key in the maximum number of phone numbers. CH.3. Try to key in yet one more phone number.	XP.1. Pass, if the program displays an error message, and if a listing doesn't contain anything from the last phone number, and if the program still executes normally.	A.RB.6.S.2.X.3. Append-phone-number-to-full-phone-number-list exception use-case

TABLE 12-95 Phonebook black-box test-cases for list-phonebook-contacts scenario

Black-box stimulus injection	Expected response with pass/fail criteria	Behavior requirement
• Use a phonebook file which has various numbers of existing contacts described below. Start the phonebook program, and list it. CH.1. Phonebook file with one contact CH.2. Phonebook file approximately half full CH.3. Phonebook file completely full	XP.1. Pass, if listing properly displays all contacts with persons' data and phone numbers in the phonebook.	A.RB.6.S.3.U.1. List-phonebook-contacts normal use-case
CH.1. Use a correct phonebook file, but without contacts.	XP.1. Pass, if listing explains that there are no contacts in the phonebook.	A.RB.6.S.3.U.2. List-empty-phonebook normal use-case

TABLE 12-96 Phonebook black-box test-cases
for delete-phonebook-contact scenario

Black-box stimulus injection	Expected response with pass/fail criteria	Behavior requirement
• Use a phonebook file which has various numbers of existing contacts described below and start up the phonebook program. CH.1. Phonebook with one contact CH.1.a. Remove this. CH.2. Phonebook with two contacts CH.2.a. Remove the first and then the last. CH.2.b. Remove the second and then the last. CH.3. Phonebook file approximately half full CH.3.a. Remove one contact in the middle. CH.4. Phonebook with one contact still unused CH.4.a. Remove the last contact and again the last. CH.4.b. Remove the second-last and then the last. CH.5. Phonebook file completely full CH.5.a. Remove the last contact and again the last. CH.5.b. Remove the second-last and then the last.	XP.1. Pass, if the removed contact is completely removed at a listing of contacts.	A.RB.6.S.4.U.1. Delete-phonebook-contact normal use-case
CH.1. Use a correct phonebook file, but without contacts. Try to remove a contact.	XP.1. Pass, if the program displays a message that there are no contacts to remove.	A.RB.6.S.4.X.2. Delete-contact-in-empty-phonebook exception use-case
• Use a phonebook file which has various numbers of existing contacts described below and start up the phonebook program. CH.1. Phonebook with one contact CH.1.a. Use a contact number one below the only existing contact number. CH.1.b. Use a contact number one above the only existing contact number. CH.2. Phonebook file approximately half full CH.2.a. Use a contact number one below the lowest existing contact number. CH.2.b. Use a contact number one above the highest existing contact number. CH.3. Phonebook file completely full CH.3.a. Use a contact number one below the lowest existing contact number. CH.3.b. Use a contact number one above the highest existing contact number.	XP.1. Pass, if the program displays an error message that the used contact number is out of existing range.	A.RB.6.S.4.X.3. Delete-contact-not-existing exception use-case

TABLE 12-97 Phonebook black-box test-cases
for terminate-phonebook scenario

Black-box stimulus injection	Expected response with pass/fail criteria	Behavior requirement
CH.1. When phonebook half filled and program is idle, terminate it.	XP.1. Pass, if a good-bye message is displayed, and if the console windows disappears and if the phonebook file is correct when the phonebook program is started again.	A.RB.6.S.5. Terminate-phonebook scenario
CH.1. Use a half full phonebook file in one file folder to read from, and save the new phonebook to another read-only file folder.	XP.1. Pass, if the program displays an error message that the phonebook to be saved cannot be created.	A.RB.6.S.5.X.2. File-not-created exception use-case
CH.1. Use a half full phonebook file in one file folder to read from, and save the new phonebook to another file folder with no free space (possibly use a small memory stick). Start the phonebook program and terminate it.	XP.1. Pass, if the program displays an error message that nothing is saved.	A.RB.6.S.5.X.3. File-not-written exception use-case
CH.1. Use a half full phonebook file in one file folder to read from, and save the new phonebook to another file folder with very limited space (possibly use a small memory stick). Start the phonebook program. Start the phonebook program and terminate it.	XP.1. Pass, if the program displays an error message that not all contacts are saved.	A.RB.6.S.5.X.4. File-incompletely-written exception use-case

12.26.4 EXAMPLE Phonebook: Verify white-box

Now over to *verification* by using all prepared *test-cases*. First, the *white-box* is verified (see Table 12-98 below).

TABLE 12-98 Phonebook white-box verification report 1

Actual outcome and pass/fail conclusion	Expected outcome and pass/fail criteria	Requirement
VR.1. Pass	XP.1.	A.RR.1.1.
VR.1. Pass	XP.1.	A.RR.2.1.
VR.1. Pass	XP.1.	A.RR.4.1.
VR.1. Pass	XP.1.	A.RR.3.1.
VR.1. Pass	XP.1.	A.RR.5.1.
VR.1. Pass	XP.1.	A.RR.5.2.
VR.1.1. Fail. At equal string length (C.2. and C.5.) or at one character more (C.3. and C.4.), the debugger gave Assertion failed (see Fig. 12-35 below).	XP.1. Pass, if oversized person data and phone numbers get truncated at their max length, and if the program continues to execute normally, and if these data keep after termination and restart.	A.RR.5.4. MAX_PERS_LE N = 128 A.RR.5.5. MAX_NUMB_L EN = 32

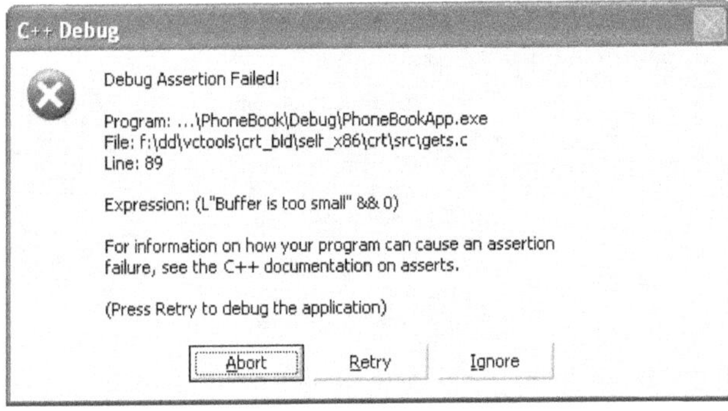

FIGURE 12-35 Test-case RR.5.4. and RR.5.5. assertion failed

12.26.5 EXAMPLE Phonebook (white-box verified): Verify black-box

Then black-box behavior is verified (see Table 12-99 below).

TABLE 12-99 Phonebook verification report 2

Actual outcome and pass/fail conclusion	Expected outcome and pass/ fail criteria	Requirement
VR.2. Pass	XP.1.	A.RB.6.S.1.U.1.
VR.2. Pass	XP.1.	A.RB.6.S.1.X.2.
VR.2. Pass	XP.1.	A.RB.6.S.1.X.3.
VR.2.1. Fail. For all cases C.a. to C.r. last contact was not read. For example, if unexpected file end occurs within first contact, then the number of contacts is displayed as -1.	XP.1. Pass, if a warning message is displayed and thereafter if a welcome message is displayed, and if all intact contacts are read and available in the phonebook.	A.RB.6.S.1.X.4. Found-file-unexpected-end exception use-case
VR.2. Pass	XP.1.	A.RB.6.S.2.U.1.
VR.2. Pass	XP.1.	A.RB.6.S.2.X.2.
VR.2. Pass	XP.1.	A.RB.6.S.2.X.3.
VR.2. Pass	XP.1.	A.RB.6.S.3.U.1.
VR.2. Pass	XP.1.	A.RB.6.S.3.U.2.
VR.2. Pass	XP.1.	A.RB.6.S.4.U.1.
VR.2. Pass	XP.1.	A.RB.6.S.4.X.2.
VR.2.2. Fail. The last contact is deleted even if a contact number one too high is used.	XP.1. Pass, if the program displays an error message that the used contact number is out of existing range.	A.RB.6.S.4.X.3. Delete-contact-not-existing exception use-case
VR.2. Pass	XP.1.	A.RB.6.S.5.U.1.
VR.2.3. Fail. Displayed failure message is correct, but despite that, the phonebook program terminates.	XP.1. Pass, if the program displays an error message that the phonebook to be saved cannot be created.	A.RB.6.S.5.X.2. File-not-created exception use-case

TABLE 12-99 Phonebook verification report 2

Actual outcome and pass/fail conclusion	Expected outcome and pass/fail criteria	Requirement
VR.2.4. Fail. Although nothing is saved, this message is not displayed.	XP.1. Pass, if the program displays an error message that nothing is saved.	A.RB.6.S.5.X.3. File-not-written exception use-case
VR.2. Pass	XP.1.	A.RB.6.S.5.X.4.

12.26.6 EXAMPLE Phonebook: Locate failure source

All failures are simple logic failures and rather easy to localize in the *program* when run in a capable debugger. The Debug Assertion Failed is localized to typing a too long string when executing the *instruction* gets_s.

12.26.7 EXAMPLE Phonebook: Evaluate optimal elimination

The gate meeting was held, and the main message was to just correct the failures detected during *verification* (see Table 12-100 below).

TABLE 12-100 Phonebook failure elimination 1

Failure location	Failure elimination plan	Actual outcome and pass/fail criteria	Expected outcome and pass/fail criteria	Restrictions and use-case requirement
FER.1.1. The safe C instruction "gets_s" takes a string length size parameter. If input exceeds, the program stops.	FER.1.1.a.Input strings are long and rarely overloaded. FER.1.1.b. A professional program should not allow runtime crashes. FER.1.1.c. Catch the failure before the program is stopped and truncate the string at maximum size. FER.1.1.d. Reverify.	VR.1.1. Fail. At equal string length (C.2. and C.5.) or at one character more (C.3. and C.4.), the debugger gave Assertion failed (see Fig. 12-35 below).	XP.1. Pass, if oversized person data and phone numbers get truncated at their max length, and if the program continues to execute normally, and if these data keep after termination and restart.	A.RR.5.4. MAX_PERS_LEN = 128 A.RR.5.5. MAX_NUMB_LEN = 32
FER.1.2. A simple logic failure when calculating number of intact contacts	FER.1.2.a.Find the failure by the debugger. FER.1.2.b. Eliminate the failure. FER.1.2.c. Reverify.	VR.2.1. Fail. For all cases C.a. to C.r. last contact was not read. For example, if unexpected file end occurs within first contact, then the number of contacts is displayed as -1.	XP.1. Pass, if a warning message is displayed and thereafter if a welcome message is displayed, and if all intact contacts are read and available in the phonebook.	A.RB.6.S.1.X.4. Found-file-unexpected-end exception use-case
FER.1.3. A simple logic failure when compared with number of last contact	FER.1.3.a.Find the failure by the debugger. FER.1.3.b. Eliminate the failure. FER.1.3.c. Reverify.	VR.2.2. Fail. The last contact is deleted even if a contact number one too high is used.	XP.1. Pass, if the program displays an error message that the used contact number is out of existing range.	A.RB.6.S.4.X.3. Delete-contact-not-existing exception use-case

TABLE 12-100 Phonebook failure elimination 1

Failure location	Failure elimination plan	Actual outcome and pass/fail criteria	Expected outcome and pass/fail criteria	Restrictions and use-case requirement
FER.1.4. The file is ordered to close, though not created.	FER.1.4.a.Change the file closing instruction to after it has first been opened. FER.1.4.b. Reverify.	VR.2.3. Fail. Displayed failure message is correct, but despite that, the phonebook program terminates.	XP.1. Pass, if the program displays an error message that the phonebook to be saved cannot be created.	A.RB.6.S.5.X.2. File-not-created exception use-case
FER.1.5. When writing to the file, the first lines are buffered and impossibility to write is not immediately detected.	FER.1.5.a.Flush the buffer after the first line is written, then the impossibility to write is discovered. FER.1.5.b. Reverify.	VR.2.4. Fail. Although nothing is saved, this message is not displayed.	XP.1. Pass, if the program displays an error message that nothing is saved.	A.RB.6.S.5.X.3. File-not-written exception use-case

A lot of exception handling is specified and implemented, protecting the phonebook file from damage, regardless of how this has happened. However, the case of the phonebook disk file containing more than the maximum number of phone-numbers or contacts has not been specified, and is probably not considered in the *program*.

If the phonebook disk file is always written on termination after running the phonebook *program*, this failure never happens. But if the file is accidentally extended, there may not be any protection against the *program's* ceasing to read past the *variable* arrays. If reading too many data and saving them beyond the *variable* array, the *program* will likely crash. It is a rare and far-fetched exception, but fatal when it happens.

The decision gate decided to include these exception cases into the *requirements*, insert the prevention in the phonebook *program*, and prepare *test-cases* to be executed in the next round of *verification*.

12.26.8 EXAMPLE Phonebook: Verify black-/white-box

The *verification* was done on the detected failures, including the newly added *requirement* and *test-case*, to conclude they are all eliminated (see Table 12-101 below).

TABLE 12-101 Phonebook verification report 3

Actual outcome and pass/fail conclusion	Expected outcome and pass/fail criteria	Requirement
VR.3. Pass	XP.1.	A.RR.5.4. A.RR.5.5.
VR.3. Pass	XP.1.	A.RB.6.S.1.X.4.
VR.3. Pass	XP.1.	A.RB.6.S.4.X.3.

TABLE 12-101 Phonebook verification report 3

Actual outcome and pass/ fail conclusion	Expected outcome and pass/fail criteria	Requirement
VR.3. Pass	XP.1.	A.RB.6.S.5.X.2.
VR.3. Pass	XP.1.	A.RB.6.S.5.X.3.
VR.3. Pass	XP.1.	New requirement A.RB.3.S.0.X.1.

12.26.9 EXAMPLE Phonebook: Manual table of contents

It is rather easy to finish the manual, because the *use-cases* are already well specified, and need only to be detailed (see Table 12-102 below).

TABLE 12-102 Phonebook manual table of contents

Manual name	Requirement / design
MA.1. Starting up the program, including exception handling	A.RB.6.S.1. Initiate-phonebook scenario
MA.2. How to append a contact to the phonebook, including exception handling	A.RB.6.S.2. Append-contact-to-phonebook scenario
MA.3. How to list all contacts, including exception handling	A.RB.6.S.3. List-phonebook-contacts scenario
MA.4. How to delete a contact, including exception handling	A.RB.6.S.4. Delete-phonebook-contact scenario
MA.5. Terminating the program, including exception handling	A.RB.6.S.5. Terminate-phonebook scenario
MA.6. Implemented phonebook-size limitations and proposed way of extension	A.RR.5.2. MAX_CONT_PER_BOOK = 128 A.RR.5.3. MAX_NUMB_PER_PERS = 16
MA.7. Implemented string length limitations	A.RR.5.4. MAX_PERS_LEN = 128 A.RR.5.4. MAX_PERS_LEN = 128
MA.8. Architecture rationale and proposed way of extension	A.Phonebook logical architecture (Fig. 9-66, p. 431)
MA.9. Object-oriented program rationale and proposed way of extension	A.Phonebook design chart (Fig. 10-58, p. 562)

12.26.10 EXAMPLE Phonebook: Proceed reading

To follow this example, these are the alternatives:

- Proceed to chapter 11.20, "EXAMPLE Phonebook environment: Realize interfaces to environment and insert/await outermost white-box," page 657.
- Trace upstream through the calc_logic example to understand how to eliminate a detected failure.

12.27 EXAMPLE Phonebook environment: Verify prototype in environment

12.27.1 EXAMPLE Phonebook environment: Process schedule to use

Pt:2:3. Tailored advanced technical schedule n = 2

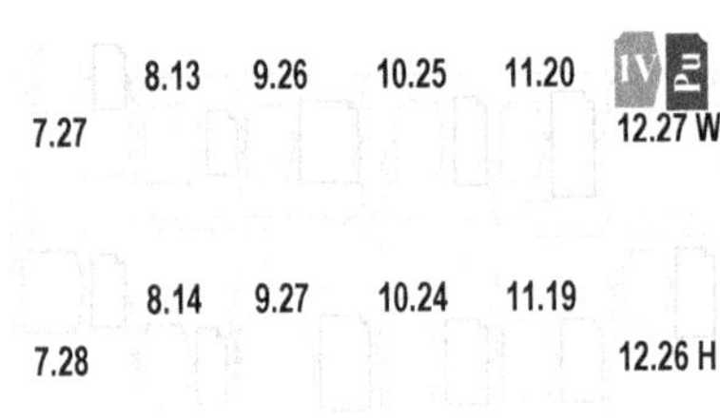

For the phonebook example, a special technical overall *schedule* is *tailored* (illustrated in Fig. 12-36, left):

- **Pt:2:3. Tailored advanced technical schedule n = 2,** page 169

In this *tailored schedule* the generic *schedule* to use is (filled-in symbols):

- **Pt.0.V. Verify prototype in environment,** page 125.

FIGURE 12-36 Position of schedule and master to use (filled in) in overall schedule

12.27.2 EXAMPLE Phonebook environment: Verify prototype in environment for evaluation

The current status of the phonebook is for laboratory evaluation only and needs not to be verified in any *target environment*, where subscribers use it in reality.

12.27.3 EXAMPLE Phonebook: Proceed reading

To follow this example, these are the alternatives:

- Trace upstream through the calc_logic example to understand how to eliminate a detected failure.

- Here the phonebook example terminates, and market validation takes place (see ch. 6.6, "P. Cpdm generic development schedule," p. 99). Open the last bottle of champagne:-). Exchange this now worn-out *Cpdm* book for a new one, buy a new case of champagne, and start a repetition of *Cpdm*.

Watch references

The reference list below may be judged too short in the academic world, and unnecessarily stringent in the industry world.

I very much appreciate theoretical discussions with skilled academic people, and I equally appreciate experience discussions with talented developers. But I want to act as a middleman to bridge between these sides, to get the academics and developers to find each other.

Nowadays, much information is kept on the Internet by such sites as Wikipedia. Countless times I have been googling to get information, and my best recommendation is that my readers do the same when looking for references.

[1] Apke, L. (2013). "Understanding the agile manifesto." Lulu Publishing.

[2] Becker, S. A., & Whittaker, J. A. (1997). "Cleanroom software engineering." Idea Group.

[3] Huzar, Z., Kuzniarz, L., Reggio, G., & Sourrouille, J. L. (2005. "Consistency problems in UML-based software development." Berlin/Heidelberg: Springer.

[4] Kruchten, P. (2003). "The rational unified process: An introduction." Addison-Wesley.

[5] Linger, R. C., Mills, H. D., & Witt, B. J. (1979). "Structured programming: Theory and practice." Addison-Wesley.

[6] Mills, H. D., Linger, R. C., & Hevner, A. R. (1986). "Principles of information systems: Analysis and design." Academic Press.

[7] Penaire, C., Edwards, M., Fernandez, A., Mancin, E., & Carroll, K. (2007). "The IBM rational unified process for System z." IBM Redbooks.

[8] Rumbaugh, J., Jaconson, I., & Booch, G. (1999). "The unified modeling language reference manual." Addison-Wesley.

[9] Rumbaugh, J., Jacobson, I., & Booch, G. (2004). "The unified modeling language reference manual (2nd. ed.)." Addison-Wesley.

Track indexes

Cornerstone words and phrases are the basic building blocks of *Cpdm* processes, explanations, and examples. Below index list shows the appearance of the most essential *Cpdm* cornerstones.

A

Ability 204–205

Abstract element 138, 345

Activity-oriented 84

Added value 66, 68–69, 73, 84–85

Address label 96, 396, 516–517, 530

Aggregate 85

Aliases 396, 414, 420, 424, 531, 533, 565, 581

Alternating activities and results 72

Any kind of items 440, 561

Architect 108, 182, 211, 249, 304, 606, 612

Architecture design 115, 117–118, 136–138, 153, 264, 289, 295, 298, 304, 359, 382, 394

Artificial complexity 21, 205, 222, 236, 596, 714

Assembler program 148–150, 219, 264, 398, 516–518, 525, 530, 550, 631, 636, 649, 767

Asynchronous interface 255, 402–403, 405

B

Basic use-case 221–224, 227, 230

Behavior requirement 130–132, 152–153, 156, 158, 185, 203–204, 206–207, 209, 217, 221, 226–227, 229, 245–246, 259, 273, 277, 285, 561

Black-/white-box hierarchy 180, 280, 345

Black-/white-box nesting level 109, 132, 177, 287, 297, 507, 524, 595

Black-box requirement 131, 134, 184, 188, 191, 196, 198, 203, 208–209, 277, 284, 297, 326, 347–348, 357, 531

Black-box test-case 152–154, 156, 160, 705, 709, 720, 733, 736

Black-box wait-state 224–225, 227, 229, 233–236, 422

Boundary interface 185, 193, 204, 211, 216, 339, 346–347, 349–352, 359, 406–407, 441, 476, 601

Business case 44, 47–48, 54

C

Capture requirements 48, 104, 130, 215, 278

Cascading results 70, 84, 337, 356, 444–445

Class 425, 512, 553–556

Class shape 559, 566, 571, 574, 577

Clean-room engineering 172–173, 345

Component 138, 143, 150, 205, 293, 297, 309, 379, 442, 472, 492–493, 507, 522, 552, 591, 642, 646

Compound requirement 131, 184, 203

Configuration management 27–28, 88, 297, 378, 443, 597, 640, 717

Connection interface 206, 226, 246, 267, 339, 346–349, 351, 357, 407, 441, 455, 469, 471, 484, 537, 559

Consistency 173, 245, 266, 339

Critical path 155

Customer 19, 28, 67–69, 176, 283, 355, 384

Customer-oriented 19, 69, 415

D

Decompose 85, 102–104, 164, 169, 344, 553

Demand 73, 89, 98

Design item 440

Design item properties 441

Designer 108, 137, 182, 191, 201, 204, 226, 264, 283–284, 308, 642, 705–706

Determined architecture ingredient 283–285

Development board 170, 533, 549–550, 643–644, 646, 649, 651–653, 761–763, 766–767

Development staffing 111, 119, 179, 185, 218, 304

Development system 146, 238, 322, 327, 519, 533, 537, 543, 546, 642–643, 646–648, 664

Development system libraries 117, 149, 322, 442, 537, 645–646

Downstream 93, 160, 600, 669, 671

E

Embedded black-/white-box 407, 596, 602, 662–664

Embedded black-box 87, 102, 128–129, 137, 139, 150, 196, 208, 309, 348, 377, 382, 390, 425, 464, 596, 668

Embedded white-box 117, 120, 139, 141, 143, 297–298, 348, 596, 601

Enfold 120, 141, 439

Enfold embedded white-box 105

Enfold outermost white-box 105

Environment architecture resumé 336–338, 342, 356, 383, 426

Environment development staffing 84, 179–180

Environment interfaces solution alternatives 112–116, 119, 287, 289, 314, 338, 347

Environment logical architecture 116, 289, 336, 410

Environment nesting level 102–103, 109, 111, 179, 182, 287, 289, 314, 336, 498, 642, 663

Environment physical architecture 289, 336

Environment restriction 177–178

Exception use-case 221, 223, 229–230, 232, 265, 738, 752, 754

Explicit verification 663

F

Failure detection 128, 154–155, 671, 704–705

Failure elimination 127, 154–155, 669, 671, 674, 676, 682, 715, 717, 746, 777

Failure localization 154, 665, 669, 723

Fake black-/white-box 123, 143, 145, 147, 166, 168, 596–598, 600

Finalize design item 439, 623

Finalize product requisite 443

Fixture 146, 152, 155–156, 656–657

Formal scenario 158, 173, 222, 230–236, 245–246, 248, 257, 265, 269–272, 275–276, 285, 419, 422, 582, 736, 752, 769–770

Formal use-case 222, 227, 229–230, 245, 273, 285, 731

Formalism 29, 168, 193, 203, 222, 224, 226–227, 250, 266, 277, 671, 704, 713–714

Full verification 596

Function shape 531, 537, 550, 559, 564, 577

Functional decomposition 285

G

Global variable 414, 531, 556, 559

Granularity 379, 499, 560–561, 579

Guru 176, 283, 559, 585, 715

H

Header file 117, 532, 537, 542, 546–547, 550–551, 559, 564, 566, 571, 574, 587

High-level program 149–150, 411, 414, 419, 559

Host 146, 152, 170, 238, 379, 519, 533, 544, 547, 642–643, 646, 733–734, 758, 761

I

Idle wait-state 231

Implicit verification 103, 145, 166, 168, 290, 298, 597, 663–664

Incorporative integration 103, 143, 162, 170, 289, 298, 594–596, 598, 624

Incremental development 28, 173–174, 278, 597, 623

In-house design item 440

In-house role item 440

Innermost black-/white-box 160, 438, 662–664, 684

Innermost black-box 166

Innermost white-box 117, 595–596, 598

Instruction 172, 353, 396, 414, 419, 513, 530–532, 535, 557, 579, 746, 750, 777

Integration report 599–600

Intellectual control 22–23, 29, 35, 283, 421, 423, 544, 623

Interaction requirement 206, 221, 225

Interrupt 401–403, 410, 442, 514, 519

Intrinsic complexity 22, 97, 138, 168, 205, 285

Invasive integration 103, 121, 143, 147, 163, 166, 168, 290, 298, 350, 439, 595–598, 606, 611–612, 615, 624, 662, 664–665, 673, 675, 677–679

J

Jump instruction 517

Justified 73, 89, 445

L

Lead time 596

Life cycle status 88

Line 39, 67, 69, 592, 623

Line management 21, 28–29, 31, 78, 83, 97, 176, 180, 186

Linkable program 145, 149–150, 171, 624, 650, 656

Local variable 414, 531–532

Low-level program 149, 396

M

Machine instruction 150–151, 395, 400, 513–517, 523, 651

Machine program 146, 150–151, 395, 437, 515–516, 566–567, 631–632, 636, 640, 645, 652, 654

Machine variable 150–151, 400, 515

Managed information 185, 301, 346, 378–379, 389

Manager 23–24, 31, 79, 83, 88, 715

Marketer 176, 278–279

Message sequence chart 380–382, 400

Meta information 80

Metrics 83

Modularize 24, 430, 513, 563, 581, 601

Module 277, 445

Module cohesion 349

Module coupling 349

Module strength 349

Monolith 585

N

Normal use-case 221, 223, 230, 232, 265, 600

Null series 58, 63–64

O

Obey 92

Object 259–260, 264, 424–425, 512, 555–559, 566–567, 569–570, 574, 577, 579

Object-oriented program 36, 149, 261, 264, 273, 277, 424–425, 429–430, 553–555, 558–559, 590

Observation post 350, 679

Off-the-shelf item 113, 121–122, 143–144, 289, 298, 440

Off-the-shelf supplier 113, 116, 119, 121–122, 134, 140, 143–144, 336, 355

Outermost black-/white-box 125, 635, 662, 710, 719

Outermost black-box 116, 129, 131, 171, 178, 196, 208, 213, 237, 287, 337, 347, 382, 384, 387, 389, 427

Outermost white-box 117–118, 120, 123, 170, 347–348, 382, 389, 407, 439, 595

P

Paradigm shift 19–21, 94, 276

Partial verification 145, 166, 168, 596, 598, 606, 609–611, 616, 620, 673

Partition 19, 23, 52, 85–86, 149, 174, 188–189, 192, 277, 282, 344, 370, 401, 425, 445, 471, 553, 560–561

Performer 79, 88

Port 117, 119, 122–124, 147, 150, 170, 322, 411, 533–534, 542, 549, 594, 636, 643–644, 646–647, 654, 712, 733, 766–768

Postcondition 222, 229

Precondition 229

Preparer 79, 88

Primordial architecture ingredient 284–285

Process management 26, 92

Product development 18, 20–23, 26–27, 30, 35–36, 38–39, 64, 66, 69, 80, 176, 261, 327, 339, 379, 421, 590

Product life cycle 19, 38–39, 41, 44, 66

Product line 43

Product management 39, 278–279

Product portfolio 41–43, 47–48, 278–279, 328, 330

Product proposition 47–48, 71–72

Product prototype 600

Product requisite 52, 89, 218, 261, 265, 289, 298, 443–444, 449, 460, 464, 481–482, 543, 587, 594, 615

Product structure 23–24, 52, 177, 354, 713

Program libraries 123, 144, 149–150, 322, 325, 327, 442, 543, 591, 640, 643, 743, 745

Progressing activities 69

Project 22, 39, 97, 132, 238, 327, 582, 592, 623, 712

Project leader 94, 132, 249, 319

Project management 21, 27, 29, 78, 111, 611

Project sponsor 218

Prototyping 45, 280, 623, 643

Provider 111, 132, 178, 249

Provider role item 440

Q

Quality 18–19, 21, 24, 27–28, 43, 59, 64, 69, 72, 78, 80, 83, 90, 92, 94, 97, 113, 120, 137, 142, 176, 205, 277–278, 284, 292, 337, 350, 422, 440, 544, 548, 592, 600, 602, 623, 661, 673, 675, 677, 713–715

Quality management 26–28, 97, 132

R

Received items deviation report 146, 599

Recycle 122, 139–140, 355, 442

Refactoring 285

Refine environment restriction 169

Refine requirement 103, 131

Regression verification 155, 717

Requirement refinement 176, 185, 188, 213, 216, 279

Restriction requirement 130–131, 152–153, 285, 561

Result 176, 179

Result-oriented 70, 84

Reusable item 122, 134, 137, 212, 238, 322, 325, 379, 411, 442, 543, 641–642, 646, 759

Reuse 35, 122, 139, 143, 155, 204–205, 238, 277, 283, 319, 322, 325, 327, 332, 351, 355, 378–380, 411, 418, 442–443, 482, 513, 542–543, 548, 641–642, 646, 708, 724, 758–759, 766

Reuse plan 43–44, 48, 51

Road map 40–42, 48, 276, 322, 325

Role 78–80, 88–89, 108, 111, 132, 142, 185–186, 335

Role item 89, 440, 560

S

Sample catalogue 113–114, 116, 119, 134, 137, 141, 286, 289, 297–298, 355

Scenario 132, 230–231, 255

Selected solutions supplier opportunities 287

Shape requirement 566–567

Solution alternatives 137, 139, 171, 282, 286, 291, 347–348, 436, 668, 670

Source file 138, 532, 537, 542, 559, 564–567, 569, 573

Source system 113, 119, 122, 134, 139–141, 143, 149, 289, 297–298, 646

Specialization 553

Specifier 108, 185–186, 211, 668

Stage 38–39, 63, 99

Stakeholder 40, 61, 111, 116, 132, 138, 154, 176–178, 188, 190, 196, 201, 205, 218, 249, 278–279, 294, 322, 325, 359, 368, 439, 660–661, 671, 684, 696, 698, 713

Stand-by wait-state 229, 231

Start-up report 124, 146, 600, 627

Stimuli history 173, 226–227, 234

Structured function call 245, 532

Structured instruction sequence 532

Structured iteration instruction 532

Structured program 149, 264, 277, 421, 529–530, 532

Structured selection instruction 532

Subroutine call 245

Supplier design item 134, 141, 143, 440

Supplier opportunities 139, 171, 304, 356

T

Terms 88, 108, 111, 132, 180, 186, 335

Test-use-case 248, 259

Thread 255, 401–402

Thread machine 255, 355, 405

Turnaround time 513, 715

Turnkey item 111, 114, 119, 123, 137, 141, 144–145, 289, 298, 304, 440

Turnkey supplier 111, 116, 119, 123, 133–135, 141–142, 144, 157, 289, 297–298, 336, 355

U

Unbiased architecture ingredient 284–285, 327, 332, 356, 419, 468

Unmanageable complexity 20

Upstream 160, 600, 665–668, 674, 678, 693, 699, 703, 712, 725, 729

Usage profile 156, 158–160, 233, 730–736, 741–742, 744, 747–748, 753–754, 756, 761, 769

Use-case 35–36, 132, 221, 227, 229–230, 232, 234, 255, 560–561, 730, 755, 779

User profile 48, 61–62, 209

Use-unit 224–225, 227, 229–233, 257, 381, 419, 421–423

V

Variability of reusable components 205

Variant 85, 87–88, 101–102, 353, 390, 407, 482, 488, 491, 493, 498, 511, 528, 623, 657

Verification report 153–154, 672, 689, 706, 753

Version 87, 99, 119, 139, 297, 345, 418, 733, 767

View 87, 101, 103, 250, 557, 601, 666, 695, 698

V-model 160

W

Wait-state explosion 401

Wait-state-machine 232–234, 733, 736–737

Waterfall model 74

White-box solution alternatives 49–50, 92, 114, 133–137, 179, 188–189, 282, 284–287, 291, 293, 295, 297–298, 301, 306, 310, 315, 332, 355–356, 359, 367

White-box supplier opportunities 287, 295, 298

White-box test-case 152–153, 705, 708, 719

Share glossary

The words and phrases below, from the commonly accepted to the self-invented to fill yet unexplored gaps, are the core building blocks in *Cpdm*, and are marked with italics below and in all running text throughout this book (yes, it constitutes overuse of italics, but it is hopefully helpful for the readers to have all of them explained below). These are also the building blocks in *process tasks*, *masters*, and *schedules*, which are mainly defined in chapter "Utilize processes," page 81.

(Disclaimer: the important thing is not what particular words are used, but rather that the reserved words throughout *Cpdm* are defined and used in a holistic, consistent, and comprehensible way).

Reserved word	Cpdm definition	Common definition
. (period)	Delimiter to denote *black-/white-box hierarchy*	Not often defined
: (colon) ; (semicolon)	Delimiter to denote *version* or *variant* (to name a file *version* in Windows, colon is not allowed in file names, but instead use semicolon)	Not often defined
- - (hyphens)	Wrapper to denote *connection interface*	Not often defined
[ ] (brackets)	Wrapper to denote *boundary interface* ( ]outside[, [inside], and]both sides])	Not often defined
Ability (of an architecture ingredient)	Often *architecture ingredients* require certain abilities, which may be specified in the *architecture resumé* or in advance as a *restriction requirements*. Also see *item*.	More vague
Abstract element	An *element* in an *architecture* encircling ingredients for the purpose of showing that they somehow belong together	Not often defined
Activity Activities	**A piece of human work (see Table 5-2, p. 68)**	Not often defined
Activity-oriented	To emphasize the importance of a *value chain's* progressing *activities*, and thereby suppressing the importance of *results*. Also see *result-oriented*.	More vague
Added value	The value of a *result* generated by an *activity*, subtracted the value of the sources used by this *activity*	About the same
Address label	**Entrance location among assembler instructions (see Table 9-27, p. 396)**	About the same
Aggregate	**Merging parts to the whole (see Table 6-3, p. 85)**	About the same
Aliases (in programs)	Assigning text names to parameters, in order to make them self-explanatory, thus easier to read and understand	About the same
Alternating activities and results	Illustrating a *value chain* as a network of interconnected alternating pairs of *activities* and *results*	Not defined
Architect	A developer *role* in designing and illustrating *architectures*	About the same
Architecture	Various kinds of *product structure* descriptions containing *architecture ingredients*. This definition includes also immaterial *program* ingredients. Also see *architecture resumé*, *logical architecture* and *physical architecture*	About the same
Architecture design	Populate *environment interfaces* or *white-boxes* with *architecture ingredients*, in such a way that its *requirements* are satisfied as well as possible	Rather similar
Architecture ingredient	*Architecture resumé* constituents, such as *users*, *black-boxes*, *white-boxes*, *elements*, and *interfaces*	Not often defined

Reserved word	Cpdm definition	Common definition
Architecture resumé	**List of decided ingredients, derived from solution alternatives or supplier samples thereof (see Table 9-8, p. 344).** Also see *logical architecture* and *physical architecture.*	Not often defined
Artifact	Something artificially developed or produced, including immaterial *artifacts*	About the same
Artificial complexity	*Complexity* caused by incompetent usage of a sufficient model, or correct usage of an insufficient model. Also see *intrinsic complexity.*	About the same
Assembler program	**Programs containing symbolic mnemonics instead of machine-dependent instructions, arithmetic calculation support, user naming of variables and address labels, which consequently need assembler and linker tools for translation to machine instructions and variables (see Table 10-73, p. 516)**	About the same
Asynchronous interface	**A connection interface with a storage buffer function, to allow receiving at another pace and rhythm than was transmitted (see Table 9-29, p. 402)**	Not often defined
Attachment (interface graph)	A graph of mechanical *connection interfaces*	Not defined
Basic use-case	**Stakeholder-expected behavior on interfaces of a black-box, specified by one interaction or a sequence of interactions (see Table 7-30, p. 221)**	More vague
Behavior requirement	**Stakeholder-expected behavior on interfaces of a black-box (see Table 7-17, p. 206)**	More vague
Black-box	**Element which conceals all of its inward content, but shows its dynamic behavior via its interfaces. Can be opened to become a white-box (see Table 9-2, p. 335).**	About the same
Black-box requirement	**Stakeholder-documented and prioritized expectations from a black-box (see Table 7-7, p. 184)**	About the same
Black-box test-case	**Concretization of behavior requirements, to be a practical base for verification of black-box behavior (see Table 12-31, p. 705)**	More vague
Black-box wait-state (or just wait-state)	**A black-box being in stand-by mode, awaiting random stimuli on any of its interfaces or from internal time-out (see Table 7-34, p. 224).** Also see *wait-state.*	Not defined (in clean room called usage state)
Black-/white-box	An autonomous *element* provided with own *requirements*	About the same
Black-/white-box hierarchy	The concept of containing *black-boxes* embedded inside each other, in order to scale up and *master* an otherwise flat, vast, and unmanageable *product*	Not often defined
Black-/white-box nesting level	The hierarchical *product structure* of *embedded black-/white-boxes*, starting at *nesting level* = 1 for the *outermost black-/white-box* embedded in the *environment*, and incrementing one *nesting level* for each *inward embedded black-/white-box*	Not often defined
Boundary interface	**The means for two or more elements to keep apart (see Table 9-10, p. 346)**	Not often defined
Business case	Estimate of any economical opportunities for a *product* (for example return on investments), to motivate the *product* to be developed	More general
Capture requirements	To refine *requirements* from an *outward black-box*, and supplement these with new stand-alone *requirements*	More vague
Cascading results	To understand and illustrate a *value chain* as a network of interconnected *results*	Only in *Cpdm*
Class	Collection of *functions, variables, objects,* and nested *objects*	Accepted
Class shape (requirement)	*Program element,* capturing appearance of *functions, variables,* and *objects,* preferably kept in separate *header files.* Also see *shape requirement* and *function shape (requirement).*	Only in *Cpdm*

Reserved word	Cpdm definition	Common definition
Clean-room engineering	An approach to apply the same "clean" *intellectual control* in *program* development as is long established in development and manufacturing of semiconductors	The same
Complex	**Something currently not sufficiently understood and thereby not predictable enough, in severe cases denoted as chaos. Paradigm shifts may accelerate understandability. (See p. 20).**	Not often defined
Component	Approved *items*, after the *product prototype* has passed *verification*, to be used (or *reused*) when manufacturing the *product*.	More vague
Compound requirement	*Requirement* specified without making any distinction in whether it is a *restriction requirement* or a *behavior requirement*.	Not often defined
Configuration management (CM)	The science of handling growth of a *modularized product* over time, by controlling changes through *version* handling, in order to judge how *product* parts are compatible with each other	About the same
Connection interface	**The means for two or more elements to influence each other (see Table 9-10, p. 346)**	More vague
Consistency	The degree to which different descriptions of a unified whole fit together. High *consistency* means that all descriptions fit together perfectly, and is achieved through heightened *formalism*.	About the same
Cpdm Complex product development model	*Cpdm* is the abbreviation of *Complex Product* Development Model, explained in this book; a model based on carefully selected world-class methods, adapted by thorough experience from the development industry	By now defined
Critical path	*Activities* impossible to execute in parallel, compressed to one sequence without time gaps, in order to estimate the shortest possible execution time of all these *activities*	About the same
Customer	The party paying for a developed *product* to satisfy a perceived need. Also see *user*	About the same
Customer-oriented	Satisfying the *customer* more than any other *stakeholder*, such as technicians or retailers	About the same
Decompose	To open identified *black-boxes* and get them populated with ingredients, which may in turn be *embedded black-/white-boxes*	About the same
Demand	How an *activity* or *value chain* ensures necessary source *result(s)*	Not defined
Designer	An engineer who provides *solution alternatives*, *architecture ingredients*, and *white-box design items*.	About the same
Design item	**Material (including linkable programs) specified with desired granularity formal enough to be orderable (see Table 10-1, p. 440)**	Vague
Design item properties	Technical characteristic of *white-box design items*, such as size, color, *quality*, reusability etc.	Not so precise
Determined architecture ingredient	*Architecture ingredient* to some extent stipulated by a *restriction requirement*, but yet chosen rather freely	Not defined
Development board	Package of tools for developing an electronics device, for example, containing a printed circuit board with a processor, memory circuits, IO circuits, sometimes an experimental area, and sometimes a *program* loader from *host development system*	The same
Development staffing	**From whom to get support for surveying the existing environment, developing the product interfaces to the environment, and developing each inward embedded white-box (see Table 7-3, p. 179)**	Sometimes defined

Reserved word	Cpdm definition	Common definition
Development system	Package of tools for development of *programs* on a *host* computer, typically containing editor, compiler, linker, built-in *program libraries*, and debugger. A developed *program* can be intended for the *host* computer or be *ported* to another computer *target*	The same
Development system libraries	Precompiled and linkable callable subroutines built into most *development systems* to be used to access the computer peripherals	About the same
Downstream	Direction of progressing development, from the original idea to the developed *product* ready for manufacturing, also see *upstream*	Not often defined
Element	**Architecture ingredients package, which has interfaces to its surroundings (see Table 9-1, p. 335)**	Many different definitions
Element (in assembler programs)	*Element* containing ingredients such as unprotected *variables* and unstructured *instructions*	About the same
Element (in structured programs)	*Element* containing *structured program* ingredients, with *functions* containing exclusively structured types of *instructions* (see *function, variable,* and *instruction*)	About the same
Element (in object-oriented programs)	*Elements* containing ingredients as in *structured programs*, but also containing *objects* (see *variable, object,* and *instruction*)	About the same
Embedded black-/white-box *Embedded white-box* *Embedded black-box*	*Black-/white-box* connected by *interfaces* to surrounding *white-box* design items	Not often defined
End user	See *user*	About the same
Enfold	To surround *white-boxes* with *white-box design items*, which consequently will belong to *outward white-boxes or environment interfaces*	Not defined
Environment	**The existing surroundings and the interfaces to the product being developed (see Table 9-2, p. 335)**	About the same
Environment architecture resumé	**List of existing relevant ingredients in the environment and newly developed interfaces to the product (see Table 9-3, p. 336)**	Not often defined
Environment interfaces	All *interfaces* between the *outermost black-box* and the *environment*	Not often defined
Environment interfaces design item	**Specified design, including technical properties, needed to realize and manufacture the product interface to environment (see Table 10-2, p. 441)**	More vague
Environment interfaces solution alternatives	**Various interface solution proposals how to best possible satisfy one or several specified environment restrictions (see Table 8-1, p. 287)**	Not often defined
Environment nesting level	Name of the *product structure* surrounding the *outermost black-/white-box*	Not often defined
Environment logical architecture	**Logical drawing of the environment ingredients, including the product black-box with its interfaces (see Table 9-3, p. 336)**	Not often defined
Environment physical architecture	**Appearance illustration of the environment surrounding the black-box of the product (see Table 9-3, p. 336)**	Not often defined

Reserved word	Cpdm definition	Common definition
Environment restriction requirement	**Limitations in the existing environment that will impact the product to be developed (see Table 7-1, p. 178)**	About the same
Exception use-case	**Unfavorable sequence of stimulus and responses interacting on interfaces to a black-box (see Table 7-29, p. 221)**	About the same
Explicit verification	*Verification* and complete *failure elimination* of a *black-/white-box* before being integrated in its *outward black-/white-box*	Not defined
Failure detection	Detecting that a *verification* outcome doesn't comply with *test-case* pass criteria	Not often defined
Failure localization	An investigation to find the exact position of a failure in the *product prototype* and/or underlying documentation. One *failure detection* may depend on several *failure localizations*	Not often defined
Failure elimination	Removal of a detected failure to get it passed in a *reverification*, either through changes in the *test-case*, changes in the *realization*, and/or changes in underlying documentation.	Not often defined
Failure elimination report	**Report of each detected failure, explaining where its source is located, and proposing how to eliminate it in the optimal way (see Table 12-6, p. 672)**	Not often defined
Fake black-/white-box	A *black-/white-box* containing *embedded black-boxes* which are (often simple) fakes, in order to make the *black-/white-box* partially verifiable. In *programs* a fake is also called stub.	Not defined
Finalize design item	Transformation of *architecture ingredients* to detailed *white-box design items*, exactly orderable from *suppliers* or in-house	Very vague
Finalize product requisite	Account for the total cost of material, work, and equipment of manufacturing, *realization*, and development.	More vague
Fixture	Temporary device for holding a work-piece (typically a *black-box*) during *verification*. Also applicable for immaterial *fixtures* such as a temporary *program* wrapping a *black-box program* to be verified.	More vague
Formal use-case	**A sequence of use-units interacting on interfaces to a black-box, starting from and ending in an explicit wait-state, with all responses in between explicitly calculated from specified stimuli histories (see Table 7-37, p. 228)**	More vague
Formal scenario	**A network of formal use-cases, assembled by attaching their wait-states (see Table 7-39, p. 230)**	Not often defined
Formalism	Complying with concepts consistently shared by colleagues	More vague
Full verification	*Verification* of a completely realized *black-/white-box*, executing a final set of *test-cases*, until all failures are eliminated, and all *test-cases* have passed.	More vague
Function	**Encapsulation of local variables and instructions (see Table 9-31, p. 414)**	About the same
Functional decomposition	A bad habit to systematically design *architecture elements* to encapsulate behavior instead of artifacts. A sign of this is that the *element* name becomes a verb instead of a noun.	About the same
Function shape (requirement)	*Program element* capturing appearance of *functions*, kept in separate *header files*. Also called "forward function declaration" or misleadingly "function prototypes". Also see *shape requirement* and *class shape (requirement)*.	Only in *Cpdm*
Global variable	**Storage of stimuli history which are accessible from any function (see Table 9-31, p. 414).** Also see *machine variable* and *local variable*.	About the same
Govern	**Documented process regulation, controlling development in reality (see Table 6-6, p. 89)**	Not defined
Guru	Developer primarily promoting own ego rather than ensuring the developing company's best interests	About the same
Granularity	Relative size of parts (compared to size of the whole)	About the same

Reserved word	Cpdm definition	Common definition
Header file	**Element exclusively containing program commands for better readability and control; never resulting in anything executable in final machine programs (see Table 10-86, p. 533)**	The same
High-level program	See *structured program* and *object-oriented program*.	About the same
Host (for development)	Capable full development system offered for convenient development of a scanty system, to later be *ported* to its final scanty *target*	About the same
Idle wait-state	The *wait-state* in a *formal scenario*, where most *use-cases* originate, and where they depart to when finished	More vague
Implicit verification	*Verification* and *failure elimination* of a *black-/white-box* after invaded in its *outward black-/white-box*	Not defined
Incorporative integration	The most preferred way to integrate, meaning that self-contained and fully verified *white-boxes* are inserted into their surroundings (also see *invasive integration*)	Not defined
Incremental development	Grouping *requirements* into increments, and ordering these for development in such a sequence that during *realization* of each increment the *product prototype* becomes useful for the *stakeholders* to evaluate	About the same
In-house design item	**Design developed by in-house workforce (see Table 10-1, p. 440)**	Not often defined
In-house role item	**Development, realization, and manufacturing in-house workforce (see Table 10-1, p. 440)**	Not often defined
Innermost white-box *Innermost black-box* *Innermost black-/white-box*	*Black-/white-boxes* not containing any *inward embedded black-/white-boxes*	Not often defined
Instruction (in programs)	Computer executable command	About the same
Instruction (in assembler programs)	**Executable limitless command (see Table 9-27, p. 396)**	About the same
Instruction (in structured and object-oriented programs)	**Executable command, exclusively consisting of the types: sequence, selection, iteration, or call to functions (see Table 9-31, p. 414)**	About the same
Integration	To integrate inside-out, see *incorporative integration, full verification* and *explicit verification*. To alternatively integrate outside-in, see *invasive integration, partial verification,* and *implicit verification*.	About the same
Integration report	**Description of fulfillments and shortcomings when joining and assembling items, like wrongly designed interfaces, items not fitting together, and so forth (see Table 11-3, p. 599)**	Not often defined
Intellectual control	A *result* so well described that developers understand that it is correct by only inspecting it (and not needing to verify it).	About the same
Interaction requirement	**Behavior requirement describing cooperation over a connection interface to a black-box or between black-boxes (see Table 7-19, p. 206)**	More vague
Interface	**The means for two or more elements to influence each other, or to keep each other apart (see Table 9-1, p. 335).** Also see *connection interface* and *boundary interface*.	Not often defined in this way

Reserved word	Cpdm definition	Common definition
Interrupt	**A signal from a port or an internal clock that causes a program to jump to a special location to perform an urgent execution before returning (see Table 9-29, p. 402)**	About the same
Intrinsic complexity	By nature being *complex*, which cannot be reduced (by models so far). Also see *artificial complexity*.	Not often defined
Invasive integration	Directly realize an *embedded white-box* to existing *environment interfaces* or within an existing *black-/white-box* (also see *incorporative integration*)	Not defined
Inward	Direction from *product environment* towards inner *nesting levels* of a *product* hierarchical structure. Also see *outward*.	Not defined
Item	**All resources needed for product development, such as material, workforce, and equipment specified with desired granularity and formal enough to be traceable (see Table 10-1, p. 440)**	Not often defined
Jump instruction	**Command to continue execution at an address label (see Table 9-27, p. 396)**	About the same
Justified	How a *result* qualifies its foregoing *task* or *value chain*	Not defined
Lead time	Calendar time between starting time and completion time, regardless of how much working time there is in between. Sum of working time between two time points is called effective time.	About the same
Life cycle status	Status of a *result*, reflecting its degree of completion during its life from birth to death	About the same
Line	The (rather) static and most often hierarchical structure of *managers* to lead a company	About the same
Line management	Organizing and leading organizations	About the same
Linkable program (object code)	*Program items* generated by assembler or compiler tools, to be integrated by a linker tool into an executable *machine program*	About the same
Local variable	**Storage of stimuli history, which are accessible only within the function (see Table 9-31, p. 414).** Also see *machine variable* and *global variable*.	About the same
Logical architecture	**Drawing illustrating how ingredients logically relate to each other (see Table 9-8, p. 344)** Also see *architecture resumé* and *physical architecture*.	More vague
Low-level program	See *machine program* and *assembler program*.	About the same
Machine program	Machine executable *program* resulting from assembler, compiler and linker tools applied on human readable *programs*. Also see *machine instruction* and *machine variable*.	About the same
Machine variable	**A read-write memory location to save and retrieve stimuli history data for further calculation (see Table 10-72, p. 515)**	About the same
Machine instruction	**Coded instruction to be decoded and executed by the controller or processor (see Table 10-72, p. 515)**	About the same
Manager	**The role of a human supervising performers and preparers in a value chain (see Table 5-4, p. 79)**	About the same
Managed information	Whenever releasing information, it must be documented, provided with unique identification, and spread to all affected people.	About the same
Marketer	A *product management role* to defend customers interest.	About the same
Master	**A documented process regulation, governing how to organize a result to justify foregoing activity (see Table 6-1, p. 84)**	Not often defined

Reserved word	Cpdm definition	Common definition
Message sequence chart (MSC)	**Chart illustrating how stimuli and responses are distributed between architectural elements, including black-/white-boxes (see Table 9-22, p. 380).**	About the same
Meta information	Information beyond *product* information needed for control of *product* development	More vague
Metrics	Measurements collected during *product* development in order to follow up the *quality* of the *process* used	About the same
Modularize	To divide an unmanageable *monolith* into interlinked self-sufficient pieces, assigned own unique identifiers	About the same
Module	Self-contained piece of a *result* (also see *modularize* and *managed information*)	About the same
Module coupling	The degree of comprehensible dependency a *module* has to its surroundings	About the same
Module strength *Module cohesion*	The degree of comprehensible purpose of a *module*	About the same
Monolith	An (often huge) *element* that can be handled only as a whole, since its content have no evident *interfaces* or its evident *interfaces* are unmanaged	Not often defined
Nesting level	Location of a *black-/white-box* in a hierarchical *product structure*, starting with *nesting level* = 0 at the *product environment*, and adding one to the *nesting level* for each *inward embedded black-/white-box*	Not defined
Normal use-case	**A favorable sequence of stimuli and responses interacting on interfaces to a black-box (see Table 7-29, p. 221)**	About the same
Null series	Incipient manufacturing of a *product* in order to finally adapt it for manufacturing and trimming the manufacturing equipment	About the same
Obey (a process)	**Development in reality performed in accordance with documented process regulation**	Not defined
Object	**Element created run-time by instantiation of a class specification. Objects may in turn contain objects, thereby enabling objects to be hierarchically organized. (See Table 9-35, p. 424).**	About the same
Object-oriented program	**Program containing classes instantiating objects, consisting of variables (can also be objects) and structured instruction functions (see Table 10-102, p. 554)**	About the same
Observation post	Design for injecting *stimuli* into, and drawing *responses* off a *connection interface* crossing a nondetachable boundary of a *black-/white-box* to make it verifiable	Not often defined
Off-the-shelf item	**Predeveloped reusable components (see Table 10-1, p. 440)**	Not often defined
Off-the-shelf supplier	*Suppliers* offering *samples* off the shelf, to be (almost) immediately delivered as specified by their *sample catalogue* with data sheets	About the same
Outermost white-box *Outermost black-box* *Outermost black-/white-box*	The *black-/white-box* having *interfaces* to the surrounding existing *environment*	Not often defined
Outward	Direction from inner *nesting levels* towards the *environment* *nesting level* of a *product* hierarchy structure. Also see *inward*.	Not defined
Paradigm shift	When a point of *complexity* is passed, where the old understanding no longer is rewarding, and a fundamental new approach must be applied to create better understanding	About the same

Reserved word	Cpdm definition	Common definition
Partial verification	Incomplete but otherwise a normal *verification* of a *black-/white-box*, because inwards *embedded white-boxes* may be missing or containing fakes.	Not defined
Partition	**Detaching parts from the whole (see Table 6-3, p. 85)**	About the same
Performer	**The role of a human participating in a value chain activity (see Table 5-4, p. 79)**	Not often defined
Phase	Distinguished periods of a *process*. The *Cpdm* total *process schedule* contains 13 *phases*, of which the 6 middle *phases* are called technical *phases*.	Not defined
Physical architecture	**Drawing illustrating how ingredients physically appear and how they combine with each other (see Table 9-8, p. 344).** Also see *architecture resumé* and *logical architecture*.	More vague
Port (a white-box)	To facilitate *integration* of a *white-box* to its *target*, it can be developed and verified in a capable *host*, and then more easily be *ported* to its scanty *target*.	About the same
Postcondition	The *wait-state* to where transitions end up when all *responses* have departed. Also see *precondition, use-units, use-cases*.	About the same
Precondition	The *wait-state* where a transition originates when *stimuli* have arrived. Also see *postcondition, use-units, use-cases*.	About the same
Preparer	**The role of a human contributing to a value chain result (see Table 5-4, p. 79)**	Not defined
Primordial architecture ingredient	Design chosen because referenced *restriction requirement* has stipulated it	Not defined
Process	Regulative documentation(e.g. directives, organizations, *metrics* and *schedules*) *governing* all aspects of *product* development	Not so strongly defined
Process management	The science of regulatory systems *governing* all aspects of *product* development, including improvement of it	About the same
Product	**An artifact created by somebody, from raw materials to finished goods for a market, to satisfy a need (see Table 4-1, p. 38)**	Same
Product development *Prototype development*	The *stage* from the time a new *product* is first identified to the launch of that particular *product* on the market. *Prototype development* is similar but lasts only until the *product prototype* is developed and verified.	Same
Product line	A set of related *products* (possibly in same *product portfolio*) that *reuse* many of their *components* from each other	About the same
Product portfolio	The overall market offering of related *products* (with possible *reuse* between them) in order to exploit a larger market by diversifying in such a way as not to cannibalize each other	About the same
Product proposition	Planned *product* based on a profitable *business case* and a market optimized *product portfolio*	More vague
Product prototype	**Realizations made during product development, for the purpose of being verified against its requirements, and to be validated against its user profiles (see Table 4-2, p. 40)**	More vague
Product life cycle	Time period from start of development of a *product* idea until the *product* is phased out of the market, including possible warranties	About the same
Product management	Ensuring that a *product* satisfies both *users'* expectation and manufacturers' return on investment	About the same
Product requisite	**Complete information about costs for all types of items, to allow accounting for development, realizations, and manufacturing (see Table 10-7, p. 443)**	About the same
Product structure	Hierarchical *partition* of a *product*, containing mechanics, electronics, and *programs*	About the same

Reserved word	Cpdm definition	Common definition
Program	*Program* intended to be machine-readable, such as *machine program* and *linkable program*, and human-readable (being the origin, sometimes called source code) such as *assembler program*, *structured program*, and *object-oriented program*	About the same
Program libraries	*Program* set of *elements* built like any *component* for *reuse*. It can be a fully changeable *source system* or precompiled to an unchangeable *linkable program*.	About the same
Progressing activities	To help understand and illustrate a *value chain* as a network of interconnected *activities*	Not defined
Project	A temporary organizational structure in a company to accomplish an assigned objective within restricted conditions (time, cost etc.)	About the same
Project leader	The *manager* of a *project*, reporting to the *project sponsor*	About the same
Project management	The science of handling *projects*, including hard facts such as preparing and following up on plans, but also soft factors such as improving attitudes and reaching specified *quality*	About the same
Project sponsor	A *line manager* owning and controlling a *project*	About the same
Prototyping	Early and finite *activity* to evaluate crucial *elements* and *interfaces*, in order to consider technical risks and opportunities not yet captured	More vague
Provider	Organization offering workforce for development, *realization*, and manufacturing	Not defined
Provider role item	**Development, realization, and manufacturing hired workforce (see Table 10-1, p. 440)**	Not often defined
Quality	The *ability* of a *product* to satisfy a *user's* explicit and implicit expectations	About the same
Quality management	The science of keeping (or improving) a specified degree of *user product* satisfaction, including indirect soft causes such as capability of the organization, people, and *processes*	More vague
Realization	A tangible construction (including *machine programs*) being built from a description of it (including human readable *programs*)	Not often defined
Received items deviation report	**Deviation report on how received items (including linkable and possibly ported programs) match placed orders and other expectations (see Table 11-1, p. 599)**	Not often defined
Recycle	Extracting *elements* from earlier developed *results*, and rework them into newly desired *elements*. Also see *reuse*.	More vague
Refactoring	Restructuring an *element* design without changing its behavior on its *interfaces*	About the same
Refine requirement	See *requirement refinement*.	Can mean whatever
Requirement refinement	To capture requirements for a *black-box* by detailing *requirements* from the *black-box* one *nesting level outwards*	Can mean whatever
Requirement summarization	Generic *requirements* being the fusion from detailed *requirements* one *nesting level inwards*	Not often defined
Regression verification	Executing *test-cases* that didn't necessarily result in failures in previous *verifications*, in order to secure that recent *failure eliminations* have not caused any new detectable failures	About the same
Requirement	See *environment restriction requirement*, *compound requirement*, *restriction requirement*, and *behavior requirement*.	About the same
Requirement management (RM)	The science of *requirements*, including how to capture and prioritize them, how to organize and store them, and how to refine them to *inward black-boxes*	About the same
Response	**Output from a black-box via connection interfaces (see Table 7-19, p. 206)**	About the same

Reserved word	Cpdm definition	Common definition
Restriction requirement	**Stakeholders' documented expectations of design inside a black-box (see Table 7-15, p. 204)**	Not often defined
Restriction (in programs)	**Limitation of any parameter in a program (see Table 9-27, p. 396)**	Not often defined
Result	**Tangible output from an activity or value chain (see Table 5-2, p. 68)**	Not often defined
Result-oriented	Emphasizing the importance of *cascading results* rather than *progressing activities* in a *value chain*. Also see *activity-oriented*.	More vague
Reuse	Develop a *component* that can easily and without significant rework be used in other developments (also see *recycle*)	About the same
Reuse plan	Based on proposed *product portfolio*, a plan can be made how these *products* can *reuse* commonalities from each other.	More vague
Reusable item	*Component* built in such a way that it is easy to *reuse*, common in mechanics and electronics, but not (yet) in *programs*	More vague
Road map	Studies of coming standards and trends, predicts new technology to be proactively tried out in-house or by preferred *suppliers*	More vague
Role	**A set of competencies, authorities, and responsibilities of a person (typically a user, an employee, or third-party contractor) (see Table 5-3, p. 79)**	About the same
Role item	**Accomplished and estimated work specified with desired granularity and formal enough to be traceable (see Table 10-1, p. 440)**	Not defined
Sample	Offering by *supplier* to match one or many *solution alternatives*. Approved *samples* become *architecture ingredients*, which, if still approved for *product prototypes*, become *white-box design items*, which, if still approved for manufacturing, become *components*.	About the same
Sample catalogue	Brochures from *suppliers* showing off-the-shelf offerings appropriate for *product prototypes*	About the same
Schedule	**A documented process regulation package of alternating tasks and masters, governing how to organize a value chain (see Table 6-2, p. 85)**	More vague
Scenario	Scenarios are interlinked use-cases. See *formal scenario*.	Many definitions exists
Scenario-thread-machine	*Wait-state-machine* awaiting a manageable number of *stimuli* of predictable sequence from one interlinked *scenario* over its synchronized *interfaces*. Also see *wait-state-machine*.	Not defined
Selected solutions supplier opportunities	**Determination of which environment solutions are part of the product, and from where (including in-house) to get appropriate samples to match these solutions (see Table 8-1, p. 287)**	Not often defined
Shape requirement	*Program element* not generating any executable *machine programs*, but specifying appearance of *functions* or *variables* (which may be *objects*), or a compound thereof. Preferably kept in separate *header files* to be referenced for inclusion in *source files*. Also see *class shape* and *function shape*.	Only in *Cpdm*
Solution alternatives	Whatever designs are found to more or less satisfy one or many *requirements*. Also see *environment interfaces solution alternatives* and *white-box solution alternatives*.	Not often defined
Source file	**Element exclusively containing program instructions, finally resulting in executable machine programs (see Table 10-86, p. 533)**	The same
Source system	A complete *element*, so well documented that this *element* can be fully incorporated and further developed in-house, like anything else developed in-house	About the same regarding programs
Specialization	**To restrict an artifact in order to make it more particular with more unique characteristics (the opposite is generalization). (See Table 10-101, p. 553).**	About the same

Reserved word	Cpdm definition	Common definition
Specifier	A developer *role* refining and capturing *requirements*	The same
Stage (of life cycle)	Period of a *product life cycle* wherein *product* development is the first *stage*	Not defined
Stakeholder	Anybody who has legitimate and relevant interest in what to expect from the *product* being developed	About the same
Stand-by wait-state	See *idle wait-state*.	More vague
Start-up report	**Report of obstacles and accomplishments when starting up a white-box or the environment interfaces (see Table 11-5, p. 600)**	Not often defined
State	Any combination of *variable* values stored in a *white-box*. Don't mix it up with *black-box wait-state*.	More vague
Stimulus *Stimuli*	**Input to a black-box via connection interfaces (see Table 7-19, p. 206)**	About the same
Stimuli history *Stimuli histories*	**Referable history of all stimuli with their arguments received on interfaces of a black-box since it started up (see Table 7-36, p. 226)**	About the same
Structured program	**Program prohibiting unstructured flow by providing only structured machine-independent instructions, exclusively being sequence, selection, iteration, and function call, needing compiler and linker tools support to be translated to machine instructions and machine variables (see Table 10-85, p. 529).** Also see *structured function call*, *structured iteration instruction*, *structured selection instruction*, and *structured instruction sequence*.	About the same
Structured function call	Command transferring execution to a *function* bringing parameters to be passed, and with a return command leaving execution back bringing parameters to be passed back. Also see *subroutine call* and *structured program*.	About the same
Structured iteration instruction	Command forming a loop of *instructions* to be repeatedly executed until a termination condition ends the loop. Also see *structured program*.	About the same
Structured selection instruction	Command selecting execution to one of several branches of *instructions*, depending on how a selection condition turned out. Also see *structured program*.	About the same
Structured instruction sequence	Commands in a sequence executed one after another. Also see *structured program*.	About the same
Subroutine call	Command transferring execution to an *address labeled* subroutine, and turning back execution after a return command, seldom with support of passing and passing back parameters. Also see *structured function call*.	About the same
Subsystem	Parts resulting from *partitioning* a system. Because of ambiguity, this word is seldom used in this book. Also see *system*.	About the same
Supplier	*Items* not developed in-house are bought from *suppliers*. In-house-developed *items* might be considered as bought from internal *suppliers*.	Accepted
Supplier design item	**Design developed by suppliers (see Table 10-1, p. 440)**	About the same
Supplier opportunities	The chance to buy *white-box design items* instead of develop them in-house	Not often defined
System	A demarcated part of the whole. Because of often seen ambiguity, this word is rarely used in this book. Also see *subsystem*.	About the same
Tailor (a process)	Adapt and concretize a generic *process* to fit a certain *product* and organization	About the same
Target	The final location where a *product* is supposed to be used (also see *host*)	About the same

Reserved word	Cpdm definition	Common definition
Task	A documented process regulation governing how to organize an activity demanding source results (see Table 6-1, p. 84)	Not often defined
Terms	Mutual governing regulation of a role between a development company and its employees, consultants, or hired contractors (see Table 6-5, p. 88)	Not often defined
Test-case	Concretization of requirements in order to be a more efficient base for performing all kinds of verification (see Table 12-2, p. 671)	More vague
Test-use-case	A use case arranged in such a way that it is easy to execute it and evaluate its correctness by the human mind	Not defined
Thread	Program elements executing apparently concurrently, but in reality getting a time slot in sequence with all other threads (see Table 9-29, p. 402)	About the same
Traceability	This means that developed *results* do reference each other, requiring that *results* are *modularized* and each *module* is assigned a unique identity.	About the same
Turnaround time	The time it takes from making a change in a finalized design until the change can be observed in the running *product prototype*	About the same
Turnkey item	Custom-made design developed to fit very well into own design (see Table 10-1, p. 440)	About the same
Turnkey supplier	*Suppliers* offering to develop *items* well fitting into design, based on own specified *requirements* and/or solution proposals	About the same
Unbiased architecture ingredient	Design chosen freely, because referenced *requirements* only specify behavior, or there are no *requirements* to reference at all	Not defined
Unmanageable complexity	Something whose further development is unpredictable, because is has turned into something that can no longer be sufficiently understood. (See p. 20).	Not often defined
Unstructured program	Program allowing various dangerous constructs, most notably the `goto` *instruction*, for example *assembler programs*, Basic, Cobol, and Fortran. Also see *structured program*.	About the same
Upstream	Reverse direction of the normal progressing development, most often applied after *verification* for the purpose of localizing and elimination of detected failures done in the past	Not often defined
Usage profile	Every stimulus causing departure from every wait-state is assigned estimated probabilities, to be used by black-box test-cases. (see Table 12-56, p. 732)	About the same
Use-case	A sequence of stimuli and responses interacting on interfaces to a black-box (see Table 7-29, p. 221)	About the same
User	Human role interacting with an element through a man-machine interface (see Table 9-1, p. 335)	The same
User profile	A *role* description of an expected or potential *user* of the *product* to be developed. Sometimes referred to as persona. (Don't mix up with *usage profile*.)	About the same
Use-unit	One stimulus arriving on an interface to a black-box being in a certain wait-state, causing zero, one, or more responses to leave the black-box before it enters its next wait-state (see Table 7-35, p. 225)	About the same
Value chain	Human work ongoing in reality for the purpose of transforming source results to added value results (see Table 5-1, p. 66)	More vague
Variable	Storage for arrived stimuli (see Table 9-27, p. 396). Also see *machine variable*, *global variable* and *local variable*.	About the same
Variability of reusable components	The degree of possibility to adapt a *reusable item* when to fit it into all sorts of different *products*	About the same

Reserved word	Cpdm definition	Common definition
Variant Version	Variants are contemporary diversifications from a common artifact. A gradually extended diversification is often called version. The opposite is unification of variants to a primal artifact (see Table 6-4, p. 87)	More vague
Verification	To confirm that a realized black-/white-box or a prototype in the environment fulfills requirements on design and behavior (see Table 12-1, p. 661)	More restricted
Verification report	Report of pass/fail conclusions from each executed test-case (see Table 12-4, p. 672)	More vague
View	Different perceptions of same artifact, depending on how the observer relates to the artifact (see Table 6-4, p. 87)	About the same
V-model	Development model, starting with outside-in development in earlier *phases*, proceeding with inside-out development in later *phases*, thus forming a V when graphically illustrated	About the same
Wait-state	Often used for *black-box wait-state*. See *black-box wait-state*, *idle wait-state*, *precondition* and *postcondition*.	Not defined
Wait-state explosion	Situation when an *element* becomes incapable of coping with too many sequences of unsynchronized *stimuli* on its *interfaces*. Also see *black-box wait-state* and *scenario-thread-machine*.	More vague
Wait-state-machine	Modeling black-box behavior by stimuli-initiated transitions between wait-states (see Table 7-40, p. 232)	The same
Waterfall model	An approach that each and every *result* needs to be fully completed before it may be *demanded* by any succeeding *activity*	More vague
White-box	Element which is an opened black-box, containing disclosed architecture ingredients, some of them possibly embedded black-boxes (see Table 9-7, p. 343)	Accepted
White-box design item	Specified design, including technical properties, needed to realize and manufacture the white-box (see Table 10-2, p. 441)	More vague
White-box solution alternatives	Various solution proposals for how to best possible satisfy one or several specified requirements (see Table 8-8, p. 296)	Not often defined
White-box supplier opportunities	Proposals of which suppliers (including in-house) can possibly provide appropriate samples to match specified solution alternatives (see Table 8-8, p. 296)	Not often defined
White-box test-case	Concretization of restriction requirements, to be an efficient base for verification of white-box design (see Table 12-31, p. 705)	More vague
White-/black-box	Same as *black-/white-box*.	About the same